Predictive Statistics

Analysis and Inference beyond Models

All scientific disciplines prize predictive success. Conventional statistical analyses, however, treat prediction as secondary, instead focusing on modeling and hence on estimation, testing, and detailed physical interpretation, tackling these tasks before the predictive adequacy of a model is established. This book outlines a fully predictive approach to statistical problems based on studying predictors; the approach does not require that predictors correspond to a model although this important special case is included in the general approach. Throughout, the point is to examine predictive performance before considering conventional inference. These ideas are traced through five traditional subfields of statistics, helping readers to refocus and adopt a directly predictive outlook. The book also considers prediction via contemporary 'blackbox' techniques and emerging data types and methodologies, where conventional modeling is so difficult that good prediction is the main criterion available for evaluating the performance of a statistical method. Well-documented open-source R code in a Github repository allows readers to replicate examples and apply techniques to other investigations.

BERTRAND S. CLARKE is Chair of the Department of Statistics at the University of Nebraska, Lincoln. His research focuses on predictive statistics and statistical methodology in genomic data. He is a fellow of the American Statistical Association, serves as editor or associate editor for three journals, and has published numerous papers in several statistical fields as well as a book on data mining and machine learning.

JENNIFER L. CLARKE is Professor of Food Science and Technology, Professor of Statistics, and Director of the Quantitative Life Sciences Initiative at the University of Nebraska, Lincoln. Her current interests include statistical methodology for metagenomics and also prediction, statistical computation, and multitype data analysis. She serves on the steering committee of the Midwest Big Data Hub and is Co-Principal Investigator on an award from the NSF focused on data challenges in digital agriculture.

CAMBRIDGE SERIES IN STATISTICAL AND PROBABILISTIC MATHEMATICS

This series of high-quality upper-division textbooks and expository monographs covers all aspects of stochastic applicable mathematics. The topics range from pure and applied statistics to probability theory, operations research, optimization, and mathematical programming. The books contain clear presentations of new developments in the field and also of the state of the art in classical methods. While emphasizing rigorous treatment of theoretical methods, the books also contain applications and discussions of new techniques made possible by advances in computational practice.

A complete list of books in the series can be found at www.cambridge.org/statistics.
Recent titles include the following:

Predictive Statistics

Analysis and Inference beyond Models

Bertrand S. Clarke
University of Nebraska, Lincoln

Jennifer L. Clarke
University of Nebraska, Lincoln

CAMBRIDGE
UNIVERSITY PRESS

CAMBRIDGE
UNIVERSITY PRESS

Shaftesbury Road, Cambridge CB2 8EA, United Kingdom

One Liberty Plaza, 20th Floor, New York, NY 10006, USA

477 Williamstown Road, Port Melbourne, VIC 3207, Australia

314–321, 3rd Floor, Plot 3, Splendor Forum, Jasola District Centre, New Delhi – 110025, India

103 Penang Road, #05–06/07, Visioncrest Commercial, Singapore 238467

Cambridge University Press is part of Cambridge University Press & Assessment,
a department of the University of Cambridge.

We share the University's mission to contribute to society through the pursuit of
education, learning and research at the highest international levels of excellence.

www.cambridge.org
Information on this title: www.cambridge.org/9781107028289

DOI: 10.1017/9781139236003

First published 2018

A catalogue record for this publication is available from the British Library

ISBN   978-1-107-02828-9   Hardback

Additional resources for this publication at www.cambridge.org/predictivestatistics

Contents

Expanded Contents

Preface

This book grew out of a nagging dissatisfaction with the various schools of thought in statistics and their increasing disjunction. Each one – frequentist, Bayes, survey sampling, information-theoretic, etc. – has its strengths and weaknesses, and comparisons amongst their different approaches to inference has energized statistical thinking. This dynamic has only grown stronger over the last decade as more challenging data types have become commonplace. Moreover, in contrasting the techniques advocated by the different schools of thought on harder problems, such as working with big data, high-dimensional data, or complex data, the nagging doubts have only become more insistent. Otherwise stated, the less data (or other information) relative to the believed complexity of the data generator that is available, the more the modeling contributes to an analysis and therefore the more the differences in schools of thought, which largely rest on modeling, become apparent.

Concisely, the era of big data, whether high-dimensional, streaming, multitype or otherwise 'big', is forcing us all to rethink statistics and its philosophy. Questions about how to measure the distance between points in high dimensions have to be addressed since that is one version of the curse of dimensionality, likewise, questions of sparsity – when it holds, when it fails, and how to deal with it in either case – and questions of data sets that have information which is not extractable within the traditional formulation of a 'random variable on a measure space' or a valid sample from a well-defined population. In these contexts, this book is a small first step to reorganize some of what we know in order to focus on predictive structure, which is one of the few properties that cuts across all the new, exciting, developments challenging us and our field.

In our field, we have relied too much on our models by not assessing them as extensively as we should. We have not looked enough at their stability. We have not, as a rule, considered a sufficient number of alternative models to be sure that the model we used was reasonable. With few exceptions, we have not done sequential searches over modeling strategies to find a reasonable model, given a certain amount of data, and then modified it in view of getting more data. Also, we have not assessed the robustness of our inferences to our modeling strategies sufficiently. The present authors are as guilty of this as anyone else. In short, we have contented ourselves with the bromide that even if the model is wrong it may be useful, in the hope that if there is a true model (and here we argue that often there isn't) we have found at least a part of it. However, that ain't necessarily so.

This book is an attempt to focus more heavily on the data than the formalism and to focus more heavily on the performance of predictors rather than the fit or physical interpretation of a model or other construct. As a consequence, testing and estimation are given short shrift. In fact, the general enterprise of inference by modeling, testing, and estimation seems premature until a lot more is known about a data generator than that it is described by a simple model that may be useful even when it's not true. In reality, the situation is usually worse than that for conventional analysis because the inferences are generated from one data set and a pile of assumptions, often dubious. Granted, in the hands of capable statisticians with enough persistence, most schools of thought will yield useful inferences. However, such success reflects the insight and doggedness of the statistician more than the efficacy of the methods. Consequently, it is hoped that one effect of focusing on the data, here via prediction, will be to energize the debate about what the central goals of statistics should be and how to go about achieving them.

An idea that recurs throughout this book is the concept of a problem class based on the relationship between the data generator and a class of predictors. The emphasis is on predictors that do not correspond to a perfect model for the data generator. In particular, cases in which the model used is only an approximation, or in which there is no true model, are frequently considered.

This book is in three parts. Part I outlines a general approach to statistics based on prediction. No claim is made that it is complete, merely that it is an alternative to various established schools of thought and deserves more attention than it has received. It is based on the prequential (predictive sequential) ideas that emerged in the early 1980s from A. P. Dawid and M. West, amongst others. There are four chapters, outlining the importance of prediction, defining a predictive paradigm, explaining why modeling, while sometimes useful, is not as good an approach as prediction, and finally looking at some familiar predictors. The view here is also more general: other schools of thought are incorporated into the predictive approach by regarding them as techniques for generating predictors that may be tested and studied. Thus, other schools of thought are not 'wrong' so much as incomplete: one school's techniques may not yield good predictors as readily as those of another.

Part II is a review of five major fields within statistics (time series, longitudinal data, survival analysis, nonparametrics, and model selection) from the predictive standpoint. The material is not new; the perspective on it is. The point of Part II is to demonstrate the feasibility of the predictive view: that it is a valid way to think about traditional branches of statistics and is computationally feasible. The five specific subfields were chosen because they are quite different from each other, suggesting the wide applicability of a general predictive view. They are also fields where the problems are so complicated that prediction is obviously important.

Part III is brings prediction up to the present. Starting with prediction in more contemporary model classes such as trees, neural nets, kernel methods, and penalized methods, it moves on to a chapter on ensemble methods, including Bayes model averaging, bagging, stacking, boosting and median methods. Even more than in previous chapters, computing is stressed to verify that the perspective advocated here is feasible. The final chapter is intended to bring predictive concepts to branches of statistics that have either recently emerged or recently changed character through, e.g., big data, changes in data collection, or new applications that have made prediction more important. Having dealt with terrestrial matters, the last chapter also indulges in some moon-gazing, speculating on which problems become more interesting when a predictive view is taken.

On the one hand this book does not require much mathematical background; a strong, determined MS student in statistics, mathematics, engineering, computer science, or other highly quantitative field should be able to follow the formal derivations. On the other hand, the book is primarily conceptual and so makes demands on the prospective reader that likely require more sophistication than a typical MS student, even a strong one, would have. Thus, our primary target audience is mid-career PhD students, practicing statisticians, and researchers in statistical fields. The authors sincerely hope that, whether or not these audiences agree with the perspective expressed in this book, they will find this perspective worth their time to understand.

For those interested in examining the R code or data used for the many examples in this text, please visit the catalog page on the Cambridge University Press website:

www.cambridge.org/predictivestatistics

This page includes a link to the github repository containing all relevant R code and data. The repository is structured so that each chapter has a branch. All code is provided under GNU Public License 3.0.

As with every book, there are people who should be thanked. First, all the people who supplied the data sets we used for examples. Second, all the people who, over the past four years put up with us obsessing over this book; we apologize for endlessly bending your ear. Third, Diana Gillooly of Cambridge University Press, with whom we had many conversations about the content, organization, and orientation of this book. Fourth, those colleagues who encouraged us in our folly. (You know who you are!) We forbear from mentioning names for fear they will regret encouraging us.

Finally, we have consistently tried to be engaging and sometimes provocative. Of course, some people will disagree with us and some errors may remain despite our best efforts. We are reminded of the (possibly mythical) story of a French physicist who, when asked about a colleague's work, pondered a few moments and finally responded: 'It's not even wrong.' In the spirit of that witticism, we apologize in advance for any errors that remain, whether technical or philosophical, hoping that they will at least be interesting.

Bertrand S. Clarke
Jennifer L. Clarke

Part I

The Predictive View

Prediction was the earliest and more prevalent form of statistical inference. This emphasis changed during the beginning of [the twentieth] century when the mathematical foundations of modern statistics emerged. Important issues such as sampling from well-defined statistical models and clarification between statistics and parameters began to dominate the attention of statisticians. This resulted in a major shift of emphasis to parametric estimation and testing.

   The purpose of this book is to correct this emphasis. The principal intent is to revive the primary purpose of statistical endeavor, namely inferring about reasonable values [that were] not observed based on values that were observed.

Preface of Geisser (1993)

The first four chapters of this book are an exposition of the centrality of prediction. For instance, estimation, testing, and classification problems can be cast in predictive terms and, conceptually, the predictive view is as philosophically justified as any other established philosophy of statistics, perhaps more so. Of particular importance is the downgrading of the role of models to settings in which they are genuinely useful, a relatively narrow set of circumstances. The alternative is falling victim to the convenience of models and the feeling of understanding they give even when they are unreliable – which is essentially all the time, outside oversimplified contexts.

1

Why Prediction?

> ... any model ... is merely a human attempt to describe or explain reality ... models are to be assessed in terms of their success at this task. It is misguided ... to believe in Nature as obeying some theory ... Even if we can find a completely successful theory, this does not mean we have identified Nature's true model – some other, distinct theory might be just as successful ... In this view, theories can only be distinguished by means of their predictions about observables ...
>
> Dawid (1992)

For centuries, perhaps millennia, people have tried all sorts of divination methods, from yarrow sticks to tarot cards, from the innards of animals to the positions of planets. With sufficiently skilled interpreters, these methods probably work a little: a tarot card reader may use the cards to evoke the frame of mind of the subject. To the extent that the future is implicit in a subject's frame of mind, the predictions may therefore be accurate. After all, if you know some one, even a little, you can predict some of their behavior. This is second nature for good salespeople, politicians, and others whose career success depends on detecting people's preferences. Arguably, this sort of procedure might even help with economic predictions that include market psychology. Note, however, that divination methods are rarely used to predict such outcomes as how much product a given chemical reaction will produce, or other outcomes that have essentially no element of human choice.

Here, by contrast, the goal is to make predictions by rules in such a way that evaluating how well the rules work will be unambiguous. The fortunate case occurs when the rules accurately reflect something about the mechanism used by the data generator (DG) to generate outcomes. This is the main goal of much of conventional science. However, there are vast classes of data where it is implausible to model the DG. As a slightly facetious example, one can treat *MacBeth* as a sequence of letters and try to predict the $(n+1)$th letter using the first n letters. A variant on this is predicting the $(n+1)$th nucleotide – or any finite sequence of nucleotides – on a chromosome, given the first n nucleotides in the chromosome. In both cases, the DG is so complex that detailed modeling for the purposes of prediction would be premature, to say the least. Indeed, if we want to make predictions, it's unlikely modeling will help much.

Worse, many DGs might not function by rules at all. The easiest way to think of this is that the outcome y_1 at a given time step is from one distribution, say Q_1, but the outcome y_2 from the next time step is from another distribution, Q_2, chosen by some agent who may not even know Q_1 or y_1 and chooses Q_2 using a hidden mechanism or no mechanism at all.

Doing this repeatedly means there is nothing stable enough to model, so we cannot, even conceptually, use models to generate prediction rules. Another way to think of this is to ask whether more data can be generated – at least in principle – that would be informationally equivalent to the data we already have. This is not the same as asking whether an experiment is repeatable in practice – many aren't, as for example in econometrics – it only asks whether in principle we could generate further data sets of the same general form. Clearly, if the answer is no (think of *MacBeth*) then there may be no rule that the DG follows and hence it will be impossible to formulate a prediction rule that matches the DG. However, even when we admit that the DG does not follow rules, we may still want a well-defined prediction rule so that we can evaluate how good it is. Crazy as it sounds, this is not entirely impossible, as we shall see.

The stance of this book, stated concisely and unabashedly, is that predictive statistics proposes an alternative formulation of the paradigm statistical problem. The central feature of predictive statistics, as opposed to other schools of thought in statistics, is to use the data to predict ahead rather than try to find out what underlies the data generator, then try to model it, and finally use the model to make predictions. In either case, predictive or not, predictions must be compared with new data for validation. The question is when this comparison is made – is it before or after 'modelling' has been attempted? In predictive statistics, modeling does not start until good prediction has been achieved. This is the reverse of the conventional approach.

One of the key arguments for predictive statistics at this time of writing, and for the foreseeable future, is that so much of the statistical world has changed. Volumes of data have massively increased, preprocessing techniques for raw data (e.g., in the 'omics world) have increased and become more diverse, multitype data is more prevalent than before (and often very hard to model), and the complexity of data streams that confront the Statistician is nearly overwhelming. Together, these features make modeling difficult, if not infeasible. Indeed, often only a small fraction of the available data can be used in an analysis. Taking a predictive approach, and hence achieving good prediction, is likely to be better in the long run than direct modeling for understanding a data generator. Even where modeling is infeasible, good prediction is the sine qua non of a good theory. The reason is that modeling requires dealing with model uncertainty and misspecification and these can be extraordinarily difficult.

1.1 Motivating the Predictive Stance

It may seem strange to ask, but it's important to answer the question 'Why is prediction so important?' First, one obvious answer is that sometimes the goal really is to know what the next outcome is likely to be: prediction may be the goal of the statistical analysis. For instance, it might be helpful to predict who will get post-traumatic stress disorder (PTSD). That way, to prevent PTSD, or minimize its effects, a physician would want to know who it is most important to treat prophylactically. Sometimes the goal is prediction even when it's not phrased predictively. For instance, one may estimate a probability of recurrence of cancer (with a standard error), but it would be more informative to give a point predictor for when a patient will get a recurrence along with an assessment of the variability of that prediction. Aside from being the information that a patient or physician wants, a prediction interval is less abstract and more intelligible than a probability, let alone a confidence or credible interval for a probability. Of course, people might want to predict the weather, the economy, the response to a new treatment, and so forth. Who hasn't wanted to know the future for some purpose, base or laudable?

A second answer (that is less obvious) is that most other goals of statistical analysis can be subsumed within prediction. What, after all, are the main goals of statistical analyses? Any list would have to include (1) model identification, (2) decision making, and (3) answering a question about a population – even if there is some overlap among these goals as stated. For instance, identifying a model may amount to making a decision: in classification, model identification amounts to identifying a classifier, and using a classifier amounts to deciding the rule by which one will assign a future subject to a class. In general, it's hard to find a statistical problem that doesn't have a direct connection with prediction.

Let's start with *model identification* – which is essentially estimation in one guise or another. This includes, among other possibilities, parametric estimation, classification, regression, nonparametric estimation, and model selection. It also includes some hypothesis testing. A simple versus simple test such as $\mathcal{H}_0 \colon P = P_1$ versus $\mathcal{H}_1 \colon P = P_2$, where P_1 and P_2 are the only two candidate probabilities for P, is an obvious example. In general, one can use a series of goodness-of-fit tests to determine models that can't be rejected. Moreover, tests such as whether a given variable should be in a regression function must also be included as part of model identification. Even though such a test does not by itself identify a model, it constitutes an effort to reduce the class of models that must be considered and is therefore a step toward model identification.

In all these cases, how can one know in reality that a model has been successfully identified without using it to generate accurate predictions? Even more, how can one know that another model with an equally good fit, possibly using different variables, can be ignored if it is not predictively discredited – for instance, on the grounds of high bias or high variance? Loosely, outside very simple problems, model identification without predictive verification is little more than conjecture. Put otherwise, whenever a model is selected, or a parameter estimated, a predictor is formed and, if it doesn't perform well, the model it came from is discredited.

Essentially this means that the search for a good model is merely a special case of the search for a good predictor. The substitution is thought worthwhile because a scientist can bring 'modeling information' into the search for a predictor. The problem is that modeling information is usually not itself predictively verified and hence is often of dubious value. Thus, taking a purely predictive view and treating modeling information as likely to be unreliable guards against the use of such suspect 'information'.

Let us now turn this around. Just as a credible model, when it exists, can often be used to generate predictions, a predictor can sometimes be used to identify a model. In the simplest case, it is assumed that there is a parametric family $\mathcal{P} = \{p(\cdot|\theta)| \ \theta \in \Omega\}$, where $\Omega \subset \mathbf{R}^d$ for some integer $d \geq 0$, equipped with a prior on Ω having a density with respect to μ, say. Then, the predictive distribution for a random variable Y_{n+1} with outcomes y_{n+1} given $Y^n = (Y_1, \ldots, Y_n) = (y_1, \ldots, y_n) = y^n$ is

$$m(y_{n+1}|y^n) = \int p(y_{n+1}|\theta)w(\theta|y^n)\mu(\mathrm{d}\theta),$$

where $w(\theta|y^n)$ is the posterior density. It is easy to verify that $m(y_{n+1}|y^n)$ is optimal in a relative entropy sense (see Clarke *et al.* (2014)) and that when $Y_i \sim P_{\theta_0}$, with density p_0, is independent and identically distributed (IID) for all i that

$$m(y_{n+1}|y^n) \to p_{\theta_0} \tag{1.1}$$

pointwise in y_{n+1} in distribution, as $n \to \infty$. So, one could fix a distance d on densities for Y_{n+1} and choose a model based on $\theta^* = \theta^*(y^n)$ satisfying

$$p_{\theta^*}(\cdot) = \arg\min_\theta d(m(\cdot|y^n), p_\theta(\cdot)).$$

Not all predictors can be so obviously converted into models, just some of the good ones. Moreover, (1.1) only holds under highly restricted conditions.

Decision making can also be subsumed into prediction. Suppose that there is a prior, a parametric family, data, and a loss function and that the task is to seek a decision rule minimizing the Bayes risk. In these cases, the challenge is to verify that the decision rule gives a good performance; at root this depends on whether some element of the parametric family matches the DG. For instance, if the decision regards which parameter value to choose then there is a model that can be used for prediction. If the decision regards which stock to buy on a given day – i.e., an action – then the gain or loss afterward gives an assessment of how good the decision is; this evaluation is disjoint from the procedure that generated the decision or action in the first place. Another common decision problem is to decide which treatment is best for a given patient or for a given patient population. Again, one is selecting an action. One can make a decision based on data, but it is only when the predictions following from that decision are tested that one can be sure the decision was the best possibile. That is, the goal in decision making is, at root, to predict the action that will be most advantageous in some sense. Otherwise stated, the merit in a given decision is determined by how well it performs, and taking empirical performance into account (since that's what's important) makes decision making merely a way to choose a predictor. Therefore, essentially, a decision problem is a one-stage prediction problem, i.e., there is one prediction, not a sequence of predictions, so there is no chance to reformulate the predictor.

As before, in some cases decision making procedures can be turned into predictors. Indeed, the decision problem may be to find a good predictor. However, even when the decision problem is not directly about prediction, it has a predictive angle. For instance, consider the frequentist test of $\mathcal{H}_0 \colon F \neq G$ versus $\mathcal{H}_1 \colon F = G$ using IID data y^n from F and z^n from G. This is sometimes called an equivalence test. The two-sample Kolmogorov–Smirnovtest would be one of the natural test statistics if the hypotheses were reversed. To address the test, choose a distance d, write $\mathcal{H}_0$ as $\mathcal{H}_0 \colon d(F, G) > \delta$, and let $\{(F, G)|\, d(F, G) > \delta\} = \cup_k S_k$, where the S_k are sets of pairs of distribution functions, $k = 1, \ldots, K$, and the diameter of S_k is small in terms of d. Then, $\mathcal{H}_0$ vs. $\mathcal{H}_1$ is equivalent to the K tests $\mathcal{H}_{0,k} \colon S_k$ vs. $\mathcal{H}_1$. If the S_k are small enough, they can be approximated by their midpoints, say s_k. This gives the approximate simple-versus-simple when testing the problems $\mathcal{H}'_{0,k} \colon (F, G) = s_k$ vs. $\mathcal{H}_1 \colon F = G$. Now, in principle these tests can be done, using a multiple comparisons correction, and a single approximate model (or a small collection of approximate models) can be given from which to make predictions. Note that this is not modeling, and in fact nonparametric approaches can be used in a decision problem to generate predictions. (This approach will arise in Sec. 1.2.2.)

Answering a question about a *population*, or, more generally, understanding the structure of a data set, is a more nebulous goal. However, it may be regarded as trying to identify some feature of a population, or of an individual within a population, that is not obviously expressible in model identification terms. As an example, imagine trying to identify which dietary supplements the residents of a city buy or determining whether two random variables are associated. The predictive angle in these cases is one of confirmation. De facto, the prediction is that residents of the city use dietary supplements from a given list. So, if a resident of the city is chosen, does the resident use one of the identified supplements or not? If predictions are made assuming the independence of two random variables, are these predictions noticeably worse than including a dependence between them? Equally important, it is relatively rare that the final end point of an analysis is the description of a data set or the answering of a question about a population. Usually, one is doing this sort of task with a greater goal in mind, such as deciding whether to offer a new supplement for sale or classifying subjects into two classes on the basis of covariates.

Since this class of problems is less well defined, it is not obvious how to convert methods from it generically into a predictive interpretation beyond what has already been discussed. It is enough to be aware of the centrality of prediction among the various statistical goals subsumed under the term population description.

For the sake of completeness, recall that there are other statistical goals such as data presentation (graphics), data summarization, and the design of experiments. These too are generally in the service of some greater goal. Data presentation may be used to explain a statistical finding to non-statisticians, but these people generally have a reason why they want the analysis and a goal that they want fulfilled, which is usually predictive. Similarly, data summarization is rarely an end in itself but a subsidiary goal towards some other presumably greater goal. The design of experiments is done before data is collected, and its primary goal is to ensure that the data collected will suffice for the analytic goal – which, as has been argued, generally has a predictive perspective even if prediction is not recognized as the main explicit goal.

A third benefit of focusing on prediction is ensuring that inferences are testable and hence that any theories they represent are testable. Testability is not the same as interpretability, but a good predictor will typically permit some, perhaps limited, interpretation. For instance, given a predictor that uses explanatory variables one can often determine which of the explanatory variables are most important for good prediction. One would expect these to be the most important for modeling as well. More generally, apart from interpretability, theories for physical phenomena that arise from estimating a model and using hypothesis tests to simplify it must be validated predictively.

It is worth noting that, heuristically, there is almost an 'uncertainty principle' between interpretability and predictive accuracy: it's as if the more interpretability one wants, the more predictive performance one must sacrifice, and conversely. After all, the best predictors are often uninterpretable (e.g., those for the Tour and Fires data in Sec. 1.2.1, the Bacterial NGS data in Sec. 1.2.2, and the Music data in Sec. 1.2.3). Moreover, interpretable predictors (typically based on models) are almost always predictively suboptimal: it is a mathematical fact that, for instance, Bayes model averaging[1] (which is difficult to interpret) is better than

---

[1] Here and elsewhere Bayesian is abbreviated to Bayes for brevity when convenient.

using any one of the models in the average (which is usually easy to interpret), at least under squared error. Also, the adaptivity of predictors to data which has little interpretability often outperforms conventional model averages or model selection; see Wong and Clarke (2004), Clarke *et al.* (2013). Thus, interpretability does not lead to good prediction and good prediction does not require interpretability – although sometimes interpretations can be derived from predictors. Indeed, relevance vector machines (RVMs) are mathematically the best predictors in some settings (reproducing kernel Hilbert spaces), but statistically they overfit and can therefore be suboptimal because of excessive variance, meaning some terms have to be dropped for improved predictive error. This does not make RVMs more interpretable – if anything it makes them more complex and hence less interpretable – but it can make them excellent predictors.

Importantly, prediction in and of itself does not require an unseen world of abstract population quantities or measure spaces. Predictors such as 'someone with high coronary artery calcium is likely to benefit from statin treatment', paraphrased from Blaha *et al.* (2011), do not require anything we have not measured or cannot measure. Similarly, 'tomorrow's weather will be the same as today's' is a purely empirical statement. We may wish to invoke the mathematical rigor of measure theory to provide a theoretical evaluation of our prediction methods under various assumptions but this is a separate task from prediction per se. Indeed, in many cases the asymptotic properties of predictors, in terms of sample size or other indices, are of interest but cannot be obtained without making assumptions that bear scant relation to reality. For instance, formally a random variable is a deterministic function on an invisible and unspecified set. Is this a reasonable way to encapsulate the concept of randomness mathematically? The answer is probably no; it's just that a better one has yet to be proposed and accepted.

A fourth reason to focus on prediction is that predictive errors automatically include the effect of uncertainty due to the data and to all the choices used for prediction. That is, when a predictor $\hat{Y}$ of Y is wrong by $|\hat{y} - y|$, the error includes not just the bias and variability of any parameters that had to be estimated to form $\hat{Y}$ but also the bias and variability due to the predictor class (or model class if models are used) itself as well as the variability in the data. This is a blessing and a curse. One of the problems with prediction is that point predictors are more variable than point estimators, so prediction intervals (PIs) are typically wider than confidence or credibility intervals (CIs). Moreover, just like CIs, model-based PIs tend to enlarge when model uncertainty is taken into account. The consequence of this is that predictive inferences tend to be weaker than parametric or other inferences about model classes. It would be natural for investigators to prefer stronger statements – even if the justification for them rests heavily on ignoring model uncertainty. However, even though inferentially weaker, point predictors and PIs have the benefit of direct testability and accurate reflection of uncertainty, which point estimators and CIs usually lack.

One of the earliest explorations of model uncertainty was by Draper (1995), who compared two ways of accounting for model uncertainty in post-model selection inference that include prediction. Draper (1995) argued that model enlargement – basically adding an extra level to a Bayesian hierarchical model – is a better solution than trying to account for the variability of model selection from criteria such as the Akaike or Bayes information criteria in terms of the sample space. He also argued that it is better to tolerate larger prediction intervals than to model uncertainty incorrectly. (As a curious note, Draper (1997) found that there are cases where correctly accounting for modeling uncertainty actually reduces predictive uncertainty.) Of course, if PIs are too large to be useful then the arguments that a modeling approach is valid are more difficult to make, and any other inferences – estimates, hypothesis tests – may be called into question. However, to quote Draper (1995): 'Which is worse – widening the bands now or missing the truth later?'

There are *two criticisms* of the predictive approach that must be answered and dispensed with. First, a criticism of the predictive approach that is used to justify direct modeling approaches is that being able to predict well does not imply that the phenomenon in question is understood. The answer to this criticism is that modeling only implies understanding when the model has been extensively validated, i.e., found to be true, and this validation is primarily predictive. So, announcing a model before doing extensive validation – as is typically done – provides only the illusion of understanding. Prediction is a step toward model building, not the reverse, and predictive evaluation is therefore more honest. While the result of this kind of validation may be a predictor that is not interpretable, it is better than having an interpretable model with poor predictive performance. It may be that the traditional concept of modeling is too restrictive to be useful, especially in complex problems.

Second, it must be admitted that in practice the predictive approach is frequently harder than modeling. It's usually easier to find a not-implausible model (based on 'modeling assumptions' that boil down to the hope that they are not too far wrong), estimate a few parameters, verify that the fit is not too bad and then use the model to make statements about the population as a whole than it is to find a model that is not just plausible but actually close enough to being correct to give good predictions for new individual members of the population. Here, 'close enough' means that the errors from model misspecification or model uncertainty are small enough, compared with those from other sources of error, that they can be ignored. The problem, however, is that there are so many plausible models that finite data sets often cannot discriminate effectively amongst them. That is, as a generality, the plausibility of a model is insufficient for good prediction because one is quite likely to have found an incorrect model that the data have not been able to discredit yet. Since models that do not give sufficiently good prediction have to be disqualified, their suitability for other inferential goals must be justified by some argument other than goodness of fit. Thus, on the one hand, the task of finding a good predictor is usually harder than the task of finding a plausible model.

On the other hand, in reality a predictive approach is easier than implementing a true, accurate, modeling approach. Truly implementing a modeling approach requires that the model be correct or at least indistinguishable from correct. Given that the true model (when it exists) is rarely knowable this is an extremely difficult task. However, finding a serviceable predictor is easier, because it asks for less: giving good predictions is an easier task than uncovering a true model because bad predictions from a model invalidate the model while failure to provide good modeling inferences does not per se invalidate a good predictor. For example, if one predicts tomorrow's weather to be the same as today's weather this predictions may be reasonably accurate even though there is no underlying model from which to make inferences. Indeed, a good predictor may correspond to a dramatic simplification of a true model such that the prediction is good but the specific modeling inferences are poor.

Taking a predictive approach also requires another shift of perspective, namely that the data to be collected in the future are extremely important. This flies in the face of modeling which focuses on a specific data set and what it says about a hypothetical population rather than what it says about future outcomes. It also flies in the face of standard scientific practice, which underweights confirmatory studies. As a gedanken (thought) experiment, imagine how scientific practice, funding decisions, and scientific publishing would change if the confirmation of studies (by different experimental teams) were weighted as highly as initial findings. It's not that prediction is against rapid scientific advancement; rather, it's that prediction is a check that the advancement is real (not based on errors, luck, or malfeasance) so that scarce resources don't get squandered on spurious results.

Despite the considerations so far, which are fairly well known, the main approach taken by statisticians has been to look at *model classes* and use them to generate predictors in cases where prediction was an acknowledged goal. Here, however, the key point is that much of traditional statistics has been done precisely backward: instead of modeling, or more generally choosing a model, and then predicting, one should propose a predictor class and find a member that performs well. Then, if model identification is desirable for some reason, in principle one can convert the predictor to a model within a class of models that are believed to be plausible. For instance, in some settings Bayes model averages yield good predictors. One can form a single model from a Bayes-model-average predictor by looking at the most important terms in the models that go into the average. In the special case of averaging linear models, one can regard this as a way to find coefficients on variables using a criterion different from least squares. (The difference is that Bayes model averaging combines models after determining coefficients rather than determining coefficients for a combined model.) As another example, one can use a kernelized method such as an RVM, take a Taylor expansion of the kernel in each term of the RVM, and take the leading terms as a model. In this way one might obtain a model that is interpretable and gives good predictions. If the predictions are not quite as good as those from the original predictor, at least one can see the cost paid to obtain interpretability.

Forthrightly, the point of this book is that the *paradigm problem of statistics is prediction*, not estimation or other forms of inference, and problems should be formulated in such a way that they can be answered by producing a predictor and examining its properties. The tendency that analysts and methodologists have toward model formulation and therefrom to estimation, decision making, and so forth is misguided and leads to unreproducible results. This often happens because model uncertainty is usually the biggest source of error in analyses, especially with complex data. Predictor uncertainty may be just as big a problem, but it is visible in the predictive error while model uncertainty is very hard to represent accurately.

Conventional statistical modeling recognizes the problem of model uncertainty in a variety of ways. Most recently, model uncertainty has been recognized in the desire for sparse models that satisfy the oracle property (they can correctly select the nonzero coefficients with high probability; see Sec. 10.5). Clever as all this is, it is merely a way to find a model that is not implausible, i.e., cannot obviously be ruled out. Indeed, shrinkage methods generally perform better predictively when they shrink less, i.e., are less sparse and distort the data less. Moreover, as they shrink less, shrinkage methods tend to improve and become competitive with the better model-averaging methods; see Clarke and Severinski (2010). As a generality, shrinkage methods frequently are a source of model misspecification since sparse models are rarely true outside narrow contexts. Indeed, the desire for sparsity is a variant on the desire for models (and small variance), since models are a way to summarize one's purported physical understanding and sparsity is a pragmatic adaptation to settings where a purported physical understanding is an unusually remote possibility.

Indeed, many people say things like 'if the model is too complex to hold in my head there's no way I can work with it'. Leaving aside the dubious utility of modeling as a way to get sparsity (rather than just bias) and the even more dubious notion that reality should be simple enough to hold in one's head, one might consider sparsity as a desideratum for predictors. There is some merit in this – but only because good predictors have a good variance–bias tradeoff. Thus, sparsity may help reduce overall error if it reduces the variance enough to overcome any increase in bias but in fact the bias is likely to increase and, moreover, sparsity becomes less desirable as sample size increases.

However, one must recognize that sparsity is rare in reality, so there is negligible merit in seeking it on esthetic grounds or insisting as doctrine that a model or predictor be sparse. Otherwise stated, most problems in the real world are complex, and if a sparse predictor or model is not so far wrong as to be discredited it is likely that gathering more data will discredit it. Aside from being an argument against standard modeling (which is generally a severe oversimplification of a physical problem), this is an argument in favor of prediction because predictors usually include terms that are useful regardless of their meaning. In short, one wants sparsity as a way to control the variance part of a variance–bias analysis. Beyond this, sparsity is of little use to prediction since bias is bad for prediction and, in the case of a model, is an indicator that the model is wrong and hence inferences from it are called into question. One must accept that the real world has little inclination to conform itself to a short, or otherwise sparse, list of quantities that we can imagine and express conveniently, model-based or not. Seeking sparsity because it pleases us or gives us the illusion of 'understanding' has a cost in terms of bias and is all too often merely a pitfall to be avoided.

The net effect of all this is that the stance of this book is predictive. That is, the fundamental task is taken to be prediction, and problems should be formulated in such a way that a predictive approach can resolve them. This book explores a large number of ways to do this and tries to elucidate some of the main properties of predictor classes, including when to use each. This means that estimators, tests, or other statistical quantities are only used when they are helpful to the predictive enterprise. For instance, in Chapter 2, the problem of estimating a mean will be recast into that of identifying point predictors for future outcomes. Instead of being concerned with standard errors or posterior variances, the focus will be on identifying prediction intervals for future outcomes. Likewise, in Parts II and III, instead of considering classes of estimators or tests, classes of predictors will be considered. The role of parameters will be mostly to form a predictor or to quantify how well a predictor performs. An overall approach to statistics can – and should – be based on prediction, and the main reason why this book was written was to demonstrate this.

1.2 Some Examples

In this section four examples of data sets will be presented where a predictive approach is either different from (and better than) a modeling approach or for which prediction is the central issue. The first two examples (in Sec. 1.2.1) are low dimensional, and we see that even in such seemingly anodyne settings a simple predictive approach is usually better than a simple modeling approach. The third example shows how a question that presents as a decision problem (hypothesis testing) with complex data is actually better interpreted as a prediction problem. The fourth example shows how a classification problem with complex data may be better conceptualized as a prediction problem and how clustering techniques might be useful to formulate a predictor. These are widely divergent data types and statistical techniques. However, taken together they show that predictive approaches are ubiquitous and frequently superior to conventional model building.

1.2.1 Prediction with Ensembles rather than Models

Let us begin by looking at two low-dimensional data sets and contrasting a conventional model building approach (linear regression) with three predictive approaches (L2-BMA, RVMs, and bagged RVMs). Once the four predictors have been described, a computational comparison of their performances will be given. The predictive approaches are better in several senses.

The first data set is from the Tour de France. The Tour de France is an annual bicycle race consisting (currently) of 21 segments raced over 23 days; two days are set aside for the cyclists to rest. The race was started in 1903 but was not held from 1915–1918 or from 1940–1946 owing to the World Wars; thus up to 2012 there are 99 data points, one for each year. The exact route changes each year but the format stays the same. There are at least two time trials and segments of the course which run through the Pyrenees and the Alps, and the finish is on the Champs-Élysées. The length has ranged from 1484 miles in 1904 to 3570 miles in 1926. While many aspects of the race were recorded for each year the race was held, here it is enough to look only at the logarithm (base e) of the average speed (LSPEED) of the winner as a function of the year (YEAR) and the length (LENGTH) of the course. Here this will be called the Tour data. Through trial and error (or more formally Tukey's Ladder, see Tukey (1977)) one can find that using the logarithm of the average speed is a good choice because it leads to better fit in linear models. Also, the data for years 1919–1926 were removed as outliers because the LSPEED values for these years are unnaturally low; this reflects the huge number of young men who died during WWI who would have otherwise been potential competitors during those years. This leaves 91 data points.

Let's describe a standard model-based analysis. First, the scatterplots of LSPEED versus YEAR and LENGTH are given in Fig. 1.1. The plot on the left shows that LSPEED increases over time, possibly because of improved training, bicycle technology, and a larger field of competitors. There are no obvious outliers and the spread of LSPEED does not appear to change much with YEAR. The plot on the right shows that as LENGTH increases the winning speed tends to decrease, possibly because the cyclists are more tired by the end of the course. There are three outliers and it appears that the spread of the scatterplot increases slightly with LENGTH. There appears to be a small amount of curvature in LSPEED as a function of both YEAR and LENGTH.

Consider finding the usual least squares fit of the second-order model

$$\text{LSPEED} = \beta_0 + \beta_1\text{YEAR} + \beta_2\text{YEAR}^2 + \beta_3\text{LENGTH} + \beta_4\text{LENGTH}^2$$
$$+ \beta_5\text{YEAR} \times \text{LENGTH} + \epsilon, \tag{1.2}$$

assuming ϵ is a mean-zero random error. Higher-order terms such as those in LENGTH^3 or YEAR^3 are so highly correlated with LENGTH^2 and YEAR^2 that they make the regression

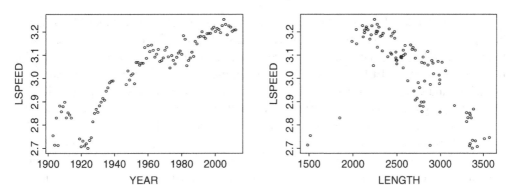

Figure 1.1 Left: Plot of LSPEED versus YEAR. Right: Plot of LSPEED versus LENGTH.

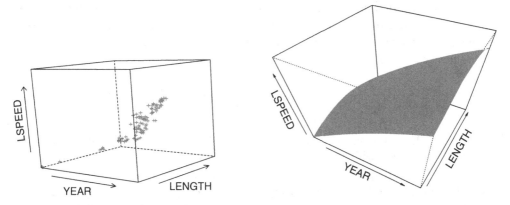

Figure 1.2 Left: Three-dimensional scatterplot of the Tour data. Right: Plot of the estimated regression function in (1.3).

worse. Moreover, a partial F-test to see whether YEAR × LENGTH is worth including, given the other included terms, shows that it can be dropped. The p-values for $\mathcal{H}_k : \beta_k = 0$ are below 0.05 except for β_2, which is 0.068. We retain the term in YEAR^2 since it is obvious that there is curvature in YEAR even if the square does not capture it well. Thus, the estimated regression function is

$$LSPEED = -41.4 + 0.04121 \, YEAR - 0.000\,009\,573 \times YEAR^2$$
$$+ 0.000\,4098 \times LENGTH - 0.000\,000\,082\,63 \times LENGTH^2, \quad (1.3)$$

and if desired one could obtain CIs, PIs, diagnostics, and so forth from standard linear model analysis. Indeed, an individual parameter estimate such as β_1 represents the change in LSPEED as a result of a unit change in YEAR and this means that parameter estimates can be converted readily into predictions.

The left-hand panel in Fig. 1.2 shows that the three-dimensional scatterplot is not very informative because the points bunch up too much for the shape to be seen in three dimensions. However, the right-hand panel shows the estimated regression function, in which the decrease of LSPEED with LENGTH and the increase in LSPEED with YEAR can be seen.

Note that a model of the form (1.2) cannot possibly be universally true; indeed, it can only be approximately valid on a range of values for LENGTH and YEAR. Indeed, if YEAR were to increase and $\beta_2 < 0$ then, beyond a certain year, LSPEED would be negative, meaning that in the limit the SPEED was zero. On the other hand, if $\beta_2 > 0$ then the speed would be increasing and sooner or later the cyclists would be traveling faster than the speed of light. If curvature is neglected, the same problem occurs with β_1. Analogous problems occur with the coefficients β_3 and β_4 on LENGTH – but it makes some sense to treat LENGTH as bounded. These criticisms, based on bias, are different from the other criticisms that one might make of the linear model. For example, the data are not independent from year to year since often the same competitors participate, and the data are not identical from year to year because the competitors and courses differ. However, one can argue that these latter criticisms represent minor deviations from the assumptions compared with the problems with bias.

This model-based analysis can be tested predictively by comparing its prediction error to the prediction error of a predictive analysis that does not even try to model LSPEED. This requires a predictive criterion to evaluate, and there are two obvious ones to consider. First is the cumulative predictive error (CPE). That is, for a predictor sequence where the elements in the sequence are formed by the same techniques we can look at

$$\text{CPE} = \sum_{i=6}^{91} \left(\hat{F}_{i-1}(Y_i, L_i) - \text{LSPEED}_i \right)^2. \tag{1.4}$$

Expression (1.4) is a sum of squared errors since the estimated linear model was found using least squares estimators with all the data. In (1.4), $\hat{F}_{i-1}$ is the predictor for LSPEED$_i$ formed from the first $i - 1$ data points. In particular, the point predictor for LSPEED$_i$ formed from fitting a linear model of the form (1.2) using the first $i - 1$ data points and evaluating $\hat{F}_{i-1}$ at $(Y_i, L_i) = (\text{YEAR}_i, \text{LENGTH}_i)$, can be used since these data points rely on data obtained before LSPEED$_i$ is measured. For the Tour data, it is natural to assume that they are ordered by YEAR, represented by $i = 1$ to 91. This means that (1.4) includes a burn-in of five data points, the minimum necessary to fit a five-term linear model. The result is that

$$\hat{F}_{90}(Y_{91}, L_{91}) \quad \text{and} \quad \hat{F}_{91}(Y_{92}, L_{92}) \tag{1.5}$$

are the predictions for the 91st and 92nd data points, but the prediction error can only be evaluated for the 91st data point.

When there is a natural order to the data as with the Tour data, expressions (1.4) and (1.5) make good sense. However, if one wants to minimize the effect of non-IID-ness, or use the same idea as the cumulative predictive error (CPE) on data that has no natural ordering, one might be led to permutations of the data. Indeed, owing to the variability of the predictive errors (the summands in (1.4)) it is natural to assess the cumulative prediction error by taking the average of expressions like (1.4) for several, say K, permutations of the data and making a prediction for the 91st data point by averaging the predictions of the K predictors found from the K permutations. That is, randomly draw permutations $\sigma_1, \ldots, \sigma_K$ of $\{1, \ldots, 91\}$ and find the average,

$$\text{CPE}(\sigma) = \frac{1}{K} \sum_{j=1}^{K} \sum_{i=6}^{91} \left(\hat{F}_{\sigma_j(i-1)}(Y_{\sigma_j(i)}, L_{\sigma_j(i)}) - \text{LSPEED}_{\sigma_j(i)} \right)^2, \tag{1.6}$$

where the five data points used for the burn-in are assumed to be $\sigma_j(1),\ldots,\sigma_j(5)$ for $j = 1,\ldots, K$. Note that different orderings of the data will give different values for the K summands; it might be informative to observe how terms in the model become more or less important, or even to find a standard error for the predictions, but that is not the point here. The predictions for the 91st and 92nd data points are

$$\frac{1}{K}\sum_{j=1}^{K}\hat{F}_{\sigma_j(90)}(Y_{91}, L_{91}) \quad \text{and} \quad \frac{1}{K}\sum_{j=1}^{K}\hat{F}_{\sigma_j(91)}(Y_{92}, L_{92}). \tag{1.7}$$

but again the predictive error can be evaluated only for the 91st data point.

Obviously, for updating the linear model, all the models used for the final prediction (at time 91) will be nearly the same (as given in (1.3)), since the coefficients are estimated by different sets of data points of size 90 and there are only 91 data points. So, the K predictions averaged in (1.7) may be similar even though along the way, e.g., for i between 26 and 90, the differences in the predictions may be larger. (For instance, this happens if a run of data points leads the estimate of a coefficient in one direction while a later run leads the estimate of the coefficient in a different direction, and then both runs may have larger than expected errors.) More generally, model reselection may be allowed at various stages of data accumulation, and this can give different models depending on the ordering of the data. This is an added level of complexity, to be discussed in Chapter 9.

While the cumulative predictive error is informative in the sense of how one arrives at a predictor, what is of more direct interest is the predictor found at the end of the process and evaluating how good it is. Thus, the final predictive error (FPE) is often a better criterion than the CPE. The FPE is the last predictive error that the data permits one to evaluate. Thus, the FPE is given by

$$\text{FPE} = \left(\hat{F}_{90}(Y_{91}, L_{91}) - \text{LSPEED}_{91} \right)^2 \tag{1.8}$$

or

$$\text{FPE}(\sigma) = \frac{1}{K}\sum_{j=1}^{K}\left(\hat{F}_{\sigma_j(90)}(Y_{\sigma_j}(91), L_{\sigma_j}(91)) - \text{LSPEED}_{\sigma_j(91)} \right)^2, \tag{1.9}$$

if there is, or is not, a natural order to the data, respectively. One benefit of using the FPE instead of the CPE is that one usually needn't correct for a burn-in. After all, correcting for burn-in can be difficult; when comparing two predictors, if one requires a much higher burn-in it will be likely to have more variability and hence a higher FPE unless the burn-in greatly improves the accuracy of predictions.

Next, consider a different predictor using the same explanatory variables as in (1.2) but based on the Bayes model average (BMA) for prediction rather than linear regression with least squares estimates. Formally, the BMA for prediction (under L^2 error) is

$$p(\text{LSPEED}_i | \mathcal{D}_{i-1}) = \sum_{k=1}^{15} p(\text{LSPEED}_i | \mathcal{D}_{i-1}, \mathcal{M}_k) w(\mathcal{M}_k | \mathcal{D}_{i-1}), \tag{1.10}$$

in which the $\mathcal{M}_k = \{p(\cdot|\beta(k)) \colon k = 1,\ldots, 15\}$ are the $2^4 - 1 = 15$ nontrivial submodels of (1.3), the vector of β-coefficients in the kth model being denoted by $\beta(k)$, $\mathcal{D}_{i-1}$ is the data set available for forming a predictor for the ith outcome, and $w(\cdot|\mathcal{D}_{i-1})$ is the posterior distribution across the models. To keep this simple, assume a uniform prior $w(\mathcal{M}_k)$ over the 15 models, independent $N(0,1)$ priors $\phi_{(0,1)}$ for the coordinates of $\beta(k)$, i.e., $w(\beta(k)|\mathcal{M}_k) \sim \prod_{u=1}^{\dim(\beta(k))} \phi_{(0,1)}$, and that the β-coefficients are estimated by their posterior means. Obviously, (1.10) can be adapted to larger models than (1.3).

Now, the marginal likelihood of LSPEED is

$$p(\text{LSPEED}|\mathcal{M}_k) = \int p(\text{LSPEED}|\beta(k)\mathcal{M}_k)p(\beta(k)|\mathcal{M}_k)d\beta(k),$$

and the posterior model probabilities are

$$p(\mathcal{M}_k|\text{LSPEED}) = \frac{p(\text{LSPEED}|\beta(k),\mathcal{M}_k)p(\beta(k)|\mathcal{M}_k)}{\sum_k p(\text{LSPEED}|\beta(k),\mathcal{M}_k)p(\beta(k)|\mathcal{M}_k)}.$$

For (1.10) we now have

$$p(\text{LSPEED}_i|\mathcal{D}_{i-1},\mathcal{M}_k) = \int p(\text{LSPEED}_i|\beta(k),\mathcal{M}_k)p(\beta(k)|\mathcal{M}_k,\mathcal{D}_{i-1})d\beta(k)$$

for fixed k, where

$$p(\beta(k)|\mathcal{M}_k,\mathcal{D}_{i-1}) = \frac{p(\beta(k)|\mathcal{M}_k)p(\mathcal{D}_{i-1}|\beta(k))}{\int p(\beta(k)|\mathcal{M}_k)p(\mathcal{D}_{i-1}|\beta(k))d\beta_k}.$$

The BMA leads to an average of point predictors on taking expectations – a procedure which is optimal in an L^2 error sense. So, applying $E(\cdot|\mathcal{D}_{i-1})$, we get the predictor L2-BMA

$$\widehat{\text{LSPEED}}_i = E(\text{LSPEED}_i|\mathcal{D}_{i-1})$$
$$= \sum_k E(\text{LSPEED}_i|\mathcal{M}_k,\mathcal{D}_{i-1})p(\mathcal{M}_k|\mathcal{D}_{i-1}). \qquad (1.11)$$

In practice, it is easiest to use the approximation

$$p(\mathcal{M}_k|\mathcal{D}_{i-1}) \approx \frac{\exp(-0.5BIC_k)}{\sum_{j\in\mathcal{M}}\exp(-0.5BIC_j)},$$

where BIC_k is the Bayes information criterion value of model k (so that $\text{Var}(\epsilon)$ is estimated by its usual empirical value) and to set

$$E(\text{LSPEED}_i|\mathcal{M}_k,\mathcal{D}_{i-1}) \approx f_k(x_i|\hat{\beta}(k)),$$

where $\hat{\beta}(k)$ is an estimator for the regression parameter $\beta(k)$ of model $\mathcal{M}_k$. Here, this is the posterior mean under the prior for the parameter.

Now, choosing K sets of $i-1$ data points using K randomly chosen permutations σ_j, $j = 1, \ldots, K$, and then using (1.11) on each set of $i-1$ data points we can form the $\hat{F}_{\sigma_j(i-1)}$, which can be used to give an FPE(σ) at stage $i = 91$ for L2-BMA, as in (1.9). Using permutations means that the natural ordering on the data is being ignored – as is reasonable since the IID assumption probably holds. Again, it is natural to use $i = 91$ since the FPE represents the predictive accuracy at the last (here 91st) data point. Note that this can be done with the same explanatory variables as in (1.3) and either method, linear models or L2-BMA, can be employed with any choice of explanatory variables that the sample size will permit.

If one really wanted a CPE, one would have to make the L2-BMA and linear models comparable by using the same burn-in with L2-BMA as for a five-term linear model, i.e., a burn-in of five data points, and take a CPE over the remaining 86. Henceforth, however, for simplicity we will focus on the FPE and neglect any burn-in for the predictors and results of this chapter.

Let us give one more technique for sequential prediction that we might use with the Tour data. There are two forms of the relevance vector machine technique that we will use here. One involves RVMs in their pure, optimal form and the other involves will be 'bagged' RVMs. Bagging – bootstrap aggregation – is a non-Bayes model-averaging technique commonly used to stabilize good but usually unstable prediction methods; see Breiman (1994, 1996b).

First, to define an RVM let $\mathcal{H}$ be a Hilbert space – basically a generalization of Euclidean space, as follows. Assume that the elements of the Hilbert space are real-valued functions on a domain $\mathcal{X}$. Then, suppose $\mathcal{H}$ has a 'reproducing kernel' k, namely, a real-valued function on $\mathcal{X} \times \mathcal{X}$ such that for any function $f \in \mathcal{H}$ we have $\langle k(\cdot, x), f(\cdot) \rangle = f(x)$, where $\langle \cdot, \cdot \rangle$ is the inner product of $\mathcal{H}$, i.e., k 'reproduces' f. We assume that k is a symmetric and positive semidefinite function on its domain, assumed compact.

Given an RKHS – a reproducing kernel Hilbert space – one can set up a minimization problem. Pick a function, say $L: (\mathcal{X} \times \mathbb{R}^2)^{i-1} \to \mathbb{R}$, and a convex penalty function, say $\Omega: [0, \infty) \to \mathbb{R}$. Let us minimize a functional J on $\mathcal{H}$ defined by

$$J(f) = L(x_1, y_1, f(x_1), \ldots, x_{i-1}, y_{i-1}, f(x_{i-1})) + \Omega(\|f\|_{\mathcal{H}}^2),$$

where $x_1, \ldots, x_{i-1} \in \mathcal{X}$, $y_1, \ldots, y_{i-1}, f(x_1), \ldots, f(x_{i-1}) \in \mathbb{R}$, and $\| \cdot \|_{\mathcal{H}}$ is the norm from the inner product on $\mathcal{H}$. The representer theorem states that there is an $f \in \mathcal{H}$ such that

$$f(x) = \arg \min_{f \in \mathcal{H}} F(f) = \sum_{j=1}^{i-1} \alpha_i k(x_j, x), \tag{1.12}$$

for some choice of $\alpha_1, \ldots, \alpha_{i-1} \in \mathbb{R}$ and, conversely, if Ω is increasing then each minimizer of $J(f)$ can be expressed in the form (1.12) . It is seen that (1.12) gives a predictor as a function of the input data, the kernel k, and $i - 1$ parameters. Loosely, an RVM is a Bayesian analysis of (1.12) treating the α_j as parameters. There are several ways to estimate the α_j. Tipping (2001) used a Bayesian approach, as did Chakraborty *et al.* (2012) and Chakraborty (2012), while Scholkopf and Smola (2002) discussed frequentist approaches.

If the α_j are estimated then (1.12) gives a third predictor, using RVMs analogous to (1.11) for L2-BMA and the predictor that one would get from finding (1.3) using $i - 1$ data points, i.e., linear models. Thus, one can use (1.8) to evaluate the final predictive error (FPE) or (1.9), by using K randomly chosen permutations σ_j, to form the $\hat{F}_{\sigma_j(i-1)}$. Setting $i = 91$ would give us the FPE for the Tour data. There is usually a tuning parameter controlling the width of the kernel function, and the value of this parameter typically has a greater effect than the shape of the kernel. This tuning parameter is usually estimated initially for the computing package being used.

Our fourth predictor is a bagged version of the estimated form of (1.12). The idea of bagging is to fix a stage i at which one wants to predict. Then, if $i < n$, choose a permutation of all the data available and select the first i data points after permutation. Next, use the first

$i - 1$ of these to predict the ith by bootstrapping. Specifically, choose $0.67(i - 1)$ of the $i - 1$ data points at random and use them to generate an estimated form of (1.12). Do this a number of times; in the computations below, 100 times gave a good performance. Then, take the predictions for the ith stage from all 100 (say) bootstrap samples and average them to get a prediction at the ith stage. To assess the FPE of this procedure one uses the analog of (1.9), because, even if there is a natural order to the data, it is disrupted by the bootstrapping. The resulting predictor is a bootstrapped RVM (BRVM). Although the RVM and BRVM are nonlinear and look quite different from linear models or L2-BMA with linear models, the predictors they give depend on which explanatory variables are included as well as which kernel function is chosen.

Let us now look at how these four predictors – linear models, L2-BMA (with linear models), RVMs, and BRVMs – actually perform for the Tour data. First, let's use the explanatory variables YEAR, YEAR^2, LENGTH, and LENGTH^2 for LSPEED. For the sake of clarity, note that it is the whole histogram formed from FPE values as σ_j varies that is really of interest. Despite this, often it is enough to look just at the mean FPE(σ) as in (1.9). However, to compare predictors, one should look at the variance of the terms in (1.9) as well, if not the whole distribution of the FPE. In the calculations below, the five-number summary of the K FPE values,

$$\left(\hat{F}_{\sigma_j(90)}(Y_{91}, L_{91}) - \text{LSPEED}_{\sigma_j(91)} \right)^2 \tag{1.13}$$

as σ_k varies, is given for the Tour data along with the mean, i.e., (1.9). This will show that the distribution of the FPE can be skewed even for sample sizes of 90 – but this obviously depends on the predictor and the random aspect of the DG, or, as it is more typically put, on the model being fitted and the true model including the error term.

Table 1.1 gives a comparison of results from the four methods described for the distribution of the FPE as σ varies for the Tour data using four nontrivial explanatory variables. The errors are found in L^1, but the results are qualitatively similar for $L^{1.5}$ and L^2. It is seen that the distribution of the FPEs, as approximated by the bootstrapping and reselection of σ, is somewhat skewed for all four predictors since the median is closer to Q1 than it is to the mean, especially for the linear predictor. Clearly, for this very simple class of predictors, the linear model gives the smallest (mean) FPE while BRVM gives the smallest median FPE. Arguably, the skewness makes the median more relevant. However, without an assessment of variability such as a standard error of the mean FPE or the interquartile range for the median FPE it is hard to decide whether the pure linear model or BRVM is the better predictor.

One reason why the linear predictor gives the smallest mean FPE, smaller in particular than the mean FPE for the L2-BMA, is that the linear model with four explanatory variables is a better predictor than any of its submodels. So, even though the posterior probability piles up on the full model, the L2-BMA puts nonzero mass on the other 14 models and thereby loses predictive power. Combining the models in the L2-BMA by adding the coefficients on their common terms means that one can regard the L2-BMA as a linear model in which the coefficients are found by a Bayesian criterion rather than using least squares estimators. So, the underperformance of the L2-BMA occurs because there is essentially no uncertainty as to which model is best (here the biggest) but the L2-BMA still puts mass on submodels.

Table 1.1 *Five-number summary plus mean for the K = 100 random choices of permutations σ for the* Tour *data. The FPEs were found using 100 bootstrap samples of proportion 0.67. For the RVM and BRVM results, the tuning parameter in the Gaussian kernel was 4. The bold numbers indicate the minimum in their column.*

	min	Q1	median	mean	Q3	max
linear	0.0017	0.1238	0.1849	**0.2287**	0.3225	0.9866
L2-BMA	0.0008	0.0931	0.2008	0.2420	0.3247	1.1750
RVM	0.0093	0.1002	0.1624	0.3949	0.3341	3.72
BRVM	0.0030	0.0719	**0.1517**	0.2571	0.2400	2.069

Table 1.2 *FPE results analogous to those in Table 1.1 but for seven rather than four nontrivial explanatory variables.*

	min	Q1	median	mean	Q3	max
linear	0.0058	0.1108	0.1795	0.2315	0.3007	1.006
L2-BMA	0.0081	0.1045	0.2001	**0.2263**	0.2993	1.0750
RVM	0.0030	0.0895	0.1601	0.3331	0.3545	3.0990
BRVM	0.0039	0.0545	**0.1557**	0.2402	0.2992	1.8950

The kernel methods do better in terms of the median FPE and worse in terms of the mean FPE than the two methods using (linear) models. This may be due to the fact that kernel methods are more flexible and can track location better, even when the median is a better assessment of location than the mean. Part of the reason why the RVM does poorly in the mean FPE may be high variance. After all, the RVM permits as many terms as there are data points (here, 90), before the prediction is made so one expects that if the variability is reduced there should be an improvement in prediction. Thus, as expected, bagging i.e., using BRVMs, improves RVM quite a lot, so that BRVM has the smallest median FPE and a smaller mean FPE than RVMs alone. So, in this case, one would want to create the full histogram of the individual FPE values (for each σ) and decide whether the skewness was small enough that one should use the linear model or large enough that one should use BRVMs. Even then, the relative sizes of the median FPEs or mean FPEs would have to be assessed, to decide whether the method with the smallest error really was the best once statistical variation is taken into account. Superficially, the skewness suggests that the median FPE is more appropriate, so the preferred predictor should be the BRVM.

For comparison with Table 1.1 it is worthwhile generating the analogous Table 1.2, using more explanatory variables. So, in addition to YEAR, $YEAR^2$, LENGTH, and $LENGTH^2$, we consider including YEAR × LENGTH, YEAR × $LENGTH^2$, and LENGTH × $YEAR^2$, all terms that would test out as unimportant under a partial F-test in a conventional linear models approach. Using seven explanatory variables means that the L2-BMA has $2^7 - 1 = 127$ terms and both arguments in the kernels, in the kernel methods, are seven dimensional.

Table 1.2 is different from Table 1.1 in several ways. First, the skewness is less: it is only seen for linear models and RVMs, where the median is closer to Q1 than to Q3. So, it is hard to decide whether the median or mean FPE is more appropriate. Second, under either form of FPE, the ensemble methods L2-BMA and BRVM have the lowest errors. Neither the linear model nor a single RVM does well. Third, in about half the cases the addition of the extra terms lowers the FPE, whether median or mean, in contrast with the entries in Table 1.1. So, some methods (chiefly the ensemble methods) are better able to make use of the extra information than the 'pure' methods. Otherwise, Table 1.2 is generally similar to Table 1.1: in the median FPE, the kernel methods do better than the linear-model-based methods and in the mean FPE the linear-model-based methods do better than the kernel-based methods. The computations using $L^{1.5}$ and L^2 were qualitatively similar. Furthermore, the BRVM does better than the RVM. Again, without an assessment of the variability of the errors it would be hard to conclude definitively that one method was performing better than another – even though the results here are suggestive.

As a way to show that the techniques and findings here are not atypical, the computations were redone on the Fires data set, available from http://archive.ics.uci.edu/ ml/datasets/Forest+Fires, first studied in Cortez and Morais (2007). The idea of the data set is to predict the burn area of a forest fire in terms of 12 explanatory variables. However, one variable (the rain) varied very little and so was dropped for the analysis here. Also, the spatial coordinates in the park were dropped as being deemed not useful in more general settings. Thus, the analysis predicts the logarithm of the area (LAREA) as a function of nine covariates. The linear model predictor is obvious: one writes LAREA as a sum of ten terms, one for each explanatory variable plus a constant. For parallelism, the L2-BMA should in principle have been formed from $2^9 - 1$ submodels but for convenience here it was formed from nine simple linear regression models, each having a single explanatory variable (plus a constant). The RVMs and BRVMs were based on nine explanatory variables, used a Gaussian kernel with sigma factor 15 (rather than 4). The sample size was 517 so, apart from permutations, the first 516 data points were used to predict the 517th.

The results are shown in Table 1.3. As for Table 1.2, it can be seen that the ensemble methods BRVM and L2-BMA give the lowest median and mean FPEs, respectively, and otherwise are competitive. The median FPEs are roughly at the midpoint between Q1 and Q3, suggesting that the mean FPE might be more appropriate. On the other hand the median FPE is very much smaller than the mean, suggesting there are high tail values that might be influential and so the median FPE may be more appropriate. It is easy to see that the single RVM predictor is very poor under either FPE criterion, and the linear model predictor, while competitive, is never best. So, one is led to either the BRVM or the L2-BMA. Since there are other explanatory variables that could be included, or a larger model list could have been used in the L2-BMA, this might have changed the conclusion. So, for the present, the conclusion must remain tentative given that an analysis of the statistical variability of the FPEs has not been done.

The implication from these three computations is clear: outside very simple cases, using predictions from relatively uninterpretable ensemble methods is preferable to using individual predictors whether based on interpretable models e.g., linear models, or uninterpretable individual predictors, e.g., RVMs. Indeed, the better the predictions, the less interpretable the predictors seem to be. Otherwise stated, unless there is overwhelming evidence that a model assumption is valid, modeling frequently leads to less predictive accuracy, thereby calling the modeling itself into question. So, predictors that do not rely heavily on modeling will be more suitable for good prediction and hence any inferences made from them have greater cogency than simple or interpretable predictors.

Table 1.3 *The FPE results for the* Fires *data parallel those in Table 1.1.*

	min	Q1	median	mean	Q3	max
linear	0.0078	0.1784	0.3463	0.5334	0.7019	3.4920
L2-BMA	0.0007	0.1586	0.3558	**0.4919**	0.5730	3.3970
RVM	0.0809	1.1750	3.5640	3.9680	5.7980	14.0900
BRVM	0.0105	0.1384	**0.2849**	0.5271	0.4100	4.714

1.2.2 Hypothesis Testing as Prediction

Although hypothesis-testing problems present themselves as decision making, in the sense of deciding which of two hypotheses to (de facto) accept, hypothesis testing, like any decision problem, is really just another technique for prediction and, like any predictive technique, its predictions must be validated on future data.

To see why this view of testing makes sense, recall that we are generally more interested in the data we will see in the future than in the data we already have. Formally, this is so because inferences are made about a population, not about a sample. So, our inferences should apply to future draws from the population. For example, the result of a hypothesis test about a parameter will tell us something about the value of the parameter. The conclusion about the parameter value is relevant to the population from which the particular sample was drawn and so should be relevant to any sample drawn from this population. Therefore the hypothesis test, and its conclusion, can be regarded as a statement about future data which such data may confirm or refute.

As a statement about the future, hypothesis testing can be regarded as a 'prediction-generating' mechanism. That is, any hypothesis test will lead to a conclusion about a parameter of interest and this conclusion represents a prediction about future samples. As a simple example, suppose that we collect data $X = (x_1, x_2, \ldots, x_n)$, where $x_i \sim N(\mu, 1), i = 1, \ldots, n$, and find that $\bar{X} = 0.4$. Consider the test

$$\mathcal{H}_0 \colon \mu \leq 0 \text{ vs. } \mathcal{H}_1 \colon \mu > 0; \tag{1.14}$$

we might decide, on the basis of the test statistic, to reject $\mathcal{H}_0$ and conclude that $\mu > 0$. Using this, we can form a prediction about the mean of the next sample or the value of the next observation from the population. Recall that a prediction interval is an interval associated with a random variable yet to be observed, with a specified probability that the random variable will lie within the interval. Under $\mathcal{H}_1$, a new individual observation would be predicted to lie in the interval $[z_\alpha, \infty)$, where z_α is the 100αth percentile of an $N(0, 1)$, with probability at least $1 - \alpha$, for some $\alpha \in (0, 1)$. Analogously, we would predict that the mean of a new sample of size n, $\bar{X}_{\text{new}}$, would be in $[z_\alpha/\sqrt{n}, \infty)$ with probability at least $1 - \alpha$. The same sort of reasoning applies to other frequentist parametric and nonparametric testing problems: the rejection – or acceptance – of a hypothesis limits the range of values that future outcomes may assume.

This type of thinking holds for Bayes testing as well. Assume that the same data as before is available and the task is to test the same hypotheses but from a Bayes perspective. Suppose there is a prior probability for each hypothesis, i.e., for $i = 0, 1$, there is a

$$w_i = P(\mathcal{H}_i) = P(\mu \in \mathcal{M}_i),$$

where $\mathcal{M}_0 = \{\mu: \mu \leq 0\}$ and $\mathcal{M}_1 = \{\mu: \mu > 0\}$. Given the prior probabilities and the likelihoods, denoted $P(x|\mathcal{H}_i)$, the posterior probability for each hypothesis can be found, namely,

$$p_i = P(\mathcal{H}_i|x) = P(\mu \in \mathcal{M}_i|x) = P(\mathcal{H}_i)P(x|\mathcal{H}_i) = w_i P(x|\mathcal{H}_i).$$

The hypotheses $\mathcal{H}_0$ and $\mathcal{H}_1$ can be compared using the Bayes factor, i.e., the ratio of the posterior odds and the prior odds; see Kass and Raftery (1995). The posterior odds in favor of $\mathcal{H}_0$ relative to $\mathcal{H}_1$ are

$$p_0/p_1 = P(\mathcal{H}_0|x)/P(\mathcal{H}_1|x),$$

and the prior odds are defined analogously.

   If the Bayes factor is sufficiently large, say > 3, it is common practice to decide in favor of $\mathcal{H}_0$ and, again, this would lead to a prediction about the next outcome or mean of the next sample. One could use prediction intervals as in the Frequentist prediction case, or, to be more consistent with the Bayes approach, one could use the predictive distribution to get a Bayes prediction interval. The latter would require not just probabilities for the hypotheses themselves but a prior distribution, say w, on $\mathcal{M}_0 \cup \mathcal{M}_1$. Then, if $\mathcal{H}_0$ were taken as true, the $(1 - \alpha)100\%$ Bayes prediction interval for the next outcome would be $[t_\alpha, \infty)$, where t_α is the lower 100αth percentile of

$$\int_{\mathcal{M}_0} p(x|\mu) \frac{w(\mu)p(x^n|\mu)}{\int_{\mathcal{M}_0} w(\mu)p(x^n|\mu)} d\mu.$$

Analogous expressions would hold if $\mathcal{H}_1$ were taken as true, if one wanted a Bayes prediction interval for a future sample mean $\bar{X}_{\text{new}}$, or other parametric or nonparametric hypotheses were tested. As in the frequentist testing case, the rejection – or acceptance – of a hypothesis limits the range of values that future outcomes may assume.

   As a more elaborate example consider the same ideas in the context of bacterial metagenomics. A DNA sample known to contain more than one strain of bacteria is called metagenomic, and a typical task is to detect which bacterial strains are present in the sample. So, the decision problem comes down to claiming the presence or absence of a given strain from a list of possible strains and hence represents a collection of hypothesis tests, one for each candidate strain. Accordingly, this will necessitate a multiple comparisons correction and will show how testing can lead to a collection of predictions.

   Suppose the DNA sample has been processed via next-generation sequencing (NGS), a complicated series of biochemical reactions that parallelizes the sequencing of DNA at the cost of generating many short DNA sequences rather than fewer longer sequences. Given the 'short reads' from a single physical sample, the task is to do the tests

$$\mathcal{H}_{0,j}: C_j \text{ is not in the population vs. } \mathcal{H}_{1,j}: C_j \text{ is in the population,} \qquad (1.15)$$

where the C_j for $j = 1, \ldots, J$ represent a collection of reference genomes for bacterial strains. That is, each C_j is a bacterial genome written as a long string of nucleotide bases. (Most bacteria have one or two circular chromosomes and a number of other smaller circular pieces of DNA called plasmids. Plasmids are the way in which bacteria exchange genetic information without sex.)

   Analyzing NGS data usually begins with alignment. Since each short read is a sequence of the nucleotides A, T, C, and G that may occur in zero, one, or more of the C_j, specialized software must be used to identify where on the C_j each short read matches (to within a certain tolerance because both the short reads and the reference genomes may be inaccurate). If a given read aligns then it is possible that the read came from the bacterial strain represented by C_j. There are some common regions across bacteria, so the alignment of some short reads may not discriminate among the bacteria. However, there are some short reads that may be unique to a specific C_j. Obviously, the efficacy of the alignment and subsequent hypothesis testing will depend on J, the coverage of the genomes, and the richness of the reference genome database across the bacterial taxa, as well as on the quality of the NGS data. Typically, J will be on the order of thousands while the number of reads will be in the hundreds of thousands to millions.

Given a sample of NGS data we can do the J tests (1.15). From a frequentist standpoint one compares the number of genomic reads which would be expected to align to a given reference C_j under $\mathcal{H}_{0,j}$ with the number of genomic reads from the sample that were observed as aligning to the same reference. So, let $r_k, k = 1, \ldots, K$, denote the sample genomic reads, i.e., finite sequences of A, T, C, and G, and let l_k be the length of r_k. Also, let X_j be the random variable representing the number of reads from C_j in the sample of size K.

Given $X_j = x_j$, the p-value $P(X_j > x_j | \mathcal{H}_{0,j})$ can be estimated by permutation testing. The basic idea behind a permutation test is to permute the data from which a test statistic is calculated, effectively creating many new data sets so that many values of the test statistic can be found. The values of the statistic are only valid under the null hypothesis because the permutation is over all the data assuming that the null is true. The collection of values of the test statistic is used to form a histogram which is an estimate of the sampling distribution of the test statistic under the null hypothesis. The p-value for the test is taken to be the area under the histogram to one side (for a one-sided test) of the actual value of the test statistic. Permutation tests can be very useful when the distribution of the test statistic under $\mathcal{H}_{0,j}$ is unknown; see Valdes *et al.* (2015).

The analog of these classical permutation tests for NGS data is to mutate the nucleotides in the sample reads $r_k, k = 1, \ldots, K$, at a fixed mutation rate q. Now, if a read r_k actually came from a specific C_j then it will be mutated, thereby simulating a sample read that might be found if $\mathcal{H}_{0,j}$ were true, i.e., if DNA from C_j were not in the sample. If this is done for all K reads then the result is a collection of K mutated reads, which can be aligned to each reference genome C_j. To quantify this, let $Y_j = Y_j(q)$ for $j = 1, \ldots, J$ be the number of mutated reads that align to C_j for a given q. Now, Y_j is an estimate of the number of reads that would be expected to align to C_j under $\mathcal{H}_{0,j}$. Note that if all the sample reads were mutated M times, the result would be M different values of Y_j, namely, $\{Y_{jm} : m = 1, \ldots, M\}$. By comparing the value of X_j with the values $\{Y_{jm}\}$ from mutated versions of the data, a p-value for testing $\mathcal{H}_{0,j}$ versus $\mathcal{H}_{1,j}$ can be found. This can be repeated for each $j \in \{1, \ldots, J\}$, giving a set of J raw p-values. After adjusting the p-values for multiple comparisons, e.g., using the Westfall–Young procedure (see Westfall and Young (1993)), since such a procedure is also based on permutations, they can be used to decide the presence of any C_j. As in Sec. 1.2.1, if this is done many times, a loss function can be chosen and a CPE or FPE found for any testing procedure.

Up to this point, this has just been a hypothesis-testing problem. However, it really is prediction: the set of adjusted p-values $P(X_j > x_j | \mathcal{H}_{0,j})$ for $j = 1, \ldots, J$ represents a prediction of which bacterial strains are present in any other sample from the same population. For instance, we may consider any adjusted p-value less than a given threshold as 'significant', reject the associated null hypothesis, and decide that the associated bacterial strain is present. Otherwise stated, this is a prediction that the strain in question will be present in a new sample from the same population. Likewise, any adjusted p-value larger than a given threshold may be considered 'not significant', so the associated null would not be rejected and the decision would be that the associated bacterial strain is not present. This is a prediction that the strain in question will not be present in a new sample from the same population. If there is a collection of samples from which to obtain a prediction, there are many ways to pool the data to get improved prediction. One way, see Vovk (2012), is to generate adjusted p-values for each sample, calculate the average p-value for each strain, and use the set of average adjusted p-values to generate predictions.

The tests (1.15) can also be done from a Bayesian perspective. Suppose the bacterial composition of the sample is modeled as an observation from a multinomial distribution, with $M + 1$ categories representing M bacterial strains and one additional category for all nonbacterial strains. Let θ_j be the proportion of the population in category j, so that $\theta_1 + \theta_2 + \cdots + \theta_{M+1} = 1$ and $\theta_j \geq 0$ for $j = 1, \ldots, M + 1$. Given a sample of K reads from the population with k_j reads aligning to category j, the likelihood function is

$$p(\mathcal{D}|\theta_1, \ldots, \theta_{M+1}) = \left(\frac{K}{k_1 k_2 \cdots k_{M+1}} \right) \theta_1^{n_1} \cdots \theta_{M+1}^{k_{M+1}},$$

where $\mathcal{D}$ is the data i.e., the set of reads. For convenience, consider a conjugate prior $w(\theta) \sim$ Dirichlet$(\alpha_1, \ldots, \alpha_{M+1})$ represented as

$$w(\theta) \propto \theta_1^{\alpha_1 - 1} \theta_2^{\alpha_2 - 1} \cdots \theta_{M+1}^{\alpha_{M+1} - 1},$$

where $\alpha_j > 0$ for $j = 1, \ldots, M + 1$; then

$$w(\theta_1, \ldots, \theta_{M+1}|\mathcal{D}) \sim \text{Dirichlet}(\alpha_1 + k_1 - 1, \alpha_2 + k_2 - 1, \ldots, \alpha_{M+1} + k_{M+1} - 1).$$

Treating the θ_j as random variables Θ_j gives the posterior expectation for $\Theta = (\theta_1, \ldots, \theta_{M+1})^T$ as

$$\hat{\theta}_j = E(\Theta_j|\mathcal{D}) = \frac{k_j + \alpha_j}{K + \sum_{j=1}^{M+1} \alpha_j},$$

which is a combination of two sources of information about the proportional composition of the sample, namely, the prior and the data; see Clarke *et al.* (2015).

In this formulation every category has a nonzero prior probability and a nonzero posterior probability. This may not accurately reflect the population if there is strong reason to believe that many bacterial strains are not present. After all, such a belief would lead to choosing a prior assigning zero mass to a category believed to be void. If the goal were solely estimation then the approach here could be modified by using either a shrinkage prior (Johndrow 2013) or a mixture prior (Morris *et al.* 2005). These analyses separate strains into abundant and scarce classes but are beyond our present scope.

If mass is assigned to the $\theta_j = 0$ by a continuous prior w such as the Dirichlet distribution then testing (1.15) can be reinterpreted (informally) as

$$\mathcal{H}_{0,j}: w(C_j|\mathcal{D}) < \epsilon \text{ vs. } \mathcal{H}_{1,j}: w(C_j|\mathcal{D}) \geq \epsilon, \tag{1.16}$$

where $\epsilon > 0$ is small enough that for all practical purposes the jth strain is not present. The associated Bayes factor for testing $\mathcal{H}_{0,j}$ versus $\mathcal{H}_{1,j}$ provides evidence about whether strain j is present in the population but would be subject to multiple comparisons problems. A common way around this is to choose one compound hypothesis, i.e., one event or set $\{\theta_1, \dots, \theta_{M+1}\}$ that corresponds to the absence of any strain C_j known to be harmful to humans using the tolerance ϵ. Then, a single test can be done to determine whether anything harmful is present in the population from which the sample was drawn.

Just as in the frequentist case, the conclusion from a Bayes test is a prediction of what is expected to be present (or absent) in the next sample from the same population, i.e., Bayes testing is a method for generating a prediction.

Note that the posterior expectation for the proportions of each bacterial strain, the $\hat{\theta}_j$, can also be regarded as a prediction for the composition of the next sample from the same population. Specifically, the $\hat{\theta}_j$ are the proportions of reads from the C_j that are expected to be in future samples. An analogous interpretation holds true for any estimation problem, parametric or nonparametric. Again, estimation and model selection can be regarded as techniques for prediction generation, as was done implicitly with linear models and L2-BMA in Sec. 1.2.1. Indeed, estimating the coefficients in an RVM has an analogous interpretation.

A further benefit of the Bayes approach is the sequential updating of the posterior. Given a posterior distribution, it can be used to generate a prediction for a future sample. Then, once the future sample is observed, it can be aggregated with the data already obtained to form a new posterior from which an updated prediction for another future sample can be obtained, and so on. Thus, again, if a loss function is chosen, the sequence of errors made by the use of this predictor sequence can be used to generate a CPE or FPE.

To complete the argument that decision problems such as hypothesis testing are really just a form of prediction, it's worth considering the standard decision-theoretic setting. Bayes testing can be shown to be the Bayes action in a formal decision-theoretic sense (under generalized zero–one loss). Moreover, even though frequentist testing emerges from the Neyman–Pearson lemma, which uses a putative non-decision-theoretic optimality criterion, standard frequentist testing can also be given a decision-theoretic interpretation. So, the reasoning already given for frequentist testing exemplifies the point that decision problems are really prediction.

Classical decision theory is based on Wald's theorem that preferences can be ordered by risk or on Savage's axioms, which can be used to show that preferences can be ordered by Bayes risk. In either case, the structure has four elements: a collection of actions $a \in \mathcal{A}$, a collection of states of Nature $\theta \in \Theta$, a distance-like function L that assigns a cost $L(a, \theta)$ to taking action a when θ is true, and a collection of distributions, either $p(\mathcal{D}|\theta)$ or $p(\theta|\mathcal{D})$, representing the current states of beliefs about Nature. Clearly, in this structure, an optimal action such as a Bayes action or maximum expected utility action is a best guess as to θ given $\mathcal{D}$. So, if θ is a parameter, the result is an estimator $\hat{\theta}$ and this identifies an element from the set of possible current states of beliefs about Nature from which predictions can be made. This four-element structure of decision theory can be adapted to prediction problems more directly by regarding θ as a future outcome, in which case a is a predictor. A lucid and comprehensive examination of the way in which decision theory, predictive model selection, and predictive model assessment are interrelated in a Bayesian context can be found in Vehtari and Ojanen (2012). In a predictive context, a loss or utility function is often called a scoring function because it is an assessment of how consistent a forecast probability is to its realized value. Gneiting (2011) provides an extensive treatment of scoring functions for point predictions; see also Parry *et al.* (2012) and Dawid *et al.* (2012).

A more general and operational description of the procedures presented here is briefly given in Sec. 8.6.3 from both the Bayesian and frequentist standpoints.

1.2.3 Predicting Classes

An example of a classification problem that may be seen as typical of a broad range of settings is provided by the analysis of data generated from musical scores. As a starting point, consider the foundational work of McKay (2004). In a wide-ranging study of the features that one might extract from pieces of music in order to identify the genre they represented, McKay identified 149 features that one could compute from musical scores. He broke them down into seven categories: dynamics (four features), instrumentation (20 features), melody (20 features), pitch (26 features), rhythm (35 features), texture (20 features), and chords (28 features). In fact, in his work, McKay only implemented 111 of these features to provide a formal categorization of all music.

Later, the music21 project built on McKay's work, developing a webtool-based on his feature selection. This webtool provides a set of computational tools to help musicologists answer quantitative questions about pieces of music on the basis of their scores. To date, music21 has implemented 70 of the features McKay identified: zero of the dynamics, 20 of the instrumentation, 19 of the melody, 26 of the pitch, 35 of the rhythm, 20 of the texture, and none of the chordal features. music21 calls these symbolic features (see http://web.mit.edu/music21/doc/html/moduleFeaturesJSymbolic.html) and adds 21 'native' features that, unlike the symbolic features, are unprocessed counts of various sorts directly obtained from the music scores (some details can be found at http://web.mit.edu/music21/doc/html/moduleFeaturesNative.html). music21 also provides a corpus of musical pieces that can be found at http://web.mit.edu/music21/doc/html/referenceCorpus.html#referencecorpus. It contains pieces of music from a variety of periods and styles. When applied to a given piece of music in the music21 corpus, the music21 software can be used along with specialized scripts to output a vector of comma-separated values of length 91. However, the values separated by commas are sometimes vectors in turn, so the real dimension of the vectors is 633, much higher than 91.

Given that this feature selection is intended to apply to all music, it's worth seeing how it performs on a binary classification problem such as determining whether a given piece of music was written by Mozart or Haydn. Mozart lived from 1756 to 1791 and Haydn lived from 1732 to 1809, so both are entirely within the period of Western music known as classical, 1730–1820. This means that, as different as Mozart's and Haydn's works are, it is reasonable to compare them and see that different classifiers give different performances.

Note that in this sort of problem a modeling approach is infeasible. It makes little sense to try to formulate a model that would accurately represent the creative process that Mozart or Haydn might have used to produce their masterpieces. Indeed, it is essentially impossible to characterize what makes a particular piece of music, or composer, great. There is just too much variety that has not been – and probably cannot be – explored. After all, if it were possible to model musical brilliance precisely it would be possible to write a 'great new music-generating algorithm' – hardly something that will be accomplished any time soon.[2] Indeed, leaving aside the perceived quality of music, even coming up with a reliable way to categorize the various genres of music reliably is by itself a formidable task.

For the present, let's use the 100 pieces of music composed by Mozart and 244 pieces composed by Haydn that are contained in the corpus. In this count, different movements from the same work are counted separately. This makes sense because, say, movement 1 from a piece by Haydn could have more in common with movement 1 from another piece by Haydn than it does with movement 2 from the same piece. Each of these 344 pieces of music was summarized by a set of features computed from the score. So, the data is neither independent nor identical and its degree of nonindependence and nonidenticality cannot realistically be assessed. The best that can be done is to test out various classification techniques and see how well they perform. In fact, this classification problem is almost a survey sampling model: the complete list of pieces by Haydn and Mozart is available. So, to make the prediction problem meaningful, half the pieces from each composer were randomly selected and put together as a training set. The other half were used as the predictive test set.

For the sake of illustration, there are three distinct ways to tackle this problem. First, one can adopt a model-based approach and use logistic regression, single trees, or support vector machines (SVMs) in an effort to model the data. Support vector machines do not actually provide an explicit model, but they are designed to give 'support vectors' which are interpreted as the boundary points between regions in a binary classification problem. Also, built into SVMs is a transformation specified by the kernel, thereby giving an implicit model. Second, one may proceed nonparametrically. In this case, the natural approach is to use k-nearest neighbors and clustering to assess whether the classes really are meaningful. Third, one can adopt a model-averaging standpoint. In this context two possible techniques are random forests and gradient boosting (the statistical formulation of Adaboost). These are purely predictive approaches since the resulting classifiers do not generally say anything directly about the DG. Since the DG in this case is in some important sense unknowable, one expects that classifiers that do not rely on finding a model for the DG ought to do better than those that do. In fact, this is more or less what is seen.

To get the data ready for analysis two tasks were performed. First, explanatory variables that did not vary enough to provide information on the response were removed, specifically, explanatory variables with a sample variance less than 0.2. This procedure removed 602 of the 633 original real variables. Second, explanatory variables that were too correlated with each other were removed. Specifically, the correlation between each pair of explanatory variables was found and if a pair had absolute correlation strictly greater than 0.9 one of the pair was removed at random and the correlations recalculated; this procedure ended up removing only six more variables, leaving a nonunique set of 25 variables.

Once this was done, the training set was found in two stages. First, since the 100 pieces by Mozart represented 29% of the total number of pieces, 344, sampling at random from the pieces by Mozart in proportion to their prevalence gave $0.5 \times 0.29 \times 344 = 49$ pieces.

---

[2] In fact, there are programs that can generate more music in the style of a given composer by mimicking some of its features. However, this is not creation *de novo* of a new sort of great music. It is more like a combination of existing great music.

Likewise, sampling at random from the pieces by Haydn gave $0.5 \times 0.71 \times 344 = 122$. Taken together this gave a total sample size of 171 for training and left 173 pieces for testing how well the classifiers performed. Second, the variance and correlation constraints that were imposed on the whole data set were imposed on the sample of size 171. This removed one more variable, which was also removed from the test set. Thus there were 24 explanatory variables, 171 samples, and each sample had a response of zero or one to indicate composition by Mozart or Haydn, respectively. Henceforth this will be called the music data and treated as a classification problem.

The results of the three model based classifiers – logistic regression, single tree, and SVM – on the test set from the music data are summarized, respectively, in the following three confusion matrices:

$$\begin{pmatrix} 24 & 25 \\ 27 & 97 \end{pmatrix}, \quad \begin{pmatrix} 19 & 13 \\ 32 & 109 \end{pmatrix}, \quad \begin{pmatrix} 8 & 0 \\ 43 & 122 \end{pmatrix}. \tag{1.17}$$

The true composer is indicated by the column (left is Mozart, right is Haydn). The predicted composer is indicated by the row (top is Mozart, bottom is Haydn). Thus, in the first matrix, logistic regression used the 24 variables to identify 24 Mozart pieces correctly and 97 Haydn pieces correctly. It misclassified 52 pieces; the upper right entry means it incorrectly identified Mozart as Haydn 25 times and the lower left entry means it incorrectly identified Haydn as Mozart 27 times. Thus, the total number of errors was $25 + 27 = 52$.

It's worth looking at the key diagnostics for this logistic regression. First, Table 1.4 shows the variables that were included, along with their estimates, SEs, and uncorrected p-values. The significant p-values are low enough that a multiple testing procedure would not make them insignificant. Note that the variables are only labeled generically. This is so because the documentation from music21 only defines the variables by their order via python code so it is hard to be sure what each one means. However, this is not essential to the analysis at this stage. It is enough to note that, aside from the intercept, only five variables, V131, V145, V425, V426, and V445, appear to be useful.

A single tree gave the middle confusion matrix in (1.17). Trees are a very rich class of models – much richer than logistic regression and consequently less stable. The contributed R package rpart was used to generate single tree. It is shown in Fig. 1.3. For this tree, the number of errors was $13 + 32 = 45$.

It is seen that the classification tree and logistic regression 'agree' that V445, V145, and V131 are important.

The standard way to assess variable importance for trees is via a permutation technique. The result is shown in Table 1.5. It is seen that a fourth variable, V425, found important by logistic regression is picked up as being important by this technique too; however, V426 is not. This sort of discrepancy is not surprising since trees are unstable: they form a large class of models that provide an overall model by fitting a series of local models using less and less data per node as the tree grows; see Breiman *et al.* (1984).

An SVM gave the right-most confusion matrix in (1.17). Support vector machines are the least model dependent of the three methods. Underlying an SVM there is a model, but it is defined by the use of a kernel which represents a transformation of the feature space and is usually only defined implicitly. Using the Gaussian kernel, the contributed R package kernlab gave an SVM classifier that made 43 errors on the test set. The SVM also has 121

Table 1.4 *This table shows the 24 variables (plus intercept term), with
their coefficients, SEs, and two-sided p-values from a logistic
regression for classifying the pieces by Mozart and Haydn in the test
set of size* 173 *from the* music *data. Bold type indicates uncorrected
p-values for significant variables.*

Variable	Estimate	Std error	*p*-value
Intercept	−50.861	15.45	**0.001**
V130	0.114	0.61	0.851
V131	−0.758	0.34	**0.026**
V135	−0.208	0.30	0.496
V145	−0.040	0.01	**0.002**
V146	−0.050	0.11	0.641
V415	−0.124	0.07	0.092
V417	0.075	0.11	0.486
V424	0.017	0.01	0.185
V425	−1.142	0.40	**0.004**
V426	0.895	0.31	**0.004**
V428	−0.285	0.67	0.669
V431	0.328	0.43	0.443
V438	−0.035	0.06	0.572
V439	−0.062	0.22	0.780
V440	0.315	0.32	0.331
V441	0.202	0.11	0.075
V442	−1.019	0.63	0.107
V443	0.031	0.11	0.783
V445	0.892	0.23	**9.86e-05**
V449	−0.105	0.08	0.193
V605	−0.130	0.08	0.107
V609	−0.004	0.01	0.732
V610	0.127	0.07	0.064
V632	−3.216	219.37	0.988

Table 1.5 *This table shows the most important variables in the* music *data as determined by the
variable importance assessment (based on permuting values) from* rpart. *Bold indicates variables
with significant p-values from logistic regression in Table 1.4.*

V445	V443	**V145**	V146	V610	V441	V417
17	11	11	8	8	8	8
V449	V609	V605	**V131**	V130	V439	V425
7	6	6	4	3	1	1

support vectors – meaning about one-third of the data is on the boundary between the two
classes. This would be considered high. Even worse, SVM did best by predicting, approximately, that all the pieces were written by Haydn. Roughly, as the model class gets richer,
i.e., the model if it exists is harder to identify, the classifier does better at prediction. Note
that even the best of these classifiers, the SVM, indicates a breakdown. (A valid criticism of
this approach is that the results used the built-in estimates of tuning parameters in the tree

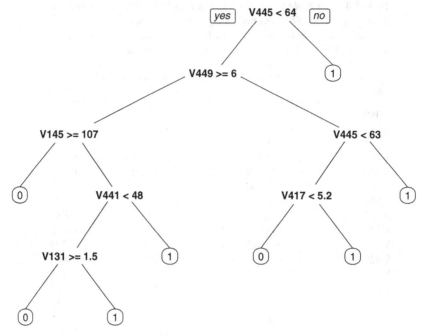

Figure 1.3 Single classification tree generated by rpart for the Mozart vs. Haydn problem and the music data. It only uses V445, V449, V417, V145, and V131.

and SVM; a more careful selection of the tuning parameters might have given better values. However, this is true of all the methods presented here. They are, in some average sense, equally disadvantaged.)

It's worth seeing whether a nonparametric approach will give better results. Using the R package RWEKA, confusion matrices for a k-nearest-neighbors classifier were generated. For the values $k = 1, 2, 3$ the confusion matrices for the music data are, respectively:

$$\begin{pmatrix} 24 & 21 \\ 27 & 101 \end{pmatrix}, \quad \begin{pmatrix} 24 & 21 \\ 27 & 101 \end{pmatrix}, \quad \begin{pmatrix} 15 & 11 \\ 36 & 111 \end{pmatrix}. \quad (1.18)$$

The first and second nearest neighbors give the same result, and the total error is 48. Including a third nearest neighbor doesn't help much – the total error is still 47. Nearest neighbor methods are known to be quite stable, so this small difference is no surprise. Overall, this method does better than logistic regression, the most restrictive model, which made 52 errors, but worse than the less restrictive tree or SVM models, which made 45 and 43 errors, respectively. It may be that the nonparametric model class is to a big for the data to overcome the model uncertainty – not to suggest that the notion of a model makes any sense in this setting. More precisely, the variability in the selection of the model from which to obtain a predictor may be hard for the data to overcome.

It is reasonable to be concerned that part of the problem with classification arises because the points in the training and test sets do not separate well. That is, the clusters of Mozart points and Haydn points in the 24-dimensional feature space are not well defined. To test this, drop the zero–one response and consider clustering on the vectors of explanatory variables. This can be done using the contributed R package pamk. For each number of clusters

Table 1.6 *The top row of this table shows the number of clusters in the* music *data found using* pamk. *The lower row shows the average silhouette distance for the clustering.*

2	3	4	5	6	7	8	9	10
0.414	0.320	0.400	0.391	0.375	0.341	0.343	0.348	0.279

k it will give a partitioning-around-medoids clustering that it can evaluate by the average silhouette distance. The results are given in Table 1.6. It is seen that $k = 2$ clusters is optimal (as 2 has the largest silhouette distance), but only weakly so. Four or even five clusters would be nearly as good. This means that the clusters are only weakly defined. Part of the problem may be that centroid-based clustering methods and even silhouette distances are better suited to convex clusterings than to more general clusterings and that if there are clusters in the 24-dimensional data then it is likely they are not convex. Nevertheless, Table 1.6 indicates that the problem is difficult. The concerns about weakly defined clusters would apply to logistic regression but less so to trees, since trees provide a rich class of models, and even less so to SVMs since they have a kernel transformation.

Turning at last to two techniques that are not at all model based, the confusion matrices for the random forests technique randomForest and the gradient boosting technique gbm applied to the music data are, respectively,

$$\begin{pmatrix} 17 & 6 \\ 34 & 110 \end{pmatrix}, \quad \begin{pmatrix} 21 & 11 \\ 30 & 11 \end{pmatrix}. \tag{1.19}$$

It is seen that the errors are 40 and 41, the smallest of the methods. The random forests technique is based on aggregating single classification trees like that found earlier. The aggregation is done via bootstrapping on the training set and *boot*strap *agg*regation is often abbreviated as 'bagging'. Thus, random forests is just bagged trees. Basically, random forests inherits all the success of individual trees but refines them by the averaging process, thereby improving the performance of the final classifier. By contrast, gradient boosting is a model-averaging technique that is a refinement of the original boosting procedure, which was called Adaboost (Friedman *et al.* 2000). The idea behind boosting is that one averages successive classifiers that are derived by putting more and more weight on the points where misclassification occurred. The refinement was to recognize that boosting could be re-expressed as fitting an additive logistic regression model by minimizing a functional; see Friedman *et al.* (2000).

It is important to note that in this problem, for which a model is unlikely to exist, the classifiers that do best are those that do not rely on model-type assumptions. Indeed, the performance of the methods roughly corresponds inversely to the strength of the modeling assumptions. For logistic regression, the modeling assumptions are very strong and it performs worst. The nearest neighbors method implicitly makes the weakest modeling assumptions and it's second worst. The trees method makes weaker modeling assumptions than logistic regression but not as weak as nearest neighbors, so its performance is intermediate between logistic and nearest neighbors. However, as the model becomes less important, as with SVMs, performance improves and, as modeling is abandoned completely in favor of relying on the data more and more, performance is improved further, as seen with random forests (which does best) and boosting (a very close second).

1.3 General Issues

The main point of the examples given in this chapter is that modeling is not as helpful as it is assumed to be, because its predictive performance can be improved rather handily by finding predictors not based on modeling. This was seen clearly in the Tour, Fires, and music data. If a worse predictor rests on modeling assumptions then, at a minimum, one is led to question the validity of those modeling assumptions. Indeed, in many realistic cases (i) models will be too complicated to formulate or (ii) simply do not exist. As noted, formulating a model for the Tour or Fires data is conceivable but practically impossible except in the most approximate sense. Also, it is unlikely that a model for the NGS data or for the music data exists at all. Even where one can imagine (with some effort) that a true model might exist, one should refer to data as coming from a data generator (DG) or source rather than a model since at best one will only be able to approximate the model, and the quality of that approximation will be hard to determine.

The better strategy for statistical analysis is to seek a predictor, not a model. The best way to find a good predictor is to look for one directly: choose a class of predictors that on the basis of experience and theory seems likely to perform well and find the best predictor in it. The optimal predictor can then be used to make predictions and to extract modeling information. For instance, one might be able to find a model that generates a predictor which approximates the optimal predictor in some sense. Or, one might be able to derive only some aspects of modeling directly from the predictor, such as identifying which variables are important even if the functional form by which they influence a response cannot be determined.

It is true that there are special, narrow, cases where modeling can be done quite accurately and the modeling gives predictions that are competitive with the best predictors. However, these cases are relatively rare in practice. When they occur, however, seeking the best predictor and seeking the best model may be equivalent, in which case it may be easier to find a predictor by obtaining it from a model. However, even though these cases exist, they are sufficiently rare that it is imprudent to build a statistical theory around the concept of a true model; better to build a statistical theory around quantities that make physical statements which can be compared with physical outcomes.

Another special case that bears mentioning is the setting where there is a true model that is extremely complex but can be well approximated in a verifiable way. This is usually the hope of subject matter specialists who seek statistical analysis. However, as much as this is a hope, it cannot be said to be realized in the absence of validation, which all too often is not even attempted. That is, it is not enough that the model 'fits' the data nor is it enough that inferences from the model be intuitively plausible. Many models will fit a given data set equally well, and if the inferences from a model are intuitively plausible that may mean only that the model has successfully encapsulated the intuition that went into it, i.e., the argument is effectively circular. If one is in the fortunate case that (i) a simple model 'fits', (ii) it gives predictions comparable with the best predictor, (iii) any inferences that one can derive from good predictors are consistent with the model, and (iv) other substantially different models that might fit well can be discarded for other reasons, then indeed one has had success. However, this is rare and mostly occurs in settings where there is so little variability that statistical analysis is not relevant.

As a generality, one approach that seems ill advised in many if not most problems is to seek a good model directly. This is so because of the likely deterioration in predictive performance resulting from using a model-based predictor rather than seeking a predictor from a more general class. Otherwise stated, model validation is likely to fail precisely because a good model was found instead of a good predictor. Moreover, to find a good model one must confront model uncertainty (and likely misspecification) or model nonexistence. The issue is therefore whether model uncertainty or the inability to specify a modeling approach is so severe as to preclude modeling as a useful approach to analysis. It may often be the case that the best modeling will result from finding a good predictor and trying to derive inferences about the DG from it.

The following chapters elaborate a predictive paradigm focusing on the properties of point prediction. The centerpiece is the goal of generating predictions on the same scale as the outcomes to be predicted. Decision-theoretic techniques for doing this have been proposed; however, this is not the same as constructing a theory of statistics centered on prediction. Comparing predictions with outcomes is different from asking whether an outcome materialized in a prediction interval, although the two are related: one can choose a scoring function (much the same as a loss function) that gives a cost of zero when the outcome is in the prediction interval (PI) and a cost of one when it isn't. So, at least in a limited sense, the use of PIs and scoring functions is conceptually included in the framework developed in the coming chapters. Likewise, probability forecasting, i.e., assigning probabilities to possible outcomes, is closely related to the approach taken here even though probabilities and outcomes are not on the same scale. In these settings one assesses the degree to which an outcome is representative of the probability of an event. Again, this is accommodated, at least in a limited sense, within the framework to be developed here.

2

Defining a Predictive Paradigm

> A philosophical and theoretical framework of a scientific school or discipline within which theories, laws, and generalizations and the experiments performed in support of them are formulated.
>
> Merriam–Webster dictionary definition of a 'paradigm'

Up to this point, four real settings in which to think about data predictively have been given. Many more are possible. As a pragmatic point, predictive approaches tend to be appropriate for settings that involve model uncertainty and complexity. Indeed, one motivation for taking a predictive approach is to back off from difficult, perhaps impossible, modeling problems since useful prediction should be easier than convincing modeling and hence more likely to be successful.

However, it is also worthwhile to abstract some key features of a predictive approach in an effort to understand what a predictive approach really means. These key features can then be compared with the key features of three other approaches – Bayesian, frequentist, and survey sampling. This will be done first in a toy problem. Then, it will be time to turn to some formalities and look at prediction in the simplest continuous cases, i.e., those with commonly occurring parametric families such as the normal distribution. This will highlight some philosophical differences among the approaches. Then, it will make sense to list and justify a set of criteria that a good predictive approach should include.

2.1 The Sunrise Problem

The Sunrise Problem is variously attributed to Thomas Bayes and Richard Price (Bayes and Price 1763) or to Pierre-Simon Laplace (Laplace 1774), amongst others, who framed it in terms of drawing lottery tickets at random, some winning prizes and some not. If there are infinitely many tickets and one has observed a given number of winners and losers it is reasonable to ask: what is the probability that the next ticket will be a winner? This question is statistically equivalent to asking what is the probability that the sun will rise tomorrow given that it has risen for a given number of days already.

In its most general form the Sunrise Problem amounts to how to evaluate the plausibility of the statement 'the sun will rise tomorrow'. The most general form leads to considerations of how notions of plausibility lead to probability in the first place from, for instance, propositional logic; see Jaynes (1995) for a good exposition. Of course, one could turn to astronomy and argue that the sun does not 'rise', the earth rotates, and one could argue that if the sun did not 'rise' then the concept of 'tomorrow' doesn't make sense. However, the point here is to use no information but the number of times a sunrise has been observed to assess probabilistically whether one should expect another sunrise. Implicitly, the approaches assume that the more sunrises that have been observed, the more one is likely to observe another. This contradicts the common sense view that things wear out: it would be preposterous, for example, to say that the more birthdays one has had, the more likely one is to have another. However, the point is to explore ideas behind prediction not to predict sunrises.

There are three main schools of thought about variability in contemporary statistics: Bayesian, frequentist, and survey sampling. So, let us see how generic versions of these three schools would conceptualize the Sunrise Problem and then contrast them with the predictive approach.

The *frequentist* conceptualization of the Sunrise Problem is based on defining the probability of an event to be the long-run relative frequency of its occurrence over many trials. So, the Frequentist is led to a 'one day, many suns' assumption. Thus, valid or not, the Frequentist imagines that there are many suns with solar systems identical to that of Earth with no interactions among them. The Frequentist chooses a sample of n of these suns and then waits 24 hours. (This is possible because since 1967 time has been measured in terms of the radiation emitted by caesium-133 atoms.) After 24 hours, the Frequentist counts the number of solar systems in which the third planet from the sun observed a sunrise. Suppose this number is k. The Frequentist would note that the 'one day, many suns' interpretation means the assumption that the binomial distribution $\text{Bin}(n, p)$ is formed from n Bernoulli(p) trials is justified and therefore gives $\hat{p} = k/n$ as a point estimate for p, the probability of a sunrise. If n were large enough, the Frequentist would then use the sampling distribution of the estimator k/n, invoke a central limit theorem for it, and announce an approximate $(1 - \alpha)100\%$ confidence interval for p, of the form

$$\frac{k}{n} \pm z_{1-\alpha/2} \sqrt{\frac{k}{n}\left(1 - \frac{k}{n}\right) \Big/ n}, \tag{2.1}$$

truncating at one or zero if necessary. The interpretation of the confidence interval would come from the repeated sampling interpretation made possible by the many suns. Thus, in terms of prediction, the Frequentist would merely issue a probability for a randomly chosen sun to rise over its third planet, i.e., a prediction problem would be answered with a probability estimate. Alternatively, the Frequentist could ignore any variability in $\hat{p}$ and approximate the random variable for whether its sun will rise over a randomly chosen third planet by a Bernoulli($\hat{p}$) random variable. This is called a *plug-in predictor* because one plugs the parameter estimate into the distribution. A more careful Frequentist might try to use the interval in (2.1) to adjust the Bernoulli($\hat{p}$) to account for the randomness in $\hat{p}$ in the plug-in predictor. In either case, the Frequentist does not have a standard way to generate an estimated distribution for the random variable representing the occurrence of the next sunrise from which to make a point prediction. Moreover, unless $k \approx 0$ or $k \approx n$, the prediction intervals obtained from a plug-in distribution will typically include zero or one, i.e., they will just be probabilistic.

By contrast, the *Bayesian* conceptualization of the Sunrise Problem is based on updating pre-experimental views encapsulated in a prior, i.e., Bayes' rule. So, the Bayesian requires a pre-experimental view of the probability of a sunrise and an updating mechanism for it. This approach can be justified by Savage's decision-theoretic axioms (Savage 1954) or by the Bernardo–Smith axiomatization (Bernardo and Smith 2000). However one justifies the approach, the Bayesian notion of probability is more abstract than the frequentist: to a Bayesian, a probability is essentially and merely a representation of a state of knowledge that is not necessarily tied to actual measurement. Conveniently, the frequentist notion of probability and the Bayesian notion of probability often coincide in reality. That is, in many problems but by no means all, the conclusions from a Bayesian analysis and from a frequentist analysis are compatible even though the reasoning is different.

In the present case, the Bayesian would adopt a 'one sun, many days' scenario. Thus, valid or not, the Bayesian treats the one sun as having given a number k of sunrises over n days, assuming there are no interactions between any of the days and that all the days are identical. So if, like Laplace, one takes the holy books of the Christians faith literally, it can be calculated that the earth is about 6000 years old, i.e., $n = 6000 \times 365$ days. Although no individual or group of individuals has recorded the occurrence of sunrises on all those days, for the sake of argument assume that $k = 6000 \times 365$ sunrises have been observed.

The Bayesian would note that the 'one sun, many days' interpretation means the assumptions for the binomial distribution $\text{Bin}(n, p)$ formed from n Bernoulli(p) trials are justified and would therefore use the binomial to update the pre-experimental views. Given p, the probability that the sun will rise tomorrow is also Bernoulli(p). However, these distributional assignments are not justified by repeated sampling, since no repeated sampling is involved; they are justified by arguing that the binomial probabilities accurately represent the available knowledge about sunrises.

It remains to identify the pre-experimental state of knowledge, the 'prior', say w, probabilistically so that the Bayesian updating rule can be applied. In the interest of objectivity, assign a Uniform[0, 1] density to p. That is, set

$$w(p) = \begin{cases} 1 & \text{if } p \in [0, 1], \\ 0 & \text{otherwise.} \end{cases} \tag{2.2}$$

This means that p is a random variable in its own right, not just a fixed but unknown number as in frequentist analysis. In fact, it is the data – the k and the n values – that are treated as fixed, known, and nonstochastic.

Now, given the distributional assignments, the past data, and the prior, Bayes rule gives that the posterior density for p is

$$w(p | X^{2\,190\,000} = 1^{2\,190\,000}) = \frac{u(p)\text{Binomial}(2\,190\,000, p)}{\int u(p)\text{Binomial}(2\,190\,000, p)\mathrm{d}p}. \tag{2.3}$$

In this expression 'Binomial' represents the binomial density with respect to counting measure and $\mathrm{d}p$ represents Lebesgue measure. In principle, one can form highest posterior density sets from (2.3) that have a pre-assigned probability, often taken as $1 - \alpha$. This is the Bayesian's analog to (2.1). These sets are usually intervals and so are called $1 - \alpha$ credibility intervals to distinguish them from $1 - \alpha$ confidence intervals. Again, there is an important interpretational difference: credibility intervals are conditional on the data. So, two different data sets can give numerically different, although typically similar, credibility sets. This is true for confidence sets as well, but this doesn't matter as much because of the repeated-sampling interpretation. The credibility interval is an actual set of parameter values that has a conditional probability given the data assigned to it. Of course, one can choose other priors such as Jeffreys' prior for a binomial parameter and get slightly different results. (In fact, Jeffreys' prior gives a higher posterior probability for a sunrise given the data, since it puts more mass at one.) The prior selection depends on what information is thought to be available pre-experimentally.

In terms of prediction, the Bayesian, like the Frequentist, answers a prediction question by estimating a probability. However, the Bayesian can go a bit further and construct the predictive distribution, an analog to the Frequentist's point prediction but better justified and more comprehensive since it's a whole distribution. The Bayesian predictive density is

$$w(X_{2\,190\,001}|X^{2\,190\,000} = 1^{2\,190\,000})$$
$$= \int \text{Bernoulli}(p)(X_{2\,190\,001})w(p|X^{2\,190\,000})\mathrm{d}p \tag{2.4}$$

and is analogous to the frequentist plug-in distribution. Prediction intervals, e.g., highest predictive density intervals from (2.4), will typically include both zero and one unless $k \approx 0$ or $k \approx n$. The expectation of (2.4) provides a point predictor.

Survey sampling provides a third conceptualization of the Sunrise Problem based on constructing a finite population from which to sample. So, suppose that there is a total of N days on which a sunrise might be observed. This would correspond to the lifespan of the solar system in days. Assume that the days are identical and noninteracting and that the number of days observed so far – which for the sake of comparison can be taken as $n = 6000 \times 365$ – is a random sample from the total of N days. Then, the observed number of sunrises is also $k = 6000 \times 365$. So far, this is the same as the 'one sun, many days' approach. However, a Survey Sampler could define N using any large but arbitrary number of suns on one day, i.e., invoke a version of the 'one day, many suns' approach. Indeed a survey sampler could also define $N = N_1 N_2$ by treating a collection of N_1 identical solar systems each with its own copy of Earth, supposing that each has N_2 days on which a sunrise might happen. Then, the $N_1 N_2$ *(sun, day)* pairs would be regarded as identical and noninteracting, invoking a 'many suns, many days' interpretation.

The point of these three cases is to emphasize that survey sampling depends directly on the actual population from which the sample is drawn. In the Bayes and frequentist conceptualizations, observations are typically regarded solely as outcomes of random variables, so that inferences are mostly about distributions rather than the actual units comprising the population. This was seen in the use of the binomial distribution – a probability law for a random variable not directly referring to population units.

To proceed with the prediction problem, the Survey Sampler's strategy is to define a function of the values of all the units in the population, say $g(y_1, \ldots, y_N)$. In the present case, one might choose the sum

$$T = g(y_1, \ldots, y_N) = \sum_{y \in S} y + \sum_{y \in S^c} y = T_1 + T_2, \tag{2.5}$$

where S is a sample of size n with complement S^c of size $N - n$ and each y_i is one or zero depending on whether a sunrise was observed on that day or not, respectively. The first term on the right-hand side of (2.5) is known: in the present case, all the observed y_i are equal to one. So, $T_1 = 6000 \times 365$. However, the second term in (2.5) is not known. It is seen that the prediction is in terms of the proportion of sunrises, which is equivalent to the total number of sunrises (since N is known), not in a forecasting sense of whether the sun will rise on the next day or not. That is, the Survey Sampler makes a statistically valid guess about the unseen y's in the sense of predicting what would be seen were sampling to be continued. Effectively the y's are fixed but unknown because they are determined for each population unit before sampling begins.

Under (2.5), predicting T_2 is equivalent to estimating T. One natural estimator for T is found by extrapolating from the sample and using $\hat{T}_s = (N-n)\sum_{y\in S} y/n$ as a predictor for T_2. It can be shown that $\hat{T}_2$ is the best linear unbiased predictor of T. Here, 'best' means that it has the smallest variance under the assumptions

$$E(Y_i) = p \quad \text{and} \quad \text{Cov}(Y_i, Y_j) = \sigma^2 \delta_{ij}$$

where $\delta_{ij} = 1$ and 0 for $i = j$ and $i \neq j$, respectively, and the y_i are regarded as outcomes of the random variables Y_i which record (essentially) whether a given unit i is included in the sample. As a proportion, $\sigma^2 = p(1-p)/n$. Here, the Survey Sampler would estimate the total number of sunrises in the population by scaling up the mean representing the proportion of days a sunrise was observed, giving

$$\hat{T} = N\left(\sum_{y\in S} y/n\right) = N,$$

since all the observed y_i are equal to one.

It is well known that the formula for the standard error of a mean in survey sampling is

$$\text{SE} = \frac{N}{n}\left(\frac{N-n}{N-1}\right)^2 \sigma \tag{2.6}$$

and the usual estimate of σ in the present case is $\sqrt{\left(\sum_{y\in S} y/n\right)\left(1 - \sum_{y\in S} y/n\right)/n}$ since the mean is actually a proportion. If one scaled up $\sum_{y\in S} y/n$ to give a predictor for T_2 directly or one wanted to estimate the proportion of sunrises, similar formulae would apply. So, to predict, the Survey Sampler, like the Frequentist or Bayesian, would give an estimation interval for the proportion, typically constructed by an analogous 'sample proportion plus normal percentile times SE' formula, scaling it up if the goal were to predict T_2. A key feature here is the factor multiplying σ in (2.6), called the finite-population correction. It explicitly takes into account the size of the sample relative to the size of the population: the larger n is relative to N, the less variability remains and the smaller the SE.

Overall these three schools of thought have different conceptual bases. Among the various differences that might be noted, one that stands out is the way in which they conceptualize variability. For a Frequentist, variability is intrinsic to the measurements: because of the repeated sampling definition of probability, the intervals are regarded as outcomes from a process that has a confidence level. The concept of confidence applies to the *process* for generating the interval, not to any interval so generated. In a sense, for the Frequentist the data retain their stochastic nature even after they are obtained while the parameter is never stochastic. The Bayesian dismisses this view partly because repeated sampling is essentially never done, but mostly because the use of a sampling distribution amounts to 'analyzing data you don't have and most likely will never have'. For the Bayesian, once the data are obtained they are no longer stochastic. The variability in the data generator is expressed in terms of how the data affect the variability of the parameter, i.e., how the initial information about the parameter, as expressed in terms of its variability and summarized in the prior, updates to a new description of the variability of the parameter. The Survey Sampler dismisses these models for variability and asserts that there is neither variability in the parameter nor in the data itself. The variability is the result of the choice of population units to be included in the sample; variability is the result of the physical act of sampling and is not due to information one believes available or to indefinitely repeated sampling.

The *predictive* conceptualization of the Sunrise Problem would be different from any of these. First, the goal is to get a predictor for the next outcome, so estimation, when it arises, only does so as a way to form predictors. If there are n observations coded as one or zero to indicate whether a sunrise was or was not observed then the data give a vector of length n, of the form $1^n = (1, \ldots, 1)$. As seen in (1.6), (1.8), and (1.9), the focus is on the FPE or perhaps the CPE. So, it is enough to specify a class of predictors. There are several that are natural. First, one might consider a sequence of functions of the form

$$\hat{Y}(n|y_1, \ldots, y_{n-1}) = \begin{cases} 1 & \text{sunrise observed,} \\ 0 & \text{sunrise not observed.} \end{cases}$$

where the structure of $\hat{Y}(n|y_1, \ldots, y_{n-1})$ is such that it depends on all the previous data. In this case, the $\hat{Y}(n|y_1, \ldots, y_{n-1})$ are full functions of $n-1$ arguments taking 2^{n-1} values, even though in practice only one value of each argument (one of the 2^{n-1} values) is observed. The functions $\hat{Y}(n|y_1, \ldots, y_{n-1})$ are much more generally defined than they really need to be, so it might make sense to use a smaller class of functions. One could do this by choosing $\hat{Y}(n|y_1, \ldots, y_{n-1}) = \hat{Y}(n)$, i.e., a function that depends on n, the time step at which one makes a prediction, but not on the data obtained. This would mean that the data do not matter, apart from their time of occurrence. An even greater reduction would be to consider exactly two predictors, $\hat{Y}(n) = 1$ and $\hat{Y}(n) = 0$, for all n, so the prediction does not even depend on the time at which it was made.

These three classes of functions lead to three forms of the FPE, namely,

$$\frac{1}{K} \sum_{k=1}^{K} \left(\hat{Y}(\sigma_k(n)|y_{\sigma_k(1)}, \ldots, y_{\sigma_k(n-1)}) - Y_{\sigma_k(n)} \right)^2, \tag{2.7}$$

$$\frac{1}{K} \sum_{k=1}^{K} \left(\hat{Y}(\sigma_k(n)) - Y_{\sigma_k(n)} \right)^2, \tag{2.8}$$

$$\frac{1}{K} \sum_{k=1}^{K} \left(\hat{Y}(n) - Y_{\sigma_k(n)} \right)^2, \tag{2.9}$$

where $\sigma_k(\cdot) = (\sigma_k(1), \sigma_k(2), \ldots, \sigma_k(n))$ is a permutation of $\{1, \ldots, n\}$ that is analogous to the permutations in (1.6) and, in each case, the predictor $\hat{Y}$ is determined from the data up to but not including the time for which the prediction is made.

Since there is only one set of observations, i.e., n ones, (2.7) should not be regarded as representative of the performance of the predictor over many samples (as we are conditioning on one entire sample). However, the other two forms are representative because the class of predictors is smaller. The expression (2.8) permits different observations to be made at different times while (2.9) requires the same prediction to be made at all times. It is important to observe that the FPE and CPE depend only on the predictions and realized values. Two predictors that gave the same predictions would be equivalent. For instance, in the present case, the predictor that took the same value all the time, $Y(n) = 1$, the predictor that predicted the next value to be the same as the last value, $Y(n) = y_n$, and the predictor that took a majority vote, $Y(n|y_1, \ldots, y_n) = \arg\max(\#0\text{'s}, \#1\text{'s})$, would all yield the same error, zero, for (2.7), (2.8), and (2.9). So, the variability of the predicted error is zero as well, however one might want to evaluate it.

The most informative of these three cases is the one that uses the smallest collection of functions since it also gives the smallest error. Thus, one is led to the constant prediction $\hat{Y} = 1$, i.e., the sun will always rise tomorrow. Loosely, one can regard predictions as actions in a decision-theoretic sense. That is, one must choose a collection of predictors to be searched for the best that it contains. This is not decision-theoretic because the evaluation is empirical and depends on the data. In a limiting sense one might try to reconstruct an expected loss or risk – but taking the limits is not realistic since n is finite. Limiting values may be useful indicators of large-sample behavior, but they necessitate assumptions about the true distribution since the limit is taken in the true distribution. In particular, one must assume that there is a true distribution and that it is approachable by limits; in many examples neither of these is true. Thus, the full specification of decision theory is not invoked either by preference relations or by expectations.

Predictions can be regarded as actions in a purely decision-theoretic sense. However, this requires expectations, loss functions, and an action space to be searched for the best predictor. While conceptually useful, beyond $\mathcal{M}$-complete cases the required formalism is not defined. Moreover, in practice, often neither the expectation nor the loss function is known to satisfactory precision. Thus, from an empirical predictive standpoint, in complex problems the mathematical form of decision theory must be replaced by data-dependent quantities, e.g., empirical risks and approximate loss functions, in which context standard decision theory is usually ineffective. The connection between the mathematical form of decision theory and the empirical form is that in a limiting sense one might try to reconstruct an expected loss, a mentioned in the previous paragraph.

Indeed, to determine a good predictor one does not even have to invoke probability theory in either its repeated sampling form or its Bayesian form. Probability only arises as we evaluate how good a predictor is expected to be and this only requires Kolmogorov's axioms, not necessarily the Bayes or frequentist justification for probability. In this sense, the predictive approach is compatible with the earlier approaches and is more closely tied to the observations than in the Bayes or frequentist case. Indeed, there is no prohibition on evaluating the quality of a predictor without using probability.

In fairness to the Bayes, frequentist, and survey sampling schools of thought, the Sunrise example was concocted to demonstrate that a predictive approach can be easier, more closely tied to observed reality, and more meaningful in the sense that it does not make extra assumptions (such as probabilities), while still giving an answer, i.e., a predictor, that can be usefully interpreted. Part of the reason why this was possible for this example was that there is essentially no variability in the observed data. However, if one were looking at a different outcome – rain tomorrow, rather than the rising of the sun – the reasoning would have been analogous and the result of the predictive analysis would have been a nontrivial function of the data that would not be very different from that found using the other schools of thought.

Finally, we will mention other schools of thought, comparatively less prevalent but conceptually important. Foremost are the fiducial and information-theoretic approaches. Roughly, the fiducial approach that originated with Fisher can be regarded as a frequentist way of generating a posterior distribution for a parameter; see Hannig (2009). From a predictive standpoint, when the fiducial approach is feasible it is merely another technique for generating a predictor that may be useful. The information-theoretic approach involves basing inferences of all sorts on codelength concepts; see Bryant and Cordero-Brana (2000) for a concise introduction. This can be regarded as an extension of the likelihood school of thought but using codelength rather than confidence or other repeated sampling notions. Aside from being another method for constructing predictors, the information-theoretic approach provides a concept of complexity in the form of codelength that may be important in predictor selection or evaluation. So, information-theoretic concepts may well have an important role to play in good prediction.

2.2 Parametric Families

Having seen how Frequentists, Bayesians, Survey Samplers, and Predictivists conceptualize their approaches in the predictive setting of the Sunrise Problem, it's time to provide some details on how Frequentists and Bayesians in particular would predict, in simple but nevertheless formal settings such as parametric families where the outcomes $Y = y$ are IID and no explanatory variables are present. The normal family will be emphasized because many of the relevant calculations can be done in closed form. The point is to see the details of prediction from the frequentist and Bayesian standpoints in a general setting.

It is important to note that the predictivist view is generally compatible with the frequentist and the Bayesian views – the Predictivist would regard Bayesian or frequentist methods primarily as techniques for generating candidate predictors whose performance would then be evaluated. Of course, the Predictivist would comfortably use any method to generate predictors that promised to give good performance in a CPE or FPE sense.

2.2.1 Frequentist Parametric Case

In the simplest prediction examples it is assumed that X is not present, i.e., there are no covariates, and that the distribution of Y is the main object of interest, since the goal is to predict outcomes from it. The predictive analog to the estimation of a population mean when the variance is known is the use of n IID observations $y_1, \ldots, y_n$ to predict Y_{n+1}. One natural predictor is $\bar{y} = (1/n) \sum y_i$. If no parametric family has been assumed, one might assume that the first and second moments of the Y_i exist, setting $\mu = EY_i$ and $\sigma^2 = \text{Var}(Y_i)$ and then use standard inequalities (Chebyshev and triangle) to obtain

$$
\begin{aligned}
P &\left(|\bar{Y} - Y_{n+1}| \geq \sqrt{\frac{\sigma^2(1 + (1/n))}{\tau}} \right) \\
&\leq \frac{\tau}{\sigma^2(1 + (1/n))} \left((E|\bar{Y} - \mu|^2) + (E|\mu - Y_{n+1}|^2) \right) \\
&\leq \tau,
\end{aligned}
\tag{2.10}
$$

for given $\tau > 0$. It is seen that the bound is nontrivial only when $\tau < 1$. For such a choice of τ, the natural frequentist prediction interval (PI) is

$$\bar{Y} \pm \sigma \sqrt{\frac{1 + (1/n)}{\tau}}, \tag{2.11}$$

provided σ is known.

In the normal case $Y_i \sim N(\mu, \sigma^2)$ and direct calculation gives $\bar{Y} - Y_{n+1} \sim N(0, \sigma^2(1 + (1/n)))$. The PI becomes $\bar{Y} \pm z_{1-\alpha/2}\sigma(1 + (1/n))^{1/2}$, where $z_{1-\alpha/2}$ is the $(1 - \alpha/2)100$th percentile of $N(0, 1)$.

Comparing the normal case with (2.11), it is easy to see that if we set $\tau = 1 - \alpha$ to make the confidence levels the same then the ratio of the widths of the two PIs is

$$\frac{\sigma z_{1-\alpha/2}(1 + (1/n))^{1/2}}{\sigma(1 + (1/n))^{1/2}/(1 - \alpha)^{1/2}} = z_{1-\alpha/2}(1 - \alpha)^{1/2} \approx 2,$$

where $\approx$ means 'approximately', i.e., roughly valid for commonly chosen values of α. This means that PIs from the general approach are around twice the width of the normal case. So, Chebyshev's inequality in (2.10) is strong enough to give the $\mathcal{O}(1/n)$ rate seen in the normal, but not strong enough to give a ratio of widths shrinking to one with increasing n.

The above bounds presume that $\bar{y}$ is the appropriate predictor, and this will usually be the case with normal data. However, if robustness were important then better choices than $\bar{y}$ for predicting Y_{n+1} would be either $\text{med}_i\, y_i$, the median of the y_i, or possibly an estimate of the mode of Y_{n+1}. We ignore this here apart from noting that, asymptotically, the mean and the median are both $\mathcal{O}(1/\sqrt{n})$ even though the mean is more efficient. The argument used to obtain (2.11) can be adapted to the median, and indeed, most other point predictors, possibly using the Hölder inequality rather than Cauchy–Schwarz.

If σ is unknown then

$$P\left(\frac{|\bar{Y} - Y_{n+1}|}{\hat{\sigma}} \geq \frac{(1 + (1/\sqrt{n}))}{\tau}\right)$$

$$\leq \frac{\tau}{(1 + (1/\sqrt{n}))} E\left(\frac{1}{\hat{\sigma}}\right) |\bar{Y} - Y_{n+1}|$$

$$\leq \frac{\tau}{(1 + (1/\sqrt{n}))} \sqrt[u]{E\left(\frac{1}{\hat{\sigma}}\right)^u} \sqrt[v]{E|\bar{Y} - Y_{n+1}|^v}, \tag{2.12}$$

where $\hat{\sigma}$ is an estimator of σ and $1/u + 1/v = 1$. When the Y_i are normal, the usual estimator for σ is $s^2 = (1/(n - 1)) \sum (Y_i - \bar{Y})^2$ and it is well known that $(n - 1)s^2/\sigma^2 \sim \chi^2_{n-1}$. So, $\sigma^2/(s^2(n - 1))$ is an inverse-chi-squared random variable with $n - 1$ degrees of freedom which has mean $1/(n - 3)$. Thus, $E(1/s^2) = (n - 1)/((n - 3)\sigma^2)$. When $u = v = 2$, (2.12) gives

$$\frac{\tau}{1 + (1/\sqrt{n})} \frac{\sqrt{n-1}}{\sqrt{n-3}\sigma} \left(\sigma\sqrt{1 + (1/n)}\right) = \tau \frac{\sqrt{n-1}}{\sqrt{n-3}} \frac{\sqrt{1 + (1/n)}}{(1 + (1/\sqrt{n}))} \tag{2.13}$$

by the triangle inequality. Thus, the bound in (2.13) is slightly increased over (2.10) owing to the extra variability from $\hat{\sigma}$. So the PIs from (2.12) become

$$\bar{Y} \pm \hat{\sigma} \frac{\sqrt{n-1}\sqrt{1+(1/n)}}{\tau\sqrt{n-3}}, \tag{2.14}$$

where the factor $(1 + 1/\sqrt{n})^{-1}$ has been absorbed into the prediction error τ, the prediction analog of confidence. An exact derivation using normality throughout rather than Chebyshev's inequality in (2.12) can be found in Geisser (1993, Chapter 2). When the underlying model is not normal, the width of the prediction interval is controlled by the two expectations in (2.12), and if the distribution of $1/\hat{\sigma}$ is too spread out or too close to zero (or $|\bar{Y} - Y_{n+1}|$ has moments that are very high) then τ would have to decrease to achieve the same prediction error. As in the case of known σ, the PI depends only on μ, σ, and the uth and vth moments.

If the data were paired, i.e., there were independent pairs (U_i, V_i) for $i = 1, \ldots, n$ and the task were to predict $U_{n+1} - V_{n+1}$ then the arguments could be applied to $Y_i = U_i - V_i$, yielding results analogous to (2.11) and (2.14).

It is worth commenting that if $\sigma = 1$ in the normal case, but upper bounds are not taken to be as in (2.10), then there are two natural ways to obtain prediction intervals and they have different properties. The first is to recognize that $(\bar{Y} - Y_{n+1})/(1 + (1/n)) \sim N(0, 1)$ and therefore a $1 - \alpha$ PI is $\bar{Y} \pm z_{1-\alpha/2}\sqrt{(n+1)/n}$. The second is to use the estimate $\hat{\mu} = \bar{y}$ to give a distribution, namely $N(\bar{y}, 1)$, that can be employed to predict Y_{n+1}. This plug-in predictor gives

$$\hat{P}(\bar{Y} - z_{1-\alpha/2} \le Y_{n+1} \le \bar{Y} + z_{1-\alpha/2}) = 1 - \alpha,$$

where $\hat{P}(\cdot)$ is the probability assigning a mass $\hat{P}(A) = N(\bar{y}, 1)(A)$ for a given set A. This means that the $1 - \alpha$ PI is $\bar{y} \pm z_{1-\alpha/2}$, slightly narrower than before as the factor of $z_{1-\alpha/2}$ is 1 rather than $\sqrt{(n+1)/n}$. The difference is that the larger interval $\bar{Y} \pm z_{1-\alpha/2}\sqrt{(n+1)/n}$ includes the variability in the estimate of μ while the interval $\bar{Y} \pm z_{1-\alpha/2}$ is conditional on the use of the data via $\bar{y}$ to identify the prediction distribution. This distinction extends to the case where σ is unknown; see Geisser (1993, example 2.2, p. 9). The normal example also extends to q-steps forward prediction; see Geisser (1993, pp. 10–11). Other examples can be derived and have similar properties, provided that a pivotal quantity with a mathematically tractable density exists so that closed-form expressions for the PIs can be derived.

Given a normal distribution with known mean, say zero, but unknown variance σ^2, it is well known that $(n - 1)s_n^2/\sigma^2 \sim \chi_{n-1}^2$, while the future observation Y_{n+1} has distribution $N(0, \sigma^2)$. Taking the ratio of the future observation Y_{n+1} and the sample standard deviation cancels σ and gives a Student's t-distribution with $n - 1$ degrees of freedom, i.e., $Y_{n+1}/s \sim t_{n-1}$. Solving for Y_{n+1} gives the prediction distribution st_{n-1}, from which PIs can be found. Notice that this prediction distribution gives slightly larger PIs than just using an $N(0, s^2)$ because the t-distributions have heavier tails than the normal distribution. This is necessary for exact interpretation of the prediction error $1 - \alpha$ in finite samples, but the two procedures give asymptotically equivalent results.

2.2.2 Bayesian Parametric Case

In the Bayesian version of Sec. 2.2.1 it is assumed that the views about a parameter Θ will be updated, from being represented by the density $w(\theta)$, by Y_i that are IID and follow a probability law drawn with density $p(\cdot|\theta)$, where Θ is a random variable with outcomes θ. Then, the Bayesian forms a posterior density $w(\theta|y^n)$ for θ given the data, i.e.,

$$w(\theta|y^n) = w(\theta)p(y^n|\theta)/m(y^n),$$

where

$$m(y^n) = \int w(\theta)p(y^n|\theta)d\theta. \tag{2.15}$$

Now, $w(\theta|y^n)$ describes the post-data information about the parameter. So the Bayesian usually writes the predictive density as

$$m(Y_{n+1} = y_{n+1}|y^n) = \frac{m(y^{n+1})}{m(y^n)} = \int p(y_{n+1}|\theta)w(\theta|y^n)d\theta \tag{2.16}$$

and uses it to form a prediction region. For instance, for given $\alpha > 0$, the highest-posterior-density predictive region is $R(\alpha) = R_{n+1}(y^n; \alpha) = \{y_{n+1}|m(y_{n+1}|y^n) \geq t_\alpha\}$, where t_α is chosen to ensure that

$$\int_{R(\alpha)} m(y_{n+1}|y^n) = 1 - \alpha. \tag{2.17}$$

It can be seen that (2.17) depends on w and uses a conditional density given the data. Closed-form expressions for (2.17) often don't exist but $R(\alpha)$ can usually be approximated quite well computationally.

Bayesian point predictors can also be found – and their theoretical performance is evaluated using P_{θ_T}. One choice for a Bayesian point predictor is

$$\hat{Y}_{n+1} = \int y_{n+1} m(y_{n+1}|y^n)dy_{n+1} = \int E_\theta(Y_{n+1})w(\theta|y^n)d\theta \rightarrow \mu = E_{\theta_T}(Y),$$

as $n \rightarrow \infty$ when θ_T is taken as the true value of θ. This leads to PIs on choosing $\tau > 0$ and, in analogy to (2.10), writing

$$\begin{aligned}
P_{\theta_T} &(|\hat{Y}_{n+1} - Y_{n+1}| > \tau) \\
&\leq \frac{1}{\tau^2} E_{\theta_T} \left(\int |Y_{n+1} - E_\theta(Y)|w(\theta|y^n)d\theta \right)^2 \\
&\leq \frac{1}{\tau^2} E_{\theta_T} \int |Y_{n+1} - E_\theta(Y)|^2 w(\theta|y^n)d\theta \\
&\leq \frac{1}{\tau^2} \sqrt{E_{\theta_T}(Y_{n+1} - \mu)^2} + \frac{1}{\tau^2} E_{\theta_T} \int |\mu - E_\theta(Y)|^2 w(\theta|y^n)d\theta \tag{2.18}
\end{aligned}$$

using Markov, the triangle inequality, and Cauchy–Schwartz. The first term in (2.18) gives σ^2/τ^2. To evaluate the second term, note that in the one-dimensional case a Taylor expansion gives

$$E_\theta(Y_{n+1}) - E_{\theta_T}(Y) = \frac{1}{2}(\theta - \theta_T)\frac{\partial E_\theta(Y)}{\partial \theta}\bigg|_{\theta=\theta_T} + o(\|\theta - \theta_T\|), \tag{2.19}$$

where the 'little-o' bound holds on a small neighborhood of θ_T where $w(\theta|y^n)$ is concentrated. Also, recall that posterior normality gives

$$w(\theta|y^n) \approx \frac{\sqrt{n}}{\sqrt{2\pi I(\theta_T)}} \exp\left(-nI(\theta_T)^{-1}(\theta - \theta_T)^2/2\right).$$

So, the second term in (2.18) is approximately

$$E_{\theta_T} \int \frac{1}{4}(\theta - \theta_T)^2 \left|\frac{\partial E_\theta(Y)}{\partial \theta}\right|^2_{\theta=\theta_T} \frac{\sqrt{n}}{\sqrt{2\pi I(\theta_T)}} \exp\left(-nI(\theta_T)^{-1}(\theta - \theta_T)^2/2\right) d\theta$$

$$= \frac{1}{4}\left|\frac{\partial E_\theta(Y)}{\partial \theta}\right|^2_{\theta=\theta_T} \frac{I(\theta_T)}{n}, \tag{2.20}$$

apart from the factor $1/\tau^2$, by a standard Laplace-approximation argument. Now (2.18) becomes

$$P_{\theta_T}\left(|\hat{Y}_{n+1} - Y_{n+1}| > \tau\right) \le \frac{1}{\tau}\left(\sigma^2 + B(\epsilon, \theta_T)\frac{I(\theta_T)}{n}\right)^{1/2} \tag{2.21}$$

where the factor $B(\epsilon, \theta_T)$ includes the absolute first derivative from (2.20) and a bound $1 + \epsilon$ from a Laplace approximation (where ϵ is the radius of the neighborhood used in the Laplace approximation). Obviously, one can assume that $\epsilon = \epsilon_n \to \infty$ as $n \to \infty$ and that the technique giving the bound in (2.21) generalizes to an analogous bound when θ is a finite-dimensional parameter in an open set with compact closure. It is seen that (2.21) is parallel to (2.10) and gives PIs of the form (2.11).

We can see from (2.17) that the key to Bayes prediction is obtaining an expression for the mixture $m(\cdot)$ of densities in (2.15). This is (relatively) easy when the prior is conjugate to the likelihood since a closed form for the posterior can then be found. The case of a normal variable with a normal prior is particularly easy. If $w(\theta)$ is $N(\mu, \tau^2)$ and $Y_i|\theta \sim N(\theta, \sigma^2)$, where μ, τ, and σ are known, then $\bar{Y}|\theta \sim N(\theta, \sigma^2/n)$ and

$$m(\bar{y}) = \frac{\sqrt{n}}{\sigma\tau\sqrt{2\pi\rho}} \exp\left(-\frac{(\mu - \bar{y})^2}{2(\sigma^2/n + \tau^2)}\right), \tag{2.22}$$

which is the density of the distribution $N(\mu, \sigma^2/n + \tau^2)$, where $\rho = 1/(\sigma^2/n) + 1/\tau^2$. So, $w(\theta|y^n)$ is the density of the distribution

$$N(E(\Theta|y^n), \text{Var}(\Theta|y^n)) = N\left(\frac{\sigma^2/n}{\sigma^2/n + \tau^2}\mu + \frac{\tau^2}{\sigma^2/n + \tau^2}\bar{y}, \frac{1}{\rho}\right).$$

That is, the posterior mean is $E(\Theta|y^n) = (\sigma^2/n)/(\sigma^2/n + \tau^2)\mu + \tau^2/(\sigma^2/n + \tau^2)\bar{y}$ and the posterior variance is $1/\rho = \tau^2\sigma^2/(n\tau^2 + \sigma^2) = \mathcal{O}(1/n)$. Now, it can be directly verified that $m(y_{n+1}|y^n)$ is the density of the distribution

$$N(E(Y_{n+1}|Y^n = y^n), \text{Var}(Y_{n+1}|Y^n = y^n))$$

and that $E(Y_{n+1}|Y^n = y^n) = E(\Theta|Y^n = y^n)$, so

$$\text{Var}(Y_{n+1}|Y^n = y^n) = \sigma^2 + \text{Var}(\Theta|Y^n = y^n).$$

So, we have that $E(\Theta|Y^n = y^n) \pm z_{1-\alpha/2}\sqrt{\text{Var}(Y_{n+1}|Y^n = y^n)}$ is a large-sample approximation for $R(\alpha)$. That is, intervals of this form have an asymptotic probability $1 - \alpha$, in the predictive distribution $M(\cdot|y^n)$, of containing Y_{n+1}, i.e.,

$$M\left(E(\Theta|Y^n = y^n) - z_{1-\alpha/2}\sqrt{\text{Var}(Y_{n+1}|Y^n = y^n)} \le Y_{n+1} \right.$$

$$\left. \le E(\Theta|Y^n = y^n) + z_{1-\alpha/2}\sqrt{\text{Var}(Y_{n+1}|Y^n = y^n)}|\, y^n \right) = 1 - \alpha.$$

Thus, if y^n is fixed, the variability in the data only affects the prediction via the normal likelihood. If one did not use $M(\cdot|y^n)$ to get prediction intervals, the properties of $E(\Theta|Y_n = y_n)$ as a predictor for Y_{n+1} would change. For instance, one could obtain PIs for Y_{n+1} based on $E(\Theta|Y_n = y_n)$ using the unconditional $M(\cdot)$ or, for Y^{n+1}, P_{θ_T}. In such a case one would expect the $1 - \alpha$ PIs to be larger, because the fixed y^n would have been replaced by the random Y^n. However, a prior often corresponds to having extra data points, effectively shrinking the lengths of the PIs. So, the tradeoff between the extra information in the prior and the extra variability in Y^n that results if one finds PIs using the unconditional distribution evident in most Bayesian procedures will remain.

Even if σ^2 is not known, calculations can be done to obtain an explicit form for $m(y_{n+1}|y^n)$. Suppose that $Y_i \sim N(\mu, \sigma^2)$ and $w(\mu, \sigma^2) \propto 1/\sigma^2$. Then

$$w(\mu, \sigma^2|\bar{y}) \propto w(\mu, \sigma^2)p(y^n|\mu, \sigma^2)$$

$$\propto \frac{1}{\sigma^2}\left(\frac{1}{2\pi\sigma^2}\right)^{n/2} \exp\left(-(1/2\sigma^2)((n-1)s^2 + n(\bar{y} - \mu)^2)\right),$$

where $s^2 = (1/(n-1))\sum_i(y_i - \bar{y})^2$. The posterior of σ^2 given $\bar{y}$ is

$$w(\sigma^2|\bar{y}) \sim \text{scaled inverse-}\chi^2(n-1, s^2)$$

$$= \text{inverse-gamma}((n-1)/2, (n-1)s^2/2)$$

and $w(\theta|\sigma^2, \bar{y})$ is the density of an $N(E(\Theta|y^n), \text{Var}(\Theta|y^n))$ distribution. Writing $\mu = E(\Theta|y^n)$ and $\tau^2 = \text{Var}(\Theta|y^n)$ in (2.16) gives

$$m(y_{n+1}|y^n) = \int p(y_{n+1}|\theta, \sigma)w(\theta|\sigma^2, \bar{y})w(\sigma^2|\bar{y})\mathrm{d}\theta\mathrm{d}\sigma^2$$

$$= \int \phi_{\mu,\sigma^2+\tau^2}(y_{n+1})w(\sigma^2|\bar{y})\mathrm{d}\sigma^2, \qquad (2.23)$$

where $\phi_{a,b}$ indicates the $N(a, b)$ density, and it can be verified that $m(y_{n+1}|y^n)$ has the density of a $t_{n-1,\bar{y},s^2(1+1/n)}$ random variable, i.e., a Student's t-distribution with degrees of freedom $n - 1$, location $\bar{y}$, and scale $s^2(1 + 1/n)$. (This is a special case of (4.29), from prediction in a linear regression context in Sec. 4.2.)

2.2.3 Interpretation

Taken together, the treatments of the Sunrise Problem (2.1) and the above derivations for PIs from the frequentist standpoint (Sec. 2.2.1) and from the Bayesian standpoint (Sec. 2.2.2) amount to showing that the standard results from binomial, normal, and more general parametric families carry over to analogs for prediction. This means that if there is strong reason to believe that a specified parametric family accurately represents the probability defining the randomness in the data there is no loss in regarding prediction as the central goal instead of parameter estimation. Even better, taking upper bounds on the probabilities of error makes the PIs somewhat independent of the specific models used, without much increase in the length of the PIs, i.e., the rate of contraction $\mathcal{O}(1/\sqrt{n})$ is unchanged even if the constant increases. However, in the presence of nontrivial model uncertainty these conclusions would not hold in general because possible misspecification of the parametric family would have to be included in the analysis and the simple parametric assumptions would have to be relaxed.

So far this chapter has focused on the construction of predictors rather than their uses. However, the reason why one wants to find a predictor is to make predictions, and the worth of a predictor is evaluated by comparing its predictions with observations over repeated uses, as with the FPE or CPE in Chapter 1. Repeated uses are important because predictors are not unique: any two predictors giving the same predictions for a data set are equivalent if one looks only at the FPE or CPE. That is, as a function, the predictor is tested on only finitely many data points and hence on only a small number of its possible arguments. This is much like function interpolation: given a finite set of points in the real plane through which one seeks an interpolant, there is no unique choice although one must invoke extra criteria to avoid getting ridiculous values at the points for which no constraint applies. Although the data set remains finite, using the FPE or CPE is a way to ensure that all the constraints that one might have from permuting the data have been imposed. While this does not give uniqueness, once a low FPE or CPE has been achieved one may proceed to impose extra criteria to avoid getting infeasible predictions.

The point is that predictors are only useful once they have been validated extensively on data. Failure to validate may be the result of a poor predictor class or a hard prediction problem. Recall that, by comparing predictions with observations, the FPE and CPE include assessments of model uncertainty – perhaps better termed *predictor uncertainty*. When one has found a well-validated predictor and ensured that it is unlikely to give ridiculous predictions in new settings, one may be confident that the amount of predictor uncertainty which it includes is not excessive. Indeed, by continually referring the predictions to the observations, the predictor choice and any downstream modeling inferences from it will be disciplined by the requirement that predictions should not deviate from the observed values too much. For instance, in the predictive version of the Sunrise Problem it is clear that the predictor means that it is reasonable to assume the sun will rise every day as part of explaining why the solar system functions in the way it does. Because of its focus on validation, the predictive approach includes model or predictor uncertainty more readily than other schools of thought.

2.3 The Abstract Version

Having seen the three predictive data analyses in Chapter 1, the gedanken examples of hypothesis testing with next-generation sequencing (NGS) data and of the Sunrise Problem, and having seen how the usual frequentist and Bayesian parametric family analyses can be re-expressed in predictive terms, it is time to pull these considerations together more abstractly. This will identify the key features of the different approaches so that they can be contrasted. The intention is that the result of this will provide the key features for the predictive approach to be regarded as a school of thought in its own right, with standing equal to – or better than – the others. Here, the comparison will include frequentism, Bayes, survey sampling, and prediction. Other schools such as the fiducial and the information-theoretic will be neglected – although notions of complexity arising from codelength will be included in Sec. 2.4.

2.3.1 Frequentism

One defining feature of frequentism is its definition of probability as a long-run average. This recurs in all aspects of frequentist theory, whether estimation, testing, prediction, or other aspects. The essential feature is that the probability of an event A is defined to be

$$P(A) = \lim_{n \to \infty} \frac{\text{\# times event } A \text{ was observed}}{\text{\# trials which could have given event } A} \tag{2.24}$$

as an empirical limit, i.e., given an $\epsilon > 0$ one cannot necessarily find an N such that $n > N$ implies that

$$\left| P(A) - \frac{\text{\# times } A \text{ was observed}}{\text{\# trials which could have given event } A} \right| < \epsilon;$$

nevertheless the limit still holds in some sense.

Frequentist estimation is conceptually easy. Suppose the probabilities that might possibly describe Y are denoted P_θ, i.e., they are indexed by a parameter θ, and, by a mild abuse of notation, the probability that might describe the nth copy of Y is denoted P_θ, too, rather than, say, P_θ^n. Then, a $1 - \alpha$ confidence region is a random set $R(Y^n)$ with the property that

$$P_\theta(\theta \in R(Y^n)) = 1 - \alpha \tag{2.25}$$

holds for any θ; essentially this means that $R(Y^n)$ must track θ by a function of the data. This function usually gives a point estimator for θ. In general the constraint (2.25) is not by itself enough to specify a sensible region uniquely. So, $R(Y^n)$ is usually based on a statistic that has a unimodal sampling distribution inherited from P_θ. In that way a uniquely defined region can be determined from the highest-density region of the sampling distribution. Then, the frequentist definition of probability (2.24) can be invoked to assert that a given set $R(y^n)$ is the result of a process with $1 - \alpha$ confidence limits, i.e., a process that produces regions having the correct θ as an interior point $(1-\alpha)\%$ of the times that any such regions are found. As a generality, θ need not have any interpretation beyond being the index of a distribution, and $R(Y^n)$ plays the role of A in (2.24).

The question is which statistics are most appropriate for a given family $\mathcal{P} = \{P_\theta\}$, bearing in mind that θ need not be a finite-dimensional real parameter. In fact, $\{P_\theta\}$ might effectively be isomorphic to the set of all distributions on a sample space. That is, if one assumes only that $Y : (\mathcal{X}, \mathcal{A}) \to (\mathbb{R}, \mathcal{B}(\mathbb{R}))$ one might set $\mathcal{P} = \{P | P \text{ is a distribution on } (\mathbb{R}, \mathcal{B}(\mathbb{R}))\}$. This is a typical nonparametric problem and one might use histograms to estimate the true distribution P_T. Regardless of whether $R(Y^n)$ is defined by statistics or whether the parameter space is nonparametric, the use of sets like $R(Y^n)$ automatically includes an assessment of variability for the estimation of θ.

To implement this in practice, $R(Y^n)$ is often defined in terms of one or more statistics, and limit theorems are proved, so that probabilities in terms of the sampling distribution as in (2.25) can be approximated easily when n is not small. In the most typical example, one sets $\theta = E(Y)$ and $\sigma = \sqrt{\text{Var}(Y)}$ and then estimates (θ, σ) by $(\bar{y}, s)$. Then the central limit theorem gives that, as $n \to \infty$,

$$P_\theta(S z_{\alpha/2} \leq \sqrt{n}(\bar{Y} - \theta) \leq S z_{1-\alpha/2}) \to 1 - \alpha.$$

So, $R(y^n) = \{s z_{\alpha/2} \leq \sqrt{n}(\bar{y} - \theta) \leq s z_{1-\alpha/2}\}$ is an asymptotic $1 - \alpha$ confidence region (interval) from which one outcome, $R(y^n) = \{s z_{\alpha/2} \leq \sqrt{n}(\bar{y} - \theta) \leq s z_{1-\alpha/2}\}$, is available. The central limit theorem is not unique: an analogous procedure can be used to approximate the sampling distribution of an extreme value such as $Y_{(n)}$ or indeed many other statistics. They key requirements are that the limit should exist in distribution (at least) and that the limiting distribution should be nontrivial. This ensures that a parameter estimator such as $\bar{Y}$ has an associated assessment of variability, e.g., S.

Frequentist hypothesis testing is the result of an optimization problem that also uses the sampling distribution. Its basic form is known as the Neyman–Pearson fundamental lemma. The statement of this result concerns the testing of a point null $\mathcal{H}_0: P = P_0$ against a point alternative $\mathcal{H}_1: P = P_1$. For fixed $\alpha > 0$ (the level) consider the class of test functions $C = \{\phi: E_0\phi(Y^n) \leq \alpha\}$. Then the element of C achieving $\arg\max P_1$ (reject $\mathcal{H}_0$) is given by

$$\phi(y^n) = \begin{cases} 1 & \text{if } p_1(y^n) > k p_0(y^n), \\ 0 & \text{if } p_1(y^n) < k p_0(y^n), \end{cases}$$

for some choice of the factor k. The hypotheses are treated asymmetrically, and the density ratio $p_1(y^n)/p_0(y^n)$ is the paradigm form of the test statistic. Thus, both C and the maximum power are defined in terms of the sampling distribution. Many extensions and special cases of this basic optimization problem, including treatments of the multiple comparison problem, have been developed.

Frequentist decision making rests on utility theory or decision theory, which are very closely related. At root is the von Neumann–Morgenstern expected utility theorem. Suppose that an agent expresses preferences as $A \succ B$, meaning option A is preferred over option B. (Formally, these options are called lotteries and correspond loosely to randomized actions with respect to a single probability.) Then, if the agent is rational, i.e., has preferences satisfying the axioms of completeness, transitivity, continuity, and independence, there is a utility function U such that the ordering of options under the preference relation $\succ$ is equivalent to the ordering of options from the size of the expected utility of the options. That is, $A \succ B \iff E(U(A)) > E(U(B))$. As before, blurring the distinction (but only a little) between a utility function and the negative of a loss function, frequentist decision making becomes Wald's decision theory, under which one chooses a loss function and considers the expected loss, or risk, of each action in the action space to find ideally either an admissible or minimax solution. Frequentist estimation can be formulated decision-theoretically; the 'decision' is which value to announce for θ given the data; however, decision theory alone will not assign a variability to a point estimator. So, this is done separately by finding the standard error. Frequentist hypothesis testing can be regarded as a form of decision making and can be formulated decision-theoretically; one loss is assigned to false rejection of the null and another loss is assigned to false rejection of the alternative. Then, the Neyman–Pearson Fundamental Lemma means that density ratio tests form a minimal complete class of decision rules; see Berger (1980).

Frequentist prediction has several forms. First, one can obtain predictors (point or interval) from estimating the parameters in a model. These plug-in predictors were seen briefly in the binomial and normal cases in Secs. 2.1 and 2.2.1; they tend to neglect variability in the plugged-in estimate. The basic idea is to identify functions $L(Y^n)$ and $U(Y^n)$ such that

$$P_\theta(L(Y^n) \leq Y_{n+1} \leq U(Y^n)) \geq 1 - \alpha \qquad (2.26)$$

for all θ. Again, this has a repeated sampling interpretation. The role of θ is of course problematic because often L or U will depend on θ. In these cases it is common to plug in an estimate of θ and regard (2.26) as holding approximately, on the grounds that the variation due to estimating θ is small and decreasing as n increases.

A typical instance of this occurs when one assumes a parametric family of densities $p(\cdot|\theta)$ for a random variable Y, with respect to a dominating measure, and indexed by a d-dimensional parameter $\theta \in \mathbb{R}^d$. One can collect an independent and identical (IID) sample of size n $(Y_1 = y_1, \ldots, Y_n = y_n)$ and use this data to give a value for an estimate $\hat{\theta} = \hat{\theta}(y^n)$ of θ, where $y^n = (y_1, \ldots, y_n)$ is a realized value of $Y^n = (Y_1, \ldots, Y_n)$ i.e., $Y^n = y^n$. Then one can make predictions for Y_{n+1} using $p(y_{n+1}|\hat{\theta})$, e.g., $E_{\hat{\theta}}(Y_{n+1})$ can be taken as a point predictor for Y_{n+1}. These are called plug-in predictors, and it can be seen that every distinct estimator leads to a distinct predictor since a prediction interval with confidence α for Y_{n+1} can be obtained from $\{y|p(y|\hat{\theta}) > t_\alpha\}$, where t_α is a threshold, to give approximately $1 - \alpha$ confidence. If predictions are made using $p(y_{n+1}|\hat{\theta})$, they will have variability due to $\hat{\theta}$ as well as due to the intrinsic variability of Y_{n+1} but the variability due to $\hat{\theta}$ usually goes to zero as $n \to \infty$.

Alternatively, one can frame the prediction problem in terms of decision theory, in which the best prediction is the optimal action. That is, suppose that the Y_i are IID, choose a loss function L and an action space $\mathcal{A}$, and examine the risks

$$R_\theta(a) = E_\theta L(Y_{n+1}, a(Y^n))$$

under θ for $a(\cdot) \in \mathcal{A}$. The optimal action, i.e., the best predictor for Y_{n+1} using Y^n, can sometimes be narrowed down or even identified. In the special case where L is the squared error loss, it is easy to show that the best point predictor is of the form $\hat{Y}_\theta = E_\theta(Y_{n+1}|Y^n = y^n)$; however, θ remains to be determined unless, as happens in some rare cases, there is an a that uniformly minimizes $R_\theta(a)$. Otherwise, θ must be estimated by 'plugging in' $\hat{\theta} = \hat{\theta}(y^n)$. However, a plug-in point estimator needn't be the result of a decision-theoretic procedure and a plug-in predictor needn't be the result of a decision-theoretic procedure. Whatever the case, decision theory typically generates only a point predictor, not a PI. So, some assessment of variability must also be given. This is sometimes done on an ad hoc basis: one might assign the variability as $\sqrt{\text{Var}_\theta(Y_{n+1}|Y^n = y^n)}$. Sometimes a variability can be assigned formally, e.g., by bootstrapping to get a range from the values $\hat{Y}_\theta = E_{\hat{\theta}(y^n)}(Y_{n+1}|Y^* = y^*)$, where y^* indicates a bootstrap sample, and then choosing the upper and lower percentiles of the histogram. And, sometimes, a PI is neglected when the goal is a narrow comparison of predictive errors.

There are numerous other methods, some which are extensions of the methods described here, that are also considered frequentist. For instance, one might select from a list of models and obtain a prediction from the model selected. This can be regarded as an estimation problem – choosing a model is often done by estimating an indexing parameter – and hence as a plug-in predictor problem. It can also be viewed as a decision problem – one decides which model to use to obtain a prediction. Indeed, weighted sums of predictions from models may also be derived from decision-theoretic procedures; see Chapter 11. These more elaborate procedures are important, and even when they are carried out within a frequentist paradigm they point to a broader view of prediction than that usually meant by frequentism.

2.3.2 Bayes Approach

Foundationally, Bayesians regard probability as a summary of information that is not necessarily associated with repeated sampling. For the Bayesian, each side of a fair die has the same probability, one-sixth, because it represents the information 'the six sides are symmetric'. One could toss a die many times, count how often each side came up and again be led to one-sixth as the probability for each side, but the argument is different. The view that probability is derived from information, or, more exactly, that under a reasonable set of postulates a collection of logical statements (in the sense of propositional logic) leads to a probability, originates in Cox (1961) and was developed further in Jaynes (1995, Chap. 2). The basic idea is that probabilities represent the plausibilities of propositions, e.g., that the side of a fair die with three dots on it will turn up has plausibility one-sixth, and this is used in prior selection. More typically, a Bayesian might reason as follows: the statement that the population mean is negative has zero plausibility, so the prior density is zero to the left of zero. Also, the statement that the population mean is bigger than $t > 0$ has plausibility that decreases exponentially. So, to the right of zero the prior density is e^{-t}. The interpretation that probabilities represent information in a logical sense recurs regularly in Bayes theory. Indeed, formal codelength arguments from information theory are often used to choose and evaluate priors.

Given this, the key features of Bayes theory have been derived axiomatically; see Bernardo and Smith (2000). Other perspectives on the Bayes approach come from: De Finetti's theorem, which represents the density of an exchangeable sequence of random variables as a mixture of IID random variables with respect to a construct that can be regarded as a prior and hence as information about how the IID experiments combine; Dutch book arguments on how to use information to post odds in gambling scenarios proposed by de Finetti and more formally established by Freedman and Purves (1969); derivations of Bayesian updating as optimal information processing (Zellner 1988). Despite the many variants and justifications of the basic Bayes approach – subjective, objective, empirical, decision-theoretic, etc. – the version presented here is intended to be generic. In particular, information and what it means via conditioning parameters on data, or data on parameters, is central to Bayes thinking in the way that repeated sampling is central to frequentist thinking.

Bayes estimation is conceptually easy. First, the parameter is regarded as a random variable in its own right, Θ with density $w(\cdot)$, not just θ. This is not because the Bayesian believes that Θ is a random variable in reality but rather that the information about the parameter has been converted into a probability. Then, the data are assumed to come from a conditional distribution $P(\cdot \,|\theta) = P(\cdot|\Theta = \theta)$ given the parameter with density $p(y|\theta)$. Given IID data $Y^n = y^n$, the Bayesian writes the joint density for (Θ, Y^n) as

$$w(\theta)p(y^n|\theta) = w(\theta|y^n)m(y^n), \tag{2.27}$$

where $m(y^n) = \int w(\theta) p(y^n|\theta) d\theta$, the marginal for the data or the mixture of densities. Rearranging gives the posterior density

$$w(\theta|y^n) = \frac{w(\theta) p(y^n|\theta)}{m(y^n)}.$$

Thus, the Bayesian follows a principle of containment: there is one big measure space within which all the operations are defined. Also, the Bayesian regards (2.27) not just as a mathematical statement about how densities factor into marginals and conditionals but as the basis for inference. That is, the Bayesian begins with a prior $w(\theta)$ and updates the prior to the posterior. Much of Bayesian foundational reasoning is intended to ensure a correspondence between what is meant by the term 'inference' and the mathematical process of going from a prior to a posterior density. In other words, Bayesians have solid arguments that justify regarding a posterior as an updated prior rather than just as a conditional density that happens to arise from factoring a joint density.

Given this, Bayesians reverse the frequentist treatment. Instead of regarding the parameter as fixed and the data as stochastic, the Bayesian regards the parameter de facto as a random variable Θ and the data y^n once obtained as no longer stochastic. Thus, Bayesians may derive interval estimates for θ from the posterior $w(\theta|y^n)$: a $1 - \alpha$ credible set $R(y^n)$ is any set of parameter values satisfying

$$W(R(y^n)|y^n) = \int_{R(y^n)} w(\theta|y^n) d\theta = 1 - \alpha. \tag{2.28}$$

This is a parallel to (2.25), but it does not invoke the frequency interpretation of probability. Also, expression (2.28) does not directly mean that $R(y^n)$ must track the true value θ_T of θ as a parameter. However, typically, the posterior is consistent, i.e., it concentrates its mass at θ_T as n increases so that $R(y^n)$ ends up tracking θ_T.

The posterior is the Bayesian's analog to the sampling distribution. So, as with (2.25), (2.28) is not by itself enough to specify a sensible region uniquely, in general. Like the frequentist, the Bayesian often chooses a region representing values of θ that have high posterior density (HPD). In this case, a credible set is unique (apart from atoms and sets of measure zero) and assumes the form

$$R(y^n) = \{\theta | w(\theta|y^n) \geq t_\alpha\}$$

for some threshold t_α. As in the frequentist case, $R(y^n)$ can be defined by statistics and these often correspond to highest-posterior-density sets. Since the posterior is the result of updating the prior by the likelihood and data, the interpretation of a credibility region R is that, conditionally on the data, $R(y^n)$ is a set that contains $1 - \alpha$ posterior probability; i.e., unlike the repeated sampling interpretation, the Bayesian gives a direct statement about θ given the data and neglects, perhaps even denies, any variability in y^n. When being more precise, a Bayesian regards the entire posterior density as the post-data summary of information about the parameter. That is, the tails are included rather than clipped off, because the rates of decay in the tails of the posterior are informative about how the credibility of large values of θ decreases.

Thus the primary statistical focus of Bayesians is on the posterior, e.g., on the posterior mean $E(\Theta|y^n)$ or posterior variance $\text{Var}(\Theta|y^n)$. So, it is reasonable to define credibility sets in terms of one or more statistics, bearing in mind that θ need not be finite dimensional. Indeed, the Bayesian, like the Frequentist, might only assume that $Y: (\mathcal{X}, \mathcal{A}) \to (\mathbb{R}, \mathcal{B}(\mathbb{R}))$ and therefore set $\mathcal{P} = \{P \,|\, P \text{ is a distribution on } (\mathbb{R}, \mathcal{B}(\mathbb{R}))\}$. The task is then to assign a reasonable prior to $\mathcal{P}$, so that the posterior can be updated by the likelihood and data to give accurate and useful information about P_T. Regardless of whether $R(Y^n)$ is defined by statistics or by HPD sets or whether the parameter space is nonparametric, the use of sets like $R(Y^n)$ automatically includes an assessment of variability for the estimation of θ inherited from the description of the post-data information provided by the posterior.

To implement this in practice, $R(y^n)$ is often approximated by using a Bayesian version of the central limit theorem to approximate the posterior. The essence of the result is that, for any set A,

$$W(A|y^n) \to \int_A \phi_{\theta_T, I(\theta_T)}(\theta) d\theta \tag{2.29}$$

in P_{θ_T}-probability. For finite-dimensional parameters θ, the focus is often on sets of the form

$$A = \{\theta \,|\, (\sqrt{n} I(\hat{\theta}))^{-1/2}(\hat{\theta} - \theta) \le t\}$$

for $t \in \mathbb{R}$, where $\hat{\theta} = \hat{\theta}(y^n)$ is a $\sqrt{n}$-consistent estimator such as the maximum likelihood estimator, the posterior mode, or the posterior mean and $I(\cdot)$ is the Fisher information. So, the Bayesian often uses approximate $1 - \alpha$ credibility sets of the form

$$R(y^n) = \{\theta \,|\, I(\hat{\theta})^{-1/2} z_{\alpha/2} \le \sqrt{n}(\theta - \hat{\theta}) \le I(\hat{\theta})^{-1/2} z_{\alpha/2}\}$$

when n is large. The posterior variance can be used in place of $I(\hat{\theta})^{-1}$ since $\mathrm{Var}(\Theta|y^n) \to I(\theta_T)^{-1}$. That is, the Bayesian central limit theorem provides an approximation to the posterior distribution, just as the frequentist central limit theorem provides an approximation to the sampling distribution of a statistic. Also, the Bayesian law of large numbers is $E(\Theta|y^n) \to E(Y)$. In this sense, the Bayesian's 'statistic' is the entire posterior.

Bayes hypothesis testing constitutes an optimization problem that also uses the posterior distribution. The paradigm case is testing $\mathcal{H}_0 : \theta \in A$ vs. $\mathcal{H}_1 : \theta \in A^c$, where A is a set in the parameter space. Testing a point null vs. a point alternative is a special case. There are two actions, a_0 meaning 'accept $\mathcal{H}_0$', and a_1, meaning 'accept $\mathcal{H}_1$'. The loss is

$$L(\theta, a_0) = \begin{cases} 0 & \text{if } \theta \in A, \\ 1 & \text{if } \theta \in A^c, \end{cases}$$

and

$$L(\theta, a_1) = \begin{cases} 0 & \text{if } \theta \in A^c, \\ 1 & \text{if } \theta \in A. \end{cases}$$

The Bayesian then finds

$$\arg\min_i E(L(\theta, a_i)|y^n),$$

the minimum posterior risk action. It can be easily derived that a_0 has a smaller posterior risk than a_1 if and only if $W(A|y^n) > W(A^c|y^n)$. The question then becomes whether any difference between $W(A|y^n)$ and $W(A^c|y^n)$ is large enough to be meaningful. To assess this, Bayesians calibrate the Bayes factor

$$\text{BF}_{0,1} = \frac{W(A|y^n)/W(A^c|y^n)}{W(A)/W(A^c)},$$

the ratio of the posterior odds to the prior odds, and usually agree that $\text{BF}_{0,1} > 3$ is a difference worth noting statistically. This choice corresponds to asking whether the post-data odds ratio has changed enough from the pre-data odds ratio that the difference in information represented by $W(A|y^n)$ and $W(A^c|y^n)$ is large enough to be worth noting. This is very different from the p-value, which, under a repeated sampling interpretation, estimates the probability under $\mathcal{H}_0$ of, in a future experiment, getting an outcome more different from $\mathcal{H}_0$ than that given by the current data. Overall, the Bayesian treats $\mathcal{H}_0$ and $\mathcal{H}_1$ symmetrically and chooses one over the other on the basis of the information that the data represent, as summarized by the posterior.

Bayesian decision making rests on utility theory or decision theory. At root is Savage's expected utility theorem. This is a Bayesian version of the von Neumann–Morgenstern expected utility theorem. If an agent expresses preferences as $A \succ B$, meaning that option A is preferred over option B, then under a set of axioms (see Savage 1954) there is a utility function U and a subjective probability measure such that the ordering of options under the preference relation $\succ$ is equivalent to the ordering of options according to the size of the expected utility of the options. That is, $A \succ B \iff E(U(A)) > E(U(B))$, but now the expectation is with respect to a subjective probability measure. Blurring the distinction between a utility function and the negative of a loss function, Bayesian decision making is based on looking at the Bayes risk in Wald's decision theory. That is, one chooses a loss and integrates the expected loss, or risk, of each action in the action space with respect to a prior. This is equivalent to ordering actions by their posterior risk. In either case, one seeks a minimum risk solution, be it Bayes or posterior.

Decision-theoretically, Bayes rules have a variety of desirable properties. First, the collection of Bayes rules forms a complete class. That is, given any non-Bayes rule there is a Bayes rule that is better than it. Also, unique Bayes rules are admissible and the limits of Bayes estimators are maximin. Under extra conditions maximin rules are minimax, too. Indeed, Bayes hypothesis testing was framed as a decision-theory problem apart from the calibration of the Bayes factor. This means that any loss function thought appropriate, not just zero–one loss, could be used to test hypotheses. Likewise, Bayes estimation can be formulated decision-theoretically; the 'decision' regards which value to announce for θ given the data. However, decision theory alone gives only a point estimate; it does not usually assign a variability to an estimate. So this would have to be done separately.

Bayesian prediction has several forms. First, one can obtain predictors (point or interval) from the predictive density

$$m(y_{n+1}|y^n) = \frac{m(y_{n+1}, y^n)}{m(y^n)}. \tag{2.30}$$

Aside the fact that (2.30) is the natural predictor from the density from $n + 1$ outcomes under the mantra 'joint density equals conditional times marginal', it is also optimal in a relative entropy sense; see Aitchison (1975). Interval predictors can be formed by identifying functions $L(Y^n)$ and $U(Y^n)$ such that

$$M(L(y^n) \leq Y_{n+1} \leq U(y^n)|y^n) \geq 1 - \alpha. \tag{2.31}$$

The functions L and U will usually be unique if an HPD region from $M(\cdot|y^n)$ is chosen. Expression (2.30) also gives the point predictor $\hat{Y}_{n+1} = E_{m(\cdot|y^n)}(Y_{n+1})$, where the subscript on E indicates the distribution in which the expectation is taken. Using this $\hat{Y}_{n+1}$ is essentially never identical to using $\hat{Y} = E_\theta(Y)$, for any θ.

More generally one can use the same procedure on a member of the class

$$m_q(y_{n+1}|y^n) = \frac{m_q(y^n, y_{n+1})}{\int m_q(y^n, y_{n+1})\mathrm{d}y_{n+1}} \tag{2.32}$$

where

$$m_q(y^n, y_{n+1}) = \left(\int w(\theta) p(y^{n+1}|\theta)^q \mathrm{d}\theta \right)^{1/q},$$

with obvious modifications if $w(\cdot)$ is replaced by $w(\cdot|y^n)$ and $p(y^{n+1}|\theta)$ is replaced by $p(y_{n+1}|\theta)$. Here, q parametrizes a class of densities and controls how much weight the modes of the densities receive relative to the tails. Analogously to the relative entropy argument in Aitchison (1975), the densities $m_{1/2}$ and m_2 are optimal under the chi-squared and Hellinger distances, respectively.

Alternatively, the Bayesian can frame the prediction problem in terms of decision theory, in which the best prediction is the optimal action. That is, suppose the Y_i are IID, choose a loss function L, and an action space $\mathcal{A}$ and examine the posterior risks

$$R(a; y^n) = \int L(a, y_{n+1}) p(y_{n+1}|y^n)\mathrm{d}y_{n+1} \tag{2.33}$$

for $a \in \mathcal{A}$. The optimal action, i.e., the best predictor for Y_{n+1} using Y^n, can sometimes be identified. For instance, when $L(a, y_{n+1}) = (a - y_{n+1})^2$ one can minimize (2.33) to obtain the posterior predictive mean $\hat{Y} = E_{m(\cdot|y^n)}(Y_{n+1})$, the mean of Y_{n+1} in the predictive distribution. In this case, the posterior risk is

$$\mathrm{Var}(Y_{n+1}|y^n) = \int (y_{n+1} - E_{m(\cdot|y^n)}(Y_{n+1}))^2 m(y_{n+1}|y^n)\mathrm{d}y_{n+1}, \tag{2.34}$$

the posterior variance. If $L(a, y_{n+1}) = |a - y_{n+1}|$ then one can minimize (2.33) to obtain $\hat{Y}_{n+1} = \mathrm{med}\ Y_{n+1}$, where med is the median of $m(y_{n+1}|y^n)$. Again, an expression of the form of the right-hand side of (2.34) is the natural choice for assessing variability.

Both point and interval predictors from either the predictive distribution or the posterior risk are conditional on the data. So, their properties are found in terms of the conditional density $m(y_{n+1}|y^n)$, not the marginal probability $M_{Y_{n+1}}$ for Y_{n+1} or any P_θ. Bayes predictors may be preferred because they may vary less owing to the conditioning on y^n. However, a different outcome $(y^n)'$ would give a different interval, i.e., one from $M(\cdot|(y^n)')$, which would be just as appropriate for Y_{n+1} as the one from $M(\cdot|y^n)$. If this variability in y^n is important, Bayes intervals might be a little too narrow to represent the variability in Y_{n+1} accurately, especially for small n.

There are numerous other methods, some of which are extensions of the methods described here, that are also considered Bayesian. For instance, one might have a list of parametric models each equipped with a prior and would therefore assign another prior over the models. Then, analogs of the procedures described here could be applied for estimation, testing, decision making, and prediction. Indeed, there will be a well-defined posterior and a marginal posterior across models. In fact the predictor from the Bayes model average, see Hoeting *et al.* (1999), under squared error loss is just a posterior predictive mean. This is an example of a weighted sum of predictions from a collection of models that can be considered Bayesian. These more elaborate procedures are important, and even when they are performed within a Bayesian paradigm they point to a broader view of prediction than that usually held by Bayesians; see Chapter 11.

2.3.3 Survey Sampling

Unlike the usual frequentist or Bayesian approach, foundationally the Survey Sampler assumes that the population is finite. The population may be very large, so that treating it as if it were infinite might give a convenient approximation, but such an approximation would require justification. Second, the Survey Sampler by and large does not assume any variability in the response. (An exception occurs where the subjects' responses include measurement error that is separately modeled.) That is, where the Frequentist or Bayesian assumes that the outcome Y is a random variable, at least before it is measured, the Survey Sampler assumes that the responses are fixed but unknown (again apart from any measurement errors on the subjects). Hence, the variability is not in the response itself but in whether the subject was selected for a sample. This means that the random variable Y represents the selection of a subject from a population, not the response of a selected subject.

Thus, the Survey Sampler starts with a population of size N, and if the plan is to select a sample of size n then there are $C(N, n)$ possible samples that one might obtain. Accordingly, Survey Samplers focus on the details of how the n subjects were selected. Simple random sampling without replacement (SRS) is the most basic procedure, and more complex sampling designs (stratified, cluster, etc.) usually incorporate SRS. One effect of a finite population, as seen, for instance, in stratified sampling, is that it necessitates the incorporation of weighting schemes to ensure that the relative contributions of outcomes to inferences match the proportions of subjects represented by the outcomes. This makes the analyses of survey data highly design dependent. So, any generic survey sampling description is less representative of survey sampling as a whole than the corresponding descriptions for frequentism or Bayes. Nevertheless, it is worth reviewing a generic survey sampling case at least for comparison to other conceptualizations of a paradigm statistical problem.

Estimation in survey sampling settings is conceptually easy and superficially almost indistinguishable from that in frequentism. For instance, given an SRS y^n of size n one can estimate the population mean $\mu = \sum_{i=1}^{N} y_i / N$ by $\bar{y}$ and assert that $E\bar{Y} = \mu$. Likewise, one can estimate the population variance $\sigma^2 = \sum_{i=1}^{N} (y_i - \mu)^2 / (N - 1)$ using $\text{Var}(\bar{Y}) = E(s^2) = (\sigma^2/n)(1 - n/N)$; in typical cases, $n/N \approx 0$ so $\text{Var}(\bar{Y}) \approx \sigma^2/n$ as in the frequentist paradigm. (The definition of σ^2 uses $N - 1$ rather than N strictly for convenience; obviously $N - 1 \approx N$ for any reasonable N.) However, the derivations of these expectations use the distribution of $\bar{Y}$ or s^2 obtained from considering all possible samples of size n. For instance, $E(\bar{Y})$ is the sum of all possible values of $\bar{Y}$ from all samples of size n weighted by the probability of choosing a sample that gives the value $\bar{Y}$; $E(S^2)$ is similar.

The question is how to get confidence intervals. One can use the usual 'mean plus factor times SE' with $\text{SE} \approx s/\sqrt{n}$ but the correct factor comes from the sampling distribution of the mean, which itself is derived from the distribution of samples of size n; this is usually inconvenient to determine. Consequently, the Survey Sampler usually employs a normal approximation, i.e., takes the factor to be a normal percentile. Thus, $(1-\alpha)100\%$ confidence intervals are approximately of the form

$$\bar{y} - z_{\alpha/2} \frac{s}{\sqrt{n}} \le \mu \le \bar{y} + z_{1-\alpha/2} \frac{s}{\sqrt{n}}. \tag{2.35}$$

Other parameters, such as proportions, percentiles, and standard deviations, can be treated analogously. Models that include measurement error such as bias, nonresponse, or correlation among sampled units require more complex analysis but still refer back to the finite population and sample. Indeed, one reason why stratified sampling gives smaller confidence intervals for a population mean is that the number of possible samples is reduced by the stratification from what it would be under SRS.

The normal approximation used to get (2.35) is parallel to the frequentist or Bayesian versions of the central limit, but of a different character. The idea is to regard the population at hand, of size N, and a sample of size n as one such pair in a sequence of populations and samples. Thus, imagine population j of size N_j and a sample of size n_j from it such that $N_j, n_j, N_j - n_j \to \infty$. With some extra conditions on the μ_j, σ_j, and the samples, one can argue that $\bar{y}$ is normal with mean μ and variance of the form σ^2/n asymptotically. Note that this imaginary sequence of populations and samples is only invoked to get intervals with approximately $1 - \alpha$ confidence over repeated sampling of the one population at hand. It is not essential to the repeated-sampling interpretation of (2.35) for the finite population. Bayesian survey sampling has also been developed but is still a minority viewpoint among Survey Samplers.

Hypothesis testing in survey sampling usually consists of one of three approaches. First, the data are categorical or discrete (say integer valued) and can be modeled by a well-known distribution such as the multinomial or hypergeometric. In this context, standard frequentist hypothesis testing is normally used. Likewise, if the data generate a contingency table, standard frequentist tests like the chi-squared test are often used. Second, the data are continuous in the sense that any real number is possible in principle. In such cases, the normal approximation is invoked and various tests, e.g., Wald's test, can be used. Again, this is indistinguishable from the frequentist approach even though the data come from a single finite population. Third, models can be proposed for the data and priors assigned to the parameters. Then Bayes tests become applicable. Overall, in survey sampling, testing is either de facto frequentist or Bayesian with the added burden of sensitivity to the sampling design. See Ghosh and Meeden (1997) for a Bayes treatment.

Decision making in a survey sampling context is relatively undeveloped. However, an example of how to use survey data to do a cost–benefit analysis for decision making can be found in Gelman *et al.* (2003).

Prediction in a survey sampling context has several forms. First, one can try to predict the next outcome, i.e., the response of the next sampled subject. This sort of procedure is relatively uncommon outside adaptive sampling but would be done in a frequentist or Bayesian way, as discussed earlier. Second, if one has a subject-level model, e.g., a linear model valid for all subjects in a population, then one can try to predict some feature of a subject from the other features of the subject. Essentially, this treats subjects as if they were from an infinite population. Both plug-in (frequentist) and Bayes estimation methods can be used. An interesting version of this involves treating response versus nonresponse as a binary variable; thus the issue of response versus nonresponse is turned into a classification problem, to which both frequentist and Bayes methods can be applied. (This could help in understanding whether a nonresponse appears to be linked to the sampling scheme or to other features of the subjects.) Third is the sort of prediction indicated in (2.5). The goal is to 'fill in' missing data, i.e., the unsampled population units, so that the value of a function on the finite population can be estimated. However, this can only be done for some functions. When it is possible, there are two ways to predict the unsampled subjects; one is exemplified in the survey sampling treatment of the Sunrise Problem, namely, the usual estimation of a total (or, more generally, the estimation of some simple function of the population by a standard approach). Alternatively, sometimes a parametrized likelihood of the form $L_\theta(S^c|S)$ for the unsampled responses S^c given the sampled responses S can be identified. If so, then the point prediction can be done in a variety of ways and in some cases a PI can be given. These methods are variations on the frequentist or Bayesian methods but they are used in a survey sampling context; see Bjørnstad (2010) for details.

There are numerous other methods, some that are extensions of the methods described here and some, that involve the shifts of emphasis that arise in survey sampling, for instance, concerns about bias and nonresponse. Indeed, there are a large number of sampling designs and each has its own form of analysis, whether it's by means of geographical designs (which use Kriging, a sort of prediction for interpolation purposes), capture–recapture sampling, adaptive sampling, designs to reduce nonresponse, designs for small-area estimation, and so forth. There are also models and hence, in principle, model selection or averaging for survey populations. However, techniques in these cases, even when they are developed, are mostly adaptations of frequentist or Bayes methods to the large-but-finite population context. Fundamentally, the survey sampling view of the paradigm statistical problem is based primarily on estimation; testing is relatively rare and prediction and decision are especially rare. So, in the absence of new developments, the survey sampling view, being tethered to finite populations, does not point to a broader view of prediction, unlike the frequentist or Bayes methods.

2.3.4 Predictivist Approach

The frequentist, Bayes, and survey sampling approaches see the paradigm statistical problem as one of estimation: that of deriving tests, decision making procedures, and predictions from the estimation framework. By contrast, the predictivist approach defines the paradigm statistical problem as that of predicting a future, or at least unseen, outcome. In forming a predictor the Predictivist may estimate parameters, use hypothesis tests, or make decisions but is unconcerned with the properties of such procedures except insofar as they lead to good prediction. Even then, if no meaningful interpretation can be given, the Predictivist is not dissatisfied. The reason is that, aside from their use in forming predictions, estimation, testing, and making decisions are generally aimed at understanding the data generator (DG). However, to a Predictivist, most real DGs are too complex to model. Whatever aspects of them can be modeled are likely to be artificial and probably too expensive to be worthwhile, outside special cases. It makes little sense to build a statistical theory around the rare cases where good modeling is feasible rather around than the typical case where good modeling is a false hope.

Essentially, the Predictivist reverses the approaches of the other three schools of thought. Instead of modeling, estimating, and then predicting, the Predictivist finds a good predictor first. Then, if examination of the predictor can generate modeling inferences, the Predictivist is happy to find them. Indeed, if a complex but extremely good predictor, e.g., a random forest, can be simplified down to a predictor that is interpretable, e.g., a single tree, the Predictivist has no objection – as long as the modeler is aware that the interpretable model is not as good predictively as the best predictor.

The Predictivist also acknowledges that Bayes methods have the flaw that, from within the Bayes paradigm, bias cannot be detected even though it is usually very important, especially with complex DGs. Recognition of this fact that has led Bayesians to 'calibrate' their inferences over the long run, essentially invoking a frequentist check on accuracy; see Dawid (1982) for one of the earliest and most salient explications. Likewise, the Predictivist admits that frequentist methods do not do as well as Bayes methods when it comes to the quantification of uncertainty – especially model uncertainty – and it must be admitted that the frequentist notion of confidence is distant from real-world experimentation. Thus, predictive performance simultaneously assesses the bias that Bayes structure does not accommodate well and the variability that frequentism does not accommodate well. As a side point, it is entertaining that frequentist hypothesis testing over-rejects the null hypothesis while, arguably, Bayes testing under-rejects the null. The Predictivist therefore eschews testing unless the hypotheses are evaluated predictively – a sort of compromise between the Bayes and frequentist philosophies.

The predictive stance is not borne of despair that modeling is so rarely feasible. Rather, it is borne of the recognition that model uncertainty or model misspecification is ubiquitous and hence a necessary component of an analysis – not one to be avoided or necessarily even minimized. Admittedly, this view is slightly extreme because there are well-controlled experiments in some fields where it is worthwhile and possible to develop precise models and test them to ensure that they cannot be discredited. However, as already noted, these cases are the exception. They are not representative of the vast majority of DGs that analysts are called to study. Indeed, as a generality, if model uncertainty cannot be eliminated – and it rarely can be – is not the main goal of modeling discredited? If so, then there is little but prediction, and interpreting predictors, that makes sense.

Above all, the goal of prediction is to make direct statements about an outcome that is measurable, not about a parameter or any other object that might be imagined. So, from the predictivist standpoint the frequentist, Bayes, and survey sampling techniques are just ways to construct predictors. Bayes' theorem is a mathematical fact that follows from factoring joint distributions and one needn't be a Bayesian to do this. Likewise, as powerful a concept as the sampling distribution is, it's abstruse and doesn't matter except possibly as a way to evaluate a predictive scheme under a set of specified assumptions. Thus, the predictivist approach doesn't deny the meanings or interpretations of many other techniques, it just ignores them. Philosophically, therefore, the predictive paradigm is a universalist approach: it doesn't matter how a predictor was found, it just matters that the predictor predicts well.

The meaning of 'Predicting well' has to be defined. The basic test of a predictor is whether it produces predictions that are close to the actual outcomes. This cannot be established theoretically except in a proof-of-concept sense because real data are produced by a physical process, the DG, not by a mathematical model that one might wish to assume. A focus on the linkage between a predictor and the real DG was already present in the computed examples of Chapter 1 in which FPEs were found. There, the assessment of the aggregate performance of a predictor was based on point prediction. The reason is that the only valid comparison is between outcomes and their predictions. As an alternative, one could compare an outcome with the probability of its occurrence as with a PI, i.e., one could ask whether an outcome was in a given prediction interval. However, this is less informative than comparing y_{n+1} with $\hat{Y}_{n+1}$ directly. Indeed, the aggregate effect of comparing predictions with their corresponding outcomes will include an assessment of variability akin to that provided by the intervals associated with point predictors.

It is obvious, however, that merely comparing predictions with outcomes is insufficient. After all, two predictors making the same set of predictions for the same data may be strikingly different away from the data points collected. This is why model uncertainty or misspecification is fundamental to Predictivists. The Predictivist would argue that the nonuniqueness of predictors equivalent under FPE or CPE means that model uncertainty is a defining feature of DGs outside rare, simple, often expensive settings. Otherwise stated, problems that have a high enough complexity invariably lead to high model uncertainty because they may have approximations that are useful only in narrow contexts or they may be so complex that good prediction (and hence modeling) is effectively impossible. As a practical point, without admitting and managing model uncertainty it is very difficult to achieve a satisfactory generalization error.

The basic predictive framework is the following. Suppose there is a sequence of outcomes y_i, $i = 1, \ldots, n$ corresponding to random variables Y_i and, for each i, there is a vector of explanatory variables $\mathbf{x}_i = (x_{i,1}, \ldots, x_{i,k})^T$. Given n data points $(x_1, y_1), \ldots, (x_n, y_n)$, the task is to find $\hat{Y}_{n+1}(\cdot)$ such that the point prediction $\hat{Y}_{n+1}(x_{n+1})$ is close to $Y_{n+1}(x_{n+1})$. This structure includes familiar signal-plus-noise models in which it is assumed there is some function $F(\cdot)$ independent of i such that, for each i, $Y_{i+1}(x_{i+1}) = F(x_{i+1}) + \epsilon_{i+1}$ where ϵ_{i+1} is the mean-zero random noise term, which is almost always assumed to have finite variance when it is permitted to be nonnormal. In these cases, it is more direct to write $\hat{Y}_{n+1}(\cdot) = \hat{F}(\cdot)$. Note that the x_i may be outcomes of random variables X_i or fixed design points and that, although the entries in the x_i are assumed to be the same for all i, this is not necessary. Different explanatory variables may be available at different times.

The simplest version of this framework, i.e., one with no explanatory variables and simple parametric families for the predictand, was used in Sec. 2.2. There, the properties of the frequentist point predictor $\hat{Y}_{n+1} = \bar{y}$ and the Bayes point predictor $\hat{Y}_{n+1} = E(Y_{n+1}|y^n)$ were given. Predictive intervals for these point predictors were also given; they were based on the probability for Y_{n+1}. However, it is immaterial whether one wishes to invoke the frequentist repeated-sampling definition of probability or to invoke the Bayes interpretation of probability as an assessment of information. After all, the goal is not to interpret probability. Rather, it is to get predictions that match future observations, and if mathematical objects satisfying Kolmogorov's axioms are useful then there is little to debate.

Whether one looks at the real data examples in Secs. 1.2.1 and 1.2.3 or the theoretical treatment of Sec. 2.2, there are some key features that can be listed. First, for a given problem one must propose a predictor or more precisely a sequence of predictors. Usually this is done by choosing a class of predictors and then letting some mix of the data and theory suggest a predictor at each time step that can be tested in terms of the FPE or CPE. That is, a key quantity to be computed for a predictor $\hat{Y}$ is its mean FPE:

$$\text{FPE} = \frac{1}{K} \sum_{k=1}^{K} L\left(\hat{Y}_{\sigma_k(1),\dots,\sigma_k(n-1)}(x_{\sigma_k(n)}), y_{\sigma_k(n)}(x_{\sigma_k(n)}) \right), \tag{2.36}$$

where L is the chosen way to assess the closeness of an outcome and predictor and σ_k is one of K randomly chosen permutations of $\{1, \dots, n\}$. However, more generally, the distribution of the FPE is of interest and can be estimated using the K values in the sum in (2.36). In this case the collection of randomly chosen permutations of the data generated a set of values from which a five-number summary of the histogram could be given. Looking at the whole distribution, or a summary of it, indicates the variability of the FPE and gives an indication of the stability of the final predictor. This is an assessment of model uncertainty as well as of generalization error. In the squared error case, it admits a variance–bias interpretation.

The issue of how to choose a suitable L and a good predictor class depends on many features of the data, including its complexity and whether one expects that a simple predictor will provide a good approximation to a complex but optimal predictor. For L, one must decide the sense in which one wants to predict well; absolute value has different properties from squared error or classification risk, for instance. As to the predictor class, experience and mathematics suggest that the more model uncertainty is permitted, the worse interpretable methods tend to perform. So, one wants as large a class of predictors as possible subject to there being the right number of constraints: not so few that the models are genuinely interpretable but not so many that the data cannot discriminate among them effectively, and one has 'overfit'. That is, the model selection problem of the Bayesian or Frequentist has been turned into a predictor class selection problem. This may not be seen initially as an improvement but it is, because it uses less information (which is usually unreliable) and because getting a good predictor is easier than getting a good model.[1] The predictor just predicts and cannot be discredited by any assertions it makes about the nature of the underlying DG, because they are so minimal.

To a Predictivist, the goal of an analysis is finding a predictor. After testing out several well-motivated predictor classes, the Predictivist would select the best of the final predictors or perhaps combine the best final predictors from each class in some way. The best predictors will usually be constructed using estimates of parameters or other conventional inference methods in their formation (e.g., trees are often built by hypothesis testing at nodes), but the evaluation of the predictors is done separately from their construction and does not require any assumptions about the DG. The overall final predictor would be announced as the key product of the analysis.

To see what sort of inferences one might derive from predictors, which might be analogous to the inferences one obtains from estimation, recall that a predictive analysis was done in Sec. 1.2.1 for three settings using four predictors; linear models, L2-BMA, RVM, and BRVM. So, if one were interested in prediction for the Tour data one would be led to L2-BMA or BRVM. In principle, L2-BMA could be converted into a linear model (but the coefficients would not be from a least squares optimization), given an interpretation in

---

[1] Here a good model is one that predicts well. It is usually easy to find a model that fits well but predicts poorly.

terms of which variables were most important and how much, say, a unit change in one of the explanatory variables would affect the response. However, BRVM, which had a lower median for its FPE distribution, would have to be, for instance, Taylor expanded to find the leading terms. The Taylor expansion would permit an interpretation again in terms of which variables were most important and how much, say, a unit change in one explanatory variable would affect the response, at least approximately.

A similar instance of this is seen with the conversion of the nonresponse problem in survey sampling into a prediction problem. In McCarthy and Jacob (2009) survey nonrespondents are predicted from their various known features by using classification trees. Once good prediction was achieved, the tree was examined to find which characteristics of the sampled units were good predictors of nonresponse. It turned out that the only variables included in this study that were helpful were indicators of how willing to participate subjects had been in the past. That is, other typically descriptive variables did not help distinguish between respondents and nonrespondents. One could have derived the same inferences, at least in principle, from a modeling approach. However, the inferences are much more convincing when there are no references to models that often do not exist or cannot even be imagined in sufficient detail to be 'meaningful'.

If desired one can use predictive methods to define an analog of point or interval estimation and hypothesis testing. Suppose that a class of predictors $F_\theta(x)$ for Y has been found to give good prediction in the sense that there is a value $\hat{\theta}$ for which $\hat{F}_{\hat{\theta}}(x)$ achieves a low FPE in comparison with the best predictors of other classes. Then, FPE in (2.36) can be written as FPE($\hat{\theta}$). So, $\hat{\theta}$ can be taken as a point estimate of θ and, holding the data fixed, one can define an interval for θ of the form

$$\{\theta | \mathrm{FPE}(\theta) - \mathrm{FPE}(\hat{\theta}) \leq t\}.$$

Here, t plays a role analogous to the normal percentile in conventional interval construction and would have to be chosen on some rational basis, e.g., how much increase in predictive error one is willing to tolerate. One choice is to set $t = \alpha\mathrm{FPE}(\hat{\theta})$, i.e., one is willing to tolerate an $\alpha\%$ increase in the FPE. This is analogous to the formation of intervals based on maximizing likelihood, but in fact there is no likelihood and one is minimizing predictive error without asserting anything about the DG except that the FPE from F_θ is low.

The situation is similar with regard to testing $\mathcal{H}_0: \theta \in A$ vs. $\mathcal{H}_1: \theta \in A^c$, as long as it can be argued that some element $F_{\hat{\theta}}$ of a predictor class F_θ gives a low FPE, so that one can choose between $\mathcal{H}_0$ and $\mathcal{H}_1$. Given data, one can simply ask whether $\hat{\theta} \in A$. If yes, A is accepted; otherwise A^c is accepted. If one wanted to test $\mathcal{H}_0: \theta = \theta_0$ vs. $\mathcal{H}_1: \theta = \theta_1$ then one merely chooses the hypothesis $i = 0, 1$ that achieves

$$\min\left(|F_{\hat{\theta}} - F_{\theta_0}|, |F_{\hat{\theta}} - F_{\theta_1}|\right).$$

In all these cases, one could just as well use a cross-validation error or CPE in place of the FPE and get similar methods. This is the way a in which hypothesis testing is done with information-theoretic statistics: The parameter leading to the shorter Shannon codelength to the data is preferred. Here, the parameter leading to the smaller FPE is preferred and would be scaled in terms of the FPE; for example, θ_0 is accepted because it caused a $q\,100\%$ increase over the minimal FPE whereas θ_1 caused a $q'\,100\%$ increase over the minimal FPE, and $q' > q$ with $q' - q$ large enough. So, one can define and calibrate predictive estimation and testing procedures. However, they only have a predictive interpretation because the predictor does not necessarily correspond to a model, true or otherwise.

Finally, for decision making purposes one can use a predictor as the best guess for what future values might be. That is, given a predictor $\hat{F}(\cdot)$ one can find the prediction $\hat{F}(x_{n+1})$ for any x_{n+1}. So, to make a decision one would use $\hat{F}$ as a reference and minimize the risk or expected loss,

$$\int L(a, \hat{F}(x_{n+1})) p(x_{n+1}) \mathrm{d}x_{n+1},$$

to find the best action a under uncertainty in x_{n+1} given that one believes that x_{n+1} would lead to $\hat{Y}_{n+1} = \hat{F}(x_{n+1})$. This necessitates assessing the distribution of X_{n+1} because otherwise there is nothing random about $\hat{F}(\cdot)$.

There are numerous other methods, some that are extensions of the methods described here and some that represent shifts of emphasis which arise in prediction; the bulk of this book consists of an exploration of them. It will be seen that many predictors arise from specific models because sometimes the model uncertainty is small enough that this is not unreasonable. However, for more complex problems the predictors that are most successful include model uncertainty in one way or another – perhaps by reselecting models at different time steps, effectively searching a collection of predictors to find the good ones, as perhaps by averaging over a collection of predictors, or in some cases by using the information in the data more thoroughly. In simple cases (to be called $\mathcal{M}$-closed in the next chapter) these correspond to the standard Bayes or frequentist approaches. Outside simple cases where models or priors are problematic (to be called $\mathcal{M}$-complete or $\mathcal{M}$-open cases in the next chapter) the resemblance to Bayes or frequentist techniques is weaker. Fundamentally, the predictive view is best suited to settings where model uncertainty is high. Roughly, when the difficulty of the statistical problem is high enough, all one can do is try to predict, and to do this any technique is fair game.

2.4 A Unified Framework for Predictive Analysis

The main point of this section is to outline a practical set of principles for predictive statistical analysis. These are more precise than the discussion of principles in Sec. 2.3.4. So, suppose that a sequence of predictors $\hat{Y}_{n+1}$ for $n \geq 1$ has been identified. There are many possible choices for such a sequence. The simplest were given in Sec. 2.2. Many more will be developed in the succeeding chapters. For the present, it is enough to consider a generic choice of predictor sequence and ask what properties it should have ideally.

In fact, it is not hard to write down a list of desirable properties for a predictor sequence. The issue is to ensure that the list is comprehensive without being too long; the list below is one possibility. Each item on it identifies an aspect of a predictor sequence that most people would agree is important to assess. The list is meant to organize the criteria for good prediction without weighting their relative importance, which is likely to be setting-specific. For instance, in a particular setting one might be willing to tolerate a larger CPE for the sake of higher robustness, in the same way that one might prefer L^1 over L^2 loss because it gives optima that are more stable even if the cost is a higher variance. The list of seven 'desiderata' offered here is one possible way to structure how one might think about the various desirable properties of a predictor sequence.

1. *The method of evaluation of a predictor $\hat{Y}(\cdot)$ should be disjoint from its method of construction, e.g., it should depend only on the predictions it makes and the future data.*
   This is the Prequential Principle, enunciated in Dawid (1984). It is a fundamental way to ensure that predictions are assessed fairly, i.e., it ensures that all predictors compete equally. For instance, modeling assumptions that may go into the formulation of a predictor play no role in its evaluation. This automatically means that once the method of comparison of predictions with outcomes has been determined, frequentist, Bayes, or decision-theoretically optimal predictors are granted no special status.

2. *The predictor $\hat{Y}_n$ should be evaluated by how close the predictions are to the outcomes using an FPE, a CPE, or a similar evaluation such as cross-validation that satisfies the prequential principle.*
   This desideratum was satisfied in the first two examples of Sec. 1.2.1 by examining the distribution of the FPE via its five-number summary. This effectively included (1.8) or (1.9). However, it would be just as valid, though different, to use the CPE as in (1.4) or (1.6) and there is no necessity to use the squared error for comparing an outcome with its prediction. This desideratum was also satisfied in the music example of Sec. 1.2.3, where a test set was used with misclassification error. The choices for the method of comparison are relatively unconstrained apart from ensuring that the comparison of the $\hat{Y}_n$ is to their respective y_n only. In fact, the use of a loss function is not necessary either; it is enough to have a sequence of values of the form $L_i = L_i(\hat{Y}_i, y_i)$ that *enable a comparison of* the $\hat{Y}_i$ with the y_i, for some functions L_i and an operation on the sequence $L_1, \ldots, L_n$ that summarizes them, provided that $\hat{Y}_i$ does not depend on y_i.

3. *The predictor $\hat{Y}_{n+1}$ should be reselected as n increases in response to the prediction errors.*
   When the value of any L_i is high, the ith prediction is poor. This may indicate that there is a problem in the formulation of $\hat{Y}_{n+1}$. If so, some component of $\hat{Y}_{n+1}$ must be changed to improve the predictive error, e.g., the model list or the predictor class. Nevertheless, there may be little to gain from changing $\hat{Y}_i$ if all the L_i are small. Indeed, if all the L_i are small, it is hard for the other desiderata to fail.
   In the data-driven examples of Secs. 1.2.1 and 1.2.3 and the theoretical examples of Sec. 2.2 the predictors were not in fact updated because there was no obvious need to do so. The FPEs were not high and, in work not shown, the CPEs do not give unusual jumps as more data is included (apart from a burn-in phase when the sample size is small). With more complicated problems, reselection is important to ensure that the predictor sequence searches a large enough space of predictors to find good ones; see Clarke and Clarke (2009).

4. *A comprehensive 'variability–accuracy' analysis for $\hat{Y}_n$ should be generated.*
   Other things being equal, the smaller is the combination of variability and accuracy, the better is $\hat{Y}_i$. Here, variability means the comprehensive variability of all random quantities that are used to form $\hat{Y}_n$ and accuracy means the collection of errors associated with these random quantities. In the case of the squared error, a variability–accuracy analysis reduces to a variance–bias analysis. The usual variance–bias decomposition of the form

$$\text{MSE}(\hat{Y}_{n+1}(x_{n+1})) = \text{Var}(\hat{Y}_{n+1}(x_{n+1})) + (E\hat{Y}_{n+1}(x_{n+1}) - F(x_{n+1}))^2$$

for the mean standard error (MSE) of a signal-plus-noise model is a quite simple analysis since the only random quantity is $\hat{F}$. Analyzing this decomposition permits the determination of whether the prediction problem stems from bias or variability. Only the latter can be improved by sample size. More generally, $\hat{F}$ is not the only quantity whose variance and bias must be assessed and the squared error is not the only way to combine assessments of the variability and accuracy of a predictor sequence. Examples of more general decompositions can be found in Domingos (2000) and Clarke (2010).

5. *A comprehensive robustness analysis should be generated for $\hat{Y}_{n+1}$.*

   In this case, there will be an optimal user-chosen level of robustness. Too much robustness would mean that the predictor sequence is insensitive to the DG; too little robustness could give unacceptable instability and/or dependence on aspects of the predictor or noise rather than on a signal. Two sorts of robustness assessments are combined in a comprehensive robustness analysis. The first applies to the fixed choices made to form $\hat{Y}_{n+1}$, such as an objective prior, a family of likelihoods, or previous data. One wants to assess the overall effect of locally varying these quantities to ensure that minor deviations do not cause excessive changes in $\hat{Y}_{n+1}$. The second applies to the arbitrary choices made to form $\hat{Y}_{n+1}$. Often the model list and the prior on the model list in a Bayes model average are arbitrary: other model lists, based, e.g., on different functions of the explanatory variables or 'features', and other choices of prior on them may make equally good sense from a predictive standpoint. In this case, stochastically varying the chosen quantities would make sense. For instance, one could assign a hyperprior over a range of model lists and priors on them to see how varying the choices according to the hyperprior affected $\hat{Y}_{n+1}$. Taken together, these two senses of robustness – fixed and stochastic – provide a comprehensive assessment of the robustness of $\hat{Y}_{n+1}$. Bayesians have long been concerned about prior sensitivity; see Berger (1994) and, more recently, Zhu *et al.* (2011) and the references therein for a recent treatment of the comprehensive robustness of Bayes procedures that could in principle extend to predictive settings. This sort of comprehensive analysis does not seem to have been conducted from a frequentist standpoint.

6. *The complexity of the DG should match the complexity of the predictor $\hat{Y}_{n+1}$.*

   This does not prescribe how the complexity of the DG or of the predictor should be measured, only that the two should be related. There are various assessments of complexity available, such as the Vapnik–Chervonenkis dimension or codelength. One expects more complex DGs to require more complex predictors. In the estimation context, minimum complexity alone can determine a consistent estimator for a distribution – see Barron and Cover (1990), Schmidt and Makalic (2012), or Schmidt (2007) – but these results require that the set of distributions should be large enough to contain the true distribution.

   As an instance, a neural net predictor may be very complex compared with a linear model predictor in a Vapnik–Chervonenkis-dimension sense, while both have similar FPEs, similar robustness, and equivalent variability–accuracy decompositions; neural nets often have higher variability and lower inaccuracy and the linear model often has lower variability and higher inaccuracy. Then, one would have to decide whether the added complexity of the neural network was worthwhile, given the complexity of the DG.

7. *In the limit of large or repeated samples the behavior of the predictor or any quantities in it calculated from the data should be acceptable.*

   This is a logical consistency requirement that considers asymptotics in both ways, namely, by repeated samples and by increasing samples. One does not want to derive properties of the predictor sequence under probability models that are reasonable but are likely to differ obviously from those of the data generator. Large-sample asymptotics are relatively common, but other cases may be more important. For instance, a parameter estimate in $\hat{Y}_n$ that depends only on, say, the ten previous data points is adaptive and has a sampling distribution but it cannot be consistent, because no sampling distribution can converge to unit mass at the true value with so few data points and repeating them will not change this. Instead, one would want to ensure that such adaptivity did not give unreasonable behavior over finite repetitions, e.g., if the parameter were the weight in a model average then one would want the model average to perform well over repeated samples of size n.

These desiderata for a sequence of predictors do not constitute a list of axioms nor are they unique; it is easy to construct other lists of desiderata. The point is to provide an organized way of coordinating the different desirable properties of a predictor sequence. The unifying theme is that predictors are the central class of objects from which to make inferences, just as the unifying theme for Bayesians is that the posterior distribution is the central object from which to make inferences and the unifying theme for Frequentists is that the sampling distribution is the central object from which to make inferences. It should also be noted that the desiderata can be implementable in practice, even if this requires a lot of work or can only be done approximately.

To conclude: the central points are (i) there is a predictive paradigm for statistical analysis, (ii) the predictive paradigm promises to be fully competitive with other paradigms for data-analytic and other inferential purposes, (iii) a predictive paradigm focuses on predictors which are more general than any other inferential class of object such as tests or estimators, (iv) the predictive paradigm is impartial in regard to how good predictors are constructed, thereby making it more general than other paradigms, and (v) a predictive paradigm is an appropriate framework within which to conceptualize statistical problems, especially those that are complex, have model uncertainty, or pose insurmountable modeling challenges.

3

What about Modeling?

'I know all models are wrong but some are useful … If I knew a model were wrong I wouldn't use it.'

– overheard at a statistics conference.

I do my own statistics.

– a physician at a medical school.

For many decades researchers have been trying to understand real phenomena and made impressive, indeed astonishing, gains. Much of this has been accomplished by modeling – extensively observing a phenomenon, positing variables that may characterize it, using physical intuition to propose models, and then testing the models. This is a procedure that works well for simple phenomena, i.e., phenomena that can be controlled so rigidly as to make it likely that a small number of 'state variables' can be identified. These state variables can then be studied and their relationship to other variables assessed. Often, the goal is to make reasonable assumptions at an individual or micro-level so that one can derive predictions for the aggregate or macro-level. Effectively, the strategy is to develop an idealized version of a phenomenon, which is exact enough in some settings that its predictions are useful, and the settings where its predictions are too far wrong to be useful are identified. This is roughly the case with much of classical science: the Carnot heat engine in thermodynamics (first outlined by Carnot in the 1820s), the ideal gas law (first stated by Clapeyron in 1834 and improved to the van der Waals equation in 1873), the Navier–Stokes fluid dynamics (dating from the mid-1840s), Maxwell's equations (early form stated in the 1860s), and Michaelis–Menten enzyme kinetics (first stated in their joint paper from 1913), among many other instances.

The same trend of explaining observed phenomena in terms of micro-level assumptions continued into twentieth-century science. The photoelectric effect was explained in terms of 'quanta' by Einstein in 1905. An effort was made to explain gravity in terms of gravitons in the 1930s but remains controversial. The field of computing was described in terms of a Turing machine in the 1930s, a logical reduction that continues to be relevant today. Probability theory as formalized in the 1930s by Kolmogorov explained an observable random variable in terms of an underlying measure space. The field of quantum electrodynamics can be regarded as the modern version of classical electromagnetism and explains interactions between charged particles in terms of photons; its foundations were established in the 1920s. Part of this theory includes explaining charged particles in terms of subnucleon particles called quarks; the last one was discovered in 1995. (It is an open question whether quarks in turn can be explained by some even lower-level theory.) The key mechanism of heredity, nuclear DNA, was explained in terms of ball-and-stick models (a reduction of the way atoms really seem to behave) by Watson and Crick in the 1950s and the subsequent development of genetics has emphasized the explanation of observed phenomena by micro-level assumptions on how atoms combine and molecules react.

Given this evident success, how could anyone be so wrongheaded as to suggest that this approach may have run its course? How could any meaningfully different approach even be imagined? The point of this chapter is to answer these questions.

Indeed, at this point, there seems to be a divergence between those sciences that have been more or less successfully model-driven, such as physics, and those fields that are distinguished by the volume and complexity of data they collect, such as biology, economics, and the behavioral sciences more generally. The latter fields seem to resist accurate models outside very narrow settings, i.e., settings where the essential complexity of the phenomenon is reduced so severely that the results do not apply with satisfactory real-world generality. As a related point, the absence of a unified field theory despite nearly a century of work may indicate a limit of the model driven approach even in physics, the result of having too much complex information to combine. (In fact, there likely is a sufficiently elaborate mathematical union of concepts from physics to produce a unified theory – but whether it is unique or could be validated would be another question.) The implication may be that, once a certain volume of data or level of complexity is reached, the conventional model-driven ways in which scientists of all types have been tackling problems may be yielding smaller and smaller returns. Consequently, how to go beyond modeling and develop an intuition for what this means may be essential.

3.1 Problem Classes for Models and Predictors

As important as model uncertainty and misspecification are in many cases, model accessibility, i.e., the degree to which a DG can be represented as an identifiable model, is probably more important. To this end, a classification of inference problems that reflects model accessibility has become accepted. It uses three classes and is described in Bernardo and Smith (2000). The importance of this is that the techniques appropriate for a given problem depend on its class. A concise summary of the classes is as follows.

The simplest case is called $\mathcal{M}$-**closed**. The idea behind this class of problems is that there is a true model; it is expressible precisely; and it is a model under consideration, e.g., it is on a model list in a model selection problem or a member of a specified parametric family. To a large extent this is the paradigm case for the frequentist and Bayes formulations. For instance, if the response $Y(x)$ is believed to have density $p_\alpha(\cdot|\theta_\alpha, x)$ for some parameter θ_α and explanatory variable x, the problem is $\mathcal{M}$-closed and frequentist analysis can proceed once data are available, provided that n is not small. Also, Bayes analysis can proceed once the Bayesian has assign a prior $w_\alpha(\cdot)$ to each θ_α and a hyperprior $w(\cdot)$ to α, again provided that n is not small. In $\mathcal{M}$-closed problems, the prior can be interpreted directly. A subjective Bayesian could interpret the priors as the summary of pre-experimental information on the parameters θ_α and α. A decision-theoretic Bayesian might choose a least favorable prior in an effort to achieve a minimax risk. Then, the prior would be an indication of the difficulty of estimation in different regions of the parameter space. A Predictivist would have a natural set of predictors to study: if one $p_\alpha(\cdot|\theta_\alpha, x)$ really is true then the collection of predictors that they generate, whether plug-in, Bayes, decision-theoretic, or based on some other principle, will typically contain an optimal predictor at least for large n. Indeed, in this case, model selection and predictor selection are essentially equivalent. If the $p_\alpha(\cdot|\theta_\alpha, x)$ were complex and n were not large then it would still be easy, at least conceptually, to define approximations to $p_\alpha(\cdot|\theta_\alpha, x)$ allowing approximately optimal inferences whether from estimation, testing, decision making, or prediction. Again, model selection and predictor selection would be essentially equivalent.

A more difficult class of problems is called $\mathcal{M}$-**complete**. The idea is that these problems are not $\mathcal{M}$-closed, in that even though a model exists it is inaccessible. It can't be written down precisely except with huge simplifications that may be hard to quantify and may not be representative of the real-world phenomenon. Bernardo and Smith (2000) refer to this as a belief model, in the sense that one can believe a true model exists – even if it is only knowable in a limited sense or not knowable at all. There is a range of inaccessibility, from slightly inaccessible to essentially completely inaccessible. An imperfect simile is the following. A model that is just barely $\mathcal{M}$-complete is like a function that has a convergent Taylor series. No finite sum actually equals the function, but summing enough terms gives an approximation to the function of verifiable accuracy. Models at the other end of the $\mathcal{M}$-complete spectrum are like complicated functions whose Taylor series, whether or not they converge, do not converge to the function value at very many, if any, points. Most of the classical theory of statistics assumes that a precise true model exists and it can be written down unambiguously, at least in principle. Arguably this is the defining feature of the classical theory of statistics. If the only 'information' available is a belief in the existence of a true model then there are not many techniques that one can use without the extra information that the true model can be identified somehow, if only by a sequence of approximations. It is in this sense that a Bayesian could assign a prior: the prior is a weight expressing pre-experimental belief about how good an approximation a given model will be.

If $\mathcal{M}$-closed techniques are used on an $\mathcal{M}$-complete problem then one cannot expect success unless the problem is only slightly inaccessible, i.e., is in a sort of 'completion' or 'closure' of the $\mathcal{M}$-closed class obtained by taking limits of the models in it. This would be a case where the $\mathcal{M}$-complete true model can be uncovered in the limit of infinite data or at least where nearly uniformly good approximations to the true model can be obtained, given enough data, however arduous the process. Indeed, the major scientific achievements listed in the introduction to this chapter are all examples of $\mathcal{M}$-closed or weakly $\mathcal{M}$-complete problems. That is, they are among the success stories for the modeling approach.

However, there are two limits. One is that there are many, many extremely important problems that are $\mathcal{M}$-complete in the sense that, even though one can imagine that a model exists, it is unrealistic to write it down in any meaningful sense. Two of these were seen in Chapter 1, namely the {Tour} and {Fires} data sets. One might argue that controlling the experimental setting adequately would permit a restricted version of the model to be written down or approximated. However, the way in which such a formulation would generalize to the broader phenomenon is usually unclear. An example would be studying a chemical reaction in a laboratory setting versus studying the same chemical reaction in a biological setting. One might be able to model the chemical reaction in a laboratory because it is such a highly controlled setting but once the reaction is in a cell all bets would be off because there would be so many other biochemical influences. In this sort of case, the real problem is

$\mathcal{M}$-complete because we can imagine that there is a true model explaining the biochemistry of a cell but the restriction of such a model to a reaction in a laboratory would be $\mathcal{M}$-closed. It may not be at all obvious how to extend the model found for the $\mathcal{M}$-closed problem to the $\mathcal{M}$-complete problem. In other words, one would not be able to discern how good an approximation a model developed in a laboratory would be to the model believed to exist for the cell.

The other limit is that, by and large, methods for the analysis of data from $\mathcal{M}$-complete DGs are relatively undeveloped. Specifically, it is unreasonable to expect $\mathcal{M}$-closed techniques to work on the less accessible end of the $\mathcal{M}$-complete model class. Two methods that have been tried are shrinkage and cross-validation. The idea behind shrinkage is that if one penalizes the squared error in a regression problem using additive models with a large number of parameters and variables, then the penalty may help give sparsity, i.e., provide reasons to eliminate extraneous variables or parameters. The smaller the penalty term is, the better these methods perform from a predictive standpoint. One of the better shrinkage methods from a predictive standpoint is from Fan and Li (2001) and is called 'smoothly clipped absolute deviation' (SCAD). The SCAD penalty is relatively small. It introduces only one tuning parameter and satisfies an 'oracle' property: in the limit of large sample size, the terms retained are correct and the parameter estimates are consistent, asymptotically normal, and efficient. Essentially this gives a good approximation in a variance–bias tradeoff sense when the problem is weakly $\mathcal{M}$-complete. Other sparsity methods seem to be similar: they frequently extend methods appropriate for the $\mathcal{M}$-closed problem class to the weakly $\mathcal{M}$-complete problem class.

Cross-validation (CV) is also intended mostly for the $\mathcal{M}$-closed or weakly $\mathcal{M}$-complete problem classes but is applicable more broadly. It requires one to specify models that can be evaluated by internal prediction, i.e., using some of the data to estimate the model and the rest of the data to test the model, and then cycling through various such partitions of the data into train and test sets. This means that one can identify models that are not too bad, at least in a weakly predictive sense. If the predictive errors are small enough, one might even be led to think of the model with the lowest CV error as being an approximation to the true model. The benefit of CV is that it does not constrain how the models to be evaluated are to be found, unlike shrinkage methods. There is no prohibition on using nonparametric models. One can even use predictors in place of the models because all that matters for CV is the input–output relationship between the training data and the predictions for the test data. While most results on CV are for the $\mathcal{M}$-closed setting, Shao (1997) showed an optimality property of CV for the $\mathcal{M}$-complete problem class. However, choosing the models or predictors whose CV error is to be evaluated is always setting-dependent. On the other hand, one could regard a model selected in this way as the closest 'wrong' model given the constraints of sample size and model class.

There is a third problem class called $\mathcal{M}$-**open** containing problems that are even less structured than $\mathcal{M}$-complete problems. The idea is that there is no model. The DG generates data but not according to any model. This is not as bizarre an idea as it may sound at first. Two examples in Chapter 1 are likely to be in this category, the analysis of whole metagenomic NGS data and the analysis of the music data. It is very hard to imagine that short pieces of DNA in a complex ecological community evolved to their current state by any statistical model, just as it is hard to imagine a 'new great-music generating' algorithm.

Probably a lot of econometric data is $\mathcal{M}$-open as well. An incorrect but easy way to think about this data is that each data point is generated by a model (that actually doesn't exist) but no two of these models have any necessary relationship to each other. That is, there is nothing stable enough to model, no matter how much data is collected. In this problem class, the only goal can be prediction. This does not mean that one cannot predict or that there are no properties which one can identify – it just means that any predictors or properties that can be identified are not going to be based on models at all. In particular, it is unclear how a Bayesian would assign a prior, apart from thinking of the prior as a weight on the believed predictive usefulness of the predictors.

Again, there are comparatively few techniques for this setting although four can be listed. First, as in the $\mathcal{M}$-open case, CV can be used with the understanding that the 'model' used in each CV error is regarded as a predictor only, not as a model. That is, the model is regarded analogously to an action in a decision theory problem: it specifies an input–output relation between explanatory variables and predictions and this input–output relationship is estimated from the previous data but otherwise has no particular meaning in the absence of further study and validation. Second, there is also a technique called 'prediction along a string', which will be discussed in Sec. 11.7.1. The idea is to make no distributional assumptions about the DG but to investigate the 'regret', i.e., how much worse than an optimal predictor a given predictor performed. These techniques have been well studied but only in narrow cases such as when a logarithmic 'score' function is used. It is not clear how broadly they generalize or how good they are.

Third, there are 'kernel methods'. Relevance vector machines and support vector machines, both seen in Chapter 1, are examples of this; a longer treatment is given in Section 10.4. However, kernels are of broader interest. One can argue that implicitly they have some sort of model assumptions because the kernel is a transformation of the explanatory variables and hence represents something about feature selection. However, the optimality properties satisfied by kernel methods in reproducing kernel Hilbert spaces (RKHSs) are mathematical optimizations to find optimal predictors (RVMs) or optimal separations (SVMs) or other quantities that are not tied to any specific model. Kernel methods can be unstable: being strictly mathematical and focused on the location of a response, the standard forms of the optimizations do not take variability into account. Hence, the desire arises to stabilize RVMs, by bagging, for instance. More interestingly, one could treat the RVM as a regression function and apply a shrinkage method to it in an effort to eliminate the smaller terms and so reduce variability (at the cost of possibly more bias).

The notion of stabilizing unstable but good methods originates with bagging, see Breiman (1996b), but as a concept applies more generally to other 'model-averaging' strategies, which constitute a fourth method for treating $\mathcal{M}$-open data. Formally, model averaging means using a probability model that is a mixture of other probability models, i.e., usually the mixing proportions are positive and sum to one. In practice, model averaging means using a convex combination of predictions from several models as a predictor that takes account of model uncertainty and hence typically gives a better performance than any individual model (unless it's actually correct, in which case the problem was not $\mathcal{M}$-open). There are numerous model-averaging techniques; the major ones are discussed in Chapter 11. An important feature of these techniques is that the models need not be compatible: the models may represent contradictory theories and this is unimportant because in $\mathcal{M}$-open problems there is no true model, so the goal is only prediction.

A fourth class of problems called $\mathcal{M}$-**mixed** was proposed by Gutierrez-Pena and Walker (2001). The idea is that one can assign a prior weight α to one set of models representing an $\mathcal{M}$-closed problem, i.e., to a fixed set of models, and a prior weight $1 - \alpha$ to the possibility that none of the models in the set is correct. This formulation essentially merges the $\mathcal{M}$-open and $\mathcal{M}$-complete classes because it does not distinguish between whether a DG has a model that is merely not in the specified class or whether there is no model at all for the DG. Essentially, one is led to decide whether or not the true model is in the specified set of models.

In terms of some intuitive notion of complexity we have that

$$\mathcal{M}\text{-closed} \prec \mathcal{M}\text{-complete} \prec \mathcal{M}\text{-open},$$

i.e., as assumptions are relaxed the problems become more complex. One consequence is that techniques for $\mathcal{M}$-open problems can be applied to $\mathcal{M}$-complete and $\mathcal{M}$-closed problems and techniques from $\mathcal{M}$-complete problems can be applied to $\mathcal{M}$-closed problems. However, one would not in general expect a technique that makes weaker assumptions to outperform techniques that make stronger assumptions, when the assumptions are satisfied. Likewise, one would not expect techniques that make assumptions that are not satisfied for a problem class to perform as well as techniques that do not make such assumptions. The techniques appropriate for a given problem depend on the problem's class.

Overall, even though these problem classes are phrased in terms of models, they are just as readily framed in terms of predictors. The $\mathcal{M}$-closed problem class consists of those problems for which a best predictor can be identified and is derived from a model class that can be said to describe the DG. The $\mathcal{M}$-complete problem class consists of those problems for which there is an optimal predictor but it is inaccessible except possibly in simple cases and then only by approximation. An inaccessible predictor corresponds to an inaccessible model but using an approximate predictor to identify an approximate model can, at best, give approximate estimation, testing, and optimal decisions. It is difficult to assess how good these approximations are. The $\mathcal{M}$-open problem class consists of those problems for which all one can do is predict and, while one might derive some weak inferences about the DG from the predictors, the inferences cannot be related to a true model for the DG, since no true model exists.

Another way to think about these three classes is in terms of validation – which is the essence of the predictive approach. In $\mathcal{M}$-closed problems it is possible to reduce model uncertainty and misspecification, or equivalently predictor uncertainty and misspecification, so much that predictions from a model can be extensively validated. In $\mathcal{M}$-complete problems, neither predictor nor model uncertainty or misspecification can be eliminated except in simple cases in a limiting sense. Even then one may have only an optimal predictor, in an FPE or CPE sense, which corresponds to a model of only indeterminate proximity to the true model. Validation therefore corresponds to reaching the situation where one cannot meaningfully reduce the FPE or CPE any further. In $\mathcal{M}$-open problems the DG has no model, therefore model misspecification is always present and model uncertainty doesn't make sense. Predictor misspecification and uncertainty may be eliminated in the sense that one has a predictor with irreducible FPE or CPE.

Since $\mathcal{M}$-complete and $\mathcal{M}$-open problems are so common, the famous statement from Box (1979) that 'all models are wrong ...but some are useful' should be amended to 'all models are wrong and mostly useless, at least as models, ... outside very narrow, controlled circumstances'.

3.2 Interpreting Modeling

In the last section we argued that the most interesting problems are $\mathcal{M}$-complete or $\mathcal{M}$-open, and these are of a character different from $\mathcal{M}$-closed problems in several ways. Basically, outside the $\mathcal{M}$-closed case modeling won't work very well because accurate, idealized, versions of the problem are difficult or impossible to formulate. Nevertheless, statisticians and other quantitatively oriented people often struggle heroically to fit models to the data and compel the models to say something important. In many cases this is done because investigators do not want to accept that models rarely encapsulate a complex physical phenomena. The dichotomy between modeling and predicting is the focus of Breiman (2001a).

Thus, there is a schizoid behavior on the part of statisticians. On the one hand there is the strong desire to model a real-world phenomenon to achieve sparsity, interpretability, and understanding. This rests on the hope that a wrong model will be useful in the Box (1979) sense. However, the admission that a model is wrong (and generally not validated) conflicts with the view that a statistician would not use a wrong model if it were known in advance that the model were wrong. It's as if, because the ways in which a model is wrong are not known, it can be used until the ways in which it is wrong have been uncovered. Even worse, statisticians and others often argue that a wrong model is a good approximation to the true model (implicitly assuming that a true model exists) without any data. That is, physical modeling arguments – suppositions from settings assumed to be related to the setting under study – are used to justify a (physically based) model even when the error of approximation to any true model cannot be meaningfully assessed. The result is a de facto over-confidence in an unvalidated model merely because it hasn't been discredited.

Despite these well-known caveats, modeling persists, not just among subject matter researchers who want to understand a phenomenon (however remote or unlikely such an understanding might be), but also among statisticians. So, if models don't represent reality very often, what are they good for? There are four obvious answers. The first is satisfying only from a classical inferential standpoint; the other three are important in their own right.

First, it must be admitted that sometimes models do turn out to be close enough to true that they enable investigators to make meaningful statements about a physical phenomenon. That is, the problem being investigated is $\mathcal{M}$-closed or just barely $\mathcal{M}$-complete and there is enough data and other information that one can identify a model and validate it. In some cases, the model may be an oversimplification of the true model that, even if its accuracy cannot be evaluated explicitly, permits some approximate inferences even after allowing for model uncertainty and misspecification.

Second, and more typically, models should be regarded as a technique for data compression or summarization. That is, a model that 'fits' a data set should be regarded as a summary statistic for the data set. Like standard summary statistics such as means or percentiles, models merely provide a feature of the data that no one would reasonably believe actually characterize the data. Many data sets may have the same mean, or be consistent with the same model, and many models may provide roughly equally good data summarization or compression. Note that data summarization or compression implies that some information will be lost. After all, given a mean, one cannot reconstruct the initial data set or even a data set guaranteed to be similar to it. Likewise, many data sets may be consistent with the same model but, in the absence of predictive validation, a model cannot be used to generate more data similar to the initial data set. Conversely, typically many model classes could be equally good for data summarization, and the summarization from the different classes would generate data sets meaningfully different from each other and from the initial data set. As a simple example, a linear model, a neural net, and a spline model may give equally good fit, but different predictions, even though any of them could be regarded as a useful summary of a data set.

Third, models serve an important role in communicating the informational content of a data set. Rather than merely plopping a data set in front of some one, it is more informative to give a model that encapsulates at least some information in the data. Again, a linear model, a neural net, and a spline model will usually provide different sorts of information that is implicit in the data, so one would not want to believe that any of these models said anything conclusive without validation. However, the three models would reveal different sorts of information contained in the data. The linear model might provide some idea about the relative importance of terms via the sizes is of the slopes. The neural network might provide some indication of the complexity of the data from the network architecture. Spline regression might provide the best smoothed representation of the data. Thus, different models permit communication about the information in the data even though it would be unreasonable to say that any of the models were true.

Fourth, models may be regarded as actions. Once its parameters have been estimated, a model is just an input–output relation between the explanatory variables and the response. The 'action' is a point prediction and could be included in the action space of a formal decision problem. In this case, the model or action has no necessary relationship with the real phenomenon.

This view – that models are tools rather than representations of objective reality – is opposed to the traditional view of inferential statistics, which regards modeling as the sine qua non of statistical analysis in one way or another. It is also opposed, though not as strongly, to the view that statistics is just a collection of search strategies. In the latter view, statistics is an intelligent way to rule out models, decisions, predictors, and so forth by finding plausible reasons to regard them as unlikely. The clearest examples of this are hypothesis testing and variable selection. In 'omics contexts, for instance, many hypothesis tests, possibly thousands, are done to identify significant hypotheses. Given the error rate for such tests, one must carry out further experimentation to validate any results of the testing procedure. Likewise, if one is doing variable selection in complex models, for instance, minimizing the MSE by simulated annealing to find a neural net, there is model uncertainty so that accepting or rejecting a node in a model has an attendant uncertainty which must be validated on new data. Obviously, statistics is a collection of search strategies – it's just that statistics can be used for far more than searching. It can be used for prediction, and sometimes other inferences, which is one step further than merely searching if not as demanding as modeling.

Again, it is not that models, even good ones, are purely figments of human imagination (although doubtless many are), it is that models have much more uncertainty or misspecification associated with them than is commonly acknowledged, even when the concept of a model makes sense. So, if one insists on using models for inferences about reality, one should regard the models as predictors and find the common features possessed by all good predictors for the phenomenon. In practice, looking at all good predictors is unrealistic, so one merely looks at a large enough class of predictors that one can assert that the robustness of the inferences about the phenomenon to the class of predictors is satisfactory. That is, for $\mathcal{M}$-complete and $\mathcal{M}$-open problems, which are ubiquitous, one should find a good predictor and then convert it to modeling information, if not an actual approximation to a model, rather than trying to model first and then predicting, i.e., validating. In short, *predict first, ask modeling questions later*. This process may provide less 'information' or fewer inferences than a modeling approach but the results should be more reliable.

3.3 The Dangers of Modeling

Although not commonly discussed, there are dangers in modeling, chiefly from overoptimistic assumptions. As noted, these stem from higher than commonly acknowledged model uncertainty and misspecification – both being at their worst when an $\mathcal{M}$-open problem is treated as an $\mathcal{M}$-closed problem. The errors made are often abstract and hence are overlooked, resulting in statistical conclusions that fail to validate except to the extent that they were already known beforehand.

There are basically three big dangers in modeling: (i) overmodeling, when modeling is valid as in $\mathcal{M}$-closed problems; (ii) circularity in reasoning about models, in cases when modeling may be helpful as in $\mathcal{M}$-closed and $\mathcal{M}$-complete problems; and (iii) using models when they are not valid, as in $\mathcal{M}$-open problems. Of course, one can undermodel as well, but that amounts to an ineffective use of information rather than a strategy likely to lead to wrong conclusions. The dangers in all three cases are worsened by over-reliance on the desire that models be 'interpretable'.

First, even though overmodeling in an $\mathcal{M}$-closed setting is a well-recognized problem, it's still done. Often models are overparameterized relative to the available data, too many hypothesis tests are done without multiple-comparisons corrections, a model class is chosen that is too narrow, or a prior is ill chosen. The net effect is to under-represent model uncertainty or misspecification. The estimated model has uncertainty or misspecification, not just due to the estimates or tests but due to the formulation. As noted above, there are many different regression models (linear, neural nets, and so forth). They may be equally justified but they are different, representing model uncertainty, or incorrect, representing model misspecification. That is, when one announces a model $p(y|\hat{\theta}, x)$ it is not just $\hat{\theta}$, and the selection of features from x (i.e., functions of the components of x), that is uncertain; it is also the functional form of $p(\cdot|\cdot)$ that is subject to uncertainty or misspecification. Even when they believe that a model has not been misspecified, often subject-matter specialists are unwilling to demand robustness on the grounds that varying $p(y|\theta, x)$ results in uninterpretable models, apparently unconcerned that if perturbations of the model destroys its conclusions then the conclusions are likely to be unreliable. Model misspecification makes this worse because misspecified models will not be stable; they typically change considerably when perturbed toward the true model, by, say, modifying the parametric family. However, even when misspecification is too small to be detected using the available data, the uncertainty is often high enough that future outcomes will not be in the PIs as often as expected. This sort of problem can be resolved simply by not overmodeling – at the risk of not being able to answer a substantive question. There is a limit to what can be inferred from limited data and other information.

Problems that are $\mathcal{M}$-complete can suffer from overmodeling as well as model misspecification. However, as long as the problem is just barely $\mathcal{M}$-complete i.e., nearly $\mathcal{M}$-closed, it will be similar to the $\mathcal{M}$-closed case. As an $\mathcal{M}$-complete problem moves away from the $\mathcal{M}$-closed class, the ability to specify a model decreases, and model misspecification will be more serious and usually has to be corrected (if it is correctable) before one worries about model uncertainty. This is so because the uncertainty around a sufficiently wrong model is not usually going to enable an analyst to capture a good approximation to the true model. The time to address model uncertainty is once one is satisfied that the model class chosen contains an adequate approximation to the true model. Requiring models to be interpretable tends to exacerbate the problems from overmodeling because it narrows the class of models so severely that it can worsen model misspecification. Overmodeling in $\mathcal{M}$-open problems is not as serious because it is easily detectable, i.e., the resulting predictors are likely to be easily outperformed by other predictive techniques.

Second, in both $\mathcal{M}$-closed and $\mathcal{M}$-complete problems, there is a tendency to bring subject-matter-specific information into the modeling process. When done correctly, this is helpful. However, all too often it leads to circular reasoning. Physical intuition is developed and a modeling strategy embodying the physical intuition is formulated. Inferences using the modeling strategy are made and found to be in conformity with the physical intuition. Therefore, the physical intuition is 'validated' because it gives conclusions that reinforce the subject-matter information. In fact, the so-called 'validation' is just a re-expression of the initial physical intuition via the model and so does not constitute a real test of the physical intuition. Only a test of the inferences on a new set of data would be able to validate the intuition, and this is rarely carried out.

In practice this circularity is harder to avoid when models are required to be 'interpretable' – as subject-matter specialists typically want – because so many of the inferences the model must give are predetermined. This narrowing of the possible models effectively means that only models that reinforce pre-experimental beliefs will end up being accepted. This is particularly so in $\mathcal{M}$-closed problems but occurs even when the phenomenon under investigation is $\mathcal{M}$-complete and far from being $\mathcal{M}$-closed. Just as in Chapter 1 we noted an uncertainty principle linking predictive accuracy and interpretability, there also seems to be an uncertainty principle between modeling accuracy and interpretability. Models simple enough to be interpretable are not likely to be very accurate. Models that are accurate tend to include quantities that do not have an obvious physical interpretation, at least not yet. It is not that subject-matter knowledge should not be brought to bear on problems. Rather, one must be certain that the knowledge is correct and this is hard to evaluate in the absence of accurate predictions.

From a predictive standpoint, modeling is at best useful only to help specify a predictor, because any physical meaning is secondary to predictive performance. In particular, for those who believe the dictum of Box (1979) that all models are wrong, it is paradoxical to get fussy about whether a model is 'true' or has a physical interpretation consistent with (often unreliable) intuition before evaluating its predictive performance.

At the risk of venturing into sociology more than statistical theory, before moving on to point (iii), consider the following. Discussions between subject-matter specialists and statisticians are often tense. The subject-matter specialist wants results to publish and has the funding. All too often this represents a power differential and its maintenance is justified by arguments that it is the subject-matter research that is important while statistics is seen as a tool to be used along the way to this goal. Thus, the pressure is to get answers, any answers that will be accepted by the subject-matter specialists. To see how this worsens the circularity problem, consider the following hypothetical (and perhaps 'worst case') but generic conversation between a subject-matter researcher (SMR) and a statistician (S). This helps show that when some people say 'I do my own statistics' it means they are really making up their own statistics.

SMR: Have you finished analyzing the data? Was the p-value less than 0.05?

S: Yes, I did a logistic regression and tested which variables should appear. I used Wald tests. The results are pretty clear; they wouldn't change if I corrected for multiple comparisons.

SMR: Was the p-value significant?

S: Overall, yes. But we can drop some of the explanatory variables.

SMR: Which ones?

S: These ones I've highlighted on the printout.

SMR: This doesn't make sense. You can't drop that variable. It's known to be important. Oh, but this one, I could have told you it didn't matter. And this one matters? I don't know what to make of that. It doesn't have any reasonable interpretation. There must be some mistake. Maybe you shouldn't include it.

Two days later:

SMR: Have you finished re-analyzing the data?

S: Yes, I tried a different analysis. You'll like the results better.

The third danger is that in $\mathcal{M}$-open problems there is no prohibition on using models; the prohibition is on using them to do anything more than provide predictions for the DG. As a generality, models, and especially sparse models, should be regarded as data summarization or data description and, in the absence of validation as being essentially useless for inference. In $\mathcal{M}$-open problems this is an even more serious consideration than for $\mathcal{M}$-closed or $\mathcal{M}$-complete problems. After all, if one has an $\mathcal{M}$-open problem and there is no model for a DG, it is possible (but unlikely) that a single model will be the best predictor. However, the lack of a model means that the model corresponding to the best predictor is not a model for the DG. One might nevertheless use the predictor to make inferences about the DG, but they would be inferences from the predictor, not from the model.

At the heart of all these considerations about the relative types of models (when they exist), and approximations to them, is a sort of generalized variance–bias argument that can also be expressed in terms of their corresponding predictors. Even when a true model exists, there will typically be biases and variances for each component in it. Models are only useful when the bias is meaningfully smaller than the other sources of variability. However, even when this happens, model uncertainty in one form or another can be large and may dominate. A predictive approach comparing different predictors takes all these sources of error into consideration because any increase in bias or variance will increase the FPE, CPE, or even CV. So, even though it is hard to determine which variances and biases it is important to reduce, focusing on the reduction of predictive error is a proxy for controlling the other sources of error.

To conclude: in $\mathcal{M}$-closed problems one can learn by proposing models and evaluating the predictors they generate. This process becomes harder as the complexity of the $\mathcal{M}$-closed problem increases. In $\mathcal{M}$-complete problems one can also learn by modeling – it just takes a lot of effort. There will be many false starts before finding anything that validates, and even then it will be, at best, approximate and the degree of approximation may be hard to evaluate. In these cases, taking a predictive approach would be likely to be more efficient. In $\mathcal{M}$-open problems it is not clear that there is any alternative to the predictive approach.

3.4 Modeling, Inference, Prediction, and Data

In this book the term modeling has been used in an extreme (and somewhat unfair) sense to mean the search for an accurate representation of a DG by combining elements from a collection of components (variables, probability densities, other mathematical forms) that have physical correlates. The hypothetical modeler, as characterized here, is obsessed with model unbiasedness – the concern that the representation should be accurate to infinite precision. (In a regression sense, this means that the response $Y = E(\hat{R}(x)) + \epsilon$, where $\hat{R}$ is the estimated regression function.) This is different from prediction, which merely seeks a $\hat{Y}(x)$ that gives values close to $Y(x)$ – a sort of empirical risk minimization (not to be confused with the theory of empirical risk minimization as elucidated in, say, Vapnik (1998)). Just as a modeler has been characterized as an extremist interested only in finding an unbiased model (bias meaning the difference between a proposed model and the DG), so the Predictivist is characterized as an extremist interested only in accurate prediction that is devoid of physical understanding, relenting only to the extent that once an ideal predictor has been found it may be studied to see whether it says anything interpretable; see Sec. 10.3.5 for one instance of making 'blackbox' neural nets partially interpretable. It is this juxtaposition of extremes that creates a conceptual tension between modeling and prediction. However, in extremes one finds clarity and the hope is that, whatever its faults, this book lays out a predictive view clearly. Obviously, in practice a compromise will often be desirable.

Indeed, in the real world of dealing with data the tension between modeling and prediction is nearly nonexistent because both approaches contribute to the enterprise of understanding the world we live in – and the problems that most statisticians address are too important to allow ideological pigheadedness to impede progress.

One aspect of this tension that has been given short shrift is what most statisticians call inference. Inference is not the same as model building, at least as defined here. Moreover, while the point of this book is that prediction should be regarded as the paradigm issue for inference, in fact there are many nonpredictive inference problems including testing and estimation. The view taken in this book is that nonpredictive inference problems receive their importance from being embedded as a component in a prediction problem. For instance, scant mention is made of testing or estimation that may be used in forming a predictor, on the grounds that the predictive performance is the overriding goal.

This limited view of nonpredictive inference is also somewhat unfair because it ignores many important statistical problems where modeling is not the main issue and it is not necessary to address prediction directly. For lack of a better term, these situations may be called 'partial modeling' since the experimenter and analyst have some modeling information that is reliable but cannot characterize a tractable class of models and the prediction problem associated with the nonpredictive inference is too distant or difficult to study directly.

This sort of situation arises in numerous fields. In biostatistics, for instance, it is common to assess risk reduction from a treatment, whether this involves drugs to prevent heart disease or lifestyle changes to prevent cancer. In these cases, there may be good reason to expect that a treatment will be effective (studies on other species, an understanding of biochemical networks, phase I/II trials, etc.). Such reasoning is insufficient to form a model but nevertheless provides a physical motivation for a trial. While a Predictivist could obstinately insist that proportional risk reduction is a nonempirical construct and hence that risk can only be said to have been lowered if a good predictor of, say, lifespan, predicts overall longer lifespans for the treatment group than the control group, this may miss the point that the construct is useful and an extension to the prediction problem would likely be a waste of resources (to say nothing of ethics).

More generally, the same sort of situation arises when the goal of experimentation and analysis can be focused on, say, a few parameters or features. This is the structure that defines semiparametric approaches, for example. In the simplest case the goal could be focused on a single parameter or variable, as is sometimes the case in clinical trials. This goal is much smaller than an exegesis of the entire DG. At the same time prediction is not the point either, even though the goal can usually be regarded as a subsidiary aspect of what is naturally a prediction problem. As with proportional risk reduction, the experimenter and analyst have partial modeling information to justify a clinical trial, for instance, and this can be used to good effect in nonpredictive inference, in which neither unbiased modeling nor prediction are immediate goals.

A related situation occurs when the goal is decision making. It is often practical to make decisions based on predicted outcomes in which the goal – decision making – is post-prediction rather than pre-prediction as in the earlier examples of this subsection. In these contexts, especially when they are sequential, regret is the issue more than prediction (although the regret can only be calculated after a decision has been made based on predictions). While decision theory, p-values, and posterior odds ratios can be used for decisions as well, one benefit of regret is that extra information can be brought in to make the next predictions and hence decisions; see Sec. 11.7.1 for an example and some details.

In many of these cases, there is an extra layer of complexity: the outcome measured may be the result of a latent trait – another sort of partial model – that is not well defined. In these cases, it is also often difficult even to define the population precisely, e.g., if the latent trait is an attitude of some sort. Taken together, this may make modeling, prediction, and decision making problematic even when regret or predictive error is, in principle, the ideal criterion to optimize. In the urgency of getting answers, nonpredictive inference – parameter estimation and testing – may be relatively feasible and resolve the problem.

Before concluding this section, it should be noted that the examples discussed here largely maintain a false dichotomy between prediction and other forms of inference. In practice, most important real-world problems are mixed, in the sense that they have a predictive component as well as a nonpredictive component. For present purposes, it is conceptually useful to separate the two components in an effort to focus on prediction since it has not previously received the attention it deserves.

As a final point it is essential to mention data quality, if mainly to emphasize its importance, especially to the reproducibility of conclusions. Although it is not always recognized by analysts, when they get analysis-ready data it is usually the result of very complicated, variable, and error-prone processes. In general, there are two major steps between experimental design and data analysis: the collection of the raw data and the preprocessing of that data into an analysis-ready format. The preprocessing often gets divided into two parts as well. The first is often to organize the data into some format amenable to computation (data cleaning, sometimes called the 'shoebox' problem) and is often something of an art since it can be highly particular to a specific experiment. The other is applying various statistical techniques to standardize the data within a batch (as in gene expression data), to upscale or downscale the data (for spatial data gathered on different scales), or to transform the data (e.g., using Tukey's ladder on a response, or using Box–Cox, etc.). Overall, the Modeler, Decision-maker, Predictivist or nonpredictive Analyst is continually at the mercy of data quality and management and must take this into account when choosing which experiment to perform and which analysis to implement; see Zozus (2017) for more on these issues.

3.5 Prequentialism

Taking a predictive approach, i.e., seeking an accurate predictor, solves the problems discussed in the earlier sections of this chapter because it avoids modeling assumptions. Instead, it relies on some of the data to form the predictor and the rest of the data for validation. Predictive errors on data not used to form the predictor automatically take account of predictor uncertainty and misspecification, regardless of the problem class. Indeed, if one chooses a function $L(\hat{Y}(x_i), y_i)$ of the predicted value $\hat{Y}(x_i)$ and the outcome $y(x_i)$ to represent the 'closeness' of these two values, then over repeated uses of $\hat{Y}_i$ it is reasonable to expect that if the predictor is unstable, i.e., there is excess uncertainty, then many $L(\hat{Y}_i, y_i)$ will tend to be large. Likewise, if the predictor class is misspecified and one gets more and more data then one expects $L(\hat{Y}_i, y_i)$ to decrease on average to a limiting value that can be reduced by choosing a better predictor class. This is the same intuition as model uncertainty and misspecification but generalized beyond $\mathcal{M}$-closed problems. *Note that the intuition for model uncertainty is based on repeated samples of size n while the intuition for model misspecification requires n to increase. This is so because model uncertainty is like variance and model misspecification is like bias.*

To be precise, it's important to define the sequential prediction problem. Henceforth, the term *prediction* will be reserved for identifying outcomes of random variables (observable or not) while *estimation* is reserved for identifying nonrandom quantities. (The Bayesian's prior is a probabilistic summary of the pre-experimental information about the true value of a parameter, assumed to be fixed but unknown). So, suppose the goal is to predict a sequence $Y_1, \ldots, Y_n, \ldots$ of random variables. To make the framework more useful, suppose there are explanatory variables $x = (x_1, \ldots, x_d)$. Then a prediction for the $(n+1)$th stage is

$$\hat{Y}_{n+1} = \hat{Y}_{n+1}(x_{n+1}), \tag{3.1}$$

where the values $x_{n+1} = (x_{1,n+1}, \ldots, x_{d,n+1})$ are available before the prediction $\hat{Y}_{n+1}$ is made and all previous data (x_i, Y_i) for $i = 1, \ldots, n$ are available to help in the construction of $\hat{Y}_{n+1}$. As stated, (3.1) is one-step-ahead prediction, but if the goal were to predict q

steps ahead then the formulation would be similar. The predictor would be a vector-valued function $\hat{Y}_{n+1,\dots,n+q} = (\hat{Y}_{n+1}, \dots, \hat{Y}_{n+q})$ and the individual functions $\hat{Y}_{n+j}$ for $j = 1, \dots, q$ would again be constructed using all the data (x_i, Y_i) for $i = 1, \dots, n$. The sequential prediction context is foundational because it permits the predictor to evolve as data accumulate and, as will be seen shortly, provides a more demanding evaluation of any prediction strategy, i.e., any choice of the sequence $\hat{Y}_{i+1}$ for $i = 2, 3, \dots$

Now, prediction error can be regarded as an empirical and more general version of the concept of generalization error. Consider an $\mathcal{M}$-closed or $\mathcal{M}$-complete problem. In the usual formulation, the *generalization error* for a predictor $\hat{Y}$ derived from a model can be written

$$E_{X,Y} L(\hat{Y}(X), Y) \tag{3.2}$$

where the expectation is over X and Y and $\hat{Y}$ is regarded as a given function, i.e., the expectation ignores the fact that $\hat{Y}$ is formed using data. The quantity $L(\hat{Y}(x), y)$ is the predictive error when the explanatory variable is $X = x$, the predictor is $\hat{Y}(\cdot)$ (possibly formed from other, earlier, data), the prediction is $\hat{Y}(x)$, and the outcome is $Y = y$. Then

$$E_{\mathcal{D}_n} E_{X,Y} L(\hat{Y}(X), Y) \tag{3.3}$$

is the *average generalization error*, where the outer expectation is taken over data sets $\mathcal{D}_n = \{(y_i, x_i)|i = 1, \dots, n\}$. If Y^* were the 'right' predictor, i.e., a predictor based on the true model, then the true generalization error would be $E_{X,Y} L(Y^*(X), Y)$ and the true average generalization error would be $E_{\mathcal{D}_n} E_{X,Y} L(Y^*(X), Y)$. If Y^* were based on a model known to be a good approximation to the true model then these statements would be approximately true. In either case, in practice these errors are not known, even asymptotically.

Two errors can be found. The *training error* is usually taken as cumulative,

$$\sum_{i=1}^{n} L(\hat{Y}(x_i), y_i), \tag{3.4}$$

where the data $\mathcal{D}_n$ are used to form $\hat{Y}$ and the terms $L(\hat{Y}(x_i), y_i)$. The *testing error* is also usually taken as cumulative,

$$\sum_{j=1}^{m} L(\hat{Y}(x_j), y_j), \tag{3.5}$$

where $\mathcal{D}_m = \{(y_j, x_j)|j = 1, \dots, m\}$ is a set of data disjoint from $\mathcal{D}_n$ and hence not used to form $\hat{Y}$. The training error typically decreases as the complexity of the predictor increases, while the testing error tends to have a minimum at the predictor with the most appropriate complexity for the problem, i.e., the testing error tends to be high for predictors that have a complexity that is too low or too high.

In practice, one wants to find a predictor $\hat{Y}$ that makes (3.2) as small as possible, i.e., equal to $E_{X,Y} L(Y^*(X), Y)$. However, the result of such a procedure might be highly dependent on $\mathcal{D}_n$ in the sense that slightly different data sets would give different $\hat{Y}$'s. Thus, for stability, one really wants a unique $\hat{Y}$ that makes (3.3) small, i.e., equal to $E_{\mathcal{D}_n} E_{X,Y} L(Y^*(X), Y)$. In fact, the set of procedures that minimizes $E_{X,Y} L(Y(X), Y)$ also minimizes $E_{\mathcal{D}_n} E_{X,Y} L(Y(X), Y)$ over Y. That is, Y^* is optimal for $\mathcal{D}_n$ if $E_{X,Y} L(Y^*(X), Y) = \arg\min_{\hat{Y}} E_{X,Y} L(Y^*(X; \mathcal{D}_n), Y)$ even if the generalization error and its expectation have different values for $\hat{Y}$'s that are not minima. So, it is enough to seek a stable predictor that minimizes the generalization error alone. Clearly, all four quantities, the generalization error, the average generalization error, the training error, and the testing error, take predictor or model uncertainty and predictor or model misspecification into account.

On the one hand the training error is a downward-biased estimate of the true generalization error because the same data are used to find the training error and to find the predictor $\hat{Y}$. (It is possible to get a training error greater than the true generalization error or true average generalization error; this only happens when one is unlucky, for instance, one has chosen an insufficiently complex class of predictors and obtained a data set that is unrepresentative of the DG.) On the other hand, the test error is an unbiased estimate of the true generalization error because the inputs required to form the predictor – the data $\mathcal{D}_n$ and the predictor class – are entirely independent of $\mathcal{D}$. Thus, the only knowable quantity that takes predictor or model uncertainty and predictor or model misspecification into account is the testing error.

Another property that sets the testing error apart from the other three errors is that the testing error, as a way to evaluate a predictor $\hat{Y}$, satisfies the Prequential Principle; see Dawid (1984) and, for a more mathematical treatment, see Dawid and Vovk (1999). One way in which the (weak) Prequential Principle can be stated is as follows: 'the method of construction of a predictor should be disjoint from the method of evaluation of the predictor's performance'. Thus, if a predictor is generated using a model then the model should not be used in the evaluation of the predictor, ruling out the generalization error and the average generalization error. The training error is ruled out by the Prequential Principle because the predictor is constructed using $\mathcal{D}_n$ and evaluated on $\mathcal{D}_n$ as well. Imposing the Prequential Principle is a way to prevent any model or overuse of data from unduly influencing how we evaluate predictors.

Adopting the Prequential Principle permits an intermediate position between the training error, which it disallows, and the testing error, which is an extreme case of what it permits. The idea is that under the Prequential Principle one can look at the sequence $L(\hat{Y}(x_i), y_i)$ for $i = 1, \ldots, n$, where it is assumed that $\hat{Y} = \hat{Y}_i = \hat{Y}(\cdot|\mathcal{D}_{i-1})$, i.e., the predictor at time i uses all previous data. In particular, the predictor can be rechosen at each time step, not just updated with more data. Two predictive errors in the sequence separated by k data points, say $L(\hat{Y}(x_i), y_i)$ and $L(\hat{Y}(x_{i+k}), y_{i+k})$, are clearly not independent but there is more information in the data than just the value of the testing error. Examining the sequence $L(\hat{Y}(x_i), y_i)$ is a natural way to try to extract information from the data beyond the testing error. So, treating data as if it had arrived in some fixed sequence (whether it did so or not) and examining the sequence of $L(\hat{Y}(x_i), y_i)$ permits an analyst to study the effect of predictor performance while taking predictor uncertainty and misspecification into account. That is, the Predictive Sequential, or Prequential, Principle tolerates interference between the predictive errors for the benefit of getting as many pieces of information out of the data, the n predictive errors, as there are data points.

The Prequential Principle precludes the use of the generalization error or average generalization error as an assessment for error because both quantities require the true model (because they involve taking an expectation). First, using the assumed model both to evaluate a predictor and construct the predictor will understate the error. This is similar to how using the same data as both training and testing data will understate the error. Second, if there is model misspecification then the assumed model will be wrong. Sometimes the degree of error is small but it can be large and it is frequently difficult to tell how far an assumed model is from the unknown true model. Thus, a predictor from any given model will routinely be suboptimal (for large enough sample sizes) relative to the true model, and the degree of suboptimality will be impossible to assess. Such a predictor may not be very suboptimal if the given model is a simplification of a complex true model and the sample size is small, but in this case our predictor will perform poorly as the sample size increases. Third, assuming a fixed true model ignores model uncertainty. Fourth, because the model is unknown, inferences should generally be stable over small neighborhoods of it. Such neighborhoods will include some models that are simpler and some that are more complex (relative to sample size) than the true model. In short, although one might use the generalization error or average generalization error to construct a predictor, using such quantities to evaluate the performance of a predictor under an assumed model is not, in general, a real test that suffices for validation on new data.

The same point can be made about other model-based criteria. In the absence of model uncertainty and model misspecification it may be reasonable to use model-based methods of evaluation. This is common in decision theory, which typically requires a fully specified model for which the risk, and variants on it, are the main criteria to be optimized. For instance, the posterior mean is the Bayes-optimal predictor under squared error loss and has good asymptotic properties when some member of the parametric model class is true; this is so because the model is used to define the optimality criterion. Criteria such as unbiasedness, minimum variance, coverage probability, and type-I and type-II error probabilities, among others, are similar in that they too are phrased in terms of the model and rely on the absence of model uncertainty or misspecification to be valid. It is easy to show examples where estimation or hypothesis testing is misleading when none of the models under consideration is true, i.e., when model misspecification or model uncertainty is present. That is, the usual decision-theoretic, estimation, and testing quantities will not usually be a good representation of the errors from real phenomena and hence will not usually lead to good prediction.

Otherwise stated, good properties from model-based criteria are only a consequence of the correctness and stability of the chosen model, not a result of validation on independent data. In general, of the criterion by which a predictor (or other inferential procedure) is judged depends on the validity of the model class, the predictor (or other inferential procedure) may break down in the presence of model uncertainty or misspecification. The breakdown can be detected by predicting on new data and finding an elevated error. In fairness to model-based criteria, empirical versions of them do usually satisfy the Prequential Principle and may help provide useful information from validation. For instance, the test error under squared error loss is an empirical form of the mean squared error. This shows how the Prequential Principle enhances validation, i.e., in the absence of modeling it makes inference rely more on the data, by using the pattern of sequential errors.

In the pure case, one looks at the data only as a single sequence. However, one can even out variations if one considers several permutations of the data and hence several sequences of the $L(\hat{Y}(x_i), y_i)$, one for each permutation, and averages over them in some way. This is what was done for the FPE (and CPE) in Chapter 1. Doing this sort of sequential bootstrapping provides enhanced validation of a predictor sequence. In addition there are other serviceable approximations to this pure case, such as various versions of cross-validation, including the use of sequential errors, which may be more convenient for obtaining combined information about model fit and generalization error, especially in linear regression. Overall, if a predictor sequence performs poorly, the way in which the predictors were chosen can be reformulated and the predictive errors re-examined. In this way data that is not sequential can be usefully treated as if it were sequential. Indeed, one would be skeptical of a predictive strategy that did not perform well under bootstrapped sequential prediction quantities such as the FPE or CPE or approximations to them.

Note that the Prequential Principle can be imposed regardless of whether a problem is in the $\mathcal{M}$-closed, $\mathcal{M}$-complete, or $\mathcal{M}$-open class and can be applied regardless of whether a predictor is derived under a Bayesian, frequentist or other inferential paradigm. It is a generally applicable condition; all that matters is how good a predictor is at predicting and that the predictor class is not improperly unconstrained. The predictor may be derived from a model, e.g., it could be a plug-in predictor, or it could be a model average or a more general input–output relation. For instance, a predictor could be adaptive in the sense that at each time step some rule is followed for rechoosing the predictor from the class, e.g., one could use the AIC and a growing list of models to rechoose at each time step a model from which to get a prediction, thereby getting different models at different times. This amounts to searching the predictor class to find good predictors.

While the Prequential Principle is central to the predictive approach in that it obviates the need for model formulation, it is not enough by itself. This was seen in Sec. 2.4, in which five principles besides Prequentialism and the use of predictive errors were listed. Obviously, for good prediction, one would update predictors by looking at how far wrong they were in predictive errors. In addition, a good predictor should be formed in such a way that the systematic error (bias) and variability (variance in the squared error case) of each ingredient in the predictor can be assessed and be satisfactorily small. For instance, one could use bootstrap samples to re-estimate a parameter in a predictor to assess its bias–variance decomposition. Also, a good predictor should be stable to perturbations of the data set that went into forming it. There are many ways to do this. One common way is to perturb the data with, say, $N(0, \sigma^2)$ noise, and rerun the procedure by which the predictor was generated, to verify that the new predictor is not unacceptably far from the original predictor. Likewise, the complexity of the DG should match the complexity of the predictor as measured by codelength arguments or possibly some notion of the Vapnik–Chervonenkis dimension (for classifiers). And, of course, one would not want the limiting behavior of a predictor to be unreasonable. Taken together, these nearly eliminate the need for model formulation for inference.

To conclude, it is only in the relatively small $\mathcal{M}$-closed problem class and in the larger (but still not very common) benign end of the $\mathcal{M}$-complete problem class that models are most useful. In a truly $\mathcal{M}$-closed problem the predictive approach will be essentially equivalent to the model-based approach because if a list of explicitly specified models can be written down then one can interconvert the model list and its corresponding natural predictor class. Away from the benign end of the $\mathcal{M}$-complete problem class, even though a true model exists, there will be ever greater problems in using its existence alone to generate good approximations to it or good predictors based on it. In short, when an $\mathcal{M}$-complete problem is not a well-behaved limit of $\mathcal{M}$-closed problems, model uncertainty and misspecification will be hard to eliminate satisfactorily, so that modeling will be less useful and a predictive approach may be the only feasible one. Of course, for $\mathcal{M}$-open problems, there is no true model, so model-based quantities such as the generalization error or comprehensive variability–accuracy analyses are inapplicable except in purely empirical form.

Consequently, if one accepts the premise that most real-world phenomena are more complex ($\mathcal{M}$-complete or $\mathcal{M}$-open) than the simple models based on limited data that are frequently used, a common mistake is underestimating the effect of model uncertainty and the effect of lack of existence of a readily approximable true model. Successful prediction centers on the Prequential Principle, which severely limits the role of models as actual models (as opposed to, say, actions) and makes model-based procedures for the evaluation of predictions or other inferences entirely inappropriate.

4

Models and Predictors: A Bickering Couple

... in most areas of statistics, inference [estimation and testing] seems to be generally carried out as if the analyst is sure that the true model is known ... In practice no-one really believes this ... when considering whether a possible additional explanatory variable is worth having in multiple regression, the question should not be 'Does it lead to a significant improvement in fit?' but 'Does it provide value for money in improving predictions?' ... The notion that there is no such thing as a true model, but rather that model building is a continual iterative search for a better model, is arguably in line with the general philosophy of science ... The idea of taking one or more confirmatory samples is a basic feature of the hard sciences, whereas statisticians seem to be primarily concerned ... with 'squeezing a single data set dry' ... Statistical inference needs to be broadened to include model formulation, but it is not clear to what extent we can formalize the steps taken by an experienced analyst during data analysis and model building ...

> Chatfield (1995) – debating model building vs. predictive
> validation in the context of model uncertainty.

Although $\mathcal{M}$-closed problems are not common, it is worthwhile to look at them from a predictive standpoint. First, it is useful to see that this is possible and is just as natural as adopting a modeling perspective and second, it is easiest to appreciate the effectiveness of the predictive approach by seeing it in some familiar contexts. This process was started in Sec. 2.2, and the predictive parts of Sec. 2.3, for parametric families where point and interval predictors were given from the frequentist and Bayesian standpoints. However, this can be taken much further. Since the model-based approach for a large number of routine statistical techniques has been amply treated in the literature, the task here is to reverse the bearings and see the same material from a predictive standpoint. We start with relatively standard techniques. So, first, frequentist and Bayes predictors are derived for some simple nonparametric settings with IID data. (Although this is actually $\mathcal{M}$-complete, it's so close to $\mathcal{M}$-closed that one finds the same rates of convergence, i.e., any reduction to a specific $\mathcal{M}$-closed problem would yield no benefit asymptotically. Also, the examples are so familiar they are often regarded as if they were $\mathcal{M}$-closed.) Then, it's time to start including explanatory variables and so turn to linear regression (Bayes and frequentist), logistic and quantile regression (Bayes and frequentist), and classification (Bayes and linear discriminants). Since details on the derivations can be found in many references, it will be enough to give and explain the forms of the predictors with minimal formal justification.

In reading this chapter and later chapters, it is important to remember that estimation and prediction provide inference about different quantities, a random variable and a parameter assuming a fixed value, respectively. An example may help to clarify this. Given IID measurements on subjects, say $y_1, \ldots, y_k$, the mean $\bar{y}$ can be used to estimate $\mu_Y = E(Y)$ or to predict Y_{k+1}. As an estimator, $\bar{Y} \to \mu_Y$ in probability (for instance) but, as a predictor, $\bar{Y} - Y_{k+1}$ converges in distribution as $k \to \infty$ to the distribution of $\mu_Y - Y$, where Y is an independent copy of any of the Y's. Loosely, estimators are meant to converge to an estimand whereas predictors are meant to track the predictand, i.e., the thing to be predicted. Thus, the analysis of a problem depends heavily on whether one takes a parameter inference perspective or a predictive perspective.

An important feature of the predictive perspective that can be seen in all the scenarios in this chapter is the abandonment of the conventional machinery for evaluating the variability of estimators. If estimators only matter insofar as they give good predictors, and the effect of an SE of an estimator is built into the predictor, the usual evaluations of estimators via SEs, MSEs, or, more generally, sampling distributions and posterior distributions are effectively pointless. In other words, the predictive approach extends the usual derivations of estimators to produce predictors and this makes the conventional inference machinery for estimation and testing irrelevant until satisfactory prediction has been achieved. Even after satisfactory prediction is achieved, it is unclear how useful the usual assessments of estimators are and also whether testing is helpful. For instance, in a linear regression setting, a high SE for a coefficient will convert into a higher variability of the predictor; the importance of a variable, as assessed by, say, the difference between including it in a predictor or not, will usually be more important for modeling than the SE of its coefficient or testing whether it equals zero. Thus, taking a predictive stance bypasses many issues that would normally be important in conventional approaches.

4.1 Simple Nonparametric Cases

As with the treatment of the mean and posterior mean in Sec. 2.2, percentiles and other fundamental properties of distributions can also be used for prediction outside the normal case. Again the goal is to provide recognizable point and interval predictors. Frequentist versions of these can be based on the empirical distribution function (EDF) $\hat{F}_n(\cdot)$. There are two versions of this. In the first one takes an interval based on $\hat{F}^{-1}$ and expands it using the uncertainty in $\hat{F}$ as bounded by the Smirnov theorem. In the second one takes endpoints for an interval using percentiles of $\hat{F}$ and enlarges it by invoking a central limit theorem for those percentiles. These results will be used in Sec. 7.1.1 to obtain estimates $\hat{S}_n$ of the survival function S.

To begin, note that a confidence level for a PI can be found by using the Kiefer–Wolfowitz Theorem 4, which establishes that for all $\epsilon > 0$ and N such that, for all $n > N$,

$$\lim_{n \to \infty} P(\sqrt{n} \sup_y |\hat{F}_n(y) - F(y)| > \epsilon) \le e^{-\alpha n}.$$

Now, let $\epsilon > 0$ and let $A_{\epsilon,n} = \{y \mid \sup_y |\hat{F}_n(y) - F(y)| < \epsilon\}$ be the 'good' set. Then, for given $\alpha > 0$, one can obtain the upper bound

$$P\left(Y_{n+1} \in [\hat{F}_n^{-1}(\alpha/2), \hat{F}_n^{-1}(1 - \alpha/2)]\right)$$
$$= P\left(F(Y_{n+1}) + (\hat{F}_n(Y_{n+1}) - F(Y_{n+1})) \in [\alpha/2, 1 - \alpha/2]\right)$$

$$= P\left(F(Y_{n+1}) \in [\alpha/2 - |\hat{F}_n(Y_{n+1}) - F(Y_{n+1})|,\right.$$
$$\left. 1 - \alpha/2 - |\hat{F}_n(Y_{n+1}) - F(Y_{n+1})|]\right)$$

$$\leq P\left(F(Y_{n+1}) \in [\alpha/2 - |\hat{F}_n(Y_{n+1}) - F(Y_{n+1})|,\right.$$
$$\left. 1 - \alpha/2 + |\hat{F}_n(Y_{n+1}) - F(Y_{n+1})|] \cap A_{\epsilon,n}\right) + P(A_{\epsilon,n}^c)$$

$$\leq P\left(F(Y_{n+1}) \in [\alpha/2 - \epsilon/\sqrt{n}, 1 - \alpha/2 + \epsilon/\sqrt{n}]\right) + \eta + e^{-\alpha n} \qquad (4.1)$$

$$\leq 1 - \alpha + \frac{2\epsilon}{\sqrt{n}} + \eta + e^{-\alpha n}, \qquad (4.2)$$

for $n \geq N$. To choose N, note that the theorem ensures that there is an N large enough that $|P(A_{\epsilon,n}) - 1| \leq \eta$ for preassigned $\eta > 0$. Hence, using $P(A_{\epsilon,n}^c) = 1 - P(A_{\epsilon,n})$ gives the bound $|P(A_{\epsilon,n}^c)| < \eta$ for $n > N$, as used in (4.1). It is seen that, as $n \to \infty$, ϵ and η can be allowed to go to zero so that the asymptotic confidence of the PI is $1 - \alpha$. (The fact that $F(Y_{n+1}) \sim \text{Unif}[0, 1]$ was also used to obtain the equality (4.2).) A lower bound follows from merely dropping $P(A_{\epsilon,n}^c)$, and hence $\eta + e^{-\alpha n}$, in (4.1) and (4.2).

A second way to look at prediction from the EDF is to observe that the rate at which the confidence limit of a PI of the form $[\hat{F}_n^{-1}(\alpha/2), \hat{F}_n^{-1}(1 - \alpha/2)]$ approaches $1 - \alpha$ is the usual $\sqrt{n}$ rate. To see this, recall the asymptotic normality of quantiles. In fact, Reiss (1989, Chap. 4, Sec. 1) shows that, for any $\alpha \in (0, 1)$,

$$\hat{F}_n^{-1}(\alpha) \to N\left(F^{-1}(\alpha), \frac{\alpha(1 - \alpha)}{n F'(F^{-1}(\alpha))^2}\right), \qquad (4.3)$$

weakly as $n \to \infty$. Now, using (4.3), asymptotic $\gamma/2 \, 100\%$ lower and $(1 - \gamma/2)100\%$ upper confidence bounds are given as

$$F^{-1}(\alpha/2) \geq F_n^{-1}(\alpha/2) + z_{\gamma/2} \frac{\sqrt{(\alpha/2)(1 - \alpha/2)}}{\sqrt{n} F'(F^{-1}(\alpha/2))}$$

and

$$F^{-1}(1 - \alpha/2) \leq F_n^{-1}(1 - \alpha/2) + z_{1-\gamma/2} \frac{\sqrt{(1 - \alpha/2)(\alpha/2)}}{\sqrt{n} F'(F^{-1}(1 - \alpha/2))},$$

where $z_{\gamma/2}$ is the $100\gamma/2$ quantile of an $N(0, 1)$ distribution. That is, we obtain a $1 - \alpha$ confidence PI for Y_{n+1} as

$$\left[F_n^{-1}(\alpha/2) + z_{\gamma/2} \frac{\sqrt{(\alpha/2)(1 - \alpha/2)}}{\sqrt{n} F'(F^{-1}(\alpha/2))}, \; F_n^{-1}(1 - \alpha/2) + z_{1-\gamma/2} \frac{\sqrt{(1 - \alpha/2)(\alpha/2)}}{\sqrt{n} F'(F^{-1}(1 - \alpha/2))}\right].$$
$$(4.4)$$

The intervals in (4.1) and (4.4) are comparable; the difference is that the constant in the $\sqrt{n}$ rate is identified in (4.4). Note that the argument leading to (4.4) treats the upper and lower bounds for $\alpha/2$ and $1 - \alpha/2$ separately. In fact, they are dependent and a joint asymptotic normality result for $(\hat{F}_n^{-1}(\alpha/2), \hat{F}_n^{-1}(1 - \alpha/2))$ can be given; see Reiss (1989). The covariances are of the same order as in (4.4) but the constants change slightly. In addition, PI bounds can also be found using the Kolmogorov–Smirnov bound for a distribution function; see DasGupta (2008, p. 395).

A third elementary frequentist nonparametric way to look at prediction is via density estimation rather than the EDF. The simplest density estimator is a histogram, but there are many others. The relationship between the data and its distribution function is different from (and closer to) the relationship between the data and its density. The histogram estimator for a density f is

$$f_n(y) = \frac{1}{n\lambda(A_{nj})} \sum_{i=1}^{n} \chi_{\{Y_i \in A_{nj}\}}(y),$$

where χ_A is the indicator function for a set A, for each n we have that $\mathcal{P}_n = \{A_{n1}, \ldots, A_{nj}, \ldots\}$ is an exhaustive and disjoint collection of sets with finite positive Lebesgue measure denoted λ. For each y there is a unique j such that $y \in A_{nj}$. At stage n, the expected value of f_n at a given $y \in A_{nj}$ is

$$g_n(y) = E(f_n(y)) = \frac{1}{\lambda(A_{nj})} \int_{A_{nj}} f(y) dy.$$

Provided that the $\mathcal{P}_n$ increase to give the Borel sigma field, i.e.,

$$\mathcal{B} = \cap_{n=1}^{\infty} \sigma \left(\cup_{m=n}^{\infty} \mathcal{P}_m \right),$$

one can verify that f_n converges in L^1 to the true density f, i.e.,

$$\int |f_n(y) - f(y)| d\lambda(y) \to 0,$$

in probability and other senses; see Devroye and Gyorfi (1985, p. 20).

Now, for a given n, one can choose percentiles $t_{\alpha/2}$ and $t_{1-\alpha/2}$ from the estimate $f_n(\cdot)$ such that

$$\int_{t_{\alpha/2}}^{t_{1-\alpha/2}} f_n(y) dy = 1 - \alpha, \tag{4.5}$$

giving an asymptotic $1 - \alpha$ PI of the form $[t_{\alpha/2}, t_{1-\alpha/2}]$ for the next outcome. These percentiles are analogous to the $(\alpha/2)$th and $(1 - \alpha/2)$th order statistics but formed from a histogram. That is, in (4.5), t_α and $t_{1-\alpha/2}$ are assumed to converge to their limits with error $o(1)$. A more exact expression for the limits of integration in (4.5) could be found if an expansion for t_α and $t_{1-\alpha/2}$ to higher order, e.g., $\mathcal{O}(1/\sqrt{n})$, were used. In the nonparametric settings considered so far, $\mathcal{M}$-closed problems and $\mathcal{M}$-complete problems are essentially being lumped together as if they were the same. Implictly the presumption is that the true distribution exists and that if the problem really is $\mathcal{M}$-complete then it can be so readily uncovered, or at least well approximated, that treating it as if if were $\mathcal{M}$-closed is not too far wrong. In low dimensions with simple data this is often reasonable.

The reasoning in Sec. 2.2 extends beyond the mean to include the median and many other point predictors. Let d be a metric and write a general 'bias–variance' bound as

$$P\left(d(\hat{Y}_{n+1}, Y_{n+1}) > \frac{1}{\epsilon} \right) \leq \epsilon \left(Ed(\hat{Y}_{n+1}, \theta) + Ed(\theta, Y_{n+1}) \right), \tag{4.6}$$

where θ is a parameter representing the limit of the predictor $\hat{Y}_{n+1}$ (assuming it exists) as an estimator. Even when the first term cannot be found for finite n it can be found asymptotically for many choices of d. Specifically, when $d(x, y) = (x - y)^q$ for some $q > 0$ and $\hat{Y}_{n+1}$

has an asymptotic distribution, e.g., one that is asymptotically normal with rate $\mathcal{O}(1/n)$, the first term becomes $E(\hat{Y}_{n+1} - \theta)^q$ and will assume a value based on the limiting distribution, provided that extra hypotheses are assumed to improve the convergence in distribution of $\hat{Y}_{n+1}$ to convergence in L^q.

For instance, one natural choice for a point predictor is the median, $\hat{F}_n^{-1}(1/2)$. It is the Bayes action under absolute loss. More generally, under the loss

$$L(\theta, a) = \begin{cases} K_1(\theta - a) & \theta > a, \\ K_2(\theta - a) & \theta < a, \end{cases} \tag{4.7}$$

the Bayes action is $\hat{F}_n^{-1}(K_1/(K_1 + K_2))$, so in principle any percentile can be an optimal point predictor.

The properties of the median (or any percentile) as a predictor follow from an analysis analogous to that given in Sec. 2.2.1 for the mean but using the absolute value in place of the squared error. Thus, apart from a factor, the right-hand side of (4.6) becomes

$$E|\hat{F}_n^{-1}(1/2) - \text{med}(Y)| + E|Y_{n+1} - \text{med}(Y)| \tag{4.8}$$

where $\theta = \text{med}(Y)$ is the median of any Y_i. Since (4.3) for $\alpha = 1/2$ gives the asymptotic distribution of the sample median, the first term in (4.8) can be evaluated asymptotically provided assumptions to ensure uniform integrability are imposed to improve the convergence in distribution in (4.3) to convergence in L^1. The result is just the expectation of the absolute value of a normal random variable and is left as an exercise for the reader. If $q = 2$ and L^2 convergence is assured then the first term in (4.8) is just the variance in (4.3). As seen in Sec. 2.2, the upper bound in (4.8) or more generally (4.6) can be bounded from above and shown to be asymptotically small.

Two other frequentist point predictors for Y_{n+1} are: (i) the trimmed mean,

$$\bar{Y}_{\alpha_1, \alpha_2} = \frac{1}{n - \lfloor n\alpha_1 \rfloor - \lfloor n\alpha_2 \rfloor} \sum_{\lfloor n\alpha_1 \rfloor + 1}^{n - \lfloor n\alpha_2 \rfloor} Y_{(i)}$$

with $\alpha_1 100\%$ and $\alpha_2 100\%$ trimming on the left and right, respectively, where $\lfloor x \rfloor$ is the largest integer less than or equal to x; and (ii) the mode. In some cases, the trimmed mean is asymptotically normal; see Shorack (1974) and more recently Wang *et al.* (2007). Results like this have also been obtained for multivariate trimmed means; see Arcones (1995). The mode of a histogram is often not uniquely defined, so, for one-dimensional data, to determine the mode of a distribution one usually takes the maximum of a density estimate,

$$\hat{Y}_{n+1} = \arg\max_x \hat{f}_n(y), \tag{4.9}$$

where $\hat{f}_n$ is an estimate of the density of Y. For instance, a kernel density estimate can be used but many other smooth functions of $y_1, \ldots, y_n$ will suffice. (A histogram does not usually give a unique value in (4.9).) Abraham *et al.* (2004) gives cases where the $\hat{Y}_{n+1}$ in (4.9) are asymptotically normal in distribution.

In a genuinely $\mathcal{M}$-closed setting with underlying parametric families, Bayesian point predictors like the posterior median of Y_{n+1}, i.e., point predictors in $m(y_{n+1}|y^n)$, converge to the true median as $n \to \infty$, as can be seen by writing

$$\text{med}_{m(y_{n+1}|y^n)}(Y_{n+1}) = \text{med}_{\int p(y_{n+1}|\theta)w(\theta|y^n)d\theta}(Y_{n+1}) \rightarrow \text{med}_{p(y_{n+1}|\theta_T)}(Y_{n+1}),$$

and again are typically asymptotically normal; see Yu and Clarke (2010).

By contrast, in the Bayes nonparametric case typical point and interval predictors for IID random variables are derived from the posterior density. This means not so much that the class of distributions (or densities) must be specified as in the frequentist case but that the prior – and its support – must be specified. The simplest case is the Dirichlet process prior; see Ferguson (1973). The Dirichlet distribution $\mathcal{D}(\alpha_1, \ldots, \alpha_K)$ with concentration parameters $\alpha_j > 0$ has density

$$\frac{\Gamma(\sum_{k=1}^K \alpha_k)}{\prod_{k=1}^K \Gamma(\alpha_k)} \prod_{k=1}^K y_k^{\alpha_k}$$

with $y_k \geq 0$ satisfying $\sum y_k = 1$, where $\Gamma(\cdot)$ is the gamma function. To extend the Dirichlet distribution to a prior on a nonparametric class of distributions, it must be converted into a process, i.e., in some sense be extended to infinitely many degrees of freedom.

To use the Dirichlet distribution to assign a prior to distribution functions (DFs) choose a base measure, say H, on the same sample space as F and a concentration parameter α. Then, F is distributed according to a Dirichlet process (DP), i.e., $F \sim \text{DP}(\alpha, H)$, if and only if given any finite measurable partition $A_1, \ldots, A_r$ we have

$$(F(A_1), \ldots, F(A_r)) \sim \mathcal{D}(\alpha H(A_1), \ldots, \alpha H(\alpha_r)),$$

i.e., a Dirichlet distribution with concentration parameters $\alpha H(A_j)$. It is not hard to show that, for any set A, $E(F(A)) = H(A)$ and $\text{Var}(F(A)) = H(A)(1 - H(A))/(\alpha + 1)$. Clearly, for any A, $F(A) \rightarrow H(A)$ in distribution as $\alpha \rightarrow \infty$ (although this does not by itself imply $F \rightarrow H$ in distribution). The support of the DP is the set of all discrete distribution functions (DFs) on the sample space, even if H is smooth; see Ghosh and Ramamoorthi (2003, Theorem 3.2.3). Although this means the support of a DP prior (DPP) is relatively small, posteriors from a DPP are generally consistent; see Ghosh and Ramamoorthi (2003, Theorem 3.2.7). Another sense in which DPPs are too small is that for different base measures they are mutually singular. That is, the DPPs do not deform continuously as a function of their base distributions; see Ghosh and Ramamoorthi (2003, Theorem 3.2.8). Fundamentally, the DP is a countable sum of point masses, as can be seen in the stick-breaking representation of a DP, see Sethuraman (1994), and so it can only represent a continuous density to the extent that it can be approximated by a countable sum of point masses.

Now, suppose that $F \sim \text{DP}(\alpha, H)$ and that $y_1, \ldots, y_n$ are drawn IID from F. To identify the posterior distribution for $(F|y^n)$ choose a partition $A_1, \ldots, A_r$ of the sample space of F and let $n_j = \#\{i|y_i \in A_j\}$. Regarding the n_j as sample sizes from a multinomial with r cells having probabilities $F(A_j)$, and recalling that the Dirichlet is conjugate to the multinomial, gives

$$(F(A_1), \ldots, F(A_r)|y^n) \sim \mathcal{D}(\alpha H(A_1) + n_1, \ldots, \alpha H(A_r) + n_r). \tag{4.10}$$

Since (4.10) is true for any finite measurable partition, the distribution $(G|y^n)$ must be a DP, too. In fact, it can be shown that

$$(G|y^n) \sim DP\left(\alpha + n, \frac{\alpha H + \sum_{i=1}^n \delta_i}{\alpha + n}\right) = DP\left(\alpha + n, \frac{\alpha}{\alpha + n}H + \frac{n}{\alpha + n}\frac{\sum_{i=1}^n \delta_i}{n}\right),$$

where δ_i is a unit mass at y_i, so that for any set A containing all the data, we have $n = \sum_{i=1}^{n} \delta_i(A)$. (A careful and accessible proof of this can be found in Susarla and van Ryzin (1976).) Now, $n_j = \sum_{i=1}^{n} \delta_i(A_j)$ and as $\alpha \to 0$ the posterior converges to the empirical distribution function (EDF). It is seen that as $n \to \infty$ the posterior also converges to the EDF, which is consistent for F_{true}; hence, the posterior concentrates at the true distribution.

For the prediction of Y_{n+1}, the Bayesian uses $(Y_{n+1}|y^n)$. That is, for measurable A,

$$P(Y_{n+1} \in A|y^n) = E(F(A)|y^n)$$

$$= \frac{1}{\alpha + n}\left(\alpha H(A) + \sum_{i=1}^{n} \delta_i(A)\right),$$

where the expectation is taken over the argument F and the last expression holds because the mean over F comes from the base distribution of the posterior. So,

$$(Y_{n+1}|y^n) \sim \frac{\alpha H + \sum_{i=1}^{n} \delta_i}{\alpha + n}, \tag{4.11}$$

i.e., the base distribution of the posterior is the predictive distribution for Y_{n+1}. Denoting the DPP distribution $W(\cdot)$ and using (4.11), a Bayes PI is of the form

$$P(L(y^n) \le Y_{n+1} \le U(y^n)|y^n),$$

for some lower and upper functions L and U of y^n.

Campbell and Hollander (1982) solved a more general problem, finding β_1 and β_2 such that

$$W(\text{at least } q \text{ of } Y_{n+1}, \ldots, Y_{n+N} \text{ are in } [\beta_1, \beta_2]|y^n) = \gamma. \tag{4.12}$$

They set up a multinomial problem with two cells, i.e., a binomial, to determine whether a future outcome is in $[\beta_1, \beta_2]$ or not. Let J be the number of Y_j in Y_{n+1}^{n+N} with values in $[\beta_1, \beta_2]$ and denote $H' = \alpha H + \sum_{i=1}^{n} \delta_i$. Then, to find a PI for q out of N future outcomes, it is enough to use (4.12) and therefore to solve

$$\sum_{j=q}^{N} P(J = j|y^n) = \sum_{j=q}^{N} \binom{N}{j} \frac{H'(\beta_1, \beta_2]^{[j]} H'(\mathbb{R} \setminus (\beta_1, \beta_2]^{[N-j]}}{H'(\mathbb{R})^{[N]}} = \gamma \tag{4.13}$$

for β_1 and β_2. In (4.13) the notation $y^{[u]} = y(y+1) \cdots (y+u-1)$ with $y^{[0]} = 1$ indicates an ascending factorial. The base case is, of course, one-step-ahead prediction with $q = N = 1$. As Campbell and Hollander (1982) noted, a unique value of $H'(\beta_1, \beta_2]$ can be found but β_1 and β_2 need not be unique without further conditions, such as requiring the PI to have the smallest length or to have endpoints corresponding to the upper and lower pth percentiles of the DF $H'(-\infty, t]/H(\mathbb{R})$, where $p = (1/2)(1 - H'(\beta_1, \beta_2])/H'(\mathbb{R})$. Indeed, for moderate sample sizes and γ's close to one, this method is likely to give PIs that are too short, with coverage lower than the nominal γ. This occurs because the point masses do not enable the prior to preserve tail areas. The point masses may also contribute to nonrobustness of the PIs; these properties seem not to be too serious when $q = 1$ but worsen as q and N increase.

Instead of getting PIs directly from the predictive density, one can obtain predictors from the posterior mean using the DPP, namely,

$$\frac{\alpha}{\alpha+n}H+\frac{n}{\alpha+n}\frac{\sum_{i=1}^{n}\delta_i}{n}.$$

Since this is the same as the predictive distribution, (4.13) still applies to give a PI. (Note that (4.13) also takes the concentration parameter $\alpha+n$ into account because the normalization factor cancels in the numerator and denominator of the right-hand side.) Of course, one could obtain other point estimators of F_{true} from $W(\cdot|y^n)$ instead.

An alternative would be to take a $1-\alpha$ neighborhood, say $R_{1-\alpha}$, satisfying

$$W(R_{1-\alpha}|y^n)=1-\alpha$$

and then form the PI

$$[\inf_{G\in R_{1-\alpha}}t_{\alpha/2}(G), \sup_{G\in R_{1-\alpha}}t_{1-\alpha/2}(G)] \tag{4.14}$$

using the smallest $(\alpha/2)$100th and largest $(1-\alpha/2)$100th percentiles from the DFs in $R_{1-\alpha}$. However, this approach does not seem to have been pursued. One could generate curves from $W(\cdot|y^n)$, plot them, and take the minimum and maximum of their $(\alpha/2)$100th and $(1-\alpha/2)$100th percentiles as in (4.14); see Mueller and Mitra (2013, Fig. 1(a)) for a setting where this would be feasible. It would probably be better to use the minimum $(\alpha/2)$100th percentile of the $(\alpha/2)$100th percentiles and the maximum $(1-\alpha/2)$100th percentile of the $(1-\alpha/2)$100th percentiles from the distributions drawn from the posterior.

To conclude our discussion about obtaining PIs using a simple nonparametric Bayes method, it's worth looking at a technique for overcoming the limitations of using the DPP for density estimation. The idea is to take a mixture of DPPs; see Antoniak (1974). If the mixing distribution is continuous then the resulting prior and posterior are also continuous, unlike single DPPs and posteriors that are not continuous and instead concentrate their mass on discrete distributions. Even though mixing will not eliminate the instability problems of having point masses in the posterior, it can improve the stability of the predictions because it adds an extra layer of 'stochasticity', as advocated by Draper (1995). Following MacEachern and Mueller (1998) (see also Mueller and Mitra (2013)), a mixture of Dirichlet processes (MDP) can be written

$$y_i\sim p_{\theta_i}(\cdot), \quad \theta_i\sim G, \quad G\sim DP(\alpha,H) \tag{4.15}$$

for $i=1,\ldots,n$. That is, the distribution of any given Y_i is

$$F(y)=\int p(y|\theta)\,G(d\theta), \quad G\sim DP(\alpha,H),$$

implicitly permitting the various y_i to originate from different θ_i. To use this, one must choose p_θ, α, and H. One common choice for p_θ is to set $\theta=(\mu,\Sigma)$, so that $p(y_i|\theta)=p(y_i|\mu,\Sigma)$ and $p(\cdot|\mu,\Sigma)$ is the density of an $N(\mu,\Sigma)$ distribution. A common choice for the distribution H is a normal-Wishart product, that is, $H(\mu,\Sigma)=N_{(m,B)}(\mu)W_{(r,R)}(\Sigma^{-1})$ (with mild abuse of notation).

As for the DPP, samples from an MDP are almost surely discrete, and explicit expressions for the prior mean and variance may be obtained. The MDP is a conjugate family in that the posterior distribution based on an MDP prior is again an MDP. (This can be seen by using a reduction to DPPs that are conjugate priors.) Hence, point predictors based on the posterior mean, median, and variance can, in principle, be obtained. For details on the MDP, and other variations on the DPP, see Ghoshal (2010, Sec. 2.3).

For the MDP, the predictive distribution for stage $n + 1$ is

$$p(y_{n+1}|y^n) = \int \left(\int p(y_{n+1}|\theta_{n+1}) p(\mathrm{d}\theta_{n+1}|\theta^n, y^n) \right) p(\mathrm{d}\theta^n, y^n). \qquad (4.16)$$

The inner integral is the density $p(y_{n+1}|\theta_{n+1})$ integrated against $p(\theta_{n+1}|\theta^n)$ because θ_{n+1} is independent of y^n. Since $p(\theta_{n+1}|\theta^n)$ is the predictive distribution for θ_{n+1} and the θ_i are drawn from a distribution itself drawn from a DP, the form of $p(\theta_{n+1}|\theta^n)$ is given by the expression in (4.11) but with θ_i in place of y_i, i.e., $(\sum_{i=1}^{n} \delta_{\theta_i} + \alpha H)/(\alpha + n)$. Moreover, the term αH may be regarded conceptually (and for computational purposes) as $\alpha \delta_{\theta^*}$, where θ^* is a new draw from H. While this cannot be evaluated in closed form, the Gibbs sampling method of MacEachern and Mueller (1998) can be used to generate estimates of the density in (4.16) from which $1 - \alpha$ PIs and point predictors can be found.

4.2 Fixed Effects Linear Regression

Another familiar example of prediction, this one involving explanatory variables, comes from linear regression models. Recall that they have a 'signal plus noise' response form:

$$Y_i = X_i \beta + \epsilon_i, \qquad (4.17)$$

where $\beta = (\beta_1, \ldots, \beta_d)^T \in \mathbb{R}^d$ is the parameter vector for the explanatory variables $X_i = (x_{1i}, \ldots, x_{di}) \in \mathbb{R}^d$, data (y_i, x_i) for $i = 1, \ldots, n$ are available, and the noise terms ϵ_i are taken to be IID $N(0, \sigma^2)$ outcomes. The n equations in (4.17) are more compactly written as

$$Y = X\beta + \epsilon, \qquad (4.18)$$

where $Y = (Y_1, \ldots, Y_n)^T$, $X = (X_1^T, \ldots, X_n^T)^T$, and $\epsilon = (\epsilon_1, \ldots, \epsilon_n)^T \sim N_n(0, \sigma^2 I_n)$ with I_n the $n \times n$ identity matrix. In the notation of (4.18), the usual least squares estimator for β is given by

$$\hat{\beta} = (X^T X)^{-1} X^T Y$$

and only requires that (i) the ϵ_i be independent and identical, but not necessarily normal, and (ii) the matrix $X^T X$ be invertible with $n > d$ (and preferably low multicollinearity). Under normality, $\hat{\beta} \sim N(\beta, \sigma^2 (X^T X)^{-1})$ and the natural point predictor for Y_{n+1} for a new design point X_{n+1} is $\hat{Y}_{n+1}(X_{n+1}) = X_{n+1}\hat{\beta}$.

To get a PI for $\hat{Y}_{n+1}$ requires an estimate for σ, unless it is known already. Denote the fitted values for Y by $\hat{Y} = X\hat{\beta}$, the ith residual by $e_i = Y_i - \hat{Y}_i$, and the residual vector by $e = (e_1, \ldots, e_n)^T = (I_n - X(X^T X)^{-1} X^T)Y$. Then, the natural estimate of σ^2 is the MSE

$$s^2 = \sum_{i=1}^{n} e_i^2/(n-d) = e \cdot e/(n-d) \sim \sigma^2 \chi_{n-d}^2/(n-d),$$

in which s^2 and $\hat{\beta}$ are independent as random variables. Since $\hat{Y}_{n+1} \sim N(X_{n+1}\beta, \sigma^2(1 + X_{n+1}^T(X^T X)^{-1}))$, a $1 - \alpha$ PI is

$$\hat{Y}_{n+1} \pm t_{1-\alpha/2, n-d} s \sqrt{1 + X_{n+1}^T(X^T X)^{-1} X_{n+1}}. \qquad (4.19)$$

If the distribution of ϵ is not normal, the form of the interval will change but the point prediction does not. In the special case $d = 2$, where $X_{1,i} = 1$ and $X_{2,i} = x_i$, so that $\bar{x} = \bar{x}_2$, the form of the interval is

$$\hat{Y}_{n+1} \pm t_{1-\alpha/2, n-2} s \sqrt{1 + \frac{1}{n} + \frac{(x_{n+1} - \bar{x})^2/n}{\sum_{i=1}^{n}(x_i - \bar{x})^2/n}}. \tag{4.20}$$

If $x_{n+1} = \bar{x}$, as may be approximately true in designed experiments, then the variance reduces to $1 + (1/n)$ as in the normal case without covariates. However, as x_{n+1} moves away from $\bar{x}$, the width of the PIs increases as $|x_{n+1} - \bar{x}|$.

If desired, the variability in s can also be included by using percentiles from the $\chi^2_{n-d}/(n-d)$. The $1 - \alpha$ lower and upper confidence limits for σ are given by $s\chi^2_{n-d,\alpha/2}/(n-d)$ and $s\chi^2_{n-d,1-\alpha/2}/(n-d)$, respectively. So, using these in place of s for the lower and upper prediction limits, respectively, will give a wider PI for Y_{n+1} than (4.19) or (4.20) does. This PI will have a probability bigger than $1 - \alpha$ and will be conservative (especially if the α in the confidence limits for σ is 1/2!).

Other choices for point predictors are possible. For instance, instead of using the squared error, one could use the least absolute deviation (LAD) or least median of squares (LMS). Estimates for β using the LAD or LMS are usually more robust than those under squared error; however, they can be nonunique. If unique solutions exist and are consistent then there are various ways to find PIs; one based on percentiles of the residuals is given in Olive (2013) and another based on bootstrapping is given in Scorbureanu (2009). Any technique that provides a good estimate of the variance of Y_i as a function of x will gives useful PIs even when the distribution of $\hat{Y}_{n+1}(x_{n+1})$ is not available in closed form.

Predictors for the model (4.17) have also been derived from a variety of Bayesian analyses. Although not necessarily the most appropriate, one natural Bayesian analysis uses the 'normal-inverse-gamma' (NIG) conjugate prior for (β, σ^2) to obtain $p(y_{n+1}|(Y, X), x_{n+1})$, the density for $Y_{n+1}(x_{n+1})$ given the design matrix X, the response vector Y, and the new design point x_{n+1}. To see this explicitly, write the normal density for β, given σ^2, as

$$w(\beta|\sigma^2) = \phi_{\mu_\beta, \sigma^2 V_\beta}(\beta), \tag{4.21}$$

where $\phi_{\mu,\Sigma}$ is the density of an $N(\mu, \Sigma)$ random variable, μ_β is the prior mean of β, and $\sigma^2 V_\beta$ is the prior variance–covariance matrix of β given σ^2. Then write the density of the inverse-gamma $IG(a, b)$ prior as

$$w(\sigma^2) = \frac{b^a}{\Gamma(a)} \left(\frac{1}{\sigma^2}\right)^{a+1} e^{-b/\sigma^2}. \tag{4.22}$$

With a mild abuse of notation the NIG$(\mu_\beta, V_\beta, a, b)$ prior for (β, σ^2) is

$$w(\beta, \sigma^2) = w(\beta|\sigma^2)w(\sigma^2) = N(\mu_\beta, \sigma^2 V_\beta)IG(a, b)$$

$$= \frac{b^a}{(2\pi)^{d/2} \det(V_\beta)^{1/2}\Gamma(a)} \left(\frac{1}{\sigma^2}\right)^{(a+d/2)+1}$$

$$\exp\left(-\frac{1}{\sigma^2}\left(b + \frac{1}{2}(\beta - \mu_\beta^T)V_\beta^{-1}(\beta - \mu_\beta)\right)\right).$$

Since the likelihood is

$$p(y|X,\beta,\sigma^2) = N(X\beta,\sigma^2 I_n) = \left(\frac{1}{2\pi\sigma^2}\right)^{n/2} \exp\left(-\frac{1}{2\sigma^2}(y-X\beta)^T(y-X\beta)\right).$$

one can find the posterior

$$w(\beta,\sigma^2|y,X,x_{n+1}) = w(\beta,\sigma^2)p(y|X,\beta,\sigma^2)/p(y|X) \tag{4.23}$$

explicitly, given an explicit expression for the marginal for the data, $p(y|X)$.

There are at least two ways to do this. Write

$$p(y|X) = \int \left(\int p(y|X,\beta,\sigma^2)w(\beta|\sigma^2)d\beta\right) w(\sigma^2)d\sigma^2. \tag{4.24}$$

It is possible to work out the two integrals in (4.24) explicitly. However, to get a useful expression for the inner integral, it is easier to proceed as follows. Suppose σ^2 is fixed. Then, the Bayesian linear regression model is hierarchical, of the form

$$Y = X\beta + \epsilon \quad \text{where} \quad \epsilon \sim N(0,\sigma^2 I_n),$$
$$\beta = \mu_\beta + \xi \quad \text{where} \quad \xi \sim N(0,\sigma^2 V_\beta),$$

in which ϵ and ξ are independent. Thus,

$$Y = X\mu_\beta + X\xi + \epsilon \sim N(X\mu_\beta, \sigma^2(I_n + XV_\beta X^T)), \tag{4.25}$$

so it is easy to identify $p(Y|X,\sigma^2,V_\beta)$. Using this in (4.24) gives

$$p(Y|X) = \int p(Y|X,\sigma^2)d\sigma^2 = \int N(X\mu_\beta, \sigma^2(I_n + XV_\beta X^T))(Y)IG(a,b)d\sigma^2$$
$$= \int \text{NIG}(X\mu_\beta, (I_n + XV_\beta X^T), a, b)(Y)d\sigma^2, \tag{4.26}$$

in which Y is random. The integral in (4.26) can be worked out generically. Using g to represent the random quantity Y with mean γ (or the random quantity β in a standard Bayesian analysis), we have

$$\int_0^\infty \text{NIG}(\gamma, V, a, b)(g,\sigma^2)d\sigma^2 = \frac{b^a}{(2\pi)^{d/2}\det(V)^{1/2}\Gamma(a)} \int \left(\frac{1}{\sigma^2}\right)^{(a+d/2)+1}$$
$$\times \exp\left(-\frac{1}{\sigma^2}\left(b + \frac{1}{2}(g-\gamma)^T V^{-1}(g-\gamma)\right)\right)d\sigma^2$$
$$= \frac{b^a}{(2\pi)^{d/2}\det(V)^{1/2}\Gamma(a)}\Gamma(a+d/2)\left(b + \frac{1}{2}(g-\gamma)^T V^{-1}(g-\gamma)\right)^{(-a+d/2)};$$

thus by transforming (starting with $u = 1/\sigma^2$), we recognize the Γ function. The result is a d-dimensional Student's t-density with v degrees of freedom that has the generic form

$$\text{MVS}t_v(\gamma,\Sigma)(g) = \frac{\Gamma((v+d)/2)}{\Gamma(v/2)\pi^{d/2}\det(v\Sigma)^{1/2}}\left(1 + \frac{(g-\gamma)^T\Sigma^{-1}(g-\gamma)}{v}\right)^{-(v+d)/2},$$

with $v = 2a$ and $\Sigma = (b/a)V$. So, it is easy to recognize that

$$p(Y|X) = \text{MVS}\, t_{2a}\left(X\mu_\beta, \frac{b}{a}(I_n + X V_\beta X^T) \right)(Y). \tag{4.27}$$

This expression can be used in (4.23) to specify the posterior. Since the posterior is proportional to the likelihood times the prior, one can use the identity

$$b + \frac{1}{2}(\beta - \mu_\beta)^T V_\beta (\beta - \mu_\beta) + \frac{1}{2}(Y - X\beta)^T (Y - X\beta) = b^* + \frac{1}{2}(\beta - \mu^*)^T (V^*)^{-1}(\beta - \mu^*),$$

in which the updated hyperparameter values are

$$\mu^* = (V_\beta^{-1} + X^T X)^{-1}(V_\beta^{-1}\mu_\beta + X^T Y),$$
$$V^* = (V_\beta^{-1} + X^T X)^{-1},$$
$$a^* = a + \frac{n}{2},$$
$$b^* = b + \frac{1}{2}\left(\mu_\beta^T V_\beta^{-1}\mu_\beta + Y^T Y - (\mu^*)^T V^{*-1}\mu^* \right).$$

So, the posterior is

$$w(\beta, \sigma^2 | Y) \propto \left(\frac{1}{\sigma^2} \right)^{a + (n+d)/2 + 1} \exp\left(-\frac{1}{\sigma^2}\left(b + \frac{1}{2}(\beta - \mu^*)^T (V_\beta^*)^{-1}(\beta - \mu^*) \right) \right), \tag{4.28}$$

i.e., an $\text{NIG}(\mu^*, V^*, a^*, b^*)$, in which (4.27) is the constant of proportionality.

Having used an NIG prior to find the NIG posterior, one can finally obtain the predictive distribution. First note that the reasoning in (4.26) and (4.27) gives

$$\int p(y|\beta, \sigma^2)w(\beta, \sigma^2)d\beta d\sigma^2 = \int N(X\beta, \sigma^2 I_n)\text{NIG}(\mu_\beta, V_\beta, a, b)d\beta d\sigma^2$$

$$= \text{MVS}\, t_{2a}\left(X\mu_\beta, \frac{b}{a}(I_n + X V_\beta X^T) \right)(Y).$$

So, using the conjugacy of the NIG priors, it can be seen that

$$p(Y_{n+1}|Y, X, x_{n+1}) = \int p(y_{n+1}|\beta, \sigma^2)w(\beta, \sigma^2 | Y, X)d\beta d\sigma^2$$

$$= \int N(x_{n+1}\beta, \sigma^2)\text{NIG}(\mu^*, V^*, a^*, b^*)d\beta d\sigma^2$$

$$= \text{MVS}\, t_{2a^*}\left(x_{n+1}\mu^*, \frac{b^*}{a^*}(1 + x_{n+1}^T V^* x_{n+1}) \right); \tag{4.29}$$

an analogous expression holds for predicting m future values of Y. Now, for the conjugate prior case, the usual Bayesian point and interval predictors for future values of Y_{n+1} can be readily found for any x_{n+1}.

In addition to the Bayes predictors that can be found using conjugate priors on (β, σ^2), Bayes predictors can be found by using noninformative or objective priors. The idea is that the priors are chosen to reflect the pre-experimental lack of information in some formal sense. In the absence of nuisance parameters, Jeffreys' prior is the 'reference prior', i.e., it is noninformative in a relative entropy sense. In the linear regression context, the Jeffreys' prior for (β, σ^2) is the product of a constant prior for β, given σ^2, and a prior $1/\sigma^2$ for σ^2. Thus, $w(\beta, \sigma^2) \propto 1/\sigma^2$, which is improper unless truncated to a compact set but often works well nevertheless. This prior can be regarded as a limit of conjugate priors by letting $V_\beta^{-1} \to 0$, $a \to -d/2$, and $b \to 0$ in an $\text{NIG}(\mu_\beta, V_\beta, a, b)$ density.

If $w(\beta, \sigma^2) \propto 1/\sigma^2$ is used then following the derivation leading to (4.28) from an $\text{NIG}(\mu_\beta, V_\beta, a, b)$ but taking the limit over the hyperparameters gives

$$\mu^* = (X^T X)^{-1} X^T Y = \hat{\beta},$$
$$V^* = (X^T X)^{-1},$$
$$a^* = \frac{n-d}{2},$$
$$b^* = \frac{(n-d)s^2}{2},$$

as the updating for the hyperparameters where $s^2 = (1/(n-d))(Y - X\hat{\beta})^T (Y - X\hat{\beta})$ is the MSE and $\hat{\beta}$ is the LSE (or MLE) for β. The MSE can also be written $s^2 = (1/(n-d))Y^T (I_n - X(X^T X)^{-1} X^T)Y$; s^2 is unbiased for σ^2. Thus, under the reference prior, the posterior for (β, σ^2) is $\text{NIG}(\mu^*, V^*, a^*, b^*)$.

It is easy to see that, in analogy to (4.29),

$$p(Y_{n+1} | Y, X, x_{n+1}) = \int p(y_{n+1} | \beta, \sigma^2) w_J(\beta, \sigma^2 | Y, X) d\beta d\sigma^2$$

$$= \int N(x_{n+1}\beta, \sigma^2) \text{NIG}(\mu^*, V^*, a^*, b^*) d\beta d\sigma^2$$

$$= \text{MVSt}_{n-d} \left(x_{n+1}\hat{\beta}, s^2 (1 + x_{n+1}^T (X^T X)^{-1} x_{n+1}) \right),$$

where w_J indicates the reference prior; again a similar expression holds for predicting m future values of Y. Now, for the Jeffreys' prior case, the usual Bayesian point and interval predictors for future values of Y_{n+1} can be readily found for any x_{n+1}. It is left as an exercise for the reader to use (4.28) to derive $w(\beta | \sigma^2, Y, X)$ and thereby verify that

$$E(Y_{n+1}(x_{n+1}) | \sigma^2, Y, X) = E \left(E(Y_{n+1}(x_{n+1}) | \beta, \sigma^2, Y, X) | \sigma^2, Y, X \right)$$

$$= E(x_{n+1}\beta | \sigma^2, Y, X) = x_{n+1}\hat{\beta}.$$

That is, the value of σ^2 does not affect the location of the predictive density; the same property can be seen in (4.29) if conjugate priors are used. For the variance, one can derive

$$\text{Var}(Y_{n+1}(x_{n+1}) | \sigma^2, Y, X) = E \left(\text{Var}(Y_{n+1}(x_{n+1}) | \beta, \sigma^2, Y, X) | \sigma^2, Y, X \right)$$

$$+ \text{Var} \left(E(Y_{n+1}(x_{n+1}) | \beta, \sigma^2, Y, X) | \sigma^2, Y, X \right)$$

$$= E(\sigma^2) + \text{Var}(x_{n+1}\beta | \sigma^2, Y, X)$$

$$= (1 + x_{n+1}(X^T X)^{-1} x_{n+1})\sigma^2,$$

meaning that, conditionally on σ, the predictive variance has two components: σ^2, representing the intrinsic variability of the response in the context of the model, and $x_{n+1}(X^T X)^{-1}\sigma^2$, representing the uncertainty due to estimating β.

The two Bayesian predictors for linear models seen so far use priors that do not depend on the explanatory variables. By contrast, Zellner's informative g-prior for β is

$$(\beta|\sigma^2, X) \sim N(\beta_0, g\sigma^2(X^T X)^{-1}), \tag{4.30}$$

(see Zellner 1986) in which g is a hyperparameter to be chosen, β_0 is often set to zero, and the prior clearly depends on X. If X is chosen so that $(X^T X)^{-1}$ is diagonal then Zellner's g-prior puts more mass close to β_0 for the components β_j of β that correspond to the smallest eigenvalues of $(X^T X)^{-1}$. Loosely, since $\text{Var}(\hat{\beta}) = \sigma^2(X^T X)^{-1}$, this means that Zellner's g-prior puts more mass on regions where the frequentist variability in $\hat{\beta}_j$ is lowest, i.e., the Zellner g-prior effectively reinforces the estimates of β that are in agreement with the prior information. However, g controls the overall prior magnitude of β and a larger g represents less prior information. (Using $X^T X$ in the variance would give an 'inverse' Zellner g-prior and might have a greater claim to objectivity or at least to compensating for regions where the data is expected to be less informative about the parameters.) See Bove and Held (2011) for further interpretation of the g-prior and variants on it.

Typically, the Jeffreys' prior $w_J(\sigma^2) \propto 1/\sigma^2$ is used for σ^2. So, to derive the predictive distribution for Y_{n+1} under the Zellner g-prior, we start by writing the posterior as

$$
\begin{aligned}
w_g(\beta, \sigma^2|X, Y) &= \frac{w_J(\sigma^2)w_g(\beta|\sigma^2)p(Y|X, \beta, \sigma^2)}{p_g(Y|X)} \\
&\propto \left(\frac{1}{\sigma^2}\right)^{-(n/2)+1} \exp\left(-\frac{1}{2g\sigma^2}(\beta - \beta_0)X^T X(\beta - \beta_0)\right) \\
&\quad \times \exp\left(-\frac{1}{2\sigma^2}(Y - X\beta)(Y - X\beta)\right) \\
&= \left(\frac{1}{\sigma^2}\right)^{-(n/2)+1} \exp\left(-\frac{1}{2g\sigma^2}(\beta - \beta_0)X^T X(\beta - \beta_0)\right) \\
&\quad \times \exp\left(-\frac{1}{2\sigma^2}s^2 - \frac{1}{2\sigma^2}(\hat{\beta} - \beta)X^T X(\hat{\beta} - \beta)\right), \tag{4.31}
\end{aligned}
$$

by adding and subtracting $X\hat{\beta}$, where $s^2 = (Y - X\hat{\beta})^T(Y - X\hat{\beta})$. By a complete-the-square argument on β one can derive the identity

$$
\begin{aligned}
(\beta - \hat{\beta})^T X^T X(\beta - \hat{\beta}) &+ \frac{1}{g}(\beta - \hat{\beta})^T X^T X(\beta - \hat{\beta}) \\
&= \frac{1+g}{g}\left(\beta - \frac{g}{1+g}(\hat{\beta} + \beta_0/g))^T X^T X(\beta - \frac{g}{1+g}(\hat{\beta} + \beta_0/g)\right) \\
&\quad + \frac{1}{1+g}(\hat{\beta} - \beta_0)X^T X(\hat{\beta} - \beta_0). \tag{4.32}
\end{aligned}
$$

Using (4.32) in (4.31) permits the identification

$$
\begin{aligned}
\beta|\sigma^2, Y, X &\sim N\left(\frac{g}{g+1}(\hat{\beta} + \beta_0/g), \frac{\sigma^2 g}{g+1}(X^T X)^{-1}\right) \\
\sigma^2|Y, X &\sim IG\left(\frac{n}{2}, \frac{s^2}{2} + \frac{1}{2(g+1)}(\hat{\beta} - \beta_0)X^T X(\hat{\beta} - \beta_0)\right). \tag{4.33}
\end{aligned}
$$

That is, the posterior for (β, σ^2) is

$$\text{NIG}\left(\frac{g}{g+1}(\hat{\beta} + \beta_0/g), \frac{g}{g+1}(X^T X)^{-1}, \frac{n}{2}, \frac{s^2}{2} + \frac{1}{2(g+1)}(\hat{\beta} - \beta_0)X^T X(\hat{\beta} - \beta_0)\right),$$

which can be used in (4.29) to give

$$P(Y_{n+1}(x_{n+1})|Y, X) = \text{MVS}\, t_n\left(x_{n+1}\frac{g}{g+1}(\hat{\beta} + \beta_0/g),\right.$$

$$\left.\frac{2}{n}\left(\frac{s^2}{2} + \frac{1}{2(g+1)}(\hat{\beta} - \beta_0)X^T X(\hat{\beta} - \beta_0)\right)\left(1 + x_{n+1}^T\frac{g}{g+1}(X^T X)^{-1}x_{n+1}\right)\right), \quad (4.34)$$

from which it is seen that as $g \to \infty$ the location of the MVS t_n distribution has mean approaching $x_{n+1}\hat{\beta}$ and variance approaching $(s^2/n)(1 + x_{n+1}(X^T X)^{-1}x_{n+1})$, the values familiar from the frequentist analysis because s^2/n is the usual estimator of σ^2. The degrees of freedom in (4.34) is n because the Zellner g-prior is informative, possibly representing information equivalent to d data points.

There is a variety of ways to choose g; Liang *et al.* (2008) provides a good discussion and empirical comparison of several. The best methods are nearly indistinguishable in their performance for model selection purposes. For predictive purposes, Liang *et al.* (2008) identified several methods for choosing g that permit consistent prediction by BMA. However, it must be remembered that, like all the techniques of this chapter, these results are of limited use because they are essentially only applicable for $\mathcal{M}$-closed problems. Indeed, outside the narrowest $\mathcal{M}$-closed problems it is easy to outperform BMA predictively; see Clarke (2003), Wong and Clarke (2004), and Clarke *et al.* (2013).

As a final Bayesian method for constructing predictors in the linear model context, consider the spike-and-slab prior for regression originally due to Mitchel and Beauchamp (1988). The idea is to permit point masses at $\beta_j = 0$, so that there is nonzero probability of eliminating some explanatory variables from the predictor. The basic formulation can be found in Yen (2011). Write

$$Y_i = \sum_{j=1}^{d} X_{ij}\gamma_j\beta_j + \epsilon_i$$

and let $\gamma = (\gamma_1, \ldots, \gamma_d)$. Spike-and-slab regression can be specified by

$$Y_i|X_i, \beta, \gamma, \sigma^2 \sim N\left(\sum_{j=1}^{d} X_{ij}\gamma_j\beta_j, \sigma^2\right) \quad i = 1, \ldots, n$$

$$\beta_j|\sigma^2, \gamma_j, \lambda \sim \gamma_j N(0, \sigma^2/\lambda) + (1 - \gamma_j)\mathbb{I}_{\beta_j=0} \quad j = 1, \ldots, d$$

$$\sigma^2|a, b \sim \text{IG}(a, b)$$

$$\gamma|\kappa \sim \text{Ber}(\kappa).$$

From this one can write down the posterior as

$$w(\beta, \gamma, \sigma^2|X, Y, \lambda, a, b, \kappa) \propto p(y|X, \beta, \gamma, \sigma^2)w(\beta|\sigma^2, \gamma, \lambda)w(\sigma^2|a, b)w(\gamma|\kappa).$$

In principle, the predictive distribution can be found by integrating the density for $Y_i|X_i, \beta, \gamma, \sigma^2$ with respect to the posterior. Closed-form expressions for prediction do not seem to have been developed but there is no impediment to doing so, apart from the effort required. Indeed, the effort required to obtain the predictive distribution numerically is probably not greater than the effort required to estimate $(\beta, \gamma, \sigma^2)$; see Yen (2011).

Another formulation of the spike-and-slab linear model is given in Ishwaran and Rao (2005). It differs from Yen (2011) in that it does not use point masses. In hierarchical form, the model is

$$Y_i | X_i, \beta, \sigma^2 \sim N(X_i \beta, \sigma^2) \quad i = 1, \ldots, n,$$
$$\beta | \gamma \sim N(0, \operatorname{diag}(\gamma_1, \ldots, \gamma_d)),$$
$$\gamma \sim w(\gamma),$$
$$\sigma^2 \sim w(\sigma^2).$$

Here, γ_j represents the concentration around zero of the prior distribution of β_j rather than the prior probability of including the jth explanatory variable. The hyperprior on the hyper-variances in γ controls the size of γ. If the concentration of the γ_j at zero in the posterior is high enough, it makes sense to drop the corresponding jth explanatory variable from the predictive distribution.

It can be seen that this version of spike-and-slab regression reduces to the NIG conjugate prior case: an NIG prior is recovered if $w(\gamma)$ were to assign unit mass to a single (diagonal) variance matrix, say, $V_\beta = \operatorname{diag}(\gamma_1, \ldots, \gamma_d)$, and the prior on σ^2 were taken as IG(a, b). Otherwise started, the sense in which this goes beyond the conjugate prior case is that V_β is treated as a hyperparameter and the prior on σ^2 can be any continuous distribution. In the general case, therefore, it may be effectively impossible to find closed-form expressions for the posterior and hence the predictive density. However, this version of spike-and-slab regression has been coded into an the R package **spikeslab**, see Ishwaran *et al.* (2010), that does prediction on test data.

4.3 Quantile Regression

The last of the regression-based predictors that will be familiar to most people is quantile regression (QR); see Koenker and Hallock (2001) for an introductory exposition and Koenker and Basset (1978) for the original contribution. Quantile Regression can be regarded as a variation on linear models under least squares optimization. Recall that if $\mu(x, \beta)$ is a smooth class of functions then finding

$$\hat{\beta} = \arg\min_{\beta \in \mathbb{R}} \sum_{i=1}^{n} (y_i - \mu(x_i, \beta))^2$$

leads to $\mu(x | \hat{\beta}) = \widehat{E}(Y | X = x)$, an approximation for the conditional mean $E(Y | X = x)$, the optimal predictor of Y given X under squared error.

In QR, rather than finding an approximation to $E(Y | X = x)$, the goal is to find an approximation to the optimal predictor for the τth conditional quantile, $\operatorname{quant}_\tau(Y | X = x)$, for some $\tau \in (0, 1)$. For $\tau = 0.5$, this is the median of Y given $X = x$, i.e., $\operatorname{med}(Y | X = x)$, and it makes sense to use the median as a location for $(Y | X)$. It does not in general make sense to use, say, the fifth percentile, $\operatorname{quant}_{\tau=0.05}(Y | X = x)$, to predict $(Y | X)$. However, the difference between the fifth and 95th percentile curves gives a 90% PI for Y given X.

To see this more formally, let $\tau \in (0, 1)$ and consider the loss function

$$\rho_\tau(u) = u(\tau - \mathbb{I}_{(-\infty, 0)}(u))$$

where $u \in \mathbb{R}$. For $u < 0$, $\rho_\tau(u) = (\tau - 1)u$, which is a straight line with slope $\tau - 1$, and, for $u > 0$, $\rho_\tau(u) = \tau u$, which is a straight line with slope τ. For continuous Y with DF F and given τ, the minimum of $E\rho_\tau(Y - a)$ over $a \in \mathbb{R}$ is the smallest value of a satisfying $F(a) = \tau$, i.e., the τth percentile. If F is the EDF, the same reasoning gives the sample τth quantile as the minimizing value of a. That is,

$$\hat{\beta}_\tau = \arg\min_\beta \sum_{i=1}^n \rho_\tau(y_i - \mu(x_i, \beta)), \tag{4.35}$$

for data Y and X, is the analog of the ordinary least squares regression estimator but under the τth quantile loss. Under various regularity conditions and hypotheses which ensure that a central limit theorem can be applied (see Koenker (2005, Chap. 4)), $\hat{\beta}_\tau$ is asymptotically normal, with an identified variance matrix, at rate $\mathcal{O}(1\sqrt{n})$. However, this is important only if it helps to ensure good prediction. More generally, QR is more robust than squared error methods and so is thought to be a good choice when, for instance, $(Y|X)$ has heavy tails, is asymmetric, or otherwise differs from light-tailed unimodality.

If μ in (4.35) is linear, i.e., $\mu(x, \beta) = x\beta$, then (4.35) can be written

$$\hat{\beta}_\tau = \arg\min_\beta \left(\sum_{i: y_i \geq x_i\beta} \tau |y_i - x_i\beta| + \sum_{i: y_i < x_i\beta} (1 - \tau)|y_i - x_i\beta| \right).$$

There are several ways to solve this optimization problem computationally, in R, Matlab, and SAS, for example, but this is not discussed here.

The point is that if $\hat{\beta}_\tau$ is found for, say, $\tau = 0.05$ and 0.95, then quantile curves $\hat{Y}_{0.05}(x) = x\hat{\beta}_{0.05}$ and $\hat{Y}_{0.95}(x) = x\hat{\beta}_{0.95}$ can be found, leading to a 90% PI for Y_{n+1} at x of the form $[x\hat{\beta}_{0.05}, x\hat{\beta}_{0.95}]$. That is, although it does not seem to have been established formally, in general it is reasonable to write

$$P(Y_{n+1}(x_{n+1}) \in [x_{n+1}\hat{\beta}_{\alpha/2}, x_{n+1}\hat{\beta}_{(1-\alpha/2)}]) \approx 1 - \alpha, \tag{4.36}$$

i.e., to regard the QR curves as a general procedure for finding PIs given x_{n+1}. Since it is routine in QR to find SEs for coefficients such as $\hat{\beta}_{\alpha/2}$ and $\hat{\beta}_{1-\alpha/2}$, it is easy to find the variance of the endpoints of the PI and the variance of the length of the PI, $x(\hat{\beta}_{1-\alpha/2} - \hat{\beta}_{\alpha/2})$, if the dependence between $\hat{\beta}_{1-\alpha/2}$ and $\hat{\beta}_{\alpha/2}$ is ignored (or otherwise handled). That is, not only can one identify a $1 - \alpha$ PI at any x_{n+1}, one can also assess, at least approximately, its variability as a function of x_{n+1}.

Of course, if a point predictor for Y_{n+1} is all that is desired, using $\hat{Y}(X_{n+1}) = X_i\hat{\beta}_{1/2}$ i.e., QR with $\tau = 1/2$, is the natural choice. Approximate $1 - \alpha$ PIs for it could be found as before, using QR curves for $\alpha/2$ and $1 - \alpha/2$. If a point predictor for a non-median quantile is desired, the process is similar.

Numerous variations on QR have been developed. In particular, Bayesian QR originated in Yu and Moyeed (2001) with the observation that loss functions of the form ρ_τ can define an exponential family. Specifically, the asymmetric Laplace density is

$$p_\tau(u) = \tau(1 - \tau)e^{-\rho_\tau(u)}.$$

So, if u is replaced by $Y_i - X_i\beta$ and n observations are assumed, the likelihood is

$$p_\tau(Y|X, \beta) = \tau^n(1-\tau)^n \exp\left(-\sum_{i=1}^n \rho_\tau(Y_i - X_i\beta)\right). \tag{4.37}$$

It is seen that, for any fixed τ, the minimization in (4.35) over β is equivalent to a maximization in (4.37) over β, i.e., the parameters in a QR problem can be regarded as the parameters in a specific parametric family.

To generate a Bayesian analysis for (4.37), priors for the d components in β must be specified. Yu and Moyeed (2001) showed that if a uniform prior is assumed on β then the resulting posterior is proper. That is, $w(\beta|Y, X)$ exists when $w(\beta) \propto 1$ and hence it exists for any proper prior. So, one may find estimates $\hat{\beta}_\tau$ for $\tau = \alpha/2$ and $1 - \alpha/2$, e.g., posterior means, and form a $1 - \alpha$ PI for Y_{n+1} of the form $[X\hat{\beta}_{\alpha/2}, X\hat{\beta}_{1-\alpha/2}]$, essentially a frequentist PI using Bayes parameter estimates that is expected to have approximate confidence limits $1 - \alpha$ as in (4.36). Alternatively, one could derive a $1 - \alpha$ PI for Y_{n+1} at X_{n+1} by taking the $(\alpha/2)100$th and $(1 - \alpha/2)100$th percentiles from the predictive distribution $M_{1/2}(Y_{n+1}(x_{n+1})|Y, X)$ when $\tau = 1/2$.

As in the frequentist case, the natural point predictor would come from a Bayesian analysis of median regression, i.e., $\hat{Y}_{n+1} = X_{n+1}\hat{\beta}_{1/2}$ where $\hat{\beta}_{1/2}$ is a posterior mean or the median from $w(\beta|Y, X)$ under a definition of median appropriate to the dimension of β – typically greater than one. If a point predictor for a non-median quantile is desired, the process is similar.

A more 'Bayesian' approach would be to derive thresholds from the posterior distributions. It should be noted, however, that these are for percentiles, not means, i.e., to use the model (4.37) to define a posterior one must remember that the curve it represents is the quantile of Y_{n+1}, not the location of Y_{n+1}. Thus, one could solve for L and U from

$$0.5 = M_{\alpha/2}(L \leq Y_{n+1}|Y, X),$$
$$0.5 = M_{1-\alpha/2}(Y_{n+1} \leq U|Y, X).$$

This means that the $(\alpha/2)100$th and $(1 - \alpha/2)100$th percentiles of Y_{n+1} at X_{n+1} each have a 50–50 chance of being above L and below U, respectively. The second is

$$\frac{1}{2} = \int_{-\infty}^U m_{1-\alpha/2}(y_{n+1}|Y, X, X_{n+1})\mathrm{d}y_{n+1}$$
$$= \int_{-\infty}^U \int p_{1-\alpha/2}(y_{n+1}|X_{n+1})w_{1-\alpha/2}(\beta|Y, X)\mathrm{d}\beta\mathrm{d}y_{n+1};$$

the first is similar. Since the posteriors can be found numerically, the result is an approximate $1 - \alpha$ PI of the form $[L, U]$.

Finally, note that QR in its frequentist form only has a model of the generic form $(Y_i|X_i, \beta)$ because the constraint to linear functions is used only to define the action space in the decision-theoretic optimization. By contrast, Bayesian QR amounts to converting a decision problem to a specific parametric model for $(Y_i|X_i, \beta)$, in which τ is effectively a hyperparameter. In both cases, the natural approach is to find QR curves to estimate the endpoints of a $1 - \alpha$ PI and, given the ability to assess variability in the parameter estimation, the variability of the PI can be assessed. A simpler approach would be just to use median regression, i.e., set $\tau = 0.5$. Because it is based on percentiles, one expects QR, whether Bayes or frequentist, to be more stable but less efficient than modeling the mean of a random variable, and this will affect the predictive performance of QR.

4.4 Comparisons: Regression

There have been many performance comparisons of the seven regression methods presented so far but few have been primarily predictive. So, here, the goal is to examine how these methods perform in a predictive setting for two examples. The first is a simulation setting; the second uses real data.

For the simulation, an $\mathcal{M}$-closed problem was defined by taking

$$Y = 1 + X_1 + X_2 + X_3 + X_4 + + X_5 + X_1^2 + X_2^2 + X_3^2 + X_5^2 + N(0, 1)$$

as a true model where the X_i are IID Unif$[-2, 2]$ and were Studentized. The sample size was $n = 200$ and a burn-in of 11 was used. For each of 100 iterations the CPE was plotted, starting at $n = 20$ and then for every ten further data points up to 200. The seven methods used were frequentist linear models (flm), Bayes linear regression with conjugate priors (blrc), Bayes linear regression with noninformative priors (blrni), Bayes linear regression with spike and slab priors (blrss), Bayes linear regression with g-priors (blrg), frequentist quantile regression (fqr, $\tau = 0.5$), and Bayes quantile regression (bqr, $\tau = 0.5$).

For all the computing, the R statistical language was used (R Core Team 2016). The details are as follows. For blrni the package BLR was used with prior error variance assigned to the inverse-χ_3^2 prior with scale parameter set to one. For blrc, BLR was used with prior error variance assigned to the scaled inverse-χ_3^2 with scale parameter set to one. Likewise, BLR was used for blrc. For blrss, the package spikeslab was used; for blrg BMS was used with g selected by empirical Bayes (local EBL prior), and for bqr the package BSquare was used with $L = 4$ Gaussian basis functions. For all the Bayes methods, a burn-in of 500 MCMC iterations was used followed by 5000 iterations for inference. Finally, for fqr, the package quantreg was used. Other parameters not specified were taken as the defaults computed internally to the packages.

The results are shown in Fig. 4.1. The left-hand panel shows that, over most of the range, blrc, blrni, and blrss did roughly equally well although, near $n = 200$, conjugate priors did a little better than noninformative priors, which in turn did a little better than spike and slab priors. The fifth Bayes linear model method, blrg, did next best. Thus, as expected, the Bayes methods did best, as they usually do for $\mathcal{M}$-closed problems such as this. The poorest performances were from bqr (worst), fqr, and flm, in that order. That is, frequentist linear models did not do as well predictively as the Bayes linear regression methods. The quantile regressions did worst, probably because they are based on percentiles, which are not as efficient as methods based on means. However, quantile regression methods are likely to be more stable. So, in situations where the data lead to unstable predictions, one expects quantile regressions to outperform the other five methods.

The right-hand panel in Fig. 4.1 shows the same curves as in the left-hand panel but under baseline adjustment. Subtracting the 'baseline' error reveals the large-sample behavior of the CPE for all the methods. Of course, for actual usage, one would not baseline-correct because how quickly a model class identifies good predictors is a meaningful assessment of

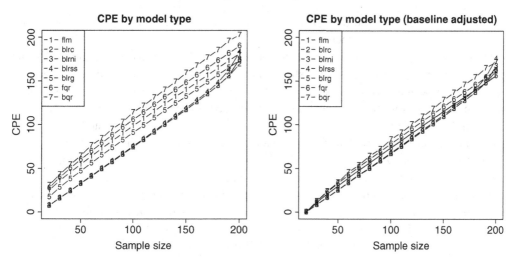

Figure 4.1 The left-hand panel shows the CPEs for seven methods, using simulated data. The right-hand panel shows the same CPEs after adjustment for the baseline error.

its usefulness. Specifically, the error at $n = 12$ data points (using $n = 11$ points as burn-in) was subtracted from each CPE so that all curves start from zero at $n = 11$. In this example it is seen that roughly similar patterns hold: the two quantile regressions have the highest CPEs. The next worst methods, over most sample sizes, were flr, blrg, and blrc, which were roughly equivalent. Over most of the range of sample sizes, the best method was blrni and the second best was blrss priors. Again, near $n = 200$, blrc, blrni, and blrss deteriorated. It appears that in a limiting sense blrg and flm would have the lowest CPEs in both panels; however, neither was best for small sample sizes.

The effect of sample size is important. The best predictor in small sample sizes, in general, is not the best predictor for large sample sizes. It is unclear whether more data permits better identification of a predictor with good limiting properties or whether it permits more complex predictors. In either case, if the goal really were model identification, one would choose the best, or one of the best, predictors at $n = 200$ and convert it back to a model that could be investigated further. The fact that there are several nonequivalent predictors with roughly equally compelling performance serves as an indicator for the degree of model uncertainty, which cannot be ignored, even in $\mathcal{M}$-closed problems.

To provide a data-driven comparison, the seven methods were applied to a real data set. The data were downloaded from `http://data.worldbank.org/topic/agricult ure-and-rural-development`, which provides precise definitions of the quantities measured. The response, Y, was chosen to be the 'food production index', which summarizes edible food crops that contain nutrients for humans. There were 20 variables in the data set that had at least 160 values, i.e., not too many missing values, not counting 'crop production' and 'food production' indices, which were thought to be more appropriate as responses than explanatory variables. We used the $n = 90$ observations for these 20 variables with enough nonmissing values. (For the sake of this analysis, the missing data was assumed noninformative.) The same seven methods as used in the simulation were applied to the data. To form the CPEs, 100 permutations of the data were randomly generated, a burn-in of 30 was used, and the permutations were used to form average CPEs for 30, 40, ..., 90 for each of the seven methods.

Unfortunately, for real data one cannot be sure that the problem is $\mathcal{M}$-closed. In the present case, given the sample size, estimates of regression parameters will have large SEs and it would be unrealistic to use a much larger model, e.g., one with more explanatory variables. All the variables and data that can be used *are* being used, making what is an $\mathcal{M}$-complete problem as close as possible to a $\mathcal{M}$-closed problem. A related point is that this is not as serious a problem for prediction since the goal is to predict a response, not to identify a model. Of course, if a model can be identified then larger sample sizes should lead to better final predictors. However, finite-sample performance, in particular, for sample sizes smaller than n is important in prediction. Indeed, when using permutations of the data to generate a CPE, the model from a given class at step n will typically be independent of the permutation of the data. So, interest focuses on the pathway $CPE(k)$ for $k \leq n$ by which $CPE(n)$ is reached. This pathway may affect $CPE(n)$.

Figure 4.2 shows the $CPE(k)$ for the seven methods. As before, the three worst methods were bqr, fqr, and flm and the two best methods were blrss and blrc. Over the range of sample sizes, it is seen that the curves for blrni and blrg cross near 77. In contrast with the left-hand panel of Fig. 4.1, the differences at burn-in are larger. So, the right-hand panel of Fig. 4.2 shows the same curves with baseline adjustment at 30. The blrss is still the best, followed by blrc (apart from the region where the curves cross around 87). Again, aside from region where the curves cross at around 80, fqr gives the worst performance. However, bqr outperforms flm and while blrni starts well it deteriorates rapidly. As before, blrc does relatively well (though not quite as well as blrss) and blrg does a little worse over most of the data range. Overall, if one had to pick a model class, for the simulated data would one choose blrc and for the real data one would choose blrss.

In this case, if the goal really were model identification, one would probably use the predictors given by blrss and blrc at $n = 90$, convert them back to models, and examine their properties. For instance, do both models put small or effectively zero weights on some explanatory variables? Do both give similar coefficients on important variables? Do both have roughly the same goodness-of-fit properties? What is the maximum discrepancy between the predictive errors of the two methods if both are given the same subset of the data on which to train? There are many other properties that could be examined and these are likely to depend on the specific application.

Taken together, these examples confirm the intuition that Bayes methods are slightly better for many $\mathcal{M}$-closed or nearly $\mathcal{M}$-closed problems. More specifically, sparsity priors such as spike-and-slab do as well as simple approaches such as conjugate priors – which are a different way to impose simplicity via the prior. The usefulness of simplicity may be important for $\mathcal{M}$-closed problems, since then a model is presumed available. To conclude this sort of analysis, the final model used for prediction from the best method would be the model chosen: it is independent of the order of the data and has the best predictive properties.

To complete the investigation of how these methods perform it is important to assess more than just minimum prediction errors. So, to argue that the separation of the curves in Figs. 4.1 and 4.2 is meaningful, Fig. 4.3 presents the SEs for each stage of the predictive schemes. The left-hand panel shows the mean CPE curves for the simulated data. The ranges indicated around each curve are one SE for each time step. That is, the 100 errors of prediction are

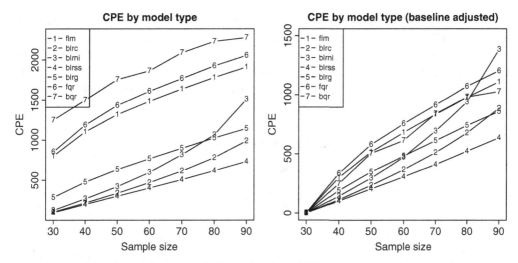

Figure 4.2 The left-hand panel shows the mean CPE curves for the seven methods for a real data set. The right-hand panel uses baseline adjustment at 30.

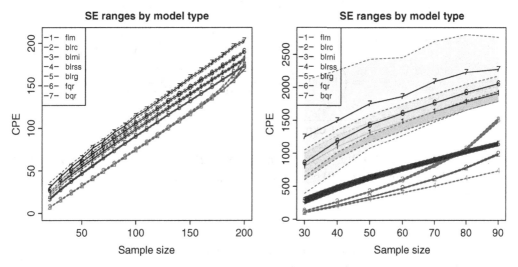

Figure 4.3 On the left, the dotted lines around the curves for the seven methods indicate the mean ± SE for the simulated data. On the right, the analogous regions are shown around the CPE lines for the real data.

taken in L^1 and the SE calculated from them gives the indicated curves. Clearly, for the simulated data, the lines separate well enough to support the conclusions. (This could be improved merely by increasing the number of iterations from 100, if desired.)

The results are not quite as clear-cut for the real data. The right-hand panel of Fig. 4.3 shows the ± SE ranges for the mean CPE curves and the best four methods are seen to separate well. However, the variability of the worst three methods, bqr, fqr, and flm, is much larger and the regions around these three curves overlap. This suggests that it is the variability of these methods that makes their predictive performance poor, and the small variability of blrss and blrc improves their performance.

In this comparison of performance, the focus was on mean CPEs and the properties of the mean CPE curves. Properties of models – or predictors – derived from consistency, asymptotic normality or other properties of solutions may be interesting from the modeling standpoint, but they are really useful only in problems that are convincingly $\mathcal{M}$-closed. From a predictive standpoint, such properties are helpful only if they lead to better prediction.

4.5 Logistic Regression

Linear regression is defined by the statement

$$Y_i \sim N(\sum_{j=0}^{d} \beta_j X_{ji}, \sigma^2), \tag{4.38}$$

which comes from writing $Y_i = X_i \beta + \epsilon_i$ and making distributional assumptions. By contrast, in logistic regression Y assumes the values zero and one and so is a Bernoulli random variable; the explanatory variables can be discrete or continuous but are treated as fixed. The basic form of logistic regression can be derived using Bayes' rule. Assume that $d = 2$, so there is one nontrivial explanatory variable X and a constant term. Then

$$
\begin{aligned}
P(Y = 0|X) &= \frac{P(Y=0)P(X|Y=0)}{P(Y=0)P(X|Y=0) + P(Y=1)P(X|Y=1)} \\
&= \left(1 + \exp\left(\ln\frac{P(Y=1)}{P(Y=0)} + \ln\frac{P(X|Y=1)}{P(X|Y=0)}\right)\right)^{-1} \\
&= \frac{1}{1 + \exp(\beta_0 + \beta_1(X))},
\end{aligned} \tag{4.39}
$$

where $\beta_0 = \ln(P(Y=1)/P(Y=0))$ and $\beta_1(X) = \ln(P(X|Y=1)/P(X|Y=0))$, if X is discrete. If X is continuous, (4.39) holds using the density p in place of P. Thus, the logistic model is

$$(Y|X) \sim \text{Ber}\big(\exp(\beta_0 + \beta_1(X))/(1 + \exp(\beta_0 + \beta_1(X)))\big), \tag{4.40}$$

which is analogous to (4.38). If X is regarded as a random variable, so as to invoke Bayes' rule, (4.39) corresponds to setting $X = x$ and treating x as a fixed value, so (4.40) remains valid, e.g., by replacing $(Y|X)$ with $(Y|X = x)$.

In the special case where $p(x|Y = k) \sim N(\mu_k, \sigma^2)$ for $k = 0, 1$, it is easy to verify that

$$\beta_1(X) = \frac{\mu_0 - \mu_1}{\sigma^2} X + \frac{\mu_1^2 - \mu_0^2}{2\sigma^2}.$$

In this case, $\beta_1(X) = \alpha_0 + \alpha_1 X$ for some choice of α_0 and α_1. Using this in (4.40) gives the usual parametrization of the logistic regression problem.

Indeed, if $d \geq 2$ and the X_j are independent then

$$\ln\frac{P(X|Y=0)}{P(X|Y=1)} = \sum_{j=1}^{d} \ln\frac{P(X_j|Y=0)}{P(X_j|Y=1)}.$$

Setting $\beta_j(X_j) = \ln P(X_j|Y=0)/P(X_j|Y=1)$ one can extend (4.39) to

$$P(Y=0|X) = \frac{1}{1 + \exp\left(\beta_0 + \sum_{j=1}^{d} \beta_j(X_j)\right)}.$$

Hence, writing $p(x_j|Y=k) \sim N(\mu_{kj}, \sigma^2)$ for $k = 0, 1$, the usual parametrization of the general logistic regression model is

$$(Y|X) \sim \text{Ber}\left(\exp\left(\beta_0 + \sum_{j=1}^{d} \beta_j X_j\right) \bigg/ \left(1 + \exp\left(\beta_0 + \sum_{j=1}^{d} \beta_j X_j\right)\right)\right).$$

Regardless of whether the X's are dichotomous, polychotomous, or continuous, logistic regression is a way to identify the distribution of a dichotomous Y as a function of X and of a parameter β, just as linear regression is a way to identify the distribution of a continuous Y as a function of X and of a (different) parameter β. If Y is polytomous, similar reasoning holds but is not presented here.

For modeling, logistic regression is often used to estimate probabilities as a function of explanatory variables X and parameters β. Often these probabilities are used to find odds ratios and relative risks. For instance, when $(X|Y=k)$ is normal, the log of the odds ratio for $(Y|X)$ is as an affine function of $X_1 = x_1$. For $d = 2$ and $Y = 1$ this is

$$\log \frac{P(Y=1|\beta_0, \beta_1, X=x)}{1 - P(Y=1|\beta_0, \beta_1, X=x)} = \beta_0 + \beta_1 x; \tag{4.41}$$

extensions to general d are similar. In the simplest case, X assumes the values zero or one. Then, (4.41) (or (4.39)) shows that β_0 and $\beta_0 + \beta_1$ are the values of the true log odds for $x_1 = 0$ and for $x_1 = 1$, respectively.

However, from a prediction standpoint, logistic regression can be used for classification and the zero and one are taken as class labels. Thus, the error term in classification is discrete, not continuous, corresponding to mistaking one class for the other. Logistic regression is one of the simplest ways to do classification. Logically, if the class predictions from a logistic regression model are poor, its usefulness as a model is called into question, implying that the odds ratios or relative risks may not represent anything physical.

So, suppose data of the form (Y_i, x_i) for $i = 1, \ldots, n$ is available. Estimates $\hat{\beta}_0$ and $\hat{\beta}_1$ can be obtained in a variety of standard ways. The most common is frequentist, deriving the likelihood and using MLEs for the β_j's; see Czepiel (2002). Typically, these estimators are consistent and asymptotically normal with rate $\mathcal{O}(1/\sqrt{n})$. Hence, the point is to use the logistic model to estimate the conditional probabilities of events such as $\{Y_{n+1} = 1\}$ by

$$\hat{P}(Y_{n+1} = 1|\hat{\beta}_0, \hat{\beta}_1, X_{n+1}) = \frac{\exp\left(\hat{\beta}_0 + \hat{\beta}_1 X_{n+1}\right)}{1 + \exp\left(\hat{\beta}_0 + \hat{\beta}_1 X_{n+1}\right)}. \tag{4.42}$$

Clearly, if $\hat{P}(Y = 1|\hat{\beta}_0, \hat{\beta}_1, x_{n+1})$ is greater than 1/2, one is led to predict $Y_{n+1} = 1$ for x_{n+1}; otherwise, to predict $Y_{n+1} = 0$. There are variety of standard ways to assess the appropriateness of the logistic regression model – Wald tests, R^2 etc. (see Hosmer and Lemeshow (2000)) – but these procedures are not the point here apart from noting that they are model-based efforts to ensure that estimates such as (4.42) will be good predictors for class membership.

Note that $P(Y_{n+1} = 1|\beta_0, \beta_1, x_{n+1})$ is not a random variable. However, Y is always a random variable and $P(Y_{n+1} = 1|\beta_0, \beta_1, X_{n+1})$ is a random variable when X_{n+1} is a random variable. In many simple settings for logistic regression, ignoring the distinction between treating X_{n+1} as fixed and treating X_{n+1} as random does not lead to problems. (Of course, generalization error, see Sec. 3.5, is an exception to this.) So, it is often reasonable to regard $\hat{P}(Y = 1|\hat{\beta}_0, \hat{\beta}_1, x_{n+1})$ as an estimator for $P(Y = 1|\beta_0, \beta_1, x_{n+1})$. Strictly speaking, one would therefore want an SE, so as to be satisfied that the predicted class is decisive. Nevertheless, the estimate of the probability leads naturally to a predictor for Y if one simply chooses the value of Y which has higher conditional probability.

Bayes logistic regression has also been developed – indeed, Bayes' rule was used to derive the frequentist logistic regression model. The Bayes formulation is more forthright because it explicitly rests only on the likelihood function for Y and the prior for the parameters. The cost is that analyzing the Bayes logistic regression model requires more computing to find the posterior for the parameters.

To see this, recall the form of the general logistic regression model and note that, for given X_i and β, the ith subject contributes a factor

$$L_i(\beta|y_i, X_i) = \pi(X_i, \beta)^{y_i}(1 - \pi(X_i, \beta))^{1-y_i}, \tag{4.43}$$

where

$$\pi(x, \beta) = \exp(\beta_0 + \beta_1 X_1 + \cdots + \beta_d X_d)/(1 + \exp(\beta_0 + \beta_1 X_1 + \cdots + \beta_d X_d)).$$

If the subjects are independent, the product of (4.43) over $i = 1, \ldots, n$ is

$$L(\beta|y_1^n, X) = \prod_{i=1}^{n} \pi(X_i, \beta)^{y_i}(1 - \pi(X_i, \beta))^{1-y_i}. \tag{4.44}$$

The parameter set is given by $\beta = (\beta_0, \ldots, \beta_d)$ and one common prior is the normal: $\forall j \; \beta_j \sim N(\mu_j, \sigma_j^2)$ for real μ_j and positive σ_j^2. Thus, the posterior density is

$$w(\beta|y^n, X) \propto \prod_{j=1}^{d} \frac{1}{\sqrt{2\pi}\sigma_j} \exp\left(-\frac{1}{2\sigma_i}(\mu_j - \beta_j)^2\right) L(\beta|y_1^n, X), \tag{4.45}$$

where the normalizing constant is the integral of the right-hand side with respect to β. Other priors may be used as well; scaled, noncentral Student's t-distributions are a common choice since their heavier tails make them less informative than normal priors with large σ_j^2. It is unusual for (4.45) to be available in closed form because it would be difficult in general to assign a prior to β that would lead to a conjugate beta prior on $\pi(X_i, \beta)$. So, in most cases, the posterior can only be found computationally. As in the frequentist setting, there are various standard ways to ensure that the Bayes logistic regression model fits well; see Christensen *et al.* (2011, Chap. 8), and Gelman *et al.* (2004) for overviews including a thorough treatment of the computational issues.

Whatever priors are chosen for β, the predictive distribution is

$$P(Y_{n+1}|y_1^n, X, X_{n+1}) = \int P(Y_{n+1}|X_{n+1}, \beta)w(\beta|y_1^n, X)d\beta. \tag{4.46}$$

This is more abstract than (4.42) and, like the posterior distribution (4.45), can usually be approximated only numerically. However, it does produce a natural predictor: predict class one when $P(Y_{n+1} = 1 | y_1^n, X, X_{n+1}) \geq 1/2$ and the other class, class zero, otherwise. Moreover, as n increases, the posterior density typically concentrates at the true β and is asymptotically normal at rate $\mathcal{O}(1\sqrt{n})$, so the predictive distribution converges to the true model, assuming that this exists. One can obtain a posterior mean and variance for $(Y_{n+1} | X_{n+1}, \beta)$ as an assessment of how decisive a given prediction is, but this is not commonly done, for good or for ill. As in the frequentist case, the primary goal is to predict the class of a new subject and make inferences about the population, assuming that the prediction is satisfactory.

4.6 Bayes Classifiers and LDA

The Bayes classifier or predictor identifies the most likely class for a response Y given a collection of observed X's. That is, instead of regarding an explanatory variable, say $x = (x_1, \ldots, x_d)$, as deterministic, it is treated as random, $X = (X_1, X_2, \ldots, X_d)$. The response Y gives the class that generated an observation X, i.e., $Y = 1, \ldots, K$ where K is the number of classes, and Bayes' theorem gives the probability of an observation falling into the kth class, i.e.,

$$P(Y = k | X_1, X_2, \ldots, X_d) = \frac{P(Y = k) P(X_1, X_2, \ldots, X_d | Y = k)}{P(X_1, X_2, \ldots, X_d)}, \qquad (4.47)$$

when X is discrete, where the denominator is given by

$$P(X_1, X_2, \ldots, X_d) = \sum_{k=1}^{K} P(Y = k) P(X_1, X_2, \ldots, X_d | Y = k). \qquad (4.48)$$

When X is continuous the same formula holds, but the density of X is used in place of its probability mass function and the probability for Y is regarded as a density with respect to counting measure. In either case, the denominator does not depend on k so the focus is on the numerator.

The distinctive aspect of Bayes classification or, more precisely, prediction of a class label, is that a prior for Y must be selected. That is, the prior is $P(Y = k)$ for $k = 1, \ldots, K$, the marginal probability of Y, and represents what are believed a priori to be the proportions of each class in the population. In the logistic regression classifier, by contrast, the prior for Y corresponds to $\beta_0 = \ln(P(Y = 1)/P(Y = 0))$, which has to be estimated. Given a prior and an outcome $X_i = x_i$, the Bayes classifier is the mode of (4.47), namely

$$\hat{Y}_i(x_i) = \arg\max_{y} P(Y = y) P(x_{1i}, \ldots, x_{di} | Y = y), \qquad (4.49)$$

and is the point predictor for $Y_i(x_i)$. The Bayes classifier minimizes the expected cost of misclassification, i.e., the Bayes risk, under zero–one loss or more general loss functions; see Domingos and Pazzani (1997) and Clarke *et al.* (2009, Sec. 5.2.2.2). Since classification by logistic regression is derived from the Bayes classifier, it too is optimal when the parametric form for the classifier is correct.

In practice, given n data points, one approximates the optimal Bayes rule by estimating the probabilities or densities in (4.47). For example, if $K = 2$, $P(Y = k | X)$ can be obtained by estimating the densities for $(X|Y = 1)$ and $(X|Y = 2)$ and $P(Y = k)$ can be obtained by using the sample proportions. These estimates can then be used in (4.48) to obtain the denominator of (4.47). Thus, the logistic regression classifiers from (4.42) or (4.46) can be regarded as Bayes classifiers in the sense that both provide approximations to (4.49) given the parametric form of the logistic regression model.

In the special case of binary classifiers, there is a natural way to visualize performance. Suppose that the two possible outcomes are, rather than zero and one, positive (+) and negative (−). There are two possible predictions for the outcome, again + and −. So, there are four possible 'prediction, outcome' pairs: $(+, +), (+, -), (-, +)$, and $(-, -)$. Clearly, $(+, +)$ and $(-, -)$ are true positive (TP) and true negative (TN); these are the cases when the prediction or classification is correct. The other two cases occur when the prediction is wrong: $(+, -)$ is false positive (FP) and $(-, +)$ is false negative (FN). The conditional probability $P(\hat{Y} = +|+$ is correct) is called the true positive rate (TPR) or *sensitivity* of a binary predictor and $P(\hat{Y} = -|-$ is correct) is called the true negative rate (TNR) or its *specificity*. The false positive rate (FPR) is $1 - $ TNR. The *conceptual* difference between sensitivity and specificity is the conditioning event.

Looking at sensitivity and specificity means summarizing a classifier as a point in the square, $[0, 1] \times [0, 1]$, corresponding to the FPR and TPR as an ordered pair. Without loss of generality it can be assumed that this point is above the $y = x$ line (otherwise the predictor could be improved by reversing its values). Thus, if a binary predictor has an FPR $P(\hat{Y} = +|-$ is correct$) = 0.1$ and a TPR $P(\hat{Y} = +|+$ is correct$) = 0.9$, it gives the point $(0.1, 0.9)$ in the unit square. The ideal predictor would have FPR $P(\hat{Y} = +|-$ is correct$) = 0$ and TPR $P(\hat{Y} = +|+$ is correct$) = 1$. That is, one wants a predictor corresponding to the point closest to $(0, 1)$. More precisely, one would choose among predictors by deciding the relative importance of sensitivity and specificity according to the purpose of the predictor and the costs of misclassification. Subject to these criteria, one would then choose the predictor corresponding to a point as close as possible to $(0, 1)$.

An extension of this representation of an individual classifier or binary predictor to a system of classifiers or binary predictors is provided by the receiver operator characteristic (ROC) curve. An ROC curve is a graph of the TPR versus the FPR for a binary classifier $\hat{Y}$, as a parameter defining the classifier varies. For instance, if Y is continuous and has been dichotomized by a threshold – say class 1 is $\{Y \geq c\}$ and class 0 is $\{Y < c\}$ for some $c \in \mathbb{R}$ – one can generate an ROC curve as c varies. In these cases, it is common to choose c, i.e., a classifier corresponding to a point in the ROC curve, again by deciding the relative importance of sensitivity and specificity, on the basis of the purpose of the predictor and the costs of misclassification. In this usage the population is fixed, the collection of classifiers varies, and it is the particular setting that is important.

A common way to produce an ROC curve is to note that $P(-$ is true$) + P(+$ is true$) = 1$ and then parametrize $P(TP)$ and $P(FP)$ by $v = P(+$ is true$)$, a property of the actual population for which the prediction is being made. Thus, an ROC plot can also be used to show how a fixed predictor or classifier $\hat{Y}$ behaves as the composition of the population (the proportion of +'s and −'s) varies; see Pepe (2003) for more details.

A further extension of the ROC concept is to compare different classifier systems. That is, there may be several classifiers each with a parameter that can be varied, and the task is first to choose a classifier system and then to choose a classifier within the chosen system.

Often the system is chosen by finding the area under the ROC curve (AUC) for the system on the premise that the higher the AUC is, the better the system of classifiers is at prediction. Once the system has been chosen, again a classifier in the system is chosen by deciding the relative importance of sensitivity and specificity according to the purpose of the predictor and the costs of misclassification. Clearly, finding a specific binary classifier that performs well (in terms of ROC curves, say) from several systems of classifiers will depend on the collection of systems. So, if none of the classifiers in any of the systems are true – in the sense of being able to represent the regions where Y is 0 or 1 perfectly – there will be the classification analog of model uncertainty and misspecification, which must be assessed. For further discussion of these issues, see for example Altman and Bland (1994) and Steyerberg *et al.* (2010).

Receiver operator characteristic curves give a cumulative evaluation of a classifier's performance. So, for instance, one could break a data set into two parts, testing and training, use the training set to produce a classifier and then generate its empirical ROC curve using the test data. Like cross-validation, this could be regarded as a cumulative assessment of predictive performance. However, one cannot really use an ROC curve approach sequentially. For instance, one cannot form a classifier, use it to predict the next Y, and then form an ROC curve; the specificity and sensitivity for a single outcome would be one or zero and not very informative. In fact ROC curves might be regarded as analogous to a CPE at a certain stage of data accumulation, and one could, in principle, generate specificity and sensitivity curves as data accumulate. Taken together these could be regarded as equivalent to a mean-CPE curve. Here, however, the focus will be on mean-CPE curves themselves since they combine the information one could get from the sequential use of an ROC approach.

One of the most important settings for Bayes classifiers is called discriminant analysis. The idea is that, given a response Y that assumes values in $1, \ldots, K$, one derives K functions $\delta_k(\cdot)$, $k = 1, \ldots, K$, such that, for each k, $\delta_k(x)$ assesses how representative class k is for the given x. Given data (Y_i, x_i), $i = 1, \ldots, n$, one estimates the δ_k by $\hat{\delta}_k$. The discriminant classifier predicts the class $\hat{Y}(x_{n+1}) = \hat{k}$ for x_{n+1}, where

$$\hat{k} = \arg\max_k \hat{\delta}_k(x_{n+1}). \tag{4.50}$$

The focus here is on the linear discriminant classifiers pioneered by Fisher in the 1930s, see e.g. Fisher (1936), now collectively referred to as linear discriminant analysis (LDA). There are several formulations of LDA; in the simplest x is taken as an outcome of X, i.e., as random, and $(X|Y = k) \sim N(\mu_k, \Sigma)$ for $k = 0, \ldots, K - 1$, so that each class has the same covariance matrix Σ, assumed to be full rank, but with possibly different means. If a uniform prior for the K classes is used then the Bayes predictor for $(Y|X = x)$ is

$$\hat{Y} = \arg\max_k P(Y = k|X = x)$$

$$= \arg\max_k \left(-\log((2\pi)^{p/2}|\Sigma|^{1/2}) - \frac{1}{2}(x - \mu_k)^T \Sigma^{-1}(x - \mu_k) \right)$$

$$= \arg\max_k \left(-\frac{1}{2}(x - \mu_k)^T \Sigma^{-1}(x - \mu_k) \right)$$

$$= \arg\max_k \left(x^T \Sigma^{-1} \mu_k - \frac{1}{2} \mu_k^T \Sigma^{-1} \mu_k - \frac{1}{2} x^T \Sigma^{-1} x \right)$$

$$= \arg\max_k \left(x^T \Sigma^{-1} \mu_k - \frac{1}{2} \mu_k^T \Sigma^{-1} \mu_k \right). \tag{4.51}$$

The expression inside the parentheses in (4.51) is Fisher's linear discriminant function (LDF), a particular choice for $\delta_k(\cdot)$. The boundary between any two classes j and l is $\{x : \delta_j(x) = \delta_l(x)\}$. Indeed, by writing (4.51) as

$$x^T w > C$$

for some d-dimensional normal vector w and $C > 0$, it is seen that Fisher's LDF specifies a hyperplane in d dimensions that partitions the space of explanatory variables. So, collectively, the $\hat{\delta}_k$ typically partition the space into regions within which each k will be predicted, when the region contains the given x.

In practice, the LDF is implemented by estimating the parameters in (4.51). Usually, the standard estimates $\hat{\Sigma}$ and $\hat{\mu}_k$ for $k = 0, \ldots, K - 1$ are used, i.e.,

$$\hat{\mu}_k = \sum_{i : y_i(x_i) = k} x_i / n_k \quad \text{and} \quad \hat{\Sigma} = \sum_{k=1}^{K} \sum_{i : y_i(x_i) = k} (x_i - \hat{\mu}_k)(x_i - \hat{\mu}_k)^T / (n - K),$$

where n_k is the number of observations in class k and n is the total number of observations. Extensions to the case where the classes have distinct variance matrices Σ_k that must be estimated as well are similar.

An alternative formulation of LDA is based on a signal-to-noise criterion. If the goal is to find a hyperplane that separates two classes then one seeks a vector w such that the projections of the x's onto the one-dimensional subspace with normal vector w are as far apart as possible for the two classes. If there are n_k explanatory vectors from the classes C_k for $k = 1, 2$ then one wants to maximize

$$|w^T (\hat{\mu}_1 - \hat{\mu}_2)| = \left| \frac{1}{n_1} \sum_{x_i \in C_1} w^T x_i - \frac{1}{n_2} \sum_{x_i \in C_2} w^T x_i \right|,$$

over w subject to a normalization that will account for the possibly different variances of the two classes. So, write $v_i = w^T x_i$ and form

$$s_k^2 = \sum_{y_i \in C_k} (v_i - w^T \hat{\mu}_k)^2.$$

Then it makes sense to find the w that maximizes $(w^T (\hat{\mu}_1 - \hat{\mu}_2))^2 / (s_1^2 + s_2^2)$. Defining the matrix

$$S_{\text{with}} = S_1 + S_2 = \sum_{x_i \in C_1} (x_i - \hat{\mu}_1)(x_i - \hat{\mu}_1)^T + \sum_{x_i \in C_2} (x_i - \hat{\mu}_2)(x_i - \hat{\mu}_2)^T,$$

one can verify that $s_k^2 = w^T S_k w$, so the denominator is $s_1^2 + s_2^2 = w^T S_{\text{with}} w$. Defining $S_B = (\hat{\mu}_1 - \hat{\mu}_2)(\hat{\mu}_1 - \hat{\mu}_2)^T$ one can verify that the numerator is $(w^T (\hat{\mu}_1 - \hat{\mu}_2))^2 = w^T S_B w$, so the objective function becomes

$$J(w) = \frac{(\hat{\mu}_1 - \hat{\mu}_2)^2}{s_1^2 + s_2^2} = \frac{w^T S_B w}{w^T S_{\text{with}} w},$$

i.e., the signal-to-noise criterion of the scaled distance between the means becomes a ratio of the 'between' subjects' variance to the 'within' subjects' variance. Setting the derivative of $J(w)$ to zero and solving gives $w = S_{\text{with}}^{-1}(\hat{\mu}_1 - \hat{\mu}_2)$, which is essentially equivalent to (4.51); see the classic Duda *et al.* (2000).

A third derivation of an LDA involves scaling the difference in means by the variance matrices Σ_k of X, given $Y = k$ for $k = 1, 2$. In this case, maximizing

$$J(w) = \frac{w^T (\mu_1 - \mu_2)}{w^T (\Sigma_1 + \Sigma_2) w}$$

over w gives $w \propto (\Sigma_1 + \Sigma_2)^{-1}(\hat{\mu}_1 - \hat{\mu}_2)$; see Clarke *et al.* (2009, Sec. 5.2.1). The same solution can be found by generalizing (4.51) to the case where class k has variance matrix Σ_k.

4.7 Nearest Neighbors

Although not based on a parametric model, nearest neighbor classification assumes that there is a nonparametric model and that it can be viewed as either a classification or regression model. The idea is that, given a value Y_i and covariates X_i, one looks at the X_j closest to X_i and forms some kind of average of their Y_j values to predict Y_i. Thus, to specify a nearest neighbor classifier (or regression function estimate) one must choose a distance on the X_i and a number of X_i to be taken as the nearest neighbors. One could use all the X_i within a fixed distance of a given X_i as well; this would be more adaptive at the cost of accepting that in some cases there might be no neighbors within the prescribed distance.

To be more formal, assume that data (y_i, x_i) for $i = 1, \ldots, n$ is available and consider a new covariate vector x_{n+1}. To predict $Y_{n+1}(x_{n+1})$ choose a metric $d(\cdot, \cdot)$ on the feature space and a number of neighbors $k < n$; in many applications, k is between one and ten. The k-nearest-neighbors $(k\text{NN})$[1] regression predictor for Y_{n+1} is

$$\hat{Y}_{n+1}(x_{n+1}) = \frac{1}{k} \sum_{i \in S(x_{n+1})} Y_i, \tag{4.52}$$

where $S(x_{n+1}) = S_k(x_{n+1}) = \{x_i | d(x_i, x_{n+1}) \leq e_{(k)}\}$ and $e_{(1)}, \ldots, e_{(n)}$ are the order statistics from the $e_i = d(x_{n+1}, x_i)$. The k-nearest-neighbors classifier is

$$\hat{Y}_{n+1}(x_{n+1}) = \arg\max_{j} \#(C_j(x_{n+1})), \tag{4.53}$$

i.e., the class to which the highest number of the k x_i that are closest to x_{n+1} belong, where

$$C_j(x_{n+1}) = \{x_u \in S(x_{n+1}) \text{ and } x_u \in \text{ class } j\}, \tag{4.54}$$

i.e., the subset of the k nearest neighbors that are in class j. For classification, k is often chosen to be odd to avoid ties. A more complete and formal treatment of nearest neighbors is given in Secs. 8.3 and 8.6.2.

---

[1] Note that, depending on the context, NNs may refer to nearest neighbors, as here, or neural nets, as later.

The central idea of nearest neighbor methods goes back at least to Fix and Hodges (1951) and there are many variants on the basic kNN idea. One of the most popular involves weighting the k nearest neighbors by their proximity to x_{n+1}. The value of k is usually chosen by cross-validation or some variant of it. Overall, nearest neighbor methods are among those most thoroughly studied; see Cunningham and Delany (2007) and Bhatia and Vandana (2010), and the references therein, for recent overviews. Devroye *et al.* (1994) provided one of the most general convergence theorems for kNN methods.

The key issue here is that kNN methods tend to be very stable since they are locally defined by percentiles and are able to capture many different shapes of classification region or regression function because they are nonparametric. They also, in a limiting sense, have a very small error under sufficiently strong conditions. However, as d increases these methods do not scale up well with n, largely because kNN methods involve essentially no data summarization. Thus, when d is not small, and the data is consequently sparse, kNN methods are usually outperformed by other methods; see Murtagh (2009) for an explanation of this phenomenon and Dutta and Ghosh (2012) for related discussion.

4.8 Comparisons: Classification

To compare LDA, logistic regression, and kNNs consider three scenarios for the binary classification of $n = 200$ IID data points. All three scenarios are $\mathcal{M}$-closed because there is a specific model that generates the data that can be uncovered. They are as follows.

I. The data come equally from $N(-2, 1)$ (class 1) and $N(2, 1)$ (class 2).
II. Begin with 200 data points (x_1, x_2) generated from a bivariate normal mean with $(0, 0)$, and variance matrix I_2. To form classes 1 and 2, partition the data into two sets depending on whether (x_1, x_2) satisfies $x_2 > x_1^2 - 1.1$. Then randomly take one out of ten points satisfying $x_2 > x_1^2 - 1.1$ and one out of ten points satisfying $x_2 < x_1^2 - 1.1$ and put them in the other set. Let class 1 be the set of points that are on average 90% in the upper set, i.e., satisfy $x_2 > x_1^2 - 1.1$, and let class 2 be the rest. Now, use (x_1, x_2) as the explanatory variables.
III. Same as II, but now use (x_1, x_1^2, x_2, x_2^2) as the explanatory variables.

The computations were done in R (R Core Team 2016). The form of LDA used was in the package MASS and is similar to (4.51) except that the prior over classes was taken as the class proportions in the training set rather than the uniform prior. Logistic regression classification was done using the command glm in the stats library (loaded by default). In the comparisons here, two versions of kNNs are used. The first was from the library class with $k = 3$ classes. The idea in using $k = 3$ was that it was odd and higher than one. The other form of kNNs was kkNN, from the library kknn, again with $k = 3$. The distance was chosen to be the squared error, and this form of kNNs weights the kNNs according to their distances from the given data point. (In this case, the computations employ the default based on using the maximum of the summed kernel densities to obtain weights for the kNNs according to their distances.) Thus, for scenarios II and III, the classifiers (the two forms of NNs, i.e., LDA and logistic regression) give different results purely because they are based on different explanatory variables. The CPE was calculated by treating the predictand as the class (1 or 2) and using the misclassification loss that counts the number of points misclassified.

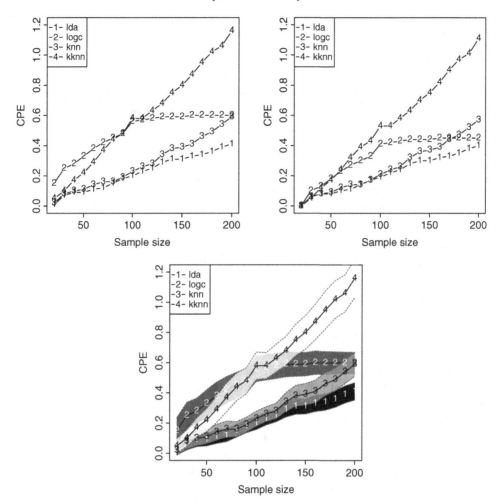

Figure 4.4 The upper left panel shows the mean CPE curves for four different classifiers for scenario I, treating the class as the predictand. The upper right panel is similar but includes a baseline adjustment. The lower panel shows the upper left panel but with ±SE regions around the CPE lines.

The upper left panel of Fig. 4.4 shows the CPE and the upper right panel shows the baseline adjusted CPE as a function of sample size for scenario I for the four class predictors, i.e., classifiers. Clearly, LDA is the best predictor and kNN with $k = 3$ is the second best; The kkNN form of NNs is seen to do poorly on both panels, while around sample size 110 the logistic regression classifier has a CPE that levels off. This suggests that for large enough sample sizes (greater than 200) it might give the best predictions, i.e., the logistic regression function may be able to 'find' the optimal boundary given 100 or so data points. However, on the range of sample sizes shown, logistic regression does not do well in a predictive sense because the setting was chosen to favor LDA. The lower panel in Fig. 4.4 contains the information shown in the upper left panel but includes the ± SE regions for the CPE curves, indicating that the separation, while not large, does permit kkNN to be seen as different from the other methods. Again, if modeling were desired, the natural approach would be to pick the best predictor found at $n = 200$ and examine its properties. One might also be led to

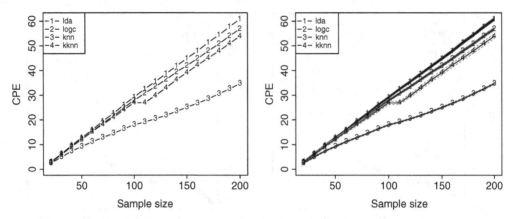

Figure 4.5 The left-hand panel shows the CPE curves for four different classifiers for scenario II, treating the class as the predictand. The right-hand panel shows the left-hand panel with ±SE regions around the CPE lines.

look at the logistic model because of its sudden leveling off. This would assess modeling uncertainty and permit further comparisons of the classifiers.

The left-hand panel of Fig. 4.5 shows the CPE as a function of sample size for scenario II, for the four class predictors, i.e., classifiers. (There is no reason to look at baseline adjustment since all four methods start with nearly the same errors.) It can be seen that kNN is the best method and the right-hand panel showing the $\pm$ SE curves confirms this. Thus, using only (x_1, x_2) as explanatory variables, neither of the model-based methods is able to find the parabolic boundary between the two classes and kkNN seems to do poorly by not weighting the kNNs equally. Indeed, scenario II was chosen to show that, when methods are unable to find a boundary between classes, methods such as kNNs can do well. In this case, although kNNs does not readily convert to a model, one can use the final kNNs expression to define regions in the plane corresponding to the two classes. These can be approximated by smooth curves or just treated as a data-dependent function. In either case, inferences would be made from the regions defined by the classifier.

The left-hand panel of Fig. 4.6 shows the CPE as a function of sample size for scenario III for the four class predictors, i.e., classifiers. (Again there is no reason to look at baseline adjustment since all four methods start with nearly the same errors.) It is seen that, for a sample size of less than around 100, LDA is a little better than the logistic regression classifier but that for sample sizes of more than about 160 the logistic regression classifier is best – and the gap between them seems to be widening. Essentially, the logistic regression classifier is able to approximate the parabolic boundary between the two regions better than LDA does, but it takes more data to see this. The two NN methods do not do well because, being nonparametric, they will be slower to converge and again it seems that unequal weighting of the neighbors does not give good results. In fact, scenario III was chosen to make these points. The right-hand panel in Fig. 4.6 is the same as the left-hand panel but includes the $\pm$SE error bars. Although the regions for the three lower curves overlap, it is likely that they would have been seen to separate had the simulation used large enough n values. In this case, logistic regression converts readily to a model that can be examined for its properties – although LDA and kNNs perform well enough for $n = 200$ that one would want to

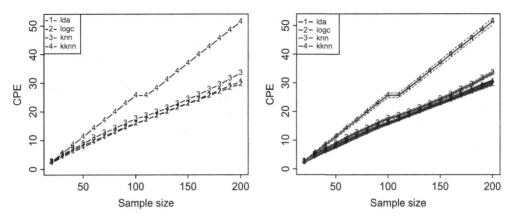

Figure 4.6 The left-hand panel shows the mean CPE curves for four different classifiers for scenario III, treating the class as the predictand. The right-hand panel shows the left-hand panel with ±SE regions around the CPE lines.

examine them too and compare the models which they generate with the logistic regression model.

To see how these classification techniques work with real data, consider the classification of colorectal cancer patients into two classes, DNA micro-satellite stable and DNA micro-satellite unstable. The difference is that patients who are micro-satellite unstable have regions of DNA that contain more errors than usual and these errors vary over the population of colorectal cancer patients. This is thought to occur because the DNA repair mechanisms are not functioning as they should; presumably this is associated with cancer-related processes. The goal of the experimentation is to discover a gene signature indicating the two classes. The data set analyzed here is from the NCBI gene expression omnibus (GEO) database, series GSE13294, and uses $n = 155$. There are 47 explanatory variables, namely, the RNA expression for selected genes; see Jorissen *et al.* (2008) for details. An earlier paper on this same topic is Kruhøffer *et al.* (2005), in which the RNA expressions for 13 selected genes were used. Entertainingly, the 13 genes used in the earlier paper were disjoint from the 47 genes used in the later paper, underscoring the uncertainty in gene selection.

The same four methods as used in the simulations were applied to the GSE13294 data, and the results are given in Fig. 4.7. In the left-hand panel it is seen that the kNN method gives the best sequential prediction and the kkNN method is second best, while the two methods that are more model-based perform worse. Moreover, the right-hand panel shows that the separation in performance is strong. That is, one would be led to convert a kNN class predictor to a model and to ignore the logistic regression and LDA classifiers.

4.9 A Look Ahead to Part II

Heuristically, it's reasonable to write

$$\text{model uncertainty} \propto \frac{1}{\text{modeling information}},$$

assuming that the modeling information is reliable. However, more is true. As the *available* modeling information decreases relative to the amount of modeling information *necessary* to

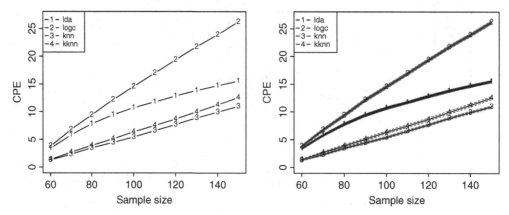

Figure 4.7 The left-hand panel shows the mean CPE curves for four different classifiers for the GSE13294 data, treating the class as the predictand. The right-hand panel shows the left-hand panel with ±SE regions around the CPE lines.

describe a DG, using the modeling information helps less and less so the model uncertainty will remain high. This is one reason why model uncertainty can remain high for complex DGs even when model information seems abundant. Indeed, as DG complexity increases for a fixed amount of modeling information, one expects the modeling information to be less and less useful. This is another way to say that high-complexity DGs correspond in practice to high model uncertainty and that modeling information or the requirement for interpretability is of limited use. Indeed, modeling information or the requirement for interpretability may well be harmful in settings that are not low complexity, because there are so many more ways to be wrong than to be right. In essence, this is one of the central points of the first three chapters of Part I.

Nevertheless, using modeling information and requiring interpretability helps researchers feel that they understand what they are studying. So, however futile or scientifically inefficient, the desire for an explanation of the mechanism of a phenomenon, as opposed to the ability to predict, will not go away. Hence the usage of models in this chapter. However, in this chapter, the models – whether parametric or nonparametric – have been regarded not as models but as the predictors that they generate. Thinking of models in terms of their predictive properties puts the focus on how well the predictions and data match rather than just on whether the model is consistent with the feelings and impressions one might have of a DG.

This is a general theme, which will be continued in Part II. Rather than relying on making assumptions and on the models they seem to justify, such assumptions will be used to formulate predictors that can be tested against the data that they are supposed to 'explain'. Note that, from a predictive standpoint, 'explain' means only 'give sufficiently good predictions'. From the modeling standpoint, 'explain' means 'provides an understanding of the DG'. This is impossible in $\mathcal{M}$-open problems and only feasible in a limited sense in $\mathcal{M}$-complete problems. However, it is, in principle, feasible for $\mathcal{M}$-closed problems and can be actually achieved for simple enough $\mathcal{M}$-closed problems (if enough experimentation and analysis is done).

Hopefully, it has been seen so far that modeling can provide predictive techniques for simple enough data, e.g., as occurs in simple nonparametrics, linear regression, and elementary classification.

In the chapters of Part II, out attention will turn to well recognized model classes, for instance, time series, where prediction is well recognized as the key criterion for models (even when the assumption of an $\mathcal{M}$-closed or even $\mathcal{M}$-complete problem is dubious). This will lead to longitudinal analysis – which also typically has an order structure. Then, it will be time to turn to data that do not have an order structure, such as survival analysis. For this setting, predictive techniques do not seem to be as well developed or prevalent. This may be due to the complexity or variability of the data (random effects add an extra layer of variability), to the nature of the quantities being estimated (estimates of density ratios in survival analysis are likely to be unstable compared with estimates of probabilities), to the fact that models that make reliably useful predictions are hard to generate (the PIs may be too wide to be useful or too hard to find readily), or simply to the over-weening fixation on modeling that leads people to ignore predictive accuracy.

These considerations may also apply to the fourth chapter of Part II, traditional nonparametrics, where the stability and generality of the methods may make good predictions hard to generate with limited data, e.g., a nonparametric function estimator may fit an irregular function with a smooth function that isn't 'too far wrong' but still misses essential features. Nonparametric techniques also, as a generality, do not scale up well to complex or high-dimensional data.

The last chapter of Part II aims to tie model selection to prediction. After all, if it is agreed that model selection is sometimes necessitated by model uncertainty, then the predictions from a selected model have to be adjusted for model uncertainty as well as parameter uncertainty within a selected model. This is an area that seems underdeveloped, or at least underused, since many experimenters essentially pick a model class and assume that the model class uncertainty will be effectively zero because they 'understand' the key features of the data generator. This position is usually too simplistic to merit refutation: model uncertainty and misspecification should be afforded at least as careful an assessment as any other aspect of uncertainty, e.g., parameter uncertainty. Such a careful assessment is afforded by the predictive approach.

Part II

Established Settings for Prediction

> The greatest plus of data modeling is that it produces a simple and understandable picture of the relationship between the input variables and responses ... different models, all of them equally good, may give different pictures of the relation between the predictor and response variables ... One reason for this multiplicity is that goodness-of-fit tests and other methods for checking fit give a yes–no answer. With the lack of power of these tests with data having more than a small number of dimensions, there will be a large number of models whose fit is acceptable. There is no way, among the yes–no methods for gauging fit, of determining which is the better model.
>
> Breiman (2001a, Sec. 5.3)

In Part II we consider the effect of viewing five subfields of statistics predictively. The first three assume the problem under investigation is $\mathcal{M}$-closed, requiring the data to have a complexity that matches the complexity of finite-dimensional parametric models even when this is far from true. Outside this class of problems, real data is more complex than finite-dimensional models can summarize well. For $\mathcal{M}$-complete problems, nonparametric methods – seen from a predictive standpoint in Chapter 8 – are sometimes adequate but often require unattainable sample sizes. Model selection from a predictive standpoint is the topic of Chapter 9.

The hope is that a statistical analysis of real data will yield something worth reporting because a true model exists. In fact, in the absence of a predictive verification, there is usually nothing to report from a modeling perspective because SEs and tests (based on fit) rarely take model uncertainty and misspecification, and hence generalization error, into account. The point of predictively examining these subfields is to reveal features of them when the data generator (DG) is more complex than the model presumes or sample size allows.

The five chapters of Part II progress from two obviously predictive subfields, time series and longitudinal data, to two subfields not usually seen as predictive, survival and nonparametrics, ending with the subfield, model selection, at the heart of model-based predictor formation.

5

Time Series

For without generalization foreknowledge is impossible. The circumstances under which one has worked will never reproduce themselves all at once. The observed action then will never recur; the only thing that can be affirmed in that under analogous circumstances an analogous action will be produced. In order to foresee, then, it is necessary to invoke at least analogy, that is to say, already then to generalize ...

Thus, thanks to generalization, each fact observed enables us to see many others; only we must not forget that the first alone is certain, that all the others are merely probable. No matter how solidly founded a prediction may appear to us, we are never *absolutely* sure that experiment will not contradict it ... It is far better to foresee even without certainty than not to foresee at all.

Science and hypothesis, Poincaré (1913, pp. 126–127).

The field of time series is vast. Here, the point is merely to see enough of the basics that the predictive perspective on time series is apparent. So, the first section reviews the well-known 'classical decomposition model', where the goal is to detrend and deseasonalize a continuous response in discrete time in such a way that the residual is well behaved. The second section presents the Box–Jenkins approach for 'SARIMA' models and also for models with a continuous response in discrete time. The third section then redoes this from a Bayesian standpoint, focusing on the 'ARMA' part since 'seasonality' and 'integration' are mostly driven by exploratory data analysis. The fourth section presents computational results to illustrate the predictive properties of the techniques covered in the first three sections. The fifth section looks at the stochastic modeling that is more appropriate to discrete-response continuous time series. In the endnotes we briefly discuss more advanced topics such as ARFIMA, ARMAX, regression with time series errors, and the dynamic linear model.

5.1 Classical Decomposition Model

The *classical decomposition model* for a time series $\langle Y_t \rangle$ is based on writing

$$Y_t = m_t + s_t + E_t, \quad t = 1, \dots, T, \tag{5.1}$$

where $Y_t = y_t$ is the observed sequence, m_t is a 'slowly changing' function of t called a trend component, s_t is a periodic function called a *seasonal* component, and E_t is a second-order stationary random error term; t and T are used in place of i and n since the observations are ordered in time. (Recall that second-order stationarity means that $E(E_t)$ exists and equals the same constant value for all E_t and that the covariance between E_t and E_{t+h} depends only on the distance h.) Here, m_t is slowly changing compared with the period of s_t and the variance of E_t; some variability in s_t may be tolerated, e.g., the period may be random but fluctuate around a fixed value with a small variance. The task is to identify $\hat{m}_t$ and $\hat{s}_t$ such that $e_t = y_t - \hat{m}_t - \hat{s}_t$ are approximately the outcomes of E_t and can be seen to represent outcomes from a second-order stationary process; sometimes a time series must be transformed in order to achieve this. Once this is done, point predictors and PIs for Y_{T+1} can be found.

If there is no seasonal component in (5.1), i.e., $s_t = 0$, then one elementary technique for detrending is to use standard least squares procedures. For instance, write $m_t = \beta_0 + \beta_1 t + \cdots + \beta_K t^K = v_t^T \beta$ as the regression function, where $v_t = (1, t, \ldots, t^K)$ and $\beta = (\beta_0, \ldots, \beta_K)^T$. (Whether T means a transpose or the length of the time series will be clear from the context.) Then, minimizing the expression

$$\sum_{t=1}^{T} (y_t - m_t)^2$$

gives the usual least squares estimators for the β_j, i.e., $\hat{\beta} = (V^T V)^{-1} V^T Y$ where $Y = (Y_1, \ldots, Y_T)^T$ and $V = [v_1, \ldots, v_T]^T$ is the $T \times K$ design matrix. The residuals are given by $Y_t - v_t^T \hat{\beta}$ for $t = 1, \ldots, T$ and should be second-order stationary.

A variant on this is to try to smooth out the seasonality by regressing on Fourier series (or other periodic) terms. Thus, one can write

$$m_t = \beta_0 + \beta_1 \cos(2\pi t) + \cdots + \beta_K \cos(2\pi K t) + \beta_{K+1} \sin(2\pi t) + \cdots + \beta_{2K} \sin(2\pi K t),$$

and again invoke least squares optimality to estimate the β's. In principle, one could combine polynomial and periodic terms in one regression function, simultaneously pulling out trend and seasonality.

A second elementary technique (also for $s_t = 0$) is to identify an m_t by smoothing the Y_t with a moving average. The idea is to ensure that random fluctuations cancel out in $\hat{m}_t$. One reasonable choice is to set

$$\hat{m}_t = \sum_{j=-q}^{} Y_j \quad \text{for some } q \geq 1.$$

Another choice is to use the one-sided moving average defined by the recursion

$$\hat{m}_t = \alpha Y_t + (1 - \alpha)\hat{m}_{t-1}, \qquad \hat{m}_t = Y_1,$$

for some $a \in (0, 1)$, sometimes called exponential smoothing. The term 'exponential' arises because the influence of Y_{t-j} on Y_t is weighted by the factor $(1 - \alpha)^j$, which decreases by a factor that is less than one as j increases and hence the factors can be regarded as exponentially decreasing. (If $\alpha = 0$ or 1 then $\hat{m}_t$ is trivial.) Other smoothing techniques can be applied as well to pull out a trend when $s_t = 0$. For instance, any nonparametric smoothing technique can be used: splines, loess, Nadaraya–Watson technique, nearest neighbor methods, etc. (Of course, if the tuning parameter is chosen to give a good fit rather than to smooth the time series, and $s_t \neq 0$, the seasonality may be smoothed out as well.)

As it stands, these techniques are merely efforts to summarize data by a model; one might examine model fit by testing formally whether the residuals look second-order stationary and try to use the model to say something about the DG. However, the goal of time series is, above all, prediction. So, when $s_t = 0$ is valid, the natural predictor for Y_{t+1} is $\hat{m}_{t+1}$ with a PI given by the apparent distribution of the residuals $e_t = y_t - \hat{m}_t$. For this purpose, only the first elementary method (least squares) can be used since smoothing the data from one to T does not provide a useful prediction for $T + 1$. An estimate of the function m_t that can be extrapolated is required, and in most settings it would be unusual to predict Y_{T+1} by $\hat{m}_t$ alone. (Actually, this may not always be as unreasonable as it looks: predicting tomorrow's temperature by today's temperature may be adequate when the temperature does not fluctuate too wildly – even though this is a case where seasonality is present).

A third elementary technique is to use differencing. Let B be a backshift operator such that $BY_t = Y_{t-1}$, i.e., the operator that converts the tth random variable in the time series to the $(t - 1)$th random variable in the time series. Letting '1' denote the identity operator, the differenced time series is $(1 - B)Y_t = Y_t - Y_{t-1}$, $(1 - B)^2 T_t = B(Y_t - Y_{t-1}) = Y_t - 2Y_{t-1} - Y_{t-2}$, etc., and the hope is that $(1 - B)^d Y_t$ will be stationary for some $d \in \mathbb{N}$. This is plausible because $1 - B$ loosely corresponds to differentiating a function, and, for many functions, e.g., polynomials, there is an order of derivative that makes them zero or nearly so. Obviously, if this is possible with a time series that has zero seasonality, one can model the trend of $\langle BY_t \rangle$ as in either of the first two elementary techniques and correct for the differencing in the predictions. A more sophisticated form of this kind of procedure is given in Sec. 5.2.

If the goal is to smooth out a trend and any seasonality, there are two cases. In the first case it is assumed that the trend can be taken as a constant in each season or period. To see this, relabel the Y_t to be of the form Y_{jk}, where $j = 1, \ldots, J$ is the number of seasons, each season has length K, and $k = 1, \ldots, K$. This requires that $T = JK$. Averaging over the K time points in the jth season gives $\hat{m}_j = (1/K) \sum_{k=1}^{K} y_{jk}$. That is, $\hat{m}_j$ is the (constant) trend for the jth season, which is unbiased in its mean. Then, $\hat{s}_k = (1/J) \sum_{j=1}^{J} (y_{jk} - \hat{m}_j)$ has the property that it sums to zero over the K time steps in a season, i.e., $\hat{s}_1 + \cdots + \hat{s}_K = 0$. The residuals are now

$$\hat{e}_{jk} = y_{jk} - \hat{m}_j - \hat{s}_k$$

and should be second-order stationary if the identification of trend and of seasonality has been successful. Obtaining predictions from this analysis is difficult unless one assumes that the trend repeats from period to period (season to season), so that the prediction for Y_{T+u} for, say, $u = 1, \ldots, T$, is based on $\hat{m}_u$, the mean of $Y_u, Y_{u+K}, Y_{u+2K}, \ldots$, with a PI derived using their variance (for instance).

The second case, which is more complex and realistic, occurs when it is not reasonable to assume that the trend is nearly constant over each season. For $t = JK$, with J seasons and K time steps in a season, begin by deseasonalizing. For K even, let

$$\hat{m}_{K/2} = \frac{1}{K} \sum_{t=1}^{K} y_t, \ldots, \hat{m}_{T-(K/2)} = \frac{1}{K} \sum_{t=T-K+1}^{T} y_t.$$

This means that there are $(K/2) - 1 Y$'s at the beginning and at the end of the time series that do not have an associated moving average. Next, let

$$w_k = \frac{1}{J} \sum_{u=0}^{J-1} (y_{k+uK} - \hat{m}_{k+uK})$$

be the weights associated with each time step within a period or season. To enforce seasonality, i.e., to ensure that the seasonal effects cancel out over each season, set

$$\hat{s}_k = w_k = \frac{1}{K} \sum_{v=1}^{K} w_v,$$

so that the s_k sum to zero. Since s_t represents period or seasonal effects it makes sense to write $\hat{s}_k = \hat{s}_{k-K}$ for $k = K+1, \ldots, 2K$ and similarly for succeeding blocks of length K. With this convention, the deseasonalized time series is

$$Y_t^* = Y_t - \hat{s}_t.$$

If the deseasonalizing was successful, $\langle Y_t^* \rangle$ can be analyzed according to either of the first two elementary techniques for detrending a time series that can be assumed deseasonalized, i.e., has $s_t = 0$. Again, if the first elementary technique, based on linear regression, is used, one can form point predictions for $t \geq T+1$ from $\hat{m}_{T+1} + s_{T+1}$, with PIs determined by using the residuals from all the first time steps in the available periods to estimate a variance (or, more generally, the DF of E_t for those time points). One can pool the residuals over all the time steps to get a better estimate if it is believed that the variance really is constant, i.e., $\langle Y_t^* \rangle$ is second-order stationary.

Whatever the specific techniques chosen, the goal is to find $\hat{m}_t$ and $\hat{s}_t$ such that the residuals are second-order stationary and then predict Y_t for $t \geq T+1$ by using $\hat{m}_t$ and $\hat{s}_t$, using the appropriate residuals to obtain a PI. That is, (5.1) is merely a class of predictors that is being searched to find a good predictor. Given that one has found a good predictor, one might inquire about its other properties but, at least initially, this is incidental to its ability for prediction.

5.2 Box–Jenkins: Frequentist SARIMA

Another broad class of predictors that one might search is derived from the collection of SARIMA$(p,d,q) \times (P,D,Q)_s$ models, i.e., seasonal auto-regressive integrated moving average models with orders $p, d, q, s, P, D, Q \in \mathbb{N}$ and a standard parametrization that involves up to four collections of real parameters. Often one talks merely about SARIMA models, leaving the parameters (integer and real) understood. The aggregation of techniques associated with SARIMA models is usually called the Box–Jenkins methodology.

To write the general form of a SARIMA model, recall the 'backshift' operator B that, for a given time series $\langle Y_t \rangle$, shifts any Y_t to Y_{t-1}. That is, $BY_t = Y_{t-1}$. Leaving aside the formal justification of this operator, it is obvious that the derived series $(1-B)Y_t$ is just the difference of successive terms and thus $(1-B^s)Y_t = Y_t - Y_{t-s}$. If s is interpreted as the number of time steps in a season then a time series is SARIMA if and only if

$$Z_t = (1-B)^d (1-B^s)^D Y_t$$

has a second-order stationary ARMA$(p,q) \times (P,Q)_s$ distribution. That is, s is the length of the season and d and D are the orders of the differencing on the time step and seasonal time scales, respectively. Given s, d, and D, an ARMA$(p, q) \times (P, Q)_s$ time series is defined by the equation

$$\phi(B)\Phi(B)Z_t = \theta(B)\Theta(B)\epsilon_t,$$

where ϕ and θ are polynomials of order p and q in B while Φ and Θ are polynomials of order P and Q in B^s. The error terms, the ϵ_t, are white-noise processes (mean zero, variance σ^2, pairwise uncorrelated). That is,

$$\phi(B)\Phi(B^s)\left((1 - B)^d(1 - D^s)^D Y_t\right) = \theta(B)\Theta(B^s)\epsilon_t \tag{5.2}$$

is the defining property of a SARIMA$(p, d, q) \times (P, D, Q)_s$ time series Y_t. Sometimes Φ and Θ are called seasonal polynomials while ϕ and θ are called nonseasonal polynomials. In contrast with (5.1), (5.2) permits the seasonal pattern to be random, i.e., to vary from one cycle to the next. (Actually, the right-hand side of (5.2) should have a mean μ added to it. Here, it has been assumed that either $\mu = 0$ or the sample mean has been subtracted from the left-hand side to ensure that $\mu = 0$ is a reasonable assumption.)

This section focuses on the Box–Jenkins (BJ) method for prediction when Y_t is SARIMA$(p, d, q) \times (P, D, Q)_s$. The BJ method for prediction has four steps: (1) predictor class identification; (2) parameter estimation; (3) internal validation; and (4) forecasting. There are several excellent references for the BJ methodology; see Brockwell and Davis (1987), Chandler and Skouras (1998) (from which the presentation here draws heavily), Shumway and Stoffer (2010), Chandler and Scott (2010), and the classic publication Box and Jenkins (1970), which focuses on ARIMA models (the factors in (5.2) involving seasonality are removed). For brevity, the discussion of many technical topics such as invertibility, stationarity, and causality is omitted since our point is not to understand time series analysis per se but to use it to generate predictors.

5.2.1 Predictor Class Identification

The first step in finding a predictor from the class of SARIMA models is to reduce to ARMA models. That is, while a SARIMA model might be nonstationary, the first task is to find an s, d, and D that will make Z_t second-order stationary, i.e., remove the seasonality and integration. Recall that second-order stationary means that the first two moments for Z_t are the same for all t and the autocovariances of all orders are the same, i.e., for any $k \geq 0$, the functions $\gamma(k) = \text{Cov}(Z_t, Z_{t+k})$ are independent of t. Thus, one can compute the $\gamma(\cdot)$ and do formal tests of stationarity for given s, d, and D. However, it is more common merely to choose an s on the basis of physical modeling ($s = 4$ if data is quarterly with an annual cycle, $s = 12$ if data is monthly with an annual cycle, $s = 24$ if measurements are taken every hour with a daily cycle, and so forth). Then, d and D are chosen so that a plot of Z_t versus t looks stationary to the eye. To achieve this, it may be necessary to transform the Z_t (or Y_t) by taking logarithms or using the Box–Cox transformation. Once this is done, Z_t can be assumed to be a SARMA$(p, q) \times (P, Q)_s$ (since $d = D = 0$ for Z_t) process. Since s is known, it remains to determine p, P, q, Q, σ^2, and the parameters associated with the four polynomials in (5.2).

To begin this process, note that a SARMA$(p, q) \times (P, Q)_s$ process is an ARMA$(p + sP, q + sQ)$ process because the products $\phi(B)\Phi(B^s)$ and $\theta(B)\Theta(B^s)$ of polynomials remain polynomials. So, in principle, it is enough to be able to choose the orders $p^* = p + sP$ and $q^* = q + sQ$ in an ARMA(p^*, q^*) process. The usual way to do this is to invoke an extra criterion such as the Akaike information criterion (AIC) or the Bayes information criterion (BIC). However, this would only identify the AR and MA orders $p + sP$ and $q + sQ$, not the individual orders p, P, q, and Q. From the predictive standpoint, this would be enough because it would identify a predictor class unambiguously, leaving only the coefficients in the polynomials to be estimated. Unfortunately, this approach does not use the fact that s is known, so the number of parameters to be estimated could be much higher than necessary, especially if s were not small. The result would be elevated SEs for parameters. This does not matter directly but could increase the generalization error and so make for a less good predictor. Even worse, there may be more than one choice of p, P, q, and Q consistent with given s, p^*, and q^*. Consequently, it may be useful to try to obtain values for P and Q before turning to p and q.

The problem is that there seems to be no standard and readily applicable method for determining P and Q given s, because using the autocorrelation function (ACF) and the partial autocorrelation function (PACF) only works very approximately unless it is known that one of the components, AR or MA, is not present; see Shumway and Stoffer (2013, Table 3.3 and Ex. 3.42). Thus, if $P, Q \neq 0$, one finds the nonseasonal ACF and PACF and hopes that one can identify a tailing off in the ACF and PACF at lags $p, p + P, \ldots, p + sP$ and $q, q + Q, \ldots, q + sQ$ so that p, q, P, and Q can be identified. (If, say, $p + sP$ and $p + (s-1)P$ are known then, given s, P can be found; Q is similar.)

For the sake of completeness, we will review the formalities of ARMA(p, q) processes and the use of the ACF, PACF, and AIC. From this it will be obvious how to extend the techniques to the SARMA$(p, q) \times (P, Q)_s$ setting. The two fundamental classes to review are the autoregressive process of order p, AR(p), and the moving average process of order q, MA(q). Given $\phi_1, \ldots \phi_p$ the AR(p) process is

$$Z_t = \nu + \phi_1 Z_{t-1} + \cdots + \phi_p Z_{t-p} + \epsilon_t,$$

where ν is a constant, here taken as zero unless otherwise noted. We assume that the roots of the polynomial

$$1 - \sum_{j=1}^{p} \phi_j x^j$$

satisfy $|x_i| > 1$ for $i = 1, \ldots, p$ to ensure stationarity. (For an AR(1) process this is equivalent to $|\phi_1| < 1$). The MA(q) process for parameters $\theta_1, \ldots, \theta_q$ is

$$Z_t - \mu + \epsilon_t - \theta_1 \epsilon_{t-1} - \cdots - \theta_q \epsilon_{t-q}.$$

In both cases, the ϵ_t are IID mean-zero with variance σ^2 and μ is the mean, here taken to be zero unless otherwise noted. We assume that the roots of the polynomial

$$1 - \sum_{j=1}^{q} \theta_j x^j$$

are outside the unit circle in $\mathbb{C}$. Taken together, an ARMA(p, q) process satisfies

$$Z_t = c + \phi_1 Z_{t-1} + \cdots + \phi_p Z_{t-p} + \epsilon_t - \theta_1 \epsilon_{t-1} - \cdots - \theta_q \epsilon_{t-q}, \tag{5.3}$$

where c is a constant, again taken to be zero unless otherwise noted.

If it is known that a given time series is MA(q), model identification means choosing q. If it is known that a given time series is AR(p), model identification means choosing p. If it is known that a given time series is ARMA(p, q) then the task is identifying p and q.

The ACF of a time series is $\rho(k) = \gamma(k)/\gamma(0)$, where

$$\gamma(k) = \frac{E\big((Z_t - \mu)(Z_{t+k} - \mu)\big)}{\sigma^2}$$

when Z_t has mean μ and variance σ^2. Clearly, $\gamma(-k) = \gamma(k)$ if both sides are defined. It can be verified that for an MA(q) process, the ACF is zero for $k > q$.

The PACF of a time series is more complicated. Given a stationary time series Z_t, write the regression function as

$$E(Z_t | Z_{t-1} = z_{t-1}, \ldots, Z_{t-k} = z_{t-k}) = \beta_{k,1} z_{t-1} + \cdots + \beta_{k,k} z_{t-k}.$$

Each $\beta_{k,t-j}$ is the linear regression coefficient of Z_t on the $Z_{t-1}, \ldots, Z_{t-k}$ treated as explanatory variables. From standard linear-model theory, the last of these is

$$\beta_{k,k} = \text{Corr}(Z_t, Z_{t-k} | Z_{t-1}, \ldots, Z_{t-k+1}),$$

giving the dependence between Z_t and Z_{t-k} that cannot be accounted for by the intervening $Z_{t-1}, \ldots, Z_{t-k+1}$. The sequence $\beta_{1,1}, \ldots, \beta_{k,k}, \ldots$ is the PACF. It can be verified that, for an AR(p) process, the PACF is zero for $k > p$.

Now, for MA(q) and AR(p), we can get estimates for q and p by looking at plots of the ACF and PACF, respectively, as functions of $k = 1, 2, \ldots$ First, the usual estimate for the ACF is

$$\hat{\rho}(k) = \frac{\sum_{t=k+1}^{T}(z_{t-k} - \bar{z})(z_t - \bar{z})}{\sum_{t=1}^{T}(z_t - \bar{z})^2}.$$

Although the $\hat{\rho}(k)$ are themselves correlated, the correlations are weak enough that in the limit of large n the approximations

$$\text{Var}(\hat{\rho}(k)) \approx \frac{1}{T} \quad \text{and} \quad \forall j \neq k, \quad \text{Corr}(\hat{\rho}(j), \hat{\rho}(k)) \approx 0$$

hold. So, for an MA(q) process, plotting the ACF values as a function of k means that the first value of k for which the ACF lands in $0 \pm 1.96/\sqrt{t}$ (and stays there) is a good estimate $\hat{q}$ of q.

Second, the PACF is used to choose p. More exactly, the ACF ends up being used again because the PACF values can be re-expressed in terms of the ACF values. This can be done either via the Yule–Walker equations or via the Levinson–Durbin recursions, which we do not show here. The net effect is that the $\hat{\rho}(k)$ can be used to generate estimates $\hat{\beta}_{k,k}$ for the PACF values. It can be shown that, for an AR(p) process, the $\hat{\beta}_{k,k}$ have mean zero and variance $1/t$. Parallel to finding $\hat{q}$, plotting the PACF values as a function of k means that the first value of k for which the PACF lands in the interval $0 \pm 1.96/\sqrt{t}$ (and stays there) is a good estimate $\hat{p}$ of p.

For ARMA(p, q) processes we *cannot* combine these two techniques (ACF and PACF) to find $\hat{p}$ and $\hat{q}$. The reason is that including both AR and MA terms often causes the ACF and PACF to have geometrically decaying patterns that do not cleanly indicate a specific order (p when $q > 0$ or q when $p > 0$). In these cases, the AIC or BIC is often used, see Sec. 5.2.3, to determine the proper choices for p and q. For a review see de Gooijer *et al.* (1985).

5.2.2 *Estimating Parameters in an ARMA(p, q) Process*

Given that $\hat{p}$ and $\hat{q}$ have been satisfactorily chosen, the next task is to estimate the θ_j in the case of an MA(q) process, the ϕ_j in the case of an AR(p) process, or both in the case of an ARMA(p, q) process. This can be done by the method of moments, i.e., equating sample moments to population moments and solving the resulting implicit equations for parameter estimates, but the estimators obtained are quite inefficient. (Actually, the method of moments estimators for AR(p) processes based on $\hat{\rho}$ are not too bad: they can be derived by solving the Yule–Walker equations $\rho(k) = \phi_1 \rho(k-1) + \cdots + \phi_p \rho(k-p)$ for $k \geq 1$ and substituting $\hat{\rho}(k)$ for $\rho(k)$ for $k = 1, \ldots, p$. The analogous procedure for MA(q) processes has much worse results.)

The two methods typically used to obtain estimators are maximum likelihood (MLE) and least squares (LSEs). To use maximum likelihood, one must specify a likelihood. This has not yet been done because only the first two moments of Z_n have been used so far. In the simplest case, the error terms are assumed to be IID $N(0, \sigma^2)$ so that any random vector $(Z_1, \ldots, Z_t)^T$ can be regarded as following a t-dimensional normal distribution. Consequently, the likelihood can be maximized – if the unobserved initial values of $\epsilon_0, \ldots, \epsilon_{-q+1}$ are properly dealt with. A simple technique for this is used below for MA(q) processes. The downside of this likelihood approach is that it requires careful checking that the normality assumption is reasonable. Nevertheless, this is the method used in many software packages.

The least squares method does not require likelihood assumptions and works in much the same way for time series as for linear regression. For instance, consider an AR(1) model,

$$\epsilon_t = Z_t - \nu - \phi_1 Z_{t-1},$$

noting that $\nu = (1 - \phi_1)\mu$ where $\mu = E(Z_t)$ for any t. The sum of squared errors is

$$S(\nu, \phi) = \sum_{t=1}^{T} (Z_t - \nu - \phi_1 Z_{t-1})^2, \tag{5.4}$$

where the default $Z_0 = 0$ is often chosen. Solving $\partial S / \partial \nu = 0$ and $\partial S / \partial \phi_1 = 0$ gives solutions $\hat{\phi}_1$ and $\hat{\nu}$ that can be found in closed form. Usually, $\hat{\nu} \approx \bar{Z}$ and $\hat{\phi}_1 \approx \hat{\rho}(1)$, i.e., the results are very similar to those for straight line regression (and the method of moments based on $\hat{\rho}$ for AR(1) processes). The parallel continues for higher-order AR processes and, asymptotically in t, the results are similar to those of linear regression. In particular, t-tests can be used to test any hypothesis of the form $\mathcal{H}_{0,j} : \phi_j = 0$ for $j = 1, \ldots, p$.

The LSE for an MA(q) process involves an extra wrinkle. If $\mu = 0$, the MA(1) model leads to a sum of squared errors of the form

$$S(\theta) = \sum_{t=1}^{T} (Z_t + \theta_1 \epsilon_{t-1})^2. \tag{5.5}$$

The problem is that the noise terms ϵ_k are unknown. One way around this is to set the white-noise process at time zero to equal its mean, i.e., set $\epsilon_0 = 0$, so that $\hat{\epsilon}_t = Z_t + \theta_1 \hat{\epsilon}_{t-1}$ for

$t = 1, \ldots, T$. Using these in (5.5) gives a new expression, say $S^*(\theta)$, that can be optimized, but not in closed form; some sort of numerical optimization procedure must be used. In this procedure, it is important to use several plausible values of $\epsilon_0, \ldots, \epsilon_{-q+1}$ to be sure that the parameter estimates are not unduly sensitive.

The same procedure can be used for higher-order MA models but, for an MA(q) process, one must set $\epsilon_0 = \cdots = \epsilon_{-q+1} = 0$. It is seen that any process with a moving average component will have a problem with the initial ϵ's and that any choice for ϵ_0 (for instance) will affect the solutions and propagate over time. Two ways around this are as follows. (i) Sensitivity analysis: choose other values for $\epsilon_0, \ldots, \epsilon_{-q+1}$ and decide whether they affect the parameter estimates overmuch; (ii) refitting: assume $\epsilon_0, \ldots, \epsilon_{-q+1}$, get estimates of $(\theta_1, \ldots, \theta_q)$, and use these with the Z_t to find updated values of $\epsilon_0, \ldots, \epsilon_{-q+1}$, which can then be used again to generate new estimates for $(\theta_1, \ldots, \theta_q)$, cycling until stable values for $\epsilon_0, \ldots, \epsilon_{-q+1}$ are found. It turns out that usually the estimates of $(\theta_1, \ldots, \theta_q)$ are fairly stable unless t is small (or the model is nearly noninvertible, a case not treated here).

The methods for (5.4) and (5.5), and their extensions to general p and q, can be combined to give LSEs for general ARMA processes. For instance, the ARMA$(1, 1)$ process is

$$\epsilon_t = Z_t - \phi_1 Z_{t-1} + \theta_1 \epsilon_{t-1},$$

leading to

$$S(\phi_1, \theta_1) = \sum_{t=1}^{T} (Z_t - \phi_1 Z_{t-1} + \theta_1 \epsilon_{t-1})^2.$$

Setting $Z_0 = \epsilon_0 = 0$ permits a numerical solution for $\hat{\phi}_1$ and $\hat{\theta}_1$, and the general case (ARMA(p, q)) would require $\epsilon_0 = \cdots = \epsilon_{-q+1} = 0$. As in the MA($q$) case, the effect of the initial values propagates over time, necessitating stability checks.

5.2.3 Validation in an ARMA(p, q) Process

There are two sorts of validation to be done: verification that the correct p or q has been found (assuming an ARMA model is correct in the first place) and, given that p and q are correct, verification that the parameters in the model have been properly estimated.

There are two standard ways to evaluate whether p and q are plausible. First, one can use several model selection techniques, such as the Akaike information criterion (AIC) or the Bayes information criterion (BIC), and compare the resulting choices for p and q. Recall that the AIC is given by

$$\text{AIC}(p, q) = -\ln L(\theta, \phi | z^T) + k$$

where $L(\theta, \phi | z^T)$ is the likelihood, $k = p + q$, and $\theta = (\theta_1, \ldots, \theta_q)$, $\phi = (\phi_1, \ldots, \phi_p)$. Finding the pair (p, q) that minimizes the AIC often provides a check of the $\hat{p}$ and $\hat{q}$ found using the ACF and PACF. The BIC, like the AIC, relies on there being a likelihood (usually normal) but has a heavier penalty because k is multiplied by the log sample size:

$$\text{BIC}(p, q) = -2\ln L(\theta, \phi | z^T) + k \ln T.$$

Because of the larger penalty on the number of parameters, the BIC tends to select smaller models than the AIC. Both the AIC and BIC can be written in terms of the residual error, which is of the form $\hat{\sigma} = s^2 = \text{SSE}/df(\text{SSE})$ in the normal-error case. That is,

$$\text{AIC} = T \ln s^2 + 2k \quad \text{and} \quad \text{BIC} = T \ln s^2 + k \ln T.$$

There have been extensive comparisons of AIC versus BIC in a wide variety of contexts; see Burnham and Anderson (2002, Chap. 6, Sec. 4) and Clarke *et al.* (2013) for summaries and Yang (2005) and van Erven *et al.* (2012) for recent contributions. Roughly, AIC identifies a model, often depending heavily on T, that is good in a variance–bias tradeoff sense. This often enables models chosen by the AIC to give better predictions when the true model is hard to specify, e.g., in an $\mathcal{M}$-complete or $\mathcal{M}$-open setting.

By contrast, BIC identifies the model, from those available, that is closest to the true model in relative entropy (AIC also has a relative entropy interpretation based on a prediction criterion). This often enables models chosen by the BIC to give better predictions when the true model is in the model class under consideration, i.e., $\mathcal{M}$-closed settings, or when the model approximation error is much smaller than any other source of error. Overall, the BIC favors smaller models (given that they are equally good); the AIC tends to favor larger models provided they are at least a little better than any smaller model.

In practice the two criteria often give similar results – at least when the sample sizes are moderate, the model sizes are not too large, and the true model is not too far from the list of models under consideration. See Hannan (1980) for a further comparison of AIC, BIC, and the Hannan–Quinn criterion, and see Brockwell and Davis (1987, Chap. 9.2) for a corrected form of the AIC. Since there are numerous information criteria, if they give similar choices for p and q then one may regard the determination of p and q as more reliable.

Another technique for satisfying oneself that $\hat{p}$ and $\hat{q}$ are reasonable is to overfit and prune back by testing. That is, if one method such as the ACF and PACF led to $\hat{p}$ and $\hat{q}$, one might look at the fit using an ARMA$(\hat{p}+1, \hat{q})$ or an ARMA$(\hat{p}, \hat{q}+1)$ and use a t-test to see whether the extra parameter can be set to zero. This approach is logical and simple to implement but ignores the highly collinear nature of the AR terms. As in linear regression one can calculate $R^2 = 1 - s^2/s_z^2$, where s_z^2 is the variance of the observations, or modifications of it such as the adjusted R^2, to compare various models. A caveat to this approach is seen in Chandler and Skouras (1998) in which it was noted that, in an AR(1) model, the process variance is $\gamma(0) = \sigma^2/(1 - \phi_1)^2$ and if the correct model were known it would give

$$R^2 = 1 - \sigma^2(1 - \phi_1)^2/\sigma^2 = 2\phi_1 - \phi_1^2.$$

So, for $\phi = 0.2$, $R^2 = 0.36$, which is commonly regarded as small but here represents the best that one could possibly do because of the intrinsic variability in the dependent process. It is also a fact that models with high values of R^2 do not necessarily provide good forecasts.

Given that a satisfactory choice of p and q has been made, the usual (and often ill-advised) sort of residual analysis is done conditionally on the model selection, i.e., ignoring the variability in the chosen $\hat{p}$ and $\hat{q}$, to ensure that the parameter estimates are not too far wrong. That is, if the model class $(\hat{p}, \hat{q})$ is taken to be correct, the residuals $e_t = z_t - \hat{z}_t$ plotted over time should have IID mean zero and constant variance. Here, the z_i are the data points and the $\hat{z}_t$ are the fitted values at time t using model $(\hat{p}, \hat{q})$ with parameter estimates $\hat{\theta}_1, \ldots, \hat{\theta}_q$ or $\hat{\phi}_1, \ldots, \hat{\phi}_p$, with obvious simplifications if $p = 0$ or $q = 0$. Of course, one should check that the histogram of the residuals looks normal as well, particularly as normality of the residuals is important in constructing predictive intervals. Then, the main remaining task is to check that the residuals are uncorrelated. This can be examined by plotting the sample autocorrelations

$$\hat{\rho}(k) = \frac{\sum_{t=1}^{T-k}(e_t - \bar{e})(e_{t+k} - \bar{e})}{\sum_{t=1}^{n}(e_t - \bar{e})^2}$$

as a function of k and making sure most of them land in the interval $\pm 1.96/\sqrt{T}$.

There are formal testing procedures for whether the sample autocorrelations are close enough to zero that they can be taken as zero. Many of these, like the Box–Pierce or Ljung–Box–Pierce statistics (i.e., portmanteau statistics), effectively compare a sum of squared errors to percentiles from a chi-squared distribution, essentially checking whether fitting the residuals with an IID model satisfies a goodness-of-fit criterion. This, as per the quote from Breiman that opened Part II, is likely to be a weak criterion at best.

5.2.4 Forecasting

Since the foregoing sections focused on finding an ARMA(p, q) model for $\langle Z_t \rangle$, it is worthwhile to approach the prediction of an ARIMA$(p, d, q) \times (P, D, Q)$ by starting with an ARMA(p, q) and then explaining how to extend the method to include the seasonality and the differencing meant by integration that may be in the original time series $\langle Y_t \rangle$.

So, given the successful completion of the first three stages, the fitted model can be used to generate predictions. Suppose that there are T observations $Z_1, \ldots, Z_T$ and the goal is to predict Z_{T+1}. One standard choice is

$$\hat{Z}_{T+1} = E(Z_{T+1}|Z_1, \ldots, Z_T), \tag{5.6}$$

the best predictor of Z_{T+1} under squared error loss. This can be computed if the parameters are known; more typically they are not known, so parameter estimates are just plugged into $E(Z_{T+1}|Z^T)$, giving a predictor that is often not too bad. Thus, since the earlier subsections enabled identification of $\hat{Z}_{T+1}$ and, indeed, $\hat{T}_{T+\ell}$ for ℓ-step-ahead predictions, point predictors can be generated readily. A PI is also desired. This requires a variance for the predictors; the previous subsections effectively generate that, too.

For an AR(1) process, (5.6) gives

$$\hat{Z}_{T+\ell} = v(1 - \phi_1^{\ell}) + \phi_1^{\ell} Z_T, \tag{5.7}$$

so that $\hat{Z}_{T+\ell} \to v \approx \bar{Z}$ as ℓ increases since $|\phi_1| < 1$. To get a point predictor for Z_{T+1}, substitute the estimates $\hat{\mu}$ and $\hat{\phi}_1$ and the observation z_T into (5.7).

For an MA(1) process, (5.6) gives

$$\hat{Z}_{T+1} = \mu - \theta_1 E(\epsilon_T|Z_1, \ldots, Z_T) = \mu - \theta_1 \epsilon_T \approx \mu - \theta_1 e_T. \tag{5.8}$$

In (5.8), the approximation $E(\epsilon_T|Z_1, \ldots, Z_T) \approx e_T$ was used; this is not likely to be too far wrong when it is reasonable to regard the e_T as IID outcomes of the noise distribution. (Ensuring this was part of the intention behind the residual analysis at the end of Sec. 5.2.3.) To get a point predictor, substitute the estimates $\hat{\mu}$ and $\hat{\theta}_1$ into (5.8). For $\ell > 1$, the formula is simpler:

$$\hat{Z}_{T+\ell} = \mu + E(\epsilon_{T+\ell}|Z_1, \ldots, Z_T) - \theta_1 E(\epsilon_{T+\ell-1}|Z_1, \ldots, Z_T) = \mu;$$

substitute $\hat{\mu}$ into it to get point predictions for $Z_{T+\ell}$.

Putting these two cases together leads to the predictor for a mean-zero ARMA(p, q) process. Clearly, using $Z_T = \sum_{k=1}^{p} \phi_k Z_{T-k} + \epsilon_T - \sum_{j=1}^{q} \theta_j \epsilon_{T-j}$ in (5.6) gives

$$\hat{Z}_{T+\ell} = \sum_{k=1}^{p} \phi_k \hat{Z}_{T+\ell-k} - \sum_{j=1}^{q} \theta_j E(\epsilon_{T+\ell-j} | Z_1, \ldots, Z_T), \qquad (5.9)$$

where $\hat{Z}_{T+j} = y_{T+j}$ for $j \leq 0$. The conditional expectations can be found using

$$E(\epsilon_{T+u} | Z_1, \ldots, Z_T) \approx \begin{cases} 0 & \text{if } u \geq 1, \\ e_{T+u} & \text{if } u \leq 0, \end{cases} \qquad (5.10)$$

in which $u = \ell - j$ and $e_{T+u} = e_T(u) = y_{T+u} - \hat{y}_{T+u}$. Thus, plugging in estimates of θ_j and ϕ_k and using the y_i gives point predictors iteratively, starting with $\ell = 1$ to get $\hat{z}_{T+1}$ and moving on to $\ell = 2, 3, \ldots$

Strictly speaking, under the Prequential Principle, to evaluate a predictor it is enough to have a sequence of observations and their corresponding point predictions. However, predictive variances remain important, partially because they are required to give PIs. So, recall the AR(1) process for which an ℓ-step-ahead point predictor has just been given. The ℓ-step-ahead 'predictual' is defined as $e_T(\ell) = z_{T+\ell} - \hat{z}_{T+\ell}$. One can derive $E(e_T(\ell)) = 0$ by using the MA(∞) representation of an AR(1) model, namely

$$Z_T - \nu = \sum_{k=0}^{\infty} \phi^k \epsilon_{T-k},$$

which follows from the recursive definition of Z_T and the expression

$$\begin{aligned} \hat{Z}_{T+\ell} = E(Z_{T+\ell} | Z_1^T) &= \nu + \phi(E(Z_{T+\ell-1} | Z_1^T) - \nu) + E(\epsilon_{T+\ell} | Z_1^T) \\ &= \nu + \phi(\hat{Z}_{T+\ell-1} - \nu) \\ &= \cdots = \nu + \phi^\ell(Z_T - \nu), \end{aligned}$$

from ℓ uses of the definition of an AR(1) process, where we set $Z_1^T = (Z_1, \ldots, Z_T)$ for brevity. Now, as a random variable, the predictual is

$$\begin{aligned} e_T(\ell) &= Z_{T+\ell} - \nu - \phi^\ell(Z_T - \nu) \\ &= \sum_{k=0}^{\infty} \phi^k \epsilon_{T+\ell-k} - \phi^\ell \sum_{k=0}^{\infty} \phi^k \epsilon_{T-k} \\ &= \epsilon_{T+\ell} + \phi \epsilon_{T+\ell-1} + \cdots + \phi^{\ell-1} \epsilon_{T+1}. \end{aligned}$$

Now, taking expectations on both sides gives $E(e_T(\ell)) = 0$, which means that $\hat{Z}_{T+\ell}$ is unbiased as a predictor for $Z_{T+\ell}$. Also, $\mathrm{Var}(e_T(\ell)) = \sigma^2(1 - \phi_1^{2\ell})/(1 - \phi_1^2)$, which tends to $\sigma^2/(1 - \phi_1^2)$, the marginal variance $\mathrm{Var}(Z_T)$ of an AR(1) process for any T, provided that $|\phi_1| < 1$. A prediction interval is formed by substituting estimates to give $\hat{Z}_{T+\ell} \pm \zeta_{1-\alpha/2} \sqrt{\widehat{\mathrm{Var}}(e_T(\ell))}$ as an approximate $1 - \alpha$ PI, where $\zeta_{1-\alpha/2}$ is the usual $(1 - \alpha/2)100$th percentile from an $N(0, 1)$, distribution.

Forming a PI for an MA(1) process is a little more complicated. However, as with an AR(1) process, it can be shown that, for MA(1) processes, $E(e_T(\ell)) = 0$ and hence

$\hat{Z}_{T+\ell} = \mu$. So, it is reasonable that we get the slightly modified prediction $\hat{Z}_{T+\ell} = \hat{\mu}$, since $\hat{\mu}$ is unbiased for μ. In addition, $\mathrm{Var}(e_T(1)) = \sigma^2$ and, for $\ell > 1$, $\mathrm{Var}(e_T(\ell)) = \mathrm{Var}(Z) = \sigma^2(1 + \theta_1^2)$. Again, PIs can be found from $\hat{Z}_{T+\ell} \pm \zeta_{1-\alpha/2}\sqrt{\widehat{\mathrm{Var}}(e_T(\ell))}$ for $\ell \geq 1$. This discussion assumes that ϵ_T is known and, in turn, $\epsilon_{T-1}, \epsilon_{T-2}$, and so forth down to ϵ_0. For a large class of AR models, however, the approximation $E(\epsilon_T | Z_1, \ldots, Z_T) \approx e_T$ will be valid when T is large. (The required condition is called invertibility but is not discussed here.) Also, the previous errors, $e_1, \ldots, e_{T-1}$, can be used as predictions of $\epsilon_1, \ldots, \epsilon_{T-1}$.

In the case of a stationary ARMA(p, q) model, parallel to the AR(1) process, it can be shown that $E(e_T(\ell)) = 0$ and $\mathrm{Var}(e_T(\ell)) = \sigma^2 \sum_{j=0}^{\ell-1} \psi_j^2$, where the ψ_j emerge from the MA(∞) representation of an ARMA model, namely, that

$$Z_T - c = \sum_{k=0}^{\infty} \psi_k \epsilon_{T-k},$$

for some sequence of weights ψ_k. For the special case where $p = q = 1$, $Z_{T+\ell} = c + \phi_1(Z_{T+\ell-1} - \mu) + \epsilon_{T+\ell} - \theta_1 \epsilon_{T+\ell-1}$ and therefore the predictor is

$$\hat{Z}_{T+\ell} = \hat{\mu} + \hat{\phi}_1(\hat{Z}_{T+\ell-1} - \hat{\mu}) + \hat{\epsilon}_{T+\ell} - \hat{\theta}_1 \hat{\epsilon}_{T+\ell-1}.$$

In this expression, $\hat{\mu}$ is the sample mean of Z_t, $t = 1, \ldots, T$ (estimating c) and $\hat{\phi}_1$ and $\hat{\theta}_1$ are LSEs (or MLEs if a likelihood can be credibly specified). The $\hat{\epsilon}$'s are the fitted values of the corresponding ϵ's from the least squares estimation in the MA(1) component of the model. More generally, in this expression, a 'hat' over a random variable indicates its conditional expectation given Z_1^n and a 'hat' over a parameter indicates an estimate of the parameter using $z_1^n = (z_1, \ldots, z_n)$. When $\ell = 1$, we have $\hat{Z}_{T+1} = \hat{\mu} + \hat{\phi}_1(Z_T - \hat{\mu}) - \hat{\theta}_1 e_T$ and this generalizes to $\ell \geq 2$, namely, $\hat{Z}_{T+\ell} = \hat{\mu} + \hat{\phi}_1^\ell(Z_T - \hat{\mu}) - \hat{\phi}_1^{\ell-1}\hat{\theta}_1 e_T$. Using the MA($\infty$) representation of an ARMA model, it can be shown that $\psi_0 = 1$ and $\psi_j = \phi_1^{j-1}(\phi_1 - \theta_1)$ for $j \geq 1$ (see Chandler and Skouras (1998)), so that in addition to $E(e_T(\ell)) = 0$ we have

$$\mathrm{Var}(e_T(\ell)) = \sigma^2 \left(1 + \frac{(\phi_1 - \theta_1)^2(1 - \phi_1^{2(\ell-1)})}{1 - \phi_1^2} \right),$$

which goes to $\sigma^2(1 + (\phi_1 - \theta_1)^2/(1 - \phi_1^2))$, the unconditional variance of the Z_t. So, θ_1 and ϕ_1 can be estimated and PIs of the form $\hat{Z}_{T+\ell} \pm \zeta_{1-\alpha/2}\sqrt{\mathrm{Var}(e_T(\ell))}$ can be found. The reasoning for other values of p and q is similar but more complicated.

Equipped with the predictors for when Z_t is an ARMA(p, q) process, consider the case where Z_t is an ARMA$(p, q) \times (P, Q)_s$ process. As discussed, one examines the ACF and PACF of Z_t at lags that are multiples of s to identify P and Q. That is, one finds P and Q values that are appropriate for the process defined by $\Phi(B^s)Z_t = \Theta(B^s)\epsilon_t$. So, one chooses P and Q to make the sample ACF of Z_t (i.e., $\hat{\rho}(ks)$ for $k = 1, 2, \ldots$) compatible with the ACF of an ARMA(P, Q) process. Then, p and q are chosen by matching $\hat{\rho}(1), \ldots, \hat{\rho}(s - 1)$ with the ACF of an ARMA(p, q) process. (This should be corroborated by the use of some criterion such as AIC and verified by goodness-of-fit testing.) In this fashion, one can identify p, q, P, and Q, so that the parameters in a SARMA$(p, q) \times (P, Q)_s$ process (including σ^2) can be estimated by an extension of the MLE or LSE procedures, as discussed in Sec. 5.2.2, to the case where seasonal polynomials are present. Now, an obvious extension of the techniques for prediction in ARMA(p,q) processes can be used for SARMA$(p,q) \times (P,Q)_s$ processes.

Finally, it is important to bring in integration. First, suppose that Z_t follows an ARIMA(p,d,q) model in which p, d, and q are known. Following Chandler and Skouras (1998), there are six steps.

1. Write the model equation for Z_t.
2. Take conditional expectations given Z_1^T.
3. Find $\hat{Z}_{T+1}$ using

$$\hat{Z}_{T+j} = Z_{T+j} \quad \text{for} \quad j \leq 0,$$
$$E(\epsilon_{T+j}|Z_1^T) = 0 \quad \text{for} \quad j \geq 1,$$
$$\hat{E}(\epsilon_{T+j}|Z_1^T) = e_{T+j} \quad \text{for} \quad j \leq 0.$$

4. Find $\hat{Z}_{T+2}$, $\hat{Z}_{T+3}$, and so forth using the same relationships and the values of $\hat{Z}_{T+j}$ already found.
5. Find the expressions for the parameters in the MA(∞) representation of the ARIMA model in terms of the ϕ's and θ's of the underlying ARMA model. (This can be done most easily using the backshift operator notation.)
6. Use $E(\epsilon_T(\ell)) = 0$ and derive Var$(e_T(\ell))$ in terms of the parameters from the MA(∞) representation of the ARIMA model.

(Technically, the MA(∞) representation does not exist for nonstationary models; however, when making forecasts exactly ℓ time steps ahead, one can proceed as if it were valid.) These six steps were followed in the earlier derivations for MA, AR and ARMA models but were not made explicit.

Implementing this procedure for an ARIMA$(0,1,1)$ model, write

$$(1 - B)Z_{T+\ell} = (1 - \theta B)\epsilon_{T+\ell}.$$

as the model equation, so that

$$E(Z_{T+\ell}|Z_1^T) = E(Z_{T+\ell-1}|Z_1^T) + E(\epsilon_{T+\ell}|Z_1^T) - \theta E(\epsilon_{T+\ell-1}|Z_1^T).$$

Now, $\hat{Z}_{T+1} = Z_T - \theta\epsilon_T$ and, for $\ell \geq 2$, $\hat{Z}_{T+\ell} = \hat{Z}_{T+\ell-1} - \theta\epsilon_{T+\ell-1}$. The MA$(\infty)$ representation of $(1-B)Z_t = (1 - \theta B)\epsilon_t$ is

$$Z_t = (1 - \theta B)(1 - B)^{-1}\epsilon_t = (1 - \theta B)\left(\sum_{j=0}^{\infty} B^j\right)\epsilon_t = \epsilon_t + \sum_{j-1}^{\infty}(1-\theta)\epsilon_{t-j},$$

on using a Taylor expansion of $1/(1 - x)$ for B. Hence, the parameters are $\psi_0 = 1$ and, for $j \geq 1$, $\psi_j = 1 - \theta$. So, $e_t(\ell) = \sum_{j=0}^{\ell-1} \psi_j \epsilon_{t+\ell-j}$ implies that $E(e_T(\ell) = 0$ and Var$(e_T(\ell)) = \sigma^2(1 + (\ell - 1)(1 - \theta^2))$, which can be estimated as above.

An analogous derivation can be done for an ARIMA$(1,1,0)$ model but is more complicated. As before, the PIs are given by $\hat{Z}_{T+\ell} \pm \zeta_{1-\alpha/2}\sqrt{\widehat{\text{Var}}(e_T(\ell))}$. As is typical of ARIMA processes, the variance of the forecast error increases without bound as $\ell \to \infty$. Essentially the same procedure extends to SARIMA models. The only difference is that the model equation is more complicated and the derivations consequently longer and harder.

5.3 Bayes SARIMA

The Bayesian approach to time series began with Zellner and Geisel (1970) and Zellner (1971); Box and Jenkins (1970) also devoted a section to the Bayesian treatment. Here, the task is to explain how to do forecasting when a Bayesian analysis is applied to a SARIMA$(p, d, q) \times (P, D, Q)_s$ model. In fact, one can regard s either as being chosen on the basis of modeling information or to be estimated. Similarly, one can regard the level of differencing (integration) (d, D) required for the nonseasonal and seasonal components as either (i) the result of a transformation of the initial time series and the requirement to choose among them as a collection of possible models, or (ii) to be estimated.

For simplicity, as in Sec. 5.2, it will be assumed that s is known from modeling and that the choice of D and d can be made either by the use of, say, a BIC or other model selection criterion or by using a prior on D and d and its mode, or even the ACF and PACF, even though this is not Bayesian. Thus, the SARIMA$(p, d, q) \times (P, D, Q)_s$ is converted to an ARMA$(p + sP, q + sQ) = $ ARMA(p^*, q^*) model apart from the differencing by D and d. So, given that D and d are chosen, one can do a Bayesian analysis of ARMA(p^*, q^*) and correct for the differencing. For instance, in a SARIMA$(p, d, q) \times (P, D, Q)_s$ model with $D = 0$ and $d = 1$, the differenced time series can be written as $D_t = Y_t - Y_{t-1}$ where the Y_t are ARMA$(p + sP, q + sQ)$. Then, the Bayesian ARMA$(p + sP, q + sQ)$ forecasts for $D_{T+\ell}$, such as $\hat{D}_{T+1} = \hat{Y}_{T+1} - Y_T$ when $\ell = 1$, give $\hat{Y}_{T+1} = \hat{D}_{T+1} + y_T$. This approach extends to $\ell \geq 2$, $d \geq 2$, and $D \geq 1$; see Chandler and Scott (2010, Sec. 5.2.1). A fully Bayes approach would use a prior on s, D, and d, but such a treatment does not seem to have been published. However, see West and Harrison (1997) for a more general Bayesian treatment, which includes an implicit presentation of a Bayesian SARIMA$(p, d, q) \times (P, D, Q)_s$ as a special case of a dynamic linear model (still assuming that s, d, and D are known).

Consequently, here it will suffice to present the predictors from the Bayes analysis of general ARMA(p, q) models. The first explicit and complete analysis is attributed to Monahan (1983); this section concludes with a (very) brief discussion of related work.

To present the method of Monahan (1983), write the ARMA(p, q) process as

$$(Y_t - \mu) - \phi_1(Y_{t-1} - \mu) - \cdots - \phi_p(Y_{t-p} - \mu) = \epsilon_t - \theta_1 \epsilon_{t-1} - \cdots - \theta_q \epsilon_{t-q},$$

in which the ϵ_t are IID $N(0, \sigma^2)$ and $t = -\infty, \ldots, -1, 0, 1, \ldots, \infty$. Centering by μ is included explicitly since $\mu = E(Y_t)$ is one of the parameters to be estimated. Suppose that a finite string $Y_1^T = (Y_1, \ldots, Y_T)$ is available; it can be shown that $Y_1^T \sim$ MVN$(\mu \mathbf{1}_T, \sigma^2 A_T)$, where A_T is the $T \times T$ matrix with (i, j)th elements given by $a_{i,j} = \text{Cov}(Y_i, Y_j) = \rho(|i - j|)$. It can be seen that ρ is a function of $\phi_1, \ldots, \phi_p$ and $\theta_1, \ldots, \theta_q$.

Now, suppose that the goal is to predict $Y_{T+1}^{T+\ell} = (Y_{T+1}, \ldots, Y_{T+\ell})^T$, i.e., to forecast ℓ steps ahead from T. The joint normal distribution for $(Y_1, \ldots, Y_{T+\ell})$ with variance matrix $A_{T+\ell}$ can be partitioned into blocks as

$$A_{T+\ell} = \begin{pmatrix} A_T & A_{12} \\ A_{21} & A_\ell \end{pmatrix}. \tag{5.11}$$

If we set $A_{T:\ell} = A_\ell - A_{21} A_T^{-1} A_{12}$, standard normal theory gives that $(Y_{T+1}^{T+\ell} | Y_1^T)$ is a ℓ-dimensional normal with mean $\mu \mathbf{1}_\ell + A_{21} A_T^{-1}(Y_1^T - \mu \mathbf{1}_T)$ and covariance matrix $A_{T:\ell}$. Provided that the time series is stationary (i.e., $1 - \sum \phi_i z^i = 0$ has roots outside the unit circle in $\mathbb{C}$) and invertible (i.e., the roots of $1 - \sum \theta_i z^i = 0$ lie on or outside the unit circle in $\mathbb{C}$, allowing an MA process to be represented as an $AR(\infty)$ process), there will be unique values μ, σ, p, $\phi_1, \ldots, \phi_p$, q and $\theta_1, \ldots, \theta_q$ that identify the true model.

Given unique parametrizations, the remaining tasks in a full Bayes analysis are to specify the prior distribution for $(p, q, \phi_1, \ldots, \phi_p, \theta_1, \ldots, \theta_q, \mu, \sigma^2)$ and to obtain the posterior. To begin the prior specification, order the possible pairs (p, q) by dictionary order on p for each fixed value of $p + q$. Thus, for $p + q = 0$, there is only one possibility, $(0, 0)$. For $p + q = 1$, there are two possibilities and the dictionary order is $(0, 1)$ and $(1, 0)$. For $p + q = 2$, dictionary order gives $(0, 2)$, $(1, 1)$, and $(2, 0)$, and so forth. In principle, any probability mass function with support equal to all pairs of nonnegative integers will permit inferences; choose one and denote it generically by $w(p, q)$ (with respect to counting measure).

Conditionally on the choice of p and q, a prior $w(\phi_1^p, \theta_1^q | p, q)$ must be specified on a subset of the ϕ_1^p and θ_1^q for which the process $\langle Y_t \rangle$ is stationary and invertible. The constraint on the values of ϕ_1^p and θ_1^q may be regarded as part of the prior information because they are necessary for the identifiability of a stable model. Given a choice for $w(\phi_1^p, \theta_1^q | p, q)$ it remains to select priors for μ and $r = 1/\sigma^2$ conditionally on p, q, ϕ_1^p, and θ_1^q. One standard choice is

$$w(r | p, q, \phi_1^p, \theta_1^q) \sim \text{gamma}(\alpha, \beta) \quad \text{and} \quad w(\mu | r, p, q, \phi_1^p, \theta_1^q) \sim N(\gamma, 1/\tau r),$$

in which α, β, γ, and τ are hyperparameters.

Now,

$$(Y_1^T | p, q, \phi_1^p, \theta_1^q, \mu, r) \sim N(\mu \mathbf{1}_T, A_T/r).$$

That is, conditionally on the ARMA process and the other parameters, Y_1^T has a multivariate normal density with a conjugate prior on the reciprocal of its variance. So, integration over μ and r should give a multivariate t-distribution. Indeed, from Monahan (1983), the result is the T-dimensional distribution

$$(Y_1^T | p, q, \phi_1^p, \theta_1^q) \sim \text{MVSt}_{2\alpha}\left(\gamma \mathbf{1}_T, (\alpha/\beta)\left(A_T + \mathbf{1}_T \mathbf{1}_T^T/\tau\right)^{-1}\right),$$

denoted $p(y_1^T | p, q, \phi_1^p, \theta_1^q)$, where the subscript 2α gives the degrees of freedom and the arguments of $MVSt$ are the location and scaling; see Sec. 4.2. Similar arguments on the conditional probability of $Y_{T+1}^{n+\ell}$ given Y_1^T yield

$$(Y_{T+1}^{T+\ell} | Y_1^T = y_1^T, p, q, \phi_1^p, \theta_1^q)$$
$$\sim \text{MVSt}_{2\alpha+T}(\gamma^* \mathbf{1}_T + A_{21} A_T^{-1}(y_1^T - \mu \mathbf{1}_T), (2\alpha + (T/2)\beta^*)\left(A_{T:\ell} + \tau^{-*} \mathbf{1}_\ell \mathbf{1}_\ell^T)^{-1}\right),$$

(5.12)

denoted $p(Y_{T+1}^{T+\ell} | y_1^T, p, q, \phi_1^p, \theta_1^q)$, where

$$\gamma^* = (\gamma \tau + \mathbf{1}_\ell^T A_T^{-1} \mathbf{1}_\ell)/\tau^*,$$
$$\tau^* = 1/\tau^{-*} = \tau + \mathbf{1}_T^T A_T^{-1} \mathbf{1}_T,$$

(note that the definition of τ^{-*} is implicit) and

$$\beta^* = \beta + (y_1^T - \gamma \mathbf{1}_T)^T (A_T + \tau^{-1} \mathbf{1}_T \mathbf{1}_T^T/\tau)^{-1}(y_1^T - \gamma \mathbf{1}_T);$$

see Monahan (1983) for statements, and Monahan (1980) for details.

Now, the Bayesian forecaster who believes that the relative entropy loss is relevant uses the predictive density (2.16); this is an easy extension of the optimality of (2.15) established in Aitchison (1975). In general, since the marginal of $Y_1^{T+\ell}$ is

$$p(y_1^T|p,q) = \int w(\phi_1^p,\theta_1^q|p,q)p(y_1^T|p,q,\phi_1^p,\theta_1^q)d\phi_1^p d\theta_1^q$$

Bayes' rule gives

$$w(\phi_1^p,\theta_1^q|p,q,y_1^T) = \frac{w(\phi_1^p,\theta_1^q|p,q)p(y_1^T|p,q,\phi_1^p,\theta_1^q)}{p(y_1^T|p,q)},$$

and

$$w(p,q|y_1^T) = \frac{w(p,q)p(y_1^T|p,q)}{\sum_{p,q} w(p,q)p(y_1^T|p,q)}.$$

Finally, the predictive density $p(y_{T+1}^{n+\ell}|y_1^T)$ equals

$$\sum_{p,q} w(p,q|y_1^T) \int p(y_{T+1}^{T+\ell}|y_1^T,p,q,\phi_1^p,\theta_1^q)w(\phi_1^p,\theta_1^q|y_1^T,p,q)d\phi_1^p d\theta_1^q. \qquad (5.13)$$

Of course, in the special case of (5.12) the dependence of the predictive density on p, q, ϕ_1^p, and θ_1^q drops out (except arguably in A_{12}) and, to the extent that explicit forms for these distributions can be derived, they are given in Monahan (1980). In general, however, (5.13) is an average over models and the uncertainty in p and q automatically affects the width of the prediction intervals.

Computational approaches to evaluating these numerically are given in Monahan (1983), but Bayesian computing has advanced much beyond then and so is able to accommodate more general prior and likelihood assignments that do not lead to closed form expressions for posterior densities. For instance, a more recent application of Bayesian SARIMA is found in Zhang *et al.* (2004). Indeed, contemporary techniques such as MCMC for order uncertainty in ARMA(p,q) models have been developed and are used but are not discussed here; see Philippe (2006) and Ehlers and Brooks (2013).

A key problem with any time series analysis is that the model fitting, i.e., estimation of the parameters, is confounded with the definition of outliers. That is, a given data point may be an outlier for one fitted model but not for another. Consequently, the identification of outliers or other influential data points is dependent on the parameter estimates that they influence. A weak form of the desired robustness is limited to the prior and was examined by Zellner and Williams (1973), who used different priors to represent different economic assumptions in order to evaluate the effect on inferences. A more general version of this was taken up by Barnett *et al.* (1997), who proposed a robust version of the Monahan procedure above based on writing $Z_t = Y_t + O_t$, where Y_t is the usual ARMA(p,q) series but only Z_t is observed. The additive term O_t is an outlier. An added feature of the setup in Barnett *et al.* (1997) is that there is a positive prior probability that some θ's and ϕ's are zero. The outliers O_t and the ϵ's in Y_t are permitted to have distributions given as finite mixtures of normals so, while the tails are not heavier, the effective range is larger.

Of course, in predicting time series one must bear in mind that there may be shocks not included in the Box–Jenkins approach or its standard extensions, as well as the fact that time series may change character after such a shock. That is, the models presented in this chapter may make good predictions over only a very short time period during which it is safe to assume there has been no change in the DG. Indeed, even if one regards the DG as including the generation of shocks that change the character of a finite sequence of data observed, it may be practically impossible to predict the shocks with much reliability let alone their consequences. A nice example is rainfall data. One can predict (with uncertainty) when a shock such as heavy rain will occur and try to build this into a time series model, at least in principle. It is much harder to predict when a scientific innovation will occur that will substantially change an economy. Aside from the difference in time scales (days versus years), one might regard the first of these problems as $\mathcal{M}$-complete and the second as $\mathcal{M}$-open.

5.4 Computed Examples

Three classes of time series techniques have been described so far from a predictive standpoint: (i) classical decomposition, (ii) frequentist SARIMA, and (iii) Bayes SARIMA. These three classes of techniques generate predictions that can be evaluated for accuracy. Following the stance of this text – that the vast majority of real problems are $\mathcal{M}$-complete, and of dubious approximability, or are genuinely $\mathcal{M}$-open and hence more amenable to a predictive approach than any other – our comparison of techniques will be on a real data set that seems to be far from the simple end of $\mathcal{M}$-complete problems.

The data set was generated by the cosmic ray soil-moisture-observing system (COSMOS), an NSF-supported project to measure soil moisture over hectometers using cosmic ray neutrons. The data is available as the CORR column at `http://cosmos.hwr .arizona.edu/Probes/StationDat/011/corcounts.txt`. The values in the time series represent corrected hourly neutron counts collected at a station. The corrections are made to account for changes in barometric pressure and incoming neutron intensity; see Zreda (2012) for details. The remaining time series values are composed mostly of soil moisture changes. The data starts on 2 June 2010 and ends on 26 November 2013. Since weather patterns have an annual cycle there are 2.5 cycles (years) of data that can be used to predict the data points between 3 June 2013 and 26 November 2013. Computing was done in R (R Core Team 2016), using the packages TTR and forecast.

The simplest analysis uses classical decomposition. First, logarithms (to the base 10) of the data were taken, then the transformed data was centered. The top panel of Fig. 5.1 shows a simple classical decomposition using a moving average with a 365-day window; this is the only moving average that is reasonable since the time series has a seasonality of one year. As a consequence, it is seen that the annual trend and residuals exist only on the middle portion of the range. Within these regions, the trend has a bump, possibly corresponding to winter, and the residuals are flat, apart from edge effects, meaning that the trend and seasonality essentially explain everything in the data. Off the mid-range the seasonal component is replicated, and it would be reasonable if the trend were replicated as well. In this way, one could generate point predictions for future time steps: just take the trend plus the seasonal effect and undo the logarithm. It is less clear how to get PIs since the process of obtaining the trend and seasonal components is independent of the error term required to get PIs.

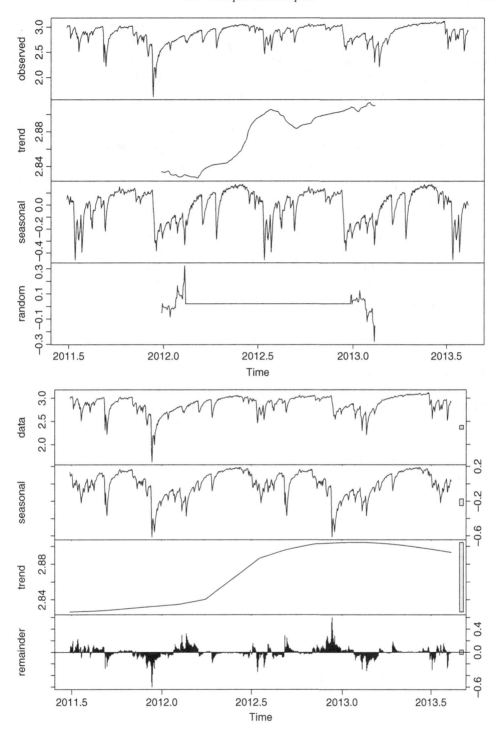

Figure 5.1 Top: Classical decomposition of the neutron count data using a 365-day moving average trend. Bottom: Classical decomposition for the neutron count data using a loess trend. Note that the plots of the seasonal and trend components are reversed in the bottom plot as compared with the top plot.

The bottom panel of Fig. 5.1 shows the result if 'loess' is used instead of a 365-day moving average. The seasonal component is found by loess-smoothing the seasonal subseries (the series of all January values, all February values, and so forth). The seasonal values are removed and the remainder smoothed to find the trend. The overall level is removed from the seasonal component and added to the trend component. This process is iterated a few times. The remainder component is the residuals from the seasonal-plus-trend fit.

The result is that there are fewer problems at the edges of the time series because a smaller window can be seen as reasonable. The seasonal effect is seen to replicate as before but the trend extends beyond a single year and the residuals are nontrivial. To make a prediction for the next time step, one would use the seasonal component plus the extrapolation of the trend component. Overall, this classical decomposition is less unsuccessful than the first.

However, in both cases – the 365-day MA and the loess-based decompositions – it is seen that the variability in the data ends up in the seasonal component; this is evidence that the model has failed. In addition, the trends may be meaningful but are hard to interpret: in the 365-day MA the trend at the end of the year is quite high while at the beginning of the next year it is very low. Indeed, the trend is continually increasing apart from an outsized bump in the center that is possibly not physically plausible. In the loess-smoothed version the trend is similar but there is no hump in the middle, just a smooth increase followed by a slight decrease. If valid, this suggests there are seasonal effects (in the statistical sense) that are much longer than one year. Overall, neither of these classical decompositions can be expected to yield predictors worth examining. Otherwise put, these models are so much less complex than the data that they are unable to represent any aspect of the DG stably enough to permit useful prediction. This is supported by Fig. 5.2, which shows the autocorrelation function from the residuals in the loess-trend classical decomposition. It is seen that the values do not fall below the bound that would indicate independence, even for relatively large lags.

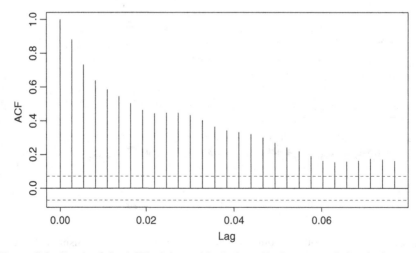

Figure 5.2 Graph of the ACF of the residuals from the loess-trend classical decomposition. The dashed lines indicate the normal cutoffs for taking the ACF as zero.

Turning next to a frequentist SARIMA predictive analysis of the same data, observe that there are a little over three years of data points. If the data are seasonally differenced once, i.e., year on year so that $D = 1$, there are two years of differenced data on which to base a predictor plus the extra data from 3 June 2013 and 26 November 2013 to use for evaluating a predictor. Differencing the two years' worth of differenced data once, i.e., day on day so that $d = 1$, means that one data point at the beginning of the seasonally differenced data is lost. The resulting graph of the doubly differenced observations looks stable, so it is not unreasonable to choose $D = d = 1$. Given the limited data available, this is fortunate. As data accumulates, however, larger values of D or d might be appropriate. One natural graph to plot is the ACF/PACF of the data for which $D = d = 1$, on the implicit assumption that the data is $ARMA(p, q)$ for some choice of p and q. The result is shown in Fig. 5.3. Although technically the ACF graph is only valid for inferring p when $q = 0$ and the PACF is only valid for inferring q when $p = 0$, these graphs are often used to get an indication of reasonable choices for p and q. The ACF suggests $p = 5$ and the PACF suggests $q = 5$.

With this in hand, one can fit an $ARMA(p, q)$ to the two years of doubly differenced data, letting (p, q) range from $(1, 1)$ to $(6, 6)$. If this is done, the AIC chooses $ARMA(3, 4)$ and the corrected AIC chooses $(3, 5)$. These models are different but as close as two different models can be in terms of ARs and MAs. Indeed, the forecast errors are also very similar. It is seen that the orders chosen by AIC are strictly less than those indicated by ACF/PACF. If the predictors from the two models, i.e., the predictors from the doubly differenced data, are found and the differencing reversed to give predictions on the original time scale,[1] the

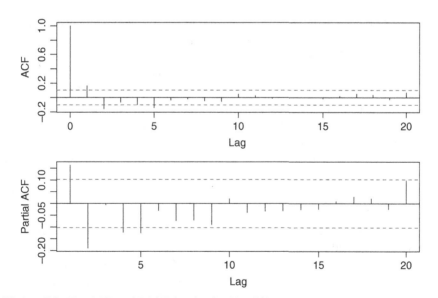

Figure 5.3 The ACF and PACF for the doubly differenced neutron count data. The dashed lines indicate the normal cutoffs for taking the function as zero.

---

[1] If d_t is the doubly differenced time series then $d_t = d_t^s - d_{t-1}^s = (y_{t+365} - y_t) - (y_{t+365-1} - y_{t-1})$, where d^s is the seasonally differenced time series. So, fitting an $ARMA(p, q)$ and getting a fitted value $\hat{d}_t$ means that the differencing can be reversed to give the predictor $\hat{y}_{t+365+1} = y_{t+365} + y_{t+1} - y_t + \hat{d}_t$ for $y_{t+365+1}$.

Residuals from ARMA(3,4)

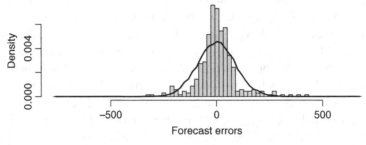

Residuals from ARMA(3,5)

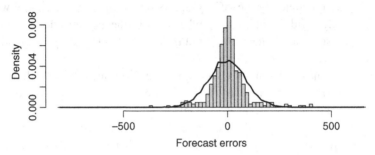

Figure 5.4 Histogram of residuals from ARMA(3, 4) and ARMA(3, 5) for the neutron count data. The curve is a kernel density estimate.

residuals can be graphed as in Fig. 5.4. The difference between the two histograms is very small, indicating that the two models fit equally well.

If predictor reselection is permitted after each time step in the period 3 June 2013 – 26 November 2013 then very different behavior is found. That is, suppose that the predictor is updated after the receipt of each new data point over 3 June 2013 – 26 November 2013, in the sense that the data are redifferenced, (p, q) is rechosen, the parameters re-estimated, the predictor on the timescale of the observations is found, and the updated predictor is used to predict the next observation. Then, Fig. 5.5 shows the results. At all stages, the BIC chooses $(p, q) = (1, 2)$. When the AIC is used to find p, p varies from 2 to 5; when the AIC is used to find q, q varies from 1 to 5. As the data accumulate, the values of (p, q) chosen by AIC converge to $(1, 2)$, the value chosen by BIC. (In this example the results of AIC and of the corrected AIC, AICc, are the same.) Thus, even though AIC has a geater variability as a model selection principle than BIC does, in a sequential predictive sense both give different models than that obtained by simply fitting the data, which gave $(p, q) = (3, 4), (3, 5)$.

The observations and point predictors from the ARMA(1, 2) predictor ($D = d = 1$) are given in the upper panel of Fig. 5.6 for the 3 June 2013 – 26 November 2013 data. The lower panel of Fig. 5.6 shows the 95% PIs around the point predictors. The point predictors and PIs track the observations, but not well: much more than 5% of the data points are outside the prediction limits.

For contrast, if an ARMA(p, q) is fitted to all the doubly differenced data, the AIC chooses ARMA(6, 6) while AICc chooses ARMA(1, 2) – the same model as that found by the BIC and prequentially (in the limit) by AIC and AICc. That is, prequentially, the

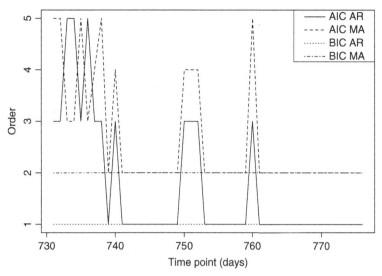

Figure 5.5 Results of sequentially refinding the predictor as the data accumulate. Both AIC and BIC converge to ARMA(1, 2) but the BIC has converged before any predictions were made.

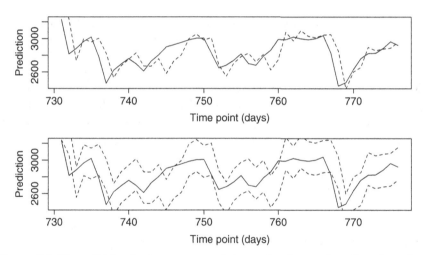

Figure 5.6 The solid line in the upper panel shows the observations from June 3 to November 26, 2013. The ARMA(1,2) $D = d = 1$ point predictors are indicated by the dashed line. The lower panel shows the observations together with the 95% PIs.

AIC finds ARMA(1, 2) but, in terms of fit, overparameterizes by finding an ARMA(6, 6). The AICc prequentially converges to an ARMA(1, 2) and this is found by using all the data as well. Only the BIC actually uncovers the limiting model quickly in a prequential sense. Figure 5.7 shows the residuals from these two models. The Box–Ljung test suggests that ARMA(6, 6) fits better but ARMA(1, 2) predicts better, suggesting that ARMA(6, 6) is an overfit.

The overall import of these results is that AIC and AICc tend to overfit on the first two years of doubly differenced data, which give ARMA(3, 4) and ARMA(3, 5). (Using

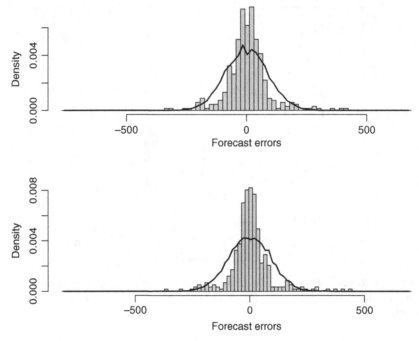

Figure 5.7 Upper panel: Residual plot from the predictor using ARMA(6, 6).
Lower panel: Residual plot from the predictor using ARMA(1, 2). The smooth line
is a kernel density estimate.

ACF/PACF gives an even worse overfit.) Also, one can find the best SARIMA class predictor using the BIC or the AICc if the sample size is large enough; AICc needs a larger sample size than BIC. The convergence of the predictors chosen by AIC, AICc, and BIC can be seen in Fig. 5.5. Even so, the predictor selected does not really perform very well. So, although the BIC converges fast, probably to the best predictor in the SARIMA class, the data are too complex to be predicted well. If all the data are used then the AIC overfit is even worse; an ARMA(6, 6) is found. By contrast, if all the data is used prequentially rather than as a batch, the AIC actually finds ARMA(1, 2). That is, if one is going to use the AIC, it may be better to proceed prequentially, i.e., use repeated reselection as data accumulate to ensure good prediction. For the AICc and BIC, proceeding prequentially did not help but does not lead to problems either.

There are several ways to do Bayesian SARIMA. One way is to regard these ways as dynamic linear models (DLMs), also called state space models. This topic is treated briefly in Sec. 5.6.2. In a DLM there are two equations, one for the observation and one for an underlying process on the parameters that are used to model the observation in terms of explanatory variables. The SARIMA models can be regarded as a subset of DLMs in the sense that they can be converted to 'state space' form. The problem is that SARIMA models are error driven, i.e., the observed variability is the result of AR and MA processes that are essentially error terms. So, SARIMA models essentially correspond to the second of the two DLM equations. One can 'fool' software such as the DLM package dlm in R to do Bayesian SARIMA by a clever choice of priors but the results are not very helpful, because SARIMAs, as a subset of DLMs, are not really natural. They do not correspond to the way analyses are done by default in dlm, because dlm (for instance) only allows a single AR term (with coefficient one) and diagonal variance matrices on the MA terms. The result is that the predictors are always simply the mean and the PIs become independent of the mean and sample size, being just normal percentiles based on how many terms are included (and the effect of the prior). In fact, one can avoid dlm and include explanatory variables in the mean of a time series; see Secs. 5.6 and 5.6.1 for a brief description, but this can be more difficult than the DLM formulation.

Here, for the sake of a simple example, the results of using a package such as dlm can be obtained more readily merely by coding the technique of Monahan (1983), as was described in detail in Sec. 5.3. So, assume the same double differencing (seasonal and day-to-day) as in the frequentist case ($D = d = 1$) to get a time series that looks stationary. Then, to use the Bayesian analysis of an ARMA(p, q) model from Sec. 5.3, it is reasonable to center the data so that $\gamma = 0$ and $\tau = \infty$ in the prior specification for μ. This means that $\gamma^* = \tau^{-*} = \mu = 0$ in (5.12). That is, the predictive distribution for the next outcome depends only on α, β, and $A_{T+\ell}$. If $A_{T+\ell}$ is estimated in the usual way (apart from setting the upper right and lower left entries to zero, as needed if one assumes that autocorrelations and partial autocorrelations are zero for large enough lags) then (5.12) is a function of the prior selection on the precision r; the dependence on p, q, ϕ_1^p, and θ_1^q drops out.

Undoing the double differencing and choosing $(\alpha, \beta) = (10, 100\,000)$ gives the top panel in Fig. 5.8 for the June 3 – November 26, 2013 data. It is seen that the bounds are unrealistically tight. If $(\alpha, \beta) = (10, 1000)$ then the result is the more reasonable bounds seen in the middle panel of Fig. 5.8. In both cases, the solid black line is the actual data and the dashed line that tracks it is the point predictor. The dotted lines indicate the 95% PI. The lowest panel in Fig. 5.8 shows the difference between the point predictor and the data. Note that the vertical scale for this panel is smaller than for the other two panels. While slightly more than 95% of the data does not fall within the 95% PIs, the difference is not too great. Even so, one might use this predictor's performance to tune the choice of gamma prior in order to achieve 95%.

If convenient priors are not used then one could employ more sophisticated computing to get the nontrivial dependence of the predictive distribution on p, q, ϕ_1^p, and θ_1^q, or more precisely their posterior densities. Unfortunately, the dependence on the prior selection could remain high if there were not enough data, and prior-driven predictions are only as good as the prior information that goes into them.

Taken together, the application of these three techniques (classical decomposition, frequentist SARIMA, Bayes SARIMA) shows that classical techniques just cannot handle data of this complexity while frequentist and Bayes techniques have to be used very gingerly. The higher penalties of the AICc and BIC seem to give improved performance. However, the AICc converges to the AIC for large sample size, and the ranges of the number of parameters and sample sizes for which AICc is better than the AIC is unclear. Indeed, the AICc, AIC, and BIC emerge from different optimality properties, and which is best depends heavily on the problem to which they are applied – but see van Erven *et al.* (2012) for a perceptive predictive resolution of the AIC–BIC dilemma. The ambiguity about which model selection procedure to use in frequentist SARIMA is mimicked by the ambiguity in the choice of prior in Bayes SARIMA. In both cases, some sort of 'calibration' is likely to be required, when choosing a model selection procedure or prior, to ensure that the results will be satisfactory.

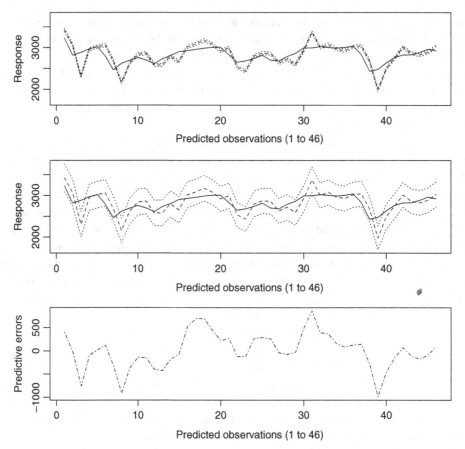

Figure 5.8 Top: The data (solid black line), point predictors (dashed line), and PIs (dotted lines) for a gamma prior on r with $(\alpha, \beta) = (10, 100\,000)$. Middle: The data, point predictors, and PIs for a gamma prior on r with $(\alpha, \beta) = (10, 1000)$ under the same conventions. Bottom: The difference between the point predictor and the stochastic data.

The notion of calibration in the predictive approach to statistical analysis has been noted several times; see Dawid (1982) and Gneiting *et al.* (2007) among others. It will be seen in the context of probability forecasting in Sec. 5.5. At root, calibration is merely the desire that a predictor should perform well as data accumulate, and so it is conceptually equivalent to looking at CPEs or FPEs and rechoosing the predictor in response to its difference from the predictand.

5.5 Stochastic Modeling

The Box–Jenkins approach to time series assumes that continuous error terms are driving the observations. However, if the time series is discrete then the error terms are not continuous, so the Box–Jenkins techniques will not apply and different techniques must be used. One way to do this is to regard a discrete time series as the outcome of a point process. Recall that a point process is a stochastic process that can be represented as a sum of Dirac δ-functions. That is, for some set of almost surely distinct random variables X_i,

$$Y = \sum_{i=1}^{N} \delta_{X_i}, \tag{5.14}$$

where N is an integer-valued random variable, the argument of Y is (ω, S), and δ is an indicator of the set $\{X_i(w) \in S\}$ in which ω is a point in the underlying measure space and S is in the Borel σ-algebra of the range of the X_i. Often the range of the X_i is $\mathbb{R}$ equipped with $\mathcal{B}(\mathcal{R})$. Thus, (5.14) is a sum of randomly many terms each of which is a random variable (being a function of the X_i). Note that, unlike in the Box–Jenkins approach, here 'time' is usually assumed to be continuous although discrete time is permitted.

The most common example of a discrete time series of the form (5.14) is a (homogeneous) Poisson process, which can be defined by three criteria; there are other equivalent definitions. Specifically, if $S = \mathbb{R}^+$ then $\langle Y(t) \rangle = \langle Y(t) \rangle_{t \geq 0}$ is a Poisson process with intensity λ if it satisfies the following.

1. *Independent increments:* For any $t_0 = 0 < t_1 < \cdots < t_k$ the random variables

$$Y(t_1) - Y(t_0), \quad Y(t_2) - Y(t_1), \quad \ldots, \quad Y(t_k) - Y(t_{k-1})$$

   are independent.
2. *Poisson probabilities:* For any $s, t \geq 0$, the distribution of $Y(s+t) - Y(s)$ is given by

$$P(Y(s+t) - Y(s) = k) = \frac{(\lambda t)e^{-\lambda t}}{k!}$$

   for any $k \in \mathbb{N}$.
3. *Initial condition:* $Y(0) = 0$.

Note that a Poisson process can be represented in the form (5.14) by defining a collection of IID exponential(λ) random variables W_i for $i = 1, 2, \ldots$, setting $S_1 = W_1, S_2 = W_1 + W_2, \ldots$ and writing $Y(t) = k \iff S_k \leq k < S_{k+1}$. Then, $Y(t) = \sum_{i=1}^{\max_j \{S_j \leq t\}} \delta_{S_i}$ and the sets are understood to be intervals of the form $[0, t]$ (a subclass of the Borel sets).

Now, consider predicting a Poisson(λ) process. First, the parameter λ must be estimated. So, suppose that the Poisson(λ) process has been observed for n (unit-length) time steps and that k_i 'arrivals' were seen in each time step. Then, from the standard theory for Poisson distributions, $\hat{\lambda} = \sum_i^n k_i / n$ is the MLE and is minimum-variance and unbiased. So, $\hat{\lambda}$ can be used to give probability forecasts, by criterion 2 (ignoring the variability in $\hat{\lambda}$). That is, the distribution of the number of events in an interval $(t, t + \ell]$ is Poisson($\ell\lambda$), so using $\hat{\lambda}$ gives a PI that can be found by using its percentiles (apart from its discreteness). If desired, a point predictor (with a variance) for the number of events in $(t, t + \ell]$ can be found by taking the mean or median of the Poisson($\ell\hat{\lambda}$). Bayesian analyses of Poisson processes and related processes do not seem to have been much developed, but see Kuo and Ghosh (2001) and Taddy and Kottas (2012).

Unfortunately, very often the Poisson process is too simple to be applied directly. One extension is to nonhomogeneous processes in which the rate is permitted to depend on t, i.e., $\lambda = \lambda(t)$. This means that one must estimate a function $\lambda(t)$, not just a parameter, a more difficult task.

An extension in a different direction is to compound Poisson processes (CPPs), in which not only are the arrival times modeled but also some feature of the event that is arriving.

For instance, the arrival times of rainfall might be modeled, but such a model would be more useful if it included the amount of rain as well as the frequency. Let $\langle Y(t) \rangle$ be a Poisson process with rate λ and consider random variables X_j assumed to be independent of each other and independent of the Poisson process. Then, $Z(t) = \sum_{j=1}^{Y(t)} X_j$ is a compound Poisson process and is continuous (so, it doesn't have a representation of the type (5.14)). That is, a compound Poisson process is the cumulative value assigned to the events that the Poisson process counts.

One classical application of the compound Poisson process is to rainfall data; see Liao *et al.* (2001), the references therein, and Alexandersson (1985), in which the compound Poisson–exponential model is developed. The idea is that the number of arrivals is chosen by the Poisson process $Y(t)$ while the magnitude of each arrival is one of the X_i, i.e., a drawn IID exponential(θ). Even though the X_i are exponential they are assumed independent of any exponential random variables that might be related to the arrivals in the Poisson process, via waiting times for instance. For a fixed T this gives

$$Z(T) = X_1 + \cdots + X_{Y(T)},$$

where $Y(\cdot)$ is the Poisson process giving the random number of IID exponential(θ) X_i that are summed. This can be written in a variation of the form of (5.14) by using factors on the δ-functions:

$$Z(t) = \sum_{i=1}^{Y(t)} x_i \delta(t - T_i),$$

where x_i is a realized value of $X_i \sim$ exponential(θ) and the T_i are the arrival time sequence of the Poisson process $Y(t)$ with rate λ (per unit time).

To predict in this context one must estimate λ and θ. The simplest way is via the method of moments, although ML methods may be used as well; see Katz (2002, Sec. 2d) for details in a related context. As noted in Alexandersson (1985), Feller (1968) obtained the moment generating function (MGF) of a CPP as $\psi_Z(t) = e^{\lambda T(\psi_X(t)-1)}$, where ψ_X is the MGF of any of the X_i. Since each X_i is exponential(θ), $\psi_X(t) = 1/(1 - \theta t)$ giving

$$\psi_Z(t) = e^{\lambda T \theta t/(1-\theta t)},$$

from which cumulants can be derived, leading to the expressions $E(Y) = \lambda T \theta$ and $\text{Var}(Y) = 2\lambda T \theta^2$ where T is known. Given the sample mean and variance of the CPP, the results are $\hat{\lambda} = 2\bar{z}^2/(T s_Z^2)$ and $\hat{\theta} = s_Z^2/(2\bar{z})$.

Using these values one can make point predictions for, say, $[T, T+1]$ simply by using the Poisson process with rate λ to predict the number of events and then drawing an outcome from an exponential($\hat{\theta}$) for each arrival time of the events. The estimate s^2 can be used to give PIs of approximate confidence. And, of course, one can use the entire process to give a predictive distribution from which probabilistic forecasts can be obtained (although it seems that this is difficult to do in a convenient closed form).

Having seen that probabilistic forecasts are natural in some contexts, the question is to evaluate how good they are. In reality, no one ever *sees* a probability since it cannot be directly measured, only indirectly inferred, so it is hard to quantify the degree of agreement between a probability forecast and a future outcome. Nevertheless, there are several ways to do this. First it is important to bear in mind the criteria of sharpness and calibration. Sharpness is merely the concept of efficiency carried over the probabilistic forecasts: one wants as narrow a predictive distribution (Bayes or not) as is realistic. Calibration is a more subtle concept: it refers to the agreement between a set of forecast distributions and the outcomes they predict. For instance, one would avoid a degree of sharpness for which the calibration of the forecast distribution and the future outcomes themselves were not compatible. One intuitive calibration check is from Dawid (1984), who proposed converting a predictive distribution for a future outcome by using the probability integral transformation (PIT), evaluating it at the future observations, and checking to what extent the resulting values represent a Unif[0, 1].

Separately from this, one often invokes the concept of a proper scoring rule. The idea is that if F is the forecast distribution and y is an outcome of the actual distribution, say G, then the penalty $S(F, x)$ is minimized when $F = G$. A strictly proper scoring rule has the property that the minimum is unique. For instance, if a forecaster announces $f(\cdot)$ as the density of a predictive distribution and the outcome is y, then $\log(1/f(y))$ is a proper scoring function: if y is representative of f, then $f(y)$ is large and $\log(1/f(x))$ is small, equaling zero when f puts all its mass at x. The log scoring rule is well motivated by codelength arguments from information theory but can be unstable unless summed over many uses. (The expectation of the log scoring rule, the entropy, is typically used in information theory, thereby reducing the instability but violating the Prequential Principle.) The obvious limitation of scoring rules is that most of them make sense only for $\mathcal{M}$-closed problems and some $\mathcal{M}$-complete problems – those for which the DG can be approximated well by a sequence of $\mathcal{M}$-closed problems. The scoring rule approach breaks down for $\mathcal{M}$-open problems and for $\mathcal{M}$-complete problems that do not have a good approximating sequence. Nevertheless, Gneiting *et al.* (2007) and Czado *et al.* (2009) gave some useful discussion of the choice of scoring rules and examples of their use, as did Gneiting (2011) – though in some cases scoring rules violate the Prequential Principle if expectations are taken under an assumed true model. Note that this work focuses on finding a good way to evaluate probabilistic predictors rather than finding good predictors in the first place. Moreover, much of this work emanates from Dawid (1982) and is more abstract; it was done before the necessary computing power was available.

To demonstrate the evaluation of probability forecasts, consider two predictors for rainfall data. First, the data was obtained from NOAA's National Climatic Data Center and consists of 58 years of rainfall data, 1954–2012, recorded at the Tucson International Airport. There was no particular reason to choose this location; however, it does have data with essentially no missing values. The annual precipitation, given in hundredths of an inch, is provided along with the number of times there was a rainfall (i.e., the number of days in each year with rainfall at least 0.1 inches). The data can be obtained from http:www.ncds.noaa.gov/cdo-web. The exact amount of rain at each rainfall is not available; it is noted on the website, however, that a different predictor using that data might be better than either of the two predictors to be examined here.

The first rainfall predictor just uses the Poisson process. Let $Y(t)$ be a Poisson process with rate λ, where t is scaled in years, and treat each of the 58 years as an independent increment. So, the data give outcomes of $Y(1), Y(2) - Y(1), Y(3) - Y(2), \ldots$, the number of rainfalls per year. If the first 30 years are used to predict the last 28 years, the natural estimate of λ is $\hat{\lambda} = (Y(1) + (Y(2) - Y(1)) + \cdots + Y(30) - Y(29))/30$. Here, $\hat{\lambda} = 36.2$ (the number of rainfall events per year). Given the total rainfall over 30 years and the number of days it rained in each of those years, one can find $\bar{a}$, the average amount of rain per rainfall. Here, $\bar{a} = 32.6$ (in hundredths of an inch). So, the forecast distribution for the number of rainfalls per year for years $31, \ldots, 58$ is the number of occurrences from a Poisson($\hat{\lambda}$) distribution and the forecast distribution for the amount of rainfall in a year is 32.6 times a draw from a Poisson(32.6). Essentially, this assumes that the rainfall events all represent the same amount of precipitation.

How good are these (very simple) probability forecasts for the latter 28 years of data? First, since the predictor is just the factor $\bar{a}$ times a Poisson(λ), the factor drops out when one finds the probability integral transform (PIT) graph and the log scoring rule value. That is, the factor can be omitted as it just relocates the amount of rain from $\mathbb{N}$ to $\bar{a}\mathbb{N}$.

The triangles on the top left panel in Fig. 5.9 give the values $\hat{DF}(y(t) - y(t-1))$ for $t = 31, \ldots, 58$, where $y(t) - y(t-1)$ is the number of rainfalls in year t and $\hat{DF}$ is the

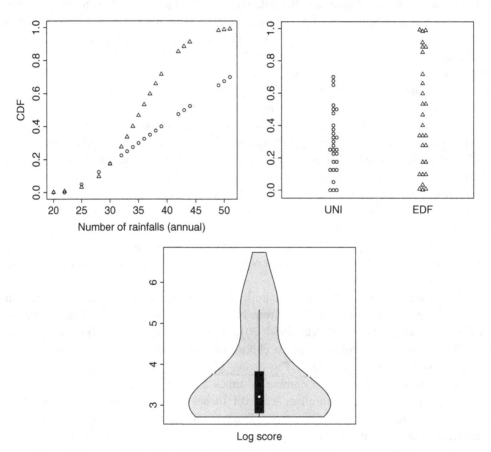

Figure 5.9 Upper left: Plot of the EDF evaluated at the amount of rainfall per year over the last 28 years and of the uniform distribution function, over the range of the number of rainfalls during the latter 28 years. Upper right: Another visualization of the PIT comparing $\hat{DF}$ evaluated at the latter 28 years of rainfall data and the uniform distribution, also over the range of rainfalls numbers. Bottom: A violin plot of the log score values using $P_{\hat{\lambda}}$.

DF of a Poisson($\hat{\lambda}$). The circles are the same except that they use a uniform DF over the range of the number of rainfalls per year in place of $\hat{DF}$. The upper right panel shows the same data points; the difference in heights in the upper right panel corresponds to the difference between the lines of triangles and circles at the right-hand side of the range on the upper left panel. The lower panel shows the results of the log scoring rule. The values $-\log P_{\hat{\lambda}}(y(t) - y(t-1))$ for $t = 31, \ldots, 58$ give the histogram depicted by the light gray region. The boxplot within the light gray region shows that over 50% of the values are under four. The white dot is the median; the mean and median over the 28 values are 3.7 and 3.2.

Now consider another predictor, based on the CPP. Let Z_t be the total rainfall during a year $t = 31, \ldots, 58$. Then, under the CPP model, $Z_t \sim \text{Erlang}(Y(t) - Y(t-1), \theta)$ where $Y(t) - Y(t-1) \sim \text{Poisson}(\lambda)$ and the $y(t) - y(t-1)$ IID outcomes are from an exponential(θ) distribution (in the mean value parametrization). Using the method of moments estimator described earlier, $\hat{\theta} = 50.5$ while the estimate for λ is $\hat{\lambda} = 23.3$, much less than the value 32.6 found for the Poisson process. With mild abuse of notation, predictions for the annual rainfall are generated from $Z_t \sim \text{Erlang}(\text{Poisson}(\hat{\lambda}), \hat{\theta})$. That is, one draws from the Poisson(23.2) distribution, uses that value as the first parameter in the Erlang distribution, and then draws from the Erlang distribution with second parameter equal to 50.5. The extra layer of variability from the Poisson adds to the variability already described by the Erlang.

Figure 5.10 shows this in practice. In the left-hand panel, the five solid lines represent the values $P_{\hat{\lambda}, \hat{\theta}}(z_t)$ for $t = 31, \ldots, 58$, for five different values drawn from the Poisson(23.3). The line of triangles represent the averages of $P_{\hat{\lambda}, \hat{\theta}}(z_t)$ for each z_t, over 1000 values drawn from the Poisson(23.3). In the right-hand panel the line of triangles is replotted along with the uniform distribution evaluated at the z_t over the range of the z_t. That is, the right-hand panel in Fig. 5.10 is analogous to the left-hand panel in Fig. 5.9, i.e., it makes more sense to compare the line of triangles (the behavior of the Erlang distribution averaged to take the Poisson variability into account) with distribution the uniform DF than to compare the DF using any one outcome that obeys the Poisson with the uniform DF.

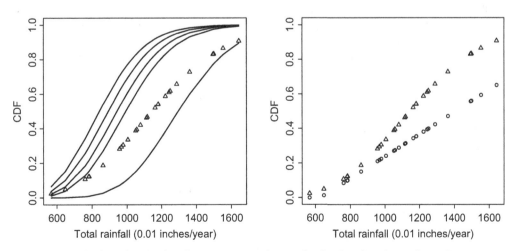

Figure 5.10 Left: The five lines represent the result of using five draws from the Poisson(23.3) to form five versions of $Z_t \sim \text{Erlang}(\text{Poisson}(23.3), 50.5)$. Many versions of these solid lines average to give the line of triangles. Right: The averaged DF using Poisson(23.3) variability (triangles) and the DF using a uniform DF for $z_{31}, \ldots, z_{58}$ over its range (circles).

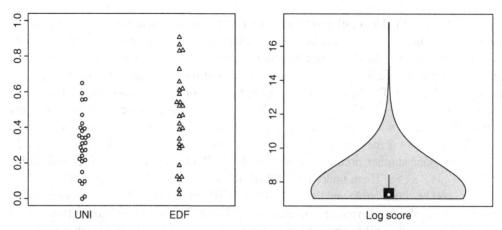

Figure 5.11 Left: Alternate visualization of the difference between the PIT for the CPP model and the uniform. Right: Violin plot of the log scores for the CPP model after averaging out over the Poisson variability.

The left-hand panel of Fig. 5.11 shows a revisualization of the right-hand panel of Fig. 5.10 (and is analogous to the top right panel of Fig. 5.9). The right-hand panel of Fig. 5.11 is analogous to the bottom panel of Fig. 5.9. In this case, the mean and median are 7.6 and 7.3.

Now, it is easy to see that in terms of the log scoring rule, where low numbers are good, that the Poisson distribution gives a better predictor than the CPP: 3.7 versus 7.6. On the other hand, visually, the PIT from the Poisson model (upper left panel of Fig. 5.9) looks further from the uniform than the PIT from the CPP (upper right panel of Fig. 5.9). That is, in one sense, the Poisson is better and in another sense, the CPP is better. A similar lack of agreement, though not as pronounced, was found in Czado *et al.* (2009).

As a generality, discrepancies between scoring rules or other ways to evaluate probability forecasts are likely to be larger in practice than discrepancies between evaluations of point predictors, because many different distributions can give the same point predictor. The implication of this is that one should use probability forecasts only when there is relatively little ambiguity in how the performance of such predictors is to be evaluated – a situation that does not seem to occur very often in practice. This general recommendation does not contradict Gneiting (2011), who investigated the variability in the selection of point predictors as a function of the sense in which pointwise errors are assessed. Again, whatever error assessment is used for point prediction, many distributions can give the same point forecast and hence there will typically be more instability in evaluating probability forecasts than in evaluating point predictors.

5.6 Endnotes: Variations and Extensions

Two further generalizations of the SARIMA framework that bear mentioning are ARMAX and ARFIMA. First, it is obvious that time series rarely occur in isolation. So, it is natural to permit the dependence of Y_n on a set of explanatory variables, say x. In the SARIMA context (and others) this is often done by replacing the constant term, e.g., the mean, whether denoted μ or ν earlier in this chapter, by Γx where Γ is a $1 \times \dim(x)$ row vector, with an obvious generalization if Y_n is, say, an r-dimensional time series (a case that has not been discussed here). This is the ARMAX model, the X standing for exogenous variables. This is clearly useful when the mean structure of a time series depends on outcomes that the time series itself does not model.

Another extension is the (S)ARFIMA model, a (seasonally) fractionally integrated ARMA. The fractional integration refers to the structure

$$\phi(B)\pi_d(B)(Y_t - \mu) = \theta(B)\epsilon_t,$$

where $d \in (-0.5, 0.5)$ and $\pi_d(B) = (1 - B)^d$ is expressed in terms of B using the Taylor series expansion for binomial expressions. Such (S)ARFIMA models are one way to incorporate long-range dependence because the Taylor series converges relatively slowly.

The rest of these endnotes look at the role of time series in linear regression but do not extend to nonlinear or other regression structures. The simplest way in which time series appear in linear regression occurs when the error term is permitted to be, say, ARMA(p, q) rather than IID or to have a dependence structure as in weighted linear regression. This is seen in the first subsection below. The second subsection looks briefly at a general regression setting from time series called dynamic linear models or state space models. Volatility models such as autoregressive conditionally heteroscedastic (ARCH) or generalized ARCH (GARCH) are beyond the present scope, though obviously of immense interest from a predictive standpoint.

5.6.1 Regression with an ARMA(p, q) Error Term

A linear model with an ARMA(p, q) error term can be given either a Bayesian or frequentist analysis. Both analyses start with

$$Y = X\beta + U, \tag{5.15}$$

where $Y = (Y_1, \ldots, Y_n)^T$, X is an $n \times k$ matrix, β is a $k \times 1$ parameter vector, and $U = (U_1, \ldots, U_n)^T$ is an $n \times 1$ vector of random outcomes from a stationary ARMA(p, q) process as in (5.3). (Note that here i rather than t indexes the sample and n rather than T is the sample size, in conformity with the usual notation for regression.) That is, one replaces the Z_i in (5.3) with the U_i in (5.15). Thus, each U_i is regarded as a linear combination of $\langle \epsilon_j \rangle_{-\infty}^i = (\epsilon_{-\infty}, \ldots, \epsilon_i)$ (using the MA(∞) representation of the AR model).

From the frequentist perspective, two estimation procedures for the parameters were studied in Zinde-Walsh and Galbraith (1991). One is an ML approach and the other is a two-stage procedure that estimates the ARMA parameters first and then uses a generalized least squares approach to estimate β. Here only their ML procedure is described. To do this requires a variance matrix for a finite string of variables in a doubly infinite stochastic vector. So, one recognizes that for a doubly infinite vector $\langle U_i \rangle|_{-\infty}^{\infty}$ the covariance matrix $\Sigma_{\infty} = E(UU^T)$ and $\langle U_i \rangle|_{-\infty}^{\infty} = \Sigma_{\infty}^{1/2} \langle \epsilon_i \rangle|_{-\infty}^{\infty}$. Since the sample size is n, an $n \times n$ block of Σ_{∞} must be extracted to represent the variance matrix of a string of n random outcomes U_i. It is enough to extract this as a block on the main diagonal of Σ_{∞}. This can be done using a doubly infinite projection matrix π_n given by the identity on the block that we want to extract with all other entries zero. Now, for a finite string $U_1, \ldots, U_n$ of the doubly infinite stochastic process, set $\Sigma_n = \pi_n \Sigma_{\infty} \pi_n$.

Then, if $\langle \epsilon_i \rangle|_{-\infty}^{\infty}$ has IID $N(0, \sigma^2)$ elements, the log-likelihood function is

$$\log L = c - \frac{1}{2} \log \det \Sigma_n - \frac{1}{2} u^T \Sigma_n^{-1} u,$$

where $u = (u_1, \ldots, u_n)^T$ and c is a constant. For fixed p and q and $\tau = (\tau_1, \ldots, \tau_{p+q}) = (\phi_1^p, \theta_1^q)$ the likelihood equations are

$$\frac{\partial \log L}{\partial \beta_k} = 0 \quad \text{and} \quad \frac{\partial \log L}{\partial \tau_m} = 0.$$

Zinde-Walsh and Galbraith (1991) showed that these are equivalent to

$$\hat{\beta}_{ML} = (X^T \Sigma_n X)^{-1} X^T \Sigma_n^{-1} y,$$

$$-\frac{1}{\det \Sigma_n} \left(\frac{\partial \det \Sigma_n}{\partial \tau_m} \right) = u_{ML}^T \frac{\partial \Sigma_n^{-1}}{\partial \tau_m} u_{ML}^T, \qquad m = 1, \ldots, p+q,$$

$$u_{ML} = y - X \hat{\beta}_{ML}. \tag{5.16}$$

For many such systems of equations, the estimates $\hat{\beta}_{ML}$ are $\sqrt{n}$-consistent, i.e., the expression $\sqrt{n}(\hat{\beta}_{ML} - \beta)$ has a nontrivial limit. When this holds, the system (5.16) is essentially a set of solvable (polynomial) constraints on the parameters β and τ. The convergence of linear estimators such as these is in mean square, i.e., L^2 on the appropriate product measure space, and hence in probability. In fact, Zinde-Walsh and Galbraith (1991) showed that a simplified form of these estimators using method-of-moments reasoning converges as well.

As in the earlier cases, a point prediction for a new input vector that is ℓ steps into the future, $X_{n+\ell}$, would be given by $\hat{Y}(X_{n+\ell}) = X_{n+\ell} \hat{\beta}_{ML}$ with prediction intervals coming from $\pm z_{1-\alpha/2} \sqrt{X_{n+\ell}^T \widehat{\text{Var}}(\hat{\beta}_{ML}) X_{n+\ell}}$ as in Sec. 4.2.

Least squares is merely one choice of loss function among many that are possible. For instance, least-absolute-deviation estimators have been studied in the ARMA(p, q) error context, and Davis and Dunsmuir (1997) established consistency and asymptotic normality for them.

A Bayesian analysis that parallels this was given in Chib and Greenberg (1994). They wrote

$$Y_i = X_i^T \beta + \epsilon_i \tag{5.17}$$

for $i = 1, \ldots, n$ where the ϵ_i follow an ARMA(p, q) as in (5.3). That is, the role of Y_n in (5.3) (see (5.2)) is played by ϵ_n and the role of ϵ_n is played by a perturbation term, say u_n, i.e.,

$$\epsilon_i = \phi_1 \epsilon_{i-1} + \cdots + u_i + \theta_1 u_{i-1} + \cdots + \theta_q u_{i-q}.$$

The error terms are sequentially dependent. The analysis in Chib and Greenberg (1994) rests on the use of a state space form described in Chandler and Scott (2010, Sec. 5.5) but adapted to a linear model. (Chib and Greenberg (1994) did not give the details of this but cited an earlier text for the result.) For the normal error case, Chib and Greenberg (1994) assumed that p and q are known. Even in this case, the likelihood cannot be written down in a closed, tractable, form. However, an expression for it can be obtained by using the conditional distributions of the predictors. Specifically,

$$p(y_i|\beta,\phi,\theta,\sigma^2;\epsilon_0,\ldots,\epsilon_{-p+1},u_0,\ldots,u_{-q+1}) = \prod_{i=1}^{n} \frac{1}{\sqrt{2\pi\sigma^2}} \exp\left(-(y_i - \hat{y}_{i|i-1})^2/2\sigma^2\right)$$

where $\hat{y}_{i|i-1}$ is the one-step-ahead prediction of Y_i, given the information up to and including step $i-1$. An explicit form was given in Chib and Greenberg (1994). Essentially, the n-dimensional normal is factored into a sequence of conditional distributions in a way that is analogous to that in (5.11) and the discussion following it. Using this, standard priors can be assigned to β, ϕ, θ, and σ^2 and conditional posterior distributions can be derived. An MCMC algorithm can be designed to generate parameter estimates. Unfortunately, Chib and Greenberg (1994) does not give the forms of the predictions that would be obtained. However, the $\hat{\beta}$ produced by their method can be used in (5.17) to get point predictions in the way one does for fixed-effect linear models. In a more Bayesian approach, PIs can be obtained from the predictive distribution using the technique in Sec. 5.3.

5.6.2 Dynamic Linear Models

In addition to numerous other generalizations, one that bears mention in closing this section is the state space model, often called the dynamic linear model (DLM) especially when its Bayesian formulation is meant. A good introductory treatment from the Bayes perspective can be found in Pole *et al.* (1994) and a more detailed treatment in West and Harrison (1997); a good introductory treatment from the frequentist perspective can be found in Shumway and Stoffer (2010, Chap. 4). The brief treatment here is a partial summary of these presentations, focusing on the predictive structure in the univariate case.

The overall DLM structure is expressed in two equations. The first is called the observation equation,

$$Y_t = x_t^T \beta_t + \epsilon_t, \tag{5.18}$$

and the second is called the system equation,

$$\beta_t = G^T \beta_{t-1} + u_t. \tag{5.19}$$

However, the system is incomplete unless initial information is specified. So, it is common to specify a value, or a distribution, for β_0. Some authors write $\mu_t = x_t^T \beta_t$ and specify $\mu_0 \sim N(m_0, C_0)$, for instance, so that the updating from time step to time step can be cleanly expressed. It is only the Y_t in (5.18) that are observed and then with error ϵ_t; the x_t are known and presumed constant in an analogy with a design matrix. The relationship between Y_t and x_t is controlled by β_t, and (5.19) means that β_t follows an autoregressive model in which the matrix G is taken as known (matrix G may be dependent on t but this case is ignored here). Usually, $\epsilon_t \sim N(0, V_t)$ and $u_t \sim N(0, W_t)$ are taken as uncorrelated, where the matrices V_t and W_t are assumed known (and often taken as $V = V_t$ and $W = W_t$). This means that the time evolution of the coefficients β_t is deterministic apart from the error u_t. Note that if $W = 0$ then we have $\beta_t = \beta$ and (5.18) is the usual model for linear regression. Indeed, any ARMAX model can be written in state space or DLM form; see Shumway and Stoffer (2010).

At least for values of t and t' that are close together, β_t and $\beta_{t'}$ are not far apart but as $t-t'$ increases, the corresponding β_t and $\beta_{t'}$ may end up far apart. Therefore, as $t-t'$ increases, the information in Y_t about $Y_{t'}$ decreases.

To see how updating, and hence forecasting, works in the DLM, suppose that

$$\beta_{t+1}|D_t \sim N(\mu_{t+1}, C_{t+1}), \qquad (5.20)$$

where D_t represents the knowledge available at time t. This includes D_0, the variances V_t, W_t for $t \geq 1$, and the values $y_1, \ldots, y_t$. Since V_t, W_t for all $t \geq 0$ are assumed known at $t = 0$ (in order to permit forecasting), $D_t = \{y_t, D_{t-1}\}$.

If $\beta_0|D_0 \sim N(\mu_0, C_0)$ and squared error loss is assumed, so that the conditional expectations given the D_t are optimal predictors, then

$$E(Y_{t+1}|D_t) = x_{t+1}^T \mu_{t+1}. \qquad (5.21)$$

Because the error terms are independent,

$$\text{Var}(Y_{t+1}|D_t) = x_{t+1}^T C_{t+1} x_{t+1} + V_{t+1}, \qquad (5.22)$$

so $(Y_{t+1}|D_t) \sim N(X_{t+1}^T \mu_{t+1}, x_{t+1}^T C_{t+1} x_{t+1} + V_{t+1})$. Forecasting two steps into the future requires an extra use of the system equation and, in general, forecasting ℓ steps into the future requires ℓ uses of the system equation to update the distribution on the β_t to $\beta_{t+\ell}$. The result is

$$\beta_{t+\ell}|D_t \sim N(\mu_t(\ell), C_t(\ell)), \qquad (5.23)$$

in which

$$\mu_t(\ell) = G^{\ell-1}\mu_{t+1}$$

and

$$C_t(\ell) = G^{\ell-1}C_{t+1}(G^{\ell-1})^T + \sum_{j=2}^{\ell} G^{\ell-j}W_{t+j}(G^{\ell-j})^T.$$

So, predictions are obtained from the observation equation for ℓ steps ahead,

$$Y_{t+\ell}|D_t \sim N(X_{t+\ell}^T \mu_i(\ell), F_{t+\ell}^T C_t(\ell)F_{t+\ell} + V_{t+\ell}). \qquad (5.24)$$

Analogous expressions for $E(Z_{t+1}^{t+k}|D_i)$ and $\text{Var}(Z_{t+1}^{t+k})$ where $Z_{t+1}^{t+k} = \sum_{j=t+1}^{t+k} Y_j$, can be derived; see West and Harrison (1997, Secs. 2.2 and 2.3), for details.

To complete the picture, note that the system equation gives the likelihood:

$$L(\beta_t|Y_t = y_t, V_t) \propto p(Y_t = y_t|\beta_t, V_t) \sim N(x_t^T\beta_t, V_t)$$

and, using this, the posterior for β_t is given by

$$p(\beta_t|D_{t-1}, y_t) = \frac{p(Y_t = y_t|\beta_i t, V_t)p(\beta_t|D_{t-1})}{p(Y_t = y_t)}. \qquad (5.25)$$

Since all three densities on the right-hand side are normal, $(\beta_t|D_{t-1}, y_i)$ is also normal and the prior on β_t can be updated to give updated forecasts for $Y_{t+\ell}$. In addition, $(\beta_{t+\ell}|D_i)$ can also be derived, to give an updated prior for $\beta_{t+\ell}$ in place of that in (5.25). See West and Harrison (1997, Chap. 4) for full details on the derivations. Note that, when nonnormal priors are used, computational techniques such as MCMC can generate predictive densities and predictors.

6

Longitudinal Data

...prediction and variable importance[1] are different goals whose optimal achievement requires the use of different tools. The objective in prediction is to specify a well defined algorithm that is capable of doing accurate predictions ... it is unnecessary [for] the intermediate calculations of the algorithm to find statistical or causal relations between the variables involved. In fact, variable importance measures defined in terms of these calculations are often inappropriate since their validity depends on correctness of the model assumed ... that may not be known. On the other hand, variable importance methods (VIM) are aimed to measure the degree to which changes in the prediction are caused by changes in each of the predictor variables ...

This difference between prediction and VIM has two main consequences. First, VIM problems are of a causal nature, whereas prediction problems are merely associational. Second, ... VIM parameters must [have] an interpretation in terms of the expected change in the outcome under a given intervention ... a meaningful interpretation can only be obtained through an intelligible characterization of [a] VIM as a statistical (or causal) parameter defined as a mapping from a honest, tenable statistical model into an Euclidean space.

Furthermore, an algorithm designed to perform well at prediction is not guaranteed to also do a good job at estimating VIM measures, because good performance is defined differently for each goal ... Performance in estimation of Euclidean parameters is assessed in terms of statistical properties like consistency and efficiency (related to bias and variance). Prediction algorithms are designed to perform well at estimating the entire regression model, resulting in an incorrect bias–variance tradeoff for each VIM measure.

Diaz *et al.* (2013, Sec. 1)

The key feature of longitudinal data is that many subjects from a population are measured repeatedly; sometimes this is called repeated measures data. This is opposed to cross-sectional data, where individuals from a population are measured at a single fixed time point. Note that although the standard terminology is based on time – repeated measures are made over time, the single measurement per subject in a cross-sectional study is made at a fixed point in time – any parameter that provided a linear order would serve just as well. For instance, the prediction of, say, pollution levels at various distances away from a source in a particular direction has the same statistical structure.

Longitudinal data can be regarded as a generalization of univariate time series data. Specifically, a single time series data set can usually be regarded as providing longitudinal data for one subject. In general, therefore, longitudinal data corresponds to many time series,

---

[1] Assessing variable importance is an important step in efforts to 'model' whether or not a model exists.

one for each of n subjects. Hence, there is a population of individuals about which one can make inferences. In time series, it is the dependence structure within the series of repeated measurements that is of primary interest. However, in longitudinal data it is usually the main effects that are of greatest interest even though in some cases the dependence structure, the population, or other aspects of the data may be important.

Longitudinal models are varied and often complex. So, since the predictive view is sometimes more difficult to implement, it is often subordinated to the goals of model fitting and checking. The hope is that inferences about the population will suffice even when predictions about next outcomes are essential for model validation. Of course, inferences about the population are typically weak – and subject to bias – because longitudinal data usually come from $\mathcal{M}$-complete and $\mathcal{M}$-open problems. Even when the goal of an analysis is prediction, practitioners usually make do with concepts of fit, i.e., the situation of being unable to disqualify a model obviously and immediately on the grounds that there is too much model uncertainty (given the sample size) to do better. In other words, in practice, when model uncertainty is high, all too often practitioners try to justify the modeling choices they have made rather than questioning the inferences. This is precisely the opposite of what reason would suggest.

One of the earliest efforts to focus on prediction in a longitudinal context was that of Rao (1987); see also the references therein. A fairly recent text on longitudinal analysis that includes a (brief) treatment of prediction is Jiang (2007); see also Fitzmaurice *et al.* (2004) and Verbeke and Molenburghs (2009). Many other authors have also thought carefully about aspects of the role of prediction in longitudinal data analysis. The goal here is to provide an overview of longitudinal prediction.

The simplest longitudinal models are based on the normal and assume all subjects are drawn from one population; later this constraint will be relaxed. For subject $i = 1, \ldots, n$, let $Y_i = (Y_{i1}, \ldots, Y_{im})$ represent $m \geq 2$ measurements and write

$$Y_i = X_i \beta + \epsilon_i \qquad \text{or} \qquad Y_i \sim N(X_i \beta, \Sigma). \tag{6.1}$$

In (6.1), X_i is an $m \times d$ matrix of explanatory variables, β is a coefficient vector of length d, and ϵ_i is the within-subject error, assumed to be IID $N(0, \Sigma)$, where Σ is an $m \times m$ covariance matrix. More generally, the model can be rewritten to permit m to depend on i, i.e., the number of measurements may depend on the subject.

Set $m_2 = m - m_1$ and partition Y_i into $Y_i = (Y'_{i1}, Y'_{i2})^T$ with $Y'_{i1} = (Y_{i1}, \ldots, Y_{im_1})^T$ and $Y'_{i2} = (Y_{im_1+1}, \ldots, Y_{im})^T$. Next, partition Σ as

$$\Sigma = \begin{pmatrix} \Sigma_{11} & \Sigma_{12} \\ \Sigma_{21} & \Sigma_{22} \end{pmatrix},$$

where Σ_{11} is $m_1 \times m_1$, Σ_{22} is $m_2 \times m_2$, and $\Sigma_{21} = \Sigma_{12}^T$ is $m_2 \times m_1$. Then, it is well known that, conditionally on $Y'_{i1} = y'_{i1}$, the distribution of Y'_{i2} is

$$(Y'_{i2} | Y'_{i1} = y'_{i1}) = N \left(\mu_{i2} - \Sigma_{21} \Sigma_{11}^{-1} (y'_{i1} - \mu_{i1}), \Sigma_{22} - \Sigma_{21} \Sigma_{11}^{-1} \Sigma_{12} \right), \tag{6.2}$$

where $\mu_{i1} = E(Y'_{i1}) = X_{i1}\beta$, $\mu_{i2} = E(Y'_{i2}) = X_{i2}\beta$, and $X_i = (X_{i1}, X_{i2})^T$, the partition of X_i into the top m_1 rows and the bottom m_2 rows. That is, if β and Σ were known, one could use the mean of the normal in (6.2) as a point predictor for Y'_{i2} (optimal in squared-error distance). If prediction regions conditional on Y'_{i1} were desired, they would be of the form

$$(y_{i2} - \mu_{i2}^{\dagger})\Sigma^{\dagger -1}(y_{i2} - \mu_{i2}^{\dagger}) \leq \chi_{m_{i2}}^2(1 - \alpha), \tag{6.3}$$

where $\mu_{i2}^{\dagger}$ and $\Sigma^{\dagger}$ are the mean and variance in (6.2). Unconditionally on Y_{i1}', the mean μ_{i2} of Y_{i2} would be used in place of $\mu^{\dagger}$ and $\Sigma^{\dagger}$ would merely be Σ_{22}.

Using (6.2) and (6.3) requires that μ_{i1}, μ_{i2} (and hence β), and Σ are known or at least can somehow be estimated well. Indeed, (6.1) reduces to the usual linear regression formulation if $\Sigma = \sigma^2 I_m$. In that case, there are m values in Y_i, p unknowns in β, and a single unknown in Σ, so $m \geq p + 1$ ensures that β can be estimated merely by pooling the linear models into one large $nm \times p$ design matrix. However, if Σ can be any covariance matrix then it has up to $m(m-1)/2$ distinct entries. So, for large enough m, we have $m \leq m(m-1)/2 + p$, meaning that one subject cannot be enough to estimate the unknown parameters. Consequently, n must be large enough that there will be enough data to fit (6.1). The basic way in which information across independent individuals $i = 1, \ldots, n$ is pooled is via the likelihood. The parameters in β and Σ can be estimated by maximizing

$$\prod_{i=1}^{n} \left(\frac{1}{2\pi}\right)^m \det(\Sigma)^{-1/2} \exp\left(\left(-\frac{1}{2}\right)(y_i - X_i\beta)^T \Sigma^{-1}(y_i - X_i\beta)\right) \tag{6.4}$$

with inference based on classical MLE theory. Essentially, one is then using Y_{i1}', for $i = 1, \ldots, n$, to estimate β and Σ for predicting one or more of the Y_{i2}'. However, note that in general (6.4) can only be maximized when Σ depends on relatively few parameters.

By contrast, the Bayesian would have at least two analogs to (6.2) and one to (6.3). The simplest would be to assume that n copies of the IID m-dimensional Y_i are distributed as $N(\mu, \Sigma)$ for some unknown μ, i.e., that the dependence of X_i on i drops out, so that $Y_i = X\beta + \epsilon_i$, most easily interpreted as the one-population case. If $b > 0$, so that $w(\mu|\Sigma) \sim N(\mu_0, (1/b)\Sigma)$ for some fixed μ_0 and $w(\Sigma) \sim \text{Wishart}^{-1}(\Psi, n_o)$ then the product is a 'normal-inverse-Wishart' $\text{NIW}(\mu_0, b, \Psi^{-1}, n_0)$. As in Gelman *et al.* (2004, p. 85) (see also Murphy (2007, Sec. 9)), the posterior is

$$\text{NIW}\left(\mu, \Sigma \left| \frac{n\bar{y} + b\mu_0}{n+b}, b+n, \Psi + nS\frac{nb}{n+b}(\bar{y} - \mu_0)(\bar{y} - \mu_0)', n+n_0\right.\right)$$

so that

$$w(\mu|\Sigma, Y^n) \sim N\left(\frac{n\bar{y} + b\mu_0}{n+b}, \frac{1}{n+b}\Sigma\right)$$

and

$$w(\Sigma|Y^n) \sim \text{Wishart}^{-1}\left(\Psi + nS + \frac{nb}{n+b}[\bar{y} - \mu_0][\bar{y} - \mu_0]', n+n_0\right), \tag{6.5}$$

where

$$\bar{y} = \frac{1}{n}\sum_{i=1}^{n} y_i \qquad \text{and} \qquad S = \frac{1}{n}\sum_{i=1}^{n}(y_i - \bar{y})(y_i - \bar{y})'.$$

Now, for the $(n+1)$th response Y_{n+1}, the predictive is a multivariate t (see Sec. 4.2 for the definition):

$$m(y_{n+1}|y_1^n) = \text{MVSt}_{n_0+n+m+1}\left(\frac{n\bar{y} + b\mu_0}{n+b}, \frac{\Psi}{(b+n)(n_0+n-m+1)}\right). \tag{6.6}$$

where $y_1^n = (y_1, \ldots, y_n)$. Note that (6.6) is for the whole vector Y_{n+1} whereas (6.2) is conditional on the past data for the individual, using the other data to estimate parameters (as in (6.4)). Now, in principle, one can marginalize out $Y_{n+1,1}$ from (6.6) to get a predictive density for $Y_{n+1,2}$ or one can find the conditional probability for $(Y_{n+1,2}|Y_{n+1,1})$ from (6.6). The same arguments as were used in Kotz and Nadarajah (2004) to obtain $Y_{n+1,1}$ lead to

$$Y_{n+1,2} \sim MVSt_{n_0+n+m_2+1} \left(\frac{n\bar{y}_2 + b\mu_{0,2}}{n+b}, \frac{\Psi}{(b+n)(n_0+n-m_2+1)} \right). \qquad (6.7)$$

However, a closed form for the conditional probability $(Y_{n+1,2}|Y_{n+1,1})$ does not seem to have been derived (except in the case where the location of Y_{n+1} is zero). One can conjecture (see Kotz and Nadarajah (2004, Chap. 1, Sec. 11)) that it would be an MVSt distribution also, with degrees of freedom $n_0 + n + m_2 + 1$, mean $(n\bar{y}_2 + b\mu_{0,2})/(n + b)$, and variance matrix $\Psi_{22}/\big((b+n)(n_0+n-m_2+1)\big)$ but this does not seem to have been formally derived. For computational purposes one would have to use the ratio of (6.6) and the marginal density for $Y_{n+1,1}$, which is analogous to the form of (6.7).

A second way in which the Bayesian might try to find the predictive density is by finding the densities in the numerator and denominator of the predictive density. If the predictive being sought is $m(y'_{21}, \ldots, y'_{2,n+1}|y'_{11}, \ldots, y'_{1,n+1}) = m(y_1, \ldots, y_{n+1})/m(y'_{12}, \ldots, y'_{1,n+1})$, meaning that the first m_1 measurements on subject Y_{n+1} are available and one can integrate out $(y'_{21}, \ldots, y'_{2n})$, effectively losing any information in it, then one can use the formulae in Murphy (2007, Sec. 9.5). These formulae give closed-form expressions for $m(y_1, \ldots, y_{n+1})$ and $m(y'_{12}, \ldots, y'_{1,n+1})$. Unfortunately, implementing this procedure seems quite hard.

A third way in which the Bayesian might try to find a predictive density starts by looking at only a single subject. That is, consider, $m(y'_{i2}|y'_{i1}) = m(y_i)/m(y'_{i1})$ and try to extend it to $n \geq 2$. Like the others, this procedure will end up giving only computational solutions, but they may be easier to implement. First note that $p(y_i|X_i, \beta, \Sigma) = N(X_i\beta, \Sigma)$ and $p(y_{i1}|X_{i1}, \beta, \Sigma_{11}) = N(X_{i1}\beta, \Sigma_{11})$. So, the numerator requires a prior on β and Σ while the denominator uses $Y_{i1} = X_{i1}\beta + \epsilon_{i1}$ and so requires a prior on β and Σ_{11}. In fact, as before it is more convenient to put a prior on $\mu_i = (\mu_{i1}, \mu_{i2})^T$, effectively absorbing the explanatory variables into the prior. That is, take $Y_{i1} \sim N(\mu_{i1}, \Sigma_{11})$ and $Y_i \sim N(\mu_i, \Sigma)$, so that the normal prior on μ_i with the Wishart prior on Σ gives the conjugate prior for (μ_i, Σ). That is, set $w(\mu_i|\Sigma) \sim N(\mu_{i0}, (1/b)\Sigma)$, i.e.,[2]

$$w \left(\left. \begin{array}{c} \mu_{i1} \\ \mu_{i2} \end{array} \right| \Sigma \right) \sim N \left(\left(\begin{array}{c} \mu_{i10} \\ \mu_{i20} \end{array} \right), \frac{1}{b} \left(\begin{array}{cc} \Sigma_{11} & \Sigma_{12} \\ \Sigma_{21} & \Sigma_{22} \end{array} \right) \right). \qquad (6.8)$$

So, the prior on μ_{i1} is $w(\mu_{i1}|(1/b)\Sigma_{11})$ To complete the prior specification, again assume $\Sigma \sim \text{Wishart}^{-1}(\Psi, n_0)$ for some prespecified degrees of freedom n_0 and an $m \times m$ matrix Ψ, and partition Ψ^{-1} in terms of the matrices Ψ_{11} ($m_1 \times m_1$), Ψ_{12} ($m_1 \times m_2$), and Ψ_{22} ($m_2 \times m_2$), in parallel to the partition of Σ in (6.8), suppressing the inverse. Loosely, Ψ corresponds to the average value of Σ^{-1}.

---

[2] To see the role of X_i explicitly, write $\beta \sim N(\mu_\beta, V_\beta)$. Then, $\mu_{i0} = X_i\mu_\beta$ and the variance of Y_i is $X_i V_\beta X_i^T + \Sigma$.

Now, with a mild abuse of notation, to see how X_i would appear in the predictive density, write it as

$$\frac{\int p(y'_{1i}, y'_{2i}|X_i\beta, \Sigma)w(X_i\beta|\Sigma)w(\Sigma)\mathrm{d}\beta\mathrm{d}\,\Sigma}{\int p(y'_{i1}|X_{i1}\beta, \Sigma_{11})w(X_{1i}\beta|\Sigma_{11})w(\Sigma_{11})\mathrm{d}\,X_{i1}\beta\mathrm{d}\Sigma_{11}}$$

$$=\frac{\int p(y'_{i1}, y'_{i2}|\mu_i, \Sigma)w(\mu_i|\Sigma)w(\Sigma)\mathrm{d}\mu_i\mathrm{d}\,\Sigma}{\int p(y'_{i1}|\mu_{i1}, \Sigma_{11})w(\mu_{i1}|\Sigma_{11})w(\Sigma_{11})\mathrm{d}\,\mu_{i1}\mathrm{d}\Sigma_{11}}$$

$$=\frac{\int p(y'_{i2}|y'_{i1}, \mu_i, \Sigma)p(y'_{i1}|\mu_i, \Sigma_{11})w(\mu_{i1}|\Sigma_{11})w(\mu_{i2}|\mu_{i1}, \Sigma)w(\Sigma)\mathrm{d}\mu_i\mathrm{d}\,\Sigma}{\int p(y'_{i1}|\mu_{i1}, \Sigma_{11})w(\mu_{i1}|\Sigma_{11})w(\Sigma_{11})\mathrm{d}\,\mu_{i1}\mathrm{d}\Sigma_{11}}, \quad (6.9)$$

using the factorization of the multivariate normal as in (6.2). It remains to factor the prior and this can be done by a series of technical arguments given in Giri (2004, p. 224–227). Specifically, Σ has three distinct parts, Σ_{11}, Σ_{12}, and Σ_{22}, and $w(\Sigma)$ can be re-expressed in terms of Σ_{11}, $\Sigma_{21}\Sigma_{11}^{-1/2}$, and $\Sigma_{22} - \Sigma_{21}\Sigma_{11}^{-1}\Sigma_{12}$:

$$w(\Sigma) = w(\Sigma_{11})w(\Sigma_{22} - \Sigma_{21}\Sigma_{11}^{-1}\Sigma_{12})w(\Sigma_{21}\Sigma_{11}^{-1/2}|\Sigma_{11})\det(\Sigma_{11})^{m_1}, \quad (6.10)$$

in which

$$w(\Sigma_{11}) \sim \text{Wishart}^{-1}(\Psi_{11}, n_0)$$
$$w(\Sigma_{22} - \Sigma_{21}\Sigma_{11}^{-1}\Sigma_{12}) \sim \text{Wishart}^{-1}(\Psi_{22} - \Psi_{21}\Psi_{11}^{-1}\Psi_{12}, n_0 - n)$$
$$w(\Sigma_{21}\Sigma_{11}^{-1/2}|\Sigma_{11}) \sim N(\Psi_{21}\Psi_{11}^{-1/2}, (\Psi_{22} - \Psi_{21}\Psi_{11}^{-1}\Psi_{12}) \otimes \Phi_{11}),$$

where $\otimes$ is the Kronecker product (used in order to give a square matrix). Recognizing that

$$w(\mu_{i1}, \Sigma_{11}|y_{i1}) = \frac{p(y'_{i1}|\mu_i, \Sigma_{11})w(\mu_{i1}|\Sigma_{11})w(\Sigma_{11})}{\int p(y'_{i1}|\mu_i, \Sigma_{11})w(\mu_{i1}|\Sigma_{11})w(\Sigma_{11})\mathrm{d}(\mu_{i1}\Sigma_{11})}$$

means that, using (6.10) in the numerator of (6.9), gives

$$\int p(y'_{i2}|y'_{i1}, \mu_i, \Sigma)w(\mu_{i2}|\mu_{i1}, \Sigma)w(\Sigma_{22} - \Sigma_{21}\Sigma_{11}^{-1}\Sigma_{12})w(\Sigma_{21}\Sigma_{11}^{-1/2}|\Sigma_{11})$$
$$\times w(\mu_{i1}, \Sigma_{11}|y_{i1})\det(\Sigma_{11})^{m_1}\mathrm{d}(\mu_{i1}\Sigma_{11})\mathrm{d}(\mu_{i2}, \Sigma_{12}, \Sigma_{22}),$$

which is parallel to (6.2). The problem is that the last expression does not obviously lead to a closed-form expression for (6.9). If such closed-form expressions are not available then PIs must be generated computationally (although one may be led to conjecture that they will be some form of MVSt).

This type of reasoning can be extended to the case where there are n independent copies of Y'_1, i.e., $Y'_{11}, \ldots, Y'_{n1}$. Then, the predictive distribution for a given Y'_{i2} is $m(y'_{i2}|y'_{11}, \ldots, y'_{n1})$. So, the parallel to (6.9) is

$$m(y'_{i2}|y'_{11}, \ldots, y'_{n1})$$
$$= \frac{p(y'_{i2}, y'_{11}, \ldots, y'_{n1}|X_{11}, \ldots, X_{n1}, X_{i2}, \beta, \Sigma)w(\beta|\Sigma)w(\Sigma)\mathrm{d}(\beta, \Sigma)}{\int p(y'_{11}, \ldots, y'_{n1}|X_{11}, \ldots, X_{n1}, \beta, \Sigma_{11})w(\beta|\Sigma_{11})w(\Sigma_{11})\mathrm{d}(\beta, \Sigma_{11})}$$
$$= \frac{p(y'_{i2}|y'_{11}, X_{i2}, \beta, \Sigma)p(y'_{11}, \ldots, y'_{1n}|X_{11}, \ldots, X_{n1}, \beta, \Sigma_{11})w(\beta|\Sigma)w(\Sigma)\mathrm{d}(\beta, \Sigma)}{\int p(y'_{11}, \ldots, y'_{n1}|X_{11}, \ldots, X_{n1}, \beta, \Sigma_{11})w(\beta|\Sigma_{11})w(\Sigma_{11})\mathrm{d}(\beta, \Sigma_{11})},$$

$$(6.11)$$

in which (6.10) can be used for $w(\Sigma)$ in the numerator. It remains to re-express $w(\beta|\Sigma)$ in the numerator in terms of $w(\beta|\Sigma_{11})$ and possibly some other factors. However, this is hard; only computational solutions can be found. A computational approach may be more feasible in cases where Σ depends only on the parameters in Σ_{11}, e.g., when Σ is of autoregressive form, so that knowledge of Σ_{11} is essentially equivalent to knowledge of Σ.

Thus, in the general frequentist or Bayesian case, sooner or later one usually arrives at predictors that do not appear to be available in closed form, necessitating either computation or simplification (which may not be justified). The result of this is to make predictive validation in longitudinal data more difficult – even though the complexity of longitudinal problems usually means that predictive validation is more important.

A simplification that is sometimes helpful is to regard time as the explanatory variable. Thus, if $Y_i = X_i\beta + \epsilon_i$, the only explanatory variable is time, and the entries of all Y_i are measured at the same times $t_1, \ldots, t_m$ then, for all i,

$$X_i = \begin{pmatrix} 1 & t_1 & t_1^2 & \cdots & t_1^{p-1} \\ \vdots & \vdots & \vdots & \vdots & \vdots \\ 1 & t_m & t_m^2 & \cdots & t_m^{p-1} \end{pmatrix}. \tag{6.12}$$

So, in this case $Y_i = X\beta + \epsilon_i$, that is, the dependence of the design matrix on i drops out. Thus, $Y_i \sim N(\mu_0, \Sigma)$ may be reasonable if the behavior of only one population of individuals is of interest since then Σ can be estimated in principle. This actually generalizes to two or more populations of individuals that are being measured repeatedly over time, as will be seen in Sec. 6.2; design matrices can be structured so that they are not dependent on i even as they reflect the population that generated i.

The remainder of this chapter continues the program of Part II, namely, looking at various important subfields of statistics – in this case longitudinal data – from a predictive standpoint. This is particularly important for longitudinal data because, even though techniques for the analysis of such data are well developed, they are usually construed in modeling terms rather than predictive terms.

The following sections review further techniques for analyzing longitudinal data. First, the traditional predictors arising from repeated measures ANOVA are presented. In some sense these are central, because they use ideas from fixed effect linear models as well as including a random-effects term. Second, general fixed effect linear models for longitudinal data will be developed predictively. Third, predictors based on generalized linear models (GLMs) will be presented. Such models generalize linear models by using a link function to identify a predictor of the mean of a response – even when the data is categorical. (For a more complete treatment of categorical longitudinal data, see Davis (2002, Chap. 7) and Agresti (2002, Chap. 11). However, in both cases their perspective is model fitting, not prediction.) It is in the GLM context that estimating equations (EEs) and generalized EEs (GEEs) will be described (although using GEEs for prediction may be problematic). Fourth, arguably the most important analytic framework currently used for analyzing longitudinal data, in which predictors are based on mixed models, will be presented. These models may be linear (LMMs), generalized linear (GLMMs), or nonlinear (NLMMs) – though the nonlinear techniques seem not to be well developed. Mixed models can accommodate much more variability than fixed-effects models; however, it is sometimes unclear whether they provide narrow enough PIs for model validation. The endnotes in Sec. 6.6 briefly present two techniques from growth curves that are important but not central to the development here.

6.1 Predictors Derived from Repeated-Measures ANOVA

In the introduction to this chapter we looked at the case of samples drawn from one normal population with the goal of predicting the next m_2 outcomes for an individual from the sample drawn from that population. It was seen that the Frequentist would use a conditional normal and pool over data to estimate parameters in order to generate a predictor, while a Bayesian would use the predictive density, conditioning on the data, to form a predictor. All effects were 'fixed' in the sense that the explanatory variables could be regarded as deterministic rather than outcomes from a random variable.

There are numerous important extensions to this basic case, and one can argue that in longitudinal analysis there are more moving parts than in other branches of statistics. First, usually there is more than one population. Second, there are numerous ways to model the covariance structure within a population. Third, there are often random effect terms that have to be dealt with differently than fixed effect terms. Fourth, there are many ways to model the data to assess variability or estimate main effects. Some of these will be seen below – but always with the goal of prediction.

So, assume that there are q populations indexed by $\ell = 1, \ldots, q$, that the times at which measurements are taken are indexed by $j = 1, \ldots, m$, and that there are n subjects indexed by $i = 1, \ldots, n$. When needed, n_ℓ will denote the number of subjects from the ℓth population, and it is assumed that $n_1 + \cdots + n_q = n$. All subjects are assumed independent and the data are assumed to be balanced, i.e., the measurements on each subject are made at the same m times for all subjects, with no missing values. Thus, each individual generates a vector Y_i of length m and the jth entry Y_{ij} in Y_i is the ith subject's response at the jth common time point. This is written as Y_{ij} or as $Y_{i(\ell)j}$ when the fact that i is in the ℓth population is being emphasized. Requiring balance is a restriction compared with the situation for real data; however, the point here is to demonstrate the derivation of predictors. The structure of the data is illustrated in Table 6.1.

Now, write the signal-plus-noise model as $Y_i = \mu_\ell + \epsilon_{i\ell}$, to mean that subject i drawn from the ℓth population has mean μ_ℓ and random component $\epsilon_{i\ell}$. The measurements over the m times on subject i are $Y_i^T = (Y_{i1}, \ldots, Y_{ij} \ldots Y_{im})$; it is understood that i is in the ℓth population. Also, $\mu_\ell = (\mu_{\ell,1}, \ldots, \mu_{\ell,m})^T$ is a vector of length m, possibly differing from population to population, and $\epsilon_{i\ell} = (\epsilon_{i\ell 1}, \ldots, \epsilon_{i\ell m})$ represents two sources of variability. One is the variability of subject i from population ℓ at time j, often regarded as 'measurement error' as in signal-plus-noise models. The second source of variability is that of subject i as a member of population ℓ, often regarded as a random effect representing its difference from other subjects in the same population. That is, if there is a distribution over the subjects in population ℓ then the second source of variation represents the 'among-units (in the population) variability', as opposed to the first source of variation, which represents the 'within-unit variability'.

So far, this description of the signal-plus-noise model is indistinguishable from a split-plot design. Indeed, the usual model for a split-plot design is

Table 6.1 *Structure of longitudinal data. There are*
$\ell = 1, \ldots, q$ *populations,* $i = 1, \ldots, n_q$ *subjects from*
population q, and $j = 1, \ldots, m$ *time points.*

Population	Subject	Time 1	...	Time m
1	1	y_{111}	...	y_{11m}
⋮	⋮	⋮	⋮	⋮
1	i	y_{i11}	...	y_{i1m}
⋮	⋮	⋮	⋮	⋮
1	n_1	y_{n_111}	...	y_{n_11m}
⋮	⋮	⋮	⋮	⋮
q	1	y_{1q1}	...	y_{1qm}
⋮	⋮	⋮	⋮	⋮
q	i	y_{iq1}	...	y_{iqm}
⋮	⋮	⋮	⋮	⋮
q	n_q	y_{n_qq1}	...	y_{n_qqm}

$$Y_{ij} = \mu + \alpha_\ell + \gamma_j + (\alpha\gamma)_{\ell j} + \epsilon_{ij}, \tag{6.13}$$

where the constraints

$$\sum_{\ell=1}^{q} \alpha_\ell = \sum_{j=1}^{m} \gamma_j = \sum_{\ell=1}^{q} (\alpha\gamma)_{\ell j} = \sum_{j=1}^{m} (\alpha\gamma)_{\ell j} = 0$$

are imposed for identifiability. That is, the whole-plot factor levels α_ℓ are analogous to the population mean of one of the q populations; the split-plot factor levels γ_j are analogous to the times or, more precisely, the mean of the measurement made at time j on subject i; and the interaction terms $(\alpha\gamma)_{\ell j}$ are analogous to the mean interaction between a member of the ℓth population and the measurement at time j. Replications over whole plots correspond to replications over subjects from the same population, and replications over the split plots correspond to repeated measures over the subjects.

The key difference between using a split-plot design for a repeated-measures ANOVA longitudinal model and using the more general linear models, as in Sec. 6.2, is in the error term. The error term in (6.13) is

$$\epsilon_{i(\ell)j} = b_{i(\ell)} + e_{i(\ell)j} \equiv N(0, \sigma_w^2) + N(0, \sigma^2), \tag{6.14}$$

in which all b_i and $\epsilon_{i(\ell)j}$ are independent. (Sometimes $b_{i(\ell)}$ is written just as b_i to emphasize its relationship to the level of the split plot.) By contrast, the parallel assumption in Sec. 6.2 will be simply that

$$\text{Var}(\epsilon_{i(\ell)}) = \Sigma \tag{6.15}$$

and is thus independent of ℓ. That is, even though $\epsilon_{i(\ell)j}$ is the sum of two terms (b_i and $e_{i(\ell)j}$) they are not separately identifiable without a lot of work, as will be seen in Sec. 6.4.

In the special case of a single population, where $q = 1$, i.e., there is only one level for the main plot, the analysis in the chapter introduction is valid. However, that treatment assumes that the prediction is for a further measurement on the same subject. Here, with two or more populations, it is worth distinguishing three prediction problems in the context of a split-plot design before moving on to other longitudinal predictors. This makes sense because, even in agricultural applications (the origin of split-plot designs), farmers are interested in the next yield of a new plot and subplot 'treatment' as much as they are interested in inference about the specific effects of a treatment on a plot.

The first prediction problem is how to predict for a new subject, say $n_\ell + 1$, from population ℓ for the jth time. The random variable is

$$Y_{n_\ell+1(\ell)j} = \mu + \alpha_\ell + \gamma_j + (\alpha\gamma)_{\ell j} + b_{n_\ell+1(\ell)} + \epsilon_{n_\ell+1(\ell)j}. \tag{6.16}$$

The variance matrix for Y_i is $\mathrm{Var}(Y_i) = \sigma^2 I_m + \sigma_w^2 \, \mathbf{1}^T \mathbf{1}$, because $\mathrm{Var}(Y_{n_\ell+1(\ell)j}) = \sigma_w^2 + \sigma^2$, $\mathrm{Cov}(Y_{n_\ell+1(\ell)j}, Y_{n_\ell+1(\ell)j'}) = \sigma^2$, and $\mathrm{Cov}(Y_{n_\ell+1(\ell)j}, Y_{i(\ell)j}) = 0$, and so it exhibits 'homogeneous compound symmetry': i.e., it is compound symmetric because all off-diagonal elements (covariances between different times) have a common value ρ and it is homogeneous because all the responses have the same variance. (Different time points or populations could have different variances.) Identifying a point predictor is not difficult: Simply take the mean $\bar{y}_{\cdot(\ell)j}$ over all n_ℓ responses at time j in population ℓ and proceed as in Sec. 2.2.1. To get a PI, simply write

$$Y_{n_\ell+1(\ell)j} \sim N(\mu + \alpha_\ell + \gamma_j + (\alpha\gamma)_{\ell j}, \sigma_w^2 + \sigma^2)$$

and use the usual estimates for σ_w^2 and σ^2, so that

$$Y_{n_\ell+1(\ell)j} - \left(\mu + \alpha_\ell + \gamma_j + (\alpha\gamma)_{\ell j}\right) \sim t_{(n_\ell-1)qm} \sqrt{\hat{\sigma}_w^2 + \hat{\sigma}^2}.$$

The usual estimates of σ_w^2 and σ^2 can be found in many books on the design of experiments, e.g., Neter *et al.* (1985, Chaps. 21, 23) or Rao (1998, Chap. 16). One unbiased estimate of $\sigma^2 + \sigma_w^2$ is from Milliken and Johnson (1992, Chap. 24):

$$\widehat{\sigma^2 + \sigma_w^2} = \frac{\mathrm{MSE(wholeplot)} + (m-1)\mathrm{MSE(subplot)}}{m},$$

in which, following Giesbrecht and Gumpertz (2004, Sec. 7.4),

$$\mathrm{MSE(wholeplot)} = \frac{m \sum_{\ell=1}^{q} \sum_{i=1}^{n_\ell} (\bar{y}_{i(\ell)\cdot} - \bar{y}_{\cdot(\ell)\cdot})}{\sum_{\ell=1}^{q} (n_\ell - 1)}$$

and

$$\mathrm{MSE(subplot)} = \frac{\sum_{\ell=1}^{q} \sum_{i=1}^{n_\ell} \sum_{j=1}^{m} (y_{i(\ell)j} - \bar{y}_{i(\ell)\cdot} - \bar{y}_{\cdot(\ell)j} + \bar{y}_{\cdot(\ell)\cdot})^2}{(m-1) \sum_{\ell=1}^{q} (n_\ell - 1)}.$$

An alternative to this simple approach is to make use of the model and other data by fixing ℓ and j and using the standard estimators $\hat{\mu}$, $\hat{\alpha}_\ell$, $\hat{\gamma}_j$, and $\widehat{(\alpha\gamma)}_{\ell j}$ for μ, α_ℓ, γ_j, and $(\alpha\gamma)_{\ell j}$ in (6.16). Likewise, the standard estimators $\hat{\sigma}_w$ and $\hat{\sigma}$ for σ_w and σ considered above can be used. Since values for the two random terms in (6.16) cannot be obtained in any general and reasonable way for data that has not been collected, it makes sense to take them as zero. Thus, following Neter *et al.* (1985, Chap. 20, p. 684), another point predictor for (6.16) is

$$\hat{Y}_{n_\ell+1(\ell)j} = \hat{\mu} + \hat{\alpha}_\ell + \hat{\gamma}_j + \widehat{(\alpha\gamma)}_{\ell j}$$

and the natural PI comes from combining the sampling distributions of $\hat{\mu}$, $\hat{\alpha}_\ell$, $\hat{\gamma}_j$, and $\widehat{(\alpha\gamma)}_{\ell j}$ with $N(0, \hat{\sigma}_w^2)$ and $N(0, \hat{\sigma}^2)$. These sampling distributions are well known (they are t-distributions with the appropriate degrees of freedom); combining them appropriately is difficult. When all the degrees of freedom are large enough, it might be not too far wrong to use normal approximations. In addition, if the dependence among the t-distributions is not too high, it would be convenient to add these normal distributions in the usual way in which independent normals are added.

The second problem is to predict for an existing subject at a new time, say $j = m + 1$, for which no data exists. Now, $Y_{i(\ell)m+1}$ is the response for the ith subject from the ℓth population at a time for which a measurement has not yet been made. The random variable is

$$Y_{i(\ell)m+1} = \mu + \alpha_\ell + \gamma_{m+1} + (\alpha\gamma)_{\ell,m+1} + b_{i(\ell)} + \epsilon_{i(\ell)m+1}, \tag{6.17}$$

and a predictor can be found only by extrapolating. Again, the usual estimates $\hat{\mu}$, $\hat{\alpha}_\ell$ for μ, α_ℓ can be used. The value of $b_{i(\ell)}$ found from the existing data can also be used; techniques for finding this value will be given in Sec. 6.4. An expedient though usually suboptimal approach is to use a $N(0, \sigma_w^2)$ variate in place of a value for $b_{i(\ell)}$. In this case, it would still make sense to take zero as the value for $\epsilon_{i(\ell)m+1}$. Even so, the problem remains that the constants γ_{m+1} and $(\alpha\gamma)_{\ell,m+1}$ cannot be directly estimated from the available data since there are no measurements at $m + 1$. Thus an extra criterion must be identified, so that γ_{m+1} and $(\alpha\gamma)_{\ell,m+1}$ can be specified using the existing data.

One way to generate estimates for γ_{m+1} and $(\alpha\gamma)_{\ell,m+1}$ is to introduce auxiliary equations. Consider writing

$$\gamma_j = \delta_0 + \delta_1 j + \eta, \tag{6.18}$$

$$(\alpha\gamma)_{\ell,j} = \delta_0' + \delta_{1\ell}' j + \eta', \tag{6.19}$$

the second equation being for each fixed $\ell = 1, \ldots, q$; it is assumed that η and η' are IID error terms $N(0, \tau_1^2)$ and $N(0, \tau_2^2)$, respectively. Since estimates for the γ_j and $(\alpha\gamma)_{\ell,j}$ are available for $j = 1, \ldots, m$, it is possible to extrapolate to $m + 1$ and get estimates for $\hat{\gamma}_{m+1}$ and the $\widehat{(\alpha\gamma)}_{\ell,m+1}$. Essentially, the error terms η and η' in (6.18) and (6.19) become components in $\epsilon_{i(\ell)m+1}$ in (6.17); the third component in $\epsilon_{i(\ell)m+1}$ would be the measurement error. So, one can either obtain a point predictor by taking the mean of $\epsilon_{i(\ell)m+1}$ or try to come up with a value for the residual. Either way, one must specify a covariance structure on the pairs $\epsilon_{i(\ell)j}$ and $\epsilon_{i(\ell)j'}$, as follows.

If the covariance is zero for all $j \neq j'$ then it makes sense to use the mean of $\epsilon_{i(\ell)m+1}$, which is zero; i.e., when the covariance structure is zero, the point predictor is

$$Y_{i(\ell)m+1} = \hat{\mu} + \hat{\alpha}_\ell + \hat{\gamma}_{m+1} + \widehat{(\alpha\gamma)}_{\ell,m+1} + b_{i(\ell)}, \tag{6.20}$$

where $b_{i(\ell)}$ is estimated from techniques to be seen in Sec. 6.4. For a PI one could, of course, use the sampling distributions for $\hat{\mu}$ and $\hat{\alpha}_\ell$ already obtained from the data. These are t-distributions with an appropriate number of degrees of freedom that, if large enough, could be taken as normal. Then, one could use the sampling distributions for $\hat{\gamma}_{m+1}$ and $\widehat{(\alpha\gamma)}_{\ell,m+1}$ obtained from fitting (6.18) and (6.19) – again, these will be t-distributions with an appropriate number of degrees of freedom $m-1$ that could be approximated as normals. Finally, one would have to use the sampling distribution of $b_{i(\ell)}$ (if this could be obtained) or merely use its actual distribution $N(0, \sigma_w^2)$, with $\hat{\sigma}_w^2$ in place of σ_w^2. Then, if the dependence were not too high, one could combine these normal distributions and get an overall distribution from which to get PIs for $Y_{i(\ell)m+1}$.

However, if the responses of subject i are systematically high relative to those of other subjects for times $j = 1, \ldots, m$, it may be likely that subject i will continue to be high compared with other subjects at time $m+1$. In this case, the zero-covariance assumption is called into question and more complicated covariance structures may be required for good prediction. Then, an extra term might be added to the point predictor, giving

$$Y_{i(\ell)m+1} = \hat{\mu} + \hat{\alpha}_\ell + \hat{\gamma}_{m+1} + \widehat{(\alpha\gamma)}_{\ell,m+1} + b_{i(\ell)} + \hat{\epsilon}_{i(\ell)m+1}, \tag{6.21}$$

where $\hat{\epsilon}_{i(\ell)m+1}$ must be derived using techniques to be presented in Sec. 6.4; see (6.65). Here, the result is that $\hat{\epsilon}_{i(\ell)m+1}$ equals

$$\text{Cov}(\epsilon_{i(\ell)m+1}, (\epsilon_{i(\ell)1}, \ldots, \epsilon_{i(\ell)m})^T) \text{Var}(\epsilon_{i(\ell)1}, \ldots, \epsilon_{i(\ell)m})^{-1} (\hat{\epsilon}_{i(\ell)1}, \ldots, \hat{\epsilon}_{i(\ell)m})^T, \tag{6.22}$$

where the last vector on the right is the vector of residuals from the first m time points for subject i. Clearly, if the covariances for different j's are zero, the first vector on the right is zero (and the variance matrix is the identity), reducing (6.21) to (6.20). However, there are numerous distinct parametrizations of the variance matrix in (6.22) that may be assumed in order to reduce the number of parameters required to identify it; compound symmetry is just one form that is convenient. It is beyond the scope of this section to indicate how to estimate the parameters in the variance matrix and the other factors in (6.22). The point is that it is even more difficult to find a distribution for the last term in (6.21) and therefore to give PIs for this point predictor in (6.22).

The third prediction problem is to predict for a new subject at a new time. Now, the random variable is

$$Y_{n_\ell+1(\ell)m+1} = \mu + \alpha_\ell + \gamma_{m+1} + (\alpha\gamma)_{\ell,m+1} + b_{n_\ell+1(\ell)} + \epsilon_{n_\ell+1(\ell)m+1}. \tag{6.23}$$

As before, $\hat{\mu}$ and $\hat{\alpha}_\ell$ can be found by analyzing the existing data and, likewise, since there is no knowledge about the $(n_\ell + 1)$th subject, $b_{n_\ell+1(\ell)}$ cannot be estimated so it is enough to revert to $N(0, \sigma_w^2)$ to describe it, and the error term can be described only as $N(0, \sigma^2)$. The remaining two terms, γ_{m+1} and $(\alpha\gamma)_{\ell,m+1}$, depend on $m+1$ and can be estimated by (6.18) and (6.19). Consequently, all the components of (6.23) have t-distributions found by estimating the variance in a normal mean-zero distribution. So, the point predictor for (6.23) is

$$Y_{n_\ell+1(\ell)m+1} = \hat{\mu} + \hat{\alpha}_\ell + \hat{\gamma}_{m+1} + \widehat{(\alpha\gamma)}_{\ell,m+1}$$

and a PI would be found by approximating the sampling distributions of the estimators (t-distributions) by normals and combining the approximating normals as usual, assuming the dependence were not too high.

There are Bayesian versions of ANOVA and split-plot designs for which predictors can be derived as well. However, the predictors derived here use a random effect which is very much at one with a Bayesian hierarchical design. Since this class of models, in general, is not particularly good or common for longitudinal data (see McCullough (2005)), there is little conceptual loss in omitting the purely Bayesian version of their predictors here.

6.2 Linear Models for Longitudinal Data

To see how the split-plot design leads to a more general linear model for longitudinal data, rewrite (6.13) using (6.15) in place of (6.14) as a linear model; thus the 'random effect' term and the 'within-subject variability' are lumped together as one error term. That is, imagine taking all the fixed effect terms (μ, α_ℓ, γ_j, $(\alpha\gamma)_{\ell j}$) and stacking them to form a parameter vector β of length d. Also, suppose that for a given subject i there is an $n \times d$ matrix X_i containing only zeroes and ones, so that $X_i\beta$ gives only the parameters appropriate for the population from which i was drawn. Now,

$$Y_i = X_i\beta + \epsilon \qquad \text{with} \qquad \text{Var}(\epsilon_i) = \Sigma \tag{6.24}$$

is a linear model that includes the ANOVA models used so far in this chapter. However, (6.24) is more general: it allows for unbalanced data and the explicit use of time and other covariates, and it permits more realistic assumptions about Σ.

As an example, recall the single-population cases that were presented in a linear model form in the introduction. Another version of the single-population case can be seen in the repeated-measures ANOVA approach given in Sec. 6.1. In some cases, however, the longitudinal behavior of a single population can be studied more cleanly by using time explicitly as a covariate and, when this is possible, it enables a straightforward extension to two or more populations. The method in the introduction to this chapter could also be extended to two or more populations; however, the complexity would become difficult to manage and closed-form expressions more difficult to obtain.

So, in this section, rather than treating responses as a collection of unrelated values over time, as in ANOVA, time itself will be used as an explanatory variable. Sometimes such approaches are called growth curve models. This means that for a single population the simplest model becomes

$$Y_{ij} = \beta_0 + \beta_1 t_j + \epsilon_{ij} \tag{6.25}$$

for an individual i at time j. When valid, this is more parsimonious than repeated-measures ANOVA. For instance, suppose there were five time points so that at a minimum there would be four γ_j (plus an overall mean) whereas (6.25) only has two parameters for the main effects. Indeed, from a predictive standpoint, the explicit use of time gives a much easier model because auxiliary equations (6.18) and (6.19) are not required – and the values of the times need not be equally spaced. However, this class of models may be too restrictive for some settings. Gathering the Y_{ij} into a vector Y_i gives

$$Y_i = \begin{pmatrix} Y_{11} \\ \vdots \\ Y_{1m} \end{pmatrix} = \begin{pmatrix} 1 & t_1 \\ \vdots & \vdots \\ 1 & t_m \end{pmatrix} \begin{pmatrix} \beta_0 \\ \beta_1 \end{pmatrix} + \begin{pmatrix} \epsilon_{11} \\ \vdots \\ \epsilon_{1m} \end{pmatrix} = X\beta + \epsilon_i$$

and again it is assumed that ϵ_i is normal mean-zero with $\text{Var}(Y_i) = \Sigma$.

To see how multiple populations can be included explicitly, it is enough to consider the case of two populations. The case of three or more populations is similar. Write two versions of (6.25),

$$Y_{i(1)j} = \beta_{01} + \beta_{11}t_j + \epsilon_{i(1)j} \quad \text{and} \quad Y_{i(2)j} = \beta_{02} + \beta_{12}t_j + \epsilon_{i(2)j}, \tag{6.26}$$

where $\ell = 1, 2$. These two linear models can be combined into a single linear model by using indicator functions. Write $\chi_1(i) = 1$ when subject i is in population $\ell = 1$ and $\chi_1(i) = 0$ when subject i is in population $\ell = 2$, with χ_2 defined analogously. Now, (6.26) is equivalent to

$$Y_{ij} = \chi_1(i)\beta_{01} + \chi_1(i)\beta_{11}t_j + \chi_2(i)\beta_{02} + \chi_2(i)\beta_{12}t_j + \epsilon_{ij},$$

which can be expressed as $Y_i = X_i\beta + \epsilon_i$ by defining

$$X_i = \begin{pmatrix} \chi_1(i) & \chi_1(i)t_1 & \chi_2(i) & \chi_2(i)t_1 \\ \vdots & \vdots & \vdots & \vdots \\ \chi_1(i) & \chi_1(i)t_m & \chi_2(i) & \chi_2(i)t_m \end{pmatrix} \quad \text{and} \quad \beta = \begin{pmatrix} \beta_{01} \\ \beta_{11} \\ \beta_{02} \\ \beta_{12} \end{pmatrix}.$$

It can be seen that in this example there are only two values of X_i, one for $\ell = 1$ and one for $\ell = 2$. If there were three or more populations, analogous X_i could be defined. Moreover, X_i can be adjusted to include higher powers of time and/or more covariates besides time, as well as to allow the set of times at which the two populations are measured to be different. Indeed, X_i can also be adjusted to permit different numbers of times for different subjects. In this case, $\text{Var}(\epsilon_i) = \Sigma_i$; however, if compound symmetry is assumed then the differently sized Σ_i depend on the same parameters. For instance, if all Y_{ij} have the same variance σ^2, for all $j \neq j'$, $\text{Cov}(Y_{ij}, Y_{ij'}) = \rho$, and Y_i has length m_i then

$$\Sigma_i = \sigma^2 \begin{pmatrix} 1 & \rho & \cdots & \rho \\ \rho & 1 & \cdots & \rho \\ \vdots & \vdots & \vdots & \vdots \\ \rho & \cdots & \rho & 1 \end{pmatrix} \tag{6.27}$$

is $m_i \times m_i$ and depends on only two parameters. Other forms of Σ_i are possible, e.g., an AR(1) assumption on the measurements on a fixed subject over time gives $\Sigma = \sigma^2 B$, where B is a banded matrix with ones on the main diagonal and powers of ρ off the main diagonal. The choice of parametrization for Σ_i is intended to model the random component in the longitudinal data parsimoniously but accurately. For ease of exposition, write $\Sigma_i = \Sigma_i(\sigma^2, \omega)$, i.e., each subject has a variance that is a function of two parameters σ and ω. In the forms of Σ_i noted above we have $\omega = \rho$, but more generally ω can be a vector of parameters.

Now, the model can be written as $Y_i \sim N(X_i\beta, \Sigma)$, where it is understood that Y_i is $n_\ell \times 1$, X_i is $n_\ell \times d$, and β is $d \times 1$. The ith design matrix contains all the (fixed effect) covariates, whether they are given as functions of time or are other measurements made at known times; these may vary from subject to subject even within the same population. It's preferred that all subjects from the same population have the same measurements because compensating for missing data, though typical, is beyond our present scope. The requirement is that the β's be common across all i's even if some design matrices must have extra columns in them to make the representation valid.

Given that all this modeling has been accomplished in some way that would not be characterized as unconvincing, it remains to estimate all the parameters so as to form the predictors. This was seen briefly in (6.4) but now can be developed for two or more populations and in greater detail. To see the general form, write the responses as a single tall column vector,

$$Y^n = (Y_1^T, \ldots, Y_{n_1}^T, \ldots, Y_{n_1+\cdots+n_{q-1}+1}^T, \ldots, Y_n^T)^T$$

(where $n = \sum_{\ell=1}^q n_\ell$) and recall that all the vectors in Y must be independent because the subjects are independent. The joint density for the vector $Y^n = y^n$ of length n is now

$$f(y^n | \beta, \sigma^2, \omega) = \prod_{\ell=1}^q \prod_{i=1}^{n_\ell} \left(\frac{1}{2\pi}\right)^{n_\ell/2} (\det \Sigma_i)^{-1/2} \exp\left(-\frac{1}{2}(y_i - X_i\beta)^T \Sigma_i^{-1}(y_i - X_i\beta)\right)$$

(6.28)

in which the unknown parameters are β, σ^2, and ω.

A rare case occurs when all the Σ_i are known or, equivalently, σ^2 and ω are known. In this case one can use calculus to solve for the MLE of β as

$$\hat{\beta} = \frac{\sum_{i=1}^n X_i^T \Sigma_i^{-1} Y_i}{(\sum_{i=1}^n X_i^n \Sigma_i^{-1} X_i)},$$

(6.29)

a sort of weighted least squares estimator generalizing the familiar 'X prime X inverse X prime Y' incantation.

Indeed, the pooling of data into large vectors as in (6.2) brings out the similarity to weighted linear regression. Combine the X_i into a large $\sum_\ell n_\ell \times d$ matrix X and the ϵ_i into a long $\sum_\ell n_\ell$ vector as in (6.2) to get

$$Y = \begin{pmatrix} X_1 \\ \vdots \\ X_n \end{pmatrix} \beta + \begin{pmatrix} \epsilon_1 \\ \vdots \\ \epsilon_n \end{pmatrix} = X\beta + \epsilon,$$

with $E(Y) = X\beta$. The variance matrix for Y assumes the form

$$\Sigma^* = \text{Var}(Y) = \begin{pmatrix} \Sigma_1 & 0 & 0 & 0 \\ 0 & \Sigma_2 & \cdots & 0 \\ \vdots & \vdots & \vdots & \vdots \\ 0 & \cdots & 0 & \Sigma_n \end{pmatrix}.$$

So, in this notation, the usual formula for the weighted least squares estimates holds and gives

$$\hat{\beta} = (X^T (\Sigma^*)^{-1} X)^{-1} X^T (\Sigma^*)^{-1} Y.$$

(6.30)

Now, it is not hard to show that $E(\hat{\beta}) = \beta$, so (6.30) is unbiased, and similar manipulations show that $\text{Var}(\hat{\beta}) = (X^T (\Sigma^*)^{-1} X)^{-1}$. Thus, given that $\hat{\beta}$ must be normal, it is easy to see that

$$\hat{\beta} \sim N(\beta, (X^T (\Sigma^*)^{-1} X)^{-1}).$$

The usual situation is that ω, and hence Σ^*, is unknown. In this case, instead of using Σ^* one must have an estimator $\hat{\omega}$ that can be used with $\hat{\sigma}$ to give an estimator $\widehat{\Sigma}_i$ of Σ_i. Given $\widehat{\Sigma}_i$ one can still obtain an MLE for β as

$$\hat{\beta} = \frac{\sum_{i=1}^n X_i^T \hat{\Sigma}_i^{-1} Y_i}{\left(\sum_{i=1}^n X_i^n \Sigma_i^{-1} X_i\right)}, \tag{6.31}$$

often called the estimated generalized least squares estimator for β. The estimator $\hat{\omega}$ can be hard to find, so usually it must be found by a numerical procedure.

Theoretically, if n is large, i.e., there are enough samples from all q populations, then the central limit theorem can be invoked to give sampling distributions for estimators. For instance, (6.29), (6.30), and (6.31) together give

$$\hat{\beta} \underset{\sim}{\infty} N\left(\beta, \left(X\Sigma^{*-1}X\right)^{-1}\right) = N\left(\beta, \left(\sum_{i=1}^n X_i^T \Sigma_i^{-1} X_i\right)^{-1}\right). \tag{6.32}$$

Sometimes it is convenient to write $\hat{V}_\beta = \left(\sum_{i=1}^n X_i^T \Sigma_i^{-1} X_i\right)^{-1}$ for brevity when discussing PIs for a given Y_i.

In practice, it may be shown that the estimators for σ^2 and ω can be biased in small sample sizes even when $\hat{\beta}$ is approximately unbiased. One way in which this is fixed (and this is the default in some computer packages) is to modify the likelihood in (6.28) by inserting an extra factor, $(\det X_i^T \Sigma_i^{-1} X_i)^{-1/2}$, giving

$$\prod_{\ell=1}^q \prod_{i=1}^{n_\ell} \left(\frac{1}{2\pi}\right)^{n_\ell/2} (\det \Sigma_i)^{-1/2} (\det X_i^T \Sigma_i^{-1} X_i)^{-1/2} \exp\left(-\frac{1}{2}(y_i - X_i\beta)^T \Sigma_i^{-1}(y_i - X_i\beta)\right) \tag{6.33}$$

as the objective function to be maximized. The estimate of β is unchanged except for a new form of $\hat{\Sigma}_i$, i.e., the estimate for the covariance changes but the dependence of the main effect parameter on the covariance does not. The result of changing the likelihood to reduce bias is often called 'restricted maximum likelihood' or REML. The details are beyond our present scope but are treated elsewhere; see Diggle *et al.* (2002).

Turning to the three prediction problems identified in Sec. 6.1, these can be stated as predicting a new observation at an existing time point, predicting an existing observation at a new time point, and predicting a new observation at a new time point.

To begin, if the new observation is for an element of population ℓ at a time j at which measurements have already been made then the marginal random variable is in general

$$Y_{n_\ell+1(\ell)j} \sim N(X_{n_\ell+1}(j)\beta, \Sigma_{n_\ell+1}(jj)), \tag{6.34}$$

where $X_{n_\ell+1}(j)$ is the jth row of $X_{n_\ell+1}$ and $\Sigma_{n_\ell+1}(jj)$ is the (j, j)th entry of $\Sigma_{n_\ell+1}$. Since the design matrix is presumed known, β and $\Sigma_{n_\ell+1}(jj)$ must be estimated from the available data. Clearly, from a frequentist standpoint one is led to use the MLE $\hat{\beta}$ estimated from all the data as a point estimator for β and to use its limiting sampling distribution to get PIs. The other parameter to be estimated is $\Sigma_{n_\ell+1}(jj)$. There are several ways to proceed depending on the assumptions made. For instance, if all populations have a common Σ, estimated by $\hat{\Sigma}$, then the (j, j)th element $\hat{\sigma}_{jj}$ of $\hat{\Sigma}$ could be used. From basic normal theory, an approximate $1 - \alpha$ PI is given by

$$Y_{n_\ell+1(\ell)j} \in X_{n_\ell+1}(j)\hat\beta \pm \frac{\hat\sigma_{jj}}{n-d} t_{n-d,1-\alpha/2}$$

if the variability in $\hat\beta$ is ignored. (If the variability in $\hat\beta$ were not ignored, an extra $\hat\sigma_{22}$ term would have to be included, as seen in Sec. 4.2.) The effectiveness of $\hat Y_{n_\ell+1,\ell,j} = X_{n_\ell+1}\hat\beta$ as a predictor can be assessed by

$$P\left(|Y_{n_\ell+1(\ell)j} - \hat Y_{n_\ell+1(\ell)j}| \ge \tau\right) \le \frac{\mathrm{Var}(Y_{n_\ell+1(\ell)j} - \hat Y_{n_\ell+1(\ell)j})}{\tau^2}$$

$$\le \frac{2}{\tau^2}\left(\mathrm{Var}(Y_{n_\ell+1(\ell)j}) + \mathrm{Var}(\hat Y_{n_\ell+1(\ell)j})\right), \quad (6.35)$$

in which

$$\mathrm{Var}(X_{n_\ell+1}(j)\hat\beta) = (X_{n_\ell+1}(j))^T \mathrm{Var}(\hat\beta) X_{n_\ell+1}(j),$$

and the variance of $\hat\beta$ from (6.32) is $\mathcal{O}(1/n)$ for n large. Thus,

$$P\left(|Y_{n_\ell+1(\ell)j} - \hat Y_{n_\ell+1(\ell)j}| \ge \tau\right) \le \frac{2}{\tau^2}\left(\frac{C}{n} + \sigma_{jj}^2\right) = \frac{2}{\tau^2}\sigma_{jj}^2\left(1 + \frac{C'}{n}\right),$$

a bound of the usual 'variance times $(1 + C/n)$' form typical of prediction problems. In principle, one could incorporate the estimate of $\hat\sigma_{jj}$ into (6.35) and therefore get a bound in terms of its sampling distribution.

More generally, predicting a collection of $Y_{n_\ell+1(\ell)j}$, say $j = 1, \ldots, m_1 < m$, is similar except that a multivariate normal PI region would be given using χ^2 rather than t_{n-d}. Thus, assuming that the populations all had the same Σ, and writing $Y_{n_\ell+1(\ell)1:\,m_1}$ to represent the m_1 time points and $X_{n_\ell+1}(1:m_1)$ to represent the corresponding rows of the design matrix,

$$P\left(\left(Y_{n_\ell+1(\ell)1:\,m_1} - X_{n_\ell+1}(1:m_1)\hat\beta\right)\Sigma_m^{-1}\left(Y_{n_\ell+1(\ell)1:\,m_1} - X_{n_\ell+1}(1:m_1)\hat\beta\right)\right.$$

$$\left. \le \chi_{m_1}^2(1-\alpha)\right) = 1-\alpha,$$

where Σ_{m_1} is the submatrix of Σ appropriate to $Y_{n_\ell+1(\ell)1:\,m_1}$. Plugging in $\hat\beta$ and $\hat\Sigma_{m_1}$ gives an approximate $1-\alpha$ PI. Inequalities for predictors such as $\hat Y_{n_\ell+1(\ell)1:\,m} = X_{n_\ell+1}(1:m)\hat\beta$ can also be derived as in (6.35) but are more complicated.

An alternative to (6.34) when partial data is available on the $(n_\ell + 1)$th subject is to treat $Y_{n_\ell+1(\ell)j}$ conditionally rather than marginally. That is, one uses the formula for $Y_{n_\ell+1(\ell)j}|Y_{n_\ell+1(\ell)\bar j}$, where $\bar j = (1, \ldots, j-1, j+1, \ldots, m)$, i.e., the times different from j where observations are available. Write $Y_{n_\ell+1,\ell} \sim N(X_{n_\ell+1}\beta, \Sigma)$, where we set $\Sigma = \Sigma_\ell$ for brevity, so that, as in (6.2), $Y_{n_\ell+1(\ell)j}|Y_{n_\ell+1(\ell)\bar j}$ has a distribution

$$N\left(EY_{n_\ell+1(\ell)j} - \Sigma_{21}\Sigma_{11}^{-1}(Y_{n_\ell+1(\ell)\bar j} - EY_{n_\ell+1(\ell)\bar j}), \Sigma_{jj} - \Sigma_{j\bar j}\Sigma_{\bar j\bar j}^{-1}\Sigma_{\bar j j}\right), \quad (6.36)$$

where the subscript ℓ's on the various submatrices of Σ_ℓ have been dropped for convenience. If fewer than $m-1$ observations on $Y_{n_\ell+1}$ are available then the formula must be adjusted appropriately. As before, the data must be pooled to estimate β, Σ_{jj}, $\Sigma_{j\bar j}$, and $\Sigma_{\bar j\bar j}$ so that conditional PIs for $Y_{n_\ell+1(\ell)j}$ can be found (assuming that the variability in the parameters can be ignored). An analysis of $P(|Y_{n_\ell+1(\ell)j} - \hat Y_{n_\ell+1(\ell)j}|)$, where $\hat Y_{n_\ell+1(\ell)j}$ is the estimated mean from (6.36), can also be carried out but is more complicated.

The Bayesian would solve these problems by finding the predictive densities assuming priors on the parameters. Thus, using (6.34) would lead to

$$m(Y_{n_\ell+1(\ell)j}|\mathcal{D}) = \int p(Y_{n_\ell+1(\ell)j}|\beta, \Sigma_{n_\ell+1}(jj))w(\beta, \Sigma_{n_\ell+1}(jj)|\mathcal{D})\mathrm{d}(\beta, \Sigma_{n_\ell+1}(jj)),$$

where $\mathcal{D}$ consists of all the data relevant to estimating the parameters in (6.34), namely, all the data from the $i = 1, \ldots, n$ subjects from the q populations at the m times, assuming that some extra structure is built into $\Sigma_{n_\ell+1}(jj)$ to ensure that the data is relevant. A simple version of this was done in closed form in Sec. 4.2; more general forms were given in (6.7) and in the discussion after (6.8). If all the Σ_ℓ are unrelated then only the data from the same population (the ℓth) would be relevant to Σ_ℓ, although it might also be relevant to estimating β; formalizing this and finding the predictive would be feasible only numerically. In this setting, it is important to recall that the predictive distribution is always a mixture of the distribution for the predictand with respect to the posterior for the parameters.

Equivalently, the Bayesian could find

$$m(Y_{n_\ell+1(\ell)1: m}|\mathcal{D}) = \frac{m(y_{1(1)1: m}, \ldots, y_{n_1(1)1: m}, y_{1(2)1: m}, \ldots, y_{n_2(2)1: m}; Y_{n_\ell+1(\ell)1: m})}{m(y_{1(1)1: m}, \ldots, y_{n_1(1)1: m}, y_{1(2)1: m}, \ldots, y_{n_2(1)1: m})}$$

and marginalize out the $\bar{j}$ entries in $Y_{n_\ell+1(\ell)1: m}$. The numerator and denominator are of the same form; for instance, the denominator is

$$\int p(y_{1(1)1: m}, \ldots, y_{n_1(1)1: m}, y_{1(2)1: m}, \ldots, y_{n_2(2)1: m}|\beta, \Sigma_1, \Sigma_2)w(\beta, \Sigma_1, \Sigma_2)\mathrm{d}(\beta, \Sigma_1, \Sigma_2).$$

Quantities like this were examined in the introduction. It seems that only computational solutions are feasible. (Note that the likelihood can be factorized, so that only data from population j is affected by Σ_j whereas all data is affected by β.)

Likewise, if a Bayesian wanted to address the conditional prediction problem in (6.36) then the predictive distribution would be

$$m(Y_{n_\ell+1(\ell)j}|\mathcal{D}, Y_{n_\ell+1(\ell)\bar{j}}).$$

This is the general $q \geq 2$ version of the n-data-point one-population predictor identified in (6.11). Since no closed form appears to be available even in the one-population setting, numerical solutions appear to be all that can be obtained.

The second sort of prediction problem would be predicting a random variable such as $Y_{i(\ell)m+1}$, where $m + 1$ represents a time for which no measurements have been made. Assuming the model still applies, we have

$$Y_{i(\ell)1: m,m+1} \sim N\left(\left(\begin{array}{c} X_i(m) \\ X_i(m+1) \end{array}\right)\beta, \left(\begin{array}{cc} \Sigma_{11} & \Sigma_{12} \\ \Sigma_{21} & \Sigma_{22} \end{array}\right)\right), \tag{6.37}$$

where the variable of interest is the last element of $Y_{i(\ell)1: m,m+1}$, in which $1: m$ denotes the first m measurements (that are available). As before, $\Sigma_{21} = \Sigma_{12}$ is the vector of covariances between the data for $j \leq m$ and time $m + 1$ and Σ_{22} is the variance for time $m + 1$, and we write $\Sigma = \Sigma_\ell$ for brevity. There is no data at any of these times to permit estimation, so extra assumptions have to be made to permit prediction. As in the second ANOVA prediction problem from Sec. 6.1, extrapolation is required: both the mean and the variance matrix must be extrapolated to time $m + 1$. This is obvious for the point predictor; one simply uses

$\hat{Y}_{i(\ell)m+1} = X_i(m+1)\hat{\beta}$ parallel to (6.18) and (6.19). Obviously, this requires that $X_i(m+1)$ is known (as a function of time or some other covariate that could be measured at time t_{m+1}). However, extrapolation of the variances may introduce extra assumptions; this will be ignored here and it will be assumed that Σ_{12} and Σ_{22} depend on the same parameters as Σ_{11}. If the values of $Y_{i(\ell)j}$ are available for $j = 1, \ldots, m$ then they would be included in the data used to form estimates of the parameters.

As with the first prediction problem there are two cases, the marginal and the conditional. The marginal is based on extracting

$$Y_{i(\ell)m+1} \sim N(X_i(m+1)\beta, \Sigma_{22}) \tag{6.38}$$

from (6.37), where $\Sigma_{22} = \sigma_{22}^2$ is a scalar. Since σ_{22} can be estimated using the same parameter estimates as are used to estimate Σ_{11}, write the estimate as $\hat{\sigma}_{22}$. Likewise, β can be estimated by $\hat{\beta}$ using the available data via the MLE or REML. Thus, a point predictor for $Y_{i(\ell)m+1}$ is $\hat{Y} = X_1(m+1)\hat{\beta}$ and the corresponding PI is of the form

$$Y_{i(\ell)m+1} \in X_i(m+1)\hat{\beta} \pm \frac{\hat{\sigma}_{22}}{n-d}t_{n-d,1-\alpha/2},$$

ignoring the variability in the $\hat{\beta}$. (As in the first prediction problem, if the variability in $\hat{\beta}$ were not ignored then, as in Sec. 4.2, an extra $\hat{\sigma}_{22}$ term would have to be included.) Analogous techniques would apply for the prediction of the next k values at times $m + 1, \ldots, m+k$.

A variant on this would be to predict using the conditional $(Y_{i(\ell)m+1}|Y_{i(\ell)1:m})$ making use of the earlier data points on subject i. From standard normal theory, $(Y_{i(\ell)m+1}|Y_{i(\ell)1:m})$ has the distribution

$$N\left(EY_{i(\ell)m+1} - \Sigma_{21}\Sigma_{11}^{-1}(Y_{i(\ell)1:m} - EY_{i(\ell)1:m}), \Sigma_{22} - \Sigma_{21}\Sigma_{11}^{-1}\Sigma_{12}\right) \tag{6.39}$$

Getting an approximate $1 - \alpha$ predictor (point or interval) clearly necessitates two extrapolation steps. First, the parameters would have to be estimated. As before, $\hat{\beta}$ could be found, so that the expectations in (6.39) could be estimated. Also, all the data that have a block Σ_{11} in the upper left corner of their covariance matrix that is the same as Σ_ℓ could be used to find a suitable $\hat{\Sigma}_{11}$ via the REML.

Next, in addition to extrapolating the mean to time point $m + 1$, the variance matrices for population ℓ would also have to be extrapolated to be $(m + 1) \times (m + 1)$. That is, extra assumptions would have to be imposed to identify $\Sigma_{21} = \Sigma_{12}$ and $\Sigma_\ell(m+1, m+1)$. As in the marginal prediction, one would be led to assume that $\Sigma_{12} = \Sigma_{21}$ and Σ_{22} from (6.37) were of the same form, e.g., of compound symmetry or AR(1), and depended on the same ω as Σ_{11}. Thus, the REML could be used to get estimates $\hat{\Sigma}_{11}$, $\hat{\Sigma}_{21}$, and $\hat{\Sigma}_{22}$. Putting this together would give an approximate point predictor and $1 - \alpha$ PI for $Y_{i(\ell)m+1}$ that is conditional on $Y_{i(\ell)1:m}$; the need for approximating would be the result of neglecting variation in the estimates $\hat{\beta}$, $\hat{\Sigma}_{11}$, $\hat{\Sigma}_{21}$, and $\hat{\Sigma}_{22}$. Prediction techniques from a more sophisticated treatment of the multivariate problem could take these sources of variation into account but are beyond our present scope.

The Bayesian would, as before, have analogs to both (6.38) and (6.39). Regarding the former, the task would be to derive the predictive distribution associated with the model (6.38), i.e., to find

$$\int p(y_{i(\ell)m+1}|\beta, \Sigma_{22})w(\beta, \Sigma_{22}|\mathcal{D})d\beta d\Sigma_{22},$$

where $\mathcal{D}$ contains all the data relevant to estimating the parameters in (6.38). In this case, the $Y_{i(\ell)1:m}$ are available and in $\mathcal{D}$. The data from populations $\bar{\ell}$ would be used to estimate β but not Σ_{22}. Priors would have to be assigned for β and the parameters of the variance matrix in (6.38). Thus, as usual, the predictive distribution would be a mixture of densities of the form (6.38) with respect to the posterior for the parameters. The Bayesian would find, in analogy to (6.39),

$$\int p(y_{i(\ell)m+1}|y_{i(\ell)1:m}, \beta, \Sigma)w(\beta, \Sigma|\mathcal{D})d\beta d\Sigma. \tag{6.40}$$

In this case, $\mathcal{D}$ would not contain $Y_{i(\ell)1:m}$. Apart from the needed extrapolation, these two cases can also be regarded as a multipopulation generalization of the one-population cases treated in the introduction to this chapter.

Note that, in contrast with the frequentist case, instead of assuming that the Σ_{ij} depend on the same parameters as Σ_{11}, a Bayesian could simply assign a prior to the unknown parameters. The predictions would then be, in part, prior-driven. As a generality, it is unclear which is worse, making assumptions about the structure of a variance matrix that are not true or having to justify prior knowledge. If the assumptions are only slightly wrong and the prior is poor, the Frequentist would be likely to have the advantage; however, if the assumptions were far wrong and the prior knowledge were accurate then the Bayesian would likely have the advantage. It makes good sense to pool all the data to estimate β, but if the matrices Σ_ℓ for $\ell = 1, \ldots, q$ are of different sizes then only some data points – those appropriate for estimating Σ_{11}, Σ_{12}, and Σ_{22} – would be included in the estimation of the entries in the covariance matrix.

The third prediction problem combines the first two: predict the outcome at time $m + 1$ for a new subject, say $n_\ell + 1$, from population ℓ. The overall model is almost the same as in (6.37), so the overall vector response can be written as

$$Y_{n_\ell+1(\ell)1:m,m+1} \sim N\left(\begin{pmatrix} X_{n_\ell+1}(m) \\ X_{n_\ell+1}(m+1) \end{pmatrix}\beta, \begin{pmatrix} \Sigma_{11} & \Sigma_{12} \\ \Sigma_{21} & \Sigma_{22} \end{pmatrix}\right), \tag{6.41}$$

in which it is the last entry of $Y_{n_\ell+1(\ell)1:m,m+1}$ that is of interest. (Predicting the first m entries of $Y_{n_\ell+1(\ell)1:m,m+1}$ was the first prediction problem.) Marginalizing out from (6.41) to get $Y_{n_\ell+1(\ell)m+1}$ gives

$$Y_{n_\ell+1(\ell)m+1} \sim N(X_{n_\ell+1}(m+1)\beta, \Sigma_{22}). \tag{6.42}$$

So, assuming there is no other data on this $(n_\ell + 1)$th subject, one would pool all the data to form $\hat{\beta}$, implicitly extrapolating the main effect to time $m + 1$. Then, one would have to assume some form for $\Sigma_{22} = \sigma_{22}$, so that it could be expressed in terms of the same ω as the other covariance matrices, use the REML to find $\hat{\sigma}_{22}$, and derive prediction intervals, possibly ignoring the variability in the parameters.

In this case, if there is no other data on the $(n_\ell + 1)$th subject, there is nothing to condition on to get a conditional PI. However, if some values were available then one could use a variant on (6.39), estimate β and the covariance matrices using whatever data are available, and so obtain a PI for $(Y_{n_\ell+1(\ell)m+1}|Y_{n_\ell+1(\ell)J})$, where the subscript J indicates the measurements on subject $n_\ell + 1$ at a subset of time points from 1 to m. Again, these methods ignore the variability due to the estimation of parameters, so the intervals they give are likely to be too narrow. More sophisticated treatments of the sampling distributions of the parameters could correct for this but are beyond our present scope. This is essentially a version of the conditional form of the second prediction problem.

The Bayes version of the third prediction problem would again have marginal and conditional predictive densities, the latter if extra observations on the $(n_\ell + 1)$th subject were available. Using (6.42), the marginal predictive for $Y_{n_\ell+1(\ell)m+1}$ is

$$m(Y_{n_\ell+1(\ell)m+1}|\mathcal{D}),$$

recalling that it is the integral of the marginal for $Y_{n_\ell+1(\ell)m+1}$ against the posterior for the parameters β and Σ_{22}. The conditional prediction would be possible only if some observations were available for subject $n_\ell + 1$. Thus, $Y_{n_\ell+1(\ell)m+1}$ would be conditional on them in an analog of (6.39) and they would be included in $\mathcal{D}$. Overall, the conditional Bayes version of the third prediction problem would resemble the conditional Bayes version of the second problem.

To conclude this section, it's worth commenting that a purely Bayesian version of the predictive form of the linear model formulation for longitudinal data from one population was developed in Brown *et al.* (1998). The problem actually solved there includes model selection. While this is important, it is beyond our present scope, so only the predictive aspect will be mentioned here. First, the model is for the outcomes of an m-variate Y using $d + 1$ explanatory variables in the context of a GLM. If the link function is the identity then this reduces to the usual regression framework; see Sec. 6.3 for some details. Write $Y = (Y_1, \ldots, Y_n)$ and $X_f = (X_1^T, \ldots, X_n^T)$ for the $m \times n$ and $m \times d$ matrices formed from their respective column vectors. Let B be a $d \times n$ matrix of coefficients. (The use of d as a dimension of B assumes that the constant is one of the coefficients.) Now write

$$Y - XB \sim \mathcal{N}(I, \Sigma).$$

The use of $\mathcal{N}$ for the normal distribution rather than N indicates that the random variable is a matrix rather than a vector; for details see Brown *et al.* (1998). The point is that conjugate priors can be assigned to B and Σ and, given data, a posterior can be derived. Thus, to predict a $k \times m$ matrix of new responses, say Z, at new design points X_f one can use the same form of model, i.e.,

$$Z - X_f \sim \mathcal{N}(I, \Sigma),$$

and use standard Bayes updating to get a point predictor $\hat{Z}$ for Z. Details are given in Brown *et al.* (1998).

6.3 Predictors Derived from Generalized Linear Models

A *generalized* linear model (GLM) – not to be confused with the general linear model – is actually a family of models, the elements of the family being determined by the nature of the data. Logistic regression for binary variables, see Sec. 4.5, is an example of a generalized linear model but there are many others. First, suppose that the Y_i are independent, even though this is unrealistic for longitudinal data; this assumption will be relaxed shortly. To begin, three features define a GLM:

- a linear predictor $\eta_i = x_i^T \beta$, where x_i is the vector of explanatory values for subject i and β is a d-dimensional parameter vector;
- a link function g specifying the relationship between $\mu_i = E(Y_i)$, the expected value of subject i at x_i, and the linear predictor, that is,

$$g(\mu_i) = g(E(Y_i)) = \eta_i = x_i^T \beta;$$

- a relationship between the conditional variance of Y_i and the covariates,

$$\text{Var}(Y_i) = \phi_i V(\mu_i),$$

where ϕ_i is a possibly subject-dependent scaling parameter that is either known or to be estimated and $V(\cdot)$ is a known variance function.

For exponential families, the link function is chosen to be a transformation of the mean so as to get a linear predictor, i.e., the link is determined by the parametric family. So, for the case of logistic regression, η_i is used with the logit function $g(u) = \log(u/(1-u))$ (see (4.41)) and $V(\mu_i) = \mu_i(1 - \mu_i)$, effectively setting all $\phi_i = 1$.

A different example arises if the goal is to model count data. Recall that for $Y \sim$ Poisson(λ), $\mu = E(Y) = \text{Var}(Y)$. If Y_i is the count for subject i with explanatory variables x_i then set $\eta_i = x_i^T \beta$ with

$$\log E(Y_i) = \log \mu_i = x_i^T \beta.$$

This can be 'derived' by putting the Poisson(λ) distribution in its natural form:

$$\frac{\lambda^x e^{-\lambda}}{x!} = e^{x \log \lambda - \lambda - \log x!} = \frac{1}{x!} e^{x \log \lambda - \lambda} = h(x) e^{\eta T(x) - a(\lambda)},$$

from which it make sense to set $\mu_i = e^{\eta_i} = e^{x_i^T \beta}$, i.e., to use a linear regression model for the log of the mean. Now, for $g = \log$, $g(\mu_i) = x^T \beta$.

The variance is simpler:

$$\text{Var}(Y_i) = \phi_i E(Y_i),$$

where $\phi_i > 1$ and $\phi_i < 1$ represent over- and underdispersion (relative to the Poisson) respectively; often a common value ϕ for the ϕ_i is assumed. Technically, if $\phi \neq 1$, the underlying distribution of Y is not Poisson, but this procedure is usually called Poisson regression anyway. Poisson regression can be generalized to log-linear models but that is beyond our present scope. The probit link, based on the inverse of the standard normal distribution function, is also used in settings where Y is binary and the explanatory variables can be taken as normal. It is sometimes regarded as quite similar to the logit link. There are numerous other link functions, many of which have been implemented in computer packages.

Provided the GLM model has been specified, one can estimate β by solving the estimating equation

$$\sum_{i=1}^{n} \left(\frac{\partial \mu_i}{\partial \beta} \right)^T \text{Var}(Y_i)^{-1}(y_i - \mu_i) = 0. \tag{6.43}$$

This equation uses only the first two moments of Y_i and so applies to any distribution with those moments, i.e., for which the distributional form for the Y's is otherwise unconstrained. The details of how to use estimating equations, and their properties, will not be given here. It is enough to observe that the estimates of β are typically consistent and asymptotically normal although not necessarily efficient. Techniques for estimating the ϕ_i (when they cannot all be taken as equal to one) are also beyond our present scope.

Turning to prediction, in principle each GLM model can be examined and a correct PI, given g, derived (but see Pan and Le (2001) for a computational approach). Since this does not seem to have been expressed in a unified mathematical framework, it's worth proceeding (with great informality) to describe a procedure that may work for any g. It rests on asymptotics; details for specific g's do not seem to have been investigated.

First consider the point predictors. Given that the parameters have been estimated, the natural point predictor for a new value x_{n+1} is $\hat{Y}_{n+1} = g^{-1}(x_{n+1}\hat{\beta})$. This was seen for logistic regression in Sec. 4.5. Obviously, $\hat{Y}_{n+1}$ is also the natural point estimator for μ_{n+1}; however, its roles in prediction and estimation are different. So, for greater accuracy, it's worth taking account of the nonlinearities in g and finding a modification of $\hat{Y}_{n+1}$.

A second-order Taylor expansion gives

$$
\begin{aligned}
Y_{n+1} &\approx g^{-1}(\eta_{n+1}) + (g^{-1}(\hat{\beta}x_{n+1}) - g^{-1}(\eta_{n+1})) \\
&\approx g^{-1}(\eta_{n+1}) + (\hat{\beta}x_{n+1} - \eta_{n+1})(g^{-1})'(\eta_{n+1}) \\
&\quad + (\hat{\beta}x_{n+1} - \eta_{n+1})^2 (g^{-1})''(\eta_{n+1})/2.
\end{aligned}
\tag{6.44}
$$

Taking expectations, the middle term on the right is zero giving

$$
\begin{aligned}
E(Y_{n+1}) &\approx g^{-1}(\eta_{n+1}) + E(\hat{\beta}x_{n+1} - \eta_{n+1})^2 (g^{-1})''(\eta_{n+1})/2 \\
&\approx g^{-1}(\eta_{n+1}) + (1/2)(g^{-1})''(\eta_{n+1})x_{n+1}^T \mathrm{Var}(\hat{\beta})x_{n+1}.
\end{aligned}
$$

So, substituting $\hat{\eta}_{n+1} = x_{n+1}\hat{\beta}$ for η_{n+1} and finding an estimate for $\mathrm{Var}(\hat{\beta})$ will give a useful point predictor $\hat{Y}_{n+1}$ for Y_{n+1} (or point estimator of $E(Y_{n+1})$), provided that the mean is a reasonable summary for the location of the distribution.

If this $\hat{Y}_{n+1}$ is to be used as an estimator for $E(Y_{n+1})$, the natural way to find an standard error is via the delta method. To see this, ignore the second-derivative term in (6.44) and take variances on both sides. Since the cross-term drops out, this gives

$$
\begin{aligned}
\mathrm{Var}(Y_{n+1}) &= \mathrm{Var}(g^{-1}(\eta_{n+1})) + \mathrm{Var}((\hat{\beta}x_{n+1} - \eta_{n+1})(g^{-1})'(\eta_{n+1})) \\
&= \mathrm{Var}(\hat{\beta}x_{n+1}(g^{-1})'(\eta_{n+1})) \\
&= ((g^{-1})'(\eta_{n+1}))^2 x_{n+1}^T \mathrm{Var}(\hat{\beta})x_{n+1}.
\end{aligned}
\tag{6.45}
$$

Now, invoking asymptotic normality gives

$$
\hat{Y}_{n+1} \pm z_{\alpha/2}\sqrt{\widehat{\mathrm{Var}(Y_{n+1})}}
\tag{6.46}
$$

as a confidence interval for $E(Y_{n+1})$.

This can be extended to PIs. Write $Y_{n+1} \sim G_{\eta_{n+1}}$ where $G_{\eta_{n+1}}$ is the distribution of Y_{n+1} given $\eta_{n+1} = g^{-1}(x_{n+1}\beta)$. For a GLM, $G_{\eta_{n+1}}$ is taken to be a known exponential family. So, in principle, expressions for $\rho_{\alpha/2,\eta_{n+1}}$ and $\rho_{1-\alpha/2,\eta_{n+1}}$, the $(\alpha/2)100$th and

$(1 - \alpha/2)100$th percentiles of $G_{\eta_{n+1}}$, respectively, can be obtained in terms of the parameters of the exponential family. It remains to estimate the percentiles, and this amounts to estimating β. Therefore, it is enough to substitute an estimate $\hat{\eta} = x_{n+1}\hat{\beta}$ for $\eta_{n+1} = x_{n+1}\beta$. Or, to be conservative, one could use a lower confidence bound on η_{n+1} in $\rho_{\alpha/2,\eta_{n+1}}$ and an upper confidence bound on η_{n+1} in $\rho_{1-\alpha/2,\eta_{n+1}}$. These can be naturally taken to be of the form $\hat{\beta}x_{n+1} - z_{\alpha/2}x_{n+1}^T \text{Var}(\hat{\beta})x_{n+1}$ and $\hat{\beta}x_{n+1} + z_{1-\alpha/2}x_{n+1}^T \text{Var}(\hat{\beta})x_{n+1}$. Furthermore, the inverse of g as in $g^{-1}(\eta_{n+1}) = E(Y_{n+1})$ can be used to transform the confidence intervals for $E(Y_{n+1})$ via (6.46) to confidence intervals on η_{n+1}, for use in $\rho_{\alpha/2,\eta_{n+1}}$ and $\rho_{1-\alpha/2,\eta_{n+1}}$.

To use GLMs for longitudinal data one extends them to generalized estimating equation models (GEEs). For $i = 1, \ldots, n$ subjects measured at $j = 1, \ldots, m$ time points the defining conditions for a GLM are $\eta_{ij} = x_{ij}^T \beta$, $E(Y_{ij}) = \mu_{ij}$, $g(\mu_{ij}) = \eta_{ij}$, and $\text{Var}(Y_{ij}) = \phi V(\mu_{ij})$. Next, one needs a 'working correlation' matrix $R_i(\theta)$, where θ indicates the parameters defining the entries in the $m \times m$ matrix R_i for the ith subject. Clearly, $R_i = I_m$, the $m \times m$ identity matrix, is unreasonable – its use would mean the observations on a given subject were uncorrelated, defeating the point of a longitudinal analysis. Sometimes $R_i(\theta)$ is taken to have all diagonal entries one and all off-diagonal entries ρ, so that any two measurements on a subject have the same correlation; other choices are possible.

To use the correlation matrices, the R_i, they must be converted into a covariance structure. By the definition of a GLM, the form $V(\cdot)$ is known; so let $A_i = \text{diag}(V(\mu_{i1}), \ldots, V(\mu_{im}))$, the $m \times m$ diagonal matrix with entries given by the variances of the Y_{ij} for $j = 1, \ldots, m$. Then, set

$$V_i(\theta) = \phi A_i^{1/2} R_i(\theta) A_i^{1/2}. \tag{6.47}$$

Now, the analog of (6.43) is

$$\sum_{i=1}^{n} \left(\frac{\partial \mu_i}{\partial \beta} \right)^T V_i(\hat{\theta})^{-1}(y_i - \mu_i) = 0, \tag{6.48}$$

in which $y_i = (y_{i1}, \ldots, y_{im})^T$, $\mu_i = (\mu_{i1}, \ldots, \mu_{im})^T$, and $\partial \mu_i / \partial \beta$ is the $m \times d$ matrix with (j,k)th entry $\partial \mu_{ij}/\partial \beta_k$ for $j = 1, \ldots, m$ and $k = 1, \ldots, d$. Expression (6.48) can be used to estimate β provided that consistent estimates $\hat{\theta}$ and $\hat{\phi}$ of θ and ϕ (for use in (6.47)) are available. The resulting estimates are often called quasi-likelihood estimators because they use only the first two moments of Y_{ij}, leaving the rest of the distribution unspecified. Roughly, one uses $\hat{\theta}$ and $\hat{\phi}$ to get $\hat{\beta}$, which is then used to obtain new estimates of θ and ϕ, cycling until convergence. The details of finding a new $\hat{\phi}$ and $\hat{\theta}$ given $\hat{\beta}$ can be found in Diggle *et al.* (2002) or Fitzmaurice *et al.* (2004), for instance.

However, a simple form for the variance of $\hat{\beta}$ can be given. Often, it is enough to use

$$\hat{V}(\hat{\beta}) = \sum_{j=}^{m} \left(\frac{\partial \mu_i}{\partial \beta} \right)^T V_i(\hat{\theta}) \left(\frac{\partial \mu_i}{\partial \beta} \right).$$

Usually better estimates can be given; see Fitzmaurice *et al.* (2009) for some more recent estimates.

Even given successful parameter estimation, prediction in the context of GEEs is problematic. The reason is that GEE models only use assumptions about the first and second moments. This is often satisfactory for parameter inference but this is not obviously so for point predictions or prediction intervals, which may depend on higher-order moments. For instance, $\hat{Y}_{i,m+1} = x_{im+1}^T \hat{\beta}$ could be adopted as a point predictor – but this requires the assumption that the tails of the distribution are relatively balanced. Getting a PI is even more difficult because it depends on the tails of a distribution of which only the first two moments are specified. (One can imagine getting ultra-conservative PIs by using Chebyshev's inequality, but they would be likely to be too weak to be useful.) Indeed, if one naively uses a GEE model as if the corresponding GLM model (with the appropriate covariance structure) were true then it is not clear how wide a class of fully specified models is being represented by the first two moments of the GLM model. More forcefully, it is not clear how far the true model could be from the GLM model while still having the same first two moments. It may be that two different parametric families that lead to the same GEE model require quite different predictors. As a generality, it seems that this issue remains to be studied.

We conclude this subsection with the observation that there seems to be little work on prediction with generalized linear models even in the case where they are treated as an extension to the general linear model, i.e., without taking a dependence structure into account as is needed for longitudinal data. There are a few notable exceptions, such as Thomas and Cook (1990) and Meyer and Laud (2002) but, overall, prediction in GLMs – and especially in GEEs – seems relatively unexplored.

6.4 Predictors Using Random Effects

A random effect in an experiment is a factor that has many possible levels (often varying over the real line) but only finitely many of the levels have been included in the data, even though interest is in all possible levels. Otherwise stated, a random effect is a draw from a population that has to be included in a model to represent the response accurately. This was seen in the split-plot design of Sec. 6.1: the effect of the subplot treatment ranged over $\mathbb{R}$ but only finitely many values could be sampled. This amounted to including an extra term in the model that behaved like a random location. At the same time, the effect of the subplot treatment could be regarded as a random variable over a population in its own right.

In this section models with both fixed effect and random effect terms are presented. These are usually called mixed models. Linear mixed models (LMMs) have been the most studied and will be the focus of this section. Indeed, the random effect term in a split-plot model is a random intercept term in LMMs. There is also much work on generalized linear mixed models (GLMMs) and even some on nonlinear mixed models (NLMMs). Some discussion of these cases will be given at the end of this section but in general the topic is still largely at the research stage.

6.4.1 Linear Mixed Models

Consider the linear mixed model for a single m_i-dimensional observation Y_i on subject i and write

$$Y_i = X_i \beta + Z_i U_i + \epsilon_i \tag{6.49}$$

for $i = 1, \ldots, n$, where $\beta = (\beta_1, \ldots, \beta_d)^T$ is a fixed effect (FE) with $m_i \times d$ design matrix X_i, $U_i \sim N_q(0, D)$ is a random effect (RE) with $m_i \times q$ design matrix Z_i, and the error term is $\epsilon_i \sim N_{m_i}(0, R_i)$. The random effects are assumed IID and the error terms are mutually independent and independent of the U_i. Here, subscripts on the normal distributions indicate the dimension; these are omitted when there will be no misunderstanding.

The model (6.49) can be regarded hierarchically. The within-subject version of (6.49) arises from setting $U_i = u_i$, where the subject-to-subject variability, i.e., the random effect, is defined by the distribution of U_i. The model (6.49) can also be regarded as a hierarchical Bayesian model in which $(Y_i | \beta, U_i, \theta)$ is specified first and $(U_i | \theta)$ is specified second. The Bayesian model would be completed by specifying a prior for (β, θ). For brevity, this section focuses on the frequentist story even though the Bayesian formulation may be useful in motivating computational techniques beyond the present scope. Whether one adopts a Bayes or frequentist view, the inclusion of RE terms provides a lot more flexibility than fixed effect terms alone. Consequently, mixed effect (ME) models like (6.49) are often able to summarize the information in the data more accurately.

The model (6.49) is used for each of n subjects, so there are n versions of it, i.e., n matrices X_i and Z_i and n random variables U_i. The different Y_i are related to each other by the FE component, specifically β, and the fact that all the U_i have a common distribution. Another way to think of this is that a longitudinal model – for one population in particular – is a collection of repeated time series data that can be pooled to estimate common features. So, for subject i, if $j = 1, \ldots, m_i$ is regarded as a series of time points, it is sometimes better to write

$$Y_{ij} = \sum_{k=1}^{d} X_i(j;k)\beta_k + \sum_{k=1}^{q} Z_i(j;k)U_{ik} + \epsilon_i(j), \qquad (6.50)$$

where $X_i(j;k)$ is the kth element of the jth row of X_i, i.e., the kth explanatory variable measured at time j on subject i. The $Z_i(j;k)$ are similar, and $\epsilon_i(j)$ is the jth element of ϵ_i.

It's worthwhile examining what (6.49) and (6.50) imply. First, taking the expectation on both sides of (6.49) gives

$$EY_i = X_i\beta \qquad \text{and} \qquad EY_{ij} = \sum_{k=1}^{p} X_i(j;k)\beta_k,$$

meaning that the random effects only matter at the subject level, not the population level. Otherwise stated, individuals with high values of $Z_i U_i$ are balanced by those with low values of $Z_i U_i$.

There are two conditional expectations, Y on U and U on Y, that are important. The first is simpler. For $k \neq i$, $E(Y_k | U_i) = X_k\beta$ since U_i and Y_k are independent. However, for $i = k$,

$$E(Y_i | U_i) = X_i\beta + Z_i U_i, \qquad (6.51)$$

because only the measurement error ϵ_i is zero in expectation. Both sides of expression (6.51) are random: individual variability is represented as U_i and the term $Z_i U_i$ represents the difference of individual i from the overall population mean $X_i\beta$. It is also easy to see that

$$\text{Cov}(Y_i | U_i) = R_i.$$

Regarding the second conditional expectation, it is possible to derive a closed-form expression for $E(U_i | Y_i)$. For given i, the model assumptions give $Y_i \sim N_{m_i}(X_i\beta, V_i)$, where

$V_i = \text{Var}(Y_i) = Z_i D Z_i^T + R_i$. So, if all the m_i are equal and the ϵ_i are identical then $Y_i \sim N(X_i\beta, V_i)$, where $V_i = \text{Var}(Y_i) = Z_i D Z_i^T + R$. A convenient simplification is to assume $R = \sigma^2 I$, where I is the identity matrix of dimension m, the common value of the m_i: this corresponds to the independence of the $\epsilon_{i,j}$. It is easy to see that the covariance between Y_i and U_i, $\text{Cov}(Y_i, U_i)$, is $D Z_i^T$. Thus, Y_i and U_i are jointly normal, with dimension $m+q$, mean $(X_i\beta, \mathbf{0}_q)$ (where $\mathbf{0}_q$ is a vector of zeroes of length q), and block covariance matrix as in

$$\begin{pmatrix} Y_i \\ U_i \end{pmatrix} \sim N\left(\begin{pmatrix} X_i\beta \\ \mathbf{0}_q \end{pmatrix}, \begin{pmatrix} V_i & Z_i D^T \\ D Z_i^T & D \end{pmatrix} \right).$$

Now, standard multivariate normal theory gives that

$$E(U_i|Y_i) \sim N_q(E(U_i|Y_i), \text{Cov}(U_i|Y_i))$$
$$= N_q(D Z_i^T V_i^{-1}(Y_i - X_i\beta), D - D Z_i^T V_i^{-1} Z_i D^T). \qquad (6.52)$$

So, (6.52) identifies a closed-form expression for $E(U_i|Y_i)$. It would be unrealistic to simplify (6.52) by setting $D = \sigma_{RE}^2 I_q$ because then the Y_{ij} would be independent even for fixed i. However, sometimes U_i is a scalar, $U_i \sim N(0, \sigma_{RE})$, e.g., as a random intercept in a split-plot design.

The best linear unbiased predictor (BLUP) for U_i is $E(U_i|Y_i)$ (in the usual squared error sense) using the generalized least squares estimate $\hat{\beta}_{GLS}$ for β in the first entry in (6.52). That is, let $\hat{U}_i = D Z_i^T V_i^{-1}(Y_i - X_i\hat{\beta}_{GLS})$. It is easy to see that $E(U_i - E(U_i|Y_i)) = E(U_i - \hat{U}_i) = 0$ but harder to verify that the variance of $U_i - \hat{U}_i$ or $E(U_i|Y_i) - \hat{U}_i$ is minimal, even with the normality assumptions we have made. See Robinson (1991) for a thorough treatment, with original references. However, the bigger problem is that (6.52) cannot be used directly because it requires D, V_i, and β to be known and usually they are not. In the rest of this subsection estimators for these parameters are given; this is important in order to find models from which to generate predictions. It's easiest to start with β and V_i.

First, re-express (6.49) more concisely in notation similar to that of Sec. 6.2. Write

$$Y = \begin{pmatrix} Y_1 \\ \vdots \\ Y_n \end{pmatrix}, \quad X = \begin{pmatrix} X_1 \\ \vdots \\ X_n \end{pmatrix}, \quad Z = \begin{pmatrix} Z_1 \\ \vdots \\ Z_n \end{pmatrix}, \quad U = \begin{pmatrix} U_1 \\ \vdots \\ U_n \end{pmatrix}, \quad \epsilon = \begin{pmatrix} \epsilon_1 \\ \vdots \\ \epsilon_n \end{pmatrix} \qquad (6.53)$$

and therefore

$$Y = X\beta + ZU + \epsilon,$$

leading to

$$Y \sim N(X\beta, V) \qquad \text{where} \qquad V = R + Z D_n Z^T,$$

in which $R = \text{diag}(R_1, \ldots, R_n)$ and $D_n = \text{diag}(D, \ldots, D)$, i.e., n copies of D (this suggests that a generalization to subject-specific D is possible but the present treatment is limited to a common D).

Now, for an individual i, the log-likelihood for Y_i is

$$\ell(V_i, \beta; y_i) = -\frac{1}{2}\left(\log|V_i| + (y_i - X_i\beta)V_i^{-1}(y_i - X_i\beta) + n_i \log(2\pi) \right). \qquad (6.54)$$

So, the MLE of β would be $\hat{\beta}_i = (X_i^T V_i^{-1} X_i)^{-1} X_i^T V_i^{-1} Y_i$, if V_i were known. However, higher efficiency in estimating β can be achieved by pooling the data over the n subjects since β is common to all subjects. Doing this gives that the log-likelihood for Y is

$$\ell(V, \beta; y) = -\frac{1}{2} \left(\log |V| + (y - X\beta) V^{-1} (y - X\beta) + n \log(2\pi) \right) + \log(2\pi) \sum_{i=1}^{n} m_i.$$

(6.55)

Now, the MLE for β using all the data is seen to be

$$\hat{\beta} = \hat{\beta}_n = (X^T V^{-1} X)^{-1} X^T V^{-1} y = \left(\sum_{i=1}^{n} X_i^T V_i^{-1} X_i \right)^{-1} \sum_{i=1}^{n} X_i^T V_i^{-1} y_i,$$

(6.56)

provided that V is known.

An analogous procedure can be used to get 'estimates' $\tilde{u}$ – actually predictions – for the U_i (as well as an estimate $\tilde{\beta}$ for β). The joint distribution for U and ϵ is

$$\propto \left| \begin{matrix} D_n & 0 \\ 0 & R \end{matrix} \right|^{-1/2} \exp\left(-\frac{1}{2} \begin{pmatrix} u \\ y - X\beta - Zu \end{pmatrix}^T \begin{pmatrix} D_n^{-1} & 0 \\ 0 & R^{-1} \end{pmatrix} \begin{pmatrix} u \\ y - X\beta - Zu \end{pmatrix} \right).$$

Maximizing this over β and U follows on minimizing the negative of the exponent. Taking derivatives with respect to the components of β and U, setting them equal to zero, and solving leads to Henderson's equations:

$$\begin{pmatrix} \tilde{\beta} \\ \tilde{u} \end{pmatrix} = \begin{pmatrix} X^T R^{-1} X & X^T R^{-1} Z \\ Z^T R^{-1} X & Z^T R^{-1} Z + D_n^{-1} \end{pmatrix} \begin{pmatrix} X^T R^{-1} y \\ Z^T R^{-1} y \end{pmatrix}$$

$$= \begin{pmatrix} (X^T V^{-1} X)^{-1} X^T V^{-1} y \\ D_n Z^T V^{-1} \left(y - X(X^T V^{-1} X)^{-1} X^T V^{-1} y \right) \end{pmatrix}.$$

(6.57)

From (6.57) it is seen that $\tilde{\beta}$ equals the pooled MLE $\hat{\beta}$. Since the second term in the factor within the lower entry of (6.57) is $\tilde{\beta}$, it is easy to see that $\tilde{u} = D_n Z^T V^{-1} (y - X\tilde{\beta})$, generalizing (6.52) from a single i to all n subjects. However, this derivation does not show that $\tilde{\beta}$ and $\tilde{u}$ are MLEs because the estimates are exhibited as a result of maximizing a joint distribution, not a true likelihood. On the other hand, the argument leading to (6.56) implies that $\tilde{\beta}$ is an MLE, and the extension of (6.52) from U_i to U gives that $\tilde{U}$ is the BLUP (for known V).

To get an estimate of β, it remains to estimate the V_i for use in any of (6.52)–(6.57). First, recall that there are $\sum_i m_i$ measurements and that the V_i represent a total of $\sum_i m_i (m_i - 1)$ values. If all $m_i = m$ then there are nm data points and $nm(m - 1)$ parameters in the V_i, an impossible situation. So, constraints must be imposed on the V_i for estimation to be feasible. For this reason, many authors write $R = R(\theta)$, where θ is the vector of components in R. This is particularly useful when it is permissible to assume $R = \sigma^2 I$ for some $\sigma > 0$. In this case, $V_i = V_i(\theta, D)$ and $V = V(\theta, D)$ on setting $\theta = \sigma$. More compactly, write $V = V(\theta)$ by incorporating D into the components of θ; see Jiang (2007). Now, the estimate $\hat{\beta}$ from (6.56) formed by initially setting all $V_i = I_{m_i}$ (say) can be substituted into (6.55) for β to give the profile likelihood $\ell(V(\theta), \hat{\beta}; y)$, which can be maximized to give the MLE $\hat{V}$. This $\hat{V}$ can be substituted back into (6.56) to find a new $\hat{\beta}$, and one can cycle until the estimates of V and β converge. (The MLE approach can also be used to find estimates for D and R_i directly, i.e., without estimating the V_i; this is discussed below but only for D. A Bayesian formulation is given in Laird and Ware (1982).)

An alternative way to find an estimate of V_i is called the restricted maximum likelihood (REML) method. This is very similar, but not identical, to the version of REML described in Sec. 6.2. As before, the REML is often better than the MLE because the latter may have unacceptably high bias in small sample sizes. The basic idea in the LMM context (see Diggle *et al.* (1996), Chap. 4) is to find a $\sum_i m_i \times \sum_i m_i$ matrix K such that $E(KY) = 0$ and then to find the log-likelihood of $Y^* = (Y_1^{*T}, \ldots, Y_n^{*T})^T = KY$ and maximize it to find estimates for the V_i. For instance, if K is block diagonal with blocks $K_i = I_{m_i} - X_i^T (X_i^T X_i)^{-1} X_i$ then $E(Y_i^*) = E(Y_i - X_i \hat{\beta}_i) = 0$, where $\hat{\beta}_i$ is the estimate of β formed by taking $V_i = I_{m_i}$. Now, Y_i^* has a multivariate normal distribution $N(0, K_i V_i K_i^T)$. Putting the Y_i^* together in a single vector Y^* and recalling that $V = V(\theta)$, the log-likelihood, given $Y^* = y^*$, is

$$\ell_R(\theta; y^*) = -\frac{1}{2}\left(\log|KVK^T| + y^{*T}(KV^{-1}K^T)^{-1}y^* + C\right), \tag{6.58}$$

where C is a constant independent of the parameters. Expression (6.58) can be differentiated with respect to θ and the derivatives set equal to zero, to give a set of equations from which $\hat{V}$ and hence the $\hat{V}_i$ can be found. Then V can be fed back into the definition of K (by the generalized least squares expression for the LSE, using $\hat{V}$ rather than the identity matrix) to yield a new Y^* and a new form of (6.58). So, the process can be iterated until $\hat{V}$ converges.

Aside from this, using $K_i = I_{m_i} - X_i^T (X_i^T X_i)^{-1} X_i$ and $Y_i^* = K_i Y_i$ one can derive the likelihood for a single y_i namely

$$\ell_R(V_i; y_i) = -\frac{1}{2}\left(\log|V_i| + \log|X_i V_i X_i| + (y_i - X_i \hat{\beta}_i)^T V_i^{-1}(y_i - X_i \hat{\beta}_i)\right).$$

This can be optimized to get $\hat{V}_i$, which can be used to define a new K_i in order to generate a new $\hat{\beta}_i$ and hence a new likelihood for y_i, from which to find a new $\hat{V}_i$, cycling until convergence.

To use (6.52) to get fitted values, D must also be estimated. This is complicated, but it can be done; see Diggle *et al.* (1996, Secs. 5.3 and 9.2). The basic idea is to use the likelihood function for (β, D) given the Y_i. That is, write

$$L(\beta, D|y_1, \ldots, y_n) = \prod_{i=1}^{n} \int \prod_{j=1}^{m_i} p(y_{ij}|u_i, \beta) p(u_i|D) du_i$$

for the likelihood given by the marginal distribution of $Y_1, \ldots, Y_n$ obtained by integrating out the U_i from (Y_i, U_i) for $i = 1, \ldots n$. The derivatives of $L(\beta, D|y_1, \ldots, y_n)$ with respect to the entries in D can be obtained and, by using a Newton–Raphson procedure, one can generate an estimate $\hat{D}$ for D. Describing this is beyond our present scope but the most standard computational procedure is explained in the documentation on PROC MIXED in SAS; see http://support.sas.com/documentation/. The EM algorithm can also be used but is less popular – and is also beyond our present scope.

When the estimates $\hat{\beta}$, $\hat{V}_i$, and $\hat{D}$ are used in the BLUP $E(U_i|Y_i)$ from (6.52) they give an empirical BLUP,

$$\hat{U}_i = \hat{D} Z_i^T \hat{V}_i^{-1}(Y_i - X_i \hat{\beta}), \tag{6.59}$$

though the adjective 'empirical' is often omitted for brevity. Moreover, different estimators may be used to form empirical BLUPs and these may have different properties. For instance, even though ML and REML estimators are usually similar, when they differ substantially REML should be less biased. It is also important to remember that U_i is a random variable, so $\hat{U}_i$ is a predictor of U even though the values of the U_i are not observed. This is prediction in the same sense that a residual can be said to predict the outcome of an error term that is not measured directly. In both cases, it is essential to verify that the predictions, as a collection, are (i) representative of the known properties of the distribution they are thought to come from, and (ii) associate the individual values from (6.59) with the observations that generated them in a credible way. If this is valid, it's possible to obtain fitted values for Y_i,

$$\hat{Y}_i = X_i\hat{\beta} + Z_i\hat{U}_i, \tag{6.60}$$

that are sometimes called empirical BLUPs for the Y_i. However, the $\hat{Y}_i$ are not predictors: Y_i goes into the 'prediction' of $\hat{U}_i$ on which $\hat{Y}_i$ depends, and all the values of Y_i go into estimating the β's.

At last, we are in a position to make a true prediction of new outcomes of the response using an LMM, as opposed to trying to recover an unobserved outcome of U. There are at least four predictive settings that are natural. The first, and simplest, is predicting the next values for an individual for whom all past observations are available. Perhaps the earliest technique for this is described in Sec. 4.4 of Rao (1987) and given in more generality in Sec. 4 of Chi and Rensel (1989). It is based on the standard formulae for extracting conditional distributions from a normal. Indeed, Rao (1987) derived the simple conditional expectation for a future outcome of a normal regression model with only a random effect term, given past outcomes from the same model. This involves the usual partitioning of a variance matrix as used earlier. Second and third, there are two other prediction cases, which differ in terms of what data is available from which to make a prediction. One is the case where data on a group of subjects is available for existing time points and the goal is to get outcome predictions for new time points. The other is where data for all time points are available on $n-1$ subjects and data for an initial set of time points are available on the nth subject, so the goal is to predict the outcomes for time points beyond the initial set for the nth subject. The fourth prediction technique is more complicated and involves basis element selection. This is analogous to, but less extensive than, the collection of prediction problems discussed in Sec. 6.2.

For the first of these four prediction problems, it makes sense to use a conditional expectation as a predictor. Recall (6.50) and assume that the error term follows an AR(1) process, a special case of which is IID (when the AR parameter is zero; see Sec. 5.2.1 or Rao (1987, Sec. 4.5)). Partition Y_i into two parts, so that $Y_i = (Y_{i1}^T, Y_{i2}^T)^T$ in which Y_{i1} corresponds to the first-stage measurements already available and Y_{i2} corresponds to the second-stage measurements to be predicted. Then X_i can be partitioned into $X_i = (X_{i1}^T, X_{i2}^T)^T$, Z_i can be partitioned into $Z_i = (Z_{i1}^T, Z_{i2}^T)^T$, and ϵ_i into $\epsilon_i = (\epsilon_{i1}^T, \epsilon_{i2}^T)^T$, to give

$$\begin{pmatrix} Y_{i,1} \\ Y_{i,2} \end{pmatrix} = \begin{pmatrix} X_{i,1} \\ X_{i,2} \end{pmatrix} \beta + \begin{pmatrix} Z_{i,1} \\ Z_{i,2} \end{pmatrix} U_i + \begin{pmatrix} \epsilon_{i,1} \\ \epsilon_{i,2} \end{pmatrix}, \tag{6.61}$$

in which β and U remain unchanged. So, $EY_i = ((X_{i,1}\beta)^T, (X_{i,2}\beta)^T)^T$. This induces a block structure on the covariance matrices. Recall that the covariance matrix of Y_i is $V_i =$

$\text{Var}(Y_i) = Z_i D Z_i^T + \sigma^2 I_{m_i}$, which can be partitioned as follows:

$$V_i = \begin{pmatrix} V_{i,11} & V_{i,12} \\ V_{i,21} & V_{i,22} \end{pmatrix} = \begin{pmatrix} Z_{i,1} D Z_{i,1}^T + \sigma^2 I_{m_i,1} & Z_{i,1} D Z_{i,2}^T \\ Z_{i,2} D Z_{i,1}^T & Z_{i,2} D Z_{i,2}^T + \sigma^2 I_{m_i,2} \end{pmatrix}, \tag{6.62}$$

where $I_{m_i,v}$ for $v = 1, 2$ is the identity matrix of dimension equal to that of the first or second part of Y_i, respectively. Since Y_i is normal, it is fully specified by its mean and covariance structure.

Suppose that there is one observation, say $y_{i,1}$, based on $X_{i,1}$ and $Z_{i,1}$, and that X_{i2} and Z_{i2} are known. Then the BLUP $\hat{Y}_{i2}$ for Y_{i2} using the outcome y_{1i} of Y_{1i} is

$$\hat{Y}_{i2} = E(Y_{i2}|Y_{i1} = y_{i1}) = X_{i2}\hat{\beta} + Z_{i2}\hat{U}_{i,1}, \tag{6.63}$$

where $\hat{\beta}$ is an estimate of β (using $\hat{V}_i$ found by REML) and $\hat{U}_{i,1}$ is a predictor of U_i formed from the model $Y_{i,1} = X_{i,1}\beta + Z_{i,1}U_i + \epsilon_{i,1}$. (The Newton–Raphson and EM algorithm methods for estimating D, needed to find $\hat{U}_{i,1}$, have been omitted here.) A more general form of (6.63), namely

$$\hat{Y}_{i2} = E(Y_{i2}|Y_{i1} = y_{i1}) = X_{i2}\hat{\beta} + Z_{i2}\hat{U}_{i,1} + \hat{\epsilon}_{i2}, \tag{6.64}$$

holds when error ϵ is AR(1); see Chi and Rensel (1989). In (6.64), $\hat{U} = E(U_i|Y_{i1} = y_{i1})$ still holds but its interpretation changes because of the AR(1) error. Also $\hat{\epsilon}_{i2} = E(\epsilon_{i2}|Y_{i1} = y_{i1})$, which is nonzero. In fact, Chi and Rensel (1989) gives

$$\hat{\epsilon}_{i2} = R_{21}R_{11}^{-1}(y_{i1} - X_{i1}\beta - Z_{i1}\hat{U}) \tag{6.65}$$

and identifies the form of $R_{21}R_{11}^{-1}$ in terms of the AR(1) parameter.

As a second predictive setting, suppose that all m_i are the same and data for a set of time points are available on all n subjects, so the goal is to predict additional time points for all n subjects. This means that all the existing data can be pooled to get improved predictors. Pooling the data gives a better estimator $\hat{\beta}$ for use in (6.63) as well as giving improved estimators of V_i and D to get a better predictor $\hat{U}_i$ of U_i. Now, (6.63) can be used n times, once for each subject, to get n predictions for the new time points of the n subjects. If predicting the outcome for new times for only one subject is of interest, the task is, of course, easier.

In a third predictive setting one imagines that complete data, i.e., for all time points, on $n - 1$ subjects, and data from an initial set of time points on the nth subject, are available. Then the goal is to use all the data to predict the outcomes for the remaining time points of the nth subject. One way to proceed is to ignore the data for the remaining time points for the first $n - 1$ subjects, so that the prediction problem reverts to the second setting. A better way to proceed is to write (6.61) and (6.62) for the first $n - 1$ subjects, i.e., for $i = 1, \ldots, n - 1$, and then use one more copy of (6.61) and (6.62) for the nth subject, to form a vector $Y = (Y_{1,1}^T, Y_{1,2}^T, \ldots, Y_{n,1}^T, Y_{n,2}^T)^T$. Then, the conditioning properties of the normal distribution are used to derive a version of (6.63) for the remaining time points of the nth subject by regarding $Y_{n,2}$ as the last component of the normally distributed vector Y.

A variant on this is to condition on $Y_{n,1}$ and integrate out the $(Y_{i,1}, Y_{i,2})$ for $i = 1, \ldots, n-1$ to find the conditional expectation $E(Y_{n,2}|Y_{n,1} = y_{n,1})$. Essentially the predictor is then the same as in the first prediction problem because the data from subjects 1 to $n - 1$ are marginalized out. However, the final predictor differs from the predictor in the first problem because one can use all the available data to estimate β, R, D (and hence V), and to find $\hat{U}_n$. This can be done by extensions of the techniques considered here. For further discussion of a related problem see Proust-Lima and Taylor (2009).

A fourth and more sophisticated setting for forecasting originates in Shi *et al.* (1996); variants were explored in James *et al.* (2000) and Rice and Wu (2001). The core idea of the Shi–Weiss–Taylor (SWT) approach is the following. For each subject i, imagine a curve $Y_i(t) = f(t) + S_i(t) + \epsilon_{it}$ in which $f(t)$ is the population mean profile, $S_i(t)$ is the difference between the population mean profile and the subject-specific profile for subject i, and ϵ_{it} is the measurement error. Represent this as

$$Y_i = X_i\beta + Z_iU_i + \epsilon_i^*, \tag{6.66}$$

where Y_i is an $m_i \times 1$ vector of measurements at $t_i = (t_{i1}, \ldots, t_{im_i})$, X_i and Z_i are, respectively, $m_i \times J$ and $m_i \times K$ matrices in which the lth columns ($l = 1, \ldots, J$ and $l = 1, \ldots, K$, respectively) are the lth B-spline basis element evaluated at the values $t_{i1}, \ldots, t_{im_i}$. So, the only difference between X_i and Z_i is that the first uses J B-spline basis elements and the second uses K B-spline basis elements. Clearly, $X_i\beta$ and Z_iU_i are approximations to $f(t)$ and $S_i(t)$, respectively. In addition, assume that $U_i \sim N(0, D_K)$, where D_K is $K \times K$ and ϵ_i^* is $N_{m_i}(0, \sigma^2 I_{m_i})$, a new error term hopefully behaving similarly to the original ϵ_{it}. Thus, making predictions from (6.66) requires the selection of J and K, the estimation of β, D, and σ, and that suitable values for the $\hat{U}_i$ should be obtained.

To simplify the problem, the SWT approach uses a principal components decomposition for D_K to reduce the number of parameters that must be estimated. Write $D_K = \Delta\Lambda\Delta^T$, where Δ is an orthogonal matrix of normalized eigenvectors Δ_k of D_K and $\Lambda = \text{diag}(\lambda_1, \ldots, \lambda_K)$, with eigenvalues in decreasing size order. Since a serviceable K must be found, it is helpful to set $\delta_{ik} = \lambda_k^{-1/2}\Delta_k^T U_i$ and $C_{ik} = \lambda_k^{1/2} Z_i \Delta_k$ (of dimension $m_i \times 1$) and represent Z_iU_i as follows:

$$Z_iU_i = \sum_{k=1}^{K}(Z_i\lambda_k^{1/2}\Delta_k)(\lambda^{-1/2}\Delta_k^T U_i) = \sum_{k=1}^{K}C_{ik}\delta_{ik}.$$

It can be seen that $\delta_{ik} \sim N(0,1)$ and the directions Δ_k corresponding to higher eigenvalues are more important to the approximation of S_i.

Looking component-wise at the vector Y_i, it is reasonable to find a K (for large enough J) such that, for any $i = 1, \ldots, n$, the real-valued function

$$Y_i(t) \approx B(t)\beta + \sum_{k=1}^{K}\delta_{ik}C_k(t) + \epsilon_{it}^* = B(t)\beta + \hat{S}_{i,1:K}(t) + \epsilon_{it}^* \tag{6.67}$$

representing a generic entry of Y_i for a generic t_{ij} (written as t), provides the best fit to the data. In (6.67), all m_i are assumed to be the same, so that $B(t)$ (with dimensions $J \times 1$) is the vector of evaluations of the first J spline functions at $t \in \mathbb{R}$, i.e., $B(t)$ is a generic row of any X_i and is independent of i. Likewise, $C_k(t)$ (with dimensions 1×1) is a component of $\lambda_k^{1/2}Z_i\Delta_k$ formed by using a generic row of Z_i and in which the spline basis elements are evaluated at a general $t \in \mathbb{R}$, i.e., $C_k(t)$ is a generic form of C_{ik} that is independent of i.

Clearly, the random effect part of $y_i(t)$ is $\hat{S}_{i,1:K}(t)$ and the appropriateness of a given K can be evaluated by examining the amount of variance that is explained. Note that the same basis elements are used in both the fixed and random effect parts of the model (though this need not be the case) and there is no harm in regarding the J variables in the fixed effect part separately from the K variables in the random-effects part. It can be seen that both $B(\cdot)$ and the $C_k(\cdot)$ depend on the number and location of knots, with each C_k contributing one random effect.

To implement this model, one can choose the number of knots via a variance–bias trade-off. More knots give a smaller bias but greater variance; fewer knots give a higher bias but a smaller variance. The location of the knots matters as well: Shi *et al.* (1996) suggested using some knots where data were collected and others where there is more curvature in the response.

There are several ways to choose J and K, and Shi *et al.* (1996) gave four: using a likelihood ratio test, using cross-validation, examining the role of the $C_{ik}(t)$ in the reduction of estimates of $\hat{\sigma}$, and, for a fixed J, choosing K by looking at the relative percentage of the variation that is explained by the random effects (this amounts to looking at an R^2 form).

Since $R = \sigma^2 I_{m_i}$, the parameters and the outcomes of the random effect terms, i.e., $\hat{\beta}$, $\hat{D}_K$, $\hat{U}_i$, and $\hat{\sigma}$, can be estimated by a variety of methods, including MLEs, REMLs, and Bayesian methods. The SWT method starts by fixing J and fitting (6.66) using $Z_i = (B(t_{i1}), \ldots, B(t_{in_i}))^T$. Now, D_K is the $K \times K$ covariance matrix of U_i and can be estimated by the MLE or by the EM algorithm. The eigenvalues and eigenvectors of D_K are combined with the B-spline basis to obtain the transformed B-splines $C_k(t)$. The second step is to do model selection (over K) using the C_k. So, refit (6.66) for $K = 1, 2, \ldots$ and use cross-validation scores, the log-likelihood, $\hat{\sigma}$ and R^2. This can be done using an EM algorithm, so that all four methods can be used to choose K.

For given K and i, the regression function can be written as

$$\hat{y}(t) = \hat{f}(t) + \hat{S}(t) = B(t)\hat{\beta} + \sum_{j=1}^{K} \hat{u}_{ij} C_j(t),$$

where the $\hat{u}_{ij}$ are predictions for the U_{ij} (which were transformed to δ_{ij}) with residuals $\hat{e}_i(t) = y_i(t) - \hat{f}(t) - \hat{S}_i(t)$. Then the natural estimate of the population curve is given by confidence bands of the form

$$\hat{f}(t) - z_{\alpha/2}\sigma_{f(t)} \le f(t) \le \hat{f}(t) + z_{\alpha/2}\sigma_{f(t)},$$

where

$$\sigma_f(t) = B^T(t) \left(\sum_{i=1}^{n} X_i^T (\hat{\sigma}^2 I_{m_i} + Z_i \hat{D}_K Z_i^T) X_i \right)^{-1} B(t).$$

In addition, Shi *et al.* (1996) identified the quantile curves associated with (6.67). If all the m_i and t_i are the same then $C_{ik} = C_k$ and the latter also can be regarded as functions of t since the J entries are based on spline functions. Writing $C(t) = (C_1(t), \ldots, C_K(t))$, the $v100\%$ quantile curve is estimated by

$$\hat{Y}_\alpha(t) = B(t)\hat{\beta} + z_\alpha \left(C^T(t)\hat{D}_K C(t) + \hat{\sigma}^2 \right)^{1/2}, \tag{6.68}$$

for any t. This gives a $(1-\alpha)100\%$ prediction interval for a future value of $Y(t)$ of the form $[\hat{Y}_{\alpha/2}, \hat{Y}_{1-\alpha/2}]$. By the symmetry of the normal, the median curve with $\alpha = 0.5$ will be the same as the conditional expectation; see (6.60) and (6.63), which are BLUPs.

6.4.2 Generalized Linear Mixed Models

Just as generalized linear models include a link function to express the conditional mean of a response given covariates in a fixed effect linear model, generalized linear mixed models (GLMMs) introduce a link function to express the conditional mean of the response given the covariates and random effects in a mixed effect linear model.

The basic idea is that some function of the conditional mean of Y_{ij}, given the random effects U_i and explanatory variables X_i, is representable as a linear mixed model and that the Y_{ij} are conditional on the U_i and are independent. So, for Y_{ij}, the jth measurement on subject i, assume that there is a vector U_i of length q such that $(Y_{ij}|U_i)$ belongs to an exponential family. A necessary assumption is that $\text{Var}(Y_{ij}|U_i)$ is a function of $E(Y_{ij}|U_i)$, so that estimating the variance is feasible. Now, for some function g, assume that

$$g(E(Y_{ij}|U_i)) = \sum_{k=1}^{p} X_i(j;k)\beta_k + \sum_{k=1}^{q} Z_i(j;k)U_{ik} \tag{6.69}$$

and that the U_i are assigned some distribution. In the LMM case, $\text{Var}(Y_{ij}|U_i) = \sigma^2$, $g \equiv 1$, and $U_i \sim N(0, D)$.

The logistic regression in Sec. 4.5 can be regarded as a GLM and is extended to a GLMM as follows. Suppose Y_{ij} is binary, taking values 0 and 1. Then, conditionally on the U_i, the Y_{ij} are Bernoulli($E(Y_{ij}|U_i)$), so that $\text{Var}(Y_{ij}|U_i) = E(Y_{ij}|U_i)(1 - E(Y_{ij}|U_i))$. The natural model is (6.69). So, $q = 1$, and $Z_i(j, 1) = 1$ gives

$$g(E(Y_{ij}|U_i)) = \sum_{k=1}^{p} X_i(j;k)\beta_k + U_{i1}$$

and, assuming that g is a logit function,

$$\log \frac{P(Y_{ij} = 1|U_i)}{P(Y_{ij} = 0|U_i)} = \sum_{k=1}^{p} X_i(j;k)\beta_k + U_{i1}.$$

Another example is Poisson regression. Suppose that $Y_{ij} \sim \text{Poisson}(\lambda_{ij})$ and that the $E(Y_{ij}|U_i)$ are independent for $j = 1, \ldots, n_i$, where U_i is the random component for subject i. Then $E(Y_{ij}|\lambda_{ij}) = \lambda_{ij}$. So, it makes sense to write

$$\log(\lambda_{ij}) = X_i(j)\beta + Z_i(j)U_i$$

and assign a distribution to the U's such as $U \sim N(0, D)$.

Estimation in GLMMs is a research topic, to a meaningful extent, although some techniques are gaining wider acceptance. Likelihood methods and least squares techniques have been developed; see Booth and Hobert (1999) for one example of the former and see Zhang (2002) for an instance of the latter.

Predicting new observations in GLMMs has also been studied; see, for instance, Booth and Hobert (1998). However, there seem to be relatively few references for the general case of predicting new outcomes even for specific choices of link function. In a spatial statistics context, Waller and Gotway (2004, Chap. 9, Sec. 7.7, p. 433) states that using a pseudo-likelihood approach for predicting new observations in a GLMM is similar to Kriging (see Sec. 12.3) with the GLM mean and variance structure and gives several references. For a Bayesian version see Zhang (2002) and the references therein.

6.4.3 Nonlinear Mixed Models

An important early contribution to NLMM methodology was that of Lindstrom and Bates (1990), who proposed a model that can be regarded as a subclass of the class to be presented below. This general class of NLMMs saw rapid methodological development in the late 1990s and early 2000s; this was well summarized in Davidian and Giltinan (2003). Using the hierarchical form of the mixed effects model they defined the individual level as

$$Y_{ij} = f(X_{ij}, \beta_i) + \epsilon_{ij},$$

for $j = 1, \ldots, m_i$, in which X_{ij} contains the explanatory variables for subject i and response j, β_i is a collection of parameters specific to individual i, and the form of f is known but may depend on β_i. It is understood that X_{ij} has two parts, t_{ij} and u_i, where t_{ij} can be regarded as time and u_i can be taken to represent other conditions (e.g., the initial dose of a drug). Moreover, at each t_{ij}, the individual deviations satisfy

$$E(\epsilon_{ij}|u_i, \beta_i) = E(Y_{ij} - f(x_{ij}, \beta_i)|u_i, \beta_i) = 0.$$

The population-level model is given by a function d, satisfying

$$\beta_i = d(a_i, \beta, U_i),$$

depending on p fixed effects β, the characteristics a_i of subject i that are independent of anything else, and q random effects in U_i. Usually, $E(U_i|a_i) = E(U_i) = 0$ and $\text{Var}(U_i|a_i) = \text{Var}(U_i) = D$. As before, it is common to take $U_i \sim N(0, D)$. In a slightly more general version of this hierarchical structure for NLMMs, Davidian and Giltinan (2003, Secs. 3.5 and 3.6) discussed Bayesian analysis and individual inference.

   An important more recent contribution was Panhard and Samson (2009). They provided an extension of the EM algorithm to multilevel NLMMs. The model class they used differs from that of Davidian and Giltinan (2003) owing to an extra layer of variability, i.e., in addition to within- and between-subject variability there may be a grouping of subjects. The details in Davidian and Giltinan (2003) are again beyond our present scope, as is the general topic of multilevel NLMMs. Altogether, prediction in these settings (GLMs, GEEs, GLMMs, and NLMMs) seems relatively unexplored.

6.5 Computational Comparisons

Numerous predictive techniques for longitudinal data have been described here. For the purpose of illustration, our attention will be limited to eight techniques, all one-time-step-forward prediction for all subjects assuming that $\ell = 1$ and that at each stage all m_i for $i = 1, \ldots, n$ are the same. That is, we will assume that there is one population of subjects and one sample of size n from it and that all subjects have been measured at the same time. The task is to predict the next measurement for all subjects in the sample, taking the covariance of observations on a given subject into account.

   One data set that can be used to evaluate predictive performance is from the Natural Resource Monitoring, Modeling, and Management (NRM3) Project, Nanyuki, Kenya, and consists of annual measurements of precipitation from a collection of sites in the central Kenyan highlands; see Franz (2007) and Franz *et al.* (2010) for a detailed description. From this data set, the annual amounts of precipitation for the 29 years from 1971 to 1999 inclusive from 29 stations were chosen since they had relatively few missing values (not more than four). The coincidence that there were 29 years and 29 stations merely reflects the maximum number of years and stations for which a nearly complete data set could be obtained. For stations missing a value for a single year which had measured values in the years before and after, linear interpolation was used to fill in the missing value. If two or more values in a row were missing for a station, the available data were used to estimate the parameters in a log-normal distribution and draws were taken from it to fill in the missing values. This is a standard technique from hydrology.

In the predictions below, the stations are treated as the subjects. For each station, its location (in planar coordinates (x, y)) and its elevation were known. These, along with the year and an intercept term, were taken as five fixed effect explanatory variables for $Y_{i,j}$, the precipitation in millimeters for the ith station ($i = 1, \ldots, 29$) and jth year ($j = 1, \ldots, 29$). The stations were treated as independent, even though they are not; this is partially justified because the elevation and location for each could be used. When a random effect term was used in the predictor it was limited to the intercept. Thus, there was at most one random effect in each of the various predictors.

The data for the 29 stations over the 29 years is plotted in Fig. 6.1. It is seen that the data is quite variable with many stations appearing to have numerous outliers; some stations have an increasing trend while others have a decreasing trend and it is impossible to ascertain whether these trends are meaningful.

For the purposes of prediction, the first 15 years at each station were used as a burn-in so that predictors could be identified. Then, one-step-ahead predictions were made for the 14 following years by each of the eight methods. For time step $j + 1$, with $j \geq 15$, all data up to and including time step j were used.

It makes sense to start with the simplest methods, to provide a baseline. Call the first Method 1 and consider using a fixed effect linear regression on each station separately, with no random effects and treating the repeated measurements at a station as if they were independent. This model is the same as that given in (6.24) but with $\Sigma = \sigma^2 I_{29}$ for each station and was more fully described in the frequentist part of Sec. 4.2. The sequential aspect of the model is given in (6.39) and described in the paragraph following that expression. The result is shown in the top panel of Fig. 6.2. For each station, the middle line of the group of three lines indicates the year on year point predictor; the upper and lower lines indicate the 95% PIs. The computations were done in R (R Core Team 2016) using the base command lm. For most stations, the point predictions do not look too bad. The worst is the fourth, Timau Marania, where the under-prediction is systematically high. In some cases, the under- or over-prediction is systematic but not so high, as in Trench Farm and Archers' Post, respectively. In cases such as these, the actual values usually remained within the PI but only in the upper or lower half of it. Thus, it is possible to match the location but not the variability of the predictions.

For contrast, the bottom panel of Fig. 6.2 shows the result of Method 2, namely, Bayes linear regression, again with no random effects and treating repeated measurements on a station as independent. The prior was chosen to be 'noninformative': the priors for the intercept, x, y, elevation, and year were all IID $N(0, 5e^8)$ and the prior on the single variance in the error term was an inverse-Wishart with precision $V = 1$ and degrees of freedom $\nu = 0.002$.

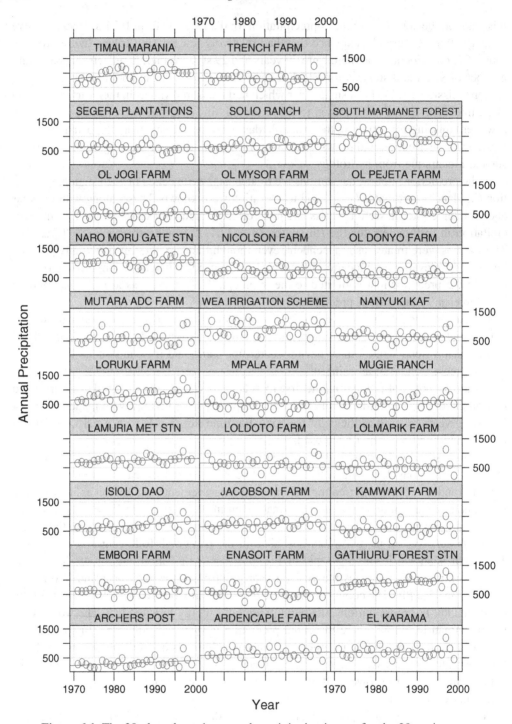

Figure 6.1 The 29 plots show the annual precipitation in mm for the 29 stations over the years 1971 to 1999. The open circles represent the amount for a given year and station; the solid line in each graph represents a simple linear regression for precipitation as a function of year.

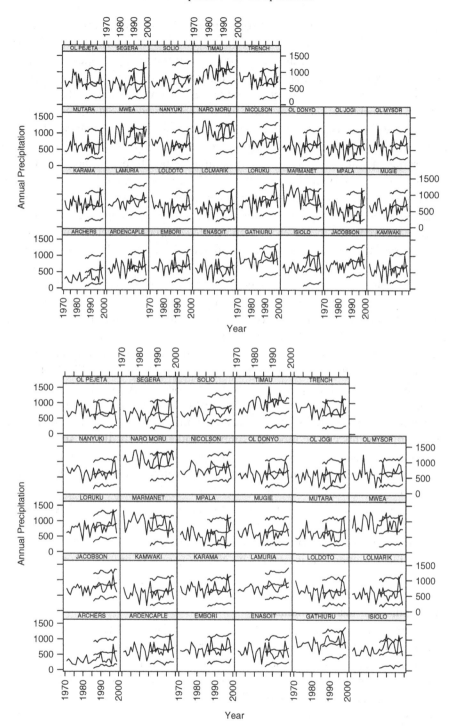

Figure 6.2 The 29 plots show Method 1, standard FE frequentist (top), and Method 2, FE Bayes linear regression with a noninformative prior (bottom), for each station separately assuming no dependence over time. In each plot the jagged line gives the observed data and the set of three curves gives the prediction with PIs above and below. Station names have been abbreviated for clarity.

This is the sequential Bayes analysis of (6.24) using $\Sigma = \sigma^2 I_{29}$ for each station and was more fully described in the Bayes discussion in Sec. 4.2. Details on the sequential aspect are given in the exposition of (6.40) that was implemented in the R package MCMCglmm. As might have been expected, the results are similar to those in the top panel. Even though both methods track the location of the response roughly equally well, visually, the Bayes results look better in the sense that there are fewer times when the actual data lie outside the 95% PIs. This suggests, but does not prove, that Bayes methods allow for more variability of the system to be encapsulated by the model. However, neither method can be called satisfactory.

Next, a series of four methods gave very similar results to each other. Method 3 was the same as Method 2 but with a conjugate prior. That is, within the R package BLR a flat prior was assigned to the five coefficients for the fixed effects and each residual variance was assigned a scaled inv-χ^2 prior with three degrees of freedom and scale $s = 1$. Method 4 used frequentist repeated measures with an AR(1) covariance structure but allowing heterogeneous variances; the form is given by modifying (6.27) to allow unequal variances on the main diagonal and powers of ρ off the main diagonal. Computations were done in the R package nlme. The AR(1) covariance structure permits a correlation structure within stations, but this is different from treating the error term as having two parts, as in (6.14); AR(1) assumes that the dependence between any two measurements at different times is a correlation with exponent equal to the distance between the observations, cf. Sec. 5.2.4. In this case, parameter estimates are found using REML; see (6.33). The technique follows the exposition on extrapolation to a future time point given around (6.37). Method 5 was the same as Method 4 (including using the R package nlme) but used a compound symmetry covariance structure in place of the AR(1) structure. Recall that a compound symmetry structure builds in an extra term, so the intrinsic variability at a time point and the correlation between two time points can be modeled simply. Mathematically, this is similar to allowing the correlation between two time points to represent the inclusion of a random effect; see the description after (6.16). Method 6 represents a different class of methods and techniques. It is a Bayesian analysis of a linear mixed model where the computational procedure is based on nested Laplace approximations, as implemented in the R package INLA. This has not been discussed earlier but the general approach can readily be imagined. A nearly flat prior is put on each coefficient of the fixed effects, namely, $N(0, 2^7)$. The distribution of the random effect (an intercept) is taken to be $N(0, \tau^2)$ and a prior is put on τ^2 by assigning a log-gamma$(1, 5e^{-5})$ prior to the precision $1/\tau$. Then, a Bayes analysis is done to obtain estimates for the parameters and the random effect and a predictive distribution is formed that can be updated as data accumulates.

The common feature among all these four methods is that they give relatively adequate point predictors but very poor PIs. For example, the results of Method 5 are shown in the top panel of Fig. 6.3. Of the four methods the graphs for Method 4 – not shown – look somewhat worse than the other three. However, qualitatively they are similar. There are several possible explanations. The most likely is that the data are over-dispersed relative to the models. That is, the ability of these models to encapsulate the actual randomness in the data is insufficient. So, the models fit the location in a minimally adequate way, but the PIs severely under-represent the variability. A second possibility is that the models are overfitting the location at the cost of poor performance on the variability. A third possibility (always present with relatively difficult computations) is that the output of the procedure is incorrect or the software implementation was unable to perform the computations. (Indeed, a Bayesian version of Method 4 failed to converge.) Whichever may be the case, it appears that these methods perform similarly.

The results of Method 7 had the reverse problem to Methods 3, 4, 5, and 6. That is, the PIs were systematically too wide to be believable rather than too narrow. This method was a Bayes analysis of linear mixed models (not presented in the text) similar to Method 6 but using conjugate priors and finding the posterior and hence the predictive distribution by MCMC using the R package MCMCglmm. In this case, the priors on the fixed effect parameters were independent $N(0, 100)$'s. The distribution on the precision of the normal error term was Inverse-Wishart(100, 0.002). The distribution of the random effect was taken to be normal with mean zero, also with an Inverse-Wishart(100, 0.002) on the precision. The results are shown in the bottom panel of Fig. 6.3. The difference between the results of Methods 6 and 7 – which are conceptually the same – may have arisen because, by implementing the Laplace approximation, Method 6 de facto treated the asymptotic case of infinite sample size, hence giving narrower but asymptotically valid PIs, while Method 7 respected the limitations of finite, indeed small, sample sizes.

Method 8 was a frequentist linear mixed model predictor using the R package lme4. This corresponds to the second predictive setting in Sec. 6.4.1, in which the data are pooled to provide better estimates for use in (6.63). The results are shown in Fig. 6.4. It can be seen that the point predictors are at least as accurate as in the best of the earlier predictors while the PIs are more accurate than any of them. Note that this method turned out best in a fair competition among the eight methods: all eight methods were judged in terms of their performance as predictors for new data from the same DG.

In fairness, there are three limitations to the results shown here. First, the prior specification in the Bayes methods was done using the defaults in the packages. Thus, it is likely that tuning the Bayes methods by prior selection would improve their performance. Second, only one model was used (the same five fixed effect terms and at most one random effect term), so improved model selection should give better results. (Of course, none of the models that could be constructed within the confines of the available data has a serious claim to veracity, so model misspecification cannot be overcome in the confines of this example.) Third, part of the reason why all methods did poorly may be that treating the stations as independent is a poor approximation. Taken together, these three points meant that the predictive errors are not small, as will be seen shortly. That is, none of the predictors did well. Nevertheless, as a series of examples to demonstrate predictive methods, the qualitative forms of the results are probably not atypical.

To conclude this example, it is instructive to look at the cumulative L^1 errors of the eight methods. Using L^1 means that the average errors are on the same scale as the measurements. These are compiled in Table 6.2. The last row of the table gives the average error per prediction in L^1 distance for each method; these are around 200 mm. Figure 6.1 suggests that the typical range for the annual precipitation is under 1000 mm, meaning that the errors are on the order of 20% or more. It is seen that the average per-prediction error of the best method, Method 8, was 171 mm. On the other hand, Method 4 was the worst, as suggested by its PIs (as noted above), meaning that whatever correlation structure is appropriate for stations it is not close to AR(1). The remaining values are relatively close to each other. Methods 1 and 2 are identical in average per-prediction error, as is no surprise in view of Fig. 6.2.

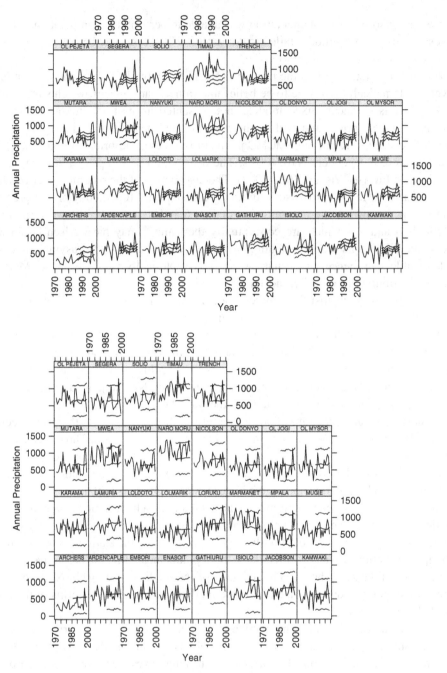

Figure 6.3 The top panel shows Method 5; the PIs shown are far too small, but are larger than the PIs for Methods 3, 4, and 6. The bottom plot shows Method 7, where the PIs appear too wide. Again, the stations were treated as if they were independent of each other even though in these methods the covariance between any two measurements on the same station was nontrivial. In each plot the jagged line gives the observed data and the set of three curves gives the prediction with PIs above and below. Station names have been abbreviated for clarity.

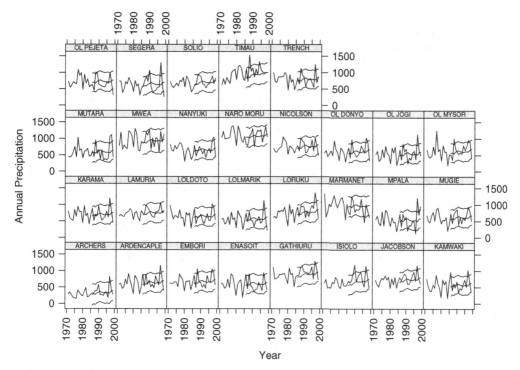

Figure 6.4 This plots here show Method 8 for the 29 stations, treated as independent but allowing dependence between any two measurements on the same station. It is seen that the point predictors from this method are better than the point predictors from Methods 3, 4, 5, 6, and 7 (and Methods 1 and 2, but this is harder to see). In addition, the width of the PIs is more representative of the actual variability than for the other methods. The jagged line gives the observed data and the set of three curves gives the prediction with PIs above and below. Station names have been abbreviated for clarity.

They are more or less in the mid-range of errors over the eight methods, suggesting that the approximation of using linear models is small compared with the other approximations that were made. Indeed, Methods 5 (using compound symmetry) and 6 (Bayes linear mixed models using Laplace's method) gave nearly the same average per prediction errors. Method 3 (conjugate prior linear models) and Method 7 (Bayes analysis of linear mixed models using conjugate priors) gave the second best results, suggesting that the use of conjugate priors may be reasonable; it is not hard to imagine tuning them to give a performance comparable to that of Method 8.

Overall, it must be admitted that even though Method 8 gave the best results, it did not outperform the other methods by very much and did not in itself perform particularly well. However, our comparison of methods is qualitatively informative and the problem is quite hard – certainly $\mathcal{M}$-complete if not $\mathcal{M}$-open – so even minimally adequate results may be as much as can be expected.

6.6 Endnotes: More on Growth Curves

The field of longitudinal data is vast and the topics covered here are elementary. Recent work on longitudinal data – for instance by David Dunson, Joe Ibrahim, and Annie Qu – is

Table 6.2 *Rows 2 through 30 of this table show the L^1 cumulative errors (in* mm*) for the eight methods for each of the 29 stations over the 14 time steps for which predictions were made. At the bottom,* SUM *is the cumulative error for each method over all 29 stations and the average per prediction sum (*PPSUM*) is* SUM*/(14 × 29).*

Station/Method	1	2	3	4	5	6	7	8
Archers Post	3139	3124	3051	3660	3139	3097	3067	1638
Ardencaple Farm	2498	2478	2357	2955	2498	2518	2358	2510
El Karama	2158	2133	2109	2505	2158	2178	2115	2265
Embori Farm	2486	2468	2301	3121	2486	2509	2263	2505
Enasoit Farm	2808	2803	2564	3349	2808	2824	2574	2586
Gathiuru Forest Stn	2390	2414	2376	1827	2390	2421	2364	2090
Isiolo Dao	3569	3576	3553	3021	3569	3599	3549	2883
Jacobson Farm	1966	1990	1780	2313	1966	1986	1778	2020
Kamwaki Farm	2489	2467	2381	3189	2489	2505	2389	2450
Lamuria Met Stn	1571	1550	1478	2073	1571	1563	1487	1527
Loldoto Farm	2147	2147	2054	2905	2147	2166	2064	2162
Lolmarik Farm	2891	2885	2624	3403	2891	2907	2584	2318
Loruku Farm	2381	2383	2285	2342	2381	2403	2271	2759
Mpala Farm	3334	3329	3272	3979	3334	3356	3266	3093
Mugie Ranch	2116	2081	2056	2395	2116	2136	2054	2193
Mutara Adc Farm	3358	3352	3277	4265	3358	3347	3309	3091
Mwea Irr. Scheme	4277	4304	4280	4242	4277	4315	4274	2344
Nanyuki Kaf	1984	1959	1894	2481	1984	2011	1925	2013
Naro Moru Gate Stn	3489	3543	3571	3269	3489	3517	3571	2369
Nicolson Farm	2314	2318	2184	2435	2314	2336	2202	2364
Ol Donyo Farm	2592	2564	2413	3183	2592	2613	2367	2492
Ol Jogi Farm	2945	2964	2747	3440	2945	2956	2805	2775
Ol Mysor Farm	2413	2457	2374	3188	2413	2408	2374	2340
Ol Pejeta Farm	2121	2128	1997	2261	2121	2143	2010	2462
Segera Plantations	3026	3061	2886	3169	3026	3029	2932	3085
Solio Ranch	2188	2131	2067	2643	2187	2179	2122	1506
S. Marmanet Ft. Stn	2904	2953	2908	2345	2907	2925	2882	2957
Timau Marania	5372	5400	5494	4590	5372	5398	5536	2652
Trench Farm 2700	2702	2712	2282	2700	2741	2685	2122	2741
SUM	79625	79666	77049	86833	79625	80082	77180	69566
PPSUM	196	196	190	214	196	197	190	171

much more complex, usually involving not just a longitudinal structure but other defining features such as missing data, survival analysis, and high-dimensional variable selection, to name but a few. Some of these topics will be treated later in this book.

An important topic not given sufficient treatment earlier in this chapter is that of growth curve models. The central idea is to predict the time evolution of some trait over time by fitting a curve. There are numerous ways to do this, sometimes as stochastic processes continuous in time but more typically as a deterministic function of time. Many models presented in this chapter may be regarded as growth curve models, even though they were presented as generic longitudinal models, simply by taking time, or powers of time, as the explanatory variable. To conclude this chapter and provide some orientation to this important field, we look briefly at two growth curve techniques.

6.6.1 A Fixed Effect Growth Curve Model

Some growth curve models can be regarded as a variant on fixed effect linear regression; see Sec. 4.2. For short time series or serial correlation, Pan and Fang (2002) suggested

$$Y = XBZ + \epsilon \tag{6.70}$$

as a useful growth curve model. In (6.70), Y is an $m \times n$ matrix, where n is the number of subjects and m is the number of measurements on each subject. The design matrix X is $m \times d$ of rank d, where d is the number of explanatory variables, here taken as $1, t, t^2, \ldots, t^{d-1}$ i.e., d is the degree of a polynomial in time t. The regression function for a subject is

$$y = \beta_0 + \beta_1 t + \beta_2 t^2 + \cdots + \beta_{d-1} t^{d-1}.$$

The matrix Z is also a design matrix, of dimension $q \times n$ and rank q, and is used to separate the parameters for the q treatment groups. The matrix B is $d \times q$ and contains the parameters. The parameters form sets, one for each of the r treatment groups. The error matrix ϵ is $m \times n$. As with Y, the columns of ϵ correspond to subjects. Usually, the columns of ϵ are assumed to be IID $N_p(0, \Sigma)$, where Σ is unknown. The structure of (6.70) therefore permits the pooling of data over groups to provide better estimation of B, primarily by better estimation of Σ since it is common to all subjects. In this sense, (6.70) can be regarded as a generalization of ANOVA.

To see what this structure means, consider the case $q = 1$, i.e., one population or treatment group. Then $Z = (1, \ldots, 1)$ of length n and B is a single column of length d. If $m = 2, d = 3$, and ϵ is ignored, the first subject gives data

$$y_{11} = \beta_0 + \beta_1 t_1 + \beta_2 t_1^2,$$
$$y_{21} = \beta_0 + \beta_1 t_2 + \beta_2 t_2^2,$$

and hence

$$y_1 = \begin{pmatrix} y_{11} \\ y_{21} \end{pmatrix} = \begin{pmatrix} 1 & t_1 & t_1^2 \\ 1 & t_2 & t_2^2 \end{pmatrix} \begin{pmatrix} \beta_0 \\ \beta_1 \\ \beta_2 \end{pmatrix}.$$

This means that in general the typical (j, k)th entry in X is t_j^{k-1}. Analogous interpretations are possible for $q = 2, 3, \ldots$

Parameter estimation in this class of models proceeds usually by a generalized least squares approach or by an MLE. For the first, one finds

$$\hat{B} = \arg\min \operatorname{tr}\big((Y - XBZ)(Y - XBZ)\big) = (X^T X)^{-1} X^T Y Z^T (ZZ^T)^{-1}$$

and

$$\hat{\Sigma} = \frac{1}{n}(Y - X\hat{B}Z)(Y - X\hat{B}Z)^T.$$

It can be shown that $\hat{B}$ is the best linear unbiased estimate of B. Prediction for new subjects or new times proceeds as in fixed-effects linear regression.

For the MLE approach to estimating B, replace the trace by a determinant. The optimization can be done, and it provides estimates for B and Σ; see Pan and Fang (2002) for details. More general growth curve models are presented in Rao (1987); Franses (2002) used a model with AR(3) error to verify that correcting for autocorrelation improves forecasts.

6.6.2 Another Fixed Effect Technique

For contrast with the SWT technique of Shi *et al.* (1996) presented at the end of Sec. 6.4.1, consider another predictive technique for longitudinal settings due to Cole and Green (1992). The Cole and Green (1992) method can be regarded as a simpler version of SWT and, while somewhat ad hoc, it can give decent results in examples (apart from boundary effects and some problems with nonuniform smoothing, as might be conjectured once the procedure is explained). At root, the approach in Cole and Green (1992) is a growth curve model but of a different form from that in Pan and Fang (2002). In fact, Cole and Green (1992) assumed one measurement per subject, so the problem is longitudinal only in the sense that it is a growth curve, i.e., it expresses Y as a function of t.

The central idea is to find curves like (6.68) that give time-dependent PIs. Again, one could take the endpoints as percentiles from the normal distributions describing the response at each time or alternatively choose the median curve as a point predictor.

For a single variable of interest Y recall the Box–Cox transformation and transform Y to X:

$$x = \frac{(y/\mu)^\lambda - 1}{\lambda}, \qquad \lambda \neq 0;$$

then $x = \ln(y/\mu)$ when $\lambda = 0$. A useful discussion of the Box–Cox transformation, including the effect of influential data, can be found in Sakia (1992). As this is a percentile-based method, μ represents the median of Y and μ maps to $X = 0$. The standard deviation (SD) of X, say σ, is the coefficient of variation of Y, and λ is chosen to minimize the SD of X. This gives

$$z = \frac{x}{\sigma} = \frac{(y/\mu)^\lambda - 1}{\lambda\sigma}, \qquad \lambda \neq 0,$$

and $z = (\ln(y/\mu))/\sigma$ for $\lambda = 0$. So, it should be safe to assume that $Z \sim N(0, 1)$.

The Box–Cox transformation can be extended to include a explanatory variable, say t. If Y depends on t then the optimal λ may depend on t and therefore so will μ and σ. Let $L(t)$, $M(t)$, and $S(t)$ represent the curves for λ, μ, and σ as t varies. Then the Box–Cox transformation takes the form

$$z(t) = \frac{(y(t)/M(t))^{L(t)} - 1}{L(t)S(t)}, \qquad L(t) \neq 0, \tag{6.71}$$

and $z(t) = S(b)^{-1} \ln(y(t)/M(t))$, when $L(t) = 0$. So, for $L(t) \neq 0$, (6.71) gives

$$\hat{Y}_\alpha(t) = M(t)(1 + L(t)S(t)z_\alpha)^{1/L(t)}, \tag{6.72}$$

and, when $L(t) = 0$,

$$\hat{Y}_\alpha = M(t)e^{S(t)z_\alpha}, \tag{6.73}$$

from which quantile-based prediction intervals can be obtained; setting $\alpha = 0.5$ gives a median point predictor.

The remaining difficulty in implementing this lies in estimating $L(t)$, $M(t)$, and $S(t)$. In (6.71), it is often safe to assume that $Z \sim N(0, 1)$ and, following Cole and Green (1992), obtain the likelihood function

$$\ell = \ell(L, M, S) = \sum_{i=1} \left(L(t_i) \ln \frac{y_i}{M(t_i)} - \ln S(t_i) - \frac{1}{2} z_i^2 \right)$$

(neglecting the constant), where $z_i = x_i / S(t_i)$ and $x_i = ((y_i / M(t_i))^{L(t_i)} - 1)/L(t_i)$ for independent observations $i = 1, \ldots, n$ (analogous to (6.71)). Now, consider the penalized likelihood

$$\ell^* = \ell - \alpha_\lambda \int (L''(t))^2 dt - \alpha_\mu \int (M''(t))^2 dt - \alpha_\sigma \int (S''(t))^2 dt, \qquad (6.74)$$

where α_λ, α_μ, and α_σ are smoothing parameters. Maximizing (6.74) leads to cubic splines with knots at distinct t_i; see Clarke *et al.* (2009, Sec. 3.2), so that only the three smoothing parameters remain to be chosen. Recall that a knot is a location where different curve segments of a spline meet; knots are often placed at a subset of the time points where responses are observed. Cole and Green (1992) used an iterative procedure to get results by choosing initial values of α_λ, α_μ, and α_σ, maximizing in (6.74), and then empirically choosing new values for α_λ, α_μ, and α_σ. The net result is three smoothing spline estimates $\hat{L}(t)$, $\hat{M}(t)$, and $\hat{S}(t)$ which can be used directly in (6.72) or (6.73). More conservatively, confidence bounds on $\hat{L}(t)$, $\hat{M}(t)$, and $\hat{S}(t)$ can be obtained and used in (6.72) or (6.73).

7

Survival Analysis

> Much of statistical analysis is concerned with making inferences about ... parameters
> ...the purpose of such inference statements is surely to convey to some second party
> information about what is likely to happen if the experiment is performed again ... It
> is surprising therefore that a greater amount of thought has not been given to the more
> direct practical type of inference where statements are required for what is likely to occur
> when future experiments are performed. Indeed, it is common practice for a statistician
> to obtain from the experimental outcome an estimate of the indexing parameter of some
> class of distributions describing an experiment and subsequently use the estimate as if
> it were the true value to allow prediction. It is paradoxical that while the folly of this
> approach is pointed out in simple situations there is an all too ready acceptance of it in
> more complicated situations.
>
> Aitchison and Dunsmore (1975, preface)

There are three distinctive features about survival data. First, the response Y to be predicted
is a 'lifetime' – a period of time with a well-defined start and end, even if the start or end
time is unavailable to us. Second, there is censoring, i.e., some values of Y are missing. This
is interpreted to mean that the lifetime exceeds the observation time. Three, although not a
regression problem in the sense of linear models, covariates X_i are available for subjects i
and can be used to improve the prediction of lifespan.

Some notation that is common to survival analyses will be defined for this chapter. First,
Y is a continuous nonnegative random variable with DF $F(y) = P(Y \le y)$ representing the
length of time (y) that a subject has survived. Here, P is the probability measure for Y with
density function $p(y)$. In practice it is the complement of the DF, $S(y) = 1 - F(y)$, that is
usually more important because interest focuses on the probability of surviving longer rather
than that of having survived up to a given time. Usually, $S(\cdot)$ is called the survival function.
As before, assume that IID outcomes $y_1, \ldots, y_n$ of Y are available. Then, it is important to
estimate $S(y)$ by, for instance, $\hat{S}(y)$, to get interval predictors for Y_{n+1} and to use features
of $\hat{S}(\cdot)$ to get point predictors for Y_{n+1}.

In the simplest cases, the EDF gives an estimate of S. That is,

$$\hat{S}(y) = \frac{\{\#y_i > y\}}{n}.$$

(Some authors use $\ge$ in place of $>$ but this usually makes little difference for continuous
data.) In this case, $\hat{S}$ looks like a step function, decreasing from one to zero. The problem
arises when the data are censored, e.g., when the lifespan y_i is not observed but partial
information, that the subject survived up to a certain time, is available. Often this is written

as $Z_i = \min(Y_i, U_i)$, for $i = 1, \ldots, n$, where U_i represents the censoring time and $Z_i = z_i$ is the last time at which the subject Y_i was known to be alive. The datum is called right censored because the true survival time is to the right of $U = u$. As a generality, if $Z_i = z_i$ then it is known whether $U_i = z_i$ or $Y_i = z_i$ occurred. So, in fact, there is another variable δ_i defined by

$$\delta_i = \begin{cases} 1 & Y_i \le U_i, \\ 0 & Y_i > U_i, \end{cases} \tag{7.1}$$

indicating which Y_i were actually observed rather than censored. In practice, the data are of the form (Z_i, δ_i) for $i = 1, \ldots, n$, i.e., for each i the observed X_i or time of censoring (and the fact that the censoring occurred).

Obtaining predictors for Y_{n+1} when some of the Y_i are censored is a key problem in survival analysis. Predictors can be obtained by estimating S nonparametrically, semi-parametrically, and parametrically; see Secs. 7.1, 7.2, 7.2.3, and 7.3, respectively. That is, representatives of all three classes will be seen in this chapter.

Another key question in survival analysis is as follows: given that a subject has survived up to time y, what is its probability of surviving longer, for instance, up to time $y + z$? This is formalized in the conditional probability $P(Y \ge y + z | Y \ge y)$. However, our interest is in z incrementally not macroscopically. So, usually it is the instantaneous change in the conditional probability of survival at y that is used. This is

$$h(y) = \lim_{\delta \to 0} \frac{P(y \le Y \le y + \delta | Y \ge y)}{\delta} = \frac{p(y)}{S(y)} = -\frac{S'(t)}{S(t)} = -\frac{d}{dy} \ln S(y) \tag{7.2}$$

and is called the hazard function. Since (7.2) is a derivative, it is natural to integrate it to obtain the cumulative hazard function,

$$H(y) = \int_0^y h(u) du = -\ln S(y) \quad \text{or} \quad H(y) = -\log S(y). \tag{7.3}$$

Note also that $h(t) = H'(t)$. From (7.2) and (7.3),

$$p(y) = h(y) \exp\left(-\int_0^y h(u) du \right),$$

i.e., the density for Y is a product of factors that depend on the hazard function. The shape of the hazard function indicates how survival up to time y affects survival in the infinitesimal interval of time following y. It is easy to imagine that survival up to y could represent 'wearing out', so that $h(y)$ is increasing, i.e., the instantaneous chance of death is increasing. It is also easy to imagine that survival up to y could represent recovering from illness, so that $h(y)$ is decreasing, i.e., the instantaneous chance of dying is decreasing. If $h(y)$ is constant, the instantaneous chance of dying does not change. All intermediate shapes are possible, too. The hazard function is central to semiparametric treatments of survival problems; see Secs. 7.2 and 7.2.3. Although not treated here, h can also be related to the nonparametrics of Sec. 7.1; see Hjort (1990).

The simplest hazard function is constant: $h(y) = \theta$ for some fixed θ. This leads to a survival function $S(y) = e^{-\theta y}$, which corresponds to the exponential(θ) distribution. In this case, if θ were known, it would be easy to find the expected lifespan $\mu = \int_0^\infty y p(y) dy$. Elementary calculus shows that $\mu = \int_0^\infty S(y) dy$ as well. (Integrate by parts using $(d/dy) S(y) =$

$-p(y)$ with $S(0) = 1$ and $S(\infty) = 0$.) Note that here the terms lifespan, survival, death, etc., are used but the mathematical structure applies to any positive random variable representing a waiting time until the occurrence of an event.

In Sec. 7.2, predictors obtained from Cox's proportional hazards (PH) model will be derived. These are semiparametric because there is a factor based on $X^T\beta$ that multiplies a 'baseline hazard', where $\beta \in \mathbb{R}^k$ (taking X_{i1} to be a constant term). This parallels the earlier derivation of predictors from regression models. Moreover, in Sec. 7.2.3 we derive predictors from the Bayes version of PH models.

Another survival function is Weibull(α, γ) with density $\alpha\gamma y^{\alpha-1}e^{-\gamma y^\alpha}$. This reduces to the exponential(θ) case when $\alpha = 1$. In this case, $S(y) = e^{-\gamma y^\alpha}$ and $h(y) = \gamma\alpha y^{\alpha-1}$, so $H(y) = \gamma t^\alpha$. These quantities can be worked out for many examples and used in a regression context. Indeed, if explanatory variables $X_i = (X_{i1}, \ldots, X_{ik})^T$, generically denoted X, are available, then Sec. 7.3 shows how to generate predictors derived from parametric models for survival data. Essentially, a location parameter is written as a linear combination of the variables X; this is the simplest context in which to introduce Bayes predictors that include explanatory variables.

The endnotes of this chapter briefly present two variations on these predictors.

As with longitudinal analysis, predictive techniques are not as commonly used with survival data as one might expect or hope. In longitudinal analysis this could be purely because of the complexity of the data sets. However, for basic survival analysis it is harder to make this argument. It may simply be that the users of results from survival analysis are so focused on modeling that prediction is ignored. Alternatively, it may be that model uncertainty, and the intrinsic variability of the populations being modeled, is high relative to the sample size, so that predictions are unreliable or PIs too wide to be useful. This is partially attributable to the fact that the relationship between the data and their distribution is tighter than the relationship of the data to their density, here the numerator of the hazard function, making for less stable inference. In the same spirit, the hazard function is a ratio and estimates for ratios are not as stable as estimates for sums.

7.1 Nonparametric Predictors of Survival

Our task is to use the n possibly censored outcomes $y_1, \ldots, y_n$ of Y, assumed IID, to predict Y_{n+1}; if an outcome y_i is censored, it is replaced by its censoring time. Prediction intervals can be found via nonparametric methods, see Sec. 4.1, where the goal is to understand the whole distribution of the lifetimes. This approach conveys more information, partly because tail behavior is so important. Prediction can also be done via point predictors (with an assessment of variability), as in Secs. 2.2.1 and 2.2.2, whether $\sigma^2 = \text{Var}(Y_i)$ is known or whether it must be estimated. The first subsection presents the main frequentist nonparametric approach to estimating S and then gives point predictors from it. The second subsection presents a nonparametric Bayes technique for estimating S and again gives point predictors from it.

7.1.1 The Kaplan–Meier predictor

The standard frequentist nonparametric estimator of S is called the Kaplan–Meier (KM) or product-limit estimator, see Kaplan and Meier (1958), and is particularly useful because it accommodates right-censored data, reducing to one minus the empirical distribution function $(1 - \text{EDF})$ when the data are not censored. To derive the KM estimator $\hat{S}(y)$, let $\tau_1 < \cdots < \tau_K$ be the distinct times of death observed over the n subjects, let d_k be the number of deaths at τ_k, and let r_k be the number of subjects 'at risk' just before τ_k, i.e., the number of subjects who have not died or been censored strictly before τ_k. (This is the number of subjects who are known to be alive at τ_k or who died at τ_k, assuming that censoring does not occur at any τ_k.) Now, d_k/r_k is the sample proportion of subjects who, having survived at least until $\tau_k - \epsilon$ for $\epsilon \to 0^+$, died at τ_k or in $[\tau_k, \tau_{k+1})$. So, $1 - d_k/r_k$ is the sample proportion of subjects who, having survived at least until $\tau_k - \epsilon$ for $\epsilon \to 0^+$, also survived the interval $[\tau_k, \tau_{k+1})$, i.e., died at τ_{k+1} or were known to be alive at τ_{k+1} if only to be censored later. More formally, for $y \geq \tau_1$, this can be written as

$$S(y) = P(Y \geq y) = \prod_{k:\,\tau_k \leq y} P(\text{survived } [\tau_k, \tau_{k+1})|\text{survived to } \tau_k)$$

$$\approx \prod_{k:\,\tau_k \leq y} (1 - d_k/r_k). \tag{7.4}$$

If $y < \tau_1$, it makes sense to set $\hat{S}(y) = 1$. That is, (7.4) can be used for $y < \tau_1$ by setting $\tau_0 = d_0 = 0$ and $r_0 = n$. Note that, for $y \in [\tau_1, \tau_2)$, (7.4) gives $\hat{S}(y) = 1 - d_1/r_1$. From (7.4) it is seen that the KM estimator does not change between events and does not change when a censoring occurs: it only changes when an event (or death) occurs. After τ_K, the KM estimator is a constant but as this is not based on any observations it is usually regarded as meaningless.

The relationship between (7.4) and the hazard function (7.2) can be seen as follows. Let $y \in (\tau_k, \tau_{k+1}]$. Then, writing

$$S(y) = P(Y \geq \tau_{k+1}) = P(Y \geq \tau_1, \ldots, Y \geq \tau_{k+1})$$

$$= P(Y \geq \tau_1) \prod_{j=1}^{k} P(Y \geq \tau_j + 1|Y \geq \tau_j)$$

$$= 1 \times \prod_{j=1}^{k} (1 - P(Y = \tau_j|Y \geq \tau_j)) = \prod_{j=1}^{k} (1 - h_j), \tag{7.5}$$

where h_j is the value of the hazard function for a discrete random variable V taking values τ_j, i.e., $h_j = P(V = \tau_j|V \geq \tau_j)$.[1] The hazard h_j can be estimated by d_j/r_j, so that (7.5) has the same form as (7.4). (Sometimes the empirical form of (7.5) is called the Nelson–Aalen estimator; fundamentally it represents an effort to estimate the cumulative hazard function $H(\cdot)$ as a way to estimate $S(\cdot)$.)

It is an exercise to verify that (7.4) can also be written

$$\hat{S}(y) = \begin{cases} 1 & \text{if } y < \tau_1, \\ \prod_{\tau_j \leq y} \left(1 - \dfrac{d_j}{n_j}\right) & \text{if } \tau_1 \leq y, \end{cases} \tag{7.6}$$

---

[1] A hazard function for a discrete random variable has the form $h(\tau_j) = P(Y = \tau_j)/P(Y \geq \tau_j)$, which equals $f(\tau_j)/S(\tau_j) = f(\tau_j)/\sum_{k \geq j} f(\tau_k)$, i.e., the proportion of deaths at or after τ_j.

where the d_j are the number of events (e.g., deaths) at time τ_j. In the absence of censoring, n_j is the number of survivors just before τ_j. If some events are right-censored then n_j is the number of survivors less the number of losses due to censoring just before τ_j. This makes sense because it is only the survivors who are still being observed who are at risk of death.

By symmetry, it is also possible to estimate the censoring distribution by

$$\hat{C}(y) = \begin{cases} 1 & \text{if } y < \tau'_1, \\ \prod_{\tau'_j \le y} \left(1 - \dfrac{d'_j}{n'_j}\right) & \text{if } \tau'_1 \le y, \end{cases} \tag{7.7}$$

where the τ'_j are the distinct censoring times, d'_j is the number censored at τ'_j, and the n'_j are the number at risk of being censored at time t. Essentially, (7.7) is the same as (7.6) but taken at the censoring points rather than at the times of death. Consequently, all the results below for estimating S consistently by $\hat{S}$ or getting an SE for it can be applied to obtain a consistent estimation of C, the survival function for the censoring process, by $\hat{C}$ and to get an SE for it.

The main $\mathcal{O}(1/\sqrt{n})$ pointwise weak consistency result with an identified covariance is due to Breslow and Crowley (1974, Secs. 5 and 6). (There are important contributions pre-dating Breslow and Crowley (1974): Greenwood's formula used below dates from 1926!) However, uniform weak consistency (with a $\mathcal{O}(1/\sqrt{n})$ rate) is attributed to Gill (1983) and strong uniform consistency, with the rate incompletely handled but decreased from $\mathcal{O}(1/\sqrt{n})$ by $((\log n)/n)^{1/2}$, is to be found in Foldes and Rejto (1981). Importantly, Chen and Lo (1997) established uniform weak and strong laws with rates $\mathcal{O}(1/n^q)$, where $q < 1/2$ is determined by the censoring.

Having defined the KM estimator, it is important to convert it to a predictive procedure. After all, in the survival analysis context, the interest in a distribution is only predictive because once a subject has died, the subject no longer plays a role. In the more general case of waiting for an 'event' to occur (not necessarily a 'death'), again, once the event has occurred interest remains only in the cases for which the event has not yet occurred.

It is easier to base a predictor on the KM estimator by writing $\hat{F}(\cdot) = 1 - \hat{S}(\cdot)$; note that $\hat{F} = \hat{F}_n$ is now a generalization of the EDF from Sec. 4.1 to permit right-censoring. So, the simpler techniques of Sec. 4.1 do not in general apply. Indeed, the effect of the censoring is to increase the uncertainty in estimating $S(\cdot)$: past the largest observed value there is no data with which to estimate the right-hand side of $\hat{S}(\cdot)$. The key statement that can be made about survival beyond τ_K is that the empirical probability of survival for $y > \tau_K$ should be bounded as

$$\frac{\#\text{ censored after } \tau_K}{n} \ge \hat{S}(y) \ge 0,$$

in which for the upper bound it is assumed that all subjects censored after τ_K would have survived beyond y and for the lower bound that all subjects at risk immediately before τ_K would have been censored at τ_K had they not died at τ_K.

To get around this, most results on the convergence of $\hat{F}$ are limited to intervals of the form $[0, T]$, where $T < \infty$. Often the value of T is influenced by the distribution G of the censoring random variables U_i, assumed IID. Write the DF of Z as F_Z with survival function S_Z. Clearly, $F_Z(t) = P(Z_i \le t) = 1 - (1 - F(t))(1 - G(t))$ for any i. In some cases, convergence results for $\hat{F}$ are on $[0, t_Z]$ where $t_Z = \sup\{t \,|\, F_Z(t) < 1\}$(Chen and Lo 1997); in other cases convergence has been proved for $[0, T]$ satisfying $\min P(Z > T), 1 - P(Z > T) \geq \delta$ for some $\delta \in (0, 1/2)$ (Foldes and Rejto 1981) or $0 \leq t \leq T$ for some $F_Z(T) < 1$ (Breslow and Crowley 1974).

The way in which G assigns mass relative to F controls the pattern of censoring. For instance, if the support of G is to the right of the support of F then $\min(Y, U) = Y$ and censoring never happens, so $Z_i = Y_i$ and $\delta_i = 1$ for all i. So, if T is the upper limit of the support of F then convergence of the KM estimator occurs on the support of $F = F_Z$, an interval of the form $[0, T]$. However, if the support of G is in a sufficiently small interval $[0, \epsilon]$ that $\min(Y, U) = U$ then essentially all data are censored, so $Z_i = U_i$ and $\delta_i = 0$ for all i. More generally, if both F and G are strictly positive on $\mathbb{R}^+$ then, in effect, $t_z = \infty$, giving convergence on compact subsets in $\mathbb{R}^+$ that can be assumed increasing. Why? Because as n increases, there will be (probabilistically) ever more points with $\delta_i = 1$, i.e., more Z_i that give a y_i rather than a u_i. In a more typical case, note that as the mass of G moves to the left relative to the mass of F then there are more censored data points. Indeed, if G corresponds to unit mass at a point, say t_m, that is the median of Y, then half the observations will be censored and

$$F_Z(\min(Y, U) \leq t) \begin{cases} = 1/2 & \text{for } t > t_m, \\ < 1/2 & \text{for } t < t_m. \end{cases}$$

Analogous statements can be made when G corresponds to other percentiles of F. In these cases, $t_z = t_{1-\alpha}$, where $t_{1-\alpha}$ is the $(1-\alpha)100$th percentile of F.

Foldes and Rejto (1981) showed that, under their assumption above on Z, for $1 > \epsilon > 12/(n\delta^4)$,

$$P(\sup_{0 \leq t \leq T} |F(t) - \hat{F}(t)| < \epsilon) \geq 1 - \frac{c_1}{\epsilon} e^{-c_2 n \epsilon^2 \delta^6},$$

where $c_1, c_2 > 0$ are constants. That is, for each ϵ, as $n \to \infty$, the left-hand side probability is $\mathcal{O}(e^{-c_3 n})$ for some $c_3 > 0$ and one can set $\epsilon = \epsilon_n \to 0$ slowly enough that the bound is maintained as n increases. (In fact, one can set $\epsilon_n = \sqrt{(\log n)/n}$.) For instance, if G concentrates its mass at, say, $t_{0.75}$ then the condition $\min[P(Z > t), 1 - P(Z > t)] \geq \delta$ for some $\delta \in (0, 1/2)$ holds; however, the Foldes and Rejto (1981) condition is much more general. Now, for prediction in a survival context, $F(0) = 0$ and one-sided PIs that correspond to the waiting time until an event happens are more important than two-sided PIs. Thus, for given $\alpha > 0$, consider the putative $1 - \alpha$ PI $[0, \hat{F}^{-1}(1-\alpha)]$. It has probability

$$P(Y_{n+1} \in [0, \hat{F}^{-1}(1-\alpha)]) = P\left(F(Y_{n+1}) + (\hat{F}(Y_{n+1}) - F(Y_{n+1})) \in [0, 1-\alpha]\right)$$

$$= P\left(F(Y_{n+1}) \in [-(\hat{F}(Y_{n+1}) - F(Y_{n+1})),\right.$$

$$\left. 1 - \alpha - (\hat{F}(Y_{n+1}) - F(Y_{n+1}))]\right). \tag{7.8}$$

If the 'good' set (locations where the predictive error is less than ϵ) is

$$A_{\epsilon,n} = \{t \,|\, \sup_{0 \leq t \leq T} |\hat{F}(t) - F(t)| < \epsilon\}$$

then (7.8) is bounded above by

$$P\left(\{F(Y_{n+1}) \in [0, 1 - \alpha + |\hat{F}(Y_{n+1}) - F(Y_{n+1})|]\} \cap A_{\epsilon,n}\right) + P(A_{\epsilon,n}^c)$$

$$\leq P\left(\{F(Y_{n+1}) \in [0, 1 - \alpha + \sup_{0 \leq t \leq T} |\hat{F}(t) - F(t)|]\} \cap A_{\epsilon,n} \cap \{Y_{n+1} \leq T\}\right)$$

$$+ P\left(\{F(Y_{n+1}) \in [0, 1 - \alpha + |\hat{F}(Y_{n+1}) - F(Y_{n+1})|]\} \cap \{Y_{n+1} > T\}\right) + P(A_{\epsilon,n}^c)$$

$$\leq P\left(\{F(Y_{n+1}) \in [0, 1 - \alpha + \epsilon]\} \cap \{F(Y_{n+1}) \leq F(T)\}\right) + P(Y_{n+1} > T) + \frac{c_1}{\epsilon} e^{-c_2 n \epsilon^2 \delta^6}$$

$$\leq 1 - \alpha + \epsilon + (1 - F(T)) + \frac{c_1}{\epsilon} e^{-c_2 n \epsilon^2 \delta^6}.$$

Analogously, (7.8) is bounded below by

$$P\left(\{F(Y_{n+1}) \in [0, 1 - \alpha + |\hat{F}(Y_{n+1}) - F(Y_{n+1})|]\} \cap A_{\epsilon,n}\right)$$

$$\geq P\left(\{F(Y_{n+1}) \in [0, 1 - \alpha - \sup_{0 \leq t \leq T} |\hat{F}(t) - F(t)|]\} \cap A_{\epsilon,n} \cap \{F(Y_{n+1}) \leq F(T)\}\right)$$

$$+ P\left(\{F(Y_{n+1}) \in [0, 1 - \alpha + |\hat{F}(Y_{n+1}) - F(Y_{n+1})|]\} \cap A_{\epsilon,n} \cap \{F(Y_{n+1}) > F(T)\}\right)$$

$$\geq \min(1 - \alpha - \epsilon, F(T))$$

$$+ P\left(\{F(Y_{n+1}) \leq 1 - \alpha + |\hat{F}(Y_{n+1}) - F(Y_{n+1})|\} \cap \{F(Y_{n+1}) > F(T)\}\right)$$

$$= \min(1 - \alpha - \epsilon, F(T)) + P\left(F(Y_{n+1}) \in [F(T), 1 - \alpha + |\hat{F}(Y_{n+1}) - F(Y_{n+1})|]\right),$$

and is seen to be less neat. (In the lower bound, $\gtrsim$ indicates that dropping $A_{\epsilon,n}$ makes little difference. Also, the closed interval in the last event may be void.) Indeed, the inequality of the upper and lower bounds reveals that the finiteness of T, or equivalently, the inability to estimate F beyond T, prevents an asymptotically tight approximation for (7.8).

In deriving these upper and lower bounds, one could have used Theorems 2.1 or 3.1 in Chen and Lo (1997) to generate essentially the same results. In this approach, conditions that depend explicitly on G can be identified, although Chen and Lo (1997) required $t_Z < \infty$. Their Theorem 3.6 also shows that a good estimation of F at t_Z is possible (although this requires estimates of the densities F' and G').

More exactly, in both the upper and lower bounds on (7.8) the limit is $1 - \alpha$, provided that $n \to \infty$, $\epsilon \to 0$, and G is such that T can be made large enough for the other terms to be small. However, for $T \to \infty$, there must effectively be no censoring. On the other hand, since it is the lower bound that is more important, one can restrict α to values making the two terms in the lower bound as large as possible. This can be done, for instance, by choosing α large enough that $\alpha \geq 1 - F(T)$ and that the event $\{\alpha \geq 1 - F(T) + |\hat{F}(Y_{n+1}) - F(Y_{n+1})|\}$ has sufficiently small probability, respectively. Again, the censoring may make these impossible to satisfy. Thus, the usefulness of $\hat{F}$ as a way to generate the predictions of greatest interest – namely, those for the right-hand tail of F – in general is limited, possibly severely, by G.

A second way to look at prediction from $\hat{F}$, parallel to the discussion in Sec. 4.1, is to find expressions for the convergence of $\hat{F}^{-1}(1 - \alpha)$, recalling that the minimal size of α is controlled by G. Since it is very difficult to find a general expression for $\hat{f}$ beyond τ_K, it must be assumed that α is big enough that a meaningful estimate of the KM-based estimate of F, $\hat{F}$, is available. Indeed, one can usually argue that even before τ_K the estimate of $\hat{F}$

is based on so few measured data points as to be unstable. Here, for the sake of deriving predictors this will be ignored – but for real data it cannot be. Now, the main goal is to find an SE for $\hat{F}^{-1}(1-\alpha)$ such that intervals of the form $[0, \hat{F}^{-1}(1-\alpha) + \phi SE]$ can be constructed, where ϕ is a factor giving a conservative fraction of the SE.

There are two ways (at least) to form such a PI. First, one might invoke Theorem 5 in Breslow and Crowley (1974). So, choose T with $F_Z(T) < 1$ and suppose that F and G are continuous. Then, for $t \in (0, T)$,

$$\sqrt{n}(\hat{F}(t) - F(t)) \to Z^*(t), \tag{7.9}$$

where $Z^*(t)$ is a mean-zero Gaussian process that, for $s \leq q$, has

$$\text{Cov}(Z^*(s), Z^*(t)) = (1 - F(s))(1 - F(t)) \int_0^s (1 - F(y))^{-2} \tilde{F}(dy),$$

where $\tilde{F}(y) = P(Y \leq y, \delta = 1) = \int_0^y (1 - G(y)) F(dy)$. Breslow and Crowley (1974) comment that if G has support $\mathbb{R}^+$ then $F(t) < 1$ for $t \in (0, \infty)$ and so their Theorem 5 probably holds on $\mathbb{R}^+$. To use (7.9) let $\hat{\xi}_{1-\alpha} = \hat{F}^{-1}(1-\alpha)$ and $\xi_{1-\alpha} = F^{-1}(1-\alpha)$ be the $(1-\alpha)100$th percentiles of $\hat{F}$ and F, respectively. Then $F(\xi_{1-\alpha}) = 1 - \alpha$. Now assume that $\hat{F}$ has been smoothed, so that it is strictly increasing and has a continuous derivative. (This assumption is very subjective but has the benefit of fixing a reasonable form for the tail behavior of $\hat{F}$; it is necessary for the present method in order to get ϕ but is not needed to locate the PI. It is well known that estimation and prediction procedures are less sensitive to scaling than to location.)

Setting $t = \xi_{1-\alpha}$ and using the function $\hat{F}^{-1}$, the δ-method gives

$$\frac{\sqrt{n}}{(\hat{F}^{-1})'(\xi_{1-\alpha})}(\hat{F}^{-1}(\hat{F}(\xi_{1-\alpha})) - \hat{F}^{-1}(F(\xi_{1-\alpha}))) \to Z^*(\xi_{1-\alpha}),$$

which, possibly with some approximation, simplifies to

$$\frac{\sqrt{n}}{(\hat{F}^{-1})'(\xi_{1-\alpha})}(\xi_{1-\alpha} - \hat{\xi}_{1-\alpha}) \to Z^*(\xi_{1-\alpha}),$$

in which, as a practical matter, it is reasonable to set $(\hat{F}^{-1})'(\xi_{1-\alpha}) = (\hat{F}^{-1})'(\hat{\xi}_{1-\alpha})$. Now, for $P(Y_{n+1} \leq \hat{F}^{-1}(1-\alpha))$, it is reasonable to use

$$\left[0, \hat{F}^{-1}(1-\alpha) + \frac{Z^*(\hat{\xi}_{1-\alpha})_{1-\alpha'}(\hat{F}^{-1})'(\hat{\xi}_{1-\alpha})}{\sqrt{n}} \right] \tag{7.10}$$

as an approximate $(1-\alpha)100\%$ PI for Y_{n+1} with confidence $1 - \alpha'$.

An alternative to invoking asymptotic normality of $\hat{F}(\cdot)$ at the beginning and then using the δ-method is to start by deriving an expression for the variance of $\hat{F}(\cdot)$ directly and then applying the δ-method to it. This gives a different form for the SE from that seen in (7.10). One expression for $\widehat{\text{Var}}(\hat{F}(\cdot))$ is called Greenwood's formula and is well known; see Collet (1994, Chap. 2) for instance.

A derivation due to Sawyer (2003) is as follows. Start with the log of the KM estimator from (7.5),

$$\log \hat{S}(y) = \sum_{j: \tau_j < y} \log(1 - \hat{h}_j),$$

and recognize that $\hat{h}_j = d_j/r_j$ where, as before, d_j is the number of deaths at τ_j and r_j is the number of survivors just before τ_j. Assuming $d_j \sim \text{binomial}(r_j, p_j)$ gives that $E(d_j) = r_j p_j$ and $\text{Var}(d_j) = r_j p_j (1 - p_j)$. Using a one-term Taylor expansion of $\log(1 - p)$ at p_j and evaluating at d_j/r_j gives

$$\log\left(1 - \frac{d_j}{r_j}\right) = \log(1 - p_j) - \frac{1}{1 - p_j}\left(\frac{d_j}{r_j} - p_j\right) + o\left(\left|p_j - \frac{d_j}{r_j}\right|\right).$$

Assuming that d_j/r_j is close enough to p_j and using this in all the terms in $\log \hat{S}$ gives

$$\log \hat{S}(y) \approx \sum_{j\,:\,\tau_j < y} \log(1 - p_j) - \sum_{j\,:\,\tau_j < y} \frac{1}{1 - p_j}\left(\frac{d_j}{r_j} - p_j\right). \tag{7.11}$$

The terms in (7.11) are not independent, since the value of d_j in any one term affects the $r_{j'}$ for $j \neq j'$. However, $E(d_1/r_1 - p_1) = 0$, $E(d_2/r_2 - p_2|d_1) = 0, \ldots, E(d_j/r_j - p_j|d_1, \ldots, d_{j-1}) = 0$, so the terms in the sum in (7.11) form a martingale. This can be used to show that the variance of the sum in (7.11) is the sum of the variances of the terms.[2] Since the constant term (in d_j and r_j) on the left-hand side of (7.11) has variance zero, using $\text{Var}(d_j/r_j|r_j) = p_j(1 - p_j)/r_j$ gives

$$\text{Var}(\log \hat{S}(t)) \approx \sum_{j\,:\,\tau_j < y} \frac{1}{(1 - p_j)^2} \frac{p_j(1 - p_j)}{r_j}$$

$$= \sum_{j\,:\,\tau_j < y} \frac{1}{r_j} \frac{p_j}{1 - p_j}$$

$$\approx \sum_{j\,:\,\tau_j < y} \frac{d_j}{r_j(r_j - d_j)}, \tag{7.12}$$

where the last approximation follows from setting $p_j = \hat{p}_j = d_j/r_j$. Note that in the first approximation $\text{Var}(d_j/r_j|r_j) = p_j(1 - p_j)/r_j$ was used, effectively ignoring the conditioning on r_j. Since $\hat{S}(y) = e^{\log(\hat{S}(y))}$ and $\text{Var}(f(X)) = f'(c)^2 \text{Var}(X)$ (the δ-method with c representing the mean of X), choosing $f(u) = e^u$ and $c = \hat{S}(y)$ leads to the estimator

$$\widehat{\text{Var}}(\hat{S}(y)) = \hat{S}^2(y) \widehat{\text{Var}}(\log \hat{S}(y)), \tag{7.13}$$

in which the second factor on the right is estimated by (7.12) and $\hat{S}$ is known from the data. That is, the Greenwood (1926) approach uses

$$\hat{S}(y) \pm z_{\alpha/2}\sqrt{\widehat{\text{Var}}(\hat{S}(y))} \quad \text{and} \quad \widehat{\text{Var}}(\hat{S}(y)) = \hat{S}(y)^2 \sum_{j\,:\,\tau_j < y} \frac{d_j}{r_j(r_j - d_j)}, \tag{7.14}$$

where $z_{\alpha/2}$ is the $(\alpha/2)100$th percentile of an $N(0, 1)$ distribution.

Ignoring the conditioning on r_j may be a problem in finite samples but, since r_j/n typically converges to a constant for any fixed interval of time, a more formal proof can be

---

[2] Recall that $\text{Var}(X + Y) = \text{Var}(X) + \text{Var}(Y) + \text{Cov}(X, Y)$ and suppose that $E(X) = 0$, $E(Y|X) = 0$. Then $\text{Cov}(X, Y) = E(XE(Y|X)) = 0$. Use this repeatedly to see that the covariances between the terms in (7.11) are zero.

established. Indeed, Meier (1975) showed $\hat{S}(y)$ to be asymptotically normal with mean $S(t)$ and variance consistently estimated by (7.14). A separate problem is that Greenwood's formula is not a good approximation in the tails of S. However, there are improvements to it, and the primary point here is to show how to derive PIs for survival, rather than to improve on estimation techniques. On the other hand, Greenwood's formula is defined in terms of the data and so is appealingly tangible. (A variation on the above reasoning can be applied to $Z(y) = \log(-\log \hat{S}(y))$, a transformation of $\hat{S}$ that, like $\log(\hat{S}(y))$, also converts the range of $\hat{S}$ from $[0,1]$ to all $\mathbb{R}$; this is sometimes regarded as giving a better estimate of $S(t)$ and its variance.)

Now, in parallel to (7.10), a PI can be given based on the SE from Greenwood's formula as in (7.14). For n large, $P(Y_{n+1} \leq \hat{\xi}_{1-\alpha}) \approx 1 - \alpha$. So, obtaining asymptotics for $\hat{\xi}_{1-\alpha}$ will give a $1 - \alpha$ PI for Y_{n+1}. In practice, Greenwood's formula is used to provide an SE for an asymptotic normal, as in

$$\hat{S}(y) - S(y) \to N(0, \mathrm{GW}(y)),$$

in distribution, approximately for n large, where $\mathrm{GW}(y)$ is Greenwood's formula at y. Observing that $S(\xi_{1-\alpha}) = \alpha$ and seting $t = \xi_{1-\alpha}$ gives

$$\hat{S}(\xi_{1-\alpha}) \to N(\alpha, \mathrm{GW}(\xi_{1-\alpha})),$$

in the same approximate asymptotic sense. Empirically, this gives

$$\hat{S}(\hat{\xi}_{1-\alpha}) \to N(\alpha, \mathrm{GW}(\hat{\xi}_{1-\alpha})).$$

Replacing $\hat{S}$ with S, using the δ-method with S^{-1}; then, replacing S^{-1} with $\hat{S}^{-1}$ gives

$$\hat{\xi}_{1-\alpha} \to N\left(\hat{S}^{-1}(\alpha), \left((\hat{S}^{-1})'(\hat{\xi}_{1-\alpha}) \right)^2 \mathrm{GW}(\hat{\xi}_{1-\alpha}) \right).$$

Using this to get an upper bound for $\xi_{1-\alpha}$ leads to a $(1-\alpha)100\%$ PI for Y_{n+1}, with confidence α' given by

$$[0, \hat{S}^{-1}(\hat{\xi}_{1-\alpha}) + z_{1-\alpha'}(\hat{S}^{-1})'(\hat{\xi}_{1-\alpha})\mathrm{GW}(\hat{\xi}_{1-\alpha})^{1/2}]. \tag{7.15}$$

This is different from (7.10) because it uses different asymptotic arguments, involving S directly rather than F.

A third way to find a PI from the Kaplan–Meier approach is to find a density estimate of F, possibly using a smoothed version of $\hat{F}$; this is beyond our present scope even though it would parallel Sec. 4.1.

A fourth way to obtain predictions is to condition on survival to a given time s. For $y \geq s$ let

$$S(y|s) = P(Y > y | Y \geq s) = \frac{S(y)}{S(s-)}.$$

In some settings there is a natural choice for the time of survival to indicate that someone has survived long enough that they are extremely unlikely to die from the same cause as the other subjects. Let this time be w. Then the conditional survival function is

$$F_y(w) = 1 - S(y + w|y),$$

which is the probability of a subject surviving at least w longer, given that the subject has already survived to y. (For the record, the SE of this can be found from the Aalen estimator $\text{SE}^2(\ln \hat{S}(y)) = \sum_{y \geq \tau_j} (d_j/r_j)^2$ as

$$\text{SE}^2(\ln \hat{F}_y(w)) = \sum_{y+w \geq \tau_j \geq y} \left(\frac{d_j}{r_j}\right)^2,$$

then inverting the logarithm to get an SE for $\hat{F}_y(w)$, the quantity of interest.) The logarithm has already been dealt with in (7.13), and this will be seen again in the discussion around (7.21) in Sec. 7.1.2. See Klein and Moeschberger (2003, Sec. 4.2) for details.

7.1.2 Median as a Predictor

Since survival distributions are often right-skewed, we will examining the median of $\hat{S}$ as a predictor for the lifetime of the next subject. This is

$$\widehat{\text{med}}(Y) = \min\{y_i | \hat{S}(y_i) \leq 0.5\} \tag{7.16}$$

with the true value denoted med(Y); other percentiles can be estimated similarly (and used to generate PIs of the form (7.15)). It is not too hard to give an approximate variance for the estimate of a percentile of $\hat{S}$ (see (7.9)), so that the results of Sec. 2.2.1 can be used. Here, however, the focus will be on Greenwood's formula and the δ-method, owing to the censoring. Indeed, Collet (1994) gave an approximation for the variance of $\widehat{\text{med}}(Y)$ in the presence of censoring based on the δ-method and Greenwood's formula. Even though the censoring distribution G will not be explicitly noted here, it is important to remember that the asymptotics given in this section assume one. This means that they are different from the asymptotics given in Reiss (1989) and used in Sec. 4.1 (see expressions (4.2) and (4.4)).

To do this, start by treating S as a real-valued function and $\widehat{\text{med}}(Y)$ as an asymptotically normal random variable. The δ-method gives

$$\text{Var}(S(\widehat{\text{med}}(Y))) = \left(\frac{dS(t)}{dt}\Big|_{\text{med}(Y)}\right)^2 \text{Var}(\widehat{\text{med}}(Y)). \tag{7.17}$$

Loosely, an empirical form can be specified by writing

$$\widehat{\text{Var}}(\hat{S}(\widehat{\text{med}}(Y))) = \left(\frac{dS(t)}{dt}\Big|_{\widehat{\text{med}(Y)}}\right)^2 \widehat{\text{Var}}(\widehat{\text{med}}(Y)), \tag{7.18}$$

from which an estimator of the last factor can be solved.

First, an estimator can be given for the left-hand side of (7.18) from Greenwood's SE in (7.14), namely,

$$\widehat{\text{Var}}(\hat{S}(\widehat{\text{med}}(Y))) = \hat{S}(\widehat{\text{med}}(Y))^2 \sum_{j:\, \tau_j < \widehat{\text{med}}(Y)} \frac{d_j}{n_j(n_j - d_j)}. \tag{7.19}$$

Second, for the first factor on the right of (7.18) write

$$\frac{dS(t)}{dt}\Big|_{\widehat{\text{med}(Y)}} = -p(\widehat{\text{med}}(Y)),$$

where $p(\cdot)$ is the density of P. So, it is enough to have an estimate $\hat{p}(\cdot)$ of the density p at $\widehat{\text{med}}(Y)$. That is, write

$$\frac{d\hat{S}(t)}{dt}\Big|_{\widehat{\text{med}}(Y)} = -\hat{p}(\widehat{\text{med}}(Y)).$$

Perhaps the simplest choice is to set

$$-\hat{p}(\widehat{\text{med}}(Y)) = \frac{\hat{S}(\hat{u}) - \hat{S}(\hat{\ell})}{\hat{\ell} - \hat{u}}, \tag{7.20}$$

where $\hat{u} = \max\{y_j | \hat{S}(y_j) \geq 1 - (q/100) + \epsilon\}$ and $\hat{\ell} = \min\{y_j | \hat{S}(y_j) \geq 1 - (q/100) - \epsilon\}$, for some small $\epsilon > 0$. Taking $q = 50$ gives an approximation to the derivative that is as close to the empirical median as possible using the data; see Collet (1994). Using (7.19) and (7.20) in (7.18) gives an estimate $\widehat{\text{Var}}(\widehat{\text{med}}(Y))$ for $\text{Var}(\widehat{\text{med}}(Y))$ from (7.17). Asymptotic normality can sometimes be invoked to give $1 - \alpha$ CIs for the median of the form $\widehat{\text{med}}(Y) \pm z_{1-\alpha/2}\sqrt{\widehat{\text{Var}}(\widehat{\text{med}}(Y))}$.

An alternative approach is to construct a CI for the median directly from the lower and upper bounds of the CI for the Kaplan–Meier estimator. An asymptotic $(1 - \alpha)100$ CI for the Kaplan–Meier estimator can be constructed, using the Greenwood variance estimate, as $\hat{S}(y) \pm z_{1-\alpha/2}\sqrt{\widehat{\text{Var}}(\hat{S}(y))}$, setting $y = \widehat{\text{med}}(Y)$. However, this interval can extend below 0 or above 1, and can have poor coverage properties; see Therneau and Grambsch (2000).

One way to avoid CIs that contain points outside $(0,1)$ is to use a transform on S such as $\log S(y)$ or $\log(-\log S(y))$ and transform the resulting CI into a CI for $\hat{S}(y)$; see Collet (1994). For instance, the log transform would lead to $(1-\alpha)100$ CIs for $\log S(y)$ of the form $\log \hat{S}(y) \pm z_{1-\alpha/2}\sqrt{\widehat{\text{Var}}(\log \hat{S}(y))}$, in which the δ-method can again be used to get an expression for the SE. So, recall (7.13) and write $\widehat{\text{Var}}(\log \hat{S}(y)) = \widehat{\text{Var}}(\hat{S}(y))/\hat{S}(y)^2$. Greenwood's formula can be used in the numerator, and the denominator is already an estimate. Inverting the logarithm, an approximate $(1 - \alpha)100$ CI for $\hat{S}(y)$ is

$$\left[\hat{S}(y)\exp\left(-z_{1-\alpha/2}\sqrt{\widehat{\text{Var}}(\hat{S}(y))} \Big/ \hat{S}(y) \right), \hat{S}(y)\exp\left(z_{1-\alpha/2}\sqrt{\widehat{\text{Var}}(\hat{S}(y))} \Big/ \hat{S}(y) \right) \right]. \tag{7.21}$$

The upper and lower limits of this confidence interval can now be used in (7.16) to get upper and lower bounds on $\widehat{\text{med}}(Y)$. Note that the results using transformations can also be used to get PIs as in (7.15).

Having examined the variance of $\widehat{\text{med}}(Y)$ as an estimator, we turn to using $\widehat{\text{med}}(Y)$ as a predictor for the survival time of the next subject, i.e., to enable the use of approaches from Sec. 2.2.1. The simplest analysis parallels (2.10). Replacing the mean with the median, invoke asymptotic normality and use the fact that the rate of convergence of $\hat{S}$ to S is $\mathcal{O}(1/\sqrt{n})$ uniformly on compact sets. Explicitly, write $\text{Var}(\widehat{\text{med}}(Y)) \approx \sigma_m^2/n$, $\text{Var}(Y_{n+1}) = \sigma^2$, and let $\tau > 0$. Then

$$P\left(|\widehat{\text{med}}(Y) - Y_{n+1}| \geq \frac{\sigma_m/\sqrt{n} + |E\,\widehat{\text{med}}(Y) - EY_{n+1}| + \sigma}{\tau} \right)$$

$$\leq \frac{\tau}{\sigma_m/\sqrt{n} + |E\,\widehat{\text{med}}(Y) - EY_{n+1}| + \sigma}\Big(E\,|\widehat{\text{med}}(Y) - E\,\widehat{\text{med}}(Y)|$$

$$+ |E\,\widehat{\text{med}}(Y) - E(Y_{n+1})| + E\,|E(Y_{n+1}) - Y_{n+1}|\Big)$$

$$\leq \frac{\tau}{\sigma_m/\sqrt{n} + |E\,\widehat{\text{med}}(Y) - EY_{n+1}| + \sigma} \left(\sqrt{\text{Var}(\widehat{\text{med}}(Y))} + |E\,\widehat{\text{med}}(Y) - E(Y_{n+1})| \right.$$
$$\left. + \sqrt{\text{Var}(Y_{n+1})} \right)$$

$$\approx \tau. \tag{7.22}$$

So, if σ, σ_m, and $E\,\widehat{\text{med}}(Y) - E(Y_{n+1})$ can be reliably estimated and τ can be chosen to be small enough that the lower bound on $|\widehat{\text{med}}(Y) - Y_{n+1}|$ is meaningful, we can obtain prediction intervals parallel to (2.11). In a similar way, (2.12), given for the mean, can be extended to the median as well. In principle, the usual variance estimate s^2 can be used for σ^2. Also, one can set $\hat{\sigma}_m^2 = n\,\widehat{\text{Var}}(\widehat{\text{med}}(Y))$ and use any estimate for the variance of a median; see (4.3), for instance. So the main remaining limitation is the difference $E\,\widehat{\text{med}}(Y) - E(Y_{n+1})$, which can be large for highly skewed distributions. If there is enough uncensored data, one can draw bootstrap samples to obtain estimates of $E\,\widehat{\text{med}}(Y)$ and $E(Y_{n+1})$ and take the difference. (This also gives an SE for the difference, if desired.)

It should be noted that the analysis here for the median can be carried out (at least in principle) for the mean and many other point predictors for Y_{n+1}.

It is not hard to imagine that, for many such probabilities P, bounds of the form (7.22) will not be as good as desired. That is, they may be too narrow to be convincing or (more likely) too wide to be useful. After all, these methods assume that the distribution F, or equivalently S, is stable enough and well enough defined to be taken as a constant and that the censoring distribution G has relatively little effect. When these assumptions fail, one way to proceed is to build variability due to P (implicitly taking into account variability from G) into the upper bound (7.22) by using a range of P's that might be valid models. Indeed, Draper (1995) explored two methods to account for model uncertainty in P. In one method, P is selected from a set of candidate P's. In the other method, the problem is enlarged so that P is a random outcome of an underlying process. Both approaches typically increase the uncertainty of downstream prediction.

For the sake of contrast, suppose that our interest is not in the survival time itself but in the probability of survival to a given time, i.e., $\hat{S}(t)$. Direct estimates of such probabilities are provided by the Kaplan–Meier estimate of the survival function, with confidence limits provided by the various SEs derived here, e.g., in (7.21) . If our interest is in the distribution of the survival time Y_{n+1} of a new patient, conditionally on the observed data, the Kaplan–Meier estimate can be used directly. This appears to work well when the sample size is large (see Chang and Schuster (1994)) but may lead to substantial cumulative error if prediction is for a group of individuals of size comparable with n. However, there are alternatives to the Kaplan–Meier estimator for this context, such as the predictive distribution proposed in Berliner and Hill (1988).

The key issue, though, is that estimating the probability of survival to a given time is different from predicting an actual survival time, and probability-based forecasting is not the focus here although an example was provided in Sec. 5.5. In fact, probability forecasting is on a different scale from predicting the next value of a random variable. If only the estimated probability of survival to year t from a group of IID observations is available then it cannot be compared with a new outcome without further assumptions. For instance, one could choose a scoring rule to enable the comparison of outcomes with probabilities. This was done in Sec. 5.5 for two scoring functions, which gave opposite conclusions. See Dawid *et al.* (2012) among others for a recent contribution on this point. Overall, using a scoring rule is not prequential because then one does not need to compare predictions directly with their outcomes. However, one could compare a predicted probability with a collection of new outcomes that gave an estimated probability; this is a calibration problem, as discussed briefly in Secs. 5.4 and 5.5.

7.1.3 Bayes Version of the Kaplan–Meier Predictor

To derive a Bayes analog of the Kaplan–Meier predictor, recall the Dirichlet process prior (DPP) $\mathcal{D}(\alpha, H)$ from Sec. 4.1, where $\alpha \in \mathbb{R}$ is a concentration parameter and H is the base measure. Rather than using this with uncensored data as before, here it will be used with censored data. Following Susarla and van Ryzin (1976), it is easier to express the results in a notation slightly different from that used in the frequentist case (see Sec. 7.1.1). Recall that the data is (Z_i, δ_i) for $i = 1, \ldots, n$ but, without loss of generality, can be partitioned into two subsets corresponding to uncensored and censored data, i.e., $(Z_1, 1), \ldots, (Z_k, 1)$ and $(Z_{k+1}, 0), \ldots, (Z_n, 0)$, respectively, for some k, where k observations have been observed and $n - k$ observations have been censored. Since $\delta_k = 1$ for $k = 1, \ldots, k$, the first k data points are $Z_1 = Y_1, \ldots, Z_k = Y_k$.

In contrast with the frequentist case, let m denote the number of distinct censoring times among $Z_{k+1}, \ldots, Z_n$ and write $\gamma_{k+1} < \cdots < \gamma_m$ to denote the ordered distinct censoring times. Let λ_j, for $j = k + 1, \ldots, m$, be the multiplicity of γ_j among $Z_{k+1}, \ldots, Z_n$. Define $N(t)$ to be the number of observations, censored or not, greater than or equal to t and define $N^+(t)$ to be the number of observations, censored or not, strictly greater than t. Thus, roughly, $N(t)$ is a parallel to r_k and $N(t) - N^+(t)$ is a parallel to d_k; see Sec. 7.1.1.

The first goal is to find the Bayes estimator under squared error loss

$$L(S^*, S) = \int_0^\infty (S^*(y) - S(y))^2 w(y) \mathrm{d}y,$$

where S^* is an action, i.e., an estimator of S, and w is a weight function. That is, the first goal is to find

$$\hat{S}_B = \arg\min E(L(S^*, S)|\mathcal{D}),$$

the estimator that minimizes the posterior risk, where $\mathcal{D}$ represents the data $\{(Z_i, \delta_i)|i = 1, \ldots, n\}$. (Since this expression uses the squared error loss, the resulting minimum posterior risk is also the posterior variance.)

The strategy for finding $\hat{S}_B$ is first to find a nonparametric Bayes estimator for F, under a DPP conditioning only on the data that was observed, and then to update this estimator by conditioning on the censoring times.

As seen in Sec. 4.1, equipping P with a DPP $\mathcal{D}(\alpha, H)$ and applying the usual Bayes updating with $(Z_1, 1), \ldots, (Z_k, 1)$ gives a posterior of the form

$$(F|z_1, \ldots, z_k) \sim DP\left(\alpha + n, \frac{\alpha + \sum_{i=1}^k \delta_i'}{\alpha + n}\right),$$

in which δ_i' is unit mass at $z_i = y_i$ so that $\delta_i'(A) = 1$ or 0 according to whether $y_i \in A$ or $y_i \notin A$. (A careful proof of this can be found in Susarla and van Ryzin (1976).)

Some further derivations (see Susarla and van Ryzin (1976)) give that, for $\gamma_j \leq y < \gamma_{j+1}$ with $j = k, \ldots, m$ and the conventions $\gamma_k = 0$ and $\gamma_{m+1} = \infty$, the Bayes estimator $\hat{S}_B(\cdot)$ of S is

$$E(S(y)|(\delta_1, Z_1), \ldots, (\delta_n, Z_n)) = \frac{\alpha H(y, \infty) + N^+(y)}{\alpha H(\mathbb{R}^+) + n}$$

$$\times \prod_{v=k+1}^{j} \frac{\alpha H([Z_{(v)}, \infty)) + N^+(Z_{(v)})}{\alpha H([Z_{(v)}, \infty)) + N(Z_{(v)}) - \lambda_v}, \qquad (7.23)$$

where the expectation is over S, and the product in (7.23) is taken to equal one for $u < Z_{(k+1)}$; $Z_{(v)}$ is the vth order statistic in the sample of size n. An extension to this result, see Susarla and van Ryzin (1976), gives the higher posterior moments of $S(t)$ as well. So, a closed-form expression for the posterior variance of $\hat{S}_B$ can be found.

Another way to write (7.23) is from Klein and Moeschberger (2003, Chap. 6). Specifically, suppose that $0 = t_0 < \cdots < t_j < \cdots < t_M < t_{M+1} = \infty$ are the M distinct times of death or censoring. For each j write r_j for the number at risk, d_j for the number of deaths, and λ_j for the number of censorings at t_j. Then, expression (7.23) becomes, for $y \in [t_j, t_{j+1})$, $j = 0, \ldots, M$,

$$\hat{S}_B(y) = \frac{\alpha H([y, \infty)) + r_{j+1}}{\alpha H(\mathbb{R}^+) + n} \times \prod_{k=1}^{j} \frac{\alpha H([t_k, \infty)) + r_{k+1} + \lambda_k}{\alpha H([t_k \infty)) + r_{k+1}}. \qquad (7.24)$$

Now, there are several ways for a Bayesian to issue PIs for Y_{n+1}. As before, write $\hat{\xi}_{1-\alpha} = \hat{F}_B^{-1}(1 - \alpha)$, where $\hat{F}_B(\cdot) = 1 - \hat{S}_B(\cdot)$. So, the simplest approximate $(1 - \alpha)100\%$ PI is of the form $[0, \hat{\xi}_{1-\alpha}]$. This follows from the approximation $P(Y_{n+1} \leq \xi_{1-\alpha}) \approx \hat{F}_B(\hat{\xi}_{1-\alpha})$. Note that an obvious Bayes point predictor for Y_{n+1} is simply the median of $\hat{F}_B$, i.e., $\hat{\xi}_{1/2} = \hat{F}_B^{-1}(1/2)$; other point predictors such as a mean under $\hat{F}_B$ could be used as well.

One can use a conservative correction to a first-order $(1 - \alpha)100\%$ PI by finding

$$\text{Var}(\hat{F}_B(y)|(Z^n, \delta^n)) = \text{Var}(\hat{F}_B(y)|(Z_1, \delta_1), \ldots, (Z_n, \delta_n)),$$

using Cor. 1 in Susarla and van Ryzin (1976). Then, it would be natural to use

$$\left[0, \hat{\xi}_{1-\alpha} + z_{1-\alpha}\sqrt{\text{Var}(\hat{F}_B(\hat{\xi}_{1-\alpha})|(Z^n, \delta^n))}\,\right]$$

as a conservative value. It is important to note that, because the posterior quantities are closed-form functions of the data, there is no need to use asymptotics. However, as long as the true DF is in the interior of the support of the DPP, the pointwise consistency of $\hat{S}_B(\cdot)$ and $\hat{F}_B(\cdot)$ is guaranteed. This is generally true of Bayes procedures; see Ghosh and Ramamoorthi (2003) and Kim and Lee (2004). Consequently, under this assumption, $\hat{\xi}_{1-\alpha} \to \xi_{1-\alpha}$.

To be more purely Bayes, in principle one should use the predictive distribution, that is, find the predictive distribution for Y_{n+1} by integrating its DF F against the posterior for F. Indeed, Theorem 5 of Susarla and van Ryzin (1976) gives the posterior distribution $(F(t)|(Z^n, \delta^n))$ as a mixture of beta distributions and identifies their parameter vectors; the expected value of $F(t)$ for Y_{n+1} under this gives (7.23) and (7.24). But, with some abuse of notation, note that

$$M(Y_{n+1} \leq y | (Z^n, \delta^n)) = \int F(y) dW(F | (Z^n, \delta^n)).$$

That is, the predictive distribution is the posterior mean of the DF, so the two are equivalent. This is not likely to be true for loss functions other than the squared error. However, such loss functions would give different Bayes actions (and hence different point and interval predictors).

As a variation one could use the hierarchical structure. Let V be a subset of DFs, ignore the evaluation at t, and find

$$M(V | (Z^n, \delta^n)) = \int_V dW(F | (Z^n, \delta^n)).$$

Then, if $V_{1-\alpha}$ were a set of DFs having posterior probability $1 - \alpha$ then

$$[0, \sup_{F \in V_{1-\alpha}} F^{-1}(1 - \alpha')]$$

could serve as a PI for Y_{n+1} for $\alpha, \alpha' > 0$. Even though many priors besides the DPP have been proposed, e.g., beta processes for cumulative hazard rates (see Hjort (1990)), it does not seem that either predictive method – finding a PI from $\hat{F}_B$ or from a well-chosen $V_{1-\alpha}$ – has been studied in detail for any of them.

7.1.4 Discrimination and Calibration

One obvious application of the sort of prediction problems discussed so far is their use in diagnostic models, e.g., as a method for predicting a binary outcome for individuals, such as 'low-risk' or 'high-risk'. That is, the real goal is to take a *prognostic* model – one which predicts the *probability* of a future event or state – and convert it into a *diagnostic* model – one which predicts an *actual* future class or state. This means defining the future classes or states, often just called categories. Usually there are many ways to do this but, once it is done, the accuracy of prognostic models is usually described in terms of two components, namely, discrimination and calibration (Cook 2008). *Discrimination* is the ability to classify individuals into their correct categories. *Calibration* is the ability to estimate the risk or probability of a future event accurately. Since some comments have already been given on calibration (see Secs. 5.4 and 5.5) and the focus here is more on the construction of predictors rather than their evaluation, it is enough to discuss discrimination, albeit briefly.

A survival distribution gives the predicted probabilities of survival and can be converted into a discriminant by classifying those with probabilities above a given cut point as belonging to one category and those with lower probabilities as belonging to another category. For such cases, sensitivity and specificity, as introduced in Sec. 4.6, can be used to choose the optimal cut point for accurate classification via a receiver operating characteristic (ROC) curve (Pepe 2003). This is a plot of sensitivity versus 1-specificity and presents all possible (sensitivity, 1-specificity) pairs attainable by dichotomizing the probabilities of survival with different cut points. It is relatively easy to identify the point on the curve that maximizes both sensitivity and specificity, bearing in mind that the optimal threshold is also a function of the relative costs of misclassifying subjects.

Often it is not clear which type of prognostic survival model will lead to the best diagnostic model. In some cases the Kaplan–Meier approach, in its frequentist or Bayes formulation, may do best. However, other prognostic methods such as proportional hazards

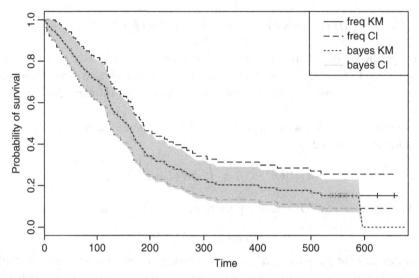

Figure 7.1 Kaplan–Meier curves for the Bayes and frequentist cases. The shaded portion indicates the 95% credibility sets for the value of the DF, pointwise in time, for the Bayes Kaplan–Meier estimate. The dashed lines indicate the 95% confidence interval for the value of the DF, pointwise in time, for the frequentist Kaplan–Meier estimate.

models, parametric survival regression models (based on, say, the Weibull or log-logistic distribution), accelerated failure time models, frailty models, competing risk models, etc. could also be used to give a diagnostic model in various settings. However, doing this is beyond our present scope.

7.1.5 Predicting with Medians

To demonstrate the use of predictive techniques in survival analysis, consider the data set uissurv available from www.umass.edu/statdata/statdata/data/ analyzed in Hosmer and Lemeshow (2008). It is a collection of times to relapse for drug abuse patients, measured in days from admission to hospital. The sample size is 628, but for illustrative purposes 80 data points were chosen at random. Of these 80, 12 were censored so there were 68 'real' observations. To visualize the data, Fig. 7.1 shows the Kaplan–Meier curves for the Bayes and frequentist cases. In the Bayes case, a DPP was used in which the concentration parameter was $\alpha = 1$ and the base measure H was taken to be the element of the exponential family having the MLE of the observed data. It is seen that the two curves nearly coincide up to time 590; their associated credibility sets and confidence intervals also nearly coincide up to time 590. After 590, the Bayes curve drops rapidly owing to the base measure in the Dirichlet prior.

Four data sets can be derived from the original by adding censoring. One kind of censoring that is illustrative is random censoring, obtained by choosing G to be a normal distribution with mean set to be the 95th or 75th percentile of the 68 observed data points and SD taken to be that of the data points below the 95th and 75th percentiles, respectively. The Kaplan–Meier curves for these two cases are shown in the top two panels of Fig. 7.2; the curves on the left are based on 56 observations (24 censored) and those on the right on 47 observations

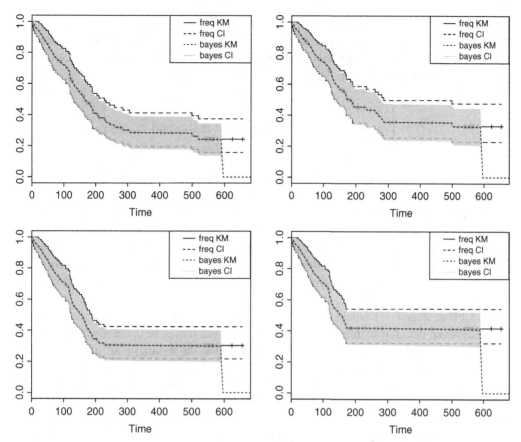

Figure 7.2 The upper two panels are the Kaplan–Meier survival probability curves
for the data set formed by further random censoring on the original data set. The
lower two panels are the Kaplan–Meier curves for the data set by truncation
censoring chosen to match the number of observations in the random censoring.
The credibility and confidence sets are indicated as in Fig. 7.1.

(33 censored). Another kind of censoring is truncation censoring. In this case, the truncation
levels were set to give 56 and 47 observations respectively, matching the random censoring.
The corresponding curves are shown in the lower two panels of Fig. 7.2.

It is seen in Fig. 7.2 that when there are fewer observations the curves rise, because the
added censoring creates ambiguity about whether the subjects in the right-hand tail survived.
Moreover, since the random censoring is spread over the whole range of the data, unlike
truncation, which affects only the larger data points, the curves for random censoring tend
to be a little lower in the right-hand tails than the curves for truncation censoring. Also, as
expected, the pointwise credibility/confidence sets increase, as there are fewer observations.

We have five data sets, and each data set generates two predictors based on the medians
of the Bayes and the frequentist Kaplan–Meier curves. So, there are a total of ten predictors
that can be evaluated in five sequential predictive contexts. For each of these, 100 random
permutations of the data were drawn and, for each fixed permutation set, the first 20 data
points were used as a burn-in to generate medians for predicting the 21st data point. Then

the first 21 data points were used to generate medians for predicting the 22nd data point, and so forth. Thus each permutation generated 60 one-step-forward predictions. The assessment of error of each prediction depended on whether the data point to be predicted was censored. For the $(i+1)$th stage, the error of predicting Y_{i+1} by $\hat{Y}_{i+1}$ was

$$\ell(\hat{y}_{i+1}, y_{i+1}) = \begin{cases} |\hat{y}_{i+1}, y_{i+1}| & \text{if } y_{i+1} \text{ observed,} \\ |\hat{y}_{i+1} - c_{i+1}| & \text{if } y_{i+1} \text{ censored at } c_{i+1} \text{ and } \hat{y}_{i+1} \leq c_{i+1}, \\ 0 & \text{if } y_{i+1} \text{ censored at } c_{i+1} \text{ and } \hat{y}_{i+1} \geq c_{i+1}. \end{cases} \quad (7.25)$$

The idea is that the absolute value distorts the distance between predictor and predictand the least. So, it should be used when the $(i+1)$th data point is observed. However, when a data point to be predicted is censored there are two cases: if $c_{i+1} \leq \hat{y}_{i+1}$ then the best one can do with a point predictor is to compare it with the censoring value by using the absolute value, and if $c_{i+1} \geq \hat{y}_{i+1}$ then the best one can do with a point predictor is to predict some number at least as great as the censoring value. Both of these cases may be regarded as under-representing the degree of error since, if y_{i+1} were known, the error could only be higher.

The CPE at stage $i+1$ is taken as the sum of the errors $\ell(\hat{y}_{i+1}, y_{i+1})$ over the permutations. However, not all possible permutations can be used. It may be that a randomly chosen permutation has so many censored values in a row that the median does not exist. Such permutations were discarded, so in fact the average is over permutations of the data for which the median predictor exists at each time step. Moreover, one could divide the CPE by its stage to get the per prediction average CPE. This is not done here but would give an equivalent result.

The top left panel in Fig. 7.3 summarizes the CPE results for all ten scenarios (Bayes and frequentist for each of the five data sets). It can be seen that the best performance (lowest CPE) occurs for the Bayes and frequentist median predictors, which are essentially identical, for the data set with the higher truncation censoring. The second best performance occurs for the Bayes and median predictors, which again are essentially identical, for the data set with the lower truncation censoring. The worst performance (highest CPE) occurs for the Bayes and median predictors, which once again are essentially identical, for the original data set. The middling performance occurs with the random censoring: all four curves are quite close but the Bayes median predictor does slightly better on the data set having the higher level of random censoring.

These results are counterintuitive: one expects better prediction with more observations. The explanation is that the median is stable, in the sense that one can censor, in principle, the 49% highest or lowest observations without changing it. By contrast, random censoring means that the values in the order statistic are further apart, making the median a little less stable. So, truncation censoring is less damaging to the median predictor than random censoring. In addition, ℓ assigns zero loss when the data point is censored and the predictor is above the censoring value. This under-representation of the error favors truncation censoring as well since it censors data at the high end only, meaning that there can be more cases where the loss is zero than when the censoring is randomly spread over the whole data set. Nevertheless, this sort of finding is at odds with the usual intuition gleaned from a modeling or other nonpredictive approach.

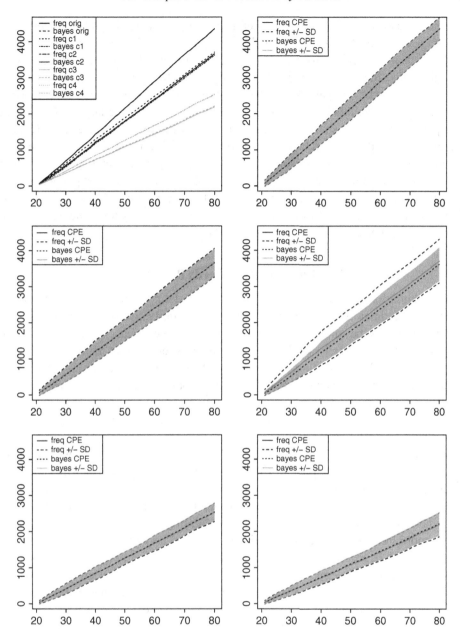

Figure 7.3 The top left panel shows the average CPE vs. sample size over 100 random permutations for the ten median predictors for the five data sets; c1 and c2 indicate random censoring at the 75th and 95th percentiles, respectively, while c3 and c4 indicate the corresponding truncation censoring down to 47 and 56 observations, respectively. The remaining panels show the CPE curves with ± SD regions indicated for the original data (top right), c2 and c1 (middle), and c4 and c3 (bottom).

The other panels in Fig. 7.3 show the average CPE curves from the Bayes and frequentist predictors seen in the first panel but with the addition of $\pm SD$ bars. It is seen that the shaded regions in the bottom two panels are slightly narrower than the other three panels, as expected, because the median is more stable under truncation than with the random censoring used in the middle two panels, i.e., the variability of the median predictor is less under truncation. Confirming this intuition, the middle two panels have the largest SDs. Also, the middle panel on the right shows the biggest difference between the Bayes and frequentist predictors; this could, however, be an artifact of the specific random permutations used and hence misleading in contrast with the other cases. The top right panel shows the average CPE curves for the original data set, for which both Bayes and frequentist predictors are the worst – this is the case where predictors get zero loss least frequently. On the other hand, getting better predictions with fewer well-chosen observations may not be unusual; the same phenomenon was noted in Wong and Clarke (2004).

7.2 Proportional Hazards Predictors

From their definitions, it is clear that $h(y)$ is only well defined when $S(y)$ is differentiable. Since the main estimators of $S(\cdot)$ are step functions with derivative zero (when it is defined), these estimators must be smoothed before being used to generate an estimate of h. However, H is easy to estimate: merely set $\hat{H}(y) = -\log \hat{S}(y)$ where $\hat{S}$ is an estimate of S such as Kaplan–Meier, the Bayes estimate, or the Nelson–Aalen estimate (not discussed here).

Although h is harder to estimate than H, it is conceptually important. For instance, when $y \geq s$, the conditional survival function is

$$S(y|s) = e^{-(H(y)-H(s))} = e^{-\int_s^y h(u)\mathrm{d}u}. \tag{7.26}$$

So, to predict survival for $y - s$ units of time beyond s, i.e., on $[s, y]$, it is enough to know h on $[s, y]$. As before, this applies to subjects that have not died during $(0, s)$. Otherwise stated, it applies to subjects that are known to be alive until, but not necessarily at, s.

Expression (7.26) shows that the qualitative properties of h determine the qualitative properties of the conditional survival function. If h is increasing then subjects with larger survival times are more likely to die (i.e., to experience an event) as y increases. If h is decreasing then subjects with larger survival times are less likely to die (i.e., to experience an event) as y increases. If h has some intermediate shape – say, increasing and then decreasing or decreasing and then increasing – the same statements hold but only for infinitesimal increments of y over which h is increasing or decreasing. Therefore, if one can overcome the difficulty in finding a useful estimator of h, the latter can be converted into an estimate for the survival function to make predictions and may permit useful understanding of the survival process.

To see this explicitly, make the simplifying assumption that any two subjects have proportional hazard functions, i.e.,

$$h_i(y) = \alpha_{ij} h_j(y)$$

for subjects i and j and some $\alpha_{ij} > 0$. Then, by integration, the cumulative hazard functions are $H_i(y) = \alpha_{ij} H_j(y)$. Now, the survival functions for the subjects are seen to be related on observing that

$$S_i(y) = e^{-H_i(y)} = e^{-\alpha_{ij} H_j(y)} = S_j(y)^{\alpha_{ij}}.$$

This means that, when the hazards are proportional, it is easy to convert one survival function to another if the α_{ij} can be found. To reduce the number of subscripts, the proportional hazards assumption on the subjects is often written as $h_i(y) = c_i h_0(y)$ for each i, where $c_i > 0$ and $h_0(t)$ is a 'baseline' hazard function; any fixed h_0 will suffice. Effectively, assuming that the hazard functions are proportional makes it relatively easy to get a good estimate of h, as will be seen.

What are the constants of proportionality c_i? Usually they are taken as $c_i = e^{X_i^T \beta}$ for a value X_i associated with subject i for some vector of explanatory variables and some vector of coefficients β that is common across subjects. Apart from the exponential, this is the usual technique of writing an unknown as a linear function of X. Thus, the proportional hazards (PH) model (i) invokes a baseline hazard function $h_0(y)$, which means that at 'baseline' the explanatory variables are all zero, and (ii) represents the factor making subject i different from baseline according to $e^{X_i^T \beta}$, which means that the explanatory variables are the X_i rather than zeros. Explicitly,

$$h_i(y) = h_i(y|X_i) = h_0(y)e^{X_i^T \beta} \qquad (7.27)$$

or, more simply, $h(y|X) = h_0(y)e^{X^T \beta}$, dropping the subject index. That is, the hazard function of a subject with explanatory variables X is the product of a baseline hazard function and factor $e^{X^T \beta}$ representing the deviation of the subject from the baseline. In terms of survival functions,

$$S(y|X) = e^{-\exp(X^T \beta)H_0(y)} = S_0(y)^{\exp(X^T \beta)}, \qquad (7.28)$$

where $H_0(y) = \int_0^y h(u)du$ is the cumulative baseline hazard function and $S_0(y) = e^{-H_0(y)}$ is the baseline survival function. So, predictors for future subjects with their values of X would be found from estimating the unknowns in (7.28).

Proportional hazards models were first proposed in Cox (1972), which also gave their foundational analysis, for which reason they are often called Cox models.

Another way to think of PH models is that they arise from a factorization criterion on the hazard function. That is, the hazard function is of the form

$$h(y|x) = \psi(x)h_0(y), \qquad (7.29)$$

where ψ is a function of the covariates that relates the baseline hazard function (with $X = 0$) to the hazard function for nonzero covariates. Often, ψ is taken to have a parametric form in which the information in x is summarized by a linear function. In these cases, (7.29) can be written as

$$h(y|x) = h_\beta(y|x) = \psi(x\beta)h_0(y). \qquad (7.30)$$

The most common form for ψ in (7.30) is an exponential, so that $\psi(x\beta) = e^{x\beta}$. This means that the ratios of hazard functions in a PH model are independent of t or, equivalently, the log ratio $\ln(h(y|x)/h_0(y))$ is independent of y. Otherwise stated, with the c_i as above, $\ln(h(y|x)/h_0(y)) = X^T \beta$. That is, the hazard functions are again seen to be proportional to each other because all are multiples of h_0. More generally, the log ratio of any two hazard functions is linear, i.e., for any covariates X and X',

$$\ln \frac{h(y|X)}{h(y|X')} = (X - X')\beta.$$

To obtain predictions from a PH model, one must find estimates $\hat{\beta}$ of β and $\hat{h}_0$ of h_0. Then $\hat{H}_0(y) = \int_0^y \hat{h}_0(s)ds$ can be used to get $\hat{S}_0(y) = e^{-\hat{H}_0(y)}$ (in place of using $H_0(y) = \int_0^y h_0(s)ds$ to get $S_0(y) = e^{-H_0(y)}$), so that (7.28) becomes

$$\hat{S}(y|X) = \hat{S}_0(y)^{\exp(X^T\hat{\beta})}, \tag{7.31}$$

for a subject with explanatory values X.

Now, suppose that data of the form (y_i, δ_i, x_i) for $i = 1, \ldots, n$ are available, where the x_i are the covariates for subject i and subject i is observed for time y_i. As before, each δ_i is an indicator function for whether the time y_i is right-censored ($\delta_i = 0$) or whether y_i really is an observation of Y ($\delta_i = 1$). Since the x_i may be regarded as outcomes of a random variable X_i, it is assumed as needed that δ_i and X_i are independent. This means that the censoring process and the explanatory variables are unrelated. Of course, this is not true in general; however, it will suffice for the derivations here.

As a final point, before discussing the estimation of β and h_0, note that the PH model is just a model even though h_0 is a nonparametric quantity. Such models have a specific definition, and, for a given selection of explanatory variables X, either a phenomenon satisfies this definition or it does not. Saying that a specific phenomenon satisfies a PH model implies that all the information that is relevant to it has been encapsulated in one way or another in X and in the structure of the model, i.e., the problem is $\mathcal{M}$-closed. Typically, this assumption cannot be made. So one merely hopes that the model misspecification and sources of variability are small enough that the model's predictions are not so far wrong as to make it useless. It remains an open question how realistic this is in many applications where PH models are used.

7.2.1 Frequentist Estimates of h_0 and β in PH Models

Normally, the MLEs for h_0 and β are used. This means that one is finding the likelihood in a form that can be readily maximized. The complete likelihood under right-censoring with exponential ψ can be derived simply. The contribution to the likelihood if y_i is an observation, i.e., $\delta_i = 1$, is $p(y_i)$. The contribution to the likelihood if y_i is a censored time, i.e., $\delta_i = 0$, is merely $S(y_i)$. So, the factor that is contributed to the likelihood by Z_i and δ_i together is $p(z_i, \delta_i) = p(y_i)^{\delta_i} S(y_i)^{1-\delta_i}$. Hence, the likelihood is

$$L(\beta, h(\cdot)|(Z^n, \delta^n)) = \prod_{i=1}^n p(y_i)^{\delta_i} S(y_i)^{1-\delta_i}$$

$$= \prod_{i=1}^n h(y_i)^{\delta_i} \exp\left(-H(y_i)\right), \tag{7.32}$$

using (7.2) when $\delta_i = 1$ and using (7.3) n times. Taking the explanatory variables into account (and suppressing (Z^n, δ^n)) this can be written as

$$L(\beta, h_0(\cdot)) = \prod_{i=1}^n h_\beta(y_i|x_i)^{\delta_i} S_\beta(y_i|x_i)$$

$$= \prod_{i=1}^n h_0(y_i)^{\delta_i} \left(\exp(x_i^T\beta)\right)^{\delta_i} \exp\left((-H_0(y_i))\exp(x_i^T\beta)\right). \tag{7.33}$$

Starting with the maximization over the $h_0(y)$, holding β fixed, suppose there are D 'deaths', i.e., values of y_i that are observed without censoring, $D \leq n$. So, (7.33) can be simplified by separating the D cases with $\delta_i = 1$ from the $n-D$ cases with $\delta_i = 0$. Relabeling the y_i as required in the first factor, this gives

$$L(\beta, h_0) = \left(\prod_{i=1}^{D} h_0(y_i) \exp\left(x_i^T \beta\right) \right) \exp\left(-\sum_{j=1}^{n} H_0(y_j) \exp\left(x_j^T \beta\right) \right). \tag{7.34}$$

From this expression, it can be seen that the likelihood is maximal (independent of β) when $h_0(y) = 0$ for $y \neq y_i$, $i = 1, \ldots, n$. This minimizes H_0. In fact, it gives

$$H_0(y_j) = \sum_{\{i \mid y_i \leq y_j\}} h_0(y_i).$$

Since this means that $L(\beta, h_0)$ depends only on the specific values $h_0(y_i)$, it is enough to maximize (7.34) over the $h_0(y_i)$ and β. Using this, (7.34) becomes

$$L(\beta, h_0(y_1), \ldots, h_0(y_D)) \propto \prod_{i=1}^{D} h_0(y_i) \exp\left(-h_0(y_i) \sum_{j \in R(y_i)} \exp\left(x_j^T \beta\right)\right), \tag{7.35}$$

in which the 'risk set' $R(y)$ for any time y is the set of subjects who are 'at risk' at time y. The idea is that, like r_k in the Kaplan–Meier or N^+ in the Bayes Kaplan–Meier, $R(y)$ groups together the set of individuals who are alive and uncensored just prior to y. (The derivation giving the proportionality in (7.35) is not obvious, but it is not hard either.) Expression (7.35) is a *partial* likelihood since D factors involving β are dropped.

Maximizing (7.35) over the $h_0(y_i)$ gives

$$\hat{h}_0(y_i) = \frac{1}{\sum_{j \in R(y_i)} e^{x_j^T \beta}} \tag{7.36}$$

and hence $\hat{H}_{0,\beta}(y) = \sum_{y_i \leq y}[\delta_i / \sum_{j \in R(y_i)} e^{x_j^T \beta}]$ as an estimate, apart from β, for the baseline cumulative hazard. Note that the role of β in (7.36) arises as an index on h_0 from the linkage between h_0 and β due to the maximization; this usage of β is different from the usage of β as a coefficient on covariates.

It remains to find $\hat{\beta}$. To express this, write $x_{(i)}$ as the covariate value associated with $y_{(i)}$, the ith order statistic among $y_1, \ldots, y_D$. Substituting (7.36) into (7.35) for $h_0(y_i)$ and restoring the factors left out in (7.35) gives

$$L^*(\beta) = \prod_{i=1}^{D} \frac{e^{x_{(i)}^T \beta}}{\sum_{j \in R(y_i)} e^{x_{(i)}^T \beta}}; \tag{7.37}$$

see Cox (1972) or Klein and Moeschberger (2003, p. 253). Cox (1972, p. 191) called L^* a conditional likelihood and Klein and Moeschberger (2003, p. 258) referred to L^* as a profile likelihood; Kalbfleisch and Prentice (1973, p. 269) called L^* a marginal likelihood.

Expression (7.37) can be maximized over β by a variety of iterative methods, including Newton–Raphson, to get a value for the MLE, a point estimate for β. A different approach to this optimization was pursued in Kalbfleisch and Prentice (1973, Sec. 4, Eqs. (7) and (8)) and leads to different estimates of $\hat{h}_0$, $\hat{H}_0$, and hence $\hat{S}_0$, and so different estimates for $\hat{h}_{\hat{\beta},x}$, $\hat{H}_{\hat{\beta},x}$, and $\hat{S}_{\hat{\beta},x}$.

The point estimate $\hat{\beta}$ can be used in (7.36) to make it purely a function of the data. Now, (7.27), (7.30) (since ψ is assumed exponential), or even more easily (7.31) can be used to get a point estimate of $h(y|x)$. First, write

$$\hat{H}_0(y) = \sum_{y_i \leq y} \left(\delta_i \Big/ \sum_{j \in R(y_i)} e^{x_j^T \hat{\beta}} \right). \tag{7.38}$$

Then $\hat{S}_0(y) = e^{-\hat{H}_0(y)}$, so (7.31) becomes explicitly

$$\hat{S}(y|X) = \hat{S}_0(y)^{e^{-X^T\hat{\beta}}}, \tag{7.39}$$

giving point estimates for the percentiles of $\hat{F}_X(y) = 1 - \hat{S}(y|X)$ for a subject with covariate value X. Indeed, the median of $\hat{F}_X$ is a point predictor for Y_{n+1}, and the $\alpha/2$ and $1 - \alpha/2$ percentiles of $\hat{F}_X$ give a $(1-\alpha)100\%$ PI for Y_{n+1}.

As convenient as this is, it is also useful to have assessments for the variability of $\hat{\beta}$ and $\hat{h}_0$. To find the Fisher information for β, first rewrite (7.37) using logs to get

$$\log L(\beta) = \sum_{i=1}^{n} \delta_i \log \frac{e^{x_i^T \beta}}{\sum_{j \in R(y_i)} e^{x_i^T \beta}}. \tag{7.40}$$

Differentiating (7.40) with respect to β once gives the vector

$$\frac{\partial}{\partial \beta} \log L(\beta) = \sum_{i=1}^{n} \delta_i \left(x_i - \frac{\sum_{j \in R(y_i)} x_j e^{x_j^T \beta}}{\sum_{j \in R(y_i)} e^{x_j^T \beta}} \right).$$

For any β, let

$$\bar{x}_i(\beta) = \frac{\sum_{j \in R(y_i)} x_j e^{x^T \beta}}{\sum_{j \in R(y_i)} e^{x^T \beta}}$$

and differentiate a second time to get the empirical Fisher information matrix as

$$\hat{I}(\beta) = \frac{\partial^2}{\partial \beta^2} \log L(\beta) = \sum_{i=1}^{n} \delta_i \frac{\sum_{j \in R(y_i)} (x_j - \bar{x}_i(\beta))(x_j - \bar{x}_i(\beta))^T e^{x_j^T \beta}}{\sum_{j \in R(y_i)} e^{x_j^T \beta}}; \tag{7.41}$$

see van Houwelingen and Putter (2012), in which (7.41) was interpreted as a weighted sum of variance matrices by the definition

$$\mathrm{Var}_i(x|\beta) = \frac{\sum_{j \in R(y_i)} (x_j - \bar{x}_i(\beta))(x_j - \bar{x}_i(\beta))^T e^{x_j^T \beta}}{\sum_{j \in R(y_i)} e^{x_j^T \beta}}.$$

Tsiatis (1981) showed that the estimate $\hat{\beta}$ obtained by maximizing (7.37) is consistent and asymptotically normal, with variance matrix n times the inverse of (7.41), i.e., asymptotically, $\sqrt{n}\hat{I}^{1/2}(\hat{\beta})(\hat{\beta} - \beta) \sim N(0, 1)$. Thus, confidence regions for β can be found readily.

It is known that $\hat{S}(y|x) \sim N(S(y|x), v^2)$, asymptotically, for some v^2. To find a form for v^2, the asymptotic variance of $\hat{S}(y|x)$, use the delta method (see (7.13)) to obtain

$$v^2 = \text{Var}(\hat{S}(y|x)) = \hat{S}^2(y|x) \text{Var}(\log \hat{S}(y|x)). \tag{7.42}$$

All that is needed is an estimate for the second factor on the right. Following van Houwelingen and Putter (2012, Chap. 3), let

$$\hat{q}(y|x) = \sum_{y_i \le y, \delta_i = 1} (x - \bar{x}_i(\hat{\beta})) \frac{e^{x^T \hat{\beta}}}{\sum_{y_i \le y, \delta_i = 1} e^{x^T \hat{\beta}}}.$$

Then

$$\sum_{y_i \le y, \delta_i = 1} \left(\frac{e^{x^T \hat{\beta}}}{\sum_{j \in R(y_i)} e^{x_j^T \hat{\beta}}} \right)^2 + \hat{q}(y|x) \hat{I}^{-1}(\hat{\beta}) \hat{q}(y|x) \tag{7.43}$$

is a consistent estimator for $\text{Var}(\log \hat{S}(y|x))$. Note that this estimator has a term for each factor on the right-hand side of $-\log \hat{S}(y|x) = \hat{H}_0(y) e^{x^T \hat{\beta}}$. Indeed, the first term in (7.43) would be the variance of $\hat{H}_0(y)$ if $\hat{\beta}$ were the correct value for β, i.e., it had no variability, and the second term is a transformation of $\hat{I}^{-1}$, the asymptotic form of the variance of $\hat{\beta}$. Together with $\hat{S}^2(y|x)$, which is known, (7.43) gives an estimator for $\text{Var}(\hat{S}(y|x))$; see also Klein and Moeschberger (2003, Sec. 8.8).

7.2.2 Frequentist PH Models as Predictors

With the results of Sec. 7.2.1 in hand, it is useful to revisit (7.31) and the discussion around it. First, recall that the choice of $\hat{\beta}$ as the maximum of (7.37) and $\hat{H}_0$ as in (7.38) gives (7.39). This is a plug-in predictor in the sense of Skouras and Dawid (1999), who proved that this procedure is efficient from the standpoint of probability forecasting. Probability forecasting is similar to the point forecasting approach mostly taken here so, in an asymptotic sense, one expects point forecasts to be efficient as well, at least when one is using finite-dimensional real parameters.

There are two more obvious predictors that one can use which take into account the variability of $\hat{\beta}$ and $\hat{H}_0$. First, one can use the conclusion of Sec. 7.2.1, namely the asymptotic normality of $\hat{S}(y|x)$, and write

$$z_{v/2} v \le S(y|x) - \hat{S}(y|x) \le z_{1-v/2} v$$

for some small v, where v is the root of the right-hand side of (7.42). Equivalently, $\hat{F}_x(y)$ is asymptotically normal with mean $F_x(y)$ and variance v^2 for each fixed y, i.e., $\hat{F}_x(y) \sim N(F_x(y), v^2)$. Using the delta method with function $\hat{F}_x$ and choosing $y = \hat{\xi}_{1-\alpha}$ to be the $(1-\alpha)100$th percentile of $\hat{F}_x$ gives $\hat{\xi}_{1-\alpha} \sim N(\xi_{1-\alpha}, (v^*)^2)$ for a $v^* > 0$ involving the derivative of $\hat{F}_x$ at $\xi_{1-\alpha}$. This statement necessitates using a smoothed version of $\hat{F}_x$ and assumes the $\sqrt{n}$ convergence of $\hat{\xi}_{1-\alpha}$ to $\xi_{1-\alpha}$. Thus, an upper confidence bound for the $(1-v)100$th percentile of $F_x(\cdot)$ is

$$\xi_{1-\alpha} \le \hat{\xi}_{1-\alpha} + z_{1-v} v^*,$$

asymptotically, giving $[0, \hat{\xi}_{1-\alpha} + z_{1-\nu}v^*]$ as a $(1-\nu)100\%$ PI for Y when $X = x$. Clearly, one can form PIs under other conditions – lower confidence-bound PIs or two-sided confidence-bound PIs – by essentially the same approach. An obvious point predictor is the median, mean, or other location derived from $\hat{F}_x$.

A different approach follows by using the SE results on $\hat{\beta}$ and $\hat{S}_0$ from Sec. 7.2.1 directly in (7.39). Obviously, $S_0(y) = S(y|x = 0)$ can be estimated by $\hat{S}(y|x = 0) = e^{-\hat{H}(y|x=0)} = e^{-\hat{H}_0(y)}$. Since the variance of $\hat{H}_0(y)$ equals the variance of $-\log S_0(y)$, which has the form (7.43) for the case $x = 0$, one can heuristically replace $\hat{S}_0(y)$ in (7.39) with $\hat{S}_0(y) \pm z_\nu v$, depending on whether upper or lower bounds are desired for a given $\nu \in (0,1)$. Then, in (7.39) it is enough to use the asymptotics for estimating $\hat{\beta}$. Specifically, the factor $e^{x^T \hat{\beta}}$ can be bounded with high probability using a confidence interval around $\hat{\beta}$ defined by $\hat{I}(\hat{\beta})$: write $x^T \hat{\beta} \sim N(x^T \beta, x^T \hat{I}(\hat{\beta})x)$ to see that

$$z_{\alpha/2} \leq \sqrt{x^T \hat{I}(\hat{\beta})x}\,(x^T \hat{\beta} - x^T \beta) \leq z_{1-\alpha/2}.$$

Now, (7.39) can be bounded by expressions of the form $\hat{S}_0(y) \pm z_\nu v$ raised to the power $x^T \hat{\beta} \pm \sqrt{x^T \hat{I}(\hat{\beta})x}\, z_{\alpha/2}$, from which PIs can be found that have natural point predictors from (7.39) merely by neglecting the variability in $\hat{\beta}$ and $\hat{h}_0$.

Overall, these predictive techniques have been derived in the same spirit as those derived earlier. The intervals here were obtained, in parallel to (4.2) or (4.4), by invoking an asymptotic normality argument; see Link (1984). Usually it is reasonable to take the median, i.e., the 50th percentile, as a natural choice for a point predictor. Alternatively, PIs from a given point predictor, such as a mean or median, can be found using $P_{\hat{\beta}, \hat{h}_0, x}$ in (2.10) or (7.22) if the variability in $\hat{\beta}$ and $\hat{h}_0$ is neglected. More carefully, one would have to incorporate the variability in $\hat{\beta}$ and $\hat{h}_0$, in analogy to (2.12).

To conclude this subsection there are two important points.

First, even though the predictors here (and elsewhere) are derived from models, it is important to remember that our point of view is not a modeling perspective; it is purely predictive and therefore a predictor has utility according to its error per prediction, whether that is low, medium, or high in relation to, say, the spread of the predictand. In other words, any two predictors can be meaningfully compared by the behavior of their predictive errors. Indeed, interpreting a model that generates a predictor as having any validity beyond producing a good (or otherwise) predictor is a mistake. For instance, it can lead to misunderstandings like the so-called 'fallacy of Greek letters',[3] which arises only from the mistake of thinking of the model as describing something real rather than merely a way to generate a predictor. The fallacy is resolved merely by noting that predictors from different models are comparable even if the models are not. This is apart from the obvious drawback of adopting a modeling perspective: it precludes comparing the coefficients for the same covariate from two different models on the grounds that these coefficients don't have the same meaning because of the presence of other terms. To see how misguided the modeling perspective is,

---

[3] In its simplest form, the so-called fallacy of Greek letters involves believing that (or at least acting as if) the parameters β_1 and β_1' in the models $Y = \beta_0 + \beta_1 X_1 + \epsilon$ and $Y = \beta_0' + \beta_1' X_1 + \beta_2' X_2 + \epsilon$ have the same meaning and can be meaningfully compared in spite of the presence of X_2 in the second model.

note that, for instance, the interpretation of a predictor from a linear model is that the factor β_i represents a change in the predicted value of Y for a unit change in X_i – a perfectly sensible interpretation that is not reliant on believing the model to be true in any meaningful sense.

In an extreme, but not indefensible, view, a model is a human construct and the 'validity' of the model exists only in the imaginations of human brains. In the absence of extraordinarily thorough testing, a model remains a fantasy in human brains. As a gedanken experiment, one may try to imagine how another species might come up with an entirely different way to explain reality on the basis of completely different principles. Given this, it's only the prediction and the outcomes that are real.

Second, as a predictive strategy, PH models have their detractors. Indeed, it is not clear that survival analysis can, as commonly used, routinely provide more than a serviceable summary of a data set in the sense of Chapter 3. Specifically, predictions often seem too weak to be useful; see Henderson *et al.* (2001) and Henderson and Keiding (2005). This may be due to the modeling being poor (e.g., the hazards are only approximately proportional at best) or difficult (e.g., the hazard function is too complex compared with the models used to approximate it) or to the high intrinsic variability of the biomedical populations to which PH models are most frequently applied. In any of these cases, PH models may require more effective validation than is commonly achieved. One further possibility is that the relationship between the data and a density such as a hazard function is much more distant than the relationship between the data and its distribution function. This may mean that reliably estimating hazard functions for populations with high intrinsic variability requires more data than are typically available.

7.2.3 Bayes PH Models

A Bayes analysis of the PH model begins by writing down the conditional survival probability, given the parameters and design matrix $X = [x_1, \ldots, x_n]$ of explanatory variables for the possibly right-censored lifetimes $y_1, \ldots, y_n$:

$$P(Y > y | X, H_0, \beta) = \exp\left(-\sum_{i=1}^{n} \exp\left(-H_0(y_i) x_i^T \beta\right)\right), \tag{7.44}$$

as in (7.27) or the simplified form following it, where $H_0(\cdot)$ is the baseline cumulative hazard function and β is the coefficient vector for the explanatory variables. The first step is rewriting (7.44) to put it in a form in which it is easy to assign priors to H_0 and β.

One way to do this is to group the data into intervals. Thus, set $0 = t_0 < t_1 < \cdots < t_m < \cdots < t_M$, where t_M is so large that it is unlikely to have a Y bigger than t_M. In particular, $y_M > y_{(n)}$, the largest of the order statistics $y_{(i)}$. This means that there are M intervals of the form $I_m = (t_{m-1}, t_j)$, for $m = 1, \ldots, J$, and that each y_i is in exactly one of those M intervals. Consequently, for each $m = 1, \ldots, M$, there is a triple (X_m, R_m, D_m) consisting of the values of the explanatory variables for the y_i in I_m, in the risk set for I_m, and in the death set for I_m. Without loss of generality, it is assumed that X has been standardized, so that

$$\sum_{j=1}^{p} x_{ij} = 0 \qquad \text{and} \qquad \sum_{j=1}^{p} x_{ij}^2 = 0.$$

The grouping approach was first used in Kalbfleisch (1978), who also introduced the gamma process used below, and was developed in Burridge (1981) to show that discretizing to intervals provided a more effective approach than regarding the data as continuous, at least in a Bayes context. The treatment here follows the more recent work in Ibrahim *et al.* (2001, Chap. 3.2) and Lee *et al.* (2011, Secs. 2 and 3) on which the R package pbscGroup is based; see R Core Team (2016).

To convert (7.44) to an interval form start by writing

$$h_m = H_0(s_m) - H_0(s_{m-1}) \qquad \text{for } m = 1, \ldots, M;$$

the h_m are the increments on which a prior will place a distribution. The definition of the h_m leads to the telescoping sum

$$\sum_{u=1}^{m} h_u = (H_0(t_1) - H_0(t_0)) + \cdots + (H_0(t_m) - H_0(t_{m-1})) = H_0(t_m) - H_0(t_0) = H_0(t_m),$$

since $H_0(t_0) = H_0(0) = 0$ is assumed. It makes mathematical sense to set $H_0(0) = 0$ because an integral over one point, zero, is zero. It also makes modeling sense because in zero units of time there is no risk of death. Now, assuming that H_0 is increasing and $m \geq 2$, use (7.44) for $n = 1$, i.e., for one generic outcome Y to get

$$
\begin{aligned}
P(Y \in I_m | X, H_0, \beta) &= P(Y \geq t_{m-1} | X, H_0, \beta) - P(Y \geq t_m | X, H_0, \beta) \\
&= \exp\left(-H_0(t_{m-1}) \exp(x_i^T \beta)\right) - \exp\left(-H_0(t_m) \exp(x_i^T \beta)\right) \\
&= \exp\left(-H_0(t_{m-1}) \exp(x_i^T \beta)\right) \left(1 - \exp\left(-H_0(t_m) \exp(x_i^T \beta) \right.\right. \\
&\qquad\qquad \left.\left. + H_0(t_{m-1}) \exp(x_i^T \beta)\right)\right) \\
&= \exp\left(-\exp(x_i^T \beta) \sum_{u=1}^{m-1} h_u\right) \left(1 - \exp\left(-h_m \exp(x_i^T \beta)\right)\right).
\end{aligned}
$$

(7.45)

When $m = 1$, the formula is simpler:

$$
\begin{aligned}
P(Y \in I_1 | X, H_0, \beta) &= P(Y \geq t_0 | X, H_0, \beta) - P(Y \geq t_1 | X, H_0, \beta) \\
&= \exp\left(-H_0(t_0) \exp(x_i^T \beta)\right) - \exp\left(-H_0(t_1) \exp(x_i^T \beta)\right) \\
&= 1 - \exp\left(-H_0(t_1) \exp(x_i^T \beta)\right).
\end{aligned}
$$

So, if $h_0 \equiv 0$, (7.45) holds for $m = 1, \ldots, M$ and the left-hand side of (7.45) can be written as $P(Y \in I_m | X, h, \beta)$, where $h = (h_1, \ldots, h_M)$, or, for n IID outcomes,

$$P(Y_1 \in I_{m_1}, \ldots, Y_n \in I_{m_n} | X, h, \beta) = \prod_{i=1}^{n} P(Y_i \in I_{m_i} | X, h, \beta).$$

Denoting $Y = (Y_1, \ldots, Y_n)$, the grouped data likelihood is written as a product of probabilities rather than as a density because the outcomes are grouped into intervals. The result is

$$L(h,\beta|X,Y) \propto \prod_{m=1}^{M} \left(\exp\left(-h_m \sum_{u \in R_m \backslash D_m}\right) \exp(x_u^T \beta) \prod_{v \in D_m} \left(1 - \exp\left(-h_m \exp(x_v^T \beta)\right)\right)\right),$$

(7.46)

where R_m and D_m are the risk and death sets at time $y_{(m)}$; see Ibrahim *et al.* (2001, Sec. 3.2.2).

To put a prior on the h_m recall the definition of a gamma process, denoted $\mathcal{GP}(H^*, c)$. The baseline cumulative hazard function H_0 has a $\mathcal{GP}(H^*, c)$ prior if and only if:

1. $H_0(0) = 0$, $H^*(0) = 0$, and H^* is increasing;
2. H_0 has independent increments for disjoint intervals; and
3. $\forall u > v$, $H_0(u) - H_0(v) \sim \text{gamma}(c(H^*(u) - H^*(v)), c)$.

In item 3, $\text{gamma}(c(H^*(u) - H^*(v)), c)$ is a gamma distribution with shape parameter $c(H^*(u) - H^*(v))$ and scale parameter c. The function H^* is the mean of the gamma process (for any time y) and c summarizes the spread around the mean. Small values of c indicate a large spread; large values of c indicate little spread. Thus, c represents the prior beliefs about how close H_0 is to H^*.

Thus, a $\mathcal{GP}(H^*, c)$ prior can be assigned to H_0 for use in (7.46) by writing

$$h_m \sim \text{gamma}(\alpha_{0m} - \alpha_{0m-1}, c),$$

(7.47)

where $\alpha_{0m} = cH^*(s_m)$, for $m = 1, \ldots, M$. Any other prior that assigns a distribution to the increments h_m could also be used, but the gamma process is one of the most common choices.

It remains to assign a prior to β. One reasonable choice is a double exponential,

$$w(\beta|\sigma^2, \lambda) = \prod_{j=1}^{p} \frac{\lambda}{2\sqrt{\sigma^2}} \exp\left(-(\lambda/\sqrt{\sigma^2})|\beta_j|\right);$$

(7.48)

Lee *et al.* (2011) showed that this guarantees unimodality of the posterior. A noninformative prior

$$w(\sigma^2) = \frac{1}{\sigma^2}$$

(7.49)

can be assigned to σ^2 and a gamma prior

$$w(\lambda^2|\delta, r) = \frac{\delta^r}{\Gamma(r)}(\lambda^2)^{r-1} \exp\left(-\delta\lambda^2\right)$$

(7.50)

can be assigned to λ^2 where the hyperparameters δ, r are chosen by the user. So, the posterior for h and β is obtained by using (7.46)–(7.50) to find

$$w(h, \beta, \sigma, \lambda^2|X, Y) \propto L(h, \beta|X, Y)w(h)w(\beta|\sigma^2, \lambda^2)w(\sigma)w(\lambda^2|\delta_0, r_0)$$

(where $w(h)$ corresponds to (7.47)) and marginalizing out σ^2, λ^2. The posterior will depend on H^*, c, δ_0, r_0, and the t_m chosen, but this is otherwise a fully Bayes solution. In the pbscGroup package, a more general setting that permits the different β_j to have different normal priors is used. This is possible because the double exponential can be represented as a scale mixture of normals with a gamma mixing density. In addition, the interval endpoints

t_m can be equipped with a prior; however, this is difficult, so often the t_m are located at data points such that $M = n$ and $t_m = y_{(m)}$ (except for t_M, which is usually chosen to be substantially larger than $y_{(M)}$).

Now, since the distribution of the Y_i has been discretized to intervals I_m, a Bayesian would predict a future Y_{n+1} by finding the predictive distribution over the same intervals,

$$P(Y_{n+1} \in I_m | X, Y) = \int P(Y_{n+1} \in I_m | X, Y, h, \beta) W(dh, d\beta | X, Y).$$

Since this is not available in closed form, one would approximate the integral by taking draws $h_1, \ldots, h_B$ and $\beta_1, \ldots, \beta_B$ of the vectors h and β from the posterior, evaluating the integrand at them, and then taking the average. That is,

$$\widehat{P}(Y \in I_m | X, Y) = \frac{1}{B} \sum_{b=1}^{B} P(Y \in I_m | X, Y, h_b, \beta_b) \qquad (7.51)$$

would be the numerical approximation to the predictive distribution over the intervals. To get a point predictor it would be natural to use the midpoint of the intervals I_m having the largest probability, recalling that the last interval is $(t_{M-1}, t_M]$ and hence bounded. An alternative point predictor would be the midpoint of the interval containing the 50th percentile of $\widehat{P}$. Of course, one could also use (7.51) to get PIs merely by taking the endpoints or midpoints of the intervals corresponding to the $(\alpha/2)100$th and $(1 - \alpha/2)100$th percentiles. It would be more accurate to use linear interpolation on the intervals containing these percentiles.

7.2.4 Continuing the Example

Recall that the data set uissurv, of times to relapse for drug abuse patients, was analyzed in Sec. 7.1.5 using Kaplan–Meier (KM) predictors. There, 80 data points were chosen at random for analysis; of these 12 were censored and 68 were real observations. Also, four other data sets were derived from those 80 data points by truncation and random censoring. Here, the same 80 data points are used but only two of the other data sets derived from them are considered, namely, those with the more severe forms of censoring. Thus, random censoring by the same normal located at the 75th percentile of the 68 observations is used, giving 47 real observations, and truncation censoring at a level to give again 47 real observations is also used. Of course, these two derived data sets do not have the same patterns of censoring because random censoring is performed over the whole data set while truncation censoring clips off the highest observed y-values.

For these three data sets, Bayes and frequentist PH models can be used to make predictions under the same loss function as before; see (7.25). That is: 100 permutations of the data sets were found for which the medians existed at all time points; a 20-data-point burn-in was used; the CPEs from the sequence of prediction errors for each permutation was summed at each time step to give an overall CPE; and these were graphed to show a (nondecreasing) CPE curve along with error bars. Thus the values being predicted and the framework within which they are being predicted are the same as in Sec. 7.1.5. The only difference is that PH predictors are used and these permit the inclusion of explanatory variables. The data set includes 10 explanatory variables, so there are 2^{10} possible predictors. However, for illustrative purposes, PH predictors based on three variables, namely, age, treatment, and site, were chosen. Despite the numerous differences between the KM predictors of Sec. 7.1 and the PH predictors in this section, it is perfectly fair to compare them. Indeed, PH predictors with other selections of explanatory variables could be included and fairly compared. It is important to bear in mind continually that the specific properties of the models are not important except insofar as they affect prediction.

To be precise, in the frequentist PH case, the predictor from the function coxph in the R library survival (R Core Team 2016) at step $i + 1$ is the median from the survival function given by the standard Cox proportional-hazards model using data up to and including step i, evaluated at x_{i+1}, the values of the explanatory variables at step $i + 1$. The Bayes PH predictor comes from the pbscGroup package with the default of treating all the explanatory variables as a single group, using $B = 2000$ MCMC draws in (7.51). Internally to pbscGroup, intervals I_m are chosen at the observed values of Y (with t_M chosen to be large). So, it is natural to use the midpoint of the interval containing the median as the Bayes predictor. That is, using the data up to step i, the approximation (7.51) is found, evaluated for x_{i+1}, its median interval found, and the midpoint of that interval taken as the prediction.

To begin the comparison, consider Fig. 7.4. It shows the CPE curves for the six cases (three data sets, Bayes and frequentist PH predictors). The top left panel shows that the frequentist PH predictor on the original data set does best. A very close second is the frequentist PH predictor with added random censoring and the third best is the frequentist PH predictor with added truncation censoring. The remaining three panels verify that the ranking of methods by predictive performance is reasonable, in that the error cones (because the CPE is not divided by n) for each method separate as the data accumulate. The fact that the lines are relatively straight suggests that the per-prediction increase in CPE stabilizes quickly for all methods.

Contrasting the performance of the KM predictors in Fig. 7.3 and the performance of the PH predictors in Fig. 7.4 is instructive. It can be seen that the vertical scales are different. In Fig. 7.3 the worst performance is for the original data and the Bayes and frequentist KM predictors are essentially equivalent, both reaching a CPE of a little over 4000 by the last data point. However, when PH predictors are used on the original data (with the selection of the three explanatory variables), performance deteriorates badly, surpassing 12 000. The difference is bigger than can be explained by taking $\pm 2SD$ bars around the averages, suggesting that the PH predictors are worse than the KM-based predictors. This may reflect the unreasonableness of the PH assumption, the instability of PH models, or the inefficacy of the selection of variables. Further comparisons of the CPEs of the KM predictors in Fig. 7.3 to their corresponding CPE curves in Fig. 7.4 reveals the same pattern. Moreover, the Bayes methods do much worse here than before, and random censoring is not as harmful as truncation censoring, the reverse of what was seen before. This may be due to the influence of the selection of priors, but there are other possible factors and further work would have to be done to explain the differences convincingly. On the plus side, since the PH predictions in this example are so poor, there is no point in carefully evaluating the parameter values (or their SEs).

As a sanity check on these conclusions, it is worthwhile to look at specific CPE paths for the two PH predictors and the three data sets. This is done in Fig. 7.5, where two representative CPE paths for each case are shown. It is seen in all cases that, for the most part,

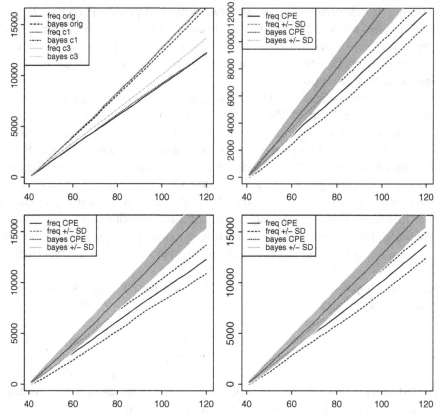

Figure 7.4 The top left panel shows the average CPE vs. sample size over 100 random permutations for the six median PH predictors for each of the three data sets; c1 indicates random censoring at the 75th percentile while c3 indicates the corresponding truncation censoring. The top right panel shows the CPE curves with ±SD curves indicated for the original data. The bottom two panels are for c1 (left) and c3 (right).

the CPE curves are close to their average. This confirms that the nearly straight lines seen in the upper left panel of Fig. 7.4 and the error bars seen in the other panels of Fig. 7.4 are reasonable.

It remains a curiosity that KM predictors perform better under truncation censoring than when the full data set is used (the final CPEs are a little over 2000 in Fig. 7.3) and that frequentist PH predictors (although the best among the PH predictors) perform so poorly even when all the available data is used (the final CPEs are a little over 12 000 in Fig. 7.3). Clearly, the interactions between the loss function, type of censoring, variable selection, prior selection, the properties of the hazard functions such as proportionality, and the nature of the predictors themselves are more complex than common practice suggests. The important thing to note here is that these comparisons are possible and natural in a predictive setting but would be theoretically impermissible in a model-based setting. Thus, model-based arguments in favor of avoiding the 'fallacy of Greek letters' amount to little more than an impediment to the comparative evaluation of statistical techniques under the overriding scientific principle of successful prediction.

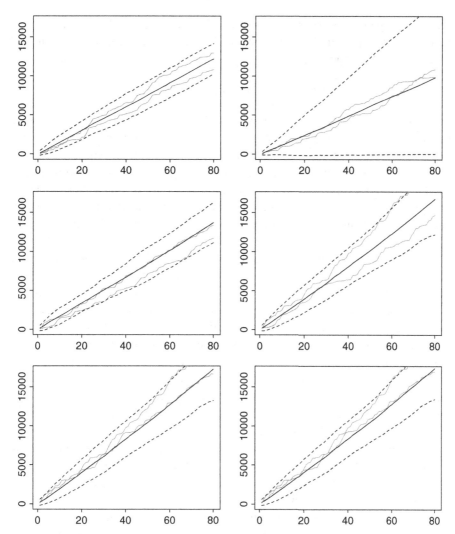

Figure 7.5 The top left panel shows the average CPE (solid black line) vs. the number of time steps, ±SD error bars (dashed lines), and two CPE paths (for two different permutations; light gray lines) for the frequentist median predictors, for the original data set. The top right and middle left panels show the same sort of graph for the random-censoring and truncation-censoring data, respectively. The middle right and bottom panels show the same graphs for the corresponding Bayes PH predictors.

7.3 Parametric Models

Having seen a purely nonparametric approach to obtaining predictors for survival by means of estimating the DF, and a semiparametric approach by means of estimating the hazard function and a parameter β, we will examine the next logical approach, namely, parametric models for survival. One of the benefits of parametric models for survival is that it is easy to introduce explanatory variables by representing a mean parameter as a linear combination of them. Another benefit is that usually the sample-size requirements for good prediction are lower for parametric models than for nonparametric models.

Given the parametric framework, one regards P as an element of the parametric family and obtains the corresponding hazard function. As seen in the introduction to this chapter, one choice for P is the exponential(θ) distribution. This gives $h(y) = \theta$, i.e., a constant, and

$$P(Y \geq y + z | Y \geq y) = P(Y \geq z),$$

i.e., exponential(θ) is 'memoryless' in the sense that having waited for y units and then a further z units is the same as having waited for z units in the first place. A generalization of the exponential(θ) distribution is the Weibull(α, θ) distribution, with survival function $S(y) = \exp(-\theta y^{\alpha})$. The extra parameter α affects the shape of the underlying density and the hazard function is $h(y) = \theta \alpha x^{\alpha-1}$, which is increasing for $\alpha > 1$, decreasing for $\alpha < 1$, and constant for $\alpha = 1$. There are numerous other distributions that have been studied in the context of lifetime data including the gamma, the log-normal, the extreme value, and the Pareto.

Whatever the parametric family chosen, the procedure for the prediction of survival times is much the same. In the frequentist case, one forms a predictor, whether a mean, median, mode, or other location, and controls the probability that it is a given distance away from the next outcome. In the Bayes case, one assigns priors to the parameters, finds their posterior, and uses it to obtain a predictive distribution. This is conceptually the same as the IID prediction scenario of Sec. 2.2, so in most cases parameters will have to be estimated and their variability assessed, in analogy to (2.12) and (2.21). If closed forms for predictive procedures do not exist, it is usually enough to substitute estimates of parameters, possibly using upper or lower confidence bounds to get conservative bounds for PIs. If one's interest focuses on probabilities of survival, these can be obtained directly from the estimated survival function, as in Secs. 7.1 and 7.2.

One step up in complexity from merely assigning a plausible parametric model is the use of explanatory variables to estimate the survival function. The idea is that the explanatory variables should enable one to get tighter bounds on S or at least get tight enough bounds that interesting questions on the difference between groups in terms of their survival characteristics can be answered. Essentially, the theoretical location parameter is represented as a regression function of the explanatory variables and its coefficients estimated. If a constant term is included in the regression function, the procedure reduces to merely assuming a parametric model for the response Y.

We will look at an example of this in detail. So, we start by looking at the no-explanatory-variables case and assume that the Y_i are distributed according to a Weibull(γ, α), i.e.,

$$Y \sim p_{\gamma,\alpha}(y) = \alpha \gamma y^{\alpha-1} \exp(-\gamma y^{\alpha}). \tag{7.52}$$

Then we use a log transform by replacing Y with $\log Y$. The survival function becomes

$$S(y) = \exp(-\gamma \exp(\alpha y)),$$

so, if the parameters are transformed by $\gamma = \exp(-\mu/\sigma)$ and $\sigma = 1/\alpha$ the density function and survival function become

$$p(y|\mu,\sigma) = \frac{1}{\sigma} \exp(((y-\mu)/\sigma) - \exp((y-\mu)/\sigma)) \tag{7.53}$$

and

$$S(y|\mu,\sigma) = \exp\left(-\exp\left((y-\mu)/\sigma\right)\right). \tag{7.54}$$

If data (y_i, δ_i, x_i) for $i = 1, \ldots, n$ (possibly right-censored) are available then (7.32) gives for the likelihood

$$
\begin{aligned}
L(\mu,\sigma) &= \prod_{i=1}^{n} p(y_i|\mu,\sigma)^{\delta_i} S(y_i|\mu,\sigma)^{1-\delta_i} \\
&= \prod_{i=1}^{n} \left(\frac{1}{\sigma} f_W\left(\frac{y_i-\mu}{\sigma}\right)\right)^{\delta_i} \left(S_W\left(\frac{y_i-\mu}{\sigma}\right)\right)^{1-\delta_i},
\end{aligned}
\tag{7.55}
$$

where $f_W(w) = \exp\left(w - \exp(-w)\right)$ and $S_W(w) = \exp\left(-\exp(w)\right)$ are the density and survival functions for a standard extreme-value distribution.

From (7.55) it is possible to obtain MLEs numerically for μ, σ, $\hat{\mu}$, and $\hat{\sigma}$. This gives $\hat{\gamma} = \exp(\hat{\mu}/\hat{\sigma})$ and $\hat{\alpha} = 1/\hat{\sigma}$. The natural choice of predictor then comes from substituting these estimates into (7.53). This means that point and interval predictors for Y_{n+1} at a given x_{n+1} can be found from the Weibull density directly, either using the formulae for the mean and variance of the Weibull or taking the median, 2.5th, and 97.5th percentiles of the estimated Weibull. This is a parallel to (7.31). Of course, it is also possible to use upper and lower confidence bounds on μ and σ to take into account the uncertainty in estimating μ and σ, obtaining slightly wider PIs as in Sec. 7.2.2.

To include explanatory variables, it is natural to write $\mu = x^T\beta$ and use $\mu_i = x_i^T\beta$ in (7.55) and again obtain numerical values for the MLEs of β and σ, which can be transformed to give $\hat{\gamma}(x)$ and $\hat{\alpha}$. (The estimates $\hat{\alpha}$ and $\hat{\sigma}$ do not depend on the x_i because α and σ are treated as real numbers.) Finally, point and interval predictors can again be obtained from substituting the estimates of γ and σ into (7.52), as in the case with no explanatory variables. A more elaborate analysis using upper or lower confidence bounds on β and σ is possible but is omitted here.

An alternative approach to obtaining estimates of β and α when explanatory variables are available was given in Klein and Moeschberger (2003). The idea is to develop a PH model by recognizing that a linear model of the form $\log Y = x^T\beta + \sigma W$, where W is a standard extreme-value distribution, leads to a PH model with a Weibull baseline hazard, $h_0(y|x) = \alpha\gamma y^{\alpha-1}\exp(x^T\beta)$. Then the techniques already developed for prediction in a PH model may be used. This approach is equivalent to ours but not used here because it seems more complicated than is minimally necessary for prediction as it uses accelerated failure time models; these are treated briefly in the endnotes to this chapter.

Bayes predictors can also be given for Weibull regression. As in Ibrahim *et al.* (2001), transform γ by setting $\lambda = \log \gamma$. Now, (7.53) becomes

$$p(y|\alpha,\lambda) = \alpha y^{\alpha-1} \exp\left(\lambda - y^\alpha \exp(\lambda)\right)$$

and (7.54) becomes

$$S(y|\alpha,\lambda) = \exp\left(-y^\alpha \exp(\lambda)\right).$$

Letting $d = \sum_i \delta_i$, the number of uncensored observations, and denoting the data by $\mathcal{D}$, (7.55) becomes

$$L(\alpha, \lambda | \mathcal{D}) = \prod_{i=1}^{n} p(y_i | \alpha, \lambda)^{\delta_i} S(y | \alpha, \lambda)^{1-\delta_i}$$

$$= \alpha^d \exp\left(d\lambda + \sum_{i=1}^{n} [\delta_i(\alpha - 1) \log y_i - y_i^{\alpha} \exp(\lambda)]\right). \qquad (7.56)$$

To complete the Bayes formulation, priors must be assigned to α and λ. As usual, the priors should be chosen to reflect the pre-experimental information believed to be available. Since there is no conjugate prior when both α and λ are assumed unknown, one convenient choice is to assign independent priors, say, gamma(α_0, κ_0) to α and Normal(μ_0, σ_0) to λ. Then, the joint posterior is

$$w(\alpha, \lambda | \mathcal{D}) \propto L(\alpha, \lambda | \mathcal{D}) w(\alpha | \alpha_0, \kappa_0) w(\mu_0, \sigma_0^2)$$

and it leads to

$$w(\alpha, \lambda | \mathcal{D}) \propto \alpha^{\alpha_0 + d - 1} \exp\left(d\lambda + \sum_{i=1}^{n} [\delta_i(\alpha - 1) \log y_i - y_i^{\alpha} \exp(\lambda)]\right)$$

$$\times \exp\left(-\kappa_0 \alpha - (\lambda - \mu_0)^2 / (2\sigma_0^2)\right).$$

Usually, this does not have a closed-form expression, so the posterior can only be found numerically.

In the absence of explanatory variables, given $w(\alpha, \lambda | \mathcal{D})$ estimates for α and λ can be found, e.g., posterior means, and substituted into the Weibull density. This means that point and interval predictors for Y_{n+1} can be found from the Weibull density directly, as in the frequentist case.

A second, and more orthodox, Bayes approach would be to find the predictive distribution defined by the predictive density

$$m(y_{n+1} | \mathcal{D}) = \int p(y_{n+1} | \alpha, \lambda) w(\alpha, \lambda | \mathcal{D}) d\alpha d\lambda$$

and obtain point predictors and PIs from it. However, since no closed form for $m(y_{n+1} | \mathcal{D})$ is available, it would have to be approximated using a sequence of draws $\alpha_1, \ldots, \alpha_B$ and $\lambda_1, \ldots, \lambda_B$ from the approximation to the posterior to find

$$\hat{m}(y_{n+1} | \mathcal{D}) = \frac{1}{B} \sum_{b=1}^{B} p(y_{n+1} | \alpha_b, \lambda_b),$$

from which point and interval predictors could be derived.

In the case where explanatory variables are present, the procedure is similar. In this parametrization, write $\lambda = x^T \beta$, i.e., as a linear regression function. Then, assigning an $N(\mu_0, \Sigma_0)$ prior to β and a gamma(α_0, κ_0) to α, the posterior $w(\beta, \alpha | \mathcal{D})$ is proportional to

$$\alpha^{\alpha_0 + d - 1} \exp\left(\sum_{i=1}^{n} [\delta_i x_i^T \beta + \delta_i(\alpha - 1) \log y_i - y_i^{\alpha} \exp(x_i^T \beta)]\right)$$

$$\times \exp\left(-\kappa_0 \alpha - (1/2)(\beta - \mu_0)^T \Sigma_0^{-1}(\beta - \mu_0)\right).$$

Again, closed forms for the posterior generally do not exist, so only numerical solutions can be found.

As before, point estimates of α and β can be found empirically from a numerical approximation of $w(\alpha, \beta | \mathcal{D})$. These estimates can be plugged into the Weibull density (use $\hat{\lambda} = \hat{\beta}^T x_{n+1}$) to generate point and interval predictors for Y_{n+1} at a given x_{n+1} directly from the Weibull distribution. Again, this can be done either using the formula for the mean and variance of the Weibull or taking the median, 2.5th, and 97.5th percentiles of the estimated Weibull. Upper and lower credibility bounds could also be used to take into account the variability of the estimates of α and β.

Again, a more orthodox Bayes approach would be to find the predictive distribution defined by the predictive density

$$m(y_{n+1}|\mathcal{D}, x_{n+1}) = \int p(y_{n+1}|\alpha, \beta, x_{n+1}) w(\alpha, \beta|\mathcal{D}) d\alpha d\beta$$

where

$$p(y_{n+1}|\alpha, \beta, x) = \alpha y^{\alpha-1} \exp\left(\beta^T x - y^\alpha \exp(\beta^T x)\right)$$

and obtain point predictors and PIs from it. However, since no closed form for $m(y_{n+1}|\mathcal{D}, x_{n+1})$ is available, it must be approximated using a sequence of draws $\alpha_1, \ldots, \alpha_B$ and $\beta_1, \ldots, \beta_B$ from the approximation to the posterior, to find

$$\hat{m}(y_{n+1}|\mathcal{D}, x_{n+1}) = \frac{1}{B} \sum_{b=1}^{B} p(y_{n+1}|\alpha_b, \beta_b, x_{n+1}), \tag{7.57}$$

from which point and interval predictors can be derived.

To conclude this section, a Weibull regression analysis of the uissurv data predictively analyzed in Sec. 7.1.5 using KM methods and in Sec. 7.2.4 using PH methods will be examined. The setting is the same three data sets and explanatory variables as in Sec. 7.2.4. The only two changes are that now (i) the frequentist and Bayes Weibull predictors are compared, and (ii) the number of permutations is reduced from 100 to 80 to save running time.

The frequentist predictors come from substituting the MLEs from(7.55) into (7.54) and finding the median. This was done in the survival package using the command survreg. The Bayes predictions were found using commands in the package LearnBayes; this is more complicated (see Albert (2009)). The computational procedure for generating a prediction is similar to the averaging in (7.57) but done on the level of survival functions rather than densities. First, a frequentist Weibull regression is performed to get MLEs that can be used as a starting point for a Laplace approximation to the posterior. (The Laplace approximation increases the independence of the downstream conclusions from the priors.) Correcting this generates a posterior mode and covariance matrix that can be used as a starting point for MCMC, i.e., a random walk of parameter values from a Metropolis algorithm that gives draws from the posterior as possible values for the Weibull regression parameters. These draws are used to generate survival functions that are then averaged. The Bayes predictor is then the median of the average.

As can be seen in the upper right panel of Fig. 7.6, the best predictions come from the frequentist predictor using all the data. The second best is the frequentist predictor with random censoring. The third best is the Bayes predictor also with random censoring. The

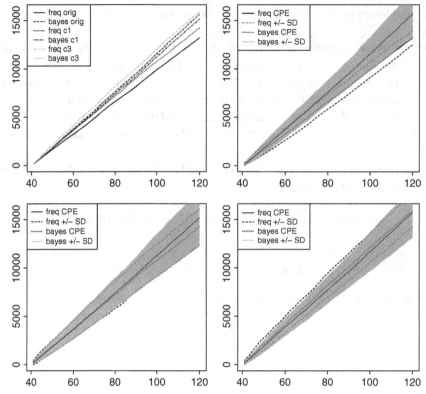

Figure 7.6 The top left panel shows the average CPE vs. sample size over 100 random permutations for the six Weibull predictors for the three data sets; c1 indicates random censoring at the 75th percentile while c3 indicates the corresponding truncation censoring. The top right panel shows the CPE curves with ± SD curves indicated for the original data. The bottom two panels are the same but for c1 (left), and c3 (right).

CPEs for all methods range from 13 000 to 15 000 units of time and so are generally worse, or at least no better, than either the KM or the PH predictors. The other three panels in Fig. 7.6 show that the methods are giving different results but that, even by the 80th step, the difference between them is small – the ±SD cones around the CPE curves do not fully separate for any pair of methods on the same data set. (Increasing the sample size would not make the SDs smaller if they were already representative.) However, the CPE lines are relatively straight so the per-prediction increase in CPE quickly becomes fairly constant, suggesting that its increments per step may be taken as representative of the actual per-prediction error.

   In any of these computed examples one could look into the details of what might be making one predictor perform better or worse than another. While this is essential for achieving good prediction, the goal here is only to provide an example comparing three different classes of predictors. Thus, this example neglects all 'model checking'. Nevertheless, it appears that the explanatory variables add little information and that the fully nonparametric approach (KM) is much better than the semiparametric approach (PH), which in turn is noticeably better than the parametric approach (Weibull regression). Whether this would be borne out in a more detailed, careful, analysis is an open question.

7.4 Endnotes: Other Models

Having seen representatives of the nonparametric, semiparametric, and parametric classes, we will discuss two further models that arise: accelerated failure time models and competing risks models. Variants on the models described here that arise less often (and so are omitted) include the models discussed from a frequentist perspective in Chapter 2 of van Houwelingen and Putter (2012) and the models discussed from a Bayes perspective in Ibrahim *et al.* (2001).

7.4.1 Accelerated Failure Time (AFT) Models

Accelerated failure time (AFT) models accelerate or decelerate the failure times Y by a factor depending on a d-dimensional covariate x, a parameter β, and a specified link function g. The general parametric case constitutes any specification of the conditional survival function $S(y|x)$ with the property that

$$S(y|x) = S_0(yg(x^T\beta)), \tag{7.58}$$

where S depends on β, S_0 is an unspecified baseline survival function, and x may be an outcome from a random variable X; see Ghosh and Ghosal (2006). Both β and S_0 must be estimated, in order to use (7.58).

In the most commonly used example of an AFT model it is assumed that

$$\ln Y = x^T\beta + \epsilon, \tag{7.59}$$

i.e., the exponential link function $g(u) = e^{-u}$ is adopted and the log of the survival time follows a linear model; the error term ϵ is often chosen to be an $N(0, \sigma^2)$, extreme value, or logistic distribution even though (7.58) does not require additive error. It is seen that (7.59) gives

$$Y = e^{x^T\beta}e^\epsilon. \tag{7.60}$$

Thus, when ϵ is $N(0, \sigma^2)$, (7.60) gives a log-normal regression model; when ϵ takes an extreme value (7.60) gives a Weibull regression model; and when ϵ is logistic, (7.60) gives a log-logistic regression model. Each of these choices of error term leads to a model class, and the classes are studied individually; see Hosmer and Lemeshow (1999) and Bedrick *et al.* (2002) among others. See Ghosh and Ghosal (2006) for a case in which S_0 is modeled as a mixture of parametric survival functions in which the mixing distribution must be specified. Using an exponential link, Ghosh and Ghosal (2006) adopted a hierarchical Bayesian model in which their equation (3) is an analog to (7.59); the survival time is taken as Weibull.

Clearly, (7.59) implies the AFT property (7.58):

$$S_{\beta,x}(y) = P_\beta(Y > y|x) = P_\beta(\sigma\epsilon > \ln y - x^T\beta|x)$$
$$= P_\beta(e^{\sigma\epsilon} > ye^{-x^T\beta}|x) = S_0(ye^{-x^T\beta}),$$

so the survival time is accelerated, relative to S_0, by $e^{-x^T\beta}$, the link function evaluated at the regression function. Other link functions give analogs of (7.59) and (7.61). If h_0 is a baseline hazard function corresponding to S_0, i.e., from $x = 0$, then an acceleration is seen on the hazard scale as well as on the survival function scale. For an exponential link this is $h_{\beta,x}(y) = h_0(ye^{-x\beta})e^{-x\beta}$, and other link functions give analogous results. So, again, the effect of the covariates is to adjust the shape of the baseline hazard rate, in effect increasing or decreasing the chance for survival given survival up to a fixed time.

Given data of the form (x_i, Y_i) for $i = 1, \ldots, n$ and an x_{n+1} for which $Y_{n+1}(x_{n+1})$ is to be predicted, denote the predictor by $\hat{y}(x_{n+1})$. If the error distribution is set to be normal$(0, \sigma^2)$, so that the variability of the future outcomes can be determined once the parameters are known and β and σ are estimated by $\hat{\beta}$ and $\hat{\sigma}$, for instance by maximum likelihood, then the survival function is

$$S_{\hat{\beta},x_{n+1}}(y) = P_{\hat{\beta}}(\hat{\sigma}\epsilon > \ln y - x_{n+1}\hat{\beta}|x_{n+1}). \tag{7.61}$$

Converting (7.61) to an estimator $\hat{F}$ of $F_{\beta,x}$ one can obtain a PI as in (4.2) or (4.4) and take the midpoint as a point predictor. Alternatively, the median survival time from $S_{\hat{\beta},x_{n+1}}(y)$ could be used as a point predictor, and its properties could be investigated as in (2.10) and (2.12). Point predictions of t-year survival can be found directly from $S_{\hat{\beta},x_{n+1}}(\cdot)$. However, none of these approaches is really satisfactory because the variability in $\hat{\beta}$ and $\hat{\sigma}$ is neglected. Including it would typically make the prediction intervals larger. On the other hand, for some predictive purposes, ignoring the variability in $\hat{\beta}$ (or $\hat{\sigma}$) may not be too damaging. Indeed, Bedrick *et al.* (2002) developed deletion diagnostics for influential data points to stabilize percentile-based predictions from AFT models with exponential link functions; this stabilization may be more important than the variability in $\hat{\beta}$.

7.4.2 Competing Risks

The idea here is that there are p possible causes for a subject's death, and the time of death from each cause is represented as T_j, $j = 1, \ldots, p$. Of course, only one Y_i can be observed. So, replace the former $\delta_i = 0, 1$, indicating that the ith subject was censored or not, with $\delta_i = j$, meaning that the time of death was $Y = T_j$, so the values of $T_{j'}$ for $j \neq j'$ were not observed. That is, $Y = \min(T_1, \ldots, T_p)$. This leads to a hazard rate for each value of j. The definition, parallel to (7.2), is

$$h_j(y|x) = \lim_{\delta \to 0} \frac{P(y \leq Y \leq y + \delta, \delta = j|Y \geq y, x)}{\delta}$$

where x denotes any explanatory variables. Since the sum of the probabilities in the numerator of the limit must be one, the overall hazard rate (as opposed to the cumulative hazard rate) is

$$h(y|x) = \sum_{j=1}^{p} h_j(y|x).$$

In terms of the cause-specific cumulative hazard functions H_j, the overall cumulative hazard function is

$$H(y|x) = \int_0^y h(u, x)du = \sum_{j=1}^{p} \int_0^y h_j(u|x)du = \sum_{j=1}^{p} H_j(y|x).$$

The overall survival function is

$$S(t|x) = e^{-H(t|x)}.$$

More generally, if the survival function is written as a function with a p-dimensional argument, $S(t_1, \ldots, t_p) = P(T_1 > t_1, \ldots, T_p > t_p)$, then h_j can be recovered by

$$h_j(y|x) = -\frac{1}{S(y, \ldots, y|x)} \frac{\partial}{\partial t_j} S(t_1, \ldots, t_p|x)|_{t_1 = \cdots = t_p = y}.$$

Indeed, one can define the function $S_j(y|x) = e^{-H_j(y|x)}$ but, for $p \geq 2$, S_j does not have a survival function interpretation. The cause-specific density of deaths at time y is

$$p_j(y|x) = \lim_{\delta \to 0} \frac{P(y \leq Y \leq y + \delta, \delta = j|x)}{\delta} = h_j(y|x)S(y|x)$$

and, since the probabilities sum to one, the overall density of deaths at y is

$$p(y|x) = \sum_{j=1}^{p} p_j(y|x).$$

Generalizing the KM estimator to the competing-risk context, consider data of the form (y_i, δ_i) for $i = 1, \ldots, n$, where $\delta_i = 0$ if the ith subject was censored and otherwise indicates which of p causes of death occurred for subject i. For each $j = 1, \ldots, p$, let $t_{j1} < t_{j2} < \cdots < t_{jk_j}$ be the K_j death times from cause j, counted without multiplicity. Let n_{ji} be the risk set just before t_{ji} and let d_{ji} be the number of deaths from cause j at time t_{ji}. The same arguments that led to the KM estimator in Sec. 7.1.1 now lead to

$$\hat{S}_j(y) = \prod_{i|t_{ji} < t} \left(1 - \frac{d_{ji}}{n_{ji}}\right),$$

where $i = 1, \ldots, n_{ji}$; see Rodriguez (2005). Moreover, $\hat{S}(y) = \prod_{j=1}^{p} \hat{S}_j(y)$ if there are no ties beween the times for two different causes of death. Once these estimates are obtained, it is straightforward to give point predictors and PIs for the death times of future subjects, using the techniques already described.

The competing-risks setting for PH models assumes data of the form $(y_i, \delta_i, x_i, d_i)$, where now $d_i = 1, \ldots, p$ indicates the cause of death for subject i (and is not defined when subject i was right-censored) and $\delta_i = 0, 1$ now indicates whether subject i was right-censored or died. Assuming the censoring process and the cause-of-death processes are independent, the likelihood function is

$$L = \prod_{i=1}^{n} h_{d_i}(y_i|x_i)^{\delta_i} S(y_i|x_i),$$

by the same reasoning as for (7.32). If the causes of death are independent then $S(y_i|x_i) = \prod_{j=1}^{p} S_j(y_i|x_i)$, so using $S_j(y_i|x_i) = e^{-H_j(y_i, |x_i)}$ gives

$$L = \prod_{i=1}^{n} \left(\prod_{j=1}^{p} h_j(y_i|x_i)^{d_{ij}} e^{-H_j(y_i|x_i)} \right)$$

(see Rodriguez (2005)), where $\delta_i = \sum_{j=1}^{p} d_{ij}$ and d_{ij} equals 1 if subject i died of cause j and equals zero otherwise.

Thus, L can be regarded as a product of p likelihoods, one for each cause of death j. Moreover, because of the presence of the d_{ij}, the likelihood for a specific cause of death j is the same as the likelihood one would get by treating all the other causes of death $j' \neq j$ as if they were censored at the time of the death from cause j'. In this sense, predictors for many competing-risk settings can be found using the techniques in Secs. 7.2.1 or 7.2.3 to estimate the $h_j(y|x)$ as long as the different causes of death do not depend on the same parameters.

8

Nonparametric Methods

...a primary criticism of using parametric methods in statistical analysis is that they oversimplify the population or process we are observing. Indeed, parametric families are not more useful because they are perfectly appropriate, rather because they are perfectly convenient. Nonparametric methods are inherently less powerful than parametric methods ...However, even when parametric assumptions hold perfectly ... nonparametric methods are only slightly less powerful than the more presumptuous statistical methods. Furthermore, if the parametric assumptions ...fail to hold, only the nonparametric method is valid ...[Also,] one of the pathways open to [confirmation bias] is the infusion of parametric assumptions into the data analysis ...Nonparametric procedures serve as a buffer against this human tendency of looking for evidence that best supports a researcher's underlying hypothesis.

Kvam and Vidakovic (2007, preface)

In contrast with the last three chapters, where parametric models were used to generate predictors, in this chapter nonparametric models will be used. In fact, it is a partial misnomer to refer to nonparametric models as 'models' because they generally do not admit of much physical interpretation and so are not intentionally modeling anything 'real'. Nonparametric models are merely a large class of estimators for an input–output relation between an explanatory variable x and a response Y. This may permit some sort of interpretation in the context of a given problem. However, this is not necessarily so.

As noted in the quote above, the intention of using nonparametric methods to generate 'models' or, more importantly, predictors, is to avoid making parametric assumptions when they can't be justified. Earlier chapters have extensively discussed the pitfalls of modeling in general and parametric modeling in particular. Here, the goal is to go beyond the rather small collection of $\mathcal{M}$-closed prediction problems and examine techniques that can be used for $\mathcal{M}$-complete prediction problems. Many of these techniques are familiar from the estimation context; here the focus is on their predictive performance.

In the nonparametric context, it is important to remember that even though the techniques are intended for $\mathcal{M}$-complete problems, a given technique may not work well (or at all) for a given problem or may have restrictions that render its performance poor. For instance, if a response $Y = R(x) + \epsilon$ has a complicated regression function R and the dimension of x is not small then it is easy to generate examples in which the sample size required for good estimation of R by kernel (or spline) methods is enormous. Even when x is unidimensional, functions with sufficiently complicated oscillations are hard to estimate using realistic sample sizes. This means that, even though a problem is $\mathcal{M}$-complete, it can be just barely

$\mathcal{M}$-complete, i.e., close to an $\mathcal{M}$-closed problem while remaining $\mathcal{M}$-complete, or nearly $\mathcal{M}$-open, i.e., close to an $\mathcal{M}$-open problem while remaining $\mathcal{M}$-complete. In short, there is a wide range of $\mathcal{M}$-complete problems. Thus, given an $\mathcal{M}$-complete problem such as choosing among a nonparametric collection of possible true functions R that are not acces-sible over their whole domain (except perhaps in a limiting sense), there may be regions within their domain on which they are accessible. Otherwise stated, an $\mathcal{M}$-complete model may be partially accessible and the hope in using nonparametric predictors is that the readily accessible parts of an $\mathcal{M}$-complete model can be found, estimated, and used to generate a predictor whose performance can be assessed.

As earlier, the stance taken here is that the appropriate test of the adequacy of nonpara-metric regression approaches is predictive and so we regard nonparametric methods from a predictive standpoint, i.e., treating the predictors arising from nonparametric estimation as the fundamental inferential quantity. This is a more honest and useful evaluation of infer-ential success than estimation-based criteria. (This applies to the traditional nonparametric methods in this chapter as well as to the more recent nonparametric methods in Part III.)

Nonparametric PIs have a long history and it is worthwhile recalling more of this than was seen in Chapter 4 before delving into nonparametric regression. One of the earliest PIs appeared in Wilks (1941) and was called a 'tolerance interval'. For IID data, $Y_1, \ldots, Y_n$, Wilks showed that $P(Y_{(j)} \leq Y \leq Y_{(n+1-j)}) = 1 - 2j/(n+1)$, where $Y_{(j)}$ is the jth order statistic in a sample of size n. Wilks' goal was to solve a sample size problem – to find the smallest n for which an interval with prespecified probability could be given. Clearly, $(y_{(j)}, y_{(n+1-j)})$ is a $100(1-\alpha)\%$ PI when $\alpha = 2j/(n+1)$. For values of α not expressible in this way, lower and upper bounds on the PI can be given. The natural point predictors include the midpoint of the interval $(y_{(k)}, y_{(k+1)})$ that has the highest probability and the midpoint of the interval that has probability as close as possible to 0.5 above it and below it. Given either of these, one can use the asymptotic approximation of the distribution of order statistics to assign an SE.

A glance at Krishnamoorthy and Mathew (2009, Sec. 1.1) will reveal that the terms 'tol-erance interval' and 'prediction interval' are nearly interchangeable. The difference, when it exists, is that a tolerance interval may have a conditional probability greater than $1 - \alpha$ for a future outcome while a PI has an unconditional probability, i.e., in P, greater than $1 - \alpha$ for a future outcome. The blending of these two terms (tolerance interval and PI) parallels the fact that the terms nonparametric and distribution-free have become essentially interchangeable.

An extension to Wilks (1941) was given in Fligner and Wolfe (1979). Consider two sam-ples, $Y_1, \ldots, Y_n$ and $Y_1', \ldots, Y_m'$, drawn IID from the same DF F and let $g = g(Y'^m)$ be a function of the second sample. Let P_F be the probability from F and denote upper and lower bounds on g by $U(Y^n)$ and $L(Y^n)$, respectively. The goal is to identify U and L for a given g and $\alpha > 0$ such that

$$P_F(L(Y^n) \leq g(Y'^m) \leq U(Y^n)) \geq 1 - \alpha$$

for any continuous F, i.e., to give a nonparametric PI for a future value $g(Y'^m)$. When g is the median of the Y_i', this is easier to do when m is odd than when m is even because the median is unique. (The results are approximately true when m is even if the midpoint of the middle interval between adjacent pairs of order statistics is taken as the median.)

As detailed in Randles and Wolfe (1979, p. 376), for any natural numbers r_1 and r_2 in $\{1, \ldots, n\}$ where $n \geq 3$ and is odd,

$$P_F(Y_{(r_1)} \leq Y'_{((m+1)/2)} \leq Y_{(r_2)}) = \sum_{i=r_1}^{r_2-1} \frac{\binom{n+(1/2)(n-1)-i}{n-i}\binom{(1/2)(m-1)+i}{i}}{\binom{n+m}{n}} \tag{8.1}$$

for any continuous F. Expression (8.1) can be regarded as an extension of the main idea in Wilks (1941). Choosing r_1, r_2, n, and m to make the right-hand side of (8.1) greater than or equal to $1 - \alpha$ gives a distribution-free $1 - \alpha$ PI for a future median. See Frey (2013) for some recent extensions and evaluations of this sort of reasoning.

As suggested by these examples, nonparametric predictors differ qualitatively from parametric predictors in that with nonparametric predictors it is more ill-advised to extrapolate than with parametric predictors. That is, if there is a valid parametric form for the regression function then in principle to estimate the parameter well is sufficient to get good predictions at any value of an explanatory variable x from the regression function. Otherwise stated, parametric predictors are typically global. However, most nonparametric predictors are local (the main exception is treated in Sec. 8.1). This means that they arise from putting together pieces of a regression function that are useful only on regions near each of the x_i in a sample. So, there is no way to extrapolate information about Y at an x_i reliably to information about Y on regions far from x_i. This was seen in the use above of order statistics: the PIs said nothing about values of Y below the first-order statistic or above the last-order statistic. As a generality, most people in the field would agree that extrapolation is riskier with nonparametric methods than with parametric methods. However, this statement assumes that the parametric method has low bias. In fact, nonparametric methods are more flexible and less prone to bias in the first place (at a potential cost in variance).

Although not usually expressed this way, the p-values of frequentist nonparametric tests are closely related to nonparametric PIs. Recall that a generic p-value for a one-sided test can be written as

$$p_T(y^n) = P_F(T(Y'^n) \geq t(y^n))$$

for some DF F, where $T(Y'^n)$ is the test statistic as a function of an independent copy Y'^n of Y^n and $t(y^n)$ is the value of T at the observed data y^n. Thus, $p_T(y^n)$ is the confidence associated with the interval $[t(y^n), \infty)$ as a PI, taking $t(y^n)$ as fixed. In frequentist hypothesis testing, the intuition is that when the probability $p_T(y^n)$ of the PI $[t(y^n), \infty)$ is small then the observations must be unrepresentative of the null hypothesis. However, from a predictive standpoint, this reasoning is incomplete because the variability in the sample y^n has been neglected. Indeed, for a lower predictive bound one would look at $P(T(Y'^n) \geq t(Y^n))$ (and insist it be large). Neglecting the variability of a p-value in frequentist testing or the variability in the posterior probability $W(\{F \in S\}|y^n)$ in Bayes testing, where the null hypothesis is the set S, say, is seen as part of the standard procedure. This is despite the fact that quantities such as $\text{Var}(p_T(Y^n))$ and $\text{Var}(W(\{F \in S\}|Y^n))$ quantify whether the p-value or posterior probability, respectively, is satisfactorily small; see Clarke (2013) and Clarke and Sun (1999), and the references therein.

In this chapter, nonparametric regression will be seen from a predictive standpoint. All the frequentist predictors will be linear, in a sense to be made precise in Sec. 8.1, which is on the use of orthonormal expansions in regression. Technically, orthonormal basis expansions introduce parameters and hence may be regarded as parametric. However, the number of parameters is unbounded. So, the class of functions leading to predictors generally contains open balls in the overall function space. Assuming such balls do not admit a finite-dimensional parametrization, the overall regression problem is nonparametric. To the extent that this problem is treated parametrically, this approach, although nonparametric, is global not local. Since it is well known that there are senses in which orthonormal basis expansions work poorly for some function-estimation problems, it is expected that orthonormal function spaces are likely to work poorly for prediction in similar cases. These problems can be corrected by looking at local nonparametric techniques, the first of which is to develop predictors from kernel density estimators and kernel regression estimators. In the regression setting, there are two cases: in the first, the explanatory variable is nonstochastic, leading to the Priestly–Chao estimator, and in the second it is treated stochastically, leading to the Nadaraya–Watson estimator. Note that, in this chapter, the term kernel is used in its local smoothing sense rather than in the sense of kernel methods, which include techniques such as relevance vector machines. (These topics are treated in Chapter 10.) Sections 8.3 and 8.4 develop predictors based respectively on nearest neighbors regression and nonparametric Bayes regression. All these techniques are well studied in an estimation context, where they are regarded as estimators of a true regression function. Here, they are presented as predictors of a future outcome in a way that is largely unconcerned with their properties as estimators except as such properties may affect prediction. Since there are few, if any, comparisons of these techniques on the basis of predictive performance, Sec. 8.5 provides examples.

The endnotes look briefly at spline predictors, nearest neighbor predictors for classification, and predictors derived from nonparametric tests.

8.1 Predictors Using Orthonormal Basis Expansions

Here, we consider the natural extension of the predictors found from linear regression (see Sec. 4.2) to more general function spaces. Actually, while technically nonparametric, it is more helpful to regard this approach as an intermediate step between predictors based on parametric regressions and predictors based on nonparametric regressions. It is 'intermediate' because it is parametric, but there may be many parameters. So, suppose that $Y(x) = f(x) + \epsilon$, where $f \in \mathcal{F}$ and ϵ has mean zero and variance σ^2. Let $\langle \phi_k \rangle|_{k=1}^\infty$ be a set of orthonormal basis functions for $\mathcal{F}$, assumed to be infinite dimensional. Now, if $\mathcal{F}$ is a Hilbert space, i.e., a complete linear space equipped with an inner product, and the inner product is of the form

$$\langle g, h \rangle = \int g(x)h(x)\lambda(\mathrm{d}x) \quad \text{for} \quad g, h \in \mathcal{F},$$

where λ is Lebesgue measure, any $f \in \mathcal{F}$ can be written as

$$f(x) = \sum_{k=1}^\infty \langle f, \phi_k \rangle \phi_k(x) \equiv \sum_{k=1}^\infty \beta_k \phi_k(x).$$

So, a natural sequence of approximations for f is $f_K(x) = \sum_{k=1}^K \beta_k \phi_k(x)$.

Suppose that all the functions in $\mathcal{F}$ have support in $[0, 1]$ and that the x_i form a regular grid, e.g., $x_i = i/n$ for unidimensional x. Assume independent data of the form $(x_i, y_i)|_{i=1}^n$; since

$$\int_0^1 E(Y_i(x)\phi_k(x))\lambda(\mathrm{d}x) = \beta_k,$$

for given k, approximating the integral by a sum over the grid points and using $\phi_k(x_i)Y_i$ as an estimate for $E(Y_i(x_i)\phi_k(x_i)) = \phi_k(x_i)E(Y_i(x_i))$ gives

$$\hat{\beta}_k = \frac{1}{n}\sum_{i=1}^n Y(x_i)\phi_k(x_i) \tag{8.2}$$

as a candidate estimator for β_k. It is easy to verify that

$$E(\hat{\beta}_k) = \frac{1}{n}\sum_{i=1}^n f(x_i)\phi_k(x_i)$$

and that $E(\hat{\beta}_k\hat{\beta}_\ell)$ equals

$$\left(\frac{1}{n}\sum_{i=1}^n f(x_i)\phi_k(x_i)\right)\left(\frac{1}{n}\sum_{i=1}^n f(x_i)\phi_\ell(x_i)\right) + \frac{\sigma^2}{n}\left(\frac{1}{n}\sum_{i=1}^n \phi_k(x_i)\phi_\ell(x_i)\right).$$

Used together, these give

$$\mathrm{Cov}(\hat{\beta}_k, \hat{\beta}_\ell) = \frac{\sigma^2}{n}\left(\frac{1}{n}\sum_{i=1}^n \phi_k(x_i)\phi_\ell(x_i)\right)$$

and

$$\mathrm{Var}(\hat{\beta}_k) = \frac{\sigma^2}{n}\left(\frac{1}{n}\sum_{i=1}^n \phi_k^2(x_i)\right)^2.$$

When the ϵ_i are $N(0, \sigma^2)$,

$$\hat{\beta}_k \sim N\left(\frac{1}{n}\sum_{i=1}^n f(x_i)\phi_k(x_i), \frac{\sigma^2}{n}\left(\frac{1}{n}\sum_{i=1}^n \phi_k^2(x_i)\right)\right) \tag{8.3}$$

marginally, although $\hat{\beta}_k$ and $\hat{\beta}_\ell$ are slightly dependent. If the x_i grid is fine enough, i.e., n is large enough, then the summations in (8.3) are nearly equal to $\langle \phi_k, f \rangle$ and unity, respectively, so $\hat{\beta}_k \sim N(\beta_k, \sigma^2/n)$ approximately. Likewise, in this limiting case, $\hat{\beta}_k$ and $\hat{\beta}_\ell$ are independent.

So, one way to derive PIs is to approximate the distribution of $\hat{\beta}_K = (\hat{\beta}_1, \ldots, \hat{\beta}_K)^T$ by $N((\beta_1, \ldots, \beta_K)^T, \sigma^2 I_n/n)$. Then,

$$E((Y(x_{\mathrm{new}}) - \hat{f}_K(x_{\mathrm{new}})) \approx f(x_{\mathrm{new}}) - f_K(x_{\mathrm{new}}) \approx 0,$$

where $\hat{f}_K(x) = \sum_{k=1}^K \hat{\beta}_k\phi_k(x)$. This means that $\hat{f}_K$ is a linear estimator since the fitted values $\hat{f} = (\hat{f}(x_1), \ldots, \hat{f}(x_n))^T$ can be written as $\hat{f} = LY$, where $Y = (y_1, \ldots, y_n)$ for some matrix L. Indeed, $L = (\phi_k(x_i))_{ik}$ for $i = 1, \ldots, n$ and $k = 1, \ldots, K$. More generally, for any x, $\hat{f}(x)$ is a linear function of Y in which L depends on the data. Moreover,

$$\text{Var}(Y(x_{\text{new}}) - \hat{f}_K(x_{\text{new}})) \approx \sigma^2 + \frac{\sigma^2}{n} \sum_{k=1}^{K} \phi_k^2(x_{\text{new}}).$$

So, for $\alpha > 0$, $(1 - \alpha)100\%$ PIs are of the form

$$\hat{f}_K(x_{\text{new}}) \pm z_{1-\alpha/2}\hat{\sigma} \sqrt{1 + \frac{1}{n} \sum_{k=1}^{K} \phi_k^2(x_{\text{new}})},$$

where $z_{1-\alpha}$ is a percentile of the $N(0, 1)$ distribution, effectively assuming ϵ is normal, and

$$\hat{\sigma}^2 = \frac{1}{n} \sum_{i=1}^{n} (Y(x_i) - \hat{f}_K(x_i))^2. \tag{8.4}$$

This form of PI can be used with orthonormal basis expansions but does not readily generalize to other methods.

Some authors have argued that (i) small $\hat{\beta}_k$ should be set to zero as a way to reduce variance without increasing bias too much, and (ii) the estimate $\hat{f}_K$ should be truncated at a bound $B = B_n$ that is allowed to increase as $n \to \infty$ as a way to reduce the mean squared error (MSE) via either bias or variance. It is clear that such extra steps can be helpful for prediction, especially when $\phi_k(x_{\text{new}})$ is large relative to the $\phi_{k'}(x_{\text{new}})$ for $k' \neq k$ or there is concern that some $\hat{\beta}_k$ are too large. For ease of exposition, these steps were omitted in the simple technique just presented and will be omitted in the two techniques below. However, these 'extras' should be borne in mind as a way to improve unsatisfactory predictors found by estimating an orthonormal basis expansion of a true regression function. On the other hand, MSE, bias, and variance are concepts from estimation that are defined in terms of the true model and so relying on them too much may lead to violation of the Prequential Principle.

Another more sophisticated way to derive PIs begins by noting that

$$\hat{f}_K(x) = \sum_{k=1}^{K} \hat{\beta}_k \phi_k(x) = \sum_{i=1}^{n} \left(\frac{1}{n} \sum_{k=1}^{K} \phi_k(x_i)\phi_k(x) \right) Y(x_i). \tag{8.5}$$

Fixing x^n and writing $\ell_i(x) = (1/n) \sum_{k=1}^{K} \phi_k(x_i)\phi_k(x)$, $\ell(x) = (\ell_1(x), \ldots, \ell_n(x))^T$, and $Y = (Y(x_1), \ldots, Y(x_n))^T$, (8.5) becomes

$$\hat{f}_K(x) = \ell(x)^T Y = \sum_{i=1}^{n} \ell_i(x) Y_i = \ell^T(x) Y = \langle \ell(x), Y \rangle, \tag{8.6}$$

i.e., $\hat{f}_K$ is a linear estimator. It is seen that

$$E(\hat{f}_K(x)) = \sum_{i=1}^{n} \ell_i(x) f(x_i) = \sum_{k=1}^{K} \left(\frac{1}{n} \sum_{i=1}^{n} \phi_k(x_i) f(x_i) \right) \phi_k(x) = \sum_{k=1}^{K} \langle \phi_k, f \rangle_n \phi_k(x)$$

where $\langle \cdot, \cdot \rangle_n$ indicates the inner product with respect to the EDF for x, in this case the values in the regular grid, so $\langle \cdot, \cdot \rangle_n$ converges to $\langle \cdot, \cdot \rangle$ and hence $E\hat{f}_K(x) \to f_K(x)$ as n increases. A similar formula can be derived for the variance of $\hat{f}_K$. Indeed, using (8.5),

$$\text{Var}(\hat{f}_K(x)) = \text{Var}(\ell^T(x)Y) = \ell^T(x)\text{Var}(Y)\ell(x) = \sigma^2 \|\ell(x)\|_2^2.$$

The fitted values are $\hat{f}_K(x_i)$ and, by linearity, satisfy

$$\begin{pmatrix} \hat{f}_K(x_1) \\ \vdots \\ \hat{f}_K(x_n) \end{pmatrix} = \begin{pmatrix} \ell(x_1)^T \\ \vdots \\ \ell(x_n)^T \end{pmatrix} Y \equiv LY,$$

where $L = L(x^n)$ is the $n \times n$ matrix formed by the $\ell_j(x_i)$. Following Loader (1999), let $\text{tr}(M)$ be the trace of matrix M and define degrees of freedom

$$\nu_1 = \text{tr}(L) \quad \text{and} \quad \nu_2 = \text{tr}(L^T L).$$

Now, since the data are assumed independent, the PIs for a new observation $Y(x_{\text{new}})$ at a point x_{new} follow from

$$\text{Var}(Y(x_{\text{new}}) - \hat{f}_K(x_{\text{new}})) = \sigma^2(1 + \|\ell(x_{\text{new}})\|^2).$$

That is, for $\alpha > 0$, $(1-\alpha)100\%$ PIs are of the form

$$\hat{f}_K(x_{\text{new}}) \pm z_{1-\alpha/2}\hat{\sigma}\sqrt{1 + \|\ell(x_{\text{new}})\|_2^2}, \tag{8.7}$$

where $z_{1-\alpha}$ is a percentile of the $N(0,1)$, assuming that ϵ is normal and

$$\hat{\sigma}^2 = \frac{1}{n - 2\nu_1 - \nu_2} \sum_{i=1}^{n} (Y(x_i) - \hat{f}_K(x_i))^2.$$

For the sake of contrast, note that a $(1-\alpha)100\%$ CI for the mean is given by

$$\hat{f}_K(x_{\text{new}}) \pm z_{1-\alpha/2}\hat{\sigma}\|\ell(x_{\text{new}})\|_2,$$

i.e., the square root factor multiplying $\hat{\sigma}$ does not need the extra '1' as the two sources of variability, in $Y(x)$ and $\hat{f}_K(x)$, are combined under the square root. This is characteristic of any method for function estimation that is linear in the sense of (8.5).

There are many other ways to obtain linear estimators. For instance, the obvious extension of least squares linear regression (see Sec. 4.2) formed by treating the first K basis functions as explanatory variables, i.e., taking the values of the explanatory variable as $(\phi_1(x_i), \ldots, \phi_K(x_i))$, leads to parameter estimates of the form $\hat{\beta}_K = (X^T X)^{-1} X^T Y$ in which the (i,k)th entry of the matrix X is $\phi_k(x_i)$. Then, the fitted values are of the form $\hat{Y} = X(X^T X)^{-1} X^T Y$ and the estimator for x_{new} is of the form

$$\hat{f}_K(x_{\text{new}}) = X(x_{\text{new}})(X^T X)^{-1} X^T = L(x_{\text{new}})Y \tag{8.8}$$

where $X(x_{\text{new}}) = (\phi_1(x_{\text{new}}), \ldots, \phi_K(x_{\text{new}}))$. So, the least squares approach to estimating the β_k is also linear in the sense of (8.5). Another linear method (see Wasserman (2006, Chap. 8)) involves the inclusion of a 'modulator' – an $\ell^2(\mathbb{N})$ sequence that puts a factor on each β_k and optimizes a risk criterion. However, it is not clear when each of these parameter estimation methods ((8.2), (8.5), least squares, and modulators) is best because predictive comparisons do not seem to have been done.

The question remains how big K should be to ensure that the bias of the predictors is sufficiently small. This has sample-size implications that are ignored here by the assumption that n is large enough. If $x \in [0, 1]$ then

$$f(x) - f_K(x) = \sum_{k=K+1}^{\infty} \beta_k \phi_k(x)$$

gives that the bias approximately satisfies

$$B^2(f, K) = \int_0^1 (f(x) - f_K(x))^2 \mathrm{d}x = \sum_{k=K+1}^{\infty} \beta_k^2, \tag{8.9}$$

and the right-hand side decreases to zero as $K \to \infty$. Sometimes, choosing a K and setting $\beta_k = 0$ for $k \geq K + 1$ and for k such that $\alpha_k \leq t$ for some threshold t is called hard thresholding. This finds the 'best' subset of ϕ_k for $k \leq K$.

One way to get a uniform bound on B^2 is to restrict the sequences of β_k for $k \geq K + 1$. Wasserman (2006) did this by using Sobolev ellipsoids. First note that the collection of sequences of β_k as used in (8.9) can be regarded as a linear subspace of $\ell^2(\mathbb{N})$. Then, define an ellipsoid with axes $a = (a_1, \ldots, a_k, \ldots)$ and radius c^2 to be the set

$$B(a, c) = \left\{ \beta = \langle \beta_k \rangle|_{k=1}^{\infty} \,\Big|\, \sum_{k=1}^{\infty} a_k^2 \beta_k^2 \leq c^2 \right\};$$

this is a direct generalization of an ellipse in finite dimensions. It is assumed that $a_k \to \infty$ as $k \to \infty$ so that $\beta_k \to 0$ as $k \to \infty$, permitting uniform bounds over $\beta \in B(a, c)$.

In the specific case where $a_k = (\pi k)^{2m}$, i.e., $a_k = \mathcal{O}(1/k^{2m})$ for some m, $B(a, c) = B_m(a, c)$ is called a Sobolev ellipsoid. Now, it can be proved that

$$\sup_{B_m(a,c)} B^2(f, K) = \mathcal{O}\left(\frac{1}{K^{2m}}\right), \tag{8.10}$$

where it is understood that the sequences in $B_m(a, c)$ start at $K + 1$. The proof of (8.10) is easy; merely observe that

$$\sum_{k=K+1}^{\infty} \beta_k^2 \leq \frac{1}{K^{2m}} \sum_{k=K+1}^{\infty} k^{2m} \beta_k^2 = \mathcal{O}\left(\frac{1}{K^{2m}}\right).$$

So, for our present purposes, if $m > 1/2$,

$$\sup_{B_m(a,c)} B^2(f, K) = o\left(\frac{1}{K}\right).$$

However, a uniform bound is not always necessary to control bias; in many, arguably most, practical settings K is chosen so large that the variance of a nonparametric predictor will dominate its bias.

The foregoing generalizes mathematically to prediction problems where X assumes values in a regular grid in $\mathbb{R}^d$. However, as d increases, estimating a regression function and using it to generate predictions gets harder and the techniques presented in this chapter do not scale up effectively. Specifically, as d gets larger, the sparsity of data in high dimensions (one version of the curse of dimensionality) becomes more pronounced. Loosely put, in higher dimensions the space between points is larger than it is in low dimensions because of the extra degrees of freedom that high dimensions permit. This makes nonparametric regression harder, in the sense that unachievably large samples will often be needed for good prediction when d is as small as 3, or even 2. Of course, the size of n needed will also increase with the complexity of the DG; the DG is the true function for $\mathcal{M}$-closed and $\mathcal{M}$-complete settings and a more general construct in the $\mathcal{M}$-open case. Further ways to predict in regression problems that may be appropriate for higher-dimensional $\mathcal{M}$-open settings will be seen in Part III, e.g., in Chapter 10.

A qualitatively different problem occurs when the values of x are not on a regular grid and in particular cannot be regarded as chosen by some nonstochastic mechanism. In this case, it may be reasonable to assume that the X_i are drawn IID from a distribution $\lambda(\cdot)$. Now, a technique analogous to the foregoing can be used to generate predictors but it is more complex – and does not scale up well to higher dimensions either. These predictors can be called empirical orthonormal basis predictors because the x-values are used to form an EDF, say $\hat{F}_n$, that is used to define an inner product and is assumed to converge to a true DF, say F. The basic idea is to use orthonormal expansions in $L^2(\hat{F}_n)$ rather than in $L^2(F)$. In the special case where F is uniform on an interval such as $[0, 1]$, the following reduces to the regular-grid case already seen because a regular grid can be regarded as a perfectly representative data set from a uniform distribution.

For any set A, let $\chi_A(x) = 1$ if $x \in A$ and zero otherwise. Then, following Gyorfi *et al.* (2002), write

$$\hat{F}_n(A) = \frac{1}{n} \sum_{i=1}^{n} \chi_A(X_i)$$

and assume that the X_i are IID F, where F is assumed to assign all its mass on $[0, 1]$. Suppose that $L^2(\hat{F}_n)$ has inner product obtained from

$$\langle g, h \rangle_n = \frac{1}{n} \sum_{i=1}^{n} f(x_i)g(x_i) = E_{\hat{F}_n}(f(X)g(X)) \quad \text{for} \quad g, h : [0, 1] \to \mathbb{R}$$

and squared norm $\|g\|_n^2 = \langle g, g \rangle$. Now it makes sense to use

$$\mathcal{F}_{n,K} = \left\{ \sum_{k=1}^{K} \beta_k \phi_{n,k} \middle| \forall k, \ \beta_k \in \mathbb{R} \right\}$$

where, for each n, the $\phi_{n,k}$ form an orthonormal basis for $\mathcal{F}_{n,K}$. Given $f \in L^2(F)$, its best approximation in $\mathcal{F}_{n,K}$ under $\| \cdot \|_n$ is

$$f_{n,K} = \sum_{k=1}^{K} \beta_{n,k} \phi_{n,K} \quad \text{where} \quad \beta_{n,k} = \langle f, \phi_{n,k} \rangle_n = \frac{1}{n} \sum_{i=1}^{n} f(x_i)\phi_{n,k}(x_i). \tag{8.11}$$

The coefficients $\beta_{n,k}$ can now be estimated by

$$\hat{\beta}_{n,k} = \frac{1}{n} \sum_{i=1}^{n} y(x_i) \phi_{n,k}(x_i), \tag{8.12}$$

using a method of moments justification. Then, the function estimator

$$\hat{f}_{n,K}(x) = \ell(x)^T Y \tag{8.13}$$

is seen to be linear, where $\ell(\cdot)$ is as in (8.6) but now for $\mathcal{F}_{n,K}$, i.e., in terms of the $\phi_{n,K}$. Alternatively, the $\beta_{n,K}$ can be estimated by

$$\hat{\beta}_n = (\hat{\beta}_{n,1}, \dots, \hat{\beta}_{n,K})^T = (X^T X)^{-1} X^T Y \tag{8.14}$$

under a least squares criterion, where the $n \times K$ matrix X now has (i,k)th entry $\phi_{n,k}(x_i)$. In view of (8.8), the estimator is

$$\hat{f}_{n,K}(x_{\text{new}}) = \left(X(x_{\text{new}})(X^T X)^{-1} X^T \right) Y.$$

This is simpler than it looks because $X = \langle \phi_{n,k}, \phi_{n,k'} \rangle_n = \delta_{k,k'}$.

Thus, with either (8.12) or (8.14), the estimate for f using $\mathcal{F}_{n,K}$ is linear and can be written

$$\hat{f}_{n,K}(x) = \sum_{k=1}^{K} \hat{\beta}_{n,K} \phi_{n,K}(x), \tag{8.15}$$

and, parallel to (8.7), generates PIs for new values of x of the form

$$\hat{f}_{n,K}(x_{\text{new}}) \pm z_{1-\alpha/2} \hat{\sigma} \sqrt{1 + \|\ell(x_{\text{new}})\|^2}, \tag{8.16}$$

where $\hat{\sigma}$ is as in (8.4), ℓ is as in (8.6) when (8.12) is used and is given by $\ell(x_{\text{new}}) = X(x_{\text{new}})(X^T X)^{-1} X^T$ when (8.14) is used, replacing $\mathcal{F}$ with $\mathcal{F}_{n,K}$ in either case. (The PIs in (8.16) may be improved by setting small $\hat{\beta}_{n,K}$ to zero or truncating the regression function.)

Gyorfi *et al.* (2002, Sec. 18.4) describe one way in which (8.15) can be constructed empirically using piecewise polynomials, at least for unidimensional x. Assuming that F – to which $\hat{F}_n$ converges – has no atoms, without loss of generality the variables $x_1, \dots, x_n$ can be assumed unique. Then, the basic idea is to partition the domain of the functions in $\mathcal{F}_{n,K}$ recursively into disjoint subintervals, so that, at each iteration, the intervals are split into two subintervals in which there are nearly the same number of x_i. On each subinterval, polynomials of order not greater than M, say, can be defined, and pieced together. Effectively, any given polynomial function of order less than or equal to M can be expressed as a linear combination of such functions, so that the different pieces are orthonormal, being defined on disjoint intervals. If no interval has more than one x_i then it is straightforward to normalize a set of basis functions that are orthogonal in $\langle \cdot, \cdot \rangle$, i.e., to use $\hat{F}_n$ to define the expectation, in order to get an orthonormal basis for $\mathcal{F}_{n,M}$ formed from the set of all polynomials in x of order at most M. (Note $\mathcal{F}_{n,M}$ is a version of $\mathcal{F}_{n,K}$ with the number of terms K replaced by the polynomial order M on the subintervals.) Thus, as $n \to \infty$, $\mathcal{F}_{n,M} \to \mathcal{F}$ and, given $f \in \mathcal{F}$, it can be approximated in a limiting sense using $f_{n,M} \in \mathcal{F}_{n,M}$, which can be estimated by $\hat{f}_{n,M}$ analogously to (8.11)–(8.13).

Finally, the estimator $\hat{f}_{n,M}$ satisfies

$$\int_0^1 |\hat{f}_{n,M}(x) - f(x)| F(\mathrm{dx}) \to 0 \quad \text{a.s.} \tag{8.17}$$

and

$$E_F \int_0^1 |\hat{f}_{n,M}(x) - f(x)| F(\mathrm{dx}) \to 0 \tag{8.18}$$

for any distribution of the pair (X, Y), provided that $EY^2 < \infty$. That is, the estimator is consistent. (This statement is a major simplification of the actual result in Gyorfi *et al.* (2002, Sec. 18.5), and omits the convergence rate in Sec. 18.6.) The estimator is also linear since it is a linear combination of orthonormal basis elements with coefficients that can be estimated as before, though this is harder to write down. Consequently, the predictor generated by this procedure will give PIs of the form (8.16) with asymptotic nominal confidence level $1 - \alpha$ though deriving this is beyond our present scope. On the other hand, the asymptotics depend on many choices so it is unclear how well this procedure will perform pre-asymptotically in a predictive sense. The point is merely to observe that function estimators based on orthonormal bases can be constructed empirically and thereby give empirical predictors. Further discussion of local polynomials is given in Sec. 8.6.1.

A related predictor arises from a partitioning estimate of f. Let $\mathcal{P} = \{A_1, \ldots, A_J\}$ be a partition of the range of X and, for $x \in A_j$, let

$$f_n(x) = \frac{\sum_{i=1}^n \chi_{X_i \in A_j}(x) y_i}{\sum_{i=1}^n \chi_{X_i \in A_j}(x)}.$$

Then, letting $W_{ij}(x) = \chi_{X_i \in A_j}(x) / \sum_{i=1}^n \chi_{X_i \in A_j}(x)$ for $x \in A_j$, it is seen that

$$f_n(x) = \left(\sum_{j=1}^J \chi_{x \in A_j} W_{1,j}(x), \ldots, \sum_{j=1}^J \chi_{x \in A_j} W_{n,j}(x) \right)^T Y, \tag{8.19}$$

meaning that f_n is a linear estimator. If the sets A_j are cuboids in d dimensions then, for sidelength h, their volume will be h^d. As in Gyorfi and Ottucsak (2012), if $h = h_n \to 0$ and $nh_n^d \to 0$, choosing $h_n = cn^{-1/(d+2)}$ for some $c > 0$ leads to

$$E\|f_n - f\|_2^2 \le Cn^{-2/(d+2)} \tag{8.20}$$

for some $C > 0$. Thus, like the previous estimators, the predictor formed from a partition estimate generates PIs as in (8.16). The cost is that $\mathcal{P}$ is subjective and must be chosen sensibly.

These last two predictors show that a predictor based on an orthonormal basis expansion can localize, i.e., be based on function estimators that accurately reflect the local behavior of the true function on all regions. This may result if the basis localizes, i.e., the individual basis functions are only nonzero, at least for all practical purposes, on regions of the domain of the function. Otherwise a basis is global, meaning that the individual basis elements are far from zero over essentially the whole domain. (Of course, it's easy to define a basis that has some elements that are local and some that are global but this would be an unusual construction.) Indeed, it is not difficult to construct examples in which a basis that does not localize can lead to poor prediction. For instance, if the convergence of the estimator is not uniform then there may be regions where the function estimator and the true function are further and further apart as more terms are included in the basis expansion. An example of this is interpolating the Runge function $y = 1/(1 + x^2)$ on $[-1, 1]$ (see Runge (1901)), on a regular grid using a polynomial basis. It can be shown that some interpolations have larger and larger oscillations near the endpoints as more terms are used. If the convergence is uniform or a localized basis is used, this problem generally disappears. However, when they work well, global bases can be better – they use fewer terms – than localizing bases. For $\mathcal{M}$-closed problems it can be determined whether a global basis will be satisfactory or whether a localizing basis must be used. For $\mathcal{M}$-complete problems, it would be unusual to get uniform convergence of a function estimator using a global basis. Hence it would be rare to get a well-localized predictor using a global basis, in the absence of assumptions that would be hard to verify in practice.

8.2 Predictors Based on Kernels

The last two predictors – the Gyorfi *et al.* (2002, Sec. 18.4) construction and (8.19), discussed in Sec. 8.1 – are almost kernel-based. In fact, (8.19) would be a kernel predictor if χ_{A_j} were replaced by a kernel function. A kernel function is any nonnegative function $K(\cdot)$ that integrates to unity – mathematically it is the same as a density but the interpretation is different. The idea is that a kernel function provides the weights for a weighted sum over the data points. The weights are scaled by an auxiliary smoothing or bandwidth parameter, usually denoted h. In the three subsections here we assume independent data. The case of dependent data is deferred to Sec. 8.3.3.

8.2.1 Kernel Density Estimation

Most kernel-based estimators are variants on a kernel density estimator, which can be defined as follows. Write

$$\hat{f}(x) = \hat{f}_h(x) = \frac{1}{nh} \sum_{i=1}^{n} K\left(\frac{X_i - x}{h}\right), \tag{8.21}$$

where $X_1, \ldots, X_n$ are the data and $h > 0$. Letting $u = (X_i - x)/h$,

$$\int \hat{f}(x)\, \mathrm{d}x = \int K(u)\mathrm{d}u = 1,$$

meaning that, when $K \geq 0$, $\hat{f}$ is a probability density and can be regarded as an estimator of $f(x)$. Thus, it can be used to generate point predictors and PIs for future outcomes, provided that K and h are specified. Expression (8.21) formalizes the idea of putting a local mass defined by K at each X_i – but letting the mass be spread out by a factor h to permit the scales of K and f to be different. It is most convenient to assume that $K(\cdot) \geq 0$ but this is not necessary. If there are regions where $K < 0$, (8.21) can be modified to

$$\hat{f}^+(x) = \frac{\hat{f}(u)\chi_{\hat{f}(u) \geq 0}}{\int \hat{f}(u)\chi_{\hat{f} \geq 0}\mathrm{d}u}.$$

Here, kernels are assumed symmetric, so their odd moments are zero. A special case of (8.21) uses the 'boxcar' kernel

$$K(x) = \begin{cases} 1/2 & |x| \leq 1, \\ 0 & \text{else}, \end{cases}$$

which assigns probability uniformly over an interval centered at zero – meaning that for each fixed x the ith term in (8.21) will be zero for $|X_i - x| > h$. Many kernels have been proposed and used; however, the difference in efficiency between the optimal Epanechnikov kernel (efficiency unity) to the boxcar kernel (efficiency about 1.075) is small. Most commonly used continuous kernels have an efficiency between the Epanechnickov and boxcar values.

To derive a specification for h in (8.21), the usual approach is to invoke a criterion such as the minimal asymptotic mean integrated squared error (AMISE). The idea is to find an asymptotic form for the MSE of $\hat{f}(x)$ pointwise in x and then integrate over x. This sort of procedure leads to a function $h = h_n$ in which the rate of decrease of h_n with increasing n is known but an unknown constant – depending on the unknown f – must be chosen.

To start, recall the that mean squared error (MSE) at a given x is the sum of the variance plus squared bias. To find the bias, write

$$E\hat{f}(x) = \int_{-\infty}^{\infty} K(u)f(x + hu)\,du.$$

Then, if the $(\nu + 1)$th derivative of f exists, the νth-order Taylor expansion of f at x can be written as

$$f(x + hu) = f(x) + f^{(1)}(x)hu + \frac{1}{2}f^{(2)}(x)h^2u^2 + \cdots + \frac{1}{\nu!}f^{(\nu)}(x)h^\nu u^\nu + o(h^\nu),$$

since it can be assumed that $h = h_n \to 0$ as $n \to 0$. If K is a νth order kernel, i.e., the jth moments of K are zero, $\kappa_j(K) = 0$ for $j \leq \nu - 1$, then using the Taylor expansion in the integral gives

$$\int_{-\infty}^{\infty} K(u)f(x + hu)\,du = f(x) + \frac{1}{\nu!}f^{(\nu)}(x)h^\nu\kappa_\nu(K) + o(h^\nu)$$

where $\kappa_\nu(K)$ is the νth moment of K and therefore

$$E(\hat{f}(x)) = f(x) + \frac{1}{\nu!}f^{(\nu)}(x)h^\nu\kappa_\nu(K) + o(h^\nu), \tag{8.22}$$

so the bias at x is given by

$$\text{bias}(\hat{f})(x) = E(\hat{f}(x) - f(x)) = \frac{1}{\nu!}f^{(\nu)}(x)h^\nu\kappa_\nu(K) + o(h^\nu),$$

an increasing function of h, meaning that, loosely, smaller h means less bias.

Turning next to the variance of $\hat{f}(x)$, note that the terms $K((X_i - x)/h)$ in $\hat{f}$ are IID. So, using 'variance equals mean square minus square mean' gives

$$\text{Var}(\hat{f}(x)) = \frac{1}{nh^2}EK\left(\frac{X_i - x}{h}\right)^2 - \frac{1}{n}\left(\frac{1}{h}EK\left(\frac{X_i - x}{h}\right)\right)^2, \tag{8.23}$$

for any i. By a change of variables, the key part of the first term is

$$
\frac{1}{h} E K \left(\frac{X_i - x}{h} \right)^2 = \int_{-\infty}^{\infty} K(u)^2 f(x + hu) du
$$

$$
\approx \int_{-\infty}^{\infty} K(u)^2 \left(f(x) + \mathcal{O}(h) \right) du
$$

$$
= f(x) \int K(u)^2 du + \mathcal{O}(h), \tag{8.24}
$$

in which $\int K^2(u) du$ can be found for any K and indicates its 'roughness'.

For the second term in (8.23),

$$
\frac{1}{h} E K \left(\frac{X_i - x}{h} \right) = \int_{-\infty}^{\infty} K(u) f(x + hu) du
$$

$$
= f(x) + o(1), \tag{8.25}
$$

by a first-order Taylor expansion. So, the second term in (8.23) is $\mathcal{O}(1/n)$.

Using (8.24) and (8.25) in (8.23) gives

$$
\mathrm{Var}(\hat{f}(x)) \approx \frac{f(x) \int K^2(u) du}{nh} + \mathcal{O}\left(\frac{1}{n} \right). \tag{8.26}
$$

The approximation error $\mathcal{O}(1/n)$ is smaller than the order of the leading term, which is $\mathcal{O}(1/nh)$ (since $1/h \to \infty$).

The asymptotic MSE of $\hat{f}(x)$ as an estimator for $f(x)$ is therefore

$$
E(\hat{f}(x) - f(x))^2 = \mathrm{Var}(\hat{f}(x)) + \mathrm{bias}^2(\hat{f}(x))
$$

$$
\approx \frac{f(x) \int K^2(u) du}{nh} + \left(\frac{\kappa_\nu(K)}{\nu!} \right)^2 (f^{(\nu)})^2(x) h^{2\nu}.
$$

Integrating over x gives the asymptotic mean integrated squared error,

$$
\mathrm{AMISE} \approx \frac{\int K^2(u) du}{nh} + \left(\frac{\kappa_\nu(K)}{\nu!} \right)^2 \int_{-\infty}^{\infty} (f^{(\nu)})^2(x) dx \, h^{2\nu}. \tag{8.27}
$$

Since one term in the AMISE is increasing in h and the other is decreasing in h, there is a tradeoff value giving a minimum AMISE. Differentiating the right-hand side of (8.27) with respect to h and setting the derivative to zero gives

$$
\frac{\int K^2(u) du}{nh^2} - 2\nu h^{2\nu-1} \left(\frac{\kappa_\nu(K)}{\nu!} \right)^2 \int_{-\infty}^{\infty} (f^{(\nu)})^2(x) dx = 0.
$$

This expression is minimized as a function of h by

$$
h_{\min} = \frac{1}{n^{1/(2\nu+1)}} \left(\frac{(\nu!)^2 \int K^2(u) du}{2\nu \kappa_\nu^2(K)} \right)^{1/(2\nu+1)} \times \left(\frac{1}{\int f^{(\nu)}(u)^2 du} \right)^{1/(2\nu+1)}, \tag{8.28}
$$

i.e., $h_{\min} \propto 1/n^{1/(2\nu+1)}$. Since K is chosen by the user, the only factor in $h_{\min}$ that remains to be identified is the last, which depends on $f^{(\nu)}(x)$. Nevertheless, one can substitute $h_{\min}$ back into the expression for the AMISE to find the minimal AMISE. Apart from not knowing

ν and $f^{(\nu)}$, the minimal AMISE is $\mathcal{O}(1/n^{2\nu/(2\nu+1)})$, i.e., $\mathcal{O}(1/n^{4/5})$ if $\nu = 2$. It is left as an exercise for the reader to verify that if the L^2 norm is denoted $\| \cdot \|_2$ then

$$\text{AMISE}_{\min} = \frac{(1+2\nu)}{n^{2\nu/(2\nu+1)}} \left(\frac{\|f^{(\nu)}\|_2^2 \|K\|_2^2 \kappa_\nu(K)^2}{(\nu!)^2 (2\nu)^{2\nu}} \right)^{1/(2\nu+1)}.$$

To finish the specification of $h_{\min}$, it is common to choose a value for $\int f^{(\nu)}(u)^2 du$ by treating f as a scale family with a normal (mean-zero) base distribution. The main justification for this is pragmatic: it gives a constant that is, hopefully, not too far wrong. So, write $f_\sigma(u) = (1/\sigma) f(u/\sigma)$. Differentiating ν times gives $f_\sigma^{(\nu)}(u) = (1/\sigma^{1+\nu}) f_\sigma^{(\nu)}(u/\sigma)$. The squared L^2 norm of $f_\sigma^{(\nu)}(u)$ is

$$\int f_\sigma^{(\nu)}(u)^2 du = \frac{1}{\sigma^{2+2\nu}} \int f_\sigma^{(\nu)}(u/\sigma) du = \frac{1}{\sigma^{1+2\nu}} \int f^{(\nu)}(u) du,$$

and hence

$$\left(\frac{1}{\int f_\sigma^{(\nu)}(u)^2 du} \right)^{1/(2\nu+1)} = \sigma \left(\frac{1}{\int f^{(\nu)}(u)^2 du} \right)^{1/(1+2\nu)}.$$

In the special case where f is a scale normal, i.e., $f_\sigma = \phi_\sigma$, we have $f = \phi$ and

$$\left(\frac{1}{\int \phi^{(\nu)}(u) du} \right)^{1/(1+2\nu)} = 2 \left(\frac{\sqrt{\pi}\nu!}{(2\nu)!} \right)^{1/(1+2\nu)}. \tag{8.29}$$

Using (8.29) and $\hat{\sigma} = \sqrt{\overline{X^2} - (\overline{X})^2}$ gives a value for the final factor in (8.28). So, in principle, a value for $h_{\min}$ can be found, once ν is chosen. More abstractly, $h_{\min}$ is given as a rate, $h_{\min} = (\hat{\sigma}C)/n^{1/(1+2\nu)}$, for some $C > 0$ depending on K (and ν). Indeed, this simplifies to $h_{\min} = \hat{\sigma}/n^5$ when a normal kernel is used. It should be noted that many other ways to select h effectively have been proposed. They often work better than the relatively simple derivation presented here, but at the cost of being more complicated; see Jones *et al.* (1996) for a review.

As was done in Sec. 8.1, it is important to comment on the case $\dim(x) \geq 2$. Suppose that $f(x) = f(x_1, \ldots, x_d)$. Then the multivariate kernel density estimator is of the form

$$\hat{f}(x) = \frac{1}{n \det H} \sum_{i=1}^n K(H^{-1}(X_i - x)) \tag{8.30}$$

where $H = \text{diag}(h_1, \ldots, h_d)$ is a $d \times d$ matrix and K is a d-dimensional kernel,

$$\int K(u) du = \int K(u_1, \ldots, u_d) du_1 \cdots du_d = 1.$$

Often, K is assumed to factorize as in $K(u) = \prod_{j=1}^d K^*(u_j)$, for some univariate kernel K^*. As in the univariate case, the ν-dimensional Taylor expansion of f can be used to derive analogous expressions for bias($\hat{f}(x)$), Var($\hat{f}(x)$) and AMISE($\hat{f}$) that depend on the multivariate h, ν, and the unknown density f. This is omitted here, but see Wand and Jones (1995) for the details.

To begin looking at (8.21) and (8.30) predictively, we will examine the pointwise convergence of $\hat{f}$ at a given x. First observe that both are a sum of independent terms and so must satisfy a central limit theorem (at least under mild hypotheses). From the form of (8.26) for $d = 1$, it is easy to see that

$$\text{Var}(\hat{f}(x)) = \frac{f(x)\left(\int K^2(u)du\right)^d}{nh_1 \cdots h_d} + \mathcal{O}\left(\frac{1}{n}\right)$$

for general d, i.e., the rate is a little slower than $\mathcal{O}(1/n)$ because $h_j \to 0$, making the first term larger than the second. Since $\hat{f}$ has a nonzero bias for finite n (even if the bias goes to zero in the limit), the correct quantity to look at is

$$\sqrt{nh_1 \cdots h_d}\left(\hat{f}(x) - E(\hat{f}(x))\right) = \frac{1}{\sqrt{n}}\sum_{i=1}^{n} Z_i,$$

where the summands

$$Z_i = \sqrt{h_1 \cdots h_d}\left(\frac{1}{\det H}K(H^{-1}(X_i - x)) - E\frac{1}{\det H}K(H^{-1}(X_i - x))\right)$$

are independent. Ignoring the error term we obtain

$$\text{Var}(Z_i) \approx f(x)\left(\int K^2(u)du\right)^d$$

and so

$$\sqrt{nh_1 \cdots h_d}\left(\hat{f}(x) - E\hat{f}(x)\right) \to N\left(0, f(x)\left(\int K^2(u)du\right)^d\right). \qquad (8.31)$$

In analogy to (8.22), the location can be written as

$$E(\hat{f}(x)) = f(x) + \frac{\kappa(K)}{\nu!}\sum_{j=1}^{d}\frac{\partial^\nu f}{\partial^\nu x_j}(x)h_j^\nu + o\left(h_1^\nu + \cdots + h_d^\nu\right),$$

using a νth-order Taylor expansion to give an alternative form for (8.31). Note that the bias, i.e., the nonzero location, is

$$\frac{\kappa(K)}{\nu!}\sum_{j=1}^{d}\frac{\partial^\nu f}{\partial^\nu x_j}(x)h_j^\nu \to 0 \quad \text{as} \quad h \to 0.$$

If $h \to 0$ faster than $\mathcal{O}(1/n^{1/(1+2\nu)})$ then the bias goes to zero faster and the location can be ignored to this level of approximation, i.e., the $E\hat{f}(x)$ term in (8.31) can be dropped. While this is appealing, choosing a smaller h means that the optimal asymptotic tradeoff between variance and bias has not been achieved; the bias may be smaller but the increase in variance is so large that the overall AMISE is increased. So, if the AMISE is a good summary of predictive performance, choosing h differently from $h_{\min}$ is not helpful. However, if there is little impact on predictive performance then choosing h smaller than $h_{\min}$ to get less bias may be sensible.

Obviously, the convergence in (8.31) leads to CIs for $f(x)$ of the form

$$\hat{f}(x) \pm z_{\alpha/2}\sqrt{\frac{\hat{f}(x)\int K^2(u)du}{nh}}$$

when x is unidimensional, with analogous intervals when x is d-dimensional. Of course, if the interval contains negative values then they would be ignored. Another sort of CI is formed by inverting a test. Suppose $\mathcal{H}_0: f(x) = f_0$ is tested using

$$t(f_0) = \frac{\hat{f}(x) - f_0}{\sqrt{nhf_0 \int K^2(u)\mathrm{d}u}}.$$

If $\mathcal{H}_0$ is rejected at level α then the values of $t(f_0)$ for which one does not reject it, i.e., $t(f_0) \leq |t_{n-1,\alpha/2}|$, give a $1 - \alpha$ CI:

$$\mathrm{CI}_{1-\alpha}(x) = \left\{ a \left| \frac{\hat{f} - a}{nha \int K^2(u)\mathrm{d}u} \right| \leq t_{n-1,\alpha/2} \right\}.$$

Turning to using the kernel density estimator for prediction, first note that $\hat{f}$ integrates to unity. So, if the sampling variability is small enough, uniformly over a large enough range of x's, naive $(1 - \alpha)100\%$ PIs can be formed using the percentiles of $\hat{f}$, i.e., there are $x_{\alpha/2} = x_{\alpha/2}(n)$ and $x_{1-\alpha/2} = x_{1-\alpha/2}(n)$ such that

$$\int_{x_{\alpha/2}}^{x_{1-\alpha/2}} \hat{f}(x)\mathrm{d}x = 1 - \alpha,$$

provided that h is known. (Recall that v is used to choose $h_{\min}$, so v must be known as well.) The dominated convergence theorem (DCT) ensures that

$$\int_I \hat{f}(u)\mathrm{d}u \to \int_I f(u)\mathrm{d}u,$$

for any interval I, since the terms in $\hat{f}$ are bounded. So, $x_{\alpha/2}(n)$ and $x_{1-\alpha/2}(n)$ converge to the $(\alpha/2)100$th and $(1 - \alpha/2)100$th percentiles of f. That is, if $\hat{P}$ is the probability measure associated with $\hat{f}$ and P is the probability measure associated with the true density f,

$$\hat{P}(X_{n+1} \in [x_{\alpha/2}(n), x_{1-\alpha/2}(n)]) \to P(X_{n+1} \in [x_{\alpha/2}(f), x_{1-\alpha/2}(f)])$$

as $n \to \infty$, implying that $[x_{\alpha/2}(n), x_{1-\alpha/2}(n)]$ is a limiting $1 - \alpha$ PI for X_{n+1}.

To obtain point predictors from $\hat{f}$ one can choose $\mathrm{med}(\hat{f})$ or the first moment of $\hat{f}$. The first moment of $\hat{f}$ is

$$\int_{-\infty}^{\infty} x\hat{f}(x)\mathrm{d}x = \frac{1}{n}\sum_{i=1}^{n} \int_{-\infty}^{\infty} (X_i + uh)K(u)\mathrm{d}u = \bar{X},$$

where $u = (X_i - x)/h$. To get a variance for this, note that the second moment of $\hat{f}$ is

$$\int_{-\infty}^{\infty} x^2\hat{f}(x)\mathrm{d}x = \frac{1}{n}\sum_{i=1}^{n} \int (X_i + uh)^2 K(u)\mathrm{d}u = \bar{X}^2 + h_n^2\kappa_2(K).$$

Taken together, we obtain

$$\mathrm{Var}\left(\int_{-\infty}^{\infty} x\hat{f}(x)\mathrm{d}x \right) = \bar{X}^2 + h^2\kappa_2(K) - \bar{X}^2 = \hat{\sigma}^2 + h_n^2\kappa_2(K), \tag{8.32}$$

where $\hat{\sigma}$ is an estimate of the SE of X_{n+1}. Thus (8.32) gives the variance of $\hat{f}(x)$ as a density in its own right, ignoring the variability from the sampling and specification of h. This is not as bad as it sounds because $\hat{f}(x) \to f(x)$ pointwise in x, and the DCT ensures that the convergence extends to intervals of x-values such that asymptotically (8.32) converges to σ^2, the variance of f.

A variation on this is to invoke the δ-method. Let $x_{\alpha/2} = x_{\alpha/2}(n)$ denote the $(\alpha/2)100$th percentile of $\hat{f}$, with limiting value $x_{\alpha/2}(f)$ under f. Then, since the DCT implies that $x_{\alpha/2} \to x_{\alpha/2}(f)$, (8.31) gives

$$\sqrt{nh_1 \cdots h_d} \left(\hat{f}(x_{\alpha/2}) - E\hat{f}(x_{\alpha/2}(f)) \right) \to N\left(0, f(x_{\alpha/2}(f))\left(\int K^2(u)du\right)^d\right).$$

Assuming that $\hat{f}^{-1}$ is monotonic and differentiable on a neighborhood of $x_{\alpha/2}$, an approximate use of the δ-method gives

$$\sqrt{nh_1 \cdots h_d} \left(x_{\alpha/2} - f^{-1}(E\hat{f}(x_{\alpha/2}(f))) \right)$$

$$\to N\left(0, ((f^{-1}(x_{\alpha/2}))')^2 f(x_{\alpha/2}(f))\left(\int K^2(u)du\right)^d\right). \tag{8.33}$$

Essentially, (8.33) means that a lower $\alpha/2$ confidence bound can be given for $x_{\alpha/2}$ and, by symmetry, an upper $1-\alpha/2$ confidence bound can be given for $x_{1-\alpha/2}$. Taken together, these two bounds give a conservative PI for X_{n+1}.

8.2.2 Kernel Regression: Deterministic Designs

As in Sec. 8.1, assume that $Y(x) = f(x) + \epsilon$, where $f \in \mathcal{F}$, ϵ has mean zero and variance σ^2, and independent data of the form $(x_i, y_i)|_{i=1}^n$ are available. For ease of exposition, suppose that the x_i are (deterministic) design points on a regular grid in an interval $[a, b]$, i.e., $x_i = i\delta$ where $\delta = (b-a)/n$. Again, let K be a square integrable kernel on $[a, b]$ that integrates to one, is assumed symmetric about zero, and has finite second moment. Let $h > 0$ be the 'bandwidth'. The Priestley–Chao estimator for f is, at any $x \in [a, b]$,

$$\hat{f}(x) = \frac{\delta}{h} \sum_{i=1}^n K\left(\frac{x - x_i}{h}\right) y_i; \tag{8.34}$$

see Priestley and Chao (1972). More generally, if the design points are not equally spaced then (8.34) becomes

$$\hat{f}(x) = \frac{1}{h} \sum_{i=1}^n (x_i - x_{i-1}) K\left(\frac{x - x_i}{h}\right) y_i.$$

To use (8.34), h must be specified. The analysis of the MSE for kernel density estimators in Sec. 8.2.1 can be adapted to choose h in the present regression-estimator setting. The core idea is to use Taylor expansions to develop an expression for the variance plus squared bias for $\hat{f}$ as a function of x and h, then to integrate out the x and optimize over h by simple calculus. The extra step required to specify h for the Priestly–Chao estimator is that (8.34) must be approximated by an integral over the x_i as well. The proof here is simplified from

Gasser and Müller (1984) and only applies for kernel functions K supported on a compact interval, here taken to be $[-1, 1]$.

First, observe that

$$E\hat{f}(x) = \frac{1}{nh} \sum_{i=1}^{n} K\left(\frac{x - x_i}{h}\right) f(x_i)$$

and that, using the $v = 0$ case of an argument given in Appendix I in Gasser and Müller (1984),

$$E\hat{f}(x) = \int_0^1 \frac{1}{h} K\left(\frac{x - u}{h}\right) f(u) du + \mathcal{O}\left(\frac{1}{n}\right)$$

$$= \int_{-(1-x)/h}^{x/h} K(u) f(x - hu) du + \mathcal{O}\left(\frac{1}{n}\right)$$

as the number of design points increases, where $t = (x - u)/h$ gives $dt = -h du$, which is then converted back to u for esthetics. Note that, since the range of the integral is taken to be $[0, 1]$, the sign of h must be preserved unlike in the derivation of (8.22).

A second-order Taylor expansion of f at x gives

$$f(x - hu) = f(x) - f^{(1)}(x)hu + \frac{1}{2}f^{(2)}(x)h^2u^2 + o(h^2), \tag{8.35}$$

as long as $x \in (h, 1 - h)$ and both $x/h \to \infty$ and $-(1 - x)/h \to -\infty$ as $h \to 0$. Using the Taylor expansion in (8.35) gives

$$E\hat{f}(x) = f(x) \int_{-1}^{1} K(u) du - hf^{(1)}(x) \int_{-1}^{1} uK(u) du$$

$$+ \frac{h^2}{2} f^{(2)}(x) \int_{-1}^{1} u^2 K(u) du + o(h^2) + \mathcal{O}\left(\frac{1}{n}\right), \tag{8.36}$$

since the domain of integration is $[-1, 1]$. From the properties of K, (8.36) is given by

$$E\hat{f}(x) = f(x) + \frac{h^2}{2} f^{(2)}(x)\kappa_2(K) + o(h^2) + \mathcal{O}\left(\frac{1}{n}\right)$$

pointwise in x as $n \to \infty$ and $h \to 0$. (Recall that $\kappa_2(K)$ is the second moment of K.) Informally, the squared bias at x is

$$\text{bias}^2(x) = \left(E\hat{f}(x) - f(x)\right)^2 \approx \frac{h^4}{4} f^{(2)}(x)^2 \kappa_2(K)^2. \tag{8.37}$$

Deriving an expression for $\text{Var}(\hat{f}(x))$ is similar. Since the x_i are deterministic and the Y_i are independent,

$$\text{Var}(\hat{f}(x)) = \frac{\sigma^2}{n^2 h^2} \sum_{i=1}^{n} K\left(\frac{x - x_i}{h}\right)^2.$$

Recognizing that, as with the bias, this sum can be approximated by an integral over the support of K gives that

$$\text{Var}(\hat{f}(x)) = \frac{\delta\sigma^2}{h} \int_0^1 \frac{1}{h} K\left(\frac{x - u}{h}\right)^2 du + \mathcal{O}\left(\frac{1}{n}\right).$$

Using the transformation $t = (x - u)/h$, and hence $dt = -h\,du$, and re-expressing in terms of u gives

$$\text{Var}(\hat{f}(x)) = \frac{\delta \sigma^2}{h} \int_{-1}^{1} K(u)^2 du + \mathcal{O}\left(\frac{1}{n}\right),$$

since the support of K is $[-1, 1]$. Informally,

$$\text{Var}(\hat{f}(x)) \approx \frac{\delta \sigma^2}{h} \int_{-1}^{1} K(u)^2 du. \tag{8.38}$$

Now, pointwise in x, (8.37) and (8.38) give

$$\text{MSE}\left(\hat{f}(x)\right) \approx \frac{\delta \sigma^2}{h} \int_{-1}^{1} K(u)^2 du + \frac{h^4}{4} f^{(2)}(x)^2 \kappa_2(K)^2, \tag{8.39}$$

where $\delta = 1/n$, $n \to \infty$, and $h \to 0$, cf. Theorem 4 in Gasser and Müller (1984). Integrating over x as in (8.27) gives that the asymptotic mean integrated squared error is

$$\begin{aligned} \text{AMISE} &= 2 \int_0^1 \text{MSE}\left(\hat{f}(x)\right) dx \\ &\approx \frac{\delta \sigma^2}{h} \int_{-1}^{1} K(u)^2 du + \frac{h^4 \kappa_2(K)^2}{4} \int_0^1 f^{(2)}(x)^2 dx. \end{aligned} \tag{8.40}$$

Differentiating and setting the derivative equal to zero gives

$$h_{\min} = \left(\frac{1}{n}\right)^{1/5} \left(\frac{\sigma^2 \int K^2(u) du}{\int_0^1 f^{(2)}(x)^2 dx \, \kappa_2(K)^2}\right)^{1/5}, \tag{8.41}$$

in which the integral of $f^{(2)}(x)^2$ and σ must be estimated. The first can be done by the same technique as used in the discussion after (8.28), and a value for $\hat{\sigma}^2$ can be found from the variance of the y_i. This value will be too large in general since the information in the x_i is not used, but one can iterate, i.e., find $\hat{\sigma}^2$, find $h_{\min}$, find $\hat{f}$ and then find the residual sum of squared errors for $\hat{f}$ to get a new value of $\hat{\sigma}^2$, and so forth.

Two comments regarding the contrast with Sec. 8.2.1 should be made. First, if a third- or higher-order Taylor expansion were used in place of (8.35), a more exact value for (8.41) could be found and would be of a form similar to (8.28). This is not usually done with the Priestley–Chao estimator unless the goal is to estimate derivatives of f, as in Gasser and Müller (1984). Second, (8.39) can be optimized pointwise in x to give $h_{\min} = h_{\min}(x)$ which is of the same form as in (8.41) apart from the integral over x. An obvious diagnostic for how reasonable a global value of h is would be the difference $h_{\min} - h_{\min}(x)$.

Multivariate forms of the Preistley–Chao estimator have been studied in detail. Their form is the obvious variant on (8.30); see Liu (2001). The asymptotic normality of $\hat{f}(x)$, pointwise in x, was established in Benedetti (1977). Here, for the sake of completeness, we consider the one-dimensional case briefly.

Write

$$\sqrt{nh}\left(\hat{f}(x) - E\hat{f}(x)\right) = \frac{1}{\sqrt{nh}} \sum_{i=1}^{n} K\left(\frac{x - x_i}{h}\right)(Y_i - f(x_i)), \tag{8.42}$$

where the summands are independent. Clearly, (8.42) is asymptotically normal with mean zero. Taking the variance, using Appendix I in Gasser and Müller (1984), Lemma 1 in Benedetti (1977), and $t = (x - u)/h$ gives

$$
\begin{aligned}
\mathrm{Var}\left(\sqrt{nh}\left(\hat{f}(x) - E\hat{f}(x)\right)\right) &= \frac{1}{nh}\sum_{i=1}^{n} K^2\left(\frac{x - x_i}{h}\right)\sigma^2 \\
&= \frac{\sigma^2}{h}\int_{-(1-x)/h}^{x/h} K^2\left(\frac{x - u}{h}\right)du + \mathcal{O}\left(\frac{1}{n}\right) \\
&= \sigma^2 \int_{-1}^{1} K^2(u)du + \mathcal{O}\left(\frac{1}{n}\right).
\end{aligned}
\tag{8.43}
$$

So, dropping the error term in (8.43) and summarizing,

$$
\sqrt{nh}\left(\hat{f}(x) - E\hat{f}(x)\right) \to N\left(0, \sigma^2 \int_{-1}^{1} K^2(u)du\right)
\tag{8.44}
$$

pointwise in x, in parallel to (8.31). This convergence gives $1 - \alpha$ CIs for $f(x)$:

$$
\hat{f}(x) \pm z_{1-\alpha/2}\sigma\sqrt{\frac{\int_{-1}^{1} K^2(u)du}{nh}},
$$

in which the natural estimator of σ is the square root of

$$
\hat{\sigma}^2 = \frac{1}{n}\sum_{i=1}^{n}(y_i - \hat{f}(x_i))^2
\tag{8.45}
$$

and h is assumed to be of the form $h_{\min}$.

Turning to using the Priestley–Chao estimator for prediction, note that, as in Sec. 8.1, $\hat{f}$ is linear. Let $Y = (y_1, \ldots, y_n)^T$ and $x \in [a, b]$. Then

$$
\hat{f}(x) = \ell(x^n)Y \quad \text{for} \quad \ell(x^n) = \frac{\delta}{h}\left(K\left(\frac{x - x_1}{h}\right), \ldots, K\left(\frac{x - x_n}{h}\right)\right),
$$

i.e., $\hat{f}(x)$ is a linear function of Y. So it is easy to derive predictors as in Sec. 8.1. Consider x_{new}. Then, $\hat{f}(x_{\text{new}})$ is the point estimator for $f(x_{\text{new}})$ and

$$
\begin{aligned}
\mathrm{Var}(Y_{\text{new}} - \hat{f}(x_{\text{new}})) &= \mathrm{Var}(Y_{\text{new}}) + \mathrm{Var}(\hat{f}(x_{\text{new}})) = \sigma^2 + \mathrm{Var}(\ell(x_{\text{new}})Y) \\
&= \sigma^2 + \ell(x_{\text{new}})^T \mathrm{Var}(Y)\ell(x_{\text{new}}) = \sigma^2 + \sigma^2\ell(x_{\text{new}})^T I_n \ell(x_{\text{new}}) \\
&= \sigma^2(1 + \|\ell(x_{\text{new}})\|^2).
\end{aligned}
\tag{8.46}
$$

So, using (8.45) in (8.46), for $\alpha \in (0, 1)$, a $(1 - \alpha)100\%$ PI at x_{new} is of the form

$$
\hat{f}(x_{\text{new}}) \pm z_{1-\alpha/2}\hat{\sigma}\sqrt{1 + \|\ell(x_{\text{new}})\|^2}.
\tag{8.47}
$$

As is typical, it is seen that one σ^2 is present because the variance of Y is σ^2 and the variance of the predictor $\hat{f}(x_{\text{new}})$ contributes another factor that has a σ^2 in it. For linear predictors this is always $\|\ell(x_{\text{new}})\|^2$, as in (8.46). Hence, even under optimal conditions such as infinite data, i.e., $n \to \infty$, *no predictor can have a smaller variance than* σ^2, although if the goal is the estimation of $f(x_{\text{new}})$ then the variance of $\hat{f}(x_{\text{new}})$ is given in (8.44). The difference is that to get (8.44) the sum of $K^2((x - x_i)/h)$ over i was approximated by the integral of K^2 while in (8.47) this approximation was not used.

8.2.3 Kernel Regression: Random Design

We will continue assuming, as in Sec. 8.2.2, that $Y(x) = f(x) + \epsilon$, where $f \in \mathcal{F}$, and that independent data of the form $(x_i, y_i)|_{i=1}^{n}$ and available. Here, the x_i are regarded as randomly generated rather than chosen design points. So, it makes sense to assume that ϵ has conditional mean zero, i.e., $E(\epsilon_i | X_i) = 0$, and hence $E(\epsilon_i) = E(E(\epsilon_i | X_i)) = 0$. Also, assume $\sigma^2(x) = E(\epsilon_i^2 | X_i = x)$. Now,

$$E(Y - g(X))^2 = E(E(Y - g(x))^2 | X = x)) \geq E(E(Y - E((Y|X))^2 | X))$$

since $\arg\min_a E(Y - a)^2 = E(Y)$. Thus, $E(Y|X)$ is the best predictor of Y under mean squared error, i.e., the goal is to estimate $E(Y|X)$ nonparametrically.

If there are many observations at a given x, it might make sense to average the values of Y that are measured, but if X is continuous then there will not in general be repeat measurements at any $X = x$. An approximation to this would be to take an average of the y_i that occurred for the x_i in a neighborhood around x. Kernels are one way to effect this local averaging. Let χ_A be the indicator function for a set A. Then, a local average around any x is of the form

$$\hat{f}(x) = \frac{\sum_{i=1}^{n} \chi_{\{|X_i - x| \leq h\}} y_i}{\sum_{i=1}^{n} \chi_{\{|X_i - x| \leq h\}}},$$

in which the denominator adjusts automatically for the number of x_i in an interval of x's of halflength h. Letting K be the boxcar kernel, i.e., unity on $[-1, 1]$ and zero elsewhere,

$$\hat{f}(x) = \frac{\sum_{i=1}^{n} K(|X_i - x|/h) y_i}{\sum_{i=1}^{n} K(|X_i - x|/h)} = \frac{\sum_{i=1}^{n} (1/h) K(|X_i - x|/h) y_i}{\sum_{i=1}^{n} (1/h) K(|X_i - x|/h)}, \qquad (8.48)$$

in which the numerator is the Priestley–Chao estimator from Sec. 8.2.2 and the denominator is the kernel density estimator from Sec. 8.2.1.

In general, the estimator $\hat{f}(x)$ in (8.48) from any kernel K and any 'bandwidth' h is called a Nadaraya–Watson (NW) estimator for f. (The NW estimator can also be motivated by writing

$$E(Y|X = x) = \frac{\int y f(x, y) dy}{f(x)},$$

using a two-dimensional kernel-density estimator for the numerator, integrating out y, and using a one-dimensional kernel-density estimator for the denominator.) This generalizes to the case where $x = (x_1, \ldots, x_d)$ is d-dimensional:

$$\hat{f}(x) = \frac{\sum_{i=1}^{n} K(H^{-1}(x_i - x)) y_i}{\sum_{i=1}^{n} K(H^{-1}(x_i - x))}.$$

Here, $H = \text{diag}(h_1, \ldots, h_d)$ is a d-dimensional matrix and K is a d-dimensional kernel, as in (8.30). (The factor $1/(n \det H)$ cancels in the numerator and denominator.)

To finish the specification of the NW estimator, and therefore to derive an NW predictor, a serviceable value of h (or H) must be chosen. In parallel to Secs. 8.2.1 and 8.2.2, h (or

H) is often found by minimizing an approximation to the AMISE formed from the sum of asymptotic expressions for a variance term and a bias term. Since the NW estimator is the ratio of the Priestly–Chao estimator and a kernel density estimator, this decomposition is more complicated than before. It will be enough to consider $d = 1$ since $d \geq 2$ is similar.

We begin by writing $\hat{f}(x) = \hat{v}(x)/\hat{w}(x)$, where

$$\hat{v}(x) = \frac{1}{nh} \sum_{i=1}^{n} K\left(\frac{X_i - x}{h}\right) y_i$$

is the PC estimator in the numerator of an NW estimator and

$$\hat{w}(x) = \frac{1}{nh} \sum_{i=1}^{n} K\left(\frac{X_i - x}{h}\right)$$

is the kernel-density estimator for the marginal $w(\cdot)$ of X that is in the denominator of an NW estimator. Let $\epsilon_i = e_i$. Then, multiplying by $(1/(nh))K((x_i - x)/h)$ and summing over $i = 1, \ldots, n$ in

$$y_i(x_i) = f(x) + (f(x_i) - f(x)) + e_i$$

gives

$$\hat{v}(x) = \hat{w}(x)f(x) + \frac{1}{nh} \sum_{i=1}^{n} K\left(\frac{X_i - x}{h}\right)(f(X_i) - f(x)) + \frac{1}{nh} \sum_{i=1}^{n} K\left(\frac{X_i - x}{h}\right)e_i$$

$$\equiv \hat{w}f(x) + \hat{m}_1(x) + \hat{m}_2(x).$$

Hence, dividing by $\hat{w}(x)$ gives

$$\hat{f}(x) = \frac{\hat{v}(x)}{\hat{w}(x)} = f(x) + \frac{\hat{m}_1(x)}{\hat{w}(x)} + \frac{\hat{m}_2(x)}{\hat{w}(x)}. \tag{8.49}$$

The first term on the right in (8.49) is not stochastic, so it is enough to examine the convergence properties of the last two terms.

Starting with the middle term on the right in (8.49), the data are IID so, for each fixed x, the mean of $\hat{m}_1(x)$ is

$$E\hat{m}_1(x) = \frac{1}{h} E K\left(\frac{X_i - x)}{h}\right)(f(X_i) - f(x))$$

$$= \frac{1}{h} \int K\left(\frac{z - x)}{h}\right)(f(z) - f(x))w(z)\mathrm{d}z$$

$$= \int K(u)(f(x + hu) - f(x))w(x + hu)\mathrm{d}u,$$

with $u = (z - x)/h$. Using a second-order Taylor expansion for f and a first-order Taylor expansion for g, both at x, in the last expression gives

$$E\hat{m}_1(x) = \int K(u)\left(uhf^{(1)}(x) + \frac{u^2 h^2}{2} f^{(2)}(x)\right)\left(w(x) + uhw^{(1)}(x)\right)\mathrm{d}u + o(h^2),$$

where the error term means that taking one more term in either Taylor expansion would give only smaller error terms. Indeed, to order $o(h^2)$, i.e., dropping the term in h^3, the last expression gives three terms:

$$E\hat{m}_1(x) = hf^{(1)}(x)w(x)\left(\int K(u)u\,du\right)$$

$$+ h^2\left(\frac{1}{2}f^{(2)}(x)w(x) + f^{(1)}(x)w^{(1)}(x)\right)\left(\int K(u)u\,du^2\right) + o(h^2).$$

The first term on the right is zero. So, writing

$$B(x) = \frac{1}{2}f^{(2)}(x) + \frac{f^{(1)}(x)w^{(1)}(x)}{w(x)}$$

gives

$$E\hat{m}_1(x) = h^2\kappa_2(K)B(x)w(x) + o(h^2). \tag{8.50}$$

Turning to the variance of $\hat{m}_1(x)$, using (8.50) gives

$$\text{Var}(\hat{m}_1(x)) \approx \frac{1}{n}E\left(\frac{1}{h}K\left(\frac{X-x}{h}\right)(f(X)-f(x)) - h^2\kappa_2 B(x)w(x)\right)^2,$$

since the data are IID. Expanding the square on the right-hand side and transforming by $u = (z-x)/h$ gives that the right-hand side is

$$\frac{1}{n}\int\frac{1}{h^2}K^2\left(\frac{z-x}{h}\right)(f(z)-f(x))^2\,w(z)dz + \frac{h^4}{n}\kappa_2^2 B(x)^2 w(x)^2$$

$$-\frac{2h^2\kappa_2 B(x)w(x)}{n}\int\frac{1}{h}K\left(\frac{z-x}{h}\right)(f(z)-f(x))\,w(z)dz$$

$$= \frac{1}{nh}\int K^2(u)(f(x+uh)-f(x))^2\,w(x+uh)du + \mathcal{O}\left(\frac{h^5}{nh}\right)$$

$$-\frac{2h\kappa_2 B(x)w(x)}{n}\int K(u)(f(x+uh)-f(x))\,w(x+uh)du$$

$$= \frac{1}{nh}\int K^2(u)\left(uhf^{(1)}+o(uh)\right)^2 w(x+uh)du + \mathcal{O}\left(\frac{h^5}{nh}\right)$$

$$-\frac{2h\kappa_2 B(x)w(x)}{n}\int K(u)\left(uhf^{(1)}(x)+o(uh)\right)w(x+uh)du$$

$$= \mathcal{O}\left(\frac{h^2}{nh}\right) + \mathcal{O}\left(\frac{h^5}{nh}\right) + \mathcal{O}\left(\frac{h^4}{nh}\right) = \mathcal{O}\left(\frac{h^2}{nh}\right).$$

Provided that $nh \to \infty$, $\mathcal{O}\left(h^2/(nh)\right)$ is of lower order than $\mathcal{O}(h^2)$. Thus,

$$\sqrt{nh}(\hat{m}_1(x) - E\hat{m}_1(x)) = \sqrt{nh}\left(\hat{m}_1(x) - h^2\kappa_2(K)B(x)w(x)\right) \to 0,$$

in probability. Since $\hat{w}(\cdot) \to w(\cdot)$ in probability as well, see Sec. 8.2.1,

$$\sqrt{nh}\left(\frac{\hat{m}_1(x)}{\hat{w}(x)} - h^2\kappa_2(K)B(x)\right) \to 0, \tag{8.51}$$

giving an asymptotic expression for the middle term on the right in (8.49).

Turning to the last term on the right in (8.49),

$$E\hat{m}_2(x) = \frac{1}{nh}\sum_{i=1}^{n} E\left(E\left(K\left(\frac{X_i - x}{h}\right)\epsilon_i | X_i\right)\right) = 0. \tag{8.52}$$

Since the data are IID, the variance of $\hat{m}_2$, $Var(\hat{m}_2(x))$, is

$$\frac{1}{nh^2} E\left(K\left(\frac{X_i - x}{h}\right)^2 E\left(\epsilon_i^2 | X_i\right)\right) = \frac{1}{nh^2}\int K\left(\frac{z - x}{h}\right)\sigma^2(z)w(z)\mathrm{d}z.$$

Transforming by $u = (z - x)/h$ and using a Taylor expansion for $\sigma(\cdot)w(\cdot)$ gives that the variance of $\hat{m}_2(x)$ is

$$\frac{1}{nh}\int K^2(u)\sigma^2(x + uh)f(x + uh)\mathrm{d}u = \frac{1}{nh}\sigma^2(x)w(x)\int K^2(u)\mathrm{d}u + o\left(\frac{1}{nh}\right).$$

Since $\hat{m}_2$ is a sum of IID random variables, see (8.42), the CLT gives

$$\sqrt{nh}\hat{m}_2(x) \to N\left(0, \sigma^2(x)w(x)\int K^2(u)\mathrm{d}u\right). \tag{8.53}$$

Taken together, (8.49)–(8.53), give

$$\sqrt{nh}\left(\hat{f}(x) - f(x) - h^2\kappa_2 B(x)\right) = \sqrt{nh}\left(\frac{\hat{m}_1(x)}{\hat{w}(x)} - h^2\kappa_2 B(x)\right) + \sqrt{nh}\left(\frac{\hat{m}_2(x)}{\hat{w}(x)}\right)$$

$$\to 0 + \frac{1}{w(x)}N\left(0, \sigma^2(x)w(x)\int K^2(u)\mathrm{d}u\right) = N\left(0, \frac{\sigma^2(x)\int K^2(u)\mathrm{d}u}{w(x)}\right),$$

in probability. A similar derivation for general d gives

$$\sqrt{n\det H}\left(\hat{f}(x) - f(x) - \kappa_2\sum_{j=1}^{d} h_j^2 B_j(x)\right) \to N\left(0, \frac{\left(\int K(u)\mathrm{d}u\right)^d \sigma^2(x)}{w(x)}\right),$$

where

$$B_j(x) = \frac{1}{2}\frac{\partial}{\partial x_j^2}f(x) + \frac{1}{w(x)}\frac{\partial}{\partial x_j}f(x)\frac{\partial}{\partial x_j}w(x).$$

Accordingly, an expression for the asymptotic MSE is

$$\mathrm{AMSE}(\hat{f}(x)) = \kappa_2^2\left(\sum_{j=1}^{d} h_j B_j(x)\right)^2 + \frac{\left(\int K(u)\mathrm{d}u\right)^d \sigma^2(x)}{n(\det H)w(x)},$$

leading to

$$\mathrm{AMISE} = \int \mathrm{AMSE}(x)w(x)\mathrm{d}x$$

$$= \kappa_2^2\int\left(\sum_{j=1}^{d} h_j B_j(x)\right)^2 w(x)\mathrm{d}x + \frac{\left(\int K(u)\mathrm{d}u\right)^d \int \sigma^2(x)\mathrm{d}x}{n\det H}, \tag{8.54}$$

where $\det H = h_1 \cdots h_d$. In the case $d = 1$, differentiating with respect to h, setting the derivative to zero, and solving gives

$$h_{\min} = \left(\frac{1}{n}\right)^{1/5} \times \left(\frac{\int K(u)du \int \sigma^2(x)dx}{4\kappa_2^2 \int B^2(x)w(x)dx}\right)^{1/5}.$$

Even though some constants on the right remain to be estimated, this can be done as discussed after (8.28). Essentially, B and w are assumed normal and the factor on $1/n^{1/5}$ is worked out. It can be hard to estimate $\sigma(x)$; however, if σ does not depend on x then the estimate in (8.45) can be used. Consequently, the NW estimator of f can be taken as fully specified, permitting the derivation of the predictor it generates.

There are other ways to choose h optimally – for instance, by cross-validation. These methods are not presented here because our present goal is merely to show that the NW estimator leads naturally to a predictor. Hence, it is enough to verify that h can be chosen optimally.

First, it is easy to see that the NW estimator is linear. Let

$$W_h(x, x_i) = \frac{K_h(x - x_i)}{\sum_{i=1}^{n} K_h(x, x_i)},$$

so that, for each x and h, $\sum_{i=1}^{n} W_h(x, x_i) = 1$. Now,

$$\hat{f}(x) = \sum_{i=1}^{n} W_h(x_i, x) y_i.$$

Loosely, for fixed h, the closer x is to an x_i, the larger the weight $W_h(x, x_i)$.

Using the reasoning at the end of Sec. 8.2.2, it is easy to derive predictors. To predict Y at x_{new}, write

$$\ell(x^n, x_{\text{new}}) = (W_h(x_1, x_{\text{new}}), \ldots, W_h(x_n, x_{\text{new}})).$$

So, the point predictor for Y_{new} at x_{new} is $\hat{Y}(x_{\text{new}}) = \ell(x^n, x_{\text{new}})Y$ and, if x^n is regarded as fixed, i.e., the variance is conditional on $X^n = x^n$, then

$$\begin{aligned}
\text{Var}(Y_{\text{new}} - \hat{Y}(x_{\text{new}})) &= \text{Var}(Y_{\text{new}}) + \text{Var}(\hat{f}(x_{\text{new}})) = \sigma^2(x_{\text{new}}) + \text{Var}(\ell(x_{\text{new}})Y) \\
&= \sigma^2(x_{\text{new}}) + \ell(x_{\text{new}})^T \text{Var}(Y) \ell(x_{\text{new}}) \\
&= \sigma^2(x_{\text{new}}) + \ell(x_{\text{new}})^T \Sigma_n \ell(x_{\text{new}}),
\end{aligned}$$

where $\Sigma_n = \text{diag}(\sigma^2(x_1), \ldots, \sigma^2(x_n))$. In the special case where $\sigma(x) = \sigma$ for all x, the last expression equals

$$\sigma^2(1 + \|\ell(x_{\text{new}})\|^2). \tag{8.55}$$

Now, (8.45) can be used in (8.55). So, for $\alpha \in (0, 1)$, a $(1 - \alpha)100\%$ PI at x_{new} is

$$\hat{f}(x_{\text{new}}) \pm z_{1-\alpha/2}\hat{\sigma}\sqrt{1 + \|\ell(x_{\text{new}})\|^2}.$$

If σ is not constant but a function of x, it may be necessary to use a nonparametric function estimate of $\sigma(x)$. As in other cases, it is seen that predictors have an irreducible variance that cannot be eliminated.

8.3 Predictors Based on Nearest Neighbors

The techniques in this section extend the brief discussion in Sec. 4.7. They are based on the idea that to get good predictions from a density, regression function, or classifier at a point x it is a good idea to look at the x_i, in a sample of size n, that are closest to x and use them to assign a value at x by some rule. Often the k values of x_i closest to x – the k nearest neighbors – are used for some fixed value of k that must be found by an auxiliary optimization analogous to determining h in the last section. The AMISE can be used, although cross-validation is more common in practice. As before, the point of obtaining these estimators here is not to estimate an unknown function but rather to convert them to predictors whose performance can be directly assessed.

As with Sec. 8.2, the first subsection focuses on density estimation. Then the second section presents the predictive view of nearest neighbors regression and the third subsection turns to the classification case, where the robustness of nearest neighbor methods make them fairly common for low-dimensional x. In these three subsections, the data are assumed IID. In the last subsection, some remarks are given on nonparametric predictors with non-IID data.

8.3.1 Nearest Neighbor Density Estimation

Nearest neighbor density estimation seems to have arisen in Fix and Hodges (1951) and to have been formalized in Loftsgaarden and Quesenberry (1965). Their results are consistent and asymptotically normal; see Devroye and Wagner (1977) and Moore and Yackel (1977), respectively. Suppose that $x_1, \ldots, x_n$ are n IID outcomes of X, a random variable assuming values in $\mathbb{R}^d$ and having bounded density f. Heuristically, the general form of a nonparametric density estimate is

$$\hat{f}(x) = \frac{k/n}{V(x)},$$

where $V(x)$ is the volume of a region having x as an interior point and $k = \#(\{x_i | x_i \in R(x)\})$. If $V = V(x)$ is independent of x, so that k is determined by the data, i.e., it is a nontrivial random variable, then the kernel-density method of Sec. 8.2.1 is recovered. When k is fixed, so that $V(x)$ is a nontrivial random variable – the volume of the region containing the k nearest neighbors for x – the result is the k-nearest-neighbors density estimator.

To be more formal, fix a distance, e.g., the Euclidean distance $\| \cdot \|$, on $\mathbb{R}^d$ and write $R_i = \|x - X_i\|$ with order statistics $R_{(i)} = R_{(i)}(x)$. So, the point x_i corresponding to $R_{(i)}$ is the ith closest data point to x based on the sample of size n. Write it as $x_{(i)}$ where the parentheses in the subscript mean that the data are ordered by the R_i. Now the k-nearest-neighbors density estimate of f at x is

$$\hat{f}(x) = \frac{1}{n R_{(k)}^d} \sum_{i=1}^{n} \frac{\chi_{\{\|x - X_i\| \leq R_{(k)}(x)\}}}{c_d} = \frac{k/n}{R_{(k)}^d c_d}, \tag{8.56}$$

where c_d is the volume of the unit ball in d dimensions,

$$c_d = \frac{\pi^{d/2}}{\Gamma((d+2)/2)}.$$

It can be seen that (8.56) is of the generic form given on the right because $R_{(k)}^d c_d$ is the volume of a ball of radius $R_{(k)}$ in d dimensions. When $R_{(k)}$ is small there are many observations near x, making $\hat{f}(x)$ large. When $R_{(k)}^d$ is large there are few observations near x, making $\hat{f}(x)$ small. Definition (8.56) can be extended to a more general form that is smooth by using a kernel, say K:

$$\hat{f}(x) = \frac{1}{n R_{(k)}^d} \sum_{i=1}^{n} K\left(\frac{\|x - X_i\|}{R_{(k)}(x)}\right), \tag{8.57}$$

where c_d has been absorbed into K and $K \geq 0$ integrates to unity over $\mathbb{R}^d$.

To finish the specification of the k-nearest-neighbors estimator, and therefore derive a k-nearest-neighbors predictor, a serviceable value of k must be chosen. In parallel to Sec. 8.2, k can be found by minimizing an approximation to the AMISE formed from the sum of asymptotic expressions for a variance term and a bias term, although k is often chosen by cross-validation. For the sake of completeness, it is worthwhile sketching the original derivation for the AMISE for this setting, established in Mack and Rosenblatt (1979). In addition to the conditions above, the key requirements are that $P(X \in B(x, r)^c) = \mathcal{O}(e^{-\alpha r})$ for some $r, \alpha > 0$ and $k = k_n \to \infty$ and $k/n \to 0$ as $n \to \infty$.

First, Mack and Rosenblatt (1979) derived some preliminary expressions. A formula for the density $h(r)$ of $R_{(k)}$ as a random variable representing the distance between x and its kth nearest neighbor can be derived by writing $G(r) = P(B(x, r))$ and noting that the derivative is the limit

$$G'(r) = \int_{\|x-t\|=r} f(t) d\sigma(t),$$

where f is the density of any X_i and σ is the surface measure of the sphere with radius $r = \|x - t\|$. This leads to

$$h(r) = n\binom{n-1}{k-1} G^{k-1}(r)(1 - G(r))^{n-k} G'(r).$$

The joint density for the x_i can be represented, for fixed x, as a density of the form $f(x_{(1)}, \ldots, x_{(k-1)}, x_{(k)}, x_{(k+1)}, \ldots, x_{(n)})$ and written with an obvious change of notation as

$$f(y_1, \ldots, y_{k-1}; v_1, \ldots, v_{n-k}; x_{(k)})$$
$$= n\binom{n-1}{k-1}\left(\prod_{j=1}^{k-1} f(y_j)\chi_{B(x,r)}(y_j)\right) \times \left(\prod_{\ell=1}^{n-k} f(v_\ell)\chi_{B(x,r)^c}(v_\ell)\right) \times f(x_{(k)}).$$

If the y's are integrated over $B(x, r)$, the v_ℓ are integrated over $B(x, r)^c$, and $f(x_{(k)})$ is integrated over $r = \|x - x_{(k)}\|$ then the result is $h(r)$ above. Hence,

$$f(y_1, \ldots, y_{k-1}; v_1, \ldots, v_{n-k}; x_{(k)}|r) = \prod_{j=1}^{k-1}\left(\frac{f(y_j)}{G(r)}\right) \times \prod_{\ell=1}^{n-k}\left(\frac{f(v_\ell)}{1 - G(r)}\right) \times \frac{f(X_{(K)})}{G'(r)}.$$

It can be seen that the Y_ℓ, V_j, and $x_{(k)}$ are conditionally independent, given $R_{(k)} = r$, with respective marginal densities $f(y)/G(r)$, $f(v)/(1 - G(r))$, and $f(x_{(k)})/G'(r)$, where $y \in B(x, r)$, $v \in B(x, r)^c$, and $x_{(k)} \in \{t | \|x - t\| = r\}$ and the conditional density of $x_{(k)}$ given $R_{(k)}$ is integrated over the surface measure on the sphere of radius r centered at x.

The derivation of the asymptotic expressions requires moments of various functions of $R_{(k)}$ to be found. Note that $R_{(k)}$ has the same distribution as T, the kth-order statistic from an IID sample of size n drawn from a Unif[0,1]. So, provided f is bounded and continuous,

$$G(r) = c_d f(x) r^d + o(r^d) \quad \text{as} \quad r \to 0^+.$$

That is, for $t = G(r)$ and $f(x) > 0$,

$$G^{-1}(t)^\lambda = r^\lambda = \left(\frac{t}{c_p f(x)} \right)^{\lambda/d} + o(t^{\lambda/d}).$$

For λ, ν, and β nonnegative integers, let

$$\phi(r) = \frac{r^\lambda}{G(r)^\nu (1 - G(r))^\beta}.$$

We have that $E(\phi(R_n))$ exists for n large enough, when $1 - G(r) = \mathcal{O}(1/r^\zeta)$ for some $\zeta > 0$. So, changing variables to $t = G(r)$,

$$E(\phi(R_{(k)})) = n \binom{n-1}{k-1} \int_0^1 \left(\left(\frac{t}{c_d f(x)} \right)^{\lambda/d} + o(t^{\lambda/d}) \right) t^{k-1-\nu} (1-t)^{n-k-\beta} dt. \quad (8.58)$$

Given these preliminaries, the variance decomposition formula gives

$$\text{Var}(\hat{f}(x)) = E \,\text{Var}(\hat{f}(x)|R_{(k)}) + \text{Var}(E(\hat{f}(x)|R_{(k)})).$$

With an obvious abuse of notation, the first term is

$$\text{Var}(\hat{f}(x)|R_{(k)}) = \frac{k-1}{n^2 R_{(k)}^{2d}} \text{Var}\left(K\left(\frac{x-Y}{R_{(k)}} \right) \bigg| R_{(k)} \right) \quad (8.59)$$

$$+ \frac{1}{n^2 R_{(k)}^{2d}} \text{Var}\left(K\left(\frac{x-X_{(k)}}{R_{(k)}} \right) \bigg| R_{(k)} \right) \quad (8.60)$$

$$+ \frac{n-k}{n^2 R_{(k)}^{2d}} \text{Var}\left(K\left(\frac{x-V}{R_{(k)}} \right) \bigg| R_{(k)} \right), \quad (8.61)$$

since, under the joint density for Y, V and $X_{(k)}$ above they are independent given $R(k)$; the Y_j are identical as are the V_j. Using the ball $B(x, R_{(k)})$, expression (8.59) becomes

$$\frac{k-1}{n^2} \left(\frac{1}{R_{(k)}^d P(B(x, R_{(k)}))} \int_{\|u\| \le 1} K^2(u) f(x - u R_{(k)}) du \right.$$

$$\left. - \frac{1}{P(B(x, R_{(k)}))^2} \left(\int_{\|u\| \le 1} K(u) f(x - u R_{(k)}) du \right)^2 \right)$$

The expectation of the first term above is

$$\frac{k-1}{n^2} f(x) \int_{\|u\| \le 1} K^2(u) du \, E\left(\frac{1}{R_{(k)} P(B(x, R_{(k)}))} \right) + \text{Error}.$$

The first term, i.e., not the Error term, is

$$\frac{c_d f(x)^2}{k} \int_{\|u\| \le 1} K^2(u) du + o\left(\frac{1}{k} \right),$$

by use of (8.58), as $k \to \infty$. It can also be shown, see Mack and Rosenblatt (1979), that Error $= o(1/k)$. A similar argument using (8.58) gives

$$\frac{k-1}{n^2} E\left(\frac{1}{P(B(x, R_{(k)}))^2} \left(\int_{\|u\| \le 1} K(u) f(x - u R_{(k)}) du\right)^2\right)$$
$$= \frac{f(x)^2}{k} \left(\int_{\|u\| \le 1} K(u) du\right) + o\left(\frac{1}{k}\right),$$

completing the argument for the $k - 1$ Y_j. For the term in (8.60) involving $X_{(k)}$, a similar argument (using (8.58)) gives

$$E\left(\frac{1}{n^2 R_{(k)}^{2d}} \mathrm{Var}\left(\frac{x - x_{(k)}}{R_{(k)}}\bigg| R_{(k)}\right)\right) = o\left(\frac{1}{k}\right).$$

The same kind of argument gives that the expectation of (8.61) is

$$E\left(\frac{n-k}{n^2 R_{(k)}^{2d}} \mathrm{Var}\left(K\left(\frac{x - V}{R_{(k)}}\right)\bigg| R_{(k)}\right)\right) = \frac{c_d f(x)^2}{k} \int_{\|u\| > 1} K^2(u) du + o\left(\frac{1}{k}\right),$$

leading to

$$E\,\mathrm{Var}(\hat{f}(x) | R_{(k)}) = \frac{c_d f(x)^2}{k} \int K^2(u) du - \frac{f(x)^2}{k} \left(\int_{\|u\| \le 1} K(u) du\right)^2 + o\left(\frac{1}{k}\right).$$

The other term in the variance decomposition formula is the variance of $E(\hat{f}(x) | R_{(k)})$, which can be expressed as the variance of

$$\frac{k-1}{n} \frac{1}{R_{(k)}^d} \frac{1}{P(B(x, R_{(k)}))} \int_{B(x, R_{(k)})} K\left(\frac{x - v}{R_{(k)}}\right) f(v) dv \tag{8.62}$$

$$+ \frac{n-k}{n} \frac{1}{R_{(k)}^d} \frac{1}{P(B(x, R_{(k)}))} \int_{B(x, R_{(k)})^c} K\left(\frac{x - v}{R_{(k)}}\right) f(v) dv \tag{8.63}$$

$$+ \frac{1}{n} \frac{1}{R_{(k)}^d} \frac{1}{P(B(x, R_{(k)}))} \int_{\partial B(x, R_{(k)})} K\left(\frac{x - v}{R_{(k)}}\right) f(v) dv \bigg/ \int_{\partial B(x, R_{(k)})} f(v) dv, \tag{8.64}$$

where $\partial B(x, R_{(k)})$ is the boundary of $B(x, R_{(k)})$. After some delicate work needed to bound the terms using expansions of the kth-order statistic $T_{(k)}$ from a uniform distribution of the form $T - (k-1)/(n-1)$, it is possible to reduce the range of the expectation by restricting to sets of the form $T - (k-1)/(n-1) \le (\alpha k)/n$; see Mack and Rosenblatt (1979) for details. Thus, it turns out that all the contributions to $\mathrm{Var}(E(\hat{f}(x) | R_{(k)}))$ except for the mean square of (8.62) are $o(1/k)$. Hence $\mathrm{Var}(E(\hat{f}(x) | R_{(k)}))$ equals

$$E\left(\frac{k-1}{n} \frac{1}{R_{(k)}^d} \frac{1}{P(B(x, R_{(k)}))} \int_{B(x, R_{(k)})} K\left(\frac{x - v}{R_{(k)}}\right) f(v) dv\right)^2 + o\left(\frac{1}{k}\right)$$
$$= \frac{f(x)^2 c_d}{k} \left(\int_{\|u\| \le 1} K(u) du\right)^2 + o\left(\frac{1}{k}\right).$$

So, the 'variance' part of the variance–bias decomposition for $\hat{f}(x)$ is

$$\text{Var}(\hat{f}(x)) = \frac{f(x)^2}{k} c_d \int K^2(x) dx + o\left(\frac{1}{k}\right)$$

and to find the AMISE of $\hat{f}$ it remains to find the squared bias.

To do this, note that $E\hat{f}(x)$ is the expectation of the sum of the three terms (8.62)–(8.64). Using a change of variables and Taylor expanding, the first term, (8.62), is

$$\frac{k-1}{nP(B(x, R_{(k)}))}\left(f(x)\int_{\|u\|\le 1} K(u) du\right.$$
$$\left. + \frac{R_{(k)}^2}{2}\int_{\|u\|\le 1}\sum_{\alpha,\beta=1}^{d}(D_\alpha D_\beta f)(x)u_\alpha u_\beta K(u) du + o(R_{(k)}^2)\right).$$

Leaving out some steps, which are given in Mack and Rosenblatt (1979), the expectation of (8.62) is

$$f(x)\int_{\|u\|\le 1} K(u) du + \frac{1}{2(c_d f(x))^{2/d}}\left(\frac{k}{n}\right)^{2/d}\int_{\|u\|\le 1}\sum_{\alpha,\beta=1}^{d}(D_\alpha D_\beta f)(x)u_\alpha u_\beta K(u) du,$$

up to order $o((k/n)^{2/d})$.

Omitting the details, an asymptotic expression for the expectation of the second term, (8.63), is

$$f(x)\int_{\|u\|>1} K(u) du + \frac{1}{2(c_d f(x))^{2/d}}\left(\frac{k}{n}\right)^{2/d}\int_{\|u\|>1}\sum_{\alpha,\beta=1}^{d}(D_\alpha D_\beta f)(x)u_\alpha u_\beta K(u) du,$$

again up to order $o((k/n)^{2/d})$, and an asymptotic expression for (8.64) is

$$\frac{c_d f(x)}{k}\int_{\|u\|=1} K(u) d\sigma(u) + o\left(\frac{1}{k}\right).$$

Taking the three asymptotic expressions together gives an asymptotic expression for the bias of the k-nearest-neighbors estimator:

$$E\hat{f}(x) = f(x) + \frac{\Gamma((d+2)/2)^{2/d}}{2\pi f(x)^{2/d}} Q(f)(x)\left(\frac{k}{n}\right)^{2/d}$$
$$+ c_d\frac{f(x)}{k}\int_{\|u\|=1} K(u) d\sigma(u) + o\left(\left(\frac{k}{n}\right)^{2/d} + \frac{1}{k}\right),$$

where $Q(f)$ indicates the quadratic term in the Taylor expansion of f, i.e., the sum of the second-order derivatives of f.

So, the AMISE for the k-nearest-neighbors estimate $\hat{f}$ is the integral of the sum of the asymptotic variance plus the asymptotic squared bias. Following Orava (2011), the expressions for the one-dimensional case are

$$\text{variance}(\hat{f})(x) = \frac{2}{k}f(x)^2 \int K(u)^2 du + o\left(\frac{1}{k}\right),$$

$$\text{bias}(\hat{f})(x) = \frac{1}{8}\left(\frac{k}{n}\right)^2 \frac{f''(x)}{f(x)^2} \int u^2 K(u) du + o\left(\left(\frac{k}{n}\right)^2 + \frac{1}{k}\right).$$

Hence, the AMISE is given by

$$\int \text{variance}(\hat{f})(x)dx + \int (\text{bias}(\hat{f})(x))^2 dx$$

$$= \frac{2}{k}\int K(u)^2 du \int f(u)^2 du + \frac{1}{64}\left(\frac{k}{n}\right)^4 \left(\int u^2 K(u)du\right)^2 \int \left(\frac{f''(u)}{f(u)^2}\right)^2 du.$$

Treating k as continuous, differentiating with respect to it, and solving for the minimizing value gives

$$k_{\min} = \text{round}\left(2n^{4/5}\left(\frac{\int K^2 \int f^2}{\int u^2 K \int (f''/f^2)}\right)\right).$$

As in the kernel methods case, finding $k_{\min}$ requires finding a serviceable value of the integral that depends on the unknown f. Discussion of how to do this in practice is given after (8.28).

A useful fact about the k-nearest-neighbors density estimator is that it is asymptotically normal. Indeed, Biau *et al.* (2011, Theorem 3.3) gives

$$\sqrt{k_n}(\hat{f}(x) - f(x)) \to N(0, f(x)^2),$$

provided that $k = k_n \to \infty$, $k_n/n^{4/(d+4)} \to 0$, and $f(x) > 0$ and differentiable. The proof is quite involved, and discussion of it is omitted here. Nevertheless, approximate $1 - \alpha$ CIs for $f(x)$ are given by

$$\hat{f}(x) \pm z_{1-\alpha/2}\frac{\hat{f}(x)}{\sqrt{k_n}};$$

intervals with negative values are truncated at zero.

Turning to the use of k-nearest-neighbors density estimators for prediction, (8.57) integrates to unity and (8.56) integrates to approximately unity since it converges pointwise to $f(x)$; this follows from its asymptotic normality but also holds more generally. Hence, approximate PIs can be found from the percentiles of $\hat{f}$ by solving for

$$\int_{x_{\alpha/2}(n)}^{x_{1-\alpha/2}(n)} \hat{f}(u)du = 1 - \alpha,$$

provided that $k = k_n$ is chosen. In fact, the pointwise consistency of $\hat{f}$ ensures that $x_{\alpha/2}(n) \to x_{\alpha/2}(f)$ and $x_{1-\alpha/2}(n) \to x_{1-\alpha/2}(f)$, the percentiles of f. Otherwise stated, if $\hat{P}$ is the estimated probability measure from $\hat{f}$, we have

$$\hat{P}(X \in [x_{\alpha/2}(n), x_{1-\alpha/2}(n)]) \to P(X \in [x_{\alpha/2}(f), x_{1-\alpha/2}(f)])$$

as $n \to \infty$, for k_n with suitable rates making $[x_{\alpha/2}(n), x_{1-\alpha/2}(n)]$ an asymptotic $1 - \alpha$ PI for X_{n+1}.

To obtain point predictors for X_{n+1} from $\hat{f}$, the simplest approach is to take the mean of $\hat{f}$:

$$E_{\hat{f}}(X) = \int x \hat{f}(x) du.$$

To get a PI, it is enough to take the variance,

$$\text{Var}_{\hat{f}}(X) = \int (x - E_{\hat{f}}(X))^2 \hat{f}(x) dx,$$

and use Chebyshev's inequality. Then, the interval

$$[E_{\hat{f}}(X) - r \text{Var}_{\hat{f}}(X), E_{\hat{f}}(X) + r \text{Var}_{\hat{f}}(X)]$$

has probability $\hat{P}$ equal to at least $1 - r^2$. This obviously gives PIs that are too wide, but may serve as a useful conservative statement. The median of $\hat{f}$ can also be a good point predictor, in which case the $(\alpha/2)100$th and $(1-\alpha/2)100$th percentiles of $\hat{f}$ as above provide a natural PI.

In parallel to (8.33) at the end of Sec. 8.2.1, the DCT and δ-method can also be used to generate PIs. From the asymptotic normality of $\hat{f}$ at $x_{\alpha/2}(f)$, we have

$$\sqrt{k}\left(x_{\alpha/2}(n) - f^{-1}(E(\hat{f}(x_{\alpha/2}(f)))) \right) \to N\left(0, f^2(x_{\alpha/2}(f))[(f^{-1})'(x_{\alpha/2}(f))]^2\right).$$

Thus, a lower $1 - \alpha/2$ confidence bound on $x_{\alpha/2}(f)$ can be given. Likewise, an upper $1 - \alpha/2$ confidence bound can be given on $x_{1-\alpha/2}(f)$. Taken together, they give an asymptotic $1 - \alpha$ PI for a future value of X.

8.3.2 Nearest Neighbor Regression

Suppose that the data follow a model of the form $Y_i = f(X_i) + \epsilon_i$ for $i = 1, \ldots, n$, where the error terms ϵ_i are IID, the explanatory variables $X_i \in \mathbb{R}^d$ are IID, and $Y_i \in \mathbb{R}$. Now, a k-nearest-neighbors regression estimator $\hat{f}(\cdot)$ for f, parallel to the k-nearest neighbors density estimator in Sec. 8.3.1, can be defined.

Given data (x_i, y_i) for $i = 1, \ldots, n$, the simplest form in which to write the k-nearest-neighbors regression estimate at some $x \in \mathbb{R}$ is

$$\hat{f}(x) = \frac{1}{k} \sum_{i \in N_k(x)} y_i,$$

cf. the definition in (4.52), where $N_k(x)$ contains the indices of the k points in $\{x_1, \ldots, x_n\}$ that are closest to x. As before, k will have to be chosen, and usually we require $k = k_n \to \infty$ and $k_n/n \to 0$ as $n \to \infty$; $k_n = \sqrt{n}$ is one common choice. Clearly, $N_k(x)$ depends on the data and plays the same kind of role as $V(x)$ at the beginning of Sec. 8.3.1.

The k-nearest-neighbors estimator is always discontinuous, as can be seen by writing $w_i(x) = 1/k$ when $x_i \in N_k(x)$ and zero otherwise. Then

$$\hat{f}(x) = \sum_{i=1}^{n} w(x_i) y_i. \tag{8.65}$$

Since the w_i are discontinuous as a function of x, $\hat{f}(x)$ is discontinuous. In the notation of Sec. 8.3.1, $\hat{f}$ can also be written as

$$\hat{f}(x) = \frac{1}{k} \sum_{i=1}^{n} \chi_{\{\|x-x_i\| \le R_{(k)}\}} y_i = \frac{\sum_{i=1}^{n} K\left(\|x - x_i\|/R_{(k)}\right) y_i}{\sum_{i=1}^{n} K\left(\|x - x_i\|/R_{(k)}\right)}$$

in which K is a kernel function and the factors $1/(nR_{(k)}^d)$ in the numerator and denominator cancel. In the simplest case, K is the indicator function for $\|x - X_i\| \le R_{(k)}$ and gives the 'pure' k-nearest-neighbors estimator. More generally, when K is smooth, so is $\hat{f}$ and the kernel permits the formation of a larger class of k-nearest-neighbors estimators. Thus, both k-nearest-neighbor estimators and NW estimators are ratios of random quantities. Below, it is assumed that K integrates to unity, has a finite second moment, and is zero outside the unit ball in d dimensions.

To finish the specification of the k-nearest-neighbors estimator and thereby derive a k-nearest-neighbors predictor, a serviceable value of k must be chosen. As done in previous sections, e.g., Sec. 8.2, one can again derive asymptotic expressions for the variance and bias of $\hat{f}$ as functions of x and hence the AMISE, as in Mack (1981). Minimizing the AMISE over k gives an asymptotic form for $k = k_n$. Cross-validation can be used as well. The derivation of the AMISE is long and complex so only a summary of the steps from Mack (1981) will be presented here. In addition to requiring $f(x) > 0$ and moment conditions on Y, the assumptions include that $P(\|x = X\| > r) = \mathcal{O}(r^{-\zeta})$ for some $\zeta > 0$ as $r \to \infty$ and that the rates of increase for n and k be related, so that $k = o(n)$ and $n = o(\log k)$.

Conditioning on $R_{(k)}$ is central to the derivation. As in Sec. 8.3.1, for any fixed x let G be the common distribution of $\|x - X_i\|$. Then, $R_{(k)}$ is the kth-order statistic[1] from n IID outcomes of $\|x - X_i\|$. So, its density is

$$h(r) = n\binom{n-1}{k-1} G(r)^{k-1}(1 - G(r))^{n-k} G'(r),$$

where

$$G'(r) = \int_{\|t-x\|=r} f(t)\sigma(dt);$$

here f is the marginal of X and σ is the surface measure for the surface of a sphere of radius r in d dimensions.

If the $k - 1$ observations inside $B(x, r)$ are denoted $((X_{(1)}, Y_{(1)}), \ldots, (X_{(k-1)}, Y_{(k-1)}))$ then, conditioned on $R_{(k)} = r$, they are conditionally independent, and

$$f((x_{(1)}, y_{(1)}), \ldots, (x_{(k-1)}, y_{(k-1)})) = \prod_{i=1}^{k-1} \left(\frac{f(X_{(i)}, Y_{(i)})}{G(r)} \right).$$

Also, as derived in Sec. 8.3.1,

$$G^{-1}(t) = r = \left(\frac{t}{c_d f(x)} \right)^{1/d} + o(t^{1/d}).$$

---

[1] In fact, as shown in Gyorfi *et al.* (2002, p. 89), $X_{(k)}(x)$, the value of the kth nearest neighbor to x, satisfies $\|X_{(k)} - x\| \to 0$ with probability unity provided that $k_n/n \to 0$ as $n \to \infty$.

Since $T = G(R_{(k)})$ is the kth-order statistic from a Unif[0, 1] distribution from a sample of size n, this last expression can be used to find moments of $R_{(k)}$.

So, write the numerator and denominator of $\hat{f}$ as

$$\hat{u}(x) = \frac{1}{nR_{(k)}^d} \sum_{i=1}^n K\left(\frac{x-X_i}{R_{(k)}}\right) Y_i \quad \text{and} \quad \hat{v}(x) = \frac{1}{nR_{(k)}^d} \sum_{i=1}^n K\left(\frac{x-X_i}{R_{(k)}}\right).$$

Recall from Mack and Rosenblatt (1979) that (as partly derived in Sec. 8.3.1) $\hat{v}$ satisfies

$$E\hat{v}(x) = v(x) + \frac{\Gamma((d+2)/2)^{2/d}}{2\pi v(x)^{2/d}} Q(v)(x) \left(\frac{k}{n}\right)^{2/d} + o\left(\left(\frac{k}{n}\right)^{2/d}\right).$$

So, letting $\pi^{-1}\Gamma((d+2)/2)^{2/d} = C_d$ we have

$$\text{Var}(\hat{v}(x)) = \frac{v(x)^2}{k} c_d \int K^2(x)\mathrm{d}x + o\left(\frac{1}{k}\right).$$

Also, $\hat{v}(x) \to v(x)$ with probability unity as $n \to \infty$. Given this notation, a final assumption is required, namely, that there is a $b \neq 0$ such that $|(1/b)\hat{v}(x) - b| < 1$ and, for some a,

$$\hat{f}(x) = \frac{\hat{u}(x)}{\hat{v}(x)} = \frac{a + \hat{u}(x) - a}{b + \hat{v}(x) - b} = \frac{a}{b} + \frac{1}{b}\left(\hat{u}(x) - a\right) - \frac{a}{b^2}\left(\hat{v}(x) - b\right)$$
$$+ \mathcal{O}\left((\hat{v}(x) - b)^2 + (\hat{u}(x) - a)(\hat{v}(x) - b)\right). \quad (8.66)$$

Asymptotic expressions for the bias and variance of $\hat{f}$ follow:

1. $E\hat{v}(x) = v(x) + \dfrac{Q(v)(x)}{2(c_d v(x))^{2/d}} \left(\dfrac{k}{n}\right)^{2/d} + o\left(\left(\dfrac{k}{n}\right)^{2/d}\right);$

2. $\text{Var}(\hat{v}(x)) = \dfrac{c_d v(x)^2}{k} \int K^2(x)\mathrm{d}x + o\left(\dfrac{1}{k}\right);$

3. $E\hat{u}(x) = f(x)v(x) + \dfrac{Q(fv)(x)}{2(c_d v(x))^{2/d}} \left(\dfrac{k}{n}\right)^{2/d} + o\left(\left(\dfrac{k}{n}\right)^{2/d}\right);$

4. $\text{Cov}(\hat{u}(x), \hat{v}(x)) = \dfrac{c_d f(x)v^2(x)}{k} \int K^2(x)\mathrm{d}x + o\left(\dfrac{1}{k}\right);$

5. $\text{Var}(\hat{u}(x)) = \dfrac{c_d v^2(x) E(Y^2 | X = x)}{k} \int K^2(x)\mathrm{d}x + o\left(\dfrac{1}{k}\right).$

Further argumentation shows that (8.66) with items 1–4 gives

$$E(\hat{f}(x)) - f(x) = \frac{Q(fv)(x) - f(x)Q(v)(x)}{2v(x)(c_d v(x))^{2/d}} \left(\frac{k}{n}\right)^{2/d} + o\left(\left(\frac{k}{n}\right)^{2/d}\right) + \mathcal{O}\left(\frac{1}{k}\right) \quad (8.67)$$

and (8.66) with items 2, 4, and 5 gives

$$\text{Var}(\hat{f}(x)) = \frac{c_d \text{Var}(Y | X = x)}{k} \int K^2(x)\mathrm{d}x + o\left(\frac{1}{k}\right). \quad (8.68)$$

Now, using (8.67) and (8.68) gives

$$\text{AMISE}(\hat{f}) = \left(\frac{k}{n}\right)^{4/d} \int \left(\frac{Q(fv)(x) - f(x)Q(v)(x)}{2v(x)(c_d v(x))^{2/d}}\right)^2 F(dx)$$

$$+ \frac{1}{k}\left(c_d \int K^2(x)dx \int \text{Var}(Y|X = x)F(dx)\right) \qquad (8.69)$$

This is of the form

$$\text{AMISE}(\hat{f}) = C_1 \left(\frac{k}{n}\right)^{4/d} + C_2 \frac{1}{k}$$

for some real C_1 and C_2. Taking k to be continuous, differentiating, and setting the derivative to zero gives the usual form of the optimal k as

$$k_{\min} = \text{round}\left(C_3 n^{4/(d+4)}\right),$$

in which C_3 is a real number that can be identified from (8.69). Again, starting values for the constants in the AMISE that depend on the unknown functions can be found in the way discussed after (8.28). Apart from constants, this is the same rate as that found in Sec. 8.3.1 although it is different from the rates in Sec. 8.2 for kernel methods where $h \propto n^{1/5}$ when x is unidimensional, as opposed to $k \propto n^{4/5}$ for nearest neighbors. Thus, the k-nearest-neighbors estimate of the regression function is fully specified.

A further result from Mack (1981) gives the asymptotic normality of $\hat{f}(x)$:

$$\sqrt{k-1}\left(\hat{f}(x) - E\hat{f}(x)\right) \to N\left(0, c_d \text{Var}(Y|X = x)\int K^2(x)dx\right)$$

in distribution; a proof can be found in Mack (1981).

Turning to using the k-nearest-neighbors estimator as a predictor, it is easy to see from (8.65) that $\hat{f}$ is linear. More formally, given any x, write

$$W_{ki}(x) = \begin{cases} 1/k & \text{if } i \in J_x, \\ 0 & \text{otherwise,} \end{cases}$$

where

$$J_x = \{i \,|\, X_i = X_{(j)} \text{ for some } j \leq k\}.$$

Now, for any k and any x,

$$\sum_{i=1}^{n} W_{ki}(x) = 1 \quad \text{and} \quad \hat{f}(x) = \sum_{i=1}^{n} W_{ki}(x)y_i.$$

So, using the reasoning at the end of Sec. 8.2.2, it is easy to derive predictors, as follows.

To predict Y at x_{new}, let $x^n = (x_1, \ldots, x_n)$ and write

$$\ell(x^n, x_{\text{new}}) = \ell_k(x^n, x_{\text{new}}) = (W_{k1}(x_n, x_{\text{new}}), \ldots, W_{kn}(x_n, x_{\text{new}})).$$

So, the point predictor for Y_{new} at x_{new} is $\hat{Y}(x_{\text{new}}) = \ell(x^n, x_{\text{new}})Y$ and, assuming x^n is held fixed, the conditional variance of $\hat{Y}_{\text{new}}$ given x^n is

$$
\mathrm{Var}(Y_{\mathrm{new}} - \hat{Y}(x_{\mathrm{new}})) = \mathrm{Var}(Y_{\mathrm{new}}) + \mathrm{Var}(\hat{f}(x_{\mathrm{new}})) = \sigma^2(x_{\mathrm{new}}) + \mathrm{Var}(\ell(x_{\mathrm{new}})Y)
$$
$$
= \sigma^2(x_{\mathrm{new}}) + \ell(x_{\mathrm{new}})^T \, \mathrm{Var}(Y) \, \ell(x_{\mathrm{new}})
$$
$$
= \sigma^2(x_{\mathrm{new}}) + \ell(x_{\mathrm{new}})^T \Sigma_n \ell(x_{\mathrm{new}}),
$$

where $\Sigma_n = \mathrm{diag}(\sigma^2(x_1), \ldots, \sigma^2(x_n))$. In the special case where $\sigma(x) = \sigma$ for all x, the last expression equals

$$
\sigma^2(1 + \|\ell(x_{\mathrm{new}})\|^2). \tag{8.70}
$$

Now, see (8.45), an estimate of σ can be used in (8.70). So, for $\alpha \in (0, 1)$, a $(1 - \alpha)100\%$ PI at x_{new} is

$$
\hat{f}(x_{\mathrm{new}}) \pm z_{1-\alpha/2} \hat{\sigma} \sqrt{1 + \|\ell(x_{\mathrm{new}})\|^2}.
$$

If σ is not constant as a function of x, it may be necessary to use a nonparametric function estimate of $\sigma(x)$. We comment here that if x^n is regarded as random, i.e., as X^n, then it would be natural here, and at the end of Sec. 8.2.3, to use either the expectation over X^n or derive a different expression that includes the variability of X^n inside the definition of the variance.

8.3.3 Beyond the Independence Case

There is an established literature that uses predictors in a sequential context, often that of time series. One general formulation is in Gyorfi and Ottucksak (2011), see also Biau *et al.* (2008b); it mirrors our use of the CPE here but the data is sequential, of the form (X_i, Y_i) for $i = 1, \ldots, n - 1$ and X_n. So, it is convenient to write $X^i = (X_1, \ldots, X_i)^T$ and $Y^i = (Y_1, \ldots, Y_i)^T$. Unlike IID data, in this case there is no benefit to considering permutations of the data. Indeed, permuting the data will be detrimental if there is a sequential structure, e.g., according to time. Let g_i be the predictor of Y_i based on X^i and Y^{i-1} and denote the average prediction error (APE) in squared error loss by

$$
\mathrm{APE}_n(g) = \frac{1}{n} \sum_{i=1}^{n-1} (g_i(X^i, Y^{i-1}) - Y_i).
$$

The established literature on sequential prediction is primarily concerned with finding predictors that are optimal in an asymptotic cumulative-risk sense, not with the properties of predictors in finite sample sizes. Here it is argued that anything that can be taken as 'true' about a DG should be derived from a predictor that has been verified to be 'good', and this includes not just the asymptotic cumulative risk but also the finite-sample variability (as has been repeatedly indicated, for instance in Fig. 7.4) and other properties such as those discussed in Sec. 2.4.

To focus on the APE, recall that Algoet (1992) showed that if (X_i, Y_i), $i = -\infty, \ldots, \infty$, is jointly stationary and ergodic then

$$
L^* = E\left(Y_0 - E(Y_0 | X_{-\infty}^0, Y_{-\infty}^{-1})\right)^2,
$$

the minimal mean squared error of prediction for Y_0 given data $X_{-\infty}^0 = (\ldots, X_0)$ and $Y_{-\infty}^{-1} = (\ldots, Y_{-1})$, is a lower bound for the liminf of the APE, i.e.,

$$
\liminf_{n \to \infty} \mathrm{APE}_n(g_n) \geq L^*
$$

for any sequence of predictors g_n, $n = 1, \ldots, \infty$. Accordingly, a prediction strategy is defined to be universally consistent with respect to a given class $\mathcal{C}$ of stationary ergodic processes if and only if

$$\lim_{n \to \infty} \text{APE}_n(g_n) = L^*$$

for each process in $\mathcal{C}$.

Various authors have established versions of the theorem that universally consistent strategies with respect to the class of all bounded, stationary, and ergodic processes exist; see Algoet (1992), Gyorfi (1984), and Morvai *et al.* (1996). These were theoretical results in which the predictors were chosen for their tractability in proving theorems; the predictors were either very complex or had a slow rate of convergence. This motivated Biau *et al.* (2008b) to establish parallel results for more standard predictors. Loosely, Biau *et al.* (2008b) showed that, under relatively reasonable hypotheses, mixtures of either kernel predictors or nearest-neighbors predictors are universally consistent for jointly stationary and ergodic processes that satisfy a fourth-moment constraint. Hence, there is a sense in which kernel and nearest-neighbors predictors that are relatively robust to outliers (compared with other predictors) are also predictively asymptotically optimal.

Note that the theorems in Biau *et al.* (2008b) use mixtures of predictors. This is no surprise: it is common for predictors based on mixtures to outperform the individual predictors used as components in the mixtures. This point will be explored in much greater detail in Chapter 11.

For the present, it is important to note that even when the X_i or the error terms are not IID, kernel-based point predictors and k-nearest-neighbors-based point predictors can be defined from their corresponding estimators and are expected not to perform too badly. For instance, Rakotomarolahy (2012) gave estimation results for k-nearest-neighbors techniques when the data arise from a strong mixing process. Politis and Vasiliev (2012) examined the sequential kernel-based estimation of a regression function with dependent observations. There is also extensive work in the use of kernel methods and k-nearest-neighbors methods in time series prediction; see Ralaivola and D'Aleché-Buc (2003) and Sasu (2012), respectively, for two examples of this literature.

8.4 Predictors from Nonparametric Bayes

A few simple cases of nonparametric Bayes predictors were presented in Sec. 4.1; these used the Dirichlet-process prior (DPP). The key result was stated in (4.11) and can be written for a set A as

$$(Y_{n+1} \in A | y^n, \alpha, H) = \frac{\alpha H(A)}{\alpha + n} + \frac{n}{\alpha + n} \hat{F}_n(A), \tag{8.71}$$

where Y_{n+1} is a future value, $y^n = (y_1, \ldots, y_n)^T$ are the data, H is the base measure, α is the concentration parameter, and $\hat{F}_n$ is the EDF. So, the predictive distribution has two obvious drawbacks: (i) it has an atomic component, and (ii) if $S = \mathbb{R} \setminus \{y_1, \ldots, y_n\}$ then $P(Y_{n+1} \in S | y^n) = \alpha H / (\alpha + n)$, so that when H is a finite measure n can be chosen large enough that $P(Y_{n+1} \in S | y^n) < \epsilon$ for any preassigned ϵ, i.e., the predictive distribution converges to zero. The predictive distribution also has the less obvious limitation that it assigns mass unity to discrete distributions since the DPP assigns mass unity to discrete distributions.

While all these features are undesirable, they are not really a surprise because this use of the DPP is a generalization of the Dirichlet distribution as a conjugate prior for a multinomial distribution. In fact, if Y assumes values in $\{1, \ldots, J\}$, the predictive distribution from a multinomial distribution with a Dirichlet prior is

$$P(Y_{n+1} = j | y^n, \alpha) = \frac{\alpha_j + n_j}{\sum_{j=1}^J \alpha_j + n} = \frac{\alpha_j}{\sum_{j=1}^J \alpha_j + n} + \frac{n}{\sum_{j=1}^J \alpha_j + n} \frac{n_j}{n}, \qquad (8.72)$$

in which n_j is an outcome of $N_j = \#(\{k | Y_i = j\})$, $\alpha = (\alpha_1, \ldots, \alpha_J)$ is the hyperparameter in the Dirichlet$(\alpha_1, \ldots, \alpha_J)$ density for $\theta = (\theta_1, \ldots, \theta_J)$, $\theta_j \geq 0$ and $\sum_{j=1}^J \theta_j = 1$ (see Sec. 1.2.2), and the multinomial is

$$P(n_1, \ldots, n_J | \theta) = \binom{n}{n_1, \ldots, n_J} \prod_{j=1}^J \theta_j^{n_j}.$$

Obviously, if H is a probability and each $\alpha_j = \alpha H(A_j)$, where the A_j are a disjoint and exhaustive set of n intervals with $y_j \in A_j$, then (8.71) and (8.72) are formally identical.

To avoid the drawbacks of the DPP, a generalization of it called a Polya tree prior can be defined; see Sec. 8.4.1. Because of its greater generality, Polya tree priors can be constructed to give probability unity to the class of continuous or absolutely continuous probabilities. On the other hand there are priors that are even more general than Polya trees – at the cost of tractability. The basic construction of a Polya tree prior is from a sequence of binary partitions of a set. For instance, $\mathbb{R}$ can be split into two sets, and each of them is split again into two sets. Each of these four sets is split again, and so on, leading to a sequence of partitions each of size 2^k for some k indicating the depth of the tree that describes the partitioning process. Then, to make the tree represent a probability, outcomes of a beta distribution are assigned as conditional probabilities to each level of the recursive partitioning. Loosely, a Polya tree prior assigns probabilities to approximations of a given probability on the basis of the partitions found by taking products of conditional probabilities as the partitions become more refined. The approximations converge to the given probability as the partition gets ever finer. The result is a prior over distributions that are measurable, at least in a limiting sense, with respect to the partitions in the sequence. The result is a prior on a nonparametric collection of DFs that gives a useful predictive distribution.

Turning to regression, rather than assigning a prior to a nonparametric collection of probabilities, it is important to assign a probability to a nonparametric collection of functions. Essentially, a real-valued function $f(x)$, $x \in \mathbb{R}$, is regarded as a vector $(\ldots, f(x_1), \ldots, f(x_2), \ldots)$ of length $\#(\mathbb{R})$ in which each element of the set $\{f(x) | x \in \mathbb{R}\}$ is taken as a parameter to be estimated. Hence, in principle, there are $\#(\mathbb{R})$ parameters. However, the continuity of f effectively reduces this to $\#(\mathbb{N})$. The most important class of priors for nonparametric function estimation are the Gaussian process priors (GPPs) for regression. These are treated in Sec. 8.4.2. Essentially, $\{f(x) | x \in \mathbb{R}\}$ is a stochastic process indexed by x and it is a Gaussian process if any finite subcollection of its variables has a multivariate normal distribution. (Frequentist and Bayes prediction in the normal case is included in Secs. 2.2.1 and 2.2.2.) The mean of the Gaussian process represents $f(x)$ and the variance matrix of the Gaussian process is generated by a covariance function $K(x, x')$ for any pair of variables indexed by $x, x' \in \mathbb{R}$; k is usually chosen by the user. The additive error gives an extra variance. Simply put, given data, future values of f at new inputs x can be predicted using a predictive distribution that is conditional on the data obtained, using the familiar rule for conditioning with normal variables.

Even though both the Polya tree priors and the GPPs require many choices that are hard to motivate (fixed partitions and covariance functions, respectively), they do provide useful nonparametric predictors.

8.4.1 Polya Tree Process Priors for Distribution Estimation

The construction of the Polya tree prior is more complex notationally than conceptually. To make it as simple as possible, consider assigning a Polya tree prior to all distributions on $S = [0, 1]$ equipped with a σ-algebra $\mathcal{B}$ – say, the Borel sets. Now, split S into left and right sets S_ℓ and S_r at some point in $(0, 1)$. Then split S_ℓ into $S_{\ell,\ell}$ and $S_{\ell,r}$ at some point in S_ℓ°, the interior of S_ℓ, and split S_r into $S_{r,\ell}$ and $S_{r,r}$ at some point in S_r°. Repeat on each of the four sets to generate eight sets, i.e., $S_{\ell,\ell,\ell}$ and $S_{\ell,\ell,r}$, where the split point is in $S_{\ell,\ell}^\circ$, and so on. Treating the initial interval $[0, 1]$ as stage zero, there are 2^k sets in the partition, say Π_k, of $[0, 1]$ at the kth stage and each set in Π_k can be indexed by a vector $v \in U_k = \{\ell, r\}^k$, the k-fold cartesian product of the set $\{\ell, r\}$. So $v = (v_1, \ldots, v_k)$, in which each $v_j \in \{\ell, r\}$, $j = 1, \ldots, k$. The first stages of this construction of nested partitions are illustrated in Fig. 8.1.

Now, given $v \in U_k$, $\exists v' \in U_{k-1}$ such that $v = (v', v_k)$. Since $v_k = \ell, r$, there are $(v', \ell), (v', r) \in U_k$ such that $S_{v',\ell} \cup S_{v',r} = S_v$ and the union is disjoint. So, $S_{v',\ell}$ and $S_{v',r}$ can be regarded as generating a (very simple) σ-field for S_v. Write the typical outcome of a Polya tree prior for the distributions on $[0, 1]$ as G. Then, the value $G(S_v)$ can be assigned by starting with the assignment

$$G(S_v | S_{v'}) = G(S_{v', v_k} | S_{v'}) = z_{v', v_k}$$

where z_{v', v_k} is an outcome from, say, a beta($\alpha_{v', \ell}, \alpha_{v', r}$). Clearly, if $v_k = \ell$ and $z_{v', \ell}$ is drawn then, for $v_k = r$, $G(S_{v', r} | S_{v'})$ is assigned the value $1 - z_{v', r}$. The reverse holds if $v_k = r$.

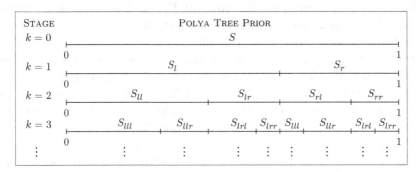

Figure 8.1 The first stages of the construction of a Polya tree prior on the set of continuous functions on $[0, 1]$. At stage k, the binary splits of the elements of the partition Π_{k-1} give the refinement Π_k.

Now, by k uses of the product rule for conditional probabilities,

$$
\begin{aligned}
G(S_{v_1,\ldots,v_k}) &= G(S_{v_1,\ldots,v_k}|S_{v_1,\ldots,v_{k-1}})G(S_{v_1,\ldots,v_{k-1}}) \\
&= z_v G(S_{v_1,\ldots,v_{k-1}}|S_{v_1,\ldots,v_{k-2}})G(S_{v_1,\ldots,v_{k-2}}) \\
&= z_{v_1,\ldots,v_k} \times \cdots \times z_{v_1,\ldots,v_j} \times \cdots \times z_{v_1} \times G([0,1]) \\
&= \prod_{j|v_j=\ell} z_{v_1,\ldots,v_j} \times \prod_{j|v_j=r}(1-z_{v_1,\ldots,v_j}),
\end{aligned}
$$

in which the left splits are gathered into the left-hand product term and the right splits are gathered into the right-hand product term.

In this formulation, a Polya tree prior has two parameters. The first is the collection of nested partitions, say, $\Pi = \{\Pi_1, \Pi_2, \ldots\}$, in which Π_k is the partition at level k, i.e., the partition of $[0,1]$ into 2^k subsets. The second is the sequence of parameters indexing the beta distributions, $\mathcal{A} = \{\alpha_\ell, \alpha_r, \alpha_{\ell,r}, \ldots\}$. Hence, a Polya tree prior is denoted $\mathrm{PT}(\Pi, \mathcal{A})$. Given the nested partition Π, the conditional probabilities for the sets from a randomly chosen $G \sim \mathrm{PT}(\Pi, \mathcal{A})$ are determined from a collection of independent beta random variables with parameters in $\mathcal{A}$. Polya trees reduce to DPPs; see Ferguson (1974) or Lavine (1994).

Obviously, conditional probabilities do not have to be randomly assigned from independent beta distributions, although this is the convention. Likewise, the base set being partitioned does not have to be $[0,1]$. Any well-defined reasonable set will do. And, of course, the split points defining the S_v can be chosen more carefully. For instance, if the goal is to center the outcomes G of a Polya tree prior $\mathrm{PT}(\Pi, \mathcal{A})$ on a base measure G_0, i.e., to have $E(G(A)) = G_0(A)$ for every set A, it is enough to choose the Π_k to be generated by the dyadic quantiles of G_0. For instance, if $S = [0,1]$ then $S_\ell = [0, G_0^{-1}(1/2)]$, $S_r = [G_0^{-1}(1/2), 1]$, $S_{\ell,\ell} = [0, G_0^{-1}(1/4)]$, $S_{\ell,r} = [G_0^{-1}(1/4), G_0^{-1}(1/2)]$, $S_{r,\ell} = [G_0^{-1}(1/2), G_0^{-1}(3/4)]$, $S_{r,r} = [G_0^{-1}(3/4), 1]$, and so on. An alternative way of arriving at the same end is to choose $\alpha_{v,\ell} \propto G_0(S_{v,\ell}|S_v)$ and $\alpha_{v,r} \propto G_0(S_{v,r}|S_v)$ for all v.

To get predictions for a new X_{n+1} given $X^n = x^n$, we start by finding the marginal for an outcome. To do this, note that because each finite union of sets $\mathcal{P}_k$ up to level k is an algebra, so it is enough to find $M(S_v)$ for any v. Since the expected value of a beta(a,b) random variable is $a/(a+b)$, for any v we have

$$
\begin{aligned}
M(X \in S_v) &= E\left(\prod_{j=2}^{k} G(S_{v_1,\ldots,v_j}|S_{v_1,\ldots,v_{j-1}})\right) \\
&= E\left(\prod_{j|v_j=\ell, j\leq k} z_{v_1,\ldots,v_{j-1}} \prod_{j|v_j=r, j\leq k}(1-z_{v_1,\ldots,v_{j-1}})\right) \\
&= \prod_{j=1}^{k} \frac{\alpha_{v_1,\ldots,v_j}}{\alpha_{v_1,\ldots,v_j\ell}+\alpha_{v_1,\ldots,v_jr}}.
\end{aligned} \tag{8.73}
$$

To derive the posterior for G, given $X_1, \ldots, X_n$ assumed IID G, consider the finite case, i.e., the unions of the partition Π_j for $j \leq k$ indexed by $U_k^* = \cup_{j\leq k}U_j$. Write the membership function $v_1(x) = \ell$ if $x \in S_\ell$ and $v_1(x) = r$ if $x \in S_\ell^c = S_r$ and, more generally, if v_k is the last element in a vector v of length k then $v_k(x) = \ell$ if $x \in S_{v_1,\dots,v_{k-1},\ell}$ and $v_k(x) = r$ if $x \in S_{v_1,\dots,v_{k-1},r}$. This is consistent because $S_{v_1,\dots,v_{k-1},r}$ is the complement of $S_{v_1,\dots,v_{k-1},\ell}$ in $S_{v_1,\dots,v_{k-1}}$.

Now the joint density of G and X, given G, is the product of the prior and the likelihood, which is, up to a constant,

$$
\left(\prod_{v \in U_k^*} G(S_{v,\ell}|S_v)^{\alpha_{v,\ell}-1}(1 - G(S_{v,\ell}|S_v))^{\alpha_{v,r}-1} \right)
$$

$$
\times \left(\prod_{j|v_j(x)=\ell} G(S_{v,\ell}|S_v) \prod_{j|v_j(x)=r} (1 - G(S_{v,\ell}|S_v)) \right)
$$

$$
= \prod_{v \in U_k^*} G(S_{v,\ell}|S_v)^{\alpha'_{v,\ell}}(1 - G(S_{v,\ell}|S_v))^{\alpha'_{v,r}},
$$

where

$$
\alpha'_v = \begin{cases} 1+\alpha_v & \text{if } x \in S_v, \\ \alpha_v & \text{otherwise.} \end{cases}
$$

From this last expression it follows that, under a Polya tree prior, the posterior distribution for G given $X_1, \dots, X_n$ is also a Polya tree prior, with parameters $\{\alpha_{v,X_1,\dots,X_n}\}$, where v ranges over U_k^*, satisfying

$$
\alpha_{v,X_1,\dots,X_n} = \alpha_v + \sum_{i=1}^n \chi_{S_v}(X_i).
$$

That is, the finite Polya tree priors are a conjugate class for distributions on $[0,1]$.

Analogous properties hold when countable unions of the Π_k and U_k are used. That is, the posterior from a Polya tree prior given X_1 with $\Pi = \cup_{k=1}^\infty \Pi_k$ and $U = \cup_{k=1}^\infty U_k$ and parameter $\mathcal{A}$ is the Polya tree distribution with parameter $\mathcal{A}_{X_1}$ in which, for all $v \in U$,

$$
v_{X_1} = v + \chi_v(X_1).
$$

In the case of n data points $X_1, \dots, X_n$, the posterior from a Polya tree prior with parameter $\mathcal{A}$ is the Polya tree distribution with parameter $\mathcal{A}_{X_1,\dots,X_n}$ in which, for all $v \in U$,

$$
v_{X_1,\dots,X_n} = v + \sum_{i=1}^n \chi_v(X_i). \tag{8.74}
$$

As in the case of DPPs, where it is possible to use mixtures of DPPs that give better performance than individual DPPs, it is again possible to use mixtures of Polya tree priors. In some respects they too offer an improvement over individual Polya trees. Both these topics are beyond our present scope, but see Lavine (1992, Sec. 3.3) for a predictive example.

Finally, the predictive distribution for X_{n+1} given $X_1, \dots, X_n$, where the X_i are IID G and G is drawn from a PT$(\Pi, \mathcal{A})$, follows from (8.73) and (8.74):

$$
M(X_{n+1} \in S_{v_1,\dots,v_k}|X_1,\dots,X_n) = \prod_{j=1}^k \frac{\alpha_{v_1,\dots,v_j} + \sum_{i=1}^n \chi_{S_{v_1,\dots,v_j}}(X_i)}{\alpha_{v_1,\dots,v_{j-1}\ell} + \alpha_{v_1,\dots,v_{j-1}r} + n_{v_1,\dots,v_{j-1}}};
$$

see Ghosh and Ramamoorthi (2003), where $n_{v_1,\ldots,v_{j-1}} = \#\{i \mid X_i \in S_{v_1,\ldots,v_{j-1}}, i = 1, \ldots, n\}$, $n_{v_0} = n$, and α_0, α_1 are α_ℓ and α_r respectively. So, provided that the partition elements shrink satisfactorily, one can get arbitrarily good approximations to $1 - \alpha$ posterior probability PIs for X_{n+1}. As in the finite-dimensional parameter case, the convergence of the posterior to the true distribution means that the predictive distribution will give valid PIs. Ghosh and Ramamoorthi (2003, Sec. 3.3) also provide conditions under which the posterior from a Polya tree prior converges.

An alternative PI follows from using the posterior mean, $\hat{G} = E(G|X_1, \ldots, X_n)$, which has the form

$$\hat{G}(S_v) = \prod_{j=1}^{k} \frac{\alpha_{v_1,\ldots,v_j} + \sum_{i=1}^{n} \chi_{S_{v_1,\ldots,v_j}}(X_i)}{\alpha_{v_1,\ldots,v_j,\ell} + \alpha_{v_1,\ldots,v_j,r} + n_{v_1,\ldots,v_{j-1}}};$$

see Lavine (1992, Sec. 2.1). Again, when the partition elements shrink satisfactorily, arbitrarily good approximations to $1 - \alpha$ PIs can be found from $\hat{G}$.

If point predictors are desired then the mean, median or other location derived from $M(\cdot|X^n)$ or $\hat{G}$ can be used. The variance of any of these locations under either distribution can be used to generate an interval for the point predictor.

The predictive distribution and the posterior mean are very similar. The difference is that the terms in the numerator of $\hat{G}$ have one fewer index than they do in the numerator of $M(\cdot|X_1, \ldots, X_n)$. Conceptually, the difference between using $\hat{G}$ and $M(\cdot|X_1, \ldots, X_n)$ is in what is being integrated with respect to the posterior. Specifically, $\hat{G}$ is the expected value of the probabilities in the support of the Polya tree prior with respect to the posterior. By contrast, $M(\cdot|X_1, \ldots, X_n)$ is the expected value of the conditional probability for X_{n+1}, given G, over G equipped with the posterior distribution. In either case, if the posterior converges to the true distribution of X as $n \to \infty$ then expectations with respect to it will converge to the value of the integrand at the true distribution. In this case, the predictive distribution converges to the correct probability and the posterior mean converges to the correct 'location', which is also the correct probability. Since these two predictors have the same limit as $n \to \infty$ (when a meaningful limit exists), the corresponding PIs are expected to be similar and well behaved.

It is possible to get analogs of variances for the convergence of these two sorts of PIs; however, the convergence of posteriors is usually expressed in terms of rates of concentration rather than numerical bounds and so they are less directly useful. (For a relatively accessible discussion of the details on rates of posterior convergence see Ghoshal *et al.* (2000) amongst others). To get around this, numerical methods can sometimes suffice.

8.4.2 Gaussian Process Priors for Regression

A Gaussian process (GP) is any stochastic process $\langle Z_m \rangle$ for $m \in I$, an index set, with the property that, for any finite set $m_1, \ldots, m_n$, the random variables $(Z_{m_1}, \ldots, Z_{m_n})^T$ have a multivariate normal distribution. A Gaussian process prior (GPP) is merely the use of a Gaussian process as a prior. In these cases, Z_m is written as $f(x)$, so that the function value $f(x)$ is regarded as a variable, with index x playing the role of m. Gaussian process priors constitute one way to put a prior on a space of functions that is too large to be usefully parametrized by a finite-dimensional real parameter. Essentially, they give a generalization of the normal regression model treated in Sec. 4.2. Again, assume a model of the form $Y = f(x) + \epsilon$ where x is unidimensional and $\epsilon \sim N(0, \sigma^2)$ and assume that n data points (x_i, y_i), $i = 1, \ldots, n$, are available. Thus, for each x, $f(x)$ is regarded as a parameter and is equipped with a normal prior. However, when f is continuous its value at x is close to its value at x' when $x - x'$ is small. This suggests that the priors on $f(x)$ and $f(x')$ should be dependent. The simplest way to reflect this is to assume that the bivariate normal assigned to $(f(x), f(x'))^T$ has a nontrivial covariance matrix.

Formally, suppose that our goal is to predict the next k values of Y at k new design points, i.e., to predict $Y_{\text{new}} = (Y_{n+1}(x_{n+1}), \ldots, Y_{n+k}(x_{n+k}))^T$. Then, providing the error terms are IID, it makes sense to write

$$
\begin{pmatrix} Y_n^1 \\ Y_{\text{new}} \end{pmatrix} \Bigg| \begin{pmatrix} x_n^1 \\ x_{n+k}^{n+1} \end{pmatrix} = \begin{pmatrix} f_n^1 \\ f_{\text{new}} \end{pmatrix} + \begin{pmatrix} \epsilon_n^1 \\ \epsilon_{\text{new}} \end{pmatrix}
$$

$$
\sim N\left(\begin{pmatrix} \mu_n^1 \\ \mu_{n+k}^{n+1} \end{pmatrix}, \begin{pmatrix} K(x_n^1, x_n^1) + \sigma^2 Id_n & K(x_n^1, x_{n+k}^{n+1}) \\ K(x_{n+k}^{n+1}, x_n^1) & K(x_{n+k}^{n+1}, x_{n+k}^{n+1}) + \sigma^2 Id_k \end{pmatrix} \right),
$$

(8.75)

where the notation Z_b^a means a vector $(Z_a, \ldots, Z_b)$, $\mu = \mu(x)$ is the mean function evaluated at $x_1, \ldots, x_{n+k}$, $f_n^1 = (f(x_1), \ldots, f(x_n))^T$, $f_{\text{new}} = (f(x_{n+1}), \ldots, f(x_{n+k}))^T$, $\epsilon_{\text{new}} = (\epsilon_{n+1}, \ldots, \epsilon_{n+k})^T$, and K is a covariance function of the form $K(x, x')$, which gives a matrix

$$
K(x_n^1, x_{n+k}^{n+1}) = \begin{pmatrix} K(x_1, x_{k+1}) & \cdots & K(x_1, x_{n+k}) \\ \vdots & \vdots & \vdots \\ K(x_n, x_{n+1}) & \cdots & K(x_n, x_{n+k}) \end{pmatrix}
$$

with $K(x_n^1, x_n^1)$, $K(x_{n+k}^{n+1}, x_n^1)$, and $K(x_{n+k}^{n+1}, x_{n+k}^{n+1})$ defined similarly and Id_k denoting the $k \times k$ identity matrix. It is assumed that, for any arguments, K gives a symmetric positive semidefinite matrix. This permits it to be a valid covariance matrix, i.e., for any x_n^1,

$$
\begin{pmatrix} f(x_1) \\ \vdots \\ f(x_n) \end{pmatrix} \sim N\left(\begin{pmatrix} \mu(x_1) \\ \vdots \\ \mu(x_n) \end{pmatrix}, \begin{pmatrix} K(x_1, x_1) & \cdots & , K(x_1, x_n) \\ \vdots & \vdots & \vdots \\ K(x_n, x_1) & \cdots & K(x_n, x_n) \end{pmatrix} \right)
$$

is a valid normal distribution. So, $K(x_i, x_j) = E\left((f(x_i) - \mu(x_i))(f(x_j) - \mu(x_j))\right)$ and $K(x_n^1, x_{n+k}^{n+1}) = K(x_{n+k}^{n+1}, x_n^1)^T$. Note that $\text{Cov}(f_n^1, f_n^1) = K(x_n^1, x_n^1)$ while $\text{Cov}(Y_n^1, Y_n^1) = K(x_n^1, x_n^1) + \sigma^2 Id_n$.

From the well-known formula for extracting the conditional distributions from a multivariate normal, it is easy to derive an expression for $Y_{\text{new}} | Y_n^1$ from (8.75):

$$
Y_{\text{new}} | Y_n^1 \sim N\Big(\mu_{n+k}^{n+1} + K(x_{n+k}^{n+1}, x_n^1)(K(x_n^1, x_n^1) + \sigma^2 Id_n)^{-1}(Y_n^1 - \mu_n^1),
$$

$$
(K(x_{n+k}^{n+1}, x_{n+k}^{n+1}) + \sigma^2 Id_k) - K(x_{n+k}^{n+1}, x_n^1)^T (K(x_n^1, x_n^1) + \sigma^2 Id_n) K(x_{n+k}^{n+1}, x_n^1) \Big).
$$

The point predictor for Y_{new} is the location of the normal distribution

$$
\hat{Y}_{\text{new}} = \mu_{n+k}^{n+1} + K(x_{n+k}^{n+1}, x_n^1)(K(x_n^1, x_n^1) + \sigma^2 Id_n)^{-1}(Y_n^1 - \mu_n^1), \tag{8.76}
$$

which is seen to be a linear function of y_n^1. Often, the mean function in the prior is taken to be zero so that this formula simplifies to $K(x_{n+k}^{n+1}, x_n^1)(K(x_n^1, x_n^1) + \sigma^2 I d_n)^{-1} Y_n^1$. The conditional variance – needed for generating PIs – is

$$
\begin{aligned}
\text{Var}(Y_{\text{new}}|X_{n+k}^{n+1}) = {} & (K(x_{n+k}^{n+1}, x_{n+k}^{n+1}) + \sigma^2 I d_k) \\
& - K(x_{n+k}^{n+1}, x_n^1)^T (K(x_n^1, x_n^1) + \sigma^2 I d_n) K(x_{n+k}^{n+1}, x_n^1).
\end{aligned} \tag{8.77}
$$

Effectively, in these two predictive formulae, the parameters f_{new} are integrated out using the GPP. In the (rare) 'noise free' setting, $\sigma = 0$ and the formulae simplify further. Note that if the goal is to predict only the function value, say $f(x_{n+1})$, rather than the measured outcome Y_{n+1}, then the σ in the lower right block of (8.75) would not appear, with corresponding changes in later formulae. The result would be the point predictor for $\hat{Y}(x_{\text{new}})$ using $E(\epsilon_{new}) = 0$.

More generally, if a loss function $L(Y_{n+1}(x_{n+1}), \hat{Y}_{n+1}(x_{n+1}))$ is available, it would be natural to choose the point predictor

$$
\arg\min_{\hat{Y}} \int L(y_{n+1}(x_{n+1}), \hat{Y}_{n+1}(x_{n+1})) p(y|x_{n+1}, X_1^n, y_1^n) \mathrm{d}y.
$$

When L is symmetric the result is as above; when L is not symmetric other predictors will be optimal and their variances would be found using $p(y|x_{n+1}, x_1^n, y_1^n)$.

What remains is to choose the covariance function K and estimate σ. Often, the form of the covariance function is chosen up to a small number of hyperparameters. One typical choice is the squared exponential

$$
K(x_1^n, z_1^n) = \sigma_K^2 \exp\left(-(x_1^n - z_1^n)^T M(x_1^n - z_1^n)/2\right) + \sigma \chi_{\{x_1^n = z_1^n\}},
$$

in which σ_K scales the exponential of the square, M is often taken as $\mathrm{diag}(\ell_1, \ldots, \ell_n)$ or $\ell \mathrm{diag}(1, \ldots, 1)$ and controls how far apart components of x must be for the corresponding function values to be uncorrelated, and the indicator function χ adds the estimate σ when $x_1^n = z_1^n$ and is zero otherwise. In this case, the hyperparameters are σ_K and the parameters in M, e.g., ℓ. Technically, σ is a parameter, not a hyperparameter, but a value for it still must be found. There are many other common choices; see Rasmussen and Williams (2006, Sec. 4.2). The choice of covariance function has a large effect on the performance of the predictive techniques; unfortunately, there seems to be little work on how to choose a class of covariance functions. One can regard the choice of covariance function as an analog to model selection. It is known that model selection requires relatively large sample sizes to be effective. Consequently, determining a covariance function by trial and error may end up overfitting the data, leading to high predictive errors.

One way to assign values to the parameters is to maximize the marginal likelihood over the parameters, that is, to write

$$
p(y_1^n|x_1^n) = \int p(y_1^n|f_1^n, x_1^n) p(f_1^n|x_1^n) \mathrm{d}f_1^n, \tag{8.78}
$$

where the integrand in the n-fold integral is the likelihood for y_1^n with respect to the prior for f_1^n. Since a GPP is assumed, $Y_1^n|f_1^n, x_1^n \sim N(f_1^n, \sigma^2 I d_n)$ and $f_1^n|x_1^n \sim N(0, K(x_1^n, x_1^n))$, in which the mean function has been set to zero, $\mu \equiv 0$. Then, suppressing the parameters in K and recognizing that the integrand is a product of normals, an explicit formula for $\log p(y_1^n | x_1^n)$ can be given, namely

$$-\frac{1}{2}(K(x_n^1, x_n^1) + \sigma^2 I d_n)^{-1} - \frac{1}{2}\log\det(K(x_n^1, x_n^1) + \sigma^2 I d_n) - \frac{n}{2}\log(2\pi); \qquad (8.79)$$

see Rasmussen and Williams (2006, p. 19 and Appendix A2). An easier derivation follows from noting that $Y_n^1 \sim N(0, K(x_n^1, x_n^1) + \sigma^2 I d_n)$. Now, in principle, the derivatives of $\log p(y_1^n | x_1^n)$ with respect to any parameter can be found such that (8.79) is maximized. This is usually done numerically. As a generality, the parameter estimates are merely substituted into the predictive distribution to generate point and interval predictions, ignoring the variability of the estimates. It is oversimplistic to ignore the variability in the estimates, but to a first approximation it is a reasonable way to specify a predictor.

In principle, a discrete prior could be put on a collection of covariance functions, with hyperpriors on their hyperparameters, and a fully Bayes analysis could be carried out to find the predictive density and point predictors. This would bypass (8.79) and be more complicated because the priors and hyperpriors would appear in the integral (8.78). This does not seem to have been done, but taking the uncertainty in the covariance function into account may give a better predictive performance.

An alternative way to assign values to parameters is to derive the leave-one-out predictive log probability

$$\log p(y_i | x_1^n, y_{-i}, \theta) = -\frac{1}{2}\log\sigma_i^2 - \frac{(y_i - \mu_i)^2}{2\sigma_i^2} - \frac{1}{2}\log(2\pi),$$

in which θ is the vector of all the parameters in K and σ, y_{-i} is the vector of length $n-1$ formed by dropping y_i from y_1^n, and μ_i and σ_i^2 are the mean and variance of y_{-i} derived from the normal in the same way as (8.76) and (8.77). Now the pseudo-likelihood is

$$L_{\mathrm{LOO}}(x_1^n, y_1^n, \theta) = \sum_{i=1}^n \log p(y_i | x_1^n, y_{-i}, \theta).$$

If explicit expressions for μ_i and σ_i are used then $L_{\mathrm{LOO}}(x_1^n, y_1^n, \theta)$ can be maximized numerically over the parameters and the estimates used in (8.76) and (8.77) to generate point and interval predictors; see Rasmussen and Williams (2006) for some details on implementing this procedure as a way to find predictors. Maximizing a pseudo-likelihood is best seen as a sort of empirical Bayes analysis or as providing an approximation to a fully Bayes analysis. That is, the predictors that arise from this are approximations to the predictive distribution.

8.5 Comparing Nonparametric Predictors

Having seen four different classes of predictors (basis expansion, kernel, k-nearest-neighbors, and Bayes) it is important to assess their relative performances. It is possible to evaluate predictors on the basis of density estimation, using scoring functions as in Sec. 5.5. However, doing so would mean that these predictors were evaluated on a probability scale while the regression-based predictors were evaluated on the scale of the response. Hence, as in earlier chapters, performance will be assessed by the cumulative predictive error (CPE) from comparing predictions with outcomes, because (i) it is easier to convert predictors based on density estimation to point predictors than to convert regression predictors to probability predictors, and (ii) point predictors can be directly related to the response. Thus, in this section, the nonparametric techniques that give densities or distributions as their primary output are converted to point predictors by taking their means. The CPEs thus computed use the absolute error, a 'flat' scale for the response.

Here, results for 14 methods will be discussed. These are: three basis expansion methods (the least asymmetric wavelets of Chebyshev, Fourier, and Daubechies), kernel methods (kernel-density estimation choosing h by AMISE and by CV and the Nadaraya–Watson (NW) method with h chosen by CV), k-nearest-neighbors methods (k-nearest-neighbors density estimation using CV and GCV to choose k and k-nearest-neighbors regressions using CV to choose k), Bayes methods (Polya trees for density estimation and GPPs for regression), and three methods that involve combining the predictive power of these first 11 methods (two model-average predictors and one model-reselection predictor). When two variants of a method gave similar results, only one is presented.

8.5.1 Description of the Data, Methods, and Results

The data analyzed here consist of measurements of soil water content collected on a 12 km × 12 km piece of land in east-central Nebraska, where the land use is predominantly agricultural. The data collection, processing, and description, are given in Franz *et al.* (2015) where a spatio-temporal analysis of the data is presented. The actual data set has about 19 000 data points, two response variables, and ten explanatory variables. Since both response variables are assessments of water content, it will be enough here to choose one, namely, the topsoil moisture, interpolated from the original measurements by Kriging and denoted SWC. There are six explanatory variables associated with SWC, namely the location of the topsoil, indicated by (x, y, z) where (x, y) indicates the location in the 12 km × 12 km grid (in meters) and z indicates the altitude of the land (in feet). The other three explanatory variables are a wetness index, WI (a standard index computed from z), the electrical resistance of the topsoil, ERT, and the electrical resistance of the subsoil, ERS. Electrical resistance is related to soil clay percentage and water holding capacity. For convenience, only 512 randomly selected data points were used in this example. Figure 8.2 provides some visualizations of the data.

Since the data are complex, and the methods of this chapter do not scale up well even to three or four variables, let alone six, regression predictors were derived from an additive structure. That is, write

$$\widehat{\text{SWC}}(x, y, z, \text{WI}, \text{ERT}, \text{ERS})$$
$$= \frac{1}{6} \left(\hat{f}(x) + \hat{f}(y) + \hat{f}(z) + \hat{f}(\text{WI}) + \hat{f}(\text{ERS}) + \hat{f}(\text{ERT}) \right) \qquad (8.80)$$

to mean that the predicted value of SWC at $(x, y, z, \text{WI}, \text{ERS}, \text{ERT})$ is the average of the predictors for SWC formed by using a nonparametric predictor $\hat{f}$ for each of the six variables individually. This may be regarded as a model-average predictor with fixed weights of $1/6$, where all the models are univariate and estimated nonparametrically. Here, however, it is regarded as a simplification, that of finding a single predictor that uses all six explanatory variables, essentially by dropping all interactions among explanatory variables. The weight $1/6$ is used because each of the six explanatory variables gives a greedy approximation to

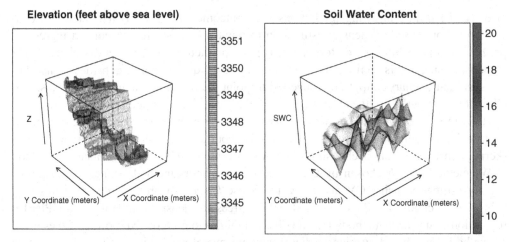

Figure 8.2 The panel on the left shows the terrain of the land, i.e., the surface $z(x, y)$. The panel on the right shows the SWC as a function of location on the 12 km × 12 km grid under study. Unsurprisingly, there is more water where the land is lower and less water where it is higher. Similar panels could be made for any triple of the seven measured variables.

SWC from the data sets (swc_i, x_i), (swc_i, y_i), etc., for $i = 1, \ldots, k$, for the kth time step, $k \leq n$.

By contrast, the three predictors based on density estimation use only the response variable, ignoring the explanatory variables.

To make all the choices for $\hat{f}$ comparable, all methods were given a burn-in of 128 data points and predictions for the next four time steps were generated. This was done because implementation of the wavelet method required groups of data to be powers of 2. That is, even for one-time-step-ahead prediction one must use a sample size that is a power of 2, e.g., 128, and predict ahead by that same number of steps – throwing away all but the first of these predictions. Hence, for convenience, predictions were taken four at a time, rather than one at a time, and compared with the corresponding four outcomes.

The average CPE for each predictor was generated over 50 runs, independently permuting the data for each run. Consequently, from these 50 runs a standard deviation (SD) at each predictive step could be found. The error bands around each CPE curve over time represent ± 2SD. The further apart these lines are, the more variable the predictor is. This is shown in Fig. 8.3.

The top row shows the CPE curves from basis expansions. For the Chebyshev basis, each of the six estimators $\hat{f}$ for the six explanatory variables were found using the first 11 basis elements (a constant term with ten nontrivial polynomials). Likewise, with the Fourier basis, the first 11 elements were used for each explanatory variable (a constant term and the first five sine–cosine pairs). For the wavelets basis, the Daubechies least asymmetric family was used and the procedure for choosing which basis elements to include was more complex. Only the first seven levels of a multiresolution analysis were used, and the procedure took the top 10% of wavelet basis elements according to their absolute correlation with the response SWC. Then, these were fed into a linear model and small coefficients were removed by individual ANOVA table tests at level 0.05. (This is different from the usual wavelet regression,

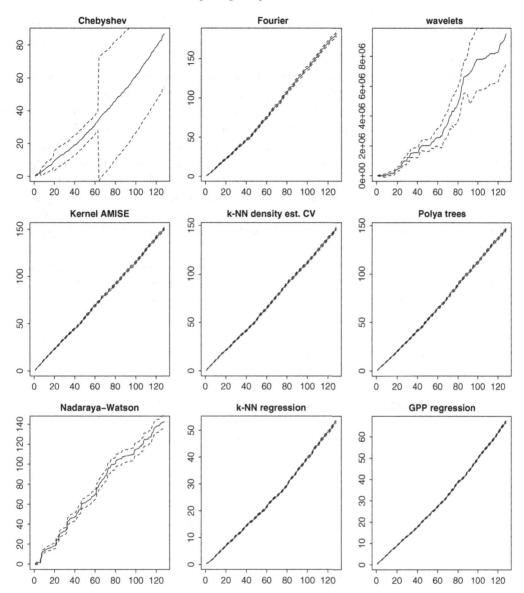

Figure 8.3 Top row: CPEs vs. predictive step for predictors from basis expansions using Chebyshev polynomials, Fourier series, and Daubechies' least asymmetric wavelets. Middle row: CPEs from predictors based on the means of density estimation using kernels, k-nearest neighbors, and Polya trees. Bottom row: CPEs from predictors based on NW, k-nearest-neighbors, and GPPs.

but the method is more comparable with the basis expansions used with the Chebyshev and Fourier bases.) It is seen that the wavelets do very poorly (high CPE), possibly because the structures they encapsulate most readily are very different from the terrain of the land as seen in Fig. 8.2. The Chebyshev expansion gives better results than the Fourier expansion, at the cost of somewhat bizarre behavior of the ± 2SD lines. The sudden increase in SD with Chebyshev around time step 62 may reflect the fact that Chebyshev polynomials can fit well up to a point determined by the number of data points (and basis elements) but, beyond that sample size, the fixed number of basis elements can rapidly become incapable of expressing the behavior of the response reliably.

The middle row of Fig. 8.3 shows the performance of three density estimation methods for forming predictors, i.e., a density estimator is formed and its mean is taken as a prediction for the next time step. (The kernel density method using unbiased CV to choose h gave the same results as using AMISE, and the k-nearest-neighbors method using GCV gave the same results as using CV. Hence neither are shown.) For the kernel and k-nearest-neighbors methods this is obvious; for the Polya tree density estimator, the mean of the predictive distribution is used. In the kernel density estimator, the Epanechnikov kernel was used. It is seen that all three methods are roughly equally good, but clearly worse than Chebyshev polyomials, the best of the basis expansion methods. The Polya tree prior was actually a mixture of Polya trees using the defaults in DPpackage; see Jara *et al.* (2011, p. 223–4), for the details.

The bottom row of Fig. 8.3 shows the performance of three regression methods: NW with the Epnechnikov kernel and h chosen by CV, k-nearest-neighbors with k chosen by CV, and GPPs with the GPP chosen to have mean zero, covariances given by the radial basis function kernel, and σ estimated internally by an empirical Bayes procedure. Clearly, the k-nearest-neighbors predictor gives the smallest average CPE and the NW predictor has the greatest variability. (The Priestley–Chao predictor might give smaller variability with little increase in its CPE because it assumes fixed design points.)

Overall, the best of the 11 methods for this data set discussed or presented so far in this section are k-nearest-neighbors regression followed by GPPs. Both of these used the explanatory variables in (8.80). The third best is a basis expansion using Chebyshev. All the density-estimation-based predictors perform comparably and worse than the best three of the regression-based predictors. The implication is that the explanatory variables are useful for explaining the response in spite of their use in (8.80). Further computational work could assign a relative importance of each explanatory variable to the response, possibly eliminating some that were not very important (assuming it had been decided that the predictor from the additive model was adequate). A related point is that all the CPEs tend to become linear in the time step with nonzero slope. This means that there is some lower bound on the average error per time step. Whether this represents an irreducible σ^2 (as in the $\mathcal{M}$-complete case) or some mixture of noise and systematically poor prediction (as in the $\mathcal{M}$-open case) is hard to determine, but see Sec. 8.5.2 for some discussion on this point.

To foreshadow some of the predictive techniques to come, consider three more predictors formed by combining the predictors already studied. There are two obvious ways to combine them: take a function of them, e.g., a mean, or reselect the predictor at each time step. In the latter case, the best predictor can change from time step to time step, when one is cycling through a subset of the available predictors.

In the upper left panel of Fig. 8.4, the predictors from the 11 methods discussed in the context of Fig. 8.3 are averaged, i.e., summed and divided by 11, to give an average predictor. It is seen that the performance is poor. However, this is largely due to the inclusion of the wavelet basis expansion. The upper right panel shows the average of the predictors omitting the wavelet basis expansion. The improvement is dramatic: the CPE is still high but much less than for all but three of the methods whose CPE curves are shown in Fig. 8.3.

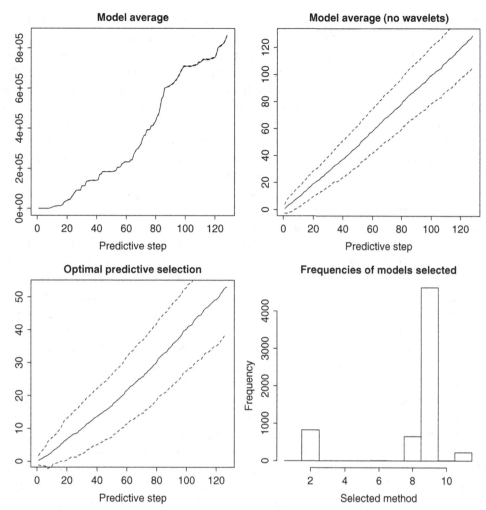

Figure 8.4 Upper row: CPEs from simple model-average predictors with and without the wavelet basis expansion predictor. Lower row: CPEs from predictors based on optimal predictor selection and a histogram showing the models selected; the horizontal axis represents the 11 methods.

The approach of model reselection, initially discussed in list item 3 in Sec. 2.4, is simple: at each time step for prediction, choose a window of data just before the time step, find the CPEs of the various methods and use the method with the smallest CPE to make the next prediction.

More precisely, it is important to separate the process of generating predictions from the process for evaluating predictions; this is the Prequential Principle. So, to form predictions, start with a burn-in of 128 data points. Then form predictions for the next four data points, 129–132. Then, form predictions for the following four data points, from 133 to 136, using all the data before and including data point 132. Continuing in this fashion generates predictions, four at a time, using all previous data points.

Next, the predictors must be evaluated. It is impossible to select a model on the basis of its predictive performance for time step 129 since the first 128 data points were used to form the predictors and there is no data point between 128 and 129 on which to predict. So, start with data point 130 and choose the predictor with the smallest predictive error on the 129th data point. For data point 131, choose the predictor with the smallest predictive error on the 129th and 130th data points. Continue in this way until the predictor is being evaluated on 64 data points. Then, when 65 data points are available, drop the earliest and use only the last 64, thus establishing a window of 64 data points on which to evaluate and hence choose predictors for the next time step. That is, the window size is the collection of previous data points back to the lower of the number of predictive steps and 64.

The lower left panel in Fig. 8.4 shows the CPE curve that results from this procedure of regular reselection of predictors. The CPE line is comparable with the best of the CPE curves for the individual methods shown in Fig. 8.3 but the SDs are generally larger for each time step. The increased variability is the direct result of rechoosing the predictor using a moving window of evaluation data points and suggests that nontrivial reselection is occurring, i.e., the predictor is not settling down to any one of the 11 that were formed. The lower right panel of Fig. 8.4 confirms this by showing a histogram of how often each method was chosen. The order of the 11 methods on the horizontal axis is wavelets, Chebychev, Fourier, kernel density estimation with AMISE and with unbiased CV, k-nearest-neighbors density estimation with GCV and with CV, NW with CV, k-nearest-neighbors regression with CV, Polya trees, and GPPs. It can be seen that four methods are chosen with nontrivial frequency – Chebyshev, NW with CV, k-nearest-neighbors regression, and GPPs. As a generality these were the better methods, as seen in Fig. 8.3. A sequence plot of which methods were chosen over time for a given run confirms that the selection of predictors does not appear to settle down; rather, the choice of best predictor at each time step cycles through the four methods – with the best method (k-nearest-neighbors regression) being chosen most often.

The implication of this is that, for complex problems, reselecting predictors may give benefits that are not achieved by fixing a specific predictor class. In particular, a reselection method or a model-averaging method (where the predictor weights are adaptive) may yield a predictor that has overall better performance than using any one fixed predictor, especially when good performance over a range of data generators is required, i.e., when the predictor must be robust to changes in the data generator. This topic will be pursued in Chapter 11.

8.5.2 $\mathcal{M}$-Complete or $\mathcal{M}$-Open?

The strength of nonparametric methods is that they do not assume any particular form for a true model. Thus, it can be argued they are useful in $\mathcal{M}$-complete problems as well as $\mathcal{M}$-closed problems. It is hard to imagine a way to treat the data in Sec. 8.5.1 as $\mathcal{M}$-closed; so, it is likely to be $\mathcal{M}$-complete or $\mathcal{M}$-open. In any event, Fig. 8.3 shows that the range of performance of nonparametric methods is wide. The best method for this data was seen to be k-nearest-neighbors – in spite of the fact that, for any finite sample size, k-nearest-neighbors gives a discontinuous 'model'. One reason for the unexpected relative performance of the methods in Fig. 8.3 is that the class of predictors was too small. The predictors were assumed to have an additive structure over the individual explanatory variables (see (8.80)), whereas almost certainly there are interactions among the explanatory variables.

In the exposition that presented the nonparametric techniques from Secs. 8.1–8.4, the focus was on the AMISE for choosing tuning parameters. However, this was for philosophical purposes; the AMISE is directly model dependent whereas CV, for instance, is not. In fact, CV is an internal (to the data set) predictive error that in the $\mathcal{M}$-closed or $\mathcal{M}$-complete cases is consistent for the expected predictive error. In the example here the various ways to choose values for k, h, or other parameters made little difference, although this will not be true in general. Indeed, if a problem is believed to be $\mathcal{M}$-open, the AMISE is not meaningful and so is unlikely to give as good results as the CV.

Up until Fig. 8.4, the strategy of this exposition had been to convert an estimator to a predictor. However, the model-average predictor used here, even though naive, does not convert to a model anyone would think of using – i.e., to greedily get 10 or 11 function estimates and then take their average – and the predictor reselection technique does not convert to a model at all. Even though the model-average predictor did not perform very well, the predictor-reselection method gave results as good, on average, as any method in spite of cycling through various of the methods. This means that the model-reselection method is likely to be robust to changes in the underlying data generator – it can predict well as long as at least one method that it selects predicts well. The cost of this robustness is the higher variability seen in the error bands of Fig. 8.4. The inference is that there is some tradeoff between robustness and variance – more of one can mean less of the other. It will be seen in later chapters that there are many predictors, like the model-reselection predictor used here, that do not correspond to a model. Mathematically, this is no surprise because every model leads to a predictor but not every predictor corresponds to a model.

Is the data set in Sec. 8.5.1 $\mathcal{M}$-complete or $\mathcal{M}$-open? This question cannot be answered unequivocally except by gathering so much data that every large class of models can be predictively discredited or validated. Even this answer must be qualified by the assumption that the data set contains all the information relevant to the description of the response. As a generality this is not true, except approximately, so the real question is whether the data at hand, if available in large enough quantities, would identify a model whose bias was as small as possible given the limitations of the data. Note that the use of bias assumes a model – even if the model is later deemed inadequate.

Since this is not a useful answer, it's worth focusing on the boundary between $\mathcal{M}$-complete and $\mathcal{M}$-open problems. It is reasonable to suppose that as an $\mathcal{M}$-complete problem moves closer to the boundary with $\mathcal{M}$-open problems it will become more and more difficult to predict: the data generator will become more complex and have fewer regularities, i.e., there will be less and less about the data stream that is stable enough to permit modeling. This is phrased generally, so it must be remembered that if a strongly $\mathcal{M}$-complete or $\mathcal{M}$-open problem is sufficiently restricted then it can become much simpler. An analogy is that even if a real-valued function on the closure of an open interval cannot be approximated uniformly by a given type of expansion, it may be very well approximated on regions of the interval. That is, the absence of a model on the whole range does not imply the absence of a model on subranges. Another analogy is that a function may have arbitrarily many small oscillations around a smooth curve on some interval, and hence be similar to an $\mathcal{M}$-complete but nearly $\mathcal{M}$-open data generator, and yet still be well approximated in supremum norm by the curve.

All this means is that, as a problem moves closer to the boundary between $\mathcal{M}$-complete and $\mathcal{M}$-open, it becomes more difficult, at least in the sense that it becomes harder and harder to give a predictively useful and interpretable solution. Indeed, it is clear that a sparse solution, one with relatively few terms, is likely to perform poorly in complex situations that tend to combine many small-to-moderate influences rather than being well summarized by a few dominant influences. This does not preclude the possibility that a model close to, but away from, the $\mathcal{M}$-complete – $\mathcal{M}$-open boundary can be unambiguously identified asymptotically; however, for finite samples, treating it as a $\mathcal{M}$-open problem may give better predictions pre-asymptotically, owing to its complexity.

Even though choosing whether a problem should be treated as $\mathcal{M}$-complete or $\mathcal{M}$-open may be impossible in practice, some guidelines can be given. If more variables are needed to characterize the state of the data generator than can be measured, there will always be nontrivial bias.[2] So, the problem is likely to be $\mathcal{M}$-open. If the data generator is complex in the sense that it is the result of innumerable small influences over time or space, the chances are that it is $\mathcal{M}$-open. Even if all relevant variables are measured, so that in principle a response can be modeled with zero bias, the response may still be $\mathcal{M}$-open if the remaining variability in the data makes the response essentially unpredictable. In this case the noise, whatever form it takes, cannot be modeled, does not wash out in any sense because it has some fleeting regularities, and yet dominates the phenomenon. It is on this basis that it has been implicitly assumed that the data in Sec. 8.5.1 is $\mathcal{M}$-open.

Another example would be human choice. How will someone react to a sale at a store next weekend? Exactly how people will react even to such mundane situations is a combination of many influences, including their formative experiences and experiences between the time one starts observing (now) and the time a choice is made (next weekend). So, outside highly restricted settings, predicting human behavior is likely to be $\mathcal{M}$-open. The same holds for complex health data and economic or financial data. On the other hand, if it is plausible that a collection of identifiable influences will characterize the data generator then the problem is likely to be $\mathcal{M}$-complete or even $\mathcal{M}$-closed. Many agronomy problems fall into this category. Even though in practice predicting crop yield, for instance, is an $\mathcal{M}$-open problem, in an experiment the range of influences will be restricted and controlled so that $\mathcal{M}$-complete techniques, e.g., sufficiently complex models, or $\mathcal{M}$-closed techniques, e.g., ANOVA, can be used to good effect. A fortiori, even if a limitless supply of data were available, a problem would still be $\mathcal{M}$-open if the data would not permit capturing enough of the regularities in the data generator to give useful predictions. In essence, this would mean the problem is just not accessible to modeling, i.e., is $\mathcal{M}$-open. See Clarke *et al.* (2013, 2014) for further discussion relating the choice of technique to the problem class.

8.6 Endnotes

There are many addenda that would be appropriate for this chapter. Here, there is only room for brief discussions of three points. The first, smoothing splines, like kernel and k-nearest-neighbors regression, is linear; once the form of the predictor has been defined prediction

---

[2] If the bias does not formally exist because there is no true model, one can replace the concept of bias by that of predictions that are systematically incorrect in some identifiable directions, possibly because they are changing over time.

is relatively easy. The second is that nearest neighbor methods are frequently used for prediction in classification. Indeed, k-nearest-neighbors may be more common in classification than any other application. Finally, it is worthwhile to pick up on a point made in Sec. 1.2.2, namely, that hypothesis tests can be converted into predictors. It is well known that collections of tests can be turned into collections of CIs and that a CI can lead to a point estimator that can be used in a parametric family, say, to choose a distribution from which to make a predictor. However, it is less obvious that the results of tests themselves can readily lead to predictors, regardless of the use of CIs or point estimates.

8.6.1 Smoothing Splines

The basic formulation of smoothing splines originated in Reinsch (1967); see also de Boor (2001). Given data (x_i, y_i) for $i = 1, \ldots, n$, assume that $x_1 < \cdots < x_n$ and find the function

$$f = \arg\min_{g | \mathcal{G}} \left(\int_a^b (g''(x))^2 dx \right),$$

where

$$\mathcal{G} = \left\{ g \left| \sum_{i=1}^n (y_i - g(x_i))^2 \leq S \quad \text{and} \quad g \in C^2[a, b] \right. \right\},$$

$S \geq 0$ is a constant, and $[a, b] \subset \mathbb{R}$. Reinsch (1967) first converted this problem to finding an f that achieves

$$\min_{g \in \mathcal{G}} \left(\sum_{i=1}^n (y_i - g(x_i))^2 + \lambda \int_a^b (g''(x))^2 dx \right),$$

by using Lagrange multipliers and the Euler–Lagrange equation. From this form, Reinsch (1967) showed that the solution is a natural cubic spline with knots at the x_i. The cubic splines of order 3 with knots at the x_i have n dimensions. So, an arbitrary element of the linear space of splines with knots at the x_i can be written as

$$g(x) = \beta_1 B_1(x) + \cdots + \beta_n B_n(x),$$

where the B_j constitute a natural basis. Let B be the $n \times n$ matrix with (i, j)th element $B_j(x_i)$ and let Ω be the $n \times n$ matrix with elements

$$\Omega_{ij} = \int_a^b B_i''(x) B_j''(x) dx.$$

Writing $\beta = (\beta_1, \ldots, \beta_n)^T$ and $y = y^n = (y_1, \ldots, y_n)^T$, the problem can be stated as finding the minimum (over β) of

$$(y - B\beta)^T (y - B\beta) + \lambda \beta^T \Omega \beta.$$

Setting the derivatives of this function with respect to the β_j to zero, the solution is

$$\hat{\beta}_\lambda = (B^T B + \lambda \Omega)^{-1} B^T y,$$

provided that the inverse exists. Consequently, the cubic spline predictor is

$$\hat{Y}(x) = \hat{\beta}_1 B_1(x) + \cdots + \hat{\beta}_n B_n(x),$$

where λ is implicit in $\hat{Y}$ and in the $\hat{\beta}_j$. There is a variety of ways to choose λ so that the predictor can be completely specified.

It is obvious that $\hat{Y}$ is linear and the point predictor at x_{new} is $\hat{Y}(x_{\text{new}})$. To obtain $1 - \alpha$ PIs the procedure at the end of Sec. 8.2.2 can be followed. Let $V(x) = (B_1(x), \ldots, B_n(x))$ and $\ell(x) = V(x)(B^T B + \hat{\lambda}\Omega)B^T$. Then

$$\hat{Y}(x_{\text{new}}) = \ell(x_{\text{new}})y$$

and $1 - \alpha$ PIs are of the form

$$\hat{Y}(x_{\text{new}}) \pm z_{1-\alpha/2}\hat{\sigma}\sqrt{1 + \|\ell(x_{\text{new}})\|^2},$$

where $\hat{\sigma}^2 = (1/n)\sum_{i=1}^{n}(y_i - \hat{Y}(x_i))^2$ and the scaling by $1/n$ is an approximation to dividing by the degrees of freedom.

8.6.2 Nearest Neighbor Classification

Although this chapter has presented k-nearest neighbors as a regression technique, it is probably more common as a classification technique. Assume data (x_i, y_i) for $i = 1, \ldots, n$ are available, where the x_i are d-dimensional vectors and the y_i are class labels drawn from $\{1, \ldots, C\}$. Then, the k-nearest-neighbors predictor $\hat{Y}$ for the class $Y(x_{\text{new}})$ at x_{new} is found by two steps: (i) find the k values among the x_i that are closest to x_{new} and note the values of the corresponding y_i, and (ii) assign $\hat{Y}(x_{\text{new}})$ the class label that has the highest frequency among the k values of x_i in step 1. One way to do this is formally expressed in the definition (4.53).

This means that three choices have to be made to formalize the predictor; the distance used on the x's, the number k, and how to combine the k values of y. The distance is chosen for convenience or is suggested by the physical nature of the problem. The most common choices are the Euclidean and L^1 distances.

The choice of k is often done by K-fold cross-validation. This is different from the AMISE but may be more appropriate for classification since the number of reasonable candidates for k is often small and K-fold cross-validation is a proxy for predictive error. Higher values of k tend to give higher variance; lower values of k tend to give higher bias. In this case, k is essentially taken as fixed. Hall *et al.* (2008) used a risk-based criterion for choosing k and this depends on the distribution by which samples from the different classes are drawn.

As a generality, there seems to be no consensus on how to choose k in the absence of extra knowledge about the data, and consequently k is often chosen empirically.

Combining the k values of y_i from the nearest neighbors of x_{new} can be done in a variety of ways. Majority vote, i.e., the most frequently occurring class among the k nearest neighbors, is typical. In some cases better results are obtained by using weights on the nearest neighbors that decrease as the distance between x_{new} and a nearest neighbor increases.

8.6.3 Test-Based Prediction

Consider testing $\mathcal{H}_0: f \in \mathcal{P}$ versus $\mathcal{H}_1: f \in \mathcal{P}^c$. Since the posterior odds ratio determines the Bayes action under generalized zero–one loss under a prior W, the posterior probability

$W(\mathcal{P}|\mathcal{D})$, where $\mathcal{D}$ is the data, is the main quantity to find in order to decide between $\mathcal{H}_0$ and $\mathcal{H}_1$. Bayes testing is symmetric in its treatment of the two hypotheses, so, without loss of generality, suppose that $\mathcal{H}_1$ is rejected. Then it is natural to base predictions on the new predictive distribution

$$m(x_{n+1}|\mathcal{D}) = \frac{1}{W(\mathcal{P}|\mathcal{D})} \int_{\mathcal{P}} f(x_{n+1}) dW(f|\mathcal{D}).$$

That is, from the decision that $\mathcal{P}^c$ is incorrect, a new predictor is justified.

In a similar manner, one could test $\mathcal{H}_0$: $f \in$ HPD versus $\mathcal{H}_1$: $f \in$ HPD^c, where HPD is a highest posterior density set, inevitably rejecting $\mathcal{H}_0$. This is permissible under the Bayes paradigm, on the simple grounds that the data once received are taken as constant and the conditional probability for the parameter (finite-dimensional or not) given the data is the mathematical quantity from which inferences are made. Using an appropriate HPD as a domain in place of $\mathcal{P}$ therefore converts a credible interval – or the result of a hypothesis test – into a prediction. Careful choice of the null would lead to regions with high posterior probability, which would be well summarized by the mixture distribution over them. Because hypothesis testing and estimation are based on the same quantity they end up giving predictors that are variations of each other. As one Bayesian put it, 'the posterior says it all'.

From a frequentist perspective the situation is more complex. Again, consider testing $\mathcal{H}_0$: $f \in \mathcal{P}$ versus $\mathcal{H}_1$: $f \in \mathcal{P}^c$, but now assume that $\mathcal{P}$ is a set of distributions for X. Denote the indicator function for a level-α rejection region S, assuming it exists, by

$$\delta(x^n) = \begin{cases} 1 & x \in S_\alpha, \\ 0 & x \in S_\alpha^c. \end{cases}$$

Then, once $X^n = x^n$ has been received, the decision to reject $\mathcal{H}_0$ or not is made. Taking the non-rejection of $\mathcal{H}_0$ to mean the acceptance of $\mathcal{H}_0$, even though the probability of type-II error has not been considered, the test amounts to decisively choosing one of the two disjoint regions $\mathcal{P}$ and $\mathcal{P}^c$ of $\mathcal{Q}$, where $\mathcal{Q} = \mathcal{P} \cup \mathcal{P}^c$. In a simple-versus-simple test, prediction is obviously from the hypothesis not rejected.

Without loss of generality suppose that $\mathcal{P}$ is the decision. The question becomes how to convert $\mathcal{P}$ into a prediction. In principle, the Frequentist can use the data to estimate which element of $\mathcal{P}$ is correct and obtain predictions from it, e.g., by some maximization of the likelihood. This may be regarded as using only a portion of the information in the test because only one of many acceptable values of $\mathcal{P}$ has been incorporated into the prediction process.

Another way in which to use the information from the test is to examine $\mathcal{P}$ and try to choose a representative element of it. In the parametric case this is obvious if the accepted hypothesis is an interval: generate predictions from the distribution indexed by the midpoint of the interval. More generally, if the accepted hypothesis has a natural 'center', e.g., the center of a ball, cuboid, or other 'regular' shape in $\mathcal{Q}$, the midpoint is a meaningful location and can be used for prediction. If the accepted region is not of a regular shape, e.g., it is the complement of a regular shape, then the natural procedure is to do another test to choose a subset of the nonregular region that is regular. In the case where the accepted region is of the form $(-\infty, a] \cup [b, \infty)$, for $a, b \in \mathbb{R}$ $a \neq b$, one of the two intervals must be chosen. If it is

$[b, \infty)$ then one must choose a value $c \in \mathbb{R}$ and test $[b, c]$ versus $[c, \infty)$ until the alternative is not accepted at a satisfactory level. Obviously, this may necessitate several tests until $[b, c]$ is satisfactory. In the nonparametric setting, where $\mathcal{P}$ is the complement of a regular region, the analog of this would be to eliminate subregions of $\mathcal{P}$ until a regular region was obtained. The center of this region could be found and used to generate predictions.

9

Model Selection

A theory is empirically adequate if what it says about observables is true ... the point of science is to provide us with accurate predictions, not tell us which theories are true. This is *instrumentalism* ... [However,] there are aspects of scientific practice that don't make sense on the model of science as the quest for truth [and] there is an alternative framework for understanding scientific inference, one that is used increasingly by scientists themselves, which says that the goal of theory evaluation is to estimate predictive accuracy. It turns out in this framework that a true theory can be less predictively accurate than a false one.

Sober (2002)

The above quote is about the use of a single probability distribution to describe a response. In practice, one rarely considers a single probability distribution, and the term 'model class' or simply 'model' typically refers to a fixed collection of probability distributions. This is a minimalist definition of a model. Usually, the probability distributions in a model are linked in some way, for instance, resulting from different values of a continuous parameter and often depending on explanatory variables. In the Bayesian setting, one probability distribution is a distribution over the other (conditional) distributions that are candidates for describing the observables. In a general sense, most conventional statistical analyses involve choosing a model or model class. For instance, every testing problem amounts to comparing two different models (one called the null and the other called the alternative) and every estimation problem can be regarded as selecting a model, in the sense that an estimator identifies a specific probability distribution to describe a response. Thus, a uniformly-most-powerful test, a maximum likelihood estimator, a Bayes test, and a Bayes estimator are model-selection procedures (MSPs) even though they are not typically regarded in that way. In fact, it's not easy to think of a standard statistical problem that cannot be regarded as model selection. This includes prediction because in practice model selection is a step along the way to finding a predictor, e.g., decay-parameter estimation in nonparametric regression.

An important feature of the quote above refers to practices that don't make sense if science is regarded as a quest for truth. The obvious one is testing a point null such as $\mathcal{H}_0 : \mu = 0$ versus $\mathcal{H}_1 : \mu \neq 0$. No one really believes that if μ has a true value it will be exactly zero to infinite precision. So, strictly, $\mathcal{H}_0$ is false. Even if either the test is more sophisticated than usual and controls the frequentist probability of type-II error or prior selection has been done carefully (e.g., a spike-and-slab prior), there can easily be cases where $\mathcal{H}_0$ is not rejected. Thus, de facto, one has 'reasoned' to a conclusion believed a priori to be false. Similarly, if a model is thought adequate, it is understood that it is wrong even though the error is too small to detect. All this is reasonable since it strains credulity to believe Nature comports itself so as to be represented by human reason and mathematics. The problem is that, again, statistical procedures are de facto generating conclusions as 'true' that are a priori believed (if not known) to be false.

A related point is that if the data generator is $\mathcal{M}$-open and so does not admit a true model then the modeling enterprise, even when deemed to be predictively successful, will necessarily fail. A less dramatic example of this occurs with $\mathcal{M}$-complete problems. If the true model for the data generator is sufficiently complicated and simplifications of it exist that give predictions that are satisfactory then such wrong models are adequate, and in many practical settings the true model will remain unknown. Indeed, estimating the true model may lead to high variance for little gain in bias, so that an approximation to it with much lower variance and only slightly increased bias may be better in a mean squared error sense. Indeed, it is possible that even if a true model is known, computing from it may be infeasible. Otherwise put, *a wrong model may be better for many purposes than the true model.*

Here, model selection will be reviewed from a predictive perspective using a narrower notion of the term model. It is assumed that the probability distributions in a model, apart from a prior, are meaningfully related to each other in some way. Often the set of probability distributions is not just the continuous image of a real parameter; the set also represents a coherent set of beliefs about the phenomenon under investigation. These beliefs are usually internally consistent, and different models correspond to different sets of beliefs. Of course, any such set of beliefs might be far wrong and a compromise position, such as combining the predictions from a collection of models, may outperform any one of them. (For details, see Chapter 11.) In addition, it will generally be assumed that the probability distributions in a model are all of the same mathematical form – linear models, survival models, etc. This is not a requirement – indeed it may unduly narrow the scope of analysis – but this assumption is traditionally made. For instance, it is unusual to have a single model class comprised of a collection of both linear and nonlinear models, or both parametric and non-parametric models, or two forms of nonlinear models such as trees and neural nets (see Sec. 10.2).

There is a lack of precision in the concept of a model class. However, this will not be a problem in practice since, to do model selection properly, the models must be precisely specified, i.e., well defined and chosen on some rational basis by the analyst. Without this, few would be satisfied by the modeling – though many might be satisfied with the predictions.

One of the key features of model-selection procedures (MSPs) is that they bring in an extra level of uncertainty, since a model class has to be defined and usually has an uncertainty associated with it (which is often ignored). So, a recurring point in this chapter will be the appropriate expression of uncertainty. In the Bayes setting, there is uncertainty about the parameter but there is also uncertainty about the model and uncertainty about how to specify the uncertainty about the parameter and the model. The same caveat pertains to frequentist methods. This is important, because overstating the uncertainty typically gives uncertainty bands that are too wide and hence conclusions that are too weak. Understating the uncertainty has the opposite problem: it leads to error bands that are too narrow and hence do not adequately hedge against the actual uncertainty. A more reified way to say this is that, if one has defined a collection of models from which to choose, the collection of models itself has an uncertainty that must also be considered.

In fact, in practice, investigators often choose a model and use it to generate predictions, trusting the model to give satisfactory margins for error on the predictions. This is an incorrect way to proceed, however. As noted more formally above, the act of selecting a specific model increases the variability of prediction in the presence of nontrivial model uncertainty. The point of this chapter is to summarize the relevant literature and use sequential prediction to evaluate the predictors that models generate. The idea is that the residuals from sequential prediction (*predictuals*) include the uncertainty of model selection (and bias) automatically and hence give a more reliable assessment of a proposed model. Hence the mistake of making predictions from a chosen model without modifying the uncertainty bands may be avoided, or at least recognized.

It is important to distinguish model selection from predictor selection even though the two are closely linked. For instance, if a parametric family is fixed, a variety of criteria can be used to choose a predictor for the next outcome. Simple examples were noted in Sec. 2.3.2; see (2.30) and (2.32). A more sophisticated treatment to finding predictors under various optimality criteria is found in Ghosh *et al.* (2008), where the optimality criteria are divergence measures. Technically this is not model selection, because only one model – the parametric family in this case – is used to form a predictor that is optimal under some criterion. More substantively, the densities used to generate predictions are essentially never elements of the parametric family unless the family is convex. (Convex sets of distributions are rare because they are often artificial.) Thus, the family is used only to construct a density outside the parametric family from which to generate predictions for the response, which is presumed to follow one of the densities in the parametric family. This is an example of predictor selection that is not model selection.

It is not hard to show that the usual predictive density under relative entropy (see (2.30) for the conditional form) is natural, in that it comes from a stochastic process, i.e., the integral over the $(n + 1)$th joint density gives the nth joint density whereas the predictors under other divergence-like distances do not, in general, have this marginalization property. Another example of predictor selection that does not correspond to model selection because it is a model average is the Shtarkov solution, described as 'prediction along a string'; see Sec. 11.7.1. It does not have the marginalization property either. However, it has a different optimality property and, like the solutions in (2.30), can be shown to converge to the natural predictor.

Using a single parametric family to construct predictors can also be regarded as model selection, in the sense that the optimality criterion leads to an estimate of the probability distribution generating the data. That is, if the individual members of the parametric families are regarded as models, the optimality criterion can be seen as achieving model selection if the selected model can be linked back to an element of the parametric family – for instance, by choosing the element of the parametric family closest to the selected model under a minimum distance criterion. An alternative model-selection interpretation involves recalling that the predictive density can be taken as an estimator of the probability distribution of the response. In this case, it is simplest to regard the problem as nonparametric (any valid distribution may be true) and the predictive density under a divergence as merely one way to estimate the true density, using the parametric family purely as an input to a predictive procedure. These latter two interpretations, while a little fanciful, are valid. So, it will be more difficult to separate model selection from predictor selection here than in earlier chapters.

Indeed, there is a well-developed theory for minimax prediction; see George *et al.* (2012) and the references therein for the normal case under relative entropy loss. The goal is to identify a distribution that can be used to predict the outcome of one normal variable given another normal variable that has the same mean and assuming that both variances are known. To wit, if $(Y|\mu) \sim N(\mu, \sigma_Y^2 I_d)$ and $(X|\mu) \sim N(\mu, \sigma_X^2 I_d)$ are independent d-dimensional normal vectors with both σ_Y^2 and σ_X^2 known then a natural problem is to find $\hat{p}(y|x)$, an estimator of $p(y|\mu)$ based on $X = x$, to use for predicting Y. The Bayes risk based on the relative entropy loss is

$$B(w, \hat{p}) = \int w(\mu) \left(\int p(x|\mu) \left(\int p(y|\mu) \left(\log \frac{p(y|\mu)}{\hat{p}(y|x)} \right) dy \right) dx \right) d\mu$$

in which the innermost integral is the loss from using $\hat{p}(y|x)$ when $p(y|\mu)$ is correct. This integrates to the risk as a function of μ and thence to the Bayes risk. As shown in Aitchison (1975), the Bayes risk is minimized by the predictive

$$\hat{p}_w(y|x) = E_w\left(p(y|\mu)|x\right) = \int p(y|\mu)w(\mu|x)d\mu. \tag{9.1}$$

George *et al.* (2012, Theorem 1) states that the sufficient conditions on mixtures are given by

$$m_w(\cdot|v) = \int p(\cdot|\mu)w(\mu)d\mu,$$

i.e., on the prior w, to ensure that (9.1) will be minimax. The sufficient conditions can be interpreted as a sort of shrinkage, see George *et al.* (2012, Sec. 4), and the analysis can be extended to include fixed effects in linear models and some nonparametric settings (see Secs. 8 and 9 of the same publication). Other divergences lead to different forms of $\hat{p}(y)$ with different conditions to ensure minimaxity; see Ghosh *et al.* (2008, Theorem 3.2). Depending on the exact definition of the problem, these procedures can be regarded as either predictor selection or model selection.

As a criterion for prediction or model selection, minimaxity is desirable but may have drawbacks. First, minimax solutions are in general not unique even though they have the same minimax risk, so different priors will give different results. Asymptotically, the performances may be the same, but pre-asymptotically the differences may be important. Second, as a generality, if used sequentially minimax procedures may not perform well in the long run as they protect most against a worst-case scenario rather than begin appropriate for the typical case. Paraphrasing Draper (1995, p. 92) from a more general decision context, if a collection of decision problems, possibly sequential, are tied together because they draw on the same finite resources for the implementation of their solutions then the combined cost of all the minimax solutions will often exceed the available resources. The implication is that some average criterion for risk that will permit allocation of a portion of the available resources to the solution of each decision problem will do better. This applies to sequential prediction problems in particular: the cost of repeatedly using minimax solutions would typically increase very rapidly, for instance, because of the cost of acquiring the information needed to choose a suitable least favorable prior from the range of possibilities. One can argue that the long-run infeasibility of minimax solutions implies that it is essential to reformulate the model in response to errors, so that minimax solutions are not used too many times before a more effective strategy is found.

Just as model selection and predictor selection are conceptually disjoint yet often coincident in practice, model selection and assessment are also conceptually disjoint yet often coincident in practice. The same is true for predictor selection and predictor assessment and, when the predictor arises from a model, the two are equivalent. The core idea is that a model or predictor chosen by one technique must be evaluated on new data to assess how well it actually performs, and these are separate procedures with possibly different criteria. As will be seen, there are many principles that can be invoked to produce candidate models or predictors, but most assessment techniques rest on some notion of how well the predictor, possibly generated from a model, performs on new data from the same population. This is called the generalization error of the predictor. The assessment guides the choice and refinement of learning methods and gives an estimate of how well the final method chosen can be expected to perform.

Philosophically, some authors have argued that in some settings, e.g., linear regression (see Shao (1993)), the focus should be on model assessment rather than model selection – at least in the sense that more data should be reserved for model assessment than for model selection. The principle seems to be that it does not much matter how the model was found as long as it performs well predictively. Indeed, the tacit goal of model assessment is model validation, or, since that is often too strong a term, then verification that the model will not easily be discredited by its predictions. This is much at one with the predictive approach, in which it makes sense to reassess predictive performance periodically so that model assessment can be used to guide model selection, even if the guidance amounts only to saying that the predictor from a chosen model performs so poorly that it must be abandoned. Overall, the intuition in Shao (1993, Theorem 1), seems compelling, and more general than stated there.

A final point to bear in mind before we turn to specific techniques is that model selection and model identification, and hence the predictions they give, are not the same conceptually. Consider an example. A model-identification task might be to identify single-nucleotide polymorphisms (SNPs) important to the onset of Alzheimer's disease. A model-selection task might be less detailed and merely be to classify which patients are most likely to get Alzheimer's so as to develop treatments for them. The model-identification task is different from the model-selection task (although the model-identification task can be regarded as a particular type of model-selection task). The difference is that the model-identification task may require one to understand an extremely complicated system at such a fine-grained level that it is impossible to do effectively, i.e., the problem is $\mathcal{M}$-open, whereas the model-selection task might be feasible, at least approximately, by not fussing over much about the impenetrable details of SNPs. Hence, the problem would be $\mathcal{M}$-complete, or at least close enough to $\mathcal{M}$-complete that real progress could be made on a diagnostically useful time line.

In the rest of this chapter the main techniques of model selection, and associated ideas of model assessment and model identification, are presented from a repeated prediction standpoint. The first topic is linear models. Here an interesting example from point prediction will be given in lieu of repeating what can already be found in many other books. Next the now-classical 'information' criteria for model selection will be presented. Third, Bayes methods will be considered. Then our attention will turn to more computationally demanding methods, including cross-validation (CV) and its variants, simulated annealing, and Markov chain Monte Carlo (MCMC); the latter is most commonly associated with Bayes analysis but can also be regarded as a stochastic search that is similar in spirit to simulated annealing. The endnotes discuss the deviance information criterion, the use of posterior predictive loss, and the minimum description length for model selection. In the last endnote scoring rules and Bayes testing are briefly compared. All these topics are addressed essentially only in $\mathcal{M}$-closed settings.

9.1 Linear Models

Recall that in Secs. 4.2 and 4.4 prediction using a single linear model was presented. In that case, only the uncertainty in the response and in the estimator of β was important for inclusion in the prediction interval (PI). The direct quantification of variability was not important for point predictors. Extending this, if there is a fixed list of linear models from which to select, uncertainty due to the selection from the list has to be included in the PI. Note that this approach will neglect uncertainty arising from the model list itself because, effectively, the PIs are formed conditionally on the model list. Unfortunately, model-list uncertainty is typically ignored even when uncertainty relative to the model list is included in the PI. See Efron (2014) for but one recent example of this naivety in an estimation context, where the estimand is not defined except as the large-sample limit of a computational procedure involving a class of models (with different model classes yielding different estimands). Dependence on the model list is an even greater problem in the predictive context because the size of a prediction and the weight given to it can be effectively unrelated to each other. In this section the focus will be only on point predictions; the development of PIs will be done in Sec. 9.2.

Various model-selection procedures (MSPs) for linear models are well known and are ably described in many undergraduate textbooks; they include forward selection, backward elimination, stepwise selection, branch-and-bound, adjusted-R^2, C_p, and best-subsets regression amongst others. So, rather than reinvent the wheel, this subsection provides a comparison of what are probably the two most popular MSPs – stepwise and best-subsets – along with a foretaste of the 'information' methods to be presented in Sec. 9.2.

The DG is

$$
\begin{aligned}
Y = {} & \ln|X_1| + \text{sgn}(X_2)\sqrt{|X_2|} + X_3 + X_4^2 \\
& + \ln|X_1|\text{sgn}(X_2)\sqrt{|X_2|} + \text{sgn}(X_2)\sqrt{|X_2|}X_3 + X_3X_4^2 + X_4^2\ln|X_1| + \epsilon. \quad (9.2)
\end{aligned}
$$

It is assumed that $\epsilon \sim N(0, 2.5^2)$ and that the X_k are generated from an $N(0, 5^2)$ distribution. So, although the data have not been Studentized to put them on the same scale, the sizes of the explanatory variables are large enough to be detected given the noise level.

Now, consider the following procedure. Take 50 data points as a burn-in to choose a model by some MSP and estimate its parameters. The model and parameter estimates are then used to make five point predictions, one for each of the next five time steps, i.e., each of the next five values of (X_1, X_2, X_3, X_4). Then, the next five Y's are measured and the sum of the squared errors between the predictions and outcomes is taken. The process is repeated using 55 data points to predict the next five Y's, and the error of the five predictions is added to the error from the first five predictions. The process repeats until fully 250 data points are used and 100 iterations have been performed. The benefit of having 100 iterations is that error bars of $\pm$ SE can be formed for each MSP and each time step so that statements can be made about how the cumulative predictive error (CPE) curves separate. (These graphs are not shown.)

To complete the specification of the example requires a model list and an MSP. In this case, the model list will be all linear models that use any number of $X_1, \ldots, X_{16}$, that is, the four variables that appear in (9.2), along with 12 decoys, so that there are 2^{16} models in all. Of course, X_1, X_2, X_3, and X_4 appear linearly in the models on the list, but they appear in (9.2) as much more complicated features (with the exception of X_3). This is a situation of high bias and is realistic in the sense that (i) the true model is tremendously more complicated than can be represented by anything on the model list, and (ii) the model list is still capable of encapsulating some features of the DG. It is also realistic in that it is not clear which model in the allowed class should be regarded as 'best'.

The five MSPs used here are stepwise, best-subset, the Akaike information criterion (AIC), the Bayes information criterion (BIC), and $\sqrt{n}$-penalty. The first two are well known and will not be described here. The AIC and the BIC will be described in Sec. 9.2. The $\sqrt{n}$-penalty is a variant on the AIC and BIC except that, where as AIC penalizes the maximized likelihood by the number of parameters and BIC penalizes it by the product of the number of parameters and $\ln n$, the $\sqrt{n}$-penalty procedure penalizes it by the product of the number of parameters and $\sqrt{n}$. To provide a bit more detail, the stepwise technique here starts with a model – the best choice is the model with four terms, one for each of X_1, X_2, X_3, and X_4, while one of the worst choices has just a constant term in it – and then uses partial F-tests for backward elimination with the p-value to exit taken as 0.05. All terms are removed until the remaining terms have p-values less than 0.05. Then the forward inclusion step considers each variable, picking those with the smallest p-values. The terms that enter are again those with p-values smaller than 0.05. The version of best-subsets used is the R function regsubsets in the package leaps (R Statistical Language 2016); it defaults to the BIC within each size of model, but other MSPs can be used as well. Given the best model of each size, leaps picks the best of them using adjusted-R^2. Thus, only stepwise requires a starting point.

Figure 9.1 shows the CPE curves for the five methods starting at the 100th time step. Clearly, best-subset selection performs worst and AIC (the lightest of the penalties) performs second worst. In the worst-case starting point for stepwise, its performance is in the middle, slightly worse than BIC, and in the best-case starting point for stepwise, it ties with the $\sqrt{n}$-penalty for which is best. The BIC is overall almost as good as the $\sqrt{n}$-penalty.

The surprisingly good performance of the $\sqrt{n}$-penalty term invites a closer examination of the terms and models that it chooses. There are two natural ways to look 'inside' the MSPs. One approach is to examine term-selection frequencies, and the second approach is to examine a representation of the sampling distributions of the MSPs. The first of these is clearly a crude version of the second, but it is easier to do and is suggestive of the general size of the models.

First, Fig. 9.2 shows the frequency of selection of terms in the final models of the 100 repetitions for the best-subsets, AIC, BIC, and $\sqrt{n}$-penalty MSPs. So, a term with frequency 100 appears in every final model. It is seen that each MSP favors a particular size of model, as indicated by number of terms. Clearly, best-subsets favors models with more terms than models with the AIC penalty, which in turn favors models with more terms than those with

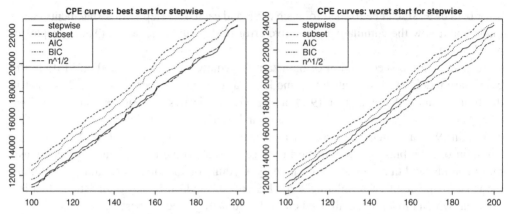

Figure 9.1 CPE curves vs. predictive step for the five model selection procedures. Left: The stepwise method is started at the best model. Right: The stepwise method is started at the worst model. Clearly, given a good starting point the stepwise method performs best and given a bad starting point it does moderately well. If a random start is used for the 100 repetitions, the corresponding CPE curves are much like those on the left, suggesting that this gives a more realistic evaluation of the stepwise method.

the BIC penalty. The $\sqrt{n}$-penalty gives even more sparsity; it concentrates more or less at a model with one term, namely X_3, the only correct linear term on the model list.

Second, Fig. 9.3 shows the same sort of histograms for stepwise regression, with three different initialization choices: the best model, containing X_1, X_2, X_3, and X_4; a random start chosen from the whole model list; and the worst model, containing only the constant term. The behavior is notably different from that seen in Fig. 9.2. In the right-hand panel, when the starting model is the worst possible, the MSP is essentially unable to find a model (apart from the constant term that is forced to occur in all models). This has been called the 'bail-out' effect: it means that the problem is so complex relative to the MSP that the MSP defaults to triviality giving the null model or the full model. That is, the MSP is unable to discriminate meaningfully over the model list. The other two cases (left-hand and middle panels) are the most similar to the BIC panel in Fig. 9.2. The left-hand panel in Fig. 9.3 has a lighter tail and the middle panel has a heavier tail. Thus the middle panel, using a random start, should be regarded as the more typical behavior, on average giving models a little larger than BIC. Of course, if prior information is available, in the form of which model is a good start, then stepwise can do better than the BIC (which does not generally specify a strong prior over models), in the sense of concentrating on the correct variables in the true model.

Overall, the best methods for prediction in the present case are (i) the $\sqrt{n}$-penalty, which reduces to the model comprising the one correct linear variable, and (ii) stepwise, provided that it starts at the best model. Other methods are generally worse from a predictive stand-point, and it is easy to see why: they hedge their bets too much by including many terms. It is worth noting that the true model, (9.2), has only one term on the model list (X_3) and that, even though X_1 and X_4 appear in terms by themselves, the fact that they are mostly of the same sign (negative for log and positive for the square) suggests that they cancel each other out or predict poorly owing to instability. The square root term in (9.2) is quite possibly too

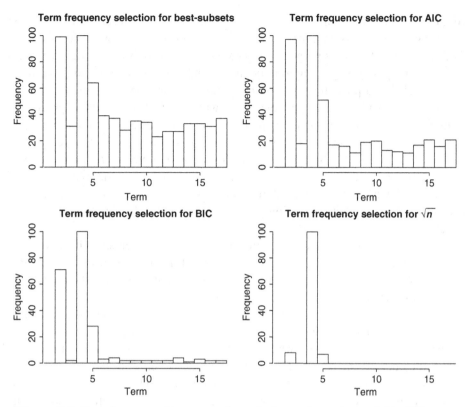

Figure 9.2 Term frequency histograms from the best-subsets (upper left), AIC (upper right), BIC (lower left), and $\sqrt{n}$-penalty (lower right) MSPs. It is seen that these four methods impose greater sparsity, i.e., generally choose fewer terms, as the penalty increases. Note that the intercept is included as the first term, so term four corresponds to X_3. The intercept occurs in all models and so is only counted when it appears alone.

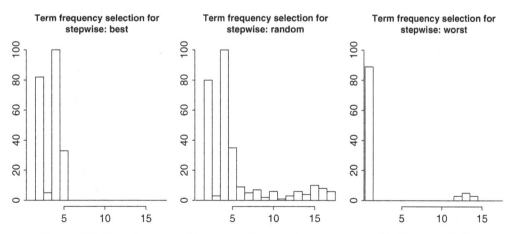

Figure 9.3 Term frequency histograms for stepwise regression with 16 terms, (9.2) as the DG, and using the best, random, and worst initializations. As in Fig. 9.2, the intercept is included as the first term so that term four corresponds to X_3. The intercept is in all models and so is only counted when it appears alone.

small relative to the other terms to make much of a difference and the cross-terms are so hard to find using the model list that they do not lead to inclusion of any X_k. Thus, even though X_3 is a poor model for many purposes (compared with, say, stepwise model selection with the best starting point), it gives a good predictor, and the MSP that led to it should probably be regarded as a predictor selection principle. This may be a typical feature of using simple model lists with complex problems: even with an $\mathcal{M}$-closed problem, prediction and estimation give unexpected results. The problems are likely to be worse for $\mathcal{M}$-complete or $\mathcal{M}$-open problems. The net effect may be that model identification is misleading, whereas at least predictors can be evaluated in comparison with other predictors (provided that they do not require information that is typically unavailable, such as an optimal starting point). A final, almost trivial, point is that different MSPs make different contributions to the variability of model and predictor selection. Hence accounting for the variability of the MSP is essential.

One limitation of the analysis just given is that it uses term-frequency histograms rather than looking directly at the sampling distributions of MSPs over the model list. Visualizing what a sampling distribution for an MSP looks like is difficult because the model list corresponds to discrete points in moderate-to-high-dimensional space. Despite this, there is a natural metric on the model list, namely, the symmetric difference between the terms that two models contain. Thus, adding or removing a term from a model creates a new model that is a distance one away from it. So, when there are K possible terms that may be included or not, the full model list can be visualized as the vertices of a cube in K or more dimensions. In the present case there are 16 terms, so the model list would be naturally regarded as corresponding to the vertices of a 16-dimensional cube. While mathematically simple, this is hard to visualize.

One way around this is to compress the points in 16 dimensions down to a two-dimensional representation. This can be done using a technique called multidimensional scaling (MDS); see Clarke *et al.* (2009, Sec. 9.10.2). The basic idea is to convert points in a higher-dimensional real space to points in a lower dimensional (e.g., two-dimensional) real space that have point-to-point distances as representative as possible to those of the initial points. A drawback of the method is that if one point is added to these in the high-dimensional space, the points in the lower-dimensional space change positions relative to each other. This technique is not the same as projection, because it's an optimization that in its simplest form resembles a chi-squared distance.

Even with MDS as a visualization technique, the number of models (i.e., points) in a high-dimensional space that can be meaningfully converted to points in $\mathbb{R}^2$ is limited, because one wants to be able to see which models are one term away from other models. Having a large number of models means that there is likely to be a large number of terms, so the resulting graph can be hard to interpret. For this reason, rather than using the earlier model list with 2^{16} terms, it's easier to use all models formed from X_1, X_2, X_3, and X_4, i.e., 16 models. Recalculating the 100 repetition CPE curves for this reduced model list using the same DG and five MSPs (and random initialization for stepwise) gives five curves that are very similar. In fact, all are roughly comparable to the BIC line and the stepwise line in the panels in Fig. 9.1.

For the sake of completeness, Fig. 9.4 gives the term-frequency histograms for the MSPs. It is enough to show three plots because the histogram for best-subsets is essentially the

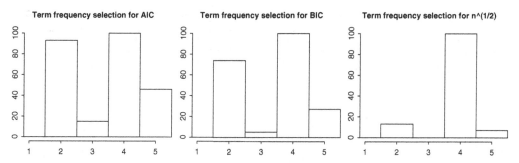

Figure 9.4 Term frequency histograms for the MSPs based on AIC, BIC, and $\sqrt{n}$-penalties with the four-variable model list and true model (9.2). The intercept-only model is included as the first term, so term k in the figure is X_{k-1} for $k = 1, 2, 3, 4$.

same as for AIC, and the histogram for stepwise (with random starts) is the same as for BIC. As suggested by the lower right panel in Fig. 9.2, the right-hand panel in Fig. 9.4 shows a concentration at the single-term model containing only X_3 (and a constant). Recall that X_3 gives the best predictive model even though some might argue that the model containing all four explanatory variables should have been chosen, since all four variables appear in the DG. Also, as the size of the penalty increases, it is seen that the frequency of including X_1, X_2, and X_4 decreases. This is possibly because they enter as $\ln|X_1|$, $\sqrt{|X_2|}$ and X_4^2, respectively, i.e., the first and third are highly nonlinear and mostly single-signed while $\sqrt{|X_2|}$ may be too small to detect in comparison with the other terms. The panels in Fig. 9.4 are very different from the panels in Figs. 9.2 and 9.3. This demonstrates the effect of model-list variability in addition to the variability due to the MSP and model uncertainty relative to a given model list. Indeed, using a larger model list by including all second-order terms would yield a further example of the effect of varying the model list.

Of greater novelty is the visualization of the sampling distributions of the MSPs. The five MDS diagrams in Figs. 9.5 and 9.6 were generated by (i) using the squared error stress, and (ii) compressing the vertices of a unit cube in four dimensions that get a strictly positive frequency, among the final models chosen over 100 repetitions of length 250. That is, the points (denoted by small circles) represent the support of the estimated sampling distribution for the respective MSP. The portions of larger circles around the small circles indicate the relative frequency of each model, labeled by the variables they include. Figure 9.5 shows sampling distributions for three of the MSPs that are nearly identical. These MSPs have modes at the model containing X_1 and X_3; the model with the second highest estimated probability in the sampling distribution contains X_1, X_3, and X_4. All the models for which the probability is positive contain the predictively optimal X_3 and usually others, e.g., X_1 and X_4. Variable X_2 was rarely included, possibly because it contributes little.

Figure 9.6 shows sampling distributions for two MSPs – best-subsets and $\sqrt{n}$-penalty – which are different from those in Fig. 9.5 and from each other. In particular, the empirical sampling distribution for best-subsets does not assign positive probability to the model containing X_3 alone; in fact, it almost always includes X_1, which is not predictively optimal. By contrast, the empirical sampling distribution for the $\sqrt{n}$-penalty concentrates at three

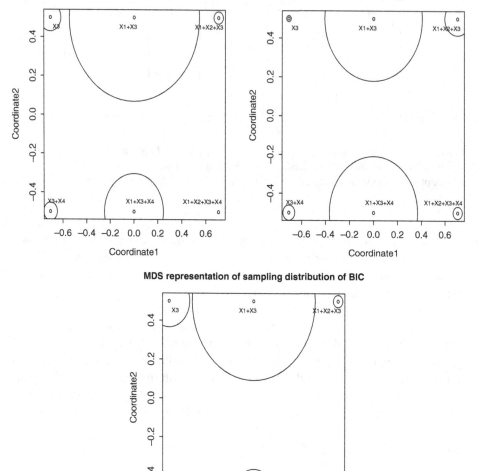

Figure 9.5 Estimated sampling distributions for stepwise with a random start AIC and BIC, with points of support found using multidimensional scaling on the vertices of a cube in four dimensions that received strictly positive frequencies of selection under their respective MSPs.

points, all of which contain X_3, and in particular concentrates at the model of X_3 that alone is predictively optimal.

If the variables on the model list had been Studentized, i.e., put on the same scale, the results might have been different. However, the point here is to visualize the sampling distribution of the MSPs. An alternative visualization in three dimensions would arise if bars with height proportional to the probability of each point were located at each point.

This example suggests that when the true model is hopelessly more complicated than any model on the model list, and hence the bias is high, better prediction results from using a

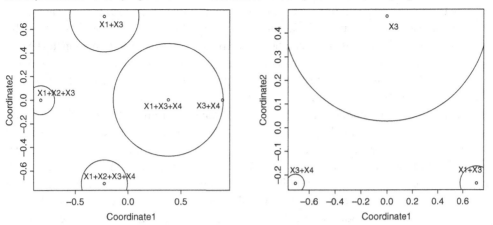

Figure 9.6 Estimated sampling distributions for best-subsets and $\sqrt{n}$-penalty MSPs with points of support found using multidimensional scaling on the vertices of a cube in four dimensions that received strictly positive frequencies of selection under their respective MSPs.

heavier penalty term. However, a lighter penalty term permits more variables to be included, but the implication of this example is that little more can be said beyond the observation that they could be included. That is, they may well be involved in the DG but the selected model may not be useful for other inferences. And, as seen in the 16-variable case, it cannot be readily inferred that the selection of a variable is a particularly strong indicator that it actually is involved in the DG. Likely, if there were extra variables in the DG that were not included in the model list, the results would be even worse from a modeling standpoint. The net effect of all this is that, outside the simplest cases, model selection is unreliable and should be abandoned in favor of prediction – if only on the grounds that the model providing the best predictor really is saying something about the DG.

To conclude, it's worth noting two other cases, in the context of linear models, where model selection and prediction diverge. First, Fan and Lv (2008) proposed and established a technique called sure independence screening (SIS) as a way to achieve effective dimension reduction. They looked at correlations between a response Y and a very large number of explanatory variables X_j; some X_j may be measured directly while others may be features constructed by a preprocessing stage that feeds into the model-selection stage. The main theorem can be informally stated as follows: with very high probability, the true (sparse) model is a subset of the models formed from explanatory variables with high univariate correlations with Y. However, SIS is a linear, marginal, procedure and may be subject in practice to the usual problems of such methods.

For instance, consider an artificial data set of the form

$$(Y, X_1, X_2) = (3, 3, 5), (4, 4, 4), (5, 5, 3), (5, 8, 10), (4, 9, 9), (3, 10, 8).$$

Then, both X_1 and X_2 have correlation with Y equal to zero even though Y appears to depend on both of them. This phenomenon can occur in any number of dimensions and sample sizes: for each X_k just make Y initially increase with X_k and then decrease as X_k

increases, or make Y initially decrease with X_k and then increase as X_k increases. Indeed, it is enough to choose nontrivial functions g_k of X_k such that $\text{Cov}(Y, g(X_k)) = 0$ for each k. As a consequence of this, if a correct X_k fails to be included owing to random variation then the theorem effectively implies that X_k can be reconstructed from other later explanatory variables which have sufficiently high correlation with Y. That is, the model that results from SIS may miss 'real' variables and so have to compensate for them by including other variables. From a predictive standpoint, this is a desirable feature (so long as the extra variables included are not too numerous). However, it represents a different notion of consistency than that commonly used in $\mathcal{M}$-complete or $\mathcal{M}$-closed problems. Indeed, there may be many different but asymptotically equivalent models, so that identifiability no longer holds.

Second, since model-list uncertainty and MSP uncertainty both contribute to the overall variability of predictions, it is quite possible that in many cases effective prediction, let alone model identification, is conceptually feasible only when a true model exists. After all, specifying a convincing collection of model lists so that model list variability can be assessed, verifying that the MSP behaves satisfactorily on the lists, and then assessing model uncertainties relative to the model lists is a long list of requirements to satisfy before making a prediction. The fact is that, as the number of candidate X_j increases, the variability due to all sources (model-list, MSP, model relative to model list) increases, thereby widening PIs. Otherwise stated, the instability may be so great as to make prediction ineffective, on the grounds that the PIs are just too wide to be useful. The problem with instability would be even worse for the more exacting problem of model identification: a specific model identified might be far wrong and a region of models in a model space might be too large to be useful. However, in practice, a single optimal model list chosen, for instance, to satisfy a variance–bias tradeoff condition (see Clarke and Fokoue (2011)) might be satisfactory, and then it might be enough to include only the variability of the MSP regarding PIs and ignore model-list variability when making a prediction. A complicating factor is that if the number of candidate X_j that should be included in a predictor increases as n increases then the variability of point predictions may increase, because the model-list uncertainty is increasing as well. This suggests that there will be cases where even though a 'good' model can be selected by a 'good' MSP, one should be satisfied with using the predictor that it generates and give up on model identification.

9.2 Information Criteria

In the last subsection, linear models were selected merely using elementary techniques and were converted to their natural point predictors. Here, the focus is on selection among finite-dimensional parametric models with d parameters using information criteria. In the context of model selection the use of information criteria means penalizing the maximized likelihood by a term dependent on n or d, but not usually on the parameter values themselves (since the goal is model selection not parameter estimation). The focus is on selecting the correct d parameters, but it is understood that the number of parameters chosen indicates the number of explanatory variables to be included, e.g., if each parameter θ_k is a factor of an X_k then not including θ_k amounts to setting $\theta_k = 0$ and hence not including X_k.

A generic form for information criteria was given and studied in Bethel and Shumway (1988). To state their criterion, fix the dimension d of the parameter $\theta = (\theta_1, \ldots, \theta_d)$. Let S

be a set of positive integers with $S \subset S_L = \{1, \ldots, L\}$ for some $L \leq d$ and let $V_S = \{\theta \in \mathbb{R}^d | \theta_k = 0$ for $k \notin S\}$. Write θ_0 for the true value of θ and $\theta_{0,S}$ and $\hat{\theta}_S$ to mean, respectively, the true value of θ and its MLE restricted to V_S. The goal is to fit the best model using some subset of the first L entries of θ. The generic criterion is

$$IC_\alpha(S) = -2 \log p(y^n | \hat{\theta}_S) + \alpha(n)\#(S), \tag{9.3}$$

where $\#(S)$ is the number of elements in S and $\alpha(n)$ satisfies

$$\alpha(n) \to \infty \quad \text{and} \quad \frac{\alpha(n)}{n} \to 0 \quad \text{as} \quad n \to \infty.$$

Clearly, the well-known BIC and the $\sqrt{n}$-penalty used in Sec. 9.1 are of this form. Other information criteria such as that given in Hannan and Quinn (1979) are also of this form ($\alpha(n) = \ln \ln n$). The Bethel–Shumway theorem, proved below, will show that any criterion within the class (9.3) is consistent.

There are other information criteria that penalize the maximized likelihood which are not in the Bethel–Shumway class. The AIC, see Akaike (1973), is probably the best known; it uses a penalty of $\#(S)$ in (9.3) independently of the sample size. A recent and clear derivation of the AIC focusing on its treatment of bias can be found in Clarke *et al.* (2014). The AIC is not consistent; however, Shibata (1981) established the optimality of the AIC in a predictive sense. The Takeuchi information criterion, see Takeuchi (1976), generalizes the AIC by penalizing by the estimated trace of a product of matrices. This was generalized in a different direction by the 'information complexity' (ICOMP, Bozdogan 1993), which penalizes by a function of the Fisher information matrix. There are many other information criteria, both in and not in the Bethel–Shumway class; the deviance information criterion (Spiegelhalter *et al.* 2002) constitutes another effort to unify the various forms of information criteria. In addition, the so-called focused information criterion (FIC, Hjort and Claeskens 2003) has a form that is not a penalized maximized likelihood and is therefore not truly an information criterion; it chooses an S on the basis of minimizing an asymptotic risk.

While in this section we will focus on the Bethel–Shumway class, and the proofs are drawn from Bethel and Shumway (1988), the predictors developed here can be used with any MSP that unambiguously ranks models in terms of their ability to summarize the data. Consistency is merely a property of an MSP that may be helpful in forming a predictor. Indeed, the point is the conversion of selected models into point and interval predictors that take account of the uncertainty in the MSP and data, even though all the statements are conditional on a model class having been fixed.

Bethel and Shumway (1988) defined a best subset of parameters, denoted by a set of integers B, as one that simultaneously minimizes $D(\theta_0 \| \theta_{0,S})$ and $\#(S)$, where $D(\theta \| \theta') = \int p(y|\theta) \ln(p(y|\theta)/p(y|\theta')) dy$ and Y with outcome y is an independent copy of any Y_i. More formally, B is a best subset if and only if, $\forall S \subset S_L$,

1. $D(\theta_0 \| \theta_{0,B}) \leq D(\theta_0 \| \theta_{0,S})$, and
2. $D(\theta_0 \| \theta_{0,B}) = D(\theta_0 \| \theta_{0,S}) \implies \#(B) \leq \#(S)$.

Thus, any best subset is the smallest subset minimizing $D(\theta_0 \| \theta_{0,S})$. Such a best subset must exist because S_L is finite and therefore has finitely many subsets.

To simplify condition 2, observe that the positivity of the relative entropy gives that, for $\theta, \theta' \in V_S$,

$$D(\theta_0 \| \theta) \leq D(\theta_0 \| \theta') \iff E \ln p(Y | \theta') \leq E \ln p(Y | \theta).$$

So, letting

$$E_S = E \ln p(Y | \theta_{0,S})$$

gives that minimizing $D(\theta_0 \| \theta)$ is equivalent to maximizing $E \ln p(Y | \theta)$. The following proposition relates best subsets to relative entropies.

Proposition 9.1 (Bethel and Shumway 1988) *Let S be a subset of S_L and let B be a best subset. Then:*

1. $E_B = E_S \iff B \subseteq S$, *and,*
2. B *is unique.*

Proof   Start by supposing that $E_B = E_S$. Since B is best and $B \subseteq B \cup S$, $E_B \geq E_{B \cup S}$. Also, from the minimization property of the relative entropy, $E_S \leq E_{B \cup S}$. So, $E_B = E_{B \cup S} = E_S$. These equalities imply $\theta_{0,S} = \theta_{0,B \cup S} = \theta_{0,B}$. So, if $\theta_{0,B,k}$, the kth component of $\theta_{0,B}$, is zero then $\theta_{0,B} \in V_{B \setminus \{k\}}$, which implies that $D(\theta_0 \| \theta_{0,B}) = D(\theta_0 \| \theta_{0,B \setminus \{k\}})$. However, this contradicts the assumption that B is a best subset because $\#(B \setminus \{k\}) < \#(B)$. Thus, $\theta_{0,B,k} \neq 0$ and hence $\theta_{0,S,k} \neq 0$ as well, implying that $k \in S$. So, the forward implication of part 1 follows.

For the reverse implication, first let $S \subseteq R$. When $\theta \in V_S$ it follows that $\theta_k = 0$ for $k \in S^c \supseteq R^c$ and hence $\theta_k = 0$ for $k \in R^c$. This gives $\theta \in V_R$ and therefore $E_R \geq E_S$. So, it follows that $B \subseteq S$. However, B is best so $E_B \geq E_S$ and therefore $E_B = E_S$, completing part 1.

The proof of part 2 is obvious. If B^* is another best subset then $E_{B^*} = E_B$ and hence $B \subseteq B^*$. By symmetry, $B^* \subseteq B$, giving $B = B^*$. $\square$

To state the three Bethel–Shumway theorems, define the following notation. For $S \subseteq \{1, \dots, L\}$ suppose that $\tilde{S}$, $\hat{S}$, and $\hat{S}_\alpha$ satisfy

$$D(\theta_0 \| \hat{\theta}_{\tilde{S}}) \leq D(\theta_0 \| \hat{\theta}_S),$$

$$\text{AIC}(\hat{S}) \leq \text{AIC}(S),$$

$$\text{IC}_\alpha(\hat{S}_\alpha) \leq \text{IC}_\alpha(S),$$

where $\text{AIC}(\cdot)$ is the usual Akaike information criterion. Now, the first of these theorems can be stated and proved.

Theorem 9.1 (Bethel and Shumway 1988) *Let B be the best subset. Then:*

$$\lim_{n \to \infty} P\left(\tilde{S} = B\right) = 1.$$

Proof Let $\delta = \min_{E_B \neq E_S}(E_B - E_S)$. Clearly, $\delta > 0$. Let

$$A_S(\theta_{0,S}) = \left(E\left(-\frac{\partial^2 \ln p(Y|\theta)}{\partial \theta_k \partial \theta_\ell} \right)\bigg|_{\theta=\theta_{0,S}} \right)_{k,\ell},$$

where k and ℓ run over the nonzero elements of S, be the Fisher information matrix, and let the standardized variate for the parameter estimate when S is a subset of parameters be

$$U_{S,n} = \left(\sqrt{n}\left(\hat{\theta}_S - \theta_{0,S}\right)^T \right) A(\theta_{0,S}) \left(\sqrt{n}\left(\hat{\theta}_S - \theta_{0,S}\right) \right).$$

Then, it is easy to verify that

$$E \ln p(Y|\hat{\theta}_S) = E \ln p(Y|\theta_{0,S}) - \frac{1}{2n} E U_{S,n} + o\left(\frac{1}{n}\right),$$

by Taylor-expanding $\ln p(y^n|\hat{\theta}_S)$ around $\theta_{0,S}$ and taking expectations. (Recall that $E \ln p(Y^n|\theta) = nE \ln p(Y|\theta)$.) Now, for any S with $E_S \neq E_B$,

$$-2E \ln p(Y|\hat{\theta}_B) + 2E \ln p(Y|\hat{\theta}_S)$$
$$\leq -2E_B + \frac{EU_{B,n}}{n} + \frac{\delta}{2} + 2E_S - \frac{EU_{S,n}}{n} + \frac{\delta}{2} \leq -\delta + \mathcal{O}\left(\frac{1}{n}\right)$$

as $n \to \infty$. Thus, for n large,

$$-2E \ln p(Y|\hat{\theta}_B) < -2E \ln p(Y|\hat{\theta}_S)$$

holds with high probability for any S with $E_S \neq E_B$. It follows that, asymptotically, $E_{\tilde{S}} = E_B$. In terms of the parameters selected,

$$\tilde{S} \in \{S|E_S = E_B\} = \{S|B \subseteq S\}.$$

However, $B \subseteq S$ implies $E_B = E_S$ and so

$$-2E \ln p(Y|\hat{\theta}_B) + 2E \ln p(Y|\hat{\theta}_S) \tag{9.4}$$

will be a sum of the expected values of χ_1^2 random variables – one for each θ_k in S that is not in B. Thus, when $B \subseteq S$, (9.4) is positive unless $B = S$. $\qquad \square$

The content of this theorem is that minimizing the relative entropy $D(\theta_0 \| \theta_S)$ selects the best subset of the θ_k consistently.

The second Bethel–Shumway theorem applies to the AIC and shows that the AIC selects a subset of the θ_k that minimizes $D(\theta_0 \| \theta_S)$ but is larger than the best subset with positive probability. The statement and proof is as follows; it subsumes much classical work on the properties of the AIC, e.g., Shibata (1976), as well as motivating more recent work.

Theorem 9.2 (Bethel and Shumway 1988) *Let B be the best subset. Then:*

1. $\lim_{n \to \infty} P(B \subseteq \hat{S}) = 1$, *and,*
2. *if* $\#(B) < L$, $\exists C > 0$ *such that* $\lim_{n \to \infty} P(\#(\hat{S}) > \#(B)) = C > 0$.

Proof Part 1: Let W be a subset of the parameters. Taylor-expanding

$$\sum_{i=1}^{n} \ln p(y_i | W)$$

at $\hat{\theta}_W$ for B and S gives

$$\text{AIC}(B) - \text{AIC}(S) = 2n\,(E_S - E_B + o(1)) + 2\,(\#(B) - \#(S)) + \mathcal{O}(1),$$

when n is large (see Bethel and Shumway (1988) for details), i.e.,

$$\text{AIC}(B) < \text{AIC}(S)$$

unless $E_B = E_S$ holds with high probability. So, Proposition 9.1 gives

$$P\left(B \subseteq \hat{S}\right) = P\left(E_{\hat{S}} = E_B\right) \to 1.$$

For part 2, note that $B \subseteq S$ gives $\theta_{B,0} = \theta_{S,0}$ so that a similar Taylor-expansion argument to that used for part 1 gives

$$\text{AIC}(B) - \text{AIC}(S) = 2\,(\#(B) - \#(S)) + U_{S,n} - U_{B,n} + o_P(1).$$

Also, recall that if $Q_{RS} = \lim\left(U_{R,n} - U_{S,n}\right)$ in probability for any sets R and S of parameters, it is possible to use Taylor-expansion arguments to show that if $S \subset R$ and $E_S = E_R$ then $Q_{RS} > 0$ with probability unity and $P(Q_{RS} > C) > 0$ for any $C > 0$; see Bethel and Shumway (1988, Lemma 3.2) for details. Essentially, this follows by grouping terms in Taylor expansions together to recognize χ^2 random variables. Setting $B = R$, if $B \subseteq S$ it is then easy to see that

$$\exists\, C > 0: \quad P(\text{AIC}(B) - \text{AIC}(S)) \to C > 0.$$

To complete part 2, note the disjoint decomposition of all possible $\hat{S}$'s:

$$\{\hat{S} \neq B\} = \left\{\#(\hat{S}) > \#(B)\right\} \cup \left\{\#(\hat{S}) < \#(B)\right\} \cup \left\{\#(\hat{S}) = \#(B), \hat{S} \neq B\right\}.$$

By the definition of best subset and part 1 of this theorem, the two sets on the right-hand side have a probability of occurrence that tends to zero as $n \to \infty$. Expressing this in terms of $\text{AIC}(\cdot)$ and rearranging gives

$$P(\#(\hat{S}) > \#(B)) = P(\hat{S} \neq B) - P(\#(\hat{S}) < \#(B)) - P(\#(\hat{S}) = \#(B), \hat{S} \neq B)$$
$$\to C > 0. \qquad \qquad \square$$

This theorem establishes that the AIC, although not consistent, does give a model of which the true model is a subset. (Although it does not seem to have been formally verified, the same is likely to be true for the TIC and ICOMP since they can be regarded as variations on the AIC.)

The third and final Bethel–Shumway theorem applies to IC_α and hence is arguably the most important because it thus applies to a class of MSPs. This theorem ensures that minimizing IC_α leads to a best subset for any given α satisfying $\alpha(n) \to \infty$ and $\alpha(n)/n \to 0$. In particular, an MSP giving much more sparsity and hence a much tighter sampling distribution than the BIC is well justified. The result is as follows; its proof is surprisingly easy.

Theorem 9.3 (Bethel and Shumway 1988) *Under the conditions on $\alpha(n)$,*

$$\lim_{n \to \infty} P(\hat{S}_\alpha = B) = 1,$$

where B is the best subset.

Proof Let S be a set of parameter indices. By Taylor-expanding,

$$\mathrm{IC}_\alpha(B) - \mathrm{IC}_\alpha(S) = -2 \sum_{i=1}^{n} \ln p(y_i | \theta_{0,B}) + 2 \sum_{i=1}^{n} \ln p(y_i | \theta_{0,S})$$
$$+ \alpha(n)(\#(B) - \#(S)) + U_{B,n} - U_{S,n} + o_P(1).$$

If $E_B - E_S > 0$ then, by the strong law of large numbers and the conditions on α,

$$\frac{\mathrm{IC}_\alpha(B) - \mathrm{IC}_\alpha(S)}{n} = -2(E_B - E_S) + o_P(1),$$

i.e., $\mathrm{IC}_\alpha(B) < \mathrm{IC}_\alpha(S)$ holds with high probability as $n \to \infty$. If $E_B = E_S$ then $B \subseteq S$, so that $\theta_{0,B} = \theta_{0,S}$. This means again that if $\#(B) < \#(S)$,

$$\mathrm{IC}_\alpha(B) - \mathrm{IC}_\alpha(S) = \alpha(n)(\#(B) - \#(S)) + \mathcal{O}_P(1) < 0.$$

So, $\#(B) = \#(S)$ and $\hat{S}_\alpha \to B$. $\qquad\square$

This theorem establishes that a large collection of MSPs are consistent and the previous theorem establishes that the AIC (and likely its variants) is (are) consistent for a model that contains the true model. So, we will now use MSPs to generate point and interval predictors. Point predictors are straightforward: merely convert the selected model into its natural and obvious predictor by estimating any parameters in it by well-behaved estimators, e.g., estimators that are consistent, asymptotically normal, and efficient, provided that this does not come at too much loss of robustness to extreme values. One can even use this point predictor to derive an interval predictor as well. However, this comes at the cost of ignoring model uncertainty and model list uncertainty. So, it is expected to be unrealistically narrow. Corrections for this problem without creating the opposite problem of PIs that are too large to be useful will be the substance of the rest of this and later sections. (A related series of results that apply only to linear regression models can be found in Müller (1992, 1993). Although not presented here, these results lead to predictors analogous to those that follow.)

There is a range of ways to generate point predictors and PIs using MSPs. Arguably, the simplest was used in Sec. 9.1, where the variability of five MSPs was seen in a linear-model context. Basically, the MSP chose a model and then a point prediction was generated from it. The distribution of the predictor $\hat{Y}$ at x_{new} was given in Sec. 4.2, where point and interval predictors were also given. While the specific formulae were particular to linear models, the ideas generalize as long as each model on the model list can be converted to a distribution for a future outcome. As in Secs. 4.2 and 9.1, if the kth parametric family $p_k(\cdot | \theta_k), k = 1, \ldots, K$, is on the list where k indexes the included variables and the parameters associated with them then a plug-in predictor follows by the estimation of θ_k by $\hat{\theta}_k$ (an MLE, posterior mean, or other well-behaved estimator) and the obtaining of point and interval predictors at x_{new} from $p_k(\cdot | \hat{\theta}_k, x_{\mathrm{new}})$. The mean of $p_k(\cdot | \hat{\theta}_k, x_{\mathrm{new}})$ is a natural point predictor, but the median or mode of $p_k(\cdot | \hat{\theta}_k, x_{\mathrm{new}})$ may work well, too. The percentiles from $p_k(\cdot | \hat{\theta}_k, x_{\mathrm{new}})$ can be used to give PIs. If the MSP and $\hat{\theta}_k$ are consistent then, in simple cases, the predictions should not be too bad. The problem with using PIs from plug-in predictors directly is that they are likely to be too narrow since they do not include the variability of the MSP (or model list). However, if the data set is reasonably large, and the CPEs are averaged over multiple permutations of the data, this may not be too bad because the variability due to the MSP is built into the repeated predictions and hence into the CPE curves.

The remainder of this section covers ways to correct the narrowness in individual one-step-ahead predictions by including variability due to the MSP.

The natural procedure for obtaining point and interval predictors is to take a convex combination of the models on the model list evaluated at x_{new} with their respective parameters estimated by the $\hat{\theta}_k$ as above. The natural choice for the weight of θ_k in the convex combination is the probability of choosing model p_k as given by the sampling distribution of the MSP. This was found in Sec. 9.1 using simulations. With real data, the analog of simulating a sampling distribution is bootstrapping, and the natural weights come from the 'worth' of a model as determined by an MSP.

So, suppose that a data set of the form $(x_i, y_i)|_{i=1}^n$ is available and that n is large enough that (i) model selection and parameter estimation will be effective and (ii) bootstrapping will give a reasonable approximation to the sampling distribution of the MSP. Let $b = 1, \ldots, B$ be the number of bootstrap samples and denote by $\mathrm{MSP}_b(p_k)$ the value that the MSP assigns to the model p_k, $k = 1, \ldots, K$, using the bth bootstrap sample. Then, the 'purist' would derive point and interval predictors from the mixture of densities

$$q_1(y|x_{new}) = \frac{1}{B} \sum_{b=1}^{B} \frac{\sum_{k=1}^{K} \mathrm{MSP}_b(p_k) p_k(y|\hat{\theta}_{k,b}, x_{new})}{\sum_{k=1}^{K} \mathrm{MSP}_b(p_k)}. \tag{9.5}$$

A variant on this is

$$q_2(y|x_{new}) = \frac{\sum_{b,k=1}^{B,K} \mathrm{MSP}_b(p_k) p_k(y|\hat{\theta}_{k,b}, x_{new})}{\sum_{b,k=1}^{B,K} \mathrm{MSP}_b(p_k)}. \tag{9.6}$$

The difference is that in (9.5) the bootstrap samples are treated independently whereas in (9.6) they are pooled. Clearly, PIs can be found from the percentiles of (9.6). If an SD were desired for a point predictor, it could be calculated in principle from the appropriate q_j.

A simplification of these that is less pure, but emphasizes the meaning of the mixture, is to start by forming bootstrap estimates of the sampling distribution of the MSP: use the MSP B times to choose a model, one for each bootstrap sample. Then each model p_k will have been chosen $m_k \geq 0$ times and $b_k = m_k/B$ is the proportion of times the MSP chose the kth model. Now, the natural density to use to form predictions is

$$q_3(y|x_{new}) = \sum_{k=1}^{K} b_k p_k(y|\hat{\theta}_k, x_{new}), \tag{9.7}$$

where $\hat{\theta}_k$ is the estimate of θ_k formed by using all the data. It is seen that if the MSP chooses one particular model p_k all the time then $b_k \approx 1$ and $q_3(y|x_{new}) \approx p_k(y|\hat{\theta}_k, x_{new})$. However, as with estimation, $b_k \approx 1$ does not mean that p_k is 'true'. It only means that, when the MSP is consistent, p_k is the best choice from the models on the list.

The three ways to derive predictors in (9.5)–(9.7) focus on forming a mixture distribution that weights each model appropriately. Essentially, this assumes that the MSP will do a good job of assessing the adequacy of a given p_k relative to the other $K - 1$ models. This is only feasible in practice when the models are quite far from each other, far enough for the MSP to provide reliable discrimination. In many cases, this is not so. There may be a few models which are relatively close together and which get nearly equivalent weights or, in an extreme case, in every region of the model space there may be numerous models that get comparable and very low weights. In the Bayes context, this second scenario is well recognized and called *dilution*; see George (1999). The problem is that the posterior model weights may all go to zero because the posterior probability is divided up among so many models, all of which may in fact be 'good'. This is likely to happen to the sampling distribution in the frequentist context, for the same reason. In these cases, weighting by the sampling distribution of the MSP (or, in the Bayes case, weighting by the posterior probabilities of the models) will give expressions that converge only vaguely to zero, i.e., the result is not a proper density. There are ways in which this problem can be eliminated, at least in principle, in the Bayes context; see George (2010).

One way to get around this problem is to shift the focus from the sampling distribution (or posterior over the models) to the predictions themselves. Write $\hat{Y}_{k,n}(x_{\text{new}})$ to mean the prediction at x_{new} from the kth model formed using n data points. This leads to the 'order statistic' $\hat{Y}_{(k),n}$ and, since each $\hat{Y}_{(k),n}$ has an associated b_k, the median value of the $\hat{Y}_{k,n}(x_{\text{new}})$ in terms of the estimated sampling distribution can be used. Then, if there are many models that are good, and roughly equally good, their values should cluster in the middle of the order statistic and one of them will be the median and chosen as the point predictor. Similarly, if K is large enough, percentiles from the order statistic (in terms of the estimated sampling distribution) can be used to get a PI.

Note that these methods assume a fixed model list – even if the model list is assumed to have some property such as being large or having many equally good models that are near each other according to some metric. Obviously, if the uncertainty in the model list is taken into account, the variability of point predictors and width of PIs would probably increase further. Uncertainty in the model list is the reason why none of the three methods given here accounts for the variability in choosing a specific model from a list by some IC_α. The uncertainty associated with any particular model depends on the other models on the list. So, the only case where the uncertainty of model selection depends only on the model selected is the extreme case, where all the other models are so unrealistic as to have approximately zero coefficients so that all the uncertainty is in the parameter estimation of the chosen model. However, in such cases, where model uncertainty is effectively zero, it is so easy to discriminate amongst the models that there is no need for formal model selection at all.

9.3 Bayes Model Selection

First consider Bayes prediction with a list of models $p_k(y|\theta_k, x_k)$, $k = 1, \ldots, K$, where $Y = y$ is an outcome, possibly from model p_k, that depends on a selection of d_k explanatory variables x_k from a collection of explanatory variables x. The parameter θ_k is d_k dimensional and $x_{k,i}$ denotes the ith value of x_k. Where no confusion will result the subscript i will be dropped. The Bayes approach would be to specify a within-model prior for each p_k, say $w_k = w_k(\theta_k)$, and then specify an across-model prior, say $W(k) = w(k)$, since the probability of a point k equals the value of the density at k. The joint parameter space is $\Omega = \bigcup_{k=1}^{K} \{k\} \times \Omega_k$, where $\Omega_k \subseteq \mathbb{R}^{d_k}$ is the parameter space of p_k for each k. Let $w(k, \theta_k) = w(k)w(\theta_k|k)$ denote the joint prior on Ω, writing $w_k(\theta_k) = w(\theta_k|k)$ for convenience. Given data $\mathcal{D}_n = (x_i, y_i)|_{i=1}^{n}$, where x_i denotes the ith value of x, the posterior density of K is

$$w(k|\mathcal{D}_n) = \frac{p_k(y^n|k, x^n)w(k)}{\sum_{k=1}^{K} p_k(y^n|k, x^n)w(k)} \tag{9.8}$$

where the terms are

$$p_k(y^n|k, x^n) = \int_{\Omega_k} w(\theta_k)p_k(y^n|k, \theta_k, x^n)\mathrm{d}\theta_k.$$

Obviously, the Bayes predictive distribution for y_{new} at x_{new} is

$$p(y_{\text{new}}|\mathcal{D}_n, x_{\text{new}}) = \sum_{k=1}^{K} w(k|\mathcal{D}_n)p_k(y_{\text{new}}|k, x_{\text{new}})$$

$$= \sum_{k=1}^{K} w(k|\mathcal{D}_n) \int_{\Omega_k} p_k(y_{\text{new}}|k, \theta_k, x_{\text{new}})w_k(\theta_k|\mathcal{D}_n)\mathrm{d}\theta_k, \tag{9.9}$$

where $w_k(\theta_k|\mathcal{D}_n)$ is the posterior for θ_k formed using the data $\mathcal{D}_n$, the model k, and the prior $w(\theta_k)$. Expression (9.9) parallels (9.7), the posterior replacing the sampling distribution, and a Bayesian would use (9.9) to find point predictors from the posterior mean, median, etc., and PIs from the percentiles. It is an important feature of the Bayes approach that model variability with respect to the fixed model list $\{p_k|k = 1, \ldots, K\}$ is automatically included in (9.9). Mathematically, a PI may involve the highest density region of more than one p_k and this reflects the posterior from the p_k as well. (The analogous observation holds for (9.5)–(9.7).)

Clearly, (9.9) gives prediction, not model selection. If a Bayesian wants to select a model, finding the posterior is not enough. The conventional approach is to use the modal model from the posterior (9.8). The justification for this is to frame the prediction problem as a decision theory problem by adding a zero–one loss. Specifically, suppose that the goal is to choose one of the K models. Then, denoting an action by a, a natural zero–one loss function is

$$L(k, a) = \begin{cases} 0 & |k - a| \le c, \\ 1 & \text{else,} \end{cases} \tag{9.10}$$

where $c \ge 0$ controls the width of the modal interval. The posterior probability $P(a - c \le k \le a + c|\mathcal{D}_n)$ is maximized if a is chosen to be the midpoint of the modal interval of length $2c$, and this gives the Bayes action. In the limit $c \to 0$, the loss is exactly unity for any wrong value of k and zero for the right value of k. Sometimes this is called classification loss. The Bayes optimal value of k achieves

$$\arg\max_{k} W(k|\mathcal{D}_n),$$

the mode of the posterior over the model list. A convenient approximate expression for the Bayes-optimal k is called the Bayes information criterion (BIC). Write

$$\text{BIC}(k) = -2\ln p_k(y^n|k, \hat{\theta}_k, x^n) + k\ln n.$$

It can be seen that BIC(k) satisfies Theorem 9.3 so taking its minimum over k is a consistent estimator for the true model if the latter is on the list; more generally it converges to the model on the list closest to the true model. A heuristic derivation for the BIC is easy. Start by writing the posterior for model k as

$$\ln W(k|\mathcal{D}) = \ln W(k) + \ln p_k(y^n|k, x^n) - \ln \sum_{k=1}^{K} W(k)p_k(y^n|k, x^n)$$

$$\approx \ln p_k(y^n|k, x^n)$$

$$= \ln \int_{\Omega_k} p_k(y^n|k, \theta_k, x^n) w_k(\theta_k|k)\mathrm{d}\theta_k, \qquad (9.11)$$

where the approximation is justified by the fact that $W(k)$ is bounded as a function of n and k (since $K < \infty$) and the last term is independent of k. Provided that there is a true model with a true value $\theta_{k,0} \in \Omega_k^0$ and p_k is smooth as a function of θ_k, Laplace's method can be applied to the integral in (9.11). The first step is to show that, for any given $\epsilon > 0$,

$$\int_{B(\theta_{k,0}, \epsilon)^c} p_k(y^n|k, \theta_k, x^n) w_k(\theta_k|k)\mathrm{d}\theta_k \leq \eta$$

for any preassigned $\eta > 0$, when n is large enough. The second step is to use a second-order Taylor expansion of $p_k(y^n|k, \theta_k, x^n)$ at $\hat{\theta}_k$, where $\hat{\theta}_k$ is the MLE from the kth model on $B(\theta_{k,0}, \epsilon)$, and then verify that, apart from lower-order terms,

$$\ln W(k|\mathcal{D}) \approx -\ln \int_{B(\theta_{k,0}, \epsilon)} p_k(y^n|k, \theta_k, x^n) w_k(\theta_k|k)\mathrm{d}\theta_k \approx p(y^n|k, \hat{\theta}_k, x^n) - \frac{d}{2}\ln n.$$

Multiplying by -2 gives the BIC. Details can be found in Clarke and Barron (1988). For a more recent and sophisticated treatment, see Berger *et al.* (2003).

In step 2, the derivation of the BIC uses the form of the true distribution to help determine which model on the list is 'true'. Being based on zero–one loss, this is really model selection. However, the AIC requires only that the true distribution exists and, technically, CV (see Sec. 9.4) does not even require this. That is, the AIC and CV are intended for prediction and the BIC predicts well when it can successfully do model selection, i.e., when model uncertainty can be neglected. For prediction in general, a Bayesian uses the predictive distribution, which can be justified by the minimization of a relative entropy; see Aitchison (1975) and (2.30). That is, the optimality criterion satisfied by Bayes model selection is different from that satisfied by Bayes prediction.

A problem with Bayes prediction is that the Bayes paradigm does not deal with bias very well. Indeed, from within the Bayes paradigm it is impossible to detect bias. Two solutions proposed for this are (i) calibration, an essentially frequentist criterion relying as it does on repeated sampling (see Dawid (1982)), and (ii) the use of scoring rules, e.g., the log score (see Sec. 5.5). However, Bayes prediction automatically takes model uncertainty into account, for which reason it can perform well as seen in Sec. 9.1. Frequentist prediction tends to be the reverse: it does not automatically take model uncertainty into account but, because of the frequentist definition of probability, tends to avoid bias better than Bayes methods do, e.g., the derivation of AIC rests on bias correction.

Another way Bayesians do model selection is by testing. Frequentists do this, too, especially in a linear-model context. However, there is little need to review frequentist methods since they are well known. Bayes testing, by contrast, is still undergoing development because selection for across-model priors remains insufficiently understood. Whether for model selection or not, a key difference between frequentist testing and Bayes testing is that the frequentist formulation suffers from multiple-comparisons problems. This is so because the definition of the probability of type-I and type-II errors involves integrals over the measure space, making tests dependent. Bayes methods involve only the posterior, which is conditional on the data, and typically there is no multiple-comparisons problem. (For instance, if the standard coherence argument due to Freedman and Purves (1969) is used then, for testing hypotheses, the posterior odds must be used. Regardless of the hypotheses, the posterior odds are found from a single posterior distribution by integration. So, if two or more tests are to be done, they do not interfere with each other because the data are fixed.)

To see how this works, consider the Zellner g-prior, $g \geq 0$. The setting is the usual linear-model framework $Y_i = x_i^T \beta + \epsilon_i$, $i = 1, \ldots, n$, where x_i is a vector giving the ith set of values of the explanatory variables, ϵ_i is the ith IID error term, usually taken to be $N(0, \sigma^2)$, and $Y_i = y_i$ is the ith outcome. The Zeller g-prior is

$$\beta | \sigma^2, X \sim w(\beta | \sigma^2, X) = N(\beta^*, g\sigma^2 (X^T X)^{-1}),$$

$$\sigma^2 \sim w(\sigma^2 | X) \propto \frac{1}{\sigma^2},$$

where $X = [x_1, \ldots, x_n]$ is the $d \times n$ matrix formed from the concatenation of the explanatory vectors x_i as columns and β^* is the prior mean for β. The hyperparameter g is interpreted as the amount of information in the prior compared with the sample. Smaller values of g indicate more information in the prior for β.

Letting $y = (y_1, \ldots, y_n)^T$, the posterior is

$$w(\beta, \sigma^2 | y, X) \propto p(y | \beta, \sigma^2, X) w(\beta, \sigma^2 | X)$$

$$\propto \left(\sigma^2\right)^{-(n/2)-1} \exp\left(-\frac{1}{2\sigma^2}(y - X\hat{\beta})^T(y - X\hat{\beta}) - \frac{1}{2\sigma^2}(\beta - \hat{\beta})^T X^T X (\beta - \hat{\beta})\right)$$

$$\times \left(\sigma^2\right)^{-k/2} \exp\left(-\frac{1}{2g\sigma^2}(\beta - \beta^*)^T X^T X (\beta - \beta^*)\right),$$

where $\hat{\beta}$ is the usual least squares estimator for β. From this, three posterior densities can be extracted:

$$\beta | \sigma^2, y, X \sim N\left(\frac{g}{g+1}\left(\frac{\beta^*}{g} + \hat{\beta}\right), \frac{\sigma^2 g}{g+1}(X^T X)^{-1}\right),$$

$$\sigma^2 | y, X \sim IG\left(\frac{n}{2}, \frac{s^2}{2} + \frac{1}{2(g+1)}(\beta^* - \hat{\beta})^T X^T X (\beta^* - \hat{\beta})\right)$$

$$\beta | y, X \sim t_d\left(n, \frac{g}{g+1}\left(\frac{\beta^*}{g} + \hat{\beta}\right), \frac{g(s^2 + (\beta^* - \hat{\beta})^T X^T X (\beta^* - \hat{\beta})/(g+1))}{n(g+1)}(X^T X)^{-1}\right),$$

see Sec. 4.2 or Marin and Robert (2007) (or their solutions manual) for the details, where s^2 is the usual MSE estimate of σ^2. The posterior $\sigma | \beta, y, X$ can also be derived but hypotheses about σ are usually less important than those about β for the purpose of model selection. Posterior means and variances can be found from these if needed. The predictive density for a future value Y_{new} at x_{new} is

$$\int w(y_{new} | \sigma^2, y, x_{new}, X) w(\sigma^2 | y, X, x_{new}) d\sigma^2.$$

The posterior mean point predictor for Y_{new} is independent of σ and is given by

$$E(Y_{new} | \sigma^2, y, X, x_{new}) = \frac{\beta^* + g\hat{\beta}}{g+1} x_{new} \tag{9.12}$$

with variance

$$\mathrm{Var}(Y_{new} | \sigma^2, y, X, x_{new}) = \sigma^2 \left(Id_d + \frac{g}{g+1} x_{new}^T (X^T X)^{-1} x_{new} \right),$$

where Id_d is the $d \times d$ identity matrix.

Now consider the null hypothesis $\mathcal{H}_0: L\beta = q$, where $q \in \mathbb{R}$ and L is a matrix of rank r. Under $\mathcal{H}_0$, the general linear model can be reduced to

$$Y | \beta_0, \sigma^2, X_0 \sim N\left(X_0 \beta_0, \sigma^2 Id_n \right),$$

where β_0 is $(d - 1 - r)$-dimensional and X_0 is the corresponding design matrix. Zellner's g-prior is now

$$\beta_0 | \sigma^2, X_0 \sim N\left(\beta_0^*, g_0 \sigma^2 (X_0^T X)^{-1} \right).$$

To derive the marginal for y under $\mathcal{H}_0$ consider the general case. From the definition of the g-prior on β we have

$$X\beta | \sigma^2, X \sim N\left(X\beta^*, g\sigma^2 X(X^T X)^{-1} X^T \right),$$

so, since $Y = X\beta + \epsilon$,

$$Y | \sigma^2, X \sim N\left(X\beta^*, \sigma^2 (Id_n + g\sigma^2 X(X^T X)^{-1} X^T) \right).$$

Integrating with respect to $w(\sigma^2)$ gives the marginal

$$p(y|X) = (g+1)^{-d/2} \pi^{n/2} \Gamma\left(\frac{n}{2}\right) \left(y^T y - \frac{g}{g+1} y^T X (X^T X)^{-1} X^T y - \frac{1}{g+1} \beta^{*T} X^T X \beta^* \right)^{-n/2} \tag{9.13}$$

for Y under the full model. Substituting β_0 for β, β_0^* for β^*, X_0 for X, and g_0 for g in (9.13) gives that the marginal $p(y|X)$ under $\mathcal{H}_0$ is

$$(g_0+1)^{-(d-r)/2} \pi^{n/2} \Gamma\left(\frac{n}{2}\right) \left(y^T y - \frac{g_0}{g_0+1} y^T X_0 (X_0^T X_0)^{-1} X_0^T y - \frac{1}{g_0+1} \beta_0^{*T} X_0^T X_0 \beta_0^* \right)^{-n/2}. \tag{9.14}$$

Taking $\mathcal{H}_1$ to be the unrestricted model implies that the ratio of (9.13) and (9.14) gives the Bayes factor $BF(1,0) = p(y|X, \mathcal{H}_1)/p(y|X, \mathcal{H}_0)$, i.e.,

$$\frac{(g_0+1)^{(d-r)/2}}{(g+1)^{d/2}} \left(\frac{y^T y - (g_0/(g_0+1))y^T X_0 (X_0^T X_0)^{-1} X_0^T y - (1/(g_0+1))\beta_0^{*T} X_0^T X_0 \beta_0^*}{y^T y - (g/(g+1))y^T X (X^T X)^{-1} X^T y - (1/(g+1))\beta^{*T} X^T X \beta^*} \right)^{n/2},$$

after some cancellation. One benefit of these Bayes factors is that they do not depend on σ because it is integrated out in both the numerator and denominator before the ratio is formed.

Bayes factors for model selection are not always well behaved even for the Zellner g-prior. For instance, a related test in the g-prior context is given in Clyde and George (2004). Consider a linear model of the form

$$M_\gamma : \mathbf{Y} = X_\gamma \beta_\gamma + \epsilon,$$

for n outcomes, where X_γ is the design matrix with columns corresponding to the γth subset of $(X_1, \ldots, X_d)$, the vector β_γ contains the regression coefficients, and $\epsilon \sim N(0, \sigma^2 I_d)$. A column vector $\mathbf{1}$ means the vector of n ones and it is easiest to think of γ as a d-dimensional vector of zeros and ones indicating which explanatory variables are in M_γ. Letting $d_\gamma = \dim(\beta_\gamma)$, assign the Zellner g-prior. Letting R_γ^2 be the usual R^2 for the γth model, we have

$$BF(\gamma, 0) = (1+g)^{(n-d_\gamma-1)/2}(1 - g(1 - R_\gamma^2))^{-(n-1)/2}.$$

(See Severinski *et al.* (2010) for a detailed derivation.) As stated in Clyde and George (2004): 'For any fixed g, the Bayes factor ... goes to $(1 + g)^{(nd_\gamma 1)/2}$ as $R_\gamma^2 \to 1$'. Thus, for a fixed n, the Bayes factor is bounded no matter how overwhelmingly the data support γ.

Another problem with Bayes factors is that they do not always work well when the dimensions of the two hypotheses are different. To see this, consider two models. The first model, M_I, is one dimensional: let $\theta \sim N(0, \sigma^2)$, and suppose that the $(X_i|\theta) \sim N(\theta, 1)$ are independent for $i = 1, 2$. The second model, M_{II}, is two dimensional: let $(\theta_1, \theta_2)^t \sim N(\mathbf{0}, \sigma^2 I d_2)$, and suppose that $(X_i|\theta) \sim N(\theta_i, 1)$ for $i = 1, 2$. Thus, in model I the two variables are tied together by a parameter while in model II they are not. Now, the form of the Bayes factor is

$$BF(M_I, M_{II}) = \frac{w(x|M_I)}{w(x|M_{II})} = \frac{\sigma^2+1}{\sqrt{\sigma^2+1}} \exp\left((\eta/2)(2x_1 x_2 - \tau(x_1^2 + x_2^2))\right),$$

where $\eta = \sigma^2/(2\sigma^2+1)$, $\tau = \nu/\eta$, and $\nu = \eta\sigma^2/(1+\sigma^2)$; see Severinski *et al.* (2010) for the details. Clearly, $BF(M_I, M_{II}) \geq 1$ and will typically favor M_I, often strongly. Indeed, when $\sigma \to 0$, $BF(M_I, M_{II}) \to \infty$, again favoring M_I. The implication is that priors selected for parameters of different dimensions may not be compatible and therefore some technique other than the naive use of BFs may be required for comparing models of different dimensions.

Taken together, this means that in some Bayes model-selection problems, as with some frequentist model-selection problems, one can use testing to arrive at a 'final' model. However, predictions from such a final model alone will only be valid if it is so representative of the data that the other models can be neglected. When this is not the case, one must revert to (9.9) for Bayes PIs or one of (9.5)–(9.7) for frequentist PIs.

From a more abstract perspective, one feature that Bayes and frequentist testing have in common, whether for model selection or not, is that both assess harm (or benefit) using optimality principles phrased in terms of expectations. The Frequentist takes expectations over the sample space via the concept of power and level; the Bayesian takes expectations over the parameter space by finding the posterior risk with respect to generalized zero–one loss (although any other type of loss would still give a risk). As noted in the introductory remarks, the cost of sequential minimax solutions can be prohibitive, and using an average-error criterion may be better. However, even if using expectations provides a better representation of the potential harm or benefit from a procedure, it may not always provide an accurate summary.

An example due to Draper (2014b) provides insight. The probabilistic argument for not buying a lottery ticket is well known: the expected gain is negative. However, if, as with some lotteries, unclaimed prizes accumulate, the expected gain can be positive even though the chance of winning is very low. The decision-theoretically optimal action is then to sell all your belongings and buy lottery tickets. However, since the chance of winning is small, the chance of losing all your net worth is high. Since the setting is sequential, in the absence of a way to recover from repeatedly going broke until you finally win, the expected gain does not summarize the likely consequences well. This represents a resource constraint of the same sort as that noted with the sequential application of minimax procedures. Also, in both the sequential minimax and expected loss cases, the problems seem to arise when the probabilities of events in either the sample space or parameter space become too small.

The use of expected risk as a way to choose a procedure rests on the von Neumann–Morgenstern expected utility theorem or Savage's axioms, which give conditions under which maximizing a preference relation can be represented by maximizing an expectation. The implication is that there is something inherent in this formulation that is more limiting than is commonly perceived. Moreover, an alternative to the theoretical structure of the expected utility theorem or Savage's axioms has been developed; see Rostek (2010). It is based on maximizing quantiles of distributions, e.g., medians, rather than the expectations that occur in means or minimax quantities. The goal of this alternative theoretical construct for decision making is to avoid unlikely but very damaging losses. Hence, there may be cases where sequential quantile maximization may outperform techniques that maximize expectations. Moreover, the scoring rule approach, see Sec. 5.5, is another theoretical construct that has advantages and disadvantages for making predictions; further comments are provided in Sec. 9.8.4. This is not to say that scoring rules or the theoretical framework of Rostek (2010) should replace conventional decision theory, merely that progress on finding predictive techniques that perform reliably better than existing hypothesis-testing methods may require a different formulation than has hitherto been proposed. In summary, from a predictive standpoint, at this time, there seems to be no standard theoretical structure that generally encapsulates the information in the CPE or other predictive quantities. That is, the discrepancies between predictions and outcomes and the properties of such discrepancies do not comfortably fit with any statistical theory yet proposed, apart from prequentialism and its extensions; see Chapter 2.

A final comment is that the BIC, having a larger penalty than the AIC, tends to parsimony, as seen by comparing Theorems 9.2 and 9.3. Thus, van Erven *et al.* (2012) proposed that the AIC be used to search for models that will give good predictors and that, once good models are found, the BIC should be used to zero-in on one model in particular. They evaluated their proposed method predictively in terms of the CPE of point predictors. However, as with many model-selection methods, extending point prediction to PIs remains difficult and relatively unexplored.

9.4 Cross-Validation

The term cross-validation (CV) comprises a collection of techniques that accomplish a statistical goal by splitting the goal into steps, using one subset of the data for each step and combining the results. The splitting of the data into subsets ensures that the steps toward the goal are made independently; the combining of the results over the subsets ensures that the inferences are valid, at least approximately, in the true distribution (assuming it exists).

The earliest reference to a procedure that is recognizably cross-validatory is in Lachenbruch and Mickey (1968): the main procedure discussed there would now be called leave-one-out CV (in a classification context). More recently, a unified treatment of CV based on decision theory was provided by Arlot and Celisse (2010). Between these two contributions, and since the second, there have been numerous other contributions to understanding how CV performs and what exactly it does. Here, it will suffice to give an overview of the main points.

Start with a data set $\mathcal{D} = (x_i, y_i)|_{i=1}^n$ and assume it can be split into, say, ℓ subsets $\mathcal{D}_s$, for $s = 1 + \cdots + \ell$, of size n_s, with $n_1 + \cdots + n_\ell = n$. Usually, $\ell = 2, 3$ but $\ell \geq 4$ is possible. As in Sec. 9.3, assume that a collection of models $p_k(\cdot | \theta_k)$ is available. A typical goal would be to make predictions for a future outcome of Y. Then, there are three steps: (i) estimate the value of k, say $\hat{k}$, and the parameters in $p_{\hat{k}}$; (ii) choose the best $\hat{k}$ and $\hat{\theta}_k$ by minimizing some criterion of fit of the models to the data, i.e., choose the model to be used in the next step; and (iii) evaluate the quality of the chosen model. Accordingly, $\ell = 3$ and $\mathcal{D}$ must be split into three disjoint subsets $\mathcal{D}_1$, $\mathcal{D}_2$, and $\mathcal{D}_3$ with one subset used for each step. There are many ways to split $\mathcal{D}$ into disjoint sets of size n_1, n_2, and n_3. Often, many splits of the data are used and the results pooled. For instance, if n_1, n_2, and $\mathcal{D}_3$ are chosen and held fixed then there are $C(n_1 + n_2, n_1)$ distinct ways to split the $n_1 + n_2$ data points. Choosing a random collection of these splits, using the $\mathcal{D}_1$ to estimate the θ_k, and using the corresponding $\mathcal{D}_2$ to evaluate the error criterion for each k gives a collection of 'CV errors' that represents how well each model k summarizes the information in the data, in a predictive sense, for each split. Averaging the resulting CV errors, i.e., averaging the empirical errors for the $\mathcal{D}_2$ from the splits, gives an overall assessment of error, so that the optimal k, $\hat{k}_{\text{opt}}$, can be found. In turn the performance of the $\hat{k}_{\text{opt}}$ model can be evaluated on $\mathcal{D}_3$. In fact, it is enough for n_3 to be held fixed so that the points in $\mathcal{D}_3$ can appear in $\mathcal{D}_1$ or $\mathcal{D}_2$. Enough random splits of $\mathcal{D}$ must be chosen that the overall result is representative of the full data set.

To be more formal, let L be a loss function, $L = L(y, f_{\hat{k}}(x | \hat{\theta}_k))$, where x and y are generic values of the explanatory variables and of the outcome, respectively, and $f_k = f_k(\cdot | \theta)$ is the predictive function determined by the density p_k. That is, $f_k(\cdot)$ is the plug-in point predictor for Y, sometimes denoted $\hat{Y}(x_{\text{new}})$, corresponding to p_k. Consider the following simplified version of an $\ell = 3$ case.

Assume that within each p_k there is a way to estimate $\hat{\theta}$ using $\mathcal{D}_1$ such that the risk is $E(L(Y, f_{\hat{k}}(X | \hat{\theta}_k)) | \mathcal{D}_1)$, where the expectation is over Y and X regarded as random variables.

The risk is seen to be an expected prediction error, i.e., an expected distance between the predictor $f_{\hat{k}}(x|\hat{\theta}_k)$ and $y(x)$. The risk can be approximated using the data in $\mathcal{D}_2$ by

$$\frac{1}{n_2} \sum_{(x_i,y_i)\in\mathcal{D}_2} L(y_i, f_{\hat{k}}(x_i|\hat{\theta}_k)). \tag{9.15}$$

The data set $\mathcal{D}_1$ is often called the training data and the data set $\mathcal{D}_2$ is often called the test set. Since (9.15) is available for each k, the value $\hat{k}$, and hence $\hat{\theta}_k$, can be chosen to minimize (9.15). For the sake of clarity, denote these by $\hat{k}_{\text{opt}}$ and $\hat{\theta}_{\hat{k}_{\text{opt}}}$. Now, holding $\mathcal{D}_1$ and $\mathcal{D}_2$ fixed, the performance of $f_{\hat{k}_{\text{opt}}}(\cdot|\hat{\theta}_{\hat{k}_{\text{opt}}})$ is assessed by

$$E(L(Y, f_{\hat{k}_{\text{opt}}}(X|\hat{\theta}_{\hat{k}_{\text{opt}}}))|\mathcal{D}_1, \mathcal{D}_2) \approx \frac{1}{n_3} \sum_{(x_i,y_i)\in\mathcal{D}_3} L(y_i, f_{\hat{k}_{\text{opt}}}(x_i|\hat{\theta}_{\hat{k}_{\text{opt}}})). \tag{9.16}$$

The left-hand side of expression (9.16) is the expected prediction error for a model chosen using $\mathcal{D}_1$ and $\mathcal{D}_2$. However, the expectation is not actually 'conditional' on $\mathcal{D}_1$ and $\mathcal{D}_2$; the conditioning notation is used only to indicate that $\mathcal{D}_1$ and $\mathcal{D}_2$ are held fixed. Thus, $\mathcal{D}_1$ is used for estimation, $\mathcal{D}_2$ is used for predictor selection, and $\mathcal{D}_3$ is used to assess the average performance of the chosen predictor. Note that, even though k is treated as discrete, the same methodology extends to continuous k.

As written, the predictor $f_{\hat{k}_{\text{opt}}}(\cdot|\hat{\theta}_{\hat{k}_{\text{opt}}})$ with performance (9.16) uses only one partition of $\mathcal{D}$ into three disjoint subsets. To see how this generates a predictor, observe that if $f_{\hat{k}_{\text{opt}}}(\cdot|\hat{\theta}_{\hat{k}_{\text{opt}}})$ comes from a model then it can be converted back to a density $p_{\hat{k}_{\text{opt}}}(\cdot|\hat{\theta}_{\hat{k}_{\text{opt}}})$ that can be used to obtain plug-in point predictors (with SDs) or PIs – at the cost of under-representing the variability due to model selection. One way to correct for this for point predictors is to use the $p_{\hat{k}_{\text{opt}}}(\cdot|\hat{\theta}_{\hat{k}_{\text{opt}}})$ and an SD from (9.16). For instance, if L is the squared error then the square root of (9.16) can be taken as an estimate of the SD of the predictor. For PIs from $p_{\hat{k}_{\text{opt}}}(\cdot|\hat{\theta}_{\hat{k}_{\text{opt}}})$, it is not clear how to correct for under-representing the variability of model selection, because only the empirically optimal model is used; other models that have relatively small values for (9.15) are not considered. (The possibility that such models could be considered is treated briefly at the end of this section by regarding CV as an MSP.) If alternative models are not going to be considered for the formation of PIs, then it may be better to put the data in $\mathcal{D}_3$ into $\mathcal{D}_1$ and $\mathcal{D}_2$, i.e., to use $\ell = 2$ instead of $\ell = 3$, so that parameter estimation and model selection will be better and then to enlarge the PI from the resulting $p_{\hat{k}_{\text{opt}}}(\cdot)$ by taking maximum and minimum percentiles using upper and lower confidence bounds on the components of $\hat{\theta}_{\hat{k}_{\text{opt}}}$.

Although it is not obvious from (9.15) and (9.16), there is no guarantee that any of the three subsets of $\mathcal{D}$ are representative of the DG, and even worse, (9.15) and (9.16) can be very large and variable. This is one main reason for using multiple splits of $\mathcal{D}$. The other main reason to use multiple disjoint subsets of the sample of size n is to correct for the fact that, for instance, the training error is much smaller than the testing error. In an extreme case, the training error can be made small merely by choosing a rich enough class of models. Analogous problems arise if the same data are used for model selection and model assessment: the model will appear to be much better than it is. Indeed, using the same data twice in the same sequence of procedures tends to give inferences that are unrealistically strong, resulting in overfit; this means that the generalization error will be large. For these two reasons, typically one averages over many splits of $\mathcal{D}$ to ensure that the final predictor and its performance are representative of the DG.

For instance, in J-fold CV,[1] $\ell = 2$ is often used when n is too small to permit choosing a reasonable value $n_3 > 0$. Then, $\mathcal{D}$ is randomly partitioned into J disjoint subsets, say $\mathcal{D}_j$, of the same size (or as close as possible). The training sets are then the $\mathcal{D}_j^c$, the complement of the $\mathcal{D}_j$ in $\mathcal{D}$, and the corresponding test sets are the $\mathcal{D}_j$ themselves. For each value of k the CV error on $\mathcal{D}_j$ is

$$Err(k, j) = \sum_{(x_i, y_i) \in \mathcal{D}_j} L(y_i, f_k(x_i | \hat{\theta}_k)),$$

giving

$$Err(k) = \frac{1}{n} \sum_{j=1}^{J} Err(k, j)$$

for each k as an analog to (9.15). The value $\hat{k}$ of k that minimizes $Err(k)$ is then chosen as the index of the best model. It can be seen that J-fold CV, like other forms of CV, can be regarded as random subsampling from the data set in much the same way as bootstrapping. The difference between CV and bootstrapping is that the subsampling for CV is done subject to the restriction that ℓ data sets will be formed and their sizes (all nonzero) will sum to n, while the subsampling for bootstrapping is only restricted to a procedure either with or without replacement. The special case of $J = n$ is called leave-one-out CV (LOOCV) because each distinct set of $n - 1$ data points is used to predict the one data point left out.

One problem with J-fold CV is that, since $\mathcal{D}_3 = \varnothing$, there is no reliable assessment of how well $f_{\hat{k}_{opt}}$ should be expected to perform predictively. However, the predictor achieves an empirical minimum so in principle it is not possible to do better. In the absence of a nonvoid set $\mathcal{D}_3$, point predictors can be given but SDs for them cannot be estimated and PIs cannot be found, unless upper and lower confidence bounds on the components of $\hat{\theta}_{\hat{k}_{opt}}$ are used.

Even though J-fold CV uses $\ell = 2$, it is easy to imagine cases where $\ell = 4$ makes sense. For instance, it is not hard to imagine reserving a subset of the data for feature selection. If a large number of functions have been proposed as 'features' of the phenomenon under investigation, their adequacy may have to be evaluated on a subset $\mathcal{D}_0$ of $\mathcal{D}$ separately from the other $\mathcal{D}_j$. One can imagine using $\ell = 5$ if a variety of model classes has been proposed, e.g., Fourier series, trees, neural nets, and kernel methods (see Sec. 10.2) and one step requires choosing amongst them. In short, ℓ increases as the number of distinct steps increases, and the latter occurs as the complexity of the inferential task increases.

Cross-validation is intuitive because it represents a within-data-set prediction, has been used extensively, and can give predictors even in the $\mathcal{M}$-open setting; see Bernardo and Smith (2000). So, it is surprising how little is really known about how well CV performs. It is

---

[1] This is usually called K-fold CV, but $k = 1, \ldots, K$ has already been used to index models p_k rather than subsets of $\mathcal{D}$.

known, however, that in many cases as much data as possible should be reserved for predictor assessment rather than predictor construction (Shao 1993). Also, CV is consistent even for comparisons involving nonparametric models and in many such cases the data reserved for predictor assessment does not have to be larger than that for predictor construction; see Yang (2007). Taken together these two statements suggest that if the model class is narrow then more data should be used for evaluation and if the model class is broad then more data should be used for predictor construction – the idea is that you put the data where it will minimize variability. On that point, on the one hand LOOCV is unstable because the data sets used to predict different y_i differ only in one data point and hence are highly dependent. On the other hand, LOOCV asymptotically satisfies an optimality property and is similar to both ' Mallows' C_p criterion and a form of CV often used for linear nonparametric regression; see Li (1987).

Overall studies of the choice of J in J-fold CV are numerous and relatively inconclusive; see, for instance, Anguita *et al.* (2012). The one heuristic that is relatively well established is that J should be larger for larger n. So, it may be best to be empirical when n is large enough: take $\ell = 3$ and use n_1 data points to form the predictors, where n_1 is large enough that the parameter estimates in f_k are thought to be reliable. The ideal is ten or more data points per parameter, but often five or six per parameter will suffice. Then use n_2 data points to form CV errors for a range of J's, where n_2 is large enough that the curve of CV errors has a discernible minimum. Then use this value of J and n_3 data points to evaluate the CV errors in choosing a predictor, ensuring that n_3 is large enough for the variability in the CV error to be below some preset threshold, e.g., use the n_3 data points to bootstrap a distribution for the CV error for the empirically optimal J. If needed, and n is large enough, n_4 data points can be reserved for assessment of the final predictor.

Despite this there is a variety of empirical and intuitive observations about CV which can be made and which may provide guidance when using CV for predictor selection. First, the CV procedure is empirical and predictive and does not require a true model to exist. The procedure can be used directly for predictor selection in the absence of a model. Consequently, CV can be applied in $\mathcal{M}$-open problems as well as $\mathcal{M}$-complete and $\mathcal{M}$-closed problems. The CV principle – matching steps in an inferential task to disjoint subsets of data and then pooling as appropriate – applies in great generality, subject primarily to the need for large enough sample sizes that the subsets of the data are in turn large enough to permit each step in the inference procedure to be carried out satisfactorily.

The key problem in all this is that the variability of the CV error can be high. For instance, if some data is reserved for bootstrapping the CV error of a predictor then the SD of the empirical distribution can be higher, for instance, than the CV error itself. In particular, if J-fold CV is used then there are J errors and the SD they generate can be very high. So, as a generality, CV does not provide strong discrimination across models. It seems that the amount of data required to do good model selection is much higher than the amount of data required to do, say, parameter estimation. In this sense, model selection is less efficient than parameter estimation. This is true of other MSPs besides CV – recall that CV is similar to Mallows' C_p, which in turn is similar to AIC – and probably arises because model classes are so much larger than the range of a real parameter. Since this seems to be inherent in CV, and other MSPs, it may be better to use MSPs in general only for comparing a small number of relatively different models. Weak discrimination across large sets of models that are not strongly differentiated from each other is the frequentist analog of the Bayesian's dilution of the posterior.

A second way to deal with the high variability of CV error is to invoke the 'one-standard-error rule': choose the most parsimonious model that has CV error not more than one SE higher than the best CV error, assuming that there is a model with this property. Otherwise, just choose the minimum-CV-error model. To follow this rule requires that the SE of each CV error, i.e., the SE of the values of (9.15) over the various splits of the data, must be found. The idea is that the minimum CV error is likely to be unrealistically small, so the model achieving it is likely to be suboptimal. Given that larger models are likely to have larger variability, one corrects for this a little by choosing a slightly smaller (more parsimonious) model that has almost as small a CV error as the minimum CV error. How well this works in practice as a correction for high variability is unclear. The one likely effect of the one-standard-error rule is that it increases the chances that the terms in the model are really present, i.e., useful for prediction, and not just an artifact of the minimization. However, whether this rule affects consistency will depend on the models on the model list that are close to the minimum-CV-error model. If there are many models on the list that have CV error near the minimum then imposing parsimony might omit important terms. If there are few or no models on the list with CV error near the minimum, the effect is likely to be small, if present at all, and any consistency properties of CV are likely to be retained. It is in the intermediate case, where the number of models with CV error near the minimum is neither too small nor too large, that the rule is likely to be most effective. The problem is that it is generally unclear how to formalize this intermediate case because it depends on the sample size and the density of the predictors on the list near the best predictor.

A third way to deal with high variability is to robustify the CV. The idea is that if a significant amount of the variability in the CV error can be attributed to an error term in a signal-plus-noise model, e.g., it has heavy tails so there are many outliers in the data, and the end effect of the outliers should be downweighted. This can be done by changing the loss function or changing the operation performed on the summands in (9.15) or (9.16). The actual summands should not be changed because they represent the comparison of the predictions with the observed values. Usually, a histogram of the summands will be positively skewed and the skewness will increase as the steepness of L increases. This suggests that truncating L so that it represents only errors of a typical size will improve the discrimination of the CV. Thus, when the true model exists and has an error term with moderately heavy tails, the Huber loss (see Huber (1964, 1973)) often gives better results than, say, the usual squared error loss. See Ronchetti *et al.* (1997) for a more recent explication of and a different justification for the Huber loss. See Lambert-Lacroix and Zwald (2011) for an extension of this to penalized methods. Overall, if the tails of the error term are light then regular CV with squared error will give the best performance for model selection. This is true even though the squared error often imposes excessive sparsity by capturing only the leading terms in a model and hence not necessarily being satisfactory for prediction. Once the tails are moderately heavy, the Huber loss, being a truncated squared error loss, outperforms the regular squared error loss for model selection and does not seem to impose oversparsity. When the tails are very heavy, as occurs with many types of complex data, it may be necessary to use some operation on the summands besides taking their average. The median has been proposed, see Yu and Clarke (2015), but it is not hard to imagine other possibilities.

Note that even in $\mathcal{M}$-closed problems there may be a series of steps that must be applied to get a useful answer in model selection. Consider the case of high-dimensional data. Some data might be used for feature selection. Then, shrinkage might be applied to find a model, necessitating a second set of data. Estimating the parameters in the model might require a third set of data. Testing whether the inclusion of parameters is useful might require a fourth set. Finally, assessing the adequacy of the reduced model would require a fifth set. Obviously, averaging the errors over splits of the data, e.g., over the data used for finding the shrinkage parameter and estimating the other parameters, should improve the estimates that go into the predictor, and averaging over the splits of the data used to assess the adequacy of the final predictor should give a better assessment of it.

When the loss function L is the squared error, the CV error is essentially an MSE and so can be expressed as a variance–bias composition. This means that J-fold CV, for instance, seeks the predictor (or model) that gives the optimal tradeoff between variance and bias. Seeking low bias favors large models (at the cost of increased variance and possibly over-fitting) whereas seeking low variance favors small models (at the cost of increased bias). As a generality, choosing J large gives smaller bias, larger variance, and a longer running time for computation while choosing J small gives larger bias, smaller variance, and a shorter running time. This sort of reasoning can be adapted to other loss functions; see Domingos (2000) and Clarke (2010). Indeed, it can be seen from (9.15) and (9.16) that, independently of the loss function, as long as the data are independent, the CV errors are consistent estimates of their expected predictive errors when a true model exists.

Finally, when the predictors being compared under CV come from models, minimizing the CV errors is similar to using information criteria in that a worth is assigned to each model. The difference is that the goal is to have a minimum CV error rather than maximum information. So, if CV is regarded as an MSP, the CV error behaves like the inverse of an IC_α. Writing MSP_k to mean the inverse of the CV error for model k, the expressions (9.5)–(9.7) can still be used to get predictors without actually doing model selection.

9.5 Simulated Annealing

In materials science, annealing is the process of heating a substance, usually metal, and then cooling it slowly so as to toughen it and make it less brittle. Simulated annealing (SA) is based on the analogy that searching for the optimal element in a set, under some criterion, is like a physical cooling process in that high values of a parameter correspond to higher perturbations of a candidate solution in the same way as higher temperatures correspond to more movement of molecules. As the parameter decreases it continues to act like a 'temperature' in that a candidate solution is perturbed less, and in the limit the solution 'freezes', i.e., is no longer perturbed. The hope is that the limiting solution is optimal and robust. Thus, SA is a directed stochastic search technique for optimizing a cost over a set.

In its abstract form, the setting for SA is the following; see Bertsimas and Tsitsiklis (1993). Start with a finite set S and a real-valued function, say $J: S \to \mathbb{R}$, to be maximized. (The procedure is much the same if J is to be minimized; it's enough to change the signs appropriately.) Given $u \in S$ there is a set $S(u) \subset S \setminus \{u\}$ called the neighbors of u. Sets of neighbors are symmetric in the sense that $v \in S(u) \Leftrightarrow u \in S(v)$. For each u there is a probability vector $q(u) = (q_{u,1}, \ldots, q_{u,\#(S(u))})^T$ that describes jumps from u to its neighbors. There is also a decreasing function $T: \mathbb{N} \to \mathbb{R}^+$ called the 'cooling schedule' that indicates how the likelihood of acceptance of a worse solution (in the sense of lower J) decreases. The value $T(m)$ is called the temperature at stage m. The SA procedure starts at a given initial value $u(0) \in S$ and generates a sequence of values $u(m)$ such that, as $m \to \infty$, $u(m)$ converges to an element of $S^* \subset S$ where $S^* = \arg\max_{u \in S} J(u)$. In principle, each run of the SA procedure results in an element of S^*, so usually many runs are done, with varying T and $u(0)$ for instance, and their results examined. Implicitly it is assumed that, although finite, $\#(S)$ is very large and that the values $J(u)$, for $u \in S$ regarded as a surface over S, are very rough. That is, J may change a lot in response to a small change in u.

The generic SA procedure produces a sequence whose last element is a candidate element of S^* as follows. Given $T(\cdot)$ and $u(0)$, choose a point $u^*(1)$ in $S(u(0))$ at random according to $q(u)$. If $J(u^*(1)) \geq J(u(0))$ then $u(1) = u^*(1)$. If, however, $J(u^*(1)) < J(u(0))$ then there are two choices, set $u(1) = u(0)$ or set $u(1) = u^*(1)$. That is, either u does not change or u takes a step down in terms of J, accepting a worse candidate in the hope that its neighbors will ultimately lead to a better solution overall, i.e., escape a local maximum. The decision to stay put or take the risk is made probabilistically:

$$u(1) = \begin{cases} u^*(1) & \text{with probability} \quad \exp\left(\dfrac{J(u(0)) - J(u^*(1))}{T(1)}\right), \\ u(0) & \text{otherwise.} \end{cases} \qquad (9.17)$$

A similar procedure is used to choose $u(2)$, $u(3)$, and so forth. The probabilistic selection of the sequence $u(m)$ may be concisely expressed as

$$P(u(m+1) = b | u(m) = a) = q_{a,b} \exp\left(\frac{\max(0, J(a) - J(b))}{T(m)}\right)$$

for $b \neq a$ and $b \in S(a)$, where it is understood that if $b \neq a$ and $b \notin S(a)$ then $P(u(m+1) = b|u(m) = a) = 0$. The quantity in the exponent is essentially the same as the acceptance probability in the Metropolis–Hastings algorithm; see Chib and Greenberg (1995). The idea is that a sequence of outcomes drawn according to these probabilities will converge to an element of S^*.

To see why this might be so, recognize that SA defines a Markov chain that is inhomogeneous (there are different transition matrices at different time steps) when T varies but for fixed values of T is homogeneous. Markov chains that are irreducible and aperiodic converge to their stationary distribution. So, when $q_{u,v} = q_{v,u}$ and T is constant the stationary distribution is

$$w_T(u) = \frac{1}{Z_T} e^{-J(u)/T} \quad \text{for} \quad u \in S^*$$

and zero otherwise, where Z_T is a normalizing constant. Moreover, as $T \to \infty$, w_T concentrates at S^*. (In fact, this is true more generally.) In principle, therefore, elements of S^* can be found, with high probability, by running the chain for many steps and taking the last step as an outcome of the stationary distribution when T is small. Since the number of steps required for the convergence of a sequence of outcomes to an element of S^* increases as $T \to 0$, SA uses a cooling schedule in the hope that homogeneous chains corresponding to stable values of T will speed the convergence of the overall inhomogeneous chain.

The main convergence result for SA is due to Hajek (1988), and its two key features are as follows. First, the usual hypotheses of aperiodicity and irreducibility for the convergence of Markov chains are in effect replaced by the condition that paths in S from any $u \in S$ to S^* exist with strictly positive probability, i.e., there is a sequence $u(m)$ such that the Markov chain has a route in S to a maximum of J. Second, points in S^* must also be reachable from any $u(0)$, in the sense that J should not become too large. Specifically, if J is evaluated at each point on a path starting from, say, $u(0)$ and ending in S^* then there is a bound on the path, of the form $J(u(0)) + h$ where $h > 0$ is uniform over all $u(0)$. Loosely, Hajek's theorem gives that SA converges if and only if $T(m) \to 0$ slowly enough. Formally:

Theorem 9.4 (Hajek 1988) *Suppose that $T(m)$ decreases to zero and that for any $u(0)$ there is a path $u(m)$ in S with endpoint in S^* and with $u(m+1) \in S(u(m))$ for all m. Also, suppose that*

$$d^* = \sup_h \{h \,|\, \forall u(0) \,\exists \text{ a path from } u(0) \text{ to an } s^* \in S^* \text{with bound } J(u(0)) + h\}$$

exists. Then, sequences in S drawn according to the SA algorithm converge to elements of S^* if and only if

$$\sum_{m=0}^{\infty} e^{-d^*/T(m)} = \infty.$$

Since the convergence of an SA sequence is only to S^*, and S^* in general is not a singleton set, the point of SA is mostly to reduce the collection of possible models S that have to be examined to a more manageable number, $\#(S^*)$. Although it is not necessary to use CV, one way to understand the role of SA is to see it as a precursor to CV when S is so large that CV cannot be effective. Recall that in CV it is assumed that all the models or predictors to be compared can be listed and their worth examined. This is not the case when the cardinality of the model list is so large that the data points are unable to provide useful discrimination over them, e.g., by interpolation of the data. In such cases one can preprocess the collection S of candidate models by SA to generate a small number of final values of finite sequences of $u(m)$'s, hopefully very close to elements of S^*. The optimality criterion J defining S^* is chosen so that the models in S^* are 'good'. Usually, 'good' refers to models that are not discredited by the data and are not so large that they merely interpolate the data, i.e., satisfy some generalized variance–bias tradeoff relationship. Examples of such model classes will be seen in Sec. 10.2. The approximations to good models produced by SA can be compared by CV (or other techniques). This is not unreasonable because CV is a predictive criterion where the prediction is internal to the data set.

However, the real test of a model or predictor selection procedure is how well it predicts future observations. So, assume that S is a set of predictors (or a set of models that can be converted to predictors) such as linear regression models with a K-dimensional vector of explanatory variables, and suppose that data $\mathcal{D} = (x_i, y_i)|_{i=1}^n$ is available. Choose an optimality criterion J on S such as IC_α (see Sec. 9.2) or the root mean squared error (RMSE) on a subset $\mathcal{D}_1$ of $\mathcal{D}$, i.e.,

$$\text{RMSE}^2 = \sum_{i \in \mathcal{D}_1} \left(\hat{Y}_i(x_i) - y_i \right)^2,$$

and a cooling schedule; $T(m) = c/(\ln m)$ for some $c > 0$ is one that is commonly used. Then, in principle, SA can be used to find a model that approximately maximizes IC_α or minimizes the RMSE. This approximation is in the sense of appearing to be sufficiently close to an optimum, and therefore one does not assume that the model generated by SA is unique. It is important to verify that the SA sequence $u(m)$ has converged to a global minimum, but diagnostics for this are not addressed here; see the discussion in Goffe *et al.* (1994), for instance.

In fact, by varying $u(0)$, T, the neighborhood structure, and any other inputs, SA can normally be used to select several distinct models. Then, the remaining data $\mathcal{D}_2 = \mathcal{D} \setminus \mathcal{D}_1$ (assuming $\ell = 2$) can be used via CV, minimally adjusted R^2, or many other criteria, to select one of the models or predictors generated by SA as a point predictor for Y_{n+1} using x_{n+1}. An added level of variability that may help to enlarge the search is to include bootstrapping over $\mathcal{D}$ to find many $\mathcal{D}_1$, so that the best models are found under a variety of similar optimality criteria. Indeed, in many cases, it may be enough to bootstrap over $\mathcal{D}_1$ alone. If S is very large, it may make sense to choose $n_1 > n_2$. However, if the range of the predictors in S is large enough then $n_2 > n_1$ may be better. Proceeding to a PI, the single model emerging from CV on the SA output would be a naive choice because it neglects the variability of predictor selection from SA and CV and their inputs. Widening it by taking the convex hull of the union of the PIs from all the models generated by SA would likewise be naive. Some procedure in between these two extremes based on $\mathcal{D}_2$ would be likely to be more accurate for setting a width.

In fact, it is unlikely that SA would be used for linear models since there are so many other good techniques; see Tenorio and Shi (1990) for a thorough treatment of a much more realistic example. Despite its oversimplicity, the example here does highlight four important features. First, it is straightforward to get point predictors. Many variants on the above procedure can be readily imagined, as follows. Use all the data for the SA optimization. Combine the SA generated predictors using weights derived from $\mathcal{D}_2$. Generate outputs from SA and then choose only those that satisfy an extra criterion based on $\mathcal{D}_2$, and so forth.

Second, it is not clear how to get PIs. Indeed, SA is an optimization procedure and its stochasticity is introduced by the user, not the data. So, it may be argued that any variability attributed to SA should come only from the fact that J is data dependent, i.e., the apparent variability of the SA procedure given J should be effectively zero even though typically $\#(S^*) > 1$. Moreover, if two procedures are used in sequence to get a point predictor, e.g., SA and CV, then the widths of the PIs should include contributions from SA and CV as well as any other sources of variation.

Third, the selection of S can strongly affect the point predictors. The richer the set S is, the stronger an optimum for J we can expect to find with SA. If multiple distinct model lists $S_1, S_2, S_3, \ldots$ are used, it is unclear how to combine them and attribute a contribution to the width of a particular PI because S is user chosen.

Fourth, there are many choices for inputs to SA (and CV if both are used), and it is important to assess the robustness of these user-specified choices. That is, it may be worthwhile to use local perturbations of the chosen $u(0)$'s, to vary the k in k-fold CV, to apply SA to a sequence of nested choices for S, or to use a range of cooling schedules, in order to see whether similar outputs are generated. Indeed, since J is usually data dependent and parameters in the predictors must be estimated, it may be useful to perturb $\mathcal{D}_1$ (adding, say, $N(0, \sigma^2)$ noise to the data points for a small value of σ) and to use various choices of $\mathcal{D}_1$ and $\mathcal{D}_2$ in order to see whether the outputs generated by SA are relatively stable. There are other important aspects of generating predictions using SA, e.g., ensuring the SA-selected predictors are parsimonious, but these seem to be as yet unexamined.

Now, to be practical, let $\hat{u}_k = \hat{u}_k(\infty)$ for $k = 1, \ldots, K'$ denote outputs from an SA run, assuming that the optimality criterion J is dependent on $\mathcal{D}_1$ and hence should be denoted J_1. If the SA run has converged then there is a $u_k \in S^*$ that is well approximated by $\hat{u}_k$ and $J_1(\hat{u}_k)$ is near $J_1(u_k)$. Thus, even though a unique optimum may be found if the computations are carried out to arbitrary precision, one finds a large number of outputs $\hat{u}_k$; it is necessary to retain only those that are within a suitable range of $\max_k J_1(\hat{u}_k)$. The reason is that all the $\hat{u}_k$ that have converged should lead to the same value of J_1. Therefore outputs $\hat{u}_k$ that are too far from the maximum must have got caught in local optima despite diagnostics that might have indicated otherwise. Note that various parameters may need to be estimated using $\mathcal{D}_1$ to form the $\hat{u}_k$; in the linear models example these would be the β's, but it is important to use SA for more complex choices of S.

To conclude this section, several ways to construct point predictors and PIs are indicated for the sake of completeness. All of these are intuitively reasonable but none can be recommended on rigorous grounds. To begin, observe that, because S is large, $J_1(\cdot)$ is rough, and possibly S^* is not small, it is safe to assume that K of the K' approximations $\hat{u}_k$ have been retained. So, it is reasonable to assign a 'worth' to each of them. Do this by using J_2, the same optimality criterion as J_1 but defined by $\mathcal{D}_2$ rather than $\mathcal{D}_1$. Write $J_{\text{sum}} = \sum_{k=1}^K J_2(\hat{u}_j)$. If the task is to predict Y_{n+1} using x_{n+1} then each prediction $\hat{u}_k(x_{n+1})$ has an associated normalized worth $J_a(\hat{u}_k(\cdot)) = J_2(\hat{u}_k(\cdot))/J_{\text{sum}}$. Let Z be the random variable, which assumes the value $\hat{u}_k(x_{n+1})$ with probability $J_a(\hat{u}_k(\cdot))$. Now med(Z) is a natural robust predictor for $Y_{n+1}(x_{n+1})$ and the $(\alpha/2)100$th and $(1-\alpha/2)100$th percentiles of Z give a natural $1-\alpha$ PI for Y_{n+1}. These predictors take into account all sources of variability ($u(0)$, T, $\mathcal{D}_1$, $\mathcal{D}_2$, SA) except for the choice of S and its corresponding S^* (which may be dependent on $\mathcal{D}_1$). This class of predictors can also be enlarged by using several splits of $\mathcal{D}$ to generate more predictions.

An alternative is to replace the bootstrapping on $\mathcal{D}_1$ by perturbing the values of the response or explanatory variables in $\mathcal{D}_1$ by $N(0, \sigma^2)$ noise. This is called the simulation–extrapolation procedure and has been studied for variable selection in several contexts using the MSE as J; see Luo *et al.* (2007) and the references therein for the linear regression and other contexts such as errors-in-variables models. The effect of this sort of perturbation is to explore the surface defined by J_1 and therefore evaluate its roughness as a function of σ (and S). In this case, σ represents the measurement error on Y or x, and including this extra variability may be useful in generating $\hat{u}_k$ that are representative of S^*.

Another alternative is to generate a small number of $\hat{u}_k$ (perhaps by selecting from a much larger set of $\hat{u}_k$) that seem particularly convincing in terms of convergence of the SA, stability under perturbations, the bootstrapping of $\mathcal{D}_1$, and other desirable criteria. A point predictor for Y_{n+1} then follows by choosing a location $\bar{u}(x_{n+1})$ to represent the $\hat{u}_k(x_{n+1})$, and a PI at x_{n+1} can be found by using $\widehat{\text{MSE}}$, the average MSE of the $\hat{u}_k$ as evaluated on $\mathcal{D}_2$, and scaling by normal percentiles, i.e., using $\bar{u}(x_{n+1}) \pm z_{1-\alpha/2}\sqrt{\widehat{\text{MSE}}}$ as an approximate $1-\alpha$ PI.

9.6 Markov Chain Monte Carlo and the Metropolis–Hastings Algorithm

Monte Carlo methods constitute a broad class of computational procedures that rely on the generation of many outcomes from a well-defined source to obtain numerical results. The Metropolis–Hastings approach originated in Metroplis *et al.* (1953) and Hastings (1970) and is a specific form of acceptance–rejection sampling such that the only outcomes that are not rejected are representative of a desired distribution. Taken together, the Markov chain Monte Carlo (MCMC) and the Metropolis–Hastings (MH) algorithms provide another computational way to do model selection – if the models are the 'outcomes', accepted or rejected – and hence obtain predictors. As seen in Sec. 9.5, SA is also a computational technique to choose models and generate predictions. The common features of SA and MCMC–MH are that (i) both are stochastic searches and (ii) both use probabilistic acceptance–rejection sampling. They differ in that SA uses a temperature where MCMC–MH uses a probability threshold and the goal of SA is to optimize a functional while the goal of MCMC–MH is to generate a representative sequence of outcomes from a partially known distribution representing a well-defined source. Thus, to use MCMC–MH for model selection, a further step is required beyond merely getting outcomes (here, models) that are representative of a distribution over outcomes (models). The extra step is needed to select, from the representative models, only those that are good in the sense of some optimality criterion.

Here, we will use discrete Markov chains (since the model space will be treated as discrete). So, let T be a transition matrix on S. That is, for $u, w \in S$, $T(u, w) = P(X_2 = w | X_1 = u)$ defines a homogeneous Markov chain on the state space S. It is well known that if a homogeneous Markov chain is irreducible and aperiodic then it has a unique stationary distribution v satisfying $Tv = v$. Much of Markov chain theory is concerned with using the outcomes of the X_j in the chain to identify v, given that usually T is at least partially known.

The presentation here is based on the lucid presentation in Chib and Greenberg (1995). As they observe, MCMC methods reverse the usual setting: T is unknown, the stationary distribution is partially known, e.g., up to a constant multiple, and it is outcomes of the stationary distribution that are desired. The term Monte Carlo indicates that many outcomes will be generated for the purposes of estimation.

A stronger sufficient condition for v to exist than $T_v = v$ is that

$$v(u)P(X_1 = w | X_0 = u) = v(w)P(X_1 = u | X_0 = w). \tag{9.18}$$

In words this means that the probability of starting at u and moving to w equals the probability of starting at w and moving to u. Sometimes this reversibility condition (9.18) is satisfied, but it's a strong hypothesis so most of the time it fails. However, MH provides a way to get around any inequality in (9.18) as long as v exists. Basically, if one side is larger than the other, a factor is introduced that will decrease it. The obvious factor to use is the ratio of the two sides. That is, if P is not known, consider the 'candidate-generating distribution' Q, where, for each u, $\sum_u Q(u, w) = 1$, and suppose that Q is not reversible but instead satisfies

$$v(u)Q(u, w) > v(w)Q(w, u). \tag{9.19}$$

This inequality corresponds to the case where the stochastic process moves from u to w too often and from w to u not often enough. To make these two conditional probabilities equal, reduce the frequency with which u moves to w by a factor, say $\alpha(u, w) < 1$, on the left-hand side of (9.19). The idea is to define a Markov chain in which transitions from u to w are made according to

$$P_{\text{MH}}(u, w) = \alpha(u, w)Q(u, w),$$

where α can be derived as follows.

Since w does not go to u often enough, $\alpha(u, w)$ should not decrease the right-hand side of (9.19). Consider the reversibility condition (9.18) and write

$$v(u)Q(u, w)\alpha(u, w) = v(w)Q(u, w)\alpha(w, u) = v(w)Q(w, u). \tag{9.20}$$

The first equality is the reversibility condition, now using α; the second equality shows how α must reduce to the right-hand side of (9.19). Solving for α from the left- and right-hand sides of (9.20) gives

$$\alpha(u, w) = \begin{cases} \min\left(1, \dfrac{v(w)Q(w, u)}{v(u)Q(u, w)}\right) & v(u)Q(u, w) > 0, \\ 1 & \text{else.} \end{cases} \tag{9.21}$$

where the value unity arises because P is bounded by unity. So, it is only the left-hand side of (9.19) that gets reduced by using P_{MH}. Note that, unlike in Chib and Greenberg (1995), the treatment here is only for the discrete case since the focus will be on model selection rather than parameter estimation. Indeed, in the model-selection case, sometimes T can be defined in such a way that $Q(u, w) = Q(w, u)$, meaning that the probability of moving from model u to model w is the same as that for moving from model w to model u. In this special case, α in (9.21) essentially depends only on $v(w)/v(u)$. In the Bayes setting this is often the ratio of the prior probabilities of models w and u, but it may be more complicated; see Hans *et al.* (2007) for an example.

Now, the MCMC–MH procedure can be summarized as follows: repeat the following steps for $i = 1, \ldots, N$, where N is large enough.

- Generate w from $Q(u_i, \cdot)$ and z from Unif[0,1].
- If $z \leq \alpha(w, y)$, set $u_{i+1} = w$.
- If $z > \alpha(w, y)$, set $u_{i+1} = u_i$.
- Obtain $\{u_1, u_2, \ldots, u_n\}$, where $n \leq N$ is the number of samples produced.

The set $U_n = \{u_1, u_2, \ldots, u_n\}$ is representative of the (partially known) stationary distribution v. More precisely, since v is a probability on S, the estimates $\hat{v}_n(u) = \#(\{i \mid u_i = u\})/n$ of $v(s)$ are consistent as $N \to \infty$. In this sense, even though v is only partially known, the probability that v assigns to subsets of S can be estimated well, and so v can be approximated to any level of accuracy desired. There are many excellent expositions of the details of the convergence of MCMC–MH chains as well as diagnostics for whether a series of MCMC–MH chains has converged. These are beyond own present scope, but see Geyer (2011) and Gelman and Shirley (2011) for accessible treatments, primarily from a Bayes parameter-estimation standpoint.

In our discussion so far, MCMC–MH has been used only to obtain a representative sample from a partially known (discrete) distribution. In fact, from a model-selection standpoint, the Markov chain is discrete, jumping from $u \in S$ to $u' \in S$ with probability given by a transition matrix T yet to be defined. There are many Markov chains that can be defined on S, and the goal is to define one that will help maximize a function of S. Using notation similar to that in Sec. 9.5, write J for an optimality criterion over a set S and let $S^* = \arg\max_{u \in S} J(u)$. As before, J is typically data dependent, but this is suppressed in the notation. When necessary, we write $S(u)$ for a neighborhood of u in $S \setminus \{u\}$.

To maximize J over S let $\sum_{u \in S} J(u) = C$, where $0 \leq C < \infty$, but C need not be known further. One way to define an MCMC–MH procedure is to write

$$T(u', u_i) = \frac{J(u')}{\sum_{u' \in S(u_i)} J(u')},$$

that is, given the starting point u_i, u' is chosen from $S(u_i)$ in proportion to how much $J(u')$ is an improvement over $J(u_i)$. Since $J(u')$ may be greater or less than $J(u_i)$ and the goal is to search all of S, it makes sense to accept u' if it gives a higher value of J than u_i does. In addition, sometimes it still makes sense to accept u' even when it gives a lower value of J than u_i does, but with a smaller probability. Thus, the MH step is to accept u' with probability

$$\alpha(u', u_i) = \min\left(1, \frac{J(u')}{J(u_i)}\right).$$

If u' is not accepted then $u_{i+1} = u_i$. Provided that the neighborhoods overlap properly, i.e., for any pair u and u' there is a sequence of $u = v_0, v_1, \ldots, v_k = u'$ for some k such that $v_j \in S(v_{j-1})$, for all $j = 1, \ldots, k$, repeating this procedure gives a sequence of outcomes $u_0, u_1, \ldots, u_n$ that is representative of outcomes from the distribution on S that assigns a probability $J(u)/C$ to each point $u \in S$.

The MCMC–MH procedure is most often used for parameter estimation in Bayes models; however, here it is used for stochastic search. That is, if a collection of models is indexed by a discrete parameter γ, the output of MCMC–MH is a sequence of parameters $\gamma_1, \gamma_2, \ldots$ that identify models representative of $Q(\gamma) = J(\gamma)/C$. Since the goal is to optimize J, it is not enough just to have a sequence of models (or γ_j) that represents a distribution based on a normalized J. It is essential to have a way to 'climb uphill' on the 'surface' $J(\gamma)$ over S to find models in S^*.

One way to do this is to let n be very large, set $J^* = \max_j J(\gamma_j)$, and define

$$\hat{S}^* = \{\gamma_j \mid J(\gamma_j) \geq J^* - \epsilon\}, \tag{9.22}$$

for some small value of $\epsilon > 0$, as an estimator of S^*. In principle, if the MCMC–MH chain is allowed to run long enough then $S^* \subset \hat{S}^*$, since the MCMC–MH chain will visit all models at least once with strictly positive probability and therefore each element of S^* will have the chance to be in $\hat{S}^*$. Of course, when S is large this is hard to ensure. So, in practice, often multiple MCMC–MH chains are run with different initial values and $\hat{J}^*$ contains the best models (in terms of J) from all of them.

There are many variations on this scheme depending on how the permitted moves in the MCMC are defined. For instance, see Hans *et al.* (2007, p. 510, (5)), for one approach that uses the BIC for J and an MH cutoff based on moves that exploit the difference between $S(u)$ and $S(u')$, where $u' \in S(u)$. An earlier example of this is found in George and McCulloch (1993), who used the Gibbs sampler, essentially a different choice of Markov chain, to implement a stochastic search for variable selection.

Again, much as with SA in Sec. 9.5, the real test of a model or predictor selection procedure is how well it predicts future observations. So, assume that S is a set of predictors (or
a set of models that can be converted to predictors), such as linear regression models with a
K-dimensional vector of explanatory variables, and suppose that data $\mathcal{D} = (x_i, y_i)|_{i=1}^{n}$ are
available. Choose an optimality criterion J on S such as IC_α (see Sec. 9.2) or the root mean
squared error (RMSE) on a subset $\mathcal{D}_1$ of $\mathcal{D}$, i.e.,

$$\text{RMSE}^2 = \sum_{i \in \mathcal{D}_1} \left(\hat{Y}_i(x_i) - y_i \right)^2,$$

and a Markov chain on S that is aperiodic and irreducible. For example, the chain could
be based on neighborhoods of models or predictors defined by the inclusion of a new variable, the exclusion of a current variable, or the interchange of a current and a new variable.
(Strictly speaking, the third move is not essential since it is a combination of the first two
moves. However, allowing more moves permits the chain or chains to search S using possibly fewer steps.) Then, in principle, MCMC–MH can be used to find an $\hat{S}^*$ whose elements
maximize IC_α or minimize RMSE to within ϵ. Then, the remaining data, $\mathcal{D}_2 = \mathcal{D} \setminus \mathcal{D}_1$, can
be used via CV, minimal adjusted R^2, or many other criteria to select one model or predictor
in $\hat{S}^*$ to give a point predictor for Y_{n+1} using x_{n+1}. Proceeding to a PI, the single model
emerging from CV on the MCMC–MH output would be a naive choice because it neglects
the variability of predictor selection from MCMC–MH and CV and their inputs. Widening it
by taking the convex hull of the union of the PIs from all the models generated by MCMC–
MH would likewise be naive. Some procedure in between these two extremes based on the
use of $\mathcal{D}_2$ to set a width would be likely to be more accurate.

Despite the variety of forms of MCMC–MH available, MCMC–MH has enough in common with SA that many of the comments in Sec. 9.5 apply to MCMC–MH also. So, to
conclude this section, several point and interval predictors will be indicated. As before,
they are intuitively reasonable but cannot be recommended on rigorous grounds. The most
obvious way to form a PI is to form the distribution

$$p(y_{n+1}|x_{n+1}) \frac{\sum_{\gamma \in \hat{S}^*} J(\gamma) p_\gamma(y_{n+1}|x_{n+1})}{\sum_{\gamma \in \hat{S}^*} J(\gamma)},$$

where the dependence on $\mathcal{D}$ is suppressed in $\hat{S}^*$. This is an average of the models in $\hat{S}^*$ for
which the weights are convex. In principle, a mean or median of $p(y_{n+1}|x_{n+1})$ can be used
as a point predictor and computationally a $1 - \alpha$ PI can be formed that is conditional on
x_{n+1}. This takes into account all the variability of the Markov chain, the initial points, and
the partition of $\mathcal{D}$ (assuming that several splits of $\mathcal{D}$ into $\mathcal{D}_1$ and $\mathcal{D}_2$ are used to find $\hat{S}^*$).

A second approach is to find a collection of p_γ that are particularly convincing in terms
of convergence, maximization of J over S, stability under random $N(0, \sigma^2)$ perturbations of
the data, bootstrapping over $\mathcal{D}$, $\mathcal{D}_1$, and $\mathcal{D}_2$, and other desirable criteria. Then, the median
(or mean) of the models identified can be taken as a point predictor and percentiles can be
obtained, using the empirical percentiles obtained from the collection of point predictions,
to give an approximate $1 - \alpha$ PI. This can also be used in SA-based prediction (as at the
end of Sec. 9.5), and the last technique mentioned at the end of Sec. 9.5 can be used with
MCMC–MH too.

9.7 Computed Examples: SA and MCMC–MH

Simulated annealing and MCMC–MH are stochastic optimization methods intended for model selection when the model list is long. By contrast, CV as an MSP is only effective when one is assessing a small number of models, i.e., the model list is short and enumerable. So, it would be unusual to find a setting where it would make sense to compare all three methods. Indeed, CV is mostly used for evaluating a small number of models that have already been chosen by some other MSP rather than for model selection in the first place. Moreover, it is sufficiently well studied that, given the discussion of Sec. 9.4, it will be enough to exemplify the use of SA and MCMC–MH predictively.

The data analyzed here are the heart data LAHEART that appeared in Afifi and Azen (1972), a subset of size $n = 200$ of the data from an epidemiological study of employees of the County of Los Angeles; see https://www.umass.edu/statdata/statdata/stat-rmult.html. A total of 17 variables were measured, but here one of them, serum cholesterol (mg) recorded to the nearest integer for each subject in 1962, is taken as the independent variable. The remaining 16 variables relate to subject health or demographics such as weight, systolic and diastolic blood pressure, age, height, and so forth. There were four predictors and ten runs were computed for each. Each of the ten runs was based on a random sample of size 150 from the initial 200 data points. That is, n was actually 150 but the data were re-chosen and randomly ordered for each run. The first 50 data points were used as a burn-in, so there were 100 predictive steps. Ensuring that the ten runs for each predictor were computed using independently drawn subsamples of size 150 meant that the results could be regarded as broadly representative of the whole data set. Using ten runs was also enough to ensure that empirical 95% PIs (generated by the ten runs) could be found at each time point. Of the four predictors implemented, two used SA and two used MCMC–MH. One of the SA methods and both of the MCMC–MH methods used linear models. The other SA method used trees or recursive partitioning; see Sec. 10.2. Trees are step functions that partition their domain according to which values of the response are associated with certain ranges of the explanatory variables.

The SA-based predictors were computed as follows. Choose the first of the ten sets of 150 data points and start with its first 50 data points. From these draw a simple bootstrap sample of size 50. Some of the first 50 data points are in the bootstrap sample and some are not. This gives two sets. The data points in the bootstrap (internal) sample are used for fitting and the data points not in the bootstrap sample (external) are used for evaluation, more specifically, to generate a root mean squared error (RMSE) that is taken as the optimality criterion. Using a random selection of the 16 variables as a starting point, the default implementation of Breiman's random forest algorithm for regression, randomForest in R (R Core Team 2016), was used with the internal data set to generate on average 50 trees, a 'random forest'. The RMSE of this random forest was calculated using the external data. From this starting point, SA was applied with 300 iterations; at each iteration variables were selected and used to form a random forest that can make predictions, so that an external RMSE can be found. The cooling schedule, i.e., the probability of acceptance when a worse value of the objective function is found, is equal to the exponential of $(\hat{J}_j - \hat{J}_{j+1})/(j\hat{J}_j)$ where $\hat{J}$ is the RMSE objective function and j is the iteration number, as suggested by (9.17). The best random forest in an RMSE sense over the 300 iterations of SA was found, and the variables in it noted. This whole procedure, starting with drawing a bootstrap sample from the first 50 data points, was repeated, giving a second set of variables. The union of the two sets of variables then served as the starting point for a third random forest and SA procedure, where the first 50 data points were used to define an internal RMSE as the objective function, and 300 iterations of SA under the same cooling schedule were done. The best random forest from this third procedure was used to make predictions for the next five data points i.e., for $i = 51, \ldots, 55$. Hence there were five errors from these five predictions from the same model. These five 'predictuals' are stored and the whole procedure is repeated for the first 55 data points rather than the first 50 data points, to make predictions for $i = 56, \ldots, 60$. That is, in these computations, data points were taken in groups of five to speed the computing. For SA with linear models, the same procedure was used except that randomForest was replaced by lm. These approaches for generating SA predictors were used on each of the ten repetitions on the ten randomly chosen subsets of size 150 from the original 200 data points.

The two MCMC–MH based predictors were computed as follows. The first is actually a Bayes model average (BMA) predictor; see Sec. 11.1 for full details. The likelihood is constructed from models of the form $Y_i = x_i^T \beta + \epsilon_i$, where the intercept is the first element of the vector β, x_i is the ith value of the 16-dimensional explanatory variable, and $\epsilon_i \sim N(0, \sigma^2)$ for $\sigma > 0$. There are $2^{16} - 1$ models indexed by j. Letting $\mathcal{D}_i = \{x_{i+1}; (y_u, x_u), u = 1, \ldots, i\}$, the predictive distribution is

$$m(y_{i+1}) = \sum_{j=1}^{2^{16}-1} W(j|\mathcal{D}_i) \int w(\beta_j|\mathcal{D}_i) p_j(y_{i+1}|\beta_j) \mathrm{d}\beta_j. \tag{9.23}$$

To evaluate (9.23) it is enough to find the integral of each model, with respect to the posterior, for its parameters and the posterior probabilities for each model.

The simplest case of (9.23) uses Zellner's g-prior as the within-model prior for the regression coefficients for each model. The prior on the constant is a flat prior over $\mathbb{R}$ and the prior over σ is $w(\sigma) \propto 1/\sigma$. The hyperprior on g is derived from the hyperprior on the 'shrinkage factor' $g/(1 + g)$ and is a beta$(1, (a/2) - 1)$ distribution in which a is chosen so that $2/a = E(g/(1 + g)) = n/(1 + n)$; see Liang *et al.* (2008) and the discussions in Secs. 9.3 and 4.2. This means there is a closed-form expression for the posterior on the parameters of any model for any sample size, in particular beyond the burn-in of 50. Hence, the integrals in (9.23) can be evaluated.

It remains to find estimates of the posterior probabilities for each model. The most straightforward approach is to use MCMC–MH. The across-model prior is a beta-binomial; see Ley and Steel (2009). For model M_j, having k_j covariates, the conditional binomial prior is

$$W(M_j|\theta) = \theta^{k_j} (1 - \theta)^{K-k_j},$$

where K is the maximal number of covariates. The parameter θ controls how many of the covariates are pre-experimentally expected to be in the true model. Since $K = 16$, it is common to choose the expected model size, set it equal to θK, and solve for θ. Ley and Steel (2009) make this fully Bayes, by putting a hyperprior on θ, namely a beta distribution. This permits more variability in model size and hence reflects more pre-experimental uncertainty.

In bms, the R package used here, the default is to set the expected model size under the hyperparameters of the beta to give an expected model size of $K/2$.

For the MCMC, the transition matrix is defined by two moves: adding or deleting a single nonconstant explanatory variable. That is, one of the 16 covariates is chosen at random. If it is already in the current model then it gets dropped in the next iteration; otherwise, it gets added in the next iteration. The MCMC used 50 000 iterations after a 10 000 iteration burn-in. The acceptance probability of a new model at a given iteration is the MH cutoff. That is, given $\mathcal{D}_i$ and 10 000 iterations in any given run, the MCMC goes from model j to model j' with probability $\min(1, W(M_{j'}|\mathcal{D}_i)/W(M_j|\mathcal{D}_i))$. (In this situation the data x_{i+1} drop out since there is no y_{i+1} in $\mathcal{D}_i$.) The burn-in values of $p(M_j|\mathcal{D}_{50})$ are used in the initial MH cutoff; they are updated as the MCMC–MH proceeds. Finally, an estimate of the posterior model probability for any M_j is given by the number of times it was chosen out of the 50 000 iterations. Now, in principle, (9.23) can be found. As with SA, the data points were taken five at a time and predictions were made five at a time to speed computation.

The second default MCMC–MH predictor from the bms package is the same as the first except that predictions are obtained from the model with the highest maximized likelihood, i.e., no averaging is used, even though the search over the model list is done by MCMC–MH. This is similar to using the BIC and is comparable to the two SA methods, in that it chooses a single best model for prediction.

The cumulative and individual predictive errors for these four predictors are shown in Fig. 9.7. The left-hand panel shows that three of the four predictors are equivalent but that the BMA predictor is substantially worse. (The BMA may be improved by Occam's window approaches; see Sec. 11.1.)

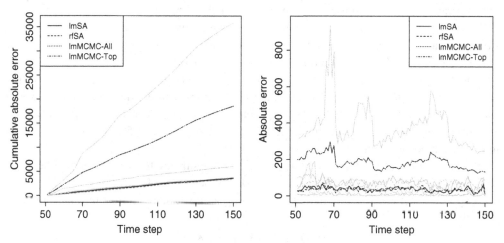

Figure 9.7 Left: Cumulative and individual prediction errors for the four predictors. SA with random forests (rfSA) or SA with linear models (lmSA) is equivalent to lmMCMC-Top, meaning that the highest maximized likelihood (Top) was used. lmMCMC-All, i.e., the BMA, was clearly the weakest predictor. Right: One-step errors from the same predictors with +2SD, confirming the conclusions from the left-hand panel. This panel suggests that the instability of the BMA is a main reason for its poor performance. Note that the error per time step is roughly constant and does not decrease.

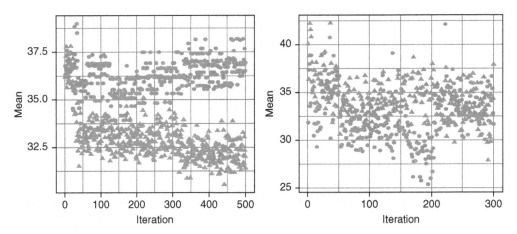

Figure 9.8 RMSEs based on internal data (triangles) and external data (circles) over 300 iterations for two different bootstrap samples for SA with trees.

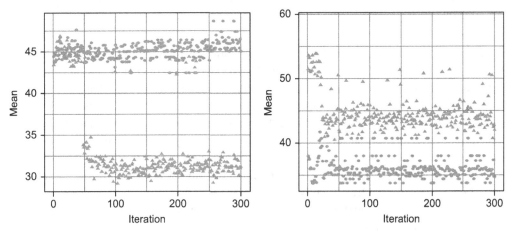

Figure 9.9 RMSEs based on internal data (triangles) and external data (circles) over 300 iterations for two different bootstrap samples for SA with linear models.

As a generality, prediction is harder than conventional inference about unknown but fixed quantities. Moreover, model selection tends to be more variable than parameter selection, even when the parameter is a whole curve as in Chapter 8. This is seen in Figs. 9.8 and 9.9. They show internal and external errors from SA over 300 iterations for the two SA-based predictors. It is no surprise that the internal errors are smaller than the external errors. The two panels in each figure represent these errors for two random samples of 150 data points from the same original data set. That they look so different is an indicator that the SA does not 'settle down' very well. Even though SA with linear models has a cumulative predictive error indistinguishable from SA with trees, these figures suggest that SA with linear models exhibits more variability. Indeed, in Fig. 9.9 the relative sizes of the internal and external errors reverse!

Analogously, it is worth looking in more detail at the MCMC–MH methods to examine various aspects of how they perform. The left-hand panel in Fig. 9.10 gives the MCMC–MH

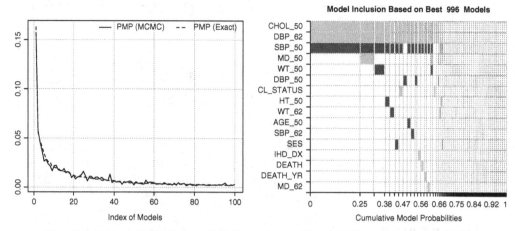

Figure 9.10 Left: Convergence of the MCMC–MH posterior probabilities of models. Right: Posterior model inclusion probabilities for the explanatory variables in the final MCMC–MH model: light gray, positive coefficient; dark gray, negative coefficient.

estimates of posterior model probabilities along with their theoretical exact values for the 100 most probable models. It is seen that the MCMC–MH has converged. So, the weak performance of the BMA cannot be attributed to convergence problems. The right-hand panel in Fig. 9.10 shows the cumulative posterior model probabilities (PMPs) and the posterior inclusion probabilities of the 16 covariates (dark gray indicates positive coefficients, light gray indicates negative regression coefficients). To ensure that, say, a model average represents at least 50% of the posterior probability requires averaging nine models representing a combination of 11 variables. The highest posterior model probability corresponds to a model with only three variables (CHOL_50, DPB_62, and SBP_50). The problem with BMA seems to be that it includes many more models than are helpful, which may contribute to instability, thereby giving poor predictive performance.

As a final examination of how the methods perform, we will examine what the final models look like. The left-hand panel of Fig. 9.11 shows how the distribution of model sizes changes for the MCMC–MH based predictors. Recall that the default **beta-binomial** prior has an expected model size equal to 8, i.e., half the number of nonconstant covariates. It is seen that the posterior distribution of models changes substantially to have its mode at size 4. This is consistent with the right-hand panel of Fig. 9.10, which shows that the top three variables are essentially always included while for the vast majority of the time at least one more variable is included.

The right-hand panel of Fig. 9.11 allows two further comparisons. Note that the comparisons do not include lmMCMC–MH–Top explicitly because it has the same values for these diagnostics as lmMCMC-MH-All or BMA; it is merely that two different predictors were chosen within the same structure. The first of these comparisons is shown in the bar chart. The bar on the right shows a posterior modal model size of 4, the same as the left-hand panel. It is seen that the average model size for SA with linear models is 7 and the average model size for SA with trees is 8. The left-hand panel of Fig. 9.7 shows that these last two procedures are essentially equivalent. So, even if trees constitute a more flexible class, the

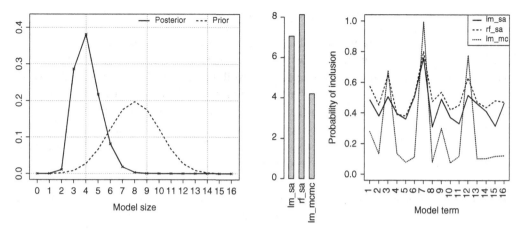

Figure 9.11 Left: Prior and posterior distributions of model sizes under MCMC–MH. Right: The bar chart gives a comparison of the average model sizes for three of the predictors, MCMC–MH with linear models and SA with linear models and trees; the graph shows the term inclusion probabilities of the three methods.

flexibility does not seem to help even when it leads to the inclusion of an extra variable. The graph on the right in Fig. 9.11 shows the term inclusion probabilities for each of the three methods. For the MCMC–MH method, these are from the estimated posterior model probabilities. For the SA methods they are from the best models found at each sample size across all runs. It is seen that all three methods track each other but the MCMC–MH method is much more variable (which is consistent with the right-hand panel of Fig. 9.10). Simulated annealing with random forests consistently includes variables with higher probabilities than SA with linear models, but again this does not seem to reduce the cumulative or individual errors, as seen in Fig. 9.7.

Overall, despite their differences, the three best methods are roughly comparable. However, it will be seen in Chapters 10 and 11 that, in fact, in some generality, the extra flexibility from using trees and random forests does provide a nontrivial improvement over linear models.

9.8 Endnotes

To conclude this chapter we provide comments on five other approaches to model and predictor selection and evaluation. The simplest approach, considered in Sec. 9.8.1, uses the deviance information criterion (DIC) and is similar in spirit to the MSPs in Sec. 9.2. The next simplest, in Sec. 9.8.2, is based on a posterior predictive decision-theoretic evaluation and proceeds by finding the model whose predictive density gives better predictions for the data gathered, in an average sense, than the predictive densities from other models. Section 9.8.3 demonstrates two information-theoretic MSPs that again represent a complexity-fit trade-off. Finally, although scoring rules were used in Sec. 5.5 and seen to be inconclusive, they are given a second look in Sec. 9.8.4, where log scoring rules and Bayes factors (BFs) are compared (even though this text advocates that neither should be used in any degree of generality).

9.8.1 DIC

The deviance information criterion (DIC) is primarily a way to compare models. However, model comparison, like testing, is one way to find predictors, by the elimination of predictors based on models that do not summarize the data well. In the parlance of Spiegelhalter *et al.* (2002, 2014), call $D(\theta) = -2\log p(y^n|\theta)$ the deviance of the likelihood $P(y|\theta)$. Since y^n is fixed, $D(\theta)$ inherits a distribution from the posterior, say, $w(\theta|y^n)$. The expectation in the posterior, $\bar{D} = E_{\Theta|y^n} D(\theta)$, can be regarded as a goodness-of-fit measure; a large value of $\bar{D}$ indicates that $p(y^n|\theta)$ is a poor summary of y^n. The idea is to trade off $\bar{D}$ against a complexity measure, because an increase in complexity should permit a decrease in $\bar{D}$ and vice versa. One measure of complexity is the effective dimension, which in some cases can be estimated by $d_{\text{eff}} = \bar{D} - D(E_{\Theta|y^n}\Theta)$. Thus, d_{eff} is the posterior mean deviance minus the deviance of the posterior mean. Now the DIC is defined as follows:

$$\text{DIC} = D(E_{\Theta|y^n}\Theta) + 2d_{\text{eff}} = \bar{D} + d_{\text{eff}}. \tag{9.24}$$

The right-hand side is a function only of y^n, and predictors formed from models with a small DIC are preferred over models with a high DIC. The predictors described at the end of Sec. 9.2 can now be applied to (9.24).

The basic DIC has some problems. First, d_{eff} can be negative and is not invariant under reparametrization. This has led to alternative choices for d_{eff}, for example, $\text{Var}(D(\theta))/2$, see Gelman *et al.* (2004, p. 182), that seem to perform better. Second, DIC has a tendency to overfit. However, Ando (2011) provided a variant on the DIC that corrects for this. On the plus side, use of the DIC does not assume there is a true model, and the quantities in it are easy to compute using MCMC runs. With careful use, the DIC may lead to good predictors. For an interesting application showing a data-driven comparison and combination of several methods (that include the risk information criterion but not the DIC) for prediction, see Foster and Stine (2004).

9.8.2 Posterior Predictive Loss

The basic idea of posterior predictive loss model selection, described in Gelfand and Ghosh (1998), is to choose a model that achieves the minimum posterior loss when predicting the data already collected. It is based on an earlier decision-theoretic formulation; see Ibrahim *et al.* (2001) and the references therein. Daniels *et al.* (2012) gave a nice summary and showed how the method can be extended to certain missing-data contexts.

Suppose data y^n are available and consider a model $p(\cdot|\theta)$. The predictive density evaluated at a copy of y^n, denoted y_{new}, is

$$m(y_{\text{new}}|y^n) = \int p(y_{\text{new}}|\theta)w(\theta|y^n)d\theta. \tag{9.25}$$

Let $\mathcal{L}(y_{\text{new}}, a; y^n)$ be the loss for guessing the action vector a when $Y_{\text{new}} = y_{\text{new}}$ is drawn and y^n is available. Then, the posterior risk is $E(\mathcal{L}(Y_{\text{new}}, a; y^n)|y^n)$, where the expectation is taken over Y_{new} using (9.25).

Next, define $\mathcal{L}$ on the ith component of y^n as

$$\mathcal{L}(y_{i,\text{new}}, a_i; Y^n = y^n) = L(y_{i,\text{new}}, a_i) + cL(y_i, a_i),$$

for some loss function L, where $c > 0$. Then, when L is the squared error, summing over $i = 1, \ldots, n$ and adding and subtracting $E(Y_{i,\text{new}}|y^n)$ in each term gives

$$\min E\mathcal{L}(Y_{\text{new}}, a; y^n)|y^n) = \sum_{i=1}^{n} \text{Var}(Y_{i,\text{new}}|y^n) + \frac{c}{c+1} \sum_{i=1}^{n} \left(E(Y_{i,\text{new}}|y^n) - y_i \right)^2. \quad (9.26)$$

Expression (9.26) can be regarded as a variance–bias decomposition. However, in the model-selection context, the sum of variances is a penalty term and the sum of 'biases' is a goodness-of-fit term. Thus, a model satisfying (9.26) provides a tradeoff between complexity and goodness of fit. In principle, this model can be used to generate point and interval predictions. However, the process for finding the optimal model neglects the variability in model selection, so it is unclear how good predictors will be that are derived from it alone. This concern can be reduced if the data are partitioned into two sets, one for model fitting and one for model validation; this amounts to a Bayes form of CV. Even if this can be done, it remains unclear how to obtain more sophisticated predictors from this approach, apart from attempting some variation on the bootstrapping described in Sec. 9.2. At this time, only the model selection properties of this approach seem to have been investigated; see Laud and Ibrahim (1995) for examples and comparisons with other methods.

9.8.3 Information-Theoretic Model Selection Procedures

As has been discussed several times now, many MSPs express a complexity–fit tradeoff. This can be seen in the two information-theoretic techniques outlined here. First, the two-stage coding MSP described in Barron and Cover (1990) can be summarized as follows (omitting many details). Consider a collection of models Γ and a codelength function $L(q)$ for $q \in \Gamma$. That is, $L(q)$ is the number of bits – zeros and ones – required to express the message q. If Γ is rich beyond a certain level, e.g., nonparametric, then the models in it may have to be discretized. Given data (Y_i, X_i) assumed IID, the joint density under $q \in \Gamma$ is $q(y^n|x^n) = \prod_i q(y_i|x_i)$. Suppressing the dependence on the x_i, the minimum-complexity density estimator is

$$\hat{p}(\cdot) = \arg\min_{q \in \Gamma} \left(L(q) + \ln \frac{1}{q(y^n)} \right) \quad (9.27)$$

where the second term on the right is, essentially, the Shannon code length of y^n under density q. The first term on the right is a complexity penalty because $L(q)$ is the description length of q and more complex q's have a longer description length. The second term is an assessment of fit: the better q summarizes the data, the smaller the code length should be. As noted in Sec. 9.8.1, the techniques at the end of Sec. 9.2, such as bootstrapping, can be applied to (9.27).

Second, Womack (2011, p. 39) proposed a posterior predictive information criterion (PPIC), similar to that of Gelfand and Ghosh (1998) in that it uses the data twice but also similar to the two-stage coding criterion of Barron and Cover (1990) in that it is based on two codelengths. The difference is that the PPIC takes the difference between two log densities (essentially the Shannon codelengths assigned to the data) and subtracts a penalty representing their average predictive discrimination. Formally, let p and q be densities with relative entropy denoted $D(p\|q)$ and write $J(p, q) = D(p\|q) - D(q\|p)$. Essentially, J is the difference in excess codelength between coding data from p using q and coding data from q using p. If there are two models $p(\cdot|\theta)$ and $q(\cdot|\phi)$ for Y^n equipped with priors $w_p(\cdot)$ and $w_q(\cdot)$ then two predictive densities $m_p(\cdot|y^n)$ and $m_q(\cdot|y^n)$ can be formed. Let

$$W_\alpha(p,q) = \ln \frac{m_p(y^n|y^n)}{m_q(y^n|y^n)} - \alpha J(m_p(\cdot|y^n), m_q(\cdot|y^n)). \qquad (9.28)$$

The log density ratio represents the relative fits of p and q while the second term is a complexity penalty. It favors models with more spread, and complex models tend to have more spread due to the integrating out of parameters. So, J limits the degree of overfitting by excess complexity. Now for the p versus q case, the PPIC rule is

$$\text{choose } p \text{ instead of } q \Leftrightarrow W_\alpha(p,q) > 0.$$

Womack (2011) proved that the PPIC is consistent for model selection.

Obviously, one can obtain PIs from any model that maximizes W_α. However, accounting for variability in model selection, or more precisely, the variability in PIs as a result of model selection, remains an open problem even though it is possible to imagine using procedures similar to those described earlier.

9.8.4 Scoring Rules and BFs Redux

To evaluate a predictor, the CPE or variants on it have been advocated here for a variety of reasons; see Sec. 1.1 for the general stance and Sec. 2.4 for key principles including the Prequential Principle. By definition, therefore, any other evaluation of a model or predictor must include information besides the data and the predicted values for it. Typically, the extra information is some sort of modeling. This is the case with scoring rules and testing, Bayes testing in particular. In Sec. 5.5, it was seen that different scoring rules can lead to different conclusions about the relative adequacy of models. In Sec. 9.3, BFs were found for several cases, showing some strengths and weaknesses. Here, some further considerations about these approaches from a predictive standpoint are given. In sum they do not lead to a better method than the CPE, but they do provide alternatives that may be worth examining if a model turns out to be sufficiently well validated. The comments here draw heavily on Draper (2014a).

First, the log scoring rule for model M_j for $j = 1, \ldots, J$ and data y^n is

$$\text{LS}_{CV}(M_j|y^n) = \frac{1}{n} \sum_{i=1}^{n} \log p(y_i|y_{-i}, M_j),$$

where y_{-i} is y^n with y_i deleted and $p(\cdot|\cdot, M_j)$ denotes the predictive density from model M_j. (Explanatory variables have been suppressed in the notation.) This is a little different from (and sometimes worse than)

$$\text{LS}_{FS}(M_j|y^n) = \frac{1}{n} \sum_{i=1}^{n} \log p(y_i|y^n, M_j),$$

the full-sample log score. A relatively large value of LS_{CV} or LS_{FS} indicates a preference for M_j over the other models. Essentially, these log scores assess how likely y_i is, given the other values of y in the context of a particular model. For continuous data, log scoring rules tend to penalize for high-dimensional parameters by spreading out the probability, thereby giving lower log scores. That is, a form of parsimony is built into log scoring rules.

From a series of examples, Draper (2014a) observed that neither BFs nor log scores uniformly dominates the other and suggested that analysts should use whichever method is better 'according to the real-world severity of the two kinds of errors they make'. As an example, BFs are often better than log scoring when false positives are a bigger problem than false negatives, as in genetic linkage studies, while log scoring should be used when a false negative is a bigger problem than false positives, as in many variable-selection problems. Draper (2014a) also argued that (i) neither BFs nor the log scoring rule is always better for decision making or for inference, and (ii) making decisions based on inferences improperly neglects the cost structure of decision problems.

In this sense, Draper (2014a) implicitly regarded decision making as the fundamental task of statistics. This is not unreasonable, since prediction can be regarded as decision making in many cases. However, good prediction is a much more general criterion, not needing models, for instance, and good predictive performance is necessary to validate any decision rule. That is, the decision-theoretic structure is a way to generate predictions, not of assessing or validating the predictors that made them.

The comparison of log scoring rules and BFs is interesting because neither is a priori invalid in $\mathcal{M}$-closed problems. However, even among such problems the performances of log scoring rules and BFs differ, and knowing which to use can be difficult. Using predictive evaluations such as the CPE seems simpler and requires less information (such as knowing the 'real-world severity' of errors). Predictive evaluations also provide reliable discrimination among predictors, understanding that reliable does not necessarily mean conclusive, since two predictors may be different but predictively equivalent.

Part III

Contemporary Prediction

Predictive modeling enables assessing the distance between theory and practice, thereby serving as a reality check to the relevance of theories. Predictive models are advantageous in terms of negative empiricism: a model either predicts accurately or it does not, and this can be observed. In contrast, explanatory models can never be confirmed and are harder to contradict ...Predictive power assessment offers a straightforward way to compare competing theories by examining the predictive power of their respective explanatory models ... The consequence of neglecting to include predictive modeling and testing alongside explanatory modeling is losing the ability to test the relevance of existing theories and to discover new causal mechanisms.

Shmueli (2010)

Part II of this text focused mostly on the use of predictors from model classes. In Part III, a purely predictive approach will be taken, i.e., the modeling aspect will be ignored. It is well known that the true model (if it exists) is not in general the best predictive model. This is a major problem for modeling because a model that makes poor predictions throws its other implications into doubt.

To avoid this problem, in the first two chapters of Part III we discuss the main 'black box' techniques and their use in forming predictors. Chapter 10 presents the most common mathematical objects used to form individual predictors. The key feature of these objects is that either they do not have an immediate physical interpretation or any immediate physical interpretation is so oversimplified or overcomplex as to give a distorted view of the physical phenomenon in question. Chapter 11 presents ways to 'ensemble' these individual predictors into a single composite predictor.

The last chapter of Part III provides a discussion of topics of current interest in prediction. It includes some techniques that have been around since before the start of the twenty-first century and are enjoying a rejuvenation. It also includes data types, e.g., as found in spatio-temporal data (remote sensing) and networks, and the predictive techniques that they require, which have emerged only comparatively recently.

10

Blackbox Techniques

...There are three possible reasons for [the] absence of predictive power. First, it is possible that the models are misspecified. Second, it is possible that the model's explanatory factors are measured at too high a level of aggregation ...Third, ...the search for statistically significant relationships may not be the strategy best suited for evaluating our model's ability to explain real world events ...the lack of predictive power is the result of too much emphasis having been placed on finding statistically significant variables, which may be overdetermined. Statistical significance is generally a flawed way to prune variables in regression models ...Statistically significant variables may actually degrade the predictive accuracy of a model ...[By using] models that are constructed on the basis of pruning undertaken with the shears of statistical significance, it is quite possible that we are winnowing our models away from predictive accuracy.

Ward *et al.* (2010), discussing a predictive analysis of an $\mathcal{M}$-open problem.

Part I explained and justified the predictive approach to data analysis. Part II drew examples from five branches of statistics to show how models can be converted to predictors. Now the task is to take a more purely predictive approach, largely ignoring interpretability until a good predictor has been found. As the quote above makes clear, there is a discrepancy between effective conventional inference and predictive power: more of the first can lead to less of the second. In fact, the reverse direction holds too: better predictive power can lead to worse conventional inference. Taken together, this means that the epistemological standing of a true model is very heavily under question.[1] As a separate point, it has been long established that informal methods – relying on past experience and training to make predictions, as opposed to using statistical prediction rules, perform so poorly that they do not merit discussion here; see Meehl (1954) for one early study that has been replicated numerous times.

Consider the following toy example drawn from Wu *et al.* (2007). Suppose that a linear model of the form $Y = X_1\beta_1 + X_2\beta_2 + \epsilon$ is true where $X_1 \in \mathbb{R}^{n \times p}$, $X_2 \in \mathbb{R}^{n \times q}$, $\beta_1 \in \mathbb{R}^p$, $\beta_2 \in \mathbb{R}^q$, $\epsilon \sim N(0, \sigma^2)$, and both X_1 and X_2 have full column rank. Writing $\beta = (\beta_1, \beta_2)^T$ and $X = (X_1, X_2) \in \mathbb{R}^{n \times (p+q)}$, the usual least squares estimator for the full model is $\hat{\beta}_F = (X^T X)^{-1} X^T Y$. It is unbiased ($E\hat{\beta}_F = \beta$) and satisfies $\mathrm{Cov}(\hat{\beta}_F) = \sigma^2 (X^T X)^{-1}$. Clearly, this problem is $\mathcal{M}$-closed.

---

[1] The results in Rissanen (1996) do not show that the true model is the unique best predictor. However, the main theorem in that paper shows that in the limit of large sample sizes the true model is one of the asymptotically best predictors. This is seen more explicitly in Skouras and Dawid (1998, Sec. 7.1). In fact, even when the true model generates the unique asymptotically optimal predictor, it is essentially never optimal or unique in finite samples.

Now consider the reduced model $Y = X_1\gamma_1 + \varepsilon$. Clearly, $\varepsilon = X_2\beta_2 + \epsilon$. The least squares estimator for γ_1 is $\hat{\gamma}_R = (X_1^T X)^{-1} X^T Y$. It is biased in general: $E\hat{\gamma}_R = \beta_1 + A_1\beta_2$, where $A_1 = (X_1^T X_1)^{-1} X_1^T X_2$ is the projection of X_1 onto X_2, but has a covariance matrix of the same form as β_F, namely, $\mathrm{Cov}(\hat{\gamma}_R) = \sigma^2 (X_1^T X_1)^{-1}$. The only cases where $\hat{\gamma}$ is not biased are (i) $\beta_2 = 0$, and (ii) X_1 and X_2 are orthogonal ($X_1^T X_2 = 0$).

The prediction from the full model at a new $Z = (Z_1, Z_2)^T$ is $\hat{Y}_F = Z_1\hat{\beta}_1 + Z_2\hat{\beta}_2$ and the prediction from the reduced model is $\hat{Y}_R = Z_1\hat{\gamma}_1$. The expectations are, respectively, $Z_1\beta_1 + Z_2\beta_2$ and $Z_1\beta_1 + Z_1 A_1\beta_2$. The unbiasedness of the parameter estimate carries over to the predictor from the full model and the bias in the parameter estimate carries over to the predictor from the reduced model. Closed-form expressions for the covariances can be derived:

$$\mathrm{Cov}(\hat{Y}_F) = \sigma^2 Z_1 (X_1^T X_1)^{-1} Z_1^T + \sigma^2 (Z_1 A_1 - Z_2)\Omega(Z_1 A_1 - Z_2)^T,$$
$$\mathrm{Cov}(\hat{Y}_R) = \sigma^2 Z_1 (X_1^T X_1)^{-1} Z_1^T,$$

where $\Omega = (X_2^T X_2)^{-1} X_1^T + A_2 (X_1^T (I_n - P_2) X_1)^{-1} A_2^T$, in which $P_2 = X_2 (X_2^2 X_2)^{-1} X_2^T$ and $A_2 = (X_2^T X_2)^{-1} X_2^T X_1$. Indeed,

$$\mathrm{Cov}(\hat{Y}_F) - \mathrm{Cov}(\hat{Y}_R) = \sigma^2 (Z_1 A_1 - Z_2)\Omega(Z_1 A_1 - Z_2)^T$$

is positive (as a matrix) since Ω is positive definite (provided that $Z_1 A_1 \neq Z_2$). Since the covariance of $\hat{Y}_R$ is less than the covariance of $\hat{Y}_F$ and the bias of $\hat{Y}_R$ is bigger than the bias of $\hat{Y}_F$ (which is zero), there is a covariance–bias tradeoff.

This can be seen more clearly by looking at the mean squared error matrices (MSEMs) of $\hat{Y}_F$ and $\hat{Y}_R$. It can be derived that

$$\mathrm{MSEM}(\hat{Y}_F) = \sigma^2 Z_1 (X_1^T X_1)^{-1} Z_1^T + \sigma^2 (Z_1 A_1 - Z_2)\Omega(Z_1 A_1 - Z_2)^T,$$
$$\mathrm{MSEM}(\hat{Y}_R) = \sigma^2 Z_1 (X_1^T X_1)^{-1} Z_1^T + (Z_1 A_1 - Z_2)\beta_2\beta_2^T (Z_1 A_1 - Z_2)^T.$$

In general, the difference is

$$\mathrm{MSEM}(\hat{Y}_F) - \mathrm{MSEM}(\hat{Y}_R) = (Z_1 A_1 - Z_2)\left(\sigma^2 (X_2^T (I_n - P_1) X_2)^{-1} - \beta_2\beta_2^T\right)(Z_1 A_1 - Z_2)^T.$$

Let

$$L = \frac{\beta_2^T Z_2^T (I_n - P_1) Z_2 \beta_2}{\sigma^2}.$$

Then it can be shown that

$$L \leq 1 \quad \Longleftrightarrow \quad \mathrm{MSEM}(\hat{Y}_R) \leq \mathrm{MSEM}(\hat{Y}_F).$$

This means that the wrong model predictor $\hat{Y}_R$ is better than the true model predictor $\hat{Y}_F$ whenever σ is large or X_1 and X_2 have high collinearity. It is also easy to see that as $n \to \infty$, L increases. So, beyond some sample size the full model will give better predictions than the reduced model.

What does this toy example suggest more generally? First, it seems to suggest that prediction is an unreliable indicator for how to choose a model. However, this is a ridiculous stance because a model that fails to validate will be discredited. In fact, what the example shows is that even when the true model is one of finitely many candidate models the notion of a true model itself is empirically problematic: even in $\mathcal{M}$-closed problems, it is a priori impossible to tell whether good predictions mean that the predictor is derived from the true model or the predictor is derived from a wrong model that just happened to make good predictions. Hence, in practice, the best that can be done is to identify a class of models or predictors that cannot be ruled out. Even so, the danger is that in an $\mathcal{M}$-complete setting (to say nothing about an $\mathcal{M}$-open setting) a true model may include many small contributions from many explanatory variables. In this case, a simpler model using explanatory variables that are related to those in the true model may perform quite well predictively but still be far wrong as a model, so that the predictor from the true model will be discredited. In short, the notion of a true model can be slippery and hard to use outside simple cases where the sample size is large enough and the explanatory variables are known to be well chosen. Otherwise stated, without knowledge that is usually not available the concept of a true model is empirically meaningless.

Second, it bears repeating that only the variance term can be reduced by increasing the sample size. The sample size does not affect the bias. Moreover, if the correct explanatory variables can themselves be readily approximated by a finite linear combination of other incorrect explanatory variables (cf. the discussion of Fan and Lv (2008) in our Sec. 9.1) then the bias may be small even when the variance is small. Essentially, in the limit of large n, a regression function may be approximated well by a collection of incorrect variables and the parameters in it may converge to fixed values at satisfactory rates – but the fixed values may be incorrect. This is another way in which the standing of the true model is ambiguous or at least unidentifiable even in $\mathcal{M}$-closed problems.

Third, the data are used primarily to eliminate candidate models or predictors. Let $\mathcal{F}$ be a collection of models or predictors and assume a data set $\mathcal{D}$ is available. Then, $\mathcal{D}$ defines a set $B(\mathcal{D})^c \subset \mathcal{F}$ of models or predictors that are implausible. As in the last example, $B(\mathcal{D})$ may contain predictors that do not correspond to the true model and give better predictions than the predictor generated by the true model. Therefore, the richer a set $\mathcal{F}$ is, the more remaining in $B(\mathcal{D})$ that cannot be discredited and can be used for predictions (or possibly models). Indeed, it is not hard to imagine settings where $B(\mathcal{D})$ is not unique even in the limit of large n: simply assume that the X's in the regression function for Y can be expressed in terms of another set of explanatory variables.

Note that if $\mathcal{F}$ were the collection of linear models used in the example above then it would be spanned by $p + q + 1$ real dimensions. If $\mathcal{F}$ were some other collection of parametric models or predictors, $\dim(\mathcal{F})$ would also have finitely many real dimensions. More generally, $\mathcal{F}$ could be a semiparametric or a nonparametric collection of models or predictors, e.g., proportional hazards, that has infinite real dimension; see Sec. 7.2 or the predictors in Chapter 8 and Sec. 7.1, respectively.

Between the truly nonparametric $\mathcal{F}$'s and the $\mathcal{F}$'s with finite real dimension, there are cases that might be called the intermediate tranche of models or predictors. These will be discussed in this chapter. The key feature they have in common is that although each model or predictor is situated in a finite-dimensional parametric family, the parametric family naturally extends to higher real dimensions, often in a variety of ways. If $\mathcal{F}$ is elaborated completely then it is usually dense in the space of continuous functions of the explanatory variables in any reasonable norm, e.g., L^1.

Most of these $\mathcal{F}$'s are 'blackbox' models or predictors. A blackbox model or predictor describes an input–output relationship between the explanatory variables and the response that the analyst does not see or need to know in order to generate predictions. In fact, this is poor terminology because a blackbox model is not a model. That is, its components usually do not have any physical meaning or interpretation. Sometimes blackbox models are called stochastic models or statistical models to emphasize that they are solely the construct of the analyst. The key benefit is the tradeoff of interpretability and predictive power, in that the latter can be increased at the cost of the former. The loss of interpretability often will not matter much because, as noted, the notion of a true model will often be empirically problematic even in $\mathcal{M}$-closed settings. A fortiori, in $\mathcal{M}$-complete or $\mathcal{M}$-open settings any loss of interpretability is even less important.

Another benefit of using blackbox techniques is that they are much more general and expressive than any fixed-dimension parametric technique and at the same time they have much more structure that can be examined than any nonparametric technique. More precisely, $B(\mathcal{D})$ from an intermediate-tranche blackbox method contains the $B(\mathcal{D})$ from any parametric method (at least in an approximate sense) but is contained in the $B(\mathcal{D})$ of any semi- or nonparametric class. This means that although a wrong model may give better predictions than the true model (and is more likely to occur for larger $B(\mathcal{D})$'s) it is better to use blackbox techniques to get good predictions, since the predictor obtained from a true model often has a problematic status. That is, the search for a true model and its concomitant predictor is often misguided from the start. Abandoning that approach and looking for good predictors, i.e., predictors in $B(\mathcal{D})$ from an intermediate tranche, can give good results that are no worse and often better than those found by searching in a finite-dimensional parametric or nonparametric family.

The next section reviews some key ideas from classical nonlinear regression, for background and motivation. Section 10.2 presents trees or recursive partitioning methods, Sec. 10.3 presents neural networks (nets) including a short survey of the latest thinking on them, called 'deep learning', Sec. 10.4 presents kernel methods, and Sec. 10.5 presents several of the more successful penalized methods. These are the main blackbox techniques – although some would argue that kernelized methods are not blackbox, just hard to visualize. It could also be argued that penalized methods are not really blackbox because they use a mathematical technique to obtain sparsity and the resulting models are, for the most part, interpretable. The problem is that the 'penalty' term makes interpretation difficult since parameter estimates can depend on it substantially. However, for practical purposes all these techniques can be seen as blackbox. As an aid to developing intuition for how they behave, Sec. 10.6 provides some simulation results and a relatively complicated real-data analysis. To conclude this chapter, the endnotes provide a brief discussion of prediction using a few less common blackbox techniques such as projection pursuit, logic trees, hidden Markov models, and errors-in-variables models.

10.1 Classical Nonlinear Regression

The motivation for the techniques that follow is much the same as the motivation for classical nonlinear regression. The core idea is that linear models of the form, say, $X\beta$ are not a rich enough class of functions from which to draw good choices for f in the signal-plus-noise model $Y = f(x) + \epsilon$. Part of the problem is that in a linear regression model the parameters enter linearly, so a multiplicative model $Y = \beta_0 x_1^{\beta_1} x_2^{\beta_2} \epsilon$ can be used only if the logarithm is taken and $\ln \epsilon$ can be regarded as essentially normal. Another problem is that there are many functions that are at best only locally approximated by a linear function, regardless of the error term or factor. For instance, any sigmoid such as $y = 1/(1 + e^{-\beta x})$ is bounded between zero and one. However, no matter how many terms there are (in x) in a linear model for y, outside a compact set any polynomial will go to $\pm \infty$. Moreover, many models are inherently nonlinear. The simplest is the exponential growth model $Y = \beta_1 e^{\beta_2 x} + \epsilon$; taking logarithms will not separate the explanatory variable from the error term. Other inherently nonlinear models occur regularly as well, such as the Weibull model $Y = \beta_1 e^{-(x/\beta_2)^{\beta_3}} + \epsilon$ (see Sec. 7.3), the Gompertz growth model $Y = \beta_1 e^{-\beta_2 \exp \beta_3 x} + \epsilon$, and any type of 'bifurcated' model such as

$$Y = \begin{cases} f(x) + \epsilon & x \geq 0, \\ g(x) + \epsilon & x < 0, \end{cases} \tag{10.1}$$

where $f \neq g$. (This last class of models introduces extra difficulty because of the possible discontinuity at zero.) The classic text Gallant (1987) is notable for the wide variety of settings in which nonlinear models commonly arise; see also Ruckstuhl (2010) for more recent examples. There are examples, such as the exponential growth model, that are linearizable in the sense that one can write

$$\ln Y = \ln \beta_0 + \beta_2 x + \epsilon.$$

However, taking the exponential gives $Y = \beta_0 e^{\beta_2 x} e^{\epsilon}$, with the error multiplicative, meaning that the initially normal error becomes log-normal and therefore distorts parameter inference when converted back to the scale of Y.

The basic theory for classical nonlinear models follows from treating the nonlinear model as if it were locally linear and adjusting for curvature. This is fundamentally different from nonlinear models, in the later sections of this chapter that belong to the intermediate tranche. Indeed, it will be seen in later sections that blackbox methods extend much more readily to complex problems than do classical nonlinear methods. To see how classical nonlinear regression models asymptotically reduce to linear models suppose that

$$Y_i = f(x_i, \theta) + \epsilon_i \tag{10.2}$$

where the x_i are d-dimensional, θ is k-dimensional, and the $\epsilon_i \sim N(0, \sigma^2)$ are IID for $i = 1, \ldots, n$. Expression (10.2) obviously includes linear models but here it is assumed that f cannot be written as any function that is linear in θ.

The simplest way to develop a predictor for use in a model such as (10.2) is to begin by defining the least squares estimator for θ by analogy with linear models:

$$\hat{\theta} = \arg \min_{\theta} SS(\theta) = \sum_{i=1}^{n} (y_i - f(x_i, \theta))^2, \tag{10.3}$$

in which the data is suppressed in the notation $SS(\theta)$. Unfortunately, explicit forms for $\hat{\theta}$ rarely exist so typically $\hat{\theta}$ is approximated by iterative numerical procedures; see Ruckstuhl (2010) for details that are not discussed here.

An explicit linear approximation of (10.2) can be found by taking derivatives. Define the $n \times k$ matrix $L(x^n, \theta) = (\ell_{i,j})_{i=1,\ldots,n, j=1,\ldots,k}$ in which

$$\ell_{ij}(\theta) = \frac{\partial f(x_i, \theta)}{\partial \theta_j}.$$

The Taylor expansion for $f(x_i, \theta)$ for θ in a small neighborhood of θ^* is

$$f(x_i, \theta) = f(x_i, \theta^*) + (\ell_{i,1}(\theta^*), \ldots, \ell_{i,k}(\theta^*))^T (\theta - \theta^*) + o\left(\|\theta - \theta^*\|\right)$$

for each i. Writing (10.2) in matrix form with $Y = (Y_1, \ldots, Y_n)^T$ and $\epsilon = (\epsilon_1, \ldots, \epsilon_n)$ and linearizing at θ^* gives the approximation

$$\tilde{Y} = L(x^n, \theta^*)\beta + \tilde{\epsilon}.$$

Clearly, approximating Y by $\tilde{Y}$ gives an 'approximation' residual of the form $\tilde{Y} = Y - E$, where $E = (f(x_1, \theta^*), \ldots, f(x_n, \theta^*))^T$. The columns of $L(x^n, \theta^*)$ are the 'explanatory variables' and the new coefficients are

$$\beta = (\beta_1, \ldots, \beta_k)^T = (\theta_1 - \theta_1^*, \ldots, \theta_k - \theta_k^*)^T;$$

the intercept term is built into the planar approximation $L(x^n, \theta^*)$. Clearly, if $\theta^* = \hat{\theta}$ and $\hat{\theta}$ is close to the true value θ_T then the linear approximation is good, at least locally around $\hat{\theta}$ (or θ_T, assuming that it exists). If $\hat{\theta}$ converges to $\theta_T = \theta^*$ then $\beta = 0$. More realistically, if the linearization is done at $\hat{\theta}$ and $\hat{\theta} \approx \theta_T$ then the covariance matrix of a good estimate $\hat{\beta}$ for β should be close to the covariance matrix of $\hat{\theta}$.

To obtain PIs for future Y's, start by obtaining CIs for θ. Let

$$V(\theta^*) = (L(x^n, \theta^*)^T L(x^n, \theta^*))^{-1},$$

so that $L(x^n, \theta^*)$ plays the role of the design matrix X in the usual linear regression setting. Then, asymptotically in n,

$$\sqrt{n}(\hat{\theta} - \theta_T) \sim N(0, \sigma^2 V(\theta_T)). \tag{10.4}$$

Since θ_T and σ are not known, $\hat{\theta}$ and $\hat{\sigma}$ are used, so that

$$\hat{V} = V(\hat{\theta}) \quad \text{and} \quad \hat{\sigma}^2 = \frac{\mathrm{SS}(\hat{\theta})}{n-k}.$$

From these asymptotics, CIs can be obtained for individual θ_j using the t-distribution, and confidence ellipsoids can be obtained for subsets of θ of size greater than unity essentially by using Hotelling's T^2 statistic, which obtains intervals using the F-distribution with degrees of freedom k and $n-k$.

Another sort of CI for $f(x, \theta_T)$ at a given x can be derived as follows. Even though the parameter is a function value, this is not the same as prediction because a different value of the function is not the same as a new value of the outcome. To see what this sort of CI looks like, recall the first-order Taylor expansion

$$f(x, \theta_T) = f(x, \hat{\theta}) + (\hat{\theta} - \theta_T)\nabla f(x, \theta)\big|_{\theta=\hat{\theta}} + o(\|\hat{\theta} - \theta_T\|); \tag{10.5}$$

so, assuming that $\hat{\theta}$ is a good estimate of θ_T, the natural point estimator of $f(x, \theta_T)$ for any x is $f(x, \hat{\theta})$. (The vector $\nabla f(x_i, \hat{\theta})$ is just the corresponding row of L if $x = x_i$ for some *i.*) Now, treating $f(x, \theta_T)$ as a parameter and using the δ-method on $\hat{\theta}$ to get asymptotic normality (essentially (10.5) and (10.4)), the nonlinear regression analog of a $1 - \alpha$ CI for the function $f(x, \beta_T) = x^T \beta_T$ in the linear regression setting is given by

$$f(x, \hat{\theta}) \pm t_{n-k, 1-\alpha/2} \hat{\sigma} \sqrt{\nabla f(x, \theta)^T \hat{V} \nabla f(x, \theta)}. \tag{10.6}$$

In the linear regression case, such CIs are narrowest at the mean of the x_i and wider as x moves away from $\bar{x}$. This is not true in general with nonlinear regression; unlike linear regression, $\nabla f(x, \theta)$ represents a curvature that typically distorts the CIs, so that their widths vary with x in more complicated ways.

Now, PIs for nonlinear regression can be stated – and contrasted with CIs from (10.4) and (10.6). To motivate the form of a PI at x_{new} write

$$Y(x_{\text{new}}) - f(x_{\text{new}}, \hat{\theta}) = (Y(x_{\text{new}}) - f(x_{\text{new}}, \theta_T)) - (f(x_{\text{new}}, \hat{\theta}) - f(x_{\text{new}}, \theta_T)). \tag{10.7}$$

The first difference on the right in (10.7) is normal and the second difference on the right is asymptotically normal (use (10.4)). The terms are independent of each other since $Y(x_{\text{new}})$ is independent of $Y(x_1), \ldots, Y(x_n)$. So, asymptotically,

$$\text{Var}(Y(x_{\text{new}}) - f(x_{\text{new}}, \hat{\theta})) \approx \sigma^2 + \sigma^2 \nabla f(x_{\text{new}}, \theta_T)^T V(\theta_T) \nabla f(x_{\text{new}}, \theta_T),$$

and hence a $1 - \alpha$ PI for $Y(x_{\text{new}})$ is given asymptotically by

$$f(x_{\text{new}}, \hat{\theta}) \pm t_{n-k, 1-\alpha/2} \hat{\sigma} \sqrt{1 + \nabla f(x_{\text{new}}, \hat{\theta})^T \hat{V} \nabla f(x_{\text{new}}, \hat{\theta})}. \tag{10.8}$$

Obviously, the expression in the square root in (10.6) (derived from (10.4)) appears in (10.8) but with the addition of an extra σ^2. The idea is that making a prediction combines the estimation of a function value with the randomness of a new outcome by taking the root of their summed variances. More precisely, the standard error (SE) that gives the width of an interval as in (10.8) represents the way in which random fluctuations in the data affect prediction. It can be seen that these random fluctuations affect $\hat{\theta}$ and therefore $f(x_{\text{new}}, \hat{\theta})$; there will be fluctuations associated with $Y(x_{\text{new}})$ as well. In principle, for large n, if the data were to change then $\hat{\theta}$ would still be approximately correct (see (10.4)) so it should remain good enough for the linearization in (10.5) that gives (10.6); analogous but simpler reasoning applies to $\hat{\sigma}$. In short, even under replication of the entire n data points, the PI for $Y(x_{\text{new}})$ should not change much.

An alternative to deriving point and interval predictors from a nonlinear model by linearization is to use the more general approach afforded by bootstrapping. There are at least two ways to do this: bootstrap on the data itself or bootstrap on the residuals from a fitted model. For the first, assume data (y_i, x_i) for $i = 1, \ldots, n$ are available and draw B bootstrap samples of size n. Denote these by $(y_{b,i}, x_{b,i})$ for $i = 1, \ldots, n$ and $b = 1, \ldots, B$. For each bootstrap sample find

$$\hat{\theta}^b = \arg\min_{\theta} \sum_{i=1}^{n} (y_{b,i} - f(x_{b,i}, \theta))^2. \tag{10.9}$$

Now, for any x_{new}, the square of the SE of $f(x_{\text{new}}, \hat{\theta})$ as an estimator for $f(x_{\text{new}}, \theta_T)$ regarded as a parameter is

$$\text{SE}^2(x_{\text{new}}) = \frac{1}{B-1} \sum_{b=1}^{B} \left(f(x_{\text{new}}, \hat{\theta}^b) - \frac{1}{B} \sum_{b=1}^{B} f(x_{\text{new}}, \hat{\theta}^b) \right)^2. \tag{10.10}$$

Assuming asymptotic normality, a $1 - \alpha$ confidence interval for $f(x_{\text{new}}, \theta_T)$ is

$$f(x_{\text{new}}, \hat{\theta}) \pm z_{\alpha/2} \text{SE}(x_{\text{new}}).$$

Using (10.7) and (10.10), this CI leads to a $1 - \alpha$ PI for $Y(x_{\text{new}})$:

$$f(x_{\text{new}}, \hat{\theta}) \pm z_{\alpha/2} \hat{\sigma} \sqrt{1 + \text{SE}^2(x_{\text{new}})}, \tag{10.11}$$

where $\hat{\sigma}$ is the estimate of σ from the original sample.

The second bootstrap approach to finding PIs is to start with residuals $r_i = Y_i - f(x_i, \hat{\theta})$ and draw B bootstrap samples from them, denoted, say, $r_{b,i}^*$ for $i = 1, \ldots, n$ and $b = 1, \ldots, B$. Let $y_i^b = f(x_i^b, \hat{\theta}) + r_i^b$. This time let

$$\hat{\theta}^b = \arg\min_{\theta} \sum_{i=1}^{n} \left(y_i^b - f(x_{b,i}, \theta) \right)^2. \tag{10.12}$$

The SE of $f(x_{\text{new}}, \hat{\theta})$ now has the same form as (10.10) even though the definition of $\hat{\theta}^b$ has changed. This leads to a $1 - \alpha$ PI of the same form as (10.11).

The two methods will give very similar results if the model $f(\cdot, \theta)$ fits well. Indeed, if the model does fit well then the r_i will be representative of the error term and bootstrapping the residuals should give an accurate PI. Likewise, when $f(\cdot, \theta)$ fits well, both (10.9) and (10.10) should be accurate and therefore give an accurate PI. However, if the model does not fit well then the r_i are likely to be inflated. So, bootstrapping over them can easily lead to PIs that are distorted, because the $\hat{\theta}^b$ depend directly on the r_i. By contrast, finding $\hat{\theta}^b$ by bootstrapping over pairs (y_i, x_i) may be less dependent on the residuals and hence more robust against model misspecification.

10.2 Trees

Trees can be used for classification or regression and both cases will be treated here. The basic idea of trees – or, more formally, recursive partitioning – is to chop the space of explanatory variables and their values into subsets and then assign a response value for each point conditionally on the subset to which the point belongs. The partitioning may be visualized for a four-dimensional space of explanatory variables as in Fig. 10.1. In the left-hand panel, the first split at the root node occurs for x_1 at the value C_1, resulting in two child nodes. Then these are split on x_2 at C_2 and on x_4 at C_4. Finally, one of the resulting nodes is split on x_3 at C_3. The process results in five 'leaves' or terminal nodes. The order of the splits can be inferred from the length of the straight line that represents the split and from the other straight lines that come from it. In the right-hand panel the sequence of splits is illustrated as a tree; the places where a split occurs are called nodes. The notation, say,

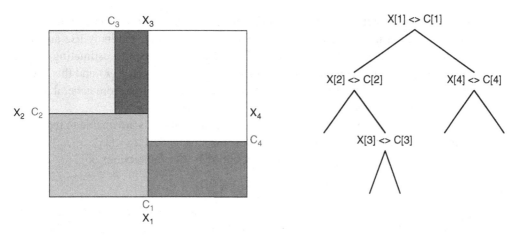

Figure 10.1 The left- and right-hand panels represent the same series of binary splits on four variables.

$X[2] <> C[2]$ means that the branch on the left corresponds to $x_2 < C_2$ and the branch on the right corresponds to $x_2 > C_2$. While all splits are binary, there is no prohibition on repeated binary splits on the same variable. That is, after the split on $X[2]$ the left-most branch could have been split again into $x_2 < C'(< C_2)$ and $C' < x_2 < C_2$; an analogous split could occur after splits on other variables as well.

The value of the response, say Y, is likely to be different on different partition elements. In a regression problem, often the value assigned to Y on a partition element is the mean of the value of the Y's whose x's are in the partition element. In a binary classification problem, the value assigned to a Y whose x's are in a given partition element might be determined by 'majority voting' – the value that the tree assigns to Y is whichever of zero and one occurs more often for the x's in the partition element.

To specify a tree mathematically, regions in the space of explanatory variables and values for the response on those regions must be specified. Write

$$Y(x) = \sum_{k=1}^{K} F_k(x) \mathbf{1}_{R_k}(x) + \epsilon, \tag{10.13}$$

where x is a d-dimensional explanatory variable, the R_k are disjoint regions in $\mathbb{R}^d$, often called leaves, F_k is the function of Y on R_K (which may have parameters that have been suppressed in the notation), $\mathbf{1}_A$ is the indicator function for a set A and ϵ represents random error. For regression trees, ϵ is often taken to be $N(0, \sigma^2)$. For classification trees, the left-hand side assumes the values 0 and 1 and therefore so must the right-hand side. Hence, ϵ can often be discrete. (A binary classification problem is a regression problem where the regression function is the indicator function for a region. In these cases the F_k assume the values 0 and 1.) Because the R_k and the F_k must be chosen, a key benefit of trees is that their degree of flexibility – often neither too small nor too large – puts them squarely in the intermediate tranche of modeling problems. More precisely, trees can encapsulate localized interactions well, so when interactions among the x_j are important they will often be more accessible to trees than to other additive models.

Consider the contrast between trees and linear models. Since linear models provide a local approximation to the true regression function, why would any one routinely use anything else? One reason is that in linear regression all the data are relevant to estimating all the parameters. That is, even if x_i is far from x it can have a big effect on $\hat{Y}(x)$ and this can be misleading for general nonlinear functions f. As an approximation to a more general model, a linear model's domain of validity may be very constrained.

This can be dramatized by observing that frequently linear models are unable to represent complex interactions among explanatory variables. For instance, if $x = (x_1, \ldots, x_d)$ and X is the design matrix, the linear model that includes all two-way interactions is

$$Y = X^T \beta + \text{vect}(\text{UT}(XX^T))\gamma + \epsilon,$$

where $\beta = (\beta_1, \ldots, \beta_d)^T$, $\gamma = (\gamma_1, \ldots, \gamma_{d(d-1)/2})$, and the second term on the right is the vector formed from the upper triangle of XX^T. So, there are $d + d(d-1)/2$ terms. Stronger nonlinearities (in the x_j) will require even bigger models. In some cases, such approximations become more oscillatory as terms are added; cf. Runge's function, Sec. 8.1. The problem is that linear models are global – they give a single predictive formula holding over the entire X-space. Trees do not localize a global model; they assemble a global model or predictor from a partition of the x-space into regions on which different models or predictors may be used. Since trees can express a relationship between a response Y and functions of explanatory variables locally, they often give better results than linear models.

What about the other extreme – nonparametric methods as seen in Chapter 8? These methods often localize well and have a useful concept of linearity too. The problem is that they don't scale up to large d. In many cases, nonparametric regression even with $d = 3$ is infeasible for commonly occurring n's. Moreover, the sensitivity to bandwidth parameters increases with d so that even when a useful solution exists it may be unstable, i.e., its predictions may have a high variance.

Of course, no method is always the best and there are some relations that are hard for trees to approximate. In particular, trees have terms representing region-dependent interactions, so functions that cut across the boundaries of the leaves R_j pose problems. These are different from the usual terms in multiple regression, which are not localized. For instance, consider the simple relation $Y = x$ with $\dim(x) = 1$. A tree would typically represent this as a staircase-like function on a partition of $\mathbb{R}$ (see Fig. 10.2) and fail to recognize the linearity while a linear model would typically find it easily. Such simple relationships are relatively rare in problems of nontrivial complexity, so trees routinely give better results in terms of prediction (and fit) than linear models in regression. In classification contexts trees – or their generalization, random forests, see Sec. 11.2 – are also highly competitive.

Usually, the regions R_k are chosen first and then, for each k, a function F_k is chosen. Loosely, the strategy is to pick one of the d variables and a value and ask a yes/no question. If $d = 73$ is chosen then one might ask whether $x_{73} < 86$? Then, the data points $x_1, \ldots, x_n$ are partitioned into two sets, those with $x_{73,i} < 86$ and those with $x_{73,i} \geq 86$. If desired, composite explanatory variables, e.g., a function of some of the x_i, can be added, giving more than d variables to search when choosing binary splits. Binary splits are chosen until the resulting R_k admit functions F_k that adequately represent the data in each R_k. Thus, there must be a reasonable way of choosing splits to test, a reasonable way to define the F_k, and a 'stop splitting' rule.

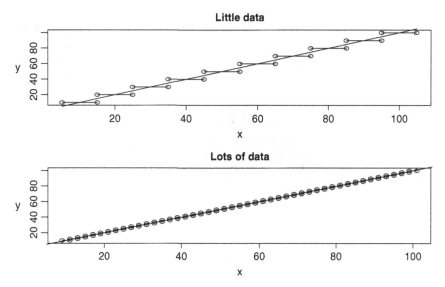

Figure 10.2 The top panel is a plot of the function that a tree would give when $Y = X$ is the correct regression function and the F_k are assumed to be constant on R_k. The bottom plot is the same but with more data.

A way in which trees are flexible is that different F_k may be different on different R_k; see Fig. 10.3. So, the functions F_k need not combine to give a continuous function on $\mathbb{R}^d$. This is bad if Y must be smooth but predictively it is rarely a problem especially for high-dimensional data. Trees are well adapted to high dimensions (but are not necessarily competitive in low dimensions) because the sparsity of the high-dimensional data permits rougher models. Moreover, for high d, the assessment of variable importance and computing in general is relatively straightforward, as will be seen. Used properly, trees provide a good method for prediction in both classification and regression settings. Below, methods for growing trees to achieve good prediction will be presented.

10.2.1 Finding a Good Tree

Any tree-growing procedure must have a rules for:

1. creating new nodes;
2. when to stop creating new nodes;
3. identifying K predictors for Y using the data in the R_k.

The first two parts are often related: a criterion by which to select splits can usually also be used to stop the selection of splits. The goal of the binary splits is to partition the training sample into increasingly 'homogeneous' groups, i.e., subsamples that can be used to generate the functions at the terminal nodes. Splitting usually stops when a satisfactorily high degree of homogeneity is achieved at the terminal nodes and, hence, on the corresponding regions of the explanatory variables. Since any procedure that will give a useful predictor at a terminal node can be used (subject to having enough data) usually the focus is on selecting splits.

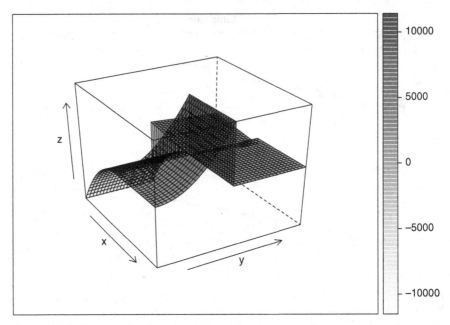

Figure 10.3 The tree function plotted has $K = 4$ regions. On two regions the tree is constant and on two regions the tree is a parabola – one opening up and one opening down.

There are three kinds of split that are commonly used at nodes to define the subsets R_k:

1. given j and $c_j \in \mathbb{R}$, split x_j at c_j;
2. given weights w_j, $j = 1, \ldots, d$, and $c \in \mathbb{R}$, split on the line $\sum_{j=1}^{d} w_j x_j \leq c$;
3. given a set U and a j, partition the x's according to $x_j \in U$ versus $x_j \in U^c$.

The first is a univariate (in x) split. The second is a linear-combination split that may arise from some criterion such as a linear discriminant; see Sec. 4.6. The third is only applicable when x_j is categorical. More complicated splits are possible, e.g., a curvilinear function of the x_j from a quadratic discriminant, but they are less common. The discussion here will be limited to the most commonly used splits.

Splitting to form more nodes tends to decrease bias, at the cost of variance. This is similar to linear regression, in which adding variables tends to decrease bias at the cost of variance. The difference is that trees can localize where they decrease variance while linear regression models cannot do so (at least not mathematically). So, there is good reason to expect that a well-chosen tree will have lower bias than a linear model. The question is how to choose a tree in the first place since poorly chosen trees can have high bias.

One natural approach to choosing splits in either a classification or a regression setting is by clustering. For a first split, find a cluster of size 2 for the data from each X_j individually, i.e., based on $x_{j,1}, \ldots, x_{j,n}$, and pick the best of these by some criterion, e.g., stability. Suppose that the best clustering uses j_0 and let its breakpoint be c_{j_0}. Then, the two half spaces $\{x | x_{j_0} \leq c_{j_0}\}$ and $\{x | x_{j_0} > c_{j_0}\}$ would give the daughter nodes from the first or 'root' node, normally taken to be $\mathbb{R}^d$. This process could be repeated in each daughter node to find further splits – hence the recursivity of the partitioning – until, for instance, the clusterings became insufficiently stable. Otherwise stated, the splitting would stop when the R_k were small enough to provide good discrimination but not so small as to lead to overfit. After generating the R_k it would remain to choose a function for the bottom nodes or 'leaves'. The simplest would be a constant that depended on the values of Y that correspond to the data points in the R_k. For regression this would be the mean; for classification this would be zero or one, depending on which had a higher frequency (majority voting). In either case, the resulting tree would be a special case of (10.13), namely

$$\hat{Y}(x) = \sum_{k=1}^{2} \hat{\beta} \mathbf{1}_{R_k}(x), \tag{10.14}$$

where $\hat{\beta}$, a function of the n data points (y_i, x_i), is either the mean of the y_i with $x_i \in R_k$ (regression) or the y-value of maximal frequency for the x_i in R_k (classification). Now, the point predictor for a new value of the explanatory variables, x_{new}, is simply $\hat{Y}(x_{\text{new}})$, using (10.14), and a PI can be found using either of the bootstrapping techniques at the end of Sec. 10.1, with bootstrapping over the data usually preferable.

There are many other ways to generate trees, and several will be seen here. However, before proceeding there are several points to make. First, at each stage the goal of splitting is to increase the homogeneity of the points in the final set of R_k. This is like nearest neighbors in the sense that leaf R_k groups together the points to be used to get a value for $\hat{Y}$ on R_k. The premise is that this will improve the predictive accuracy. Second, the choice of R_k is unstable unless the data are very accurate (not much variability) and group naturally into the R_k. When this happens, good functions F_k can usually be found. More typically, the data are highly variable and do not group decisively into leaves R_k. In these cases the trees are unstable in the sense that multiple distinct trees may provide essentially equally good predictors. Moreover, most procedures for finding the R_k are greedy, so that if a bad splitting rule is chosen at one node the effects of the poor choice propagate down to all its daughter nodes. Third, as long as the splits can be interpreted, the tree itself is interpretable. However, for complex data, the instability that trees often have means that even if a given tree can be interpreted physically, the interpretation is not reliable (although a tree can be useful for identifying explanatory variables in general, if not their specific roles). Also, bigger trees with more splits are harder to interpret – even if each individual split can be interpreted. So, it is usually best to regard trees merely as a useful set of input–output relations and ignore interpretability. Fourth, if all available data were used to find a tree then even if the data in the leaves were homogeneous and well described by the same F_k, there would be no guarantee that out-of-sample prediction would be good. This can arise because no data were held in reserve to validate the tree. Finally, in practice, if a complex tree is found, it is often hard to decide whether there would be another tree (or other model) that is close to it. This is another sense in which trees can be unstable.

An alternative to the use of clustering to choose split points c_j is to invoke an optimality criterion. When trees are being used to find a regression function, the squared error is the most popular choice of criterion because minimizing it corresponds to minimizing the within-R_k variance. When trees are being used for classification, the misclassification rate is the most popular choice because minimizing it corresponds to getting good predictions for class labels. These two cases will be treated in sequence.

Let T denote a tree, i.e., a function of the form of the first term on the right in (10.13) and suppose that a data set (y_i, x_i) for $i = 1, \ldots, n$ is available. In the regression setting, the goal is to find T such that

$$\text{SSE}(T) = \sum_{k=1}^{K} \sum_{x_i \in R_k} (y_i - F_k(x_i))^2$$

is as small as possible. If the F_k are constants then $\hat{F}_k$ is the kth within-leaf mean and this is the point predictor at x_{new} in R_k.

Often a PI is desired. The within-leaf means and variances are

$$\hat{F}_k = \bar{y}_k = \frac{1}{\#(R_k)} \sum_{x_i \in R_k} y_i \quad \text{and} \quad \widehat{\text{Var}}_k(Y) = \frac{1}{\#(R_k) - 1} \sum_{x_i \in R_k} (y_i - \bar{y}_k)^2.$$

Obviously, the within-leaf variance gives a within-leaf SE, so PIs conditional on the R_k can be given. More generally, if the F_k are linear models then the usual least squares theory (see Sec. 4.2) applies, so that PIs can be given conditionally on the R_k. The limitation of this approach is that it is naive: conditioning on the leaves ignores the variability in the prediction for $Y(x_{\text{new}})$ due to the architecture of the tree. That is, if a slightly different data set had been drawn, or the splits used to form T had been chosen slightly differently, the whole tree could have changed – discontinuously, because the architecture of the tree varies over a discrete set of possibilities. Using the bootstrap to generate PIs automatically accounts for both the within-leaf variability and the architecture variability.

One way to think about what recursively choosing splits means in a simple context is the following. Suppose that $\dim(x) = 1$. Then, $\hat{F}$ must be of the form $\hat{F}(x) = \hat{\beta}_\ell \mathbf{1}_{\{x < \hat{t}\}} + \hat{\beta}_r \mathbf{1}_{\{x \geq \hat{t}\}}$, where $\hat{t}$ is a threshold for the upper and lower leaves obtained from

$$(\hat{\beta}_\ell, \hat{\beta}_r, \hat{t}) = \arg\min_{\beta_\ell, \beta_r, t} \sum_{i=1}^{n} \left(y_i - \beta_\ell \mathbf{1}_{\{x_i < t\}} - \beta_r \mathbf{1}_{\{x_i \geq t\}} \right)^2,$$

which are estimates of the true values

$$(\beta_\ell, \beta_r, t) = \arg\min E \left(Y_i - \hat{\beta}_\ell \mathbf{1}_{\{x < \hat{t}\}} - \hat{\beta}_r \mathbf{1}_{\{x \geq \hat{t}\}} \right)^2.$$

That is, at a given node, recursion can be regarded as an extension of standard least squares regression to include the threshold. Once this optimization has been determined, at least approximately, the process is repeated on the leaves $\{x < \hat{t}\}$ and $\{x \geq \hat{t}\}$ where the expectations are conditional on the previous split. The process continues until a stopping criterion is met; the process is obviously harder to write down when $\dim(x) \geq 2$.

Because of concerns about bias for trees, it is most reasonable to bootstrap over the data as in Sec. 10.1 (although it is possible to bootstrap over residuals instead). The procedure is to draw B bootstrap samples for the n original data points and generate B trees. In principle, as in Sec. 9.1, the sampling distribution of T as a random variable over the set of possible trees can be estimated. Each tree gives a point prediction for $Y(x_{\text{new}})$ and this is an estimate of the regression function in (10.13) evaluated at x_{new}. The SD of these estimates is an estimate of the SE of the $\hat{Y}(x_{\text{new}})$; call the result $\text{SE}(\hat{T})(x_{\text{new}})$. As before, it is important to distinguish between the estimate of $T(x_{\text{new}})$ and its prediction. In parallel to (10.7), write

$$Y(x_{\text{new}}) - \hat{T}(x_{\text{new}}) = (Y(x_{\text{new}}) - T_0(x_{\text{new}})) + (T_0(x_{\text{new}}) - \hat{T}(x_{\text{new}})), \tag{10.15}$$

where $\hat{T}$ is the point predictor from a tree T and T_0 is the 'true' tree regression function. The two differences are independent and the variance of the first term on the right is σ^2. The natural estimate for σ^2 has the same form as in (10.3), namely,

$$\hat{\sigma}^2 = \min_T \text{SSE}(T).$$

Since $\text{SE}(\hat{T})(x_{\text{new}})$ is an estimate of the variance of the second term on the right in (10.15), a $1 - \alpha$ PI for $Y(x_{\text{new}})$ is

$$\hat{T}(x_{\text{new}}) \pm z_{\alpha/2} \sqrt{\hat{\sigma}^2 + \text{SE}^2(\hat{T})(x_{\text{new}})}. \tag{10.16}$$

The problem with (10.16) is that in (10.15) it is assumed that the true regression function exists and can be expressed effectively up to infinite precision as a tree. This is only valid for $\mathcal{M}$-closed problems. For $\mathcal{M}$-complete problems, an extra term summarizes the bias. Let G be the DG and let T_0 be the tree within a collection of trees that is closest to G in some sense, such as the mean integrated squared error. Then adding and subtracting gives

$$Y(x_{\text{new}}) - \hat{T}(x_{\text{new}}) = (Y(x_{\text{new}}) - G(x_{\text{new}})) + (G(x_{\text{new}}) - T_0(x_{\text{new}})) + (T_0(x_{\text{new}}) - \hat{T}(x_{\text{new}})). \tag{10.17}$$

In principle, a bootstrap estimate of the bias $G(x_{\text{new}}) - T_0(x_{\text{new}})$ or approximation error can be given: simply use bootstrap samples to produce a nonparametric estimator $\hat{G}_B(x_{\text{new}})$ for $G(x_{\text{new}})$ and $\hat{T}_{0,B}$ for T_0. If the difference between $\hat{G}_B$ and $\hat{T}_{0,B}$ is large relative to $\hat{\sigma}$ then the point prediction $\hat{T}(x_{\text{new}})$ would become $\hat{T}(x_{\text{new}}) - (\hat{G}_B(x_{\text{new}}) - \hat{T}_{0,B})$ and a second level of bootstrapping would be necessary to get an estimator for the variance of $(\hat{G}_B(x_{\text{new}}) - \hat{T}_{0,B})$. Provided that the data used for this level of bootstrapping were disjoint from the other bootstrap estimates, it would be independent of the other terms in the variances in (10.16) and would lead to adding an $\text{SE}^2(\text{bias})$ for the bias estimate under the square root sign. To do the bootstrapping in such a way that the estimates in the PIs are valid it is necessary to invoke the CV principle noted in Sec. 9.4 requiring independent data for each stage of construction of the PI.

For $\mathcal{M}$-open problems, tree methods may give useful point predictors, but the concept of a probabilistic PI no longer applies.

Given this, it is straightforward to propose a procedure for generating a tree that should be close to optimal in the sense of being close to G ($\mathcal{M}$-closed) or close to T_0 ($\mathcal{M}$-complete) when the R_k are formed by univariate splits on individual x_j in x. (Because trees are dense in the space of continuous functions, T_0 may be made as close to G as desired by choosing large enough collections of trees and data. Hence, the problem would be $\mathcal{M}$-complete if G existed.) There are two limitations that must be imposed on any procedure for it to be effective. First, once the splits have been decided there must be enough data in each leaf to specify the $\hat{F}_k$; suppose that a minimum number of data points per leaf m has been chosen. Second, to avoid having to make too many splits (and hence increasing variability) there must be a threshold $\delta > 0$ for the decrease in SSE for the acceptance of a split. That is, a chosen split will be accepted only if the SSE decreases by at least δ. Assume that the models F_k in the leaves have been chosen, e.g., linear models. The steps are as follows.

□ Start at the root node containing all the data. Set $K = 1$ and use all the data to estimate $\hat{F}_1$. Then find the SSE for $\hat{F}_1$ as a predictor for Y using the y_i.

□ If all the y's in the terminal nodes have the same value as their respective $\hat{F}_k$ then stop. Otherwise, search over all univariate splits in all terminal nodes that have nonzero SSEs and add the split that most reduces the overall SSE, provided that there will be at least m data points in each daughter node and the reduction in SSE is at least δ.

□ Repeat the last step until no splits reduce the overall SSE by more than δ and leave at least m data points in each leaf.

Clearly, the 'root node' method is simply an iterated form of classical nonlinear regression. So, tree-based regression can be regarded as an extension of it. Second, any time the SSE at a node drops below δ, that node can be dropped from further splitting. When d is large this procedure can be ineffective: in such cases, often the splits are chosen to be tested on randomly chosen x_j. Another problem that can arise is the inclusion of spurious variables. Since this method of constructing trees relies on finding $\hat{F}_k$ that replicate the data, there is no principle that forces the F_k to contain x_j that are in any way relevant to Y. Often, to get around this the data are split initially into n_1 training points and n_2 testing points, effectively permitting a CV approach. The tree is grown using the n_1 data points but then grown out to one data point per leaf (i.e., $\delta = 0$ and $m = 1$) and after that pruned back using the test set. The pruning is done by simply removing a final split and asking whether the SSE on the test data changes much, i.e., increases above a threshold. The idea behind this sort of approach is to find the optimal tradeoff between excess variance (a large tree) and bias (a small tree).

Trees are also used for classification, and are useful for classification for much the same reason that they are useful in regression. Arguably, trees are used even more in classification than regression because of the success of random forests (see Sec. 11.2). Again, a tree provides interpretable (if unrealistic) predictors, even where many variables interact in complicated and often nonlinear ways. The procedure for finding classification trees is similar to that for finding regression trees, but there are some noteworthy differences.

The key question in finding good classification trees is what to use in place of the squared error since the output of the trees is binary, not continuous. It is otherwise easy enough to use the last procedure. That is, to start at a node, possibly the root node, and find the binary split that provides the best fit to the observed classes. This binary split leads to two nodes on which the splitting process can be repeated until a stop-splitting rule is reached. Usually this will result in a tree that is too large and has to be pruned back in one way or another. As opposed to the classification methods in Sec. 4.6, which are piecewise linear, and those in Sec. 8.6.2, which are nonparametric, classification trees based on binary splits of variables are relatively easy to interpret even when n is large because usually the sets they produce have boundaries parallel to the axes of the variables.

The choice of splits for classification is different from that for regression. One way to decide whether to split at a given value of an explanatory variable is to look at the subsequent increase in 'purity' of the Y values at the relevant node. That is, does the split reduce the misclassification of all the points at the mother node by an amount that is large enough to justify adding the daughter nodes? One traditional way to answer this question is with the Gini index. For a sample of size n representing C classes, with n_c the number of data points assuming the value c, $c = 1, \cdots, C$, the Gini index is $\mathcal{G}(\mathcal{D}) = 1 - (n_1/n)^2 - \cdots - (n_k/n)^2$ for the data set $\mathcal{D}$. If the question is whether to split $\mathcal{D}$ into disjoint sets $\mathcal{D}_1$ and $\mathcal{D}_2$ of size N_1 and N_2, respectively, then the optimal split $(\mathcal{D}_1, \mathcal{D}_2)$ minimizes

$$\mathcal{G}(\mathcal{D}_1, \mathcal{D}_2) = \frac{N_1}{n}\mathcal{G}(\mathcal{D}_1) + \frac{N_2}{n}\mathcal{G}(\mathcal{D}_2).$$

This optimal split should be adopted provided that N_1 and N_2 are large enough to give a reasonable prediction for Y with explanatory variables in $\mathcal{D}_1$ and $\mathcal{D}_2$, respectively, and provided that the decrease from $\mathcal{G}$ to $\mathcal{G}(\mathcal{D}_1, \mathcal{D}_2)$ is large enough. Note that these characteristics play the same role in classification as m and δ did for regression. If the resulting tree is too large, it can again be pruned back by holding some data in reserve as a test set. Apart from variability and intermediate cases, the Gini index tends to pull off big distinct clusters of similar points first, leaving clusters of smaller or less distinct points to be pulled off by downstream nodes in the tree.

The Gini index is not the only measure of impurity that one might seek to reduce via splits. Another popular measure is the entropy. In this context, the entropy is another function of a potential split at a node. At a given node v, such as the root node, one considers splits of $\mathcal{D}$ into disjoint data sets $\mathcal{D}_1$ and $\mathcal{D}_2$ that minimize

$$H(v) = -\sum_{c=1}^{C} \hat{P}(c|v)\ln \hat{P}(v|v),$$

where $\hat{P}(c|v)$ is the estimate of the relative frequency of class c at node v. The goal now is to minimize the entropy because a small entropy corresponds to a small average codelength and hence has relatively less variability.

The classification error is another form of impurity. The task is to minimize

$$\text{Error}(v) = 1 - \max_j \hat{P}(j|v).$$

This is the most familiar misclassification error because it is zero when one of the $\hat{P}(j|v)$'s is unity, meaning that all the data points at a node are in the same class. It is easy to see that minimizing Error means finding the splits (they are usually not unique) that have as many data points as possible with $Y = 1$ in one child node and as many data points as possible with $Y = 0$ in the other child node. This is precisely the sense in which an impurity measure represents impurity; the Gini index and entropy have the same properties but this is not obvious. If a graph is plotted showing $P(Y = 1) = p \in [0, 1]$ on the x-axis and the value of the impurity on the y-axis, the Error curve looks like a tent function with a maximum of 0.5 at $p = 0.5$. The Gini index is a semicircular curve above the Error curve and also has a maximum of 0.5 at $p = 0.5$. The entropy curve is above the Gini curve and has a maximum of 1 at 0.5. It looks like a parabola that opens downward. All three curves are 0 at 0 and 1 at 1; see Figure 4.13 of Tan *et al.* (2005). The shape of the curves suggests that the entropy provides the best discrimination, if this is what is wanted, but its meaning is the most remote from actual classification.

Another way to think of the splits in a classification tree is information theoretically. Implicitly this assumes that the explanatory variables are discrete or can be discretized in such a way that sets such as U for the categorical case can be identified. (Most information-theoretic quantities can be defined for continuous variables, but here it is simplest to limit the discussion to discrete variables since Y only assumes the values 0 and 1, for instance.) Let A denote a random variable such as $\mathbf{1}_A$ taking values $\{a\}$, where A is a set defined in terms of the explanatory variables. Using A for an indicator and for the set indicated will not make for confusion since it should be obvious which is meant. Now, the Shannon mutual information (SMI) $I(Y; A)$ between Y and A is

$$I(Y; A) = H(Y) - H(Y|A) = H(Y) - \sum_a P(A = a)H(Y|A = a),$$

where $H(Y)$ is the entropy of Y and $H(Y|A = a)$ is the conditional entropy of Y given $A = a$. In terms of Shannon coding theory, $I(Y, A)$ is the rate of transmission of information (in bits per letter) across the channel $p(y|a)$. Statistically, $I(Y; A)$ is a measure of the dependence between Y and A. In either case, $I(Y; A)$ should be large for good classification.[2] It is easy to think of A and A^c as representing the possible answers to a question of interest regarding the outcome. An example would be 'Is the condition serious enough to warrant an operation?' and the answer is likely to be nuanced, meaning that A could be quite complicated. In this sense A plays the role of U in the third kind of common split.

In this formulation, if $I(Y; A)$ is sufficiently smaller than $H(Y)$, the split into A and A^c would be accepted. Then, at the node $A = a$ (and $A \neq a$) the process would be repeated. If B were the random variable $\mathbf{1}_B$ taking values $\{b\}$ then the comparison between $I(Y; B|A = a)$ and $I(Y; B|A = a)$ would determine whether the new split into B and B^c after $A = a$ would be retained. This process stops when the increase in SMI or the number of points at a node is too small.

However the classification tree is found, it is used primarily to give probability predictions that are immediately turned into point predictions. That is, given a classification tree $T(\cdot)$, a new set of covariates X_{new} is run down the tree until it ends in a leaf. If the node has 60% 1's and 40 % 0's then the probability prediction is the value in the leaf that has 60% probability. The point predictor simply predicts which value (1 or 0) has a higher probability. In the special case of an exact 50–50 split, one would not be satisfied with the tree because that would represent no information.

One use of classification trees is that the probabilities can be used in downstream calculations. Suppose that a classifier gives class probabilities $P(Y = i|X = x)$. Then there is an expected cost for each misclassification. Let $L_{i,j}$ be the cost of classifying a subject who is actually in class i as being in class j. Then, if the $P(Y = i|X)$ are known (or estimated), the conditional risk (cost) of predicting Y to be j when $X = x$ is

$$R(i|x) = \sum_j L_{i,j} P(Y = j|X = x).$$

---

[2] The notation $I(Y; A = a) = H(Y) - H(Y|A = a)$ would mean the decrease in transmission codelength due to knowing a; this is usually reinterpreted to mean a reduction in the variability of Y as a consequence of knowing a.

Typically $L_{i,i} = L_{j,j} = 0$ and one of $L_{0,1}$ and $L_{1,0}$ is much higher than the other. Think of disease detection: write $i = 1$ for diseased and $i = 0$ for not diseased. Then, for a severe disease, if the cost of further testing is not too burdensome for a given population of patients it may be more costly to get a false negative than a false positive, in which case $L_{1,0}$ may be much higher than $L_{0,1}$. If a new data point x_{new} is run through the tree then the $P(Y = i|X = x)$ are at the leaf nodes, and in each leaf the expected cost of predicting 'disease' is

$$R(i = 1|x) = L_{0,1}P(Y = 0|X = x) + L_{1,1}P(Y = 1|X = x),$$

i.e., it equals the cost of a false positive and a true positive. The expected cost of not predicting the presence of disease is

$$R(i = 0|x) = L_{1,0}P(Y = 1|X = x) + L_{0,0}P(Y = 0|X = x),$$

the cost of a false negative and a true negative. So, if $L_{1,0}$ exceeds $L_{0,1}$ by a large enough amount, the expected cost of saying someone does not have the disease when they do have it will be higher than saying that they do have it when they do not. Note that this analysis applies to all x_{new} that land in the same R_k. The cost analysis above is done for the tree as a whole. It can also be done on a subtree formed by all splits past a specific node. The results would then be conditional on the rest of the tree.

Whatever splitting rule is used, the goal in both classification and regression is to increase the homogeneity (or decrease the impurity) of the data associated with each F_k. In regression, the homogeneity is related to how well the data fit F_k; in classification, the degree of homogeneity refers to how many data points R_k have the same Y value, i.e., are in the same class. In both cases, splitting continues until a tree is produced with as much homogeneity at each terminal node as possible subject to having enough data points in the terminal node to make the homogeneity non-trivial.

10.2.2 Pruning and Selection

Having seen numerous ways to grow a tree for regression or classification purposes it is important to be able to choose among them sensibly. The intuition is that it is usually optimal to grow a larger tree than is justifiable and then prune it back. The reason for this is that splitting rules only look one step ahead. Otherwise stated, stopping rules tend to underfit, meaning that even if a rule stops at a split for which the next candidate splits give little improvement, it may be that splitting the nodes one layer further down will give a large improvement. This is a separate issue from the recursivity of splitting rules, which means that a poor split which is accepted at one stage is locked in and not rejected at a later stage – although a longer tree might compensate for it (at the cost of more variability because the tree is larger). In many cases a maximal tree is grown, i.e., $m = 1$ is used and each leaf has one data point in it. This ignores the fit in a leaf, but the true function value in a leaf is in principle unbiased for its value on the leaf.

One pragmatic way to put this intuition into effect is to consider a nested sequence of trees from an initial large (maximal) tree and apply weakest-link cutting. That is, prune off all the child nodes of a fixed nonterminal node that seems to add little to the discriminatory power of the tree. There are two basic ways to do this, namely, cost-complexity pruning and CV, although the two are often used together.

Formally, the cost–complexity measure for a tree T is

$$C(T; \alpha) = \text{Cost}(T) + \alpha |T|.$$

In this expression, $|T|$ denotes the number of terminal nodes in T, and $\alpha \in \mathbb{R}^+$ is a parameter representing the tradeoff between $|T|$, the complexity of T, and the cost of the tree, $\text{Cost}(T)$. The cost of a regression tree T is the SSE from using the estimated regression function $\hat{g}$ (so that $\text{Cost}(T) = \text{Cost}(\hat{g})$) and the cost of a classification tree is the overall misclassification rate of the function $\hat{g}$ it represents. More precisely, respectively,

$$\hat{g} = \sum_{k=1}^{K} \hat{F}_k \mathbf{1}_{\hat{R}_k}(x) \quad \text{and} \quad R(t) = \sum_{v \in T_e} P(v) R(v)$$

where T_e is the set of terminal nodes of T, $P(v)$ is the proportion of data at node v, and $R(v)$ is the misclassification rate at node $v \in T_e$. From its structure, $C(T; \alpha)$ chooses trees with fewer and fewer terminal nodes as α increases; conversely, it permits more and more nodes as α goes to zero, up to the maximum value of Cost. However, the goal is to find trees that achieve small values of $C(T; \alpha)$.

More operationally, large α's penalize large trees heavily, making small trees optimal. Small values of α permit larger trees that may be susceptible to overfitting. Thus, in the limit of large α, the one-node tree consisting of just the root is optimal; in the limit of small α, the maximal tree T is optimal. Note that pruning means that regions R_j (in a regression or classification tree) are joined so that they now have a common node function. By letting α increase from zero, a sequence of subtrees is generated. Specifically, let

$$T_\alpha = \arg \min_{T' \subset T} C(T'; \alpha)$$

be a sequence of trees as α ranges from 0 to infinity. It can be shown that $T_{\alpha_1} \subset T_{\alpha_2}$ when $\alpha_1 \geq \alpha_2$ and that the sequence itself is nested; see Breiman *et al.* (1984). Thus, there is a sequence of nested trees T_{α_j} corresponding to functions $\hat{g}_j$ for $j = 1, \ldots, J$. The structure of the $\hat{g}_j$ represents the leaves of the trees in the sequence, and it is possible to select an optimal element $\hat{g}_{\alpha_j*}$ by cross-validation. Hopefully, by choosing an optimal subtree, the resulting α_j will be a serviceable compromise between complexity and fit that has low MSE and good predictive power. To achieve this, some standard pruning rules are often followed. First, if a node provides little improvement in impurity or SSE, it is generally pruned off along with its daughter nodes. Second, if two nodes are approximately equal, in the sense that pruning off either of them, and their child nodes, reduces the cost-complexity by approximately the same amount, the usual preference is to prune off the node with the larger number of child nodes. The result of such a series of pruning steps is a sequence of ever smaller subtrees, which is good from the standpoint of parsimony and hence the variance term in an MSE. Of course, if some data are held in reserve, a CV approach can also be used to guide the pruning by $C(T; \alpha)$, removing the nodes that have higher CV errors.

There is some debate as to how much splitting criteria matter because in some contexts large differences between trees generated by different criteria, particularly in their performance, have been noticed only infrequently. As noted earlier, when data sets are small and highly accurate, trees can be generated easily and the particular splitting and stop-splitting rules do not seem to matter much. Outside these cases, however, obtaining a good answer is genuinely difficult and there is evidence that splitting rules lead to different trees. The reason is that different splitting rules favor different tree architectures. The Gini coefficient seems to favor long straggly trees while in some cases the squared error favors short bushy trees. The problem in some sense is one of identifiability: two trees may be predictively very similar in terms of their values but mathematically quite different in structure. Choosing a splitting rule is somewhat like choosing a prior. The splitting rule (or stop-splitting rule) favors trees of some forms over trees of other forms. It is unclear how much the data can overwhelm the splitting rule because the tree and the performance of the splitting rule depend on the underlying distribution of Y (outside the $\mathcal{M}$-open case).

Aside from $C(T; \alpha)$ and the various forms of errors discussed so far, there are several ways to evaluate how well a tree performs or what it says about Y. Three of these are the concept of out-of-bag (OOB) error estimates, variable importance factors (VIFs), and assessments of stability.

The OOB error estimate is the result of forming trees using, say, B bootstrap samples. The result is B trees T_b for $b = 1, \ldots, B$. All but one of the possible bootstrap samples of size n from a collection of n data points leaves out at least one of the original n data points. More typically, a bootstrap sample leaves out somewhat more than a third of the data. The OOB error estimate for a classification tree is the average of the misclassification errors of the B bootstrap-sample-generated trees as assessed on the hold-out or OOB observations. As such this error estimate is not a property of any of the trees constructed but a property of the *method* by which they were constructed. In this regard it is more like an SE than an error assessment. (In Sec. 11.2 this will become more natural.) The same property can be used for regression trees but this time the OOB error is the average of the SSEs formed from the data not drawn in the bootstrap samples. Again, the OOB error is more like an SE for a combination of the B bootstrapped trees, but if the OOB error is low or high then this may suggest that the method for generating the tree is respectively good or poor.

The VIF was seen in the example in Sec. 1.2.3. Again, the idea comes from bootstrapping. The idea is to work out the effect of a chosen explanatory variable on the average OOB error for a collection of B classification or regression trees formed by bootstrapping. This is done by randomly permuting the values of the chosen variable in each bootstrap sample. Thus, it is expected that the average OOB error will increase and, the greater the increase, the more important the variable is with respect to Y, at least in an average sense. Unlike the earlier techniques, this does not require the data to be $\mathcal{M}$-closed or $\mathcal{M}$-complete. For instance, no probabilities or expectations are assumed to be relevant. This is a purely predictive technique applicable even to $\mathcal{M}$-open data. (The earlier techniques can be applied to $\mathcal{M}$-complete data as well and they do give answers. However, it is a judgement call as to how meaningful the results are and the predictions would of course have to be validated.) Again, one can argue that this result is more appropriate to an aggregation of trees (as is used in Sec. 10.2.3), but a tree's size still indicates whether the procedure for generating it from the data is likely to be successful. One way in which VIFs can be used is to run them on all variables and then retain only the variables that have the highest VIFs. This is not theoretically justified but makes good intuitive sense and may be a pragmatic way to find good predictors.

Out-of-bag errors can be used in $\mathcal{M}$-closed and $\mathcal{M}$-complete problems – provided that the bias has been either verified to be small or corrected for, as in (10.17). In these cases, the average OOB error can be regarded as an SE for the tree found using all the data, where the

SE is taken in the sampling distribution of the collection of trees in some large collection of trees. One reason why this is valid is that, because the bootstrap samples are drawn independently, the average SE is a good approximation to an actual SE. The result for regression trees is that a $1 - \alpha$ CI for $Y(x_{\text{new}})$ can be given and modified as in (10.15) or (10.17) to give a $1 - \alpha$ PI.

It has been remarked several times that trees are highly variable and that trees with very different architectures can be mathematically very close or statistically close, in the sense that they give seemingly equally good but incompatible inferences. Specifically, single trees are sensitive to small changes in the input data. Pruning tends to reduce the overfitting problem but the choice of variable to split on and its split point strongly depend on the specific observations in the training sample. The effect is that the whole architecture of the tree can be changed dramatically if the first splitting variable, or even just the first split point, is chosen differently due to a small change in the training data. See Strobl *et al.* (2009) for a recent example of exactly this phenomenon; see Ciampi *et al.* (1988) for an earlier discussion and references. The main way in which this is counteracted is by the ensemble methods to be seen in Chapter 11. However, there are other techniques such as that in Hothorn *et al.* (2004) (hypothesis testing) or that in Kitsantas *et al.* (2007) (cost functions).

While hypothesis testing and cost functions are formal and provide good assessments of stability or sensitivity, a simpler computational approach called simulation extrapolation (SIMEX) can also be used to good effect. This was invented by Cook and Stefanski (1994) in the context of parametric errors-in-variables models but has obvious extensions; see Stefanski and Cook (1995) and, more recently, Luo *et al.* (2007). The basic idea for regression or classification trees is to add $N(0, \sigma^2)$ noise, $\sigma^2 > 0$, to each $x_{j,i}$ and regenerate the tree model to see how much it changes in response to controlled amounts of variability. If the architecture is held fixed, i.e., the variables on which to split are fixed, then this can provide a useful evaluation of the parameter selection – i.e., the split points and any parameters in the leaves. As σ increases, the increase in parameter uncertainty can be tracked and may reveal a point σ_0 beyond which the tree is unreliable. This approach can be adapted to categorical variables by changing the distribution of the perturbing noise $N(0, \sigma^2)$ to a discrete variable, possibly a Bernoulli(p) for flipping the value of a categorical variable.

This same procedure can be extended to the architecture of the tree. The whole tree can be regenerated for the perturbed data and, in principle, for small σ the architecture should not change. The first value of σ, σ_1, for which nontrivial change emerges is an assessment of the sensitivity of the tree to the data. If σ_1 is unacceptably small then the architecture is unstable and the analyst is well advised to seek a more stable tree structure, possibly by using other procedures for generating a tree or by changing the pruning strategy. As yet, it is unclear how well stability methods work on individual trees (despite their apparent interpretability) because it is model-averaging methods that have proved more successful at prediction.

To conclude this subsection we consider two points. First, trees can also be generated by hybrid methods that switch splitting criteria as they move down the decision tree. In small data sets these techniques may not lead to substantially different trees, at least after pruning, but, as the sample size increases, trees tend to show greater variability as discussed above. So, hybrid approaches may yield significant improvements.

Second, we note a recurring structure that will continue to reappear. Pruning by using $C(T; \alpha)$ rests on a tradeoff between fit and complexity. This is the same principle as

undergirds MSE (if bias is regarded as a measure of complexity) in Nadaraya–Watson estimators and linear regression. It will also be seen in neural networks (Sec. 10.3), kernelized
methods (Sec. 10.4), projection pursuit (Sec. 10.7.1), and penalized methods (Sec. 10.5),
amongst others. That is, the notion of a tradeoff among various sources of error is central.
The goal is to find an optimal tradeoff by focusing on minimizing the worst errors; cf. Sec.
2.4, principles 3 and 4.

10.2.3 Bayes Trees

The key difficulty is to write the tree model (10.13) in such a way that it is amenable to prior
specification. One way to do this is to be more abstract. First, fix a tree T. The leaves of T
are the R_k and the function in leaf k is $F_k(x|\theta)$. Then the model for T is

$$Y(x) = \sum_{k=1}^{K} F_k(x|\theta_k)\mathbf{1}_{R_k}(x) + \epsilon = T(x;\theta) + \epsilon, \tag{10.18}$$

where T is the regression function implemented by the tree, $\epsilon \sim N(0, \sigma^2)$, and the within-
leaf parameters are the θ_k concatenated into a single parameter θ. The architecture of T is
also a parameter and in (10.18) the symbol T means the architecture of T as well as the
function T implements (the distinction will be clear from the context). The first task is to
assign priors

$$w(\sigma^2, \theta, T) = w(\sigma^2)w(\theta|\sigma^2)w(T|\theta, \sigma^2).$$

There are many obvious ways to assign $w(\sigma^2)$ and $w(\theta|\sigma^2)$. Assigning a prior to T is harder
but can be done. Here the procedure of Chipman *et al.* (1998) is presented but see Denison
et al. (1998) for an alternative.

For regression trees, the basic strategy is to separate the choice of the architecture from
the parameters and to represent the architecture as a result of two processes: a probabilistic process to choose which node to split and, conditionally on it, a separate probabilistic
process to make the split. So, for example:

☐ Start with the root node; initially it represents all the data; denote it by v.
☐ Let $P_s(T, v)$ be the probability of splitting node v. One choice is

$$P_s(v, T) = \alpha(1 + d_v)^{-\beta}, \tag{10.19}$$

where d_v is the depth in the tree of node v, $\alpha \in (0, 1)$, and $\beta > 0$ indicates how fast the
probability of a split decreases as d_v increases.
☐ Let $P_r(\rho|v, T)$ be the probability of using rule ρ to split node v. A 'rule' consists of an
explanatory variable and a split point. For instance, the uniform rule chooses a value of j
from the discrete uniform distribution on $\{1, \ldots, d\}$, where d is the number of explanatory
variables, and then chooses a split point from $x_{j,i}$, $i = 1, \ldots, n$, the observed values of
x_j, again using a discrete uniform.
☐ If the node splits, write T for the resulting tree and repeat the procedure.

Clearly, this procedure is incomplete because it omits an instruction about how to choose the
hyperparameters α and β (empirically) and does not have a stopping rule (one choice is to

stop before an empty node arises). These questions were addressed in Chipman *et al.* (1998) along with alternatives to the choices for P_s and P_r noted above.

The usual choices for priors on the real parameters apply when the leaf functions are constant. These are

$$\theta_\ell | \sigma \sim N\left(\mu_0, \frac{\sigma^2}{a}\right) \quad \text{and} \quad \sigma^2 \sim \text{IG}\left(\frac{b}{2}, \frac{b\lambda}{2}\right),$$

where μ_0, a, b, and λ are hyperparameters and $\ell = 1, \ldots, \gamma = \dim(\theta)$. In this case, $\gamma \equiv K$ is the number of terminal nodes and the likelihood is

$$f(Y|X, T) = \frac{ca^{b/2}}{\prod_{\ell=1}^{\gamma}(n_\ell + a)^{1/2}}\left(\sum_{\ell=1}^{\gamma}(s_\ell + t_\ell) + b\lambda\right)^{(n+b)/2}, \quad (10.20)$$

where c is a constant, X is the design matrix, Y is the vector of outcomes, n_ℓ is the number of data points in node ℓ, s_ℓ is the SSE from node ℓ, and $t_\ell = (n_\ell a/(n_\ell + a))(\bar{y}_i - \mu_0)$; see Chipman *et al.* (1998).

Following the usual Bayes prescription, the posterior is

$$f(T|X, Y) \propto f(Y|X, T)w(T), \quad (10.21)$$

which can be approximated by an MCMC–MH approach described in Chipman *et al.* (1998). The result leads to an approximation of the predictive distribution for Y_{new} at x_{new} from

$$f(Y_{\text{new}}|X, Y, x_{\text{new}}) = \int f(Y_{\text{new}}|x_{\text{new}}, T)f(T|X, Y)\text{d}(\theta, T) \quad (10.22)$$

where the notation $\text{d}(\theta, T)$ indicates the dominating measure for the joint continuous and discrete random variables.

For classification trees, the process for growing the trees is similar, so the prior on T does not change. However, the y_i belong to one of K classes denoted C_k, $k = 1, \ldots, K$, and follow a generalized Bernoulli distribution. Write n_k to mean the number of values $y_{i,k}$, for $\ell = 1, \ldots, n_k$, i.e., the number of outcomes of Y that land in C_k, so that $n_1 + \cdots + n_K = n$. Then, the likelihood is

$$f(y^n|\theta, T) = \prod_{\ell=1}^{n_k}\prod_{k=1}^{K}p_{uk}^{I(y_i \in C_k)}, \quad (10.23)$$

where $\theta = (\theta_1, \ldots, \theta_\eta)$, with η is the number of nodes and $\theta_u = (p_{u1}, \ldots, p_{uK})$ for $u = 1, \ldots, \eta$; here $p_{uk} \geq 0$, for each u, $p_{u1} + \cdots + p_{uK} = 1$, and $I(y_i \in C_k)$ is the indicator function for y_i to be in C_k.

The prior assignment changes because the θ_u are now restricted to the $K - 1$ simplex in K dimensions just defined. Chipman *et al.* (1998) suggested a Dirichlet prior with hyperparameter $\alpha = (\alpha_1, \ldots, \alpha_K)$ and all $\alpha_k > 0$. That is, for each u, $u = 1, \ldots, K$, the θ_u are IID and

$$(p_{u1}, \ldots, p_{uK})|T \sim \text{Dirichlet}(p|\alpha) \propto p_1^{\alpha_1 - 1} \cdots p_K^{\alpha_K - 1},$$

where $p = (p_1, \ldots, p_\eta)$ is the 'center' of the Dirichlet distribution. In parallel to (10.20), using (10.23) and this prior, the likelihood is

$$f(Y|X, T) = \left(\frac{\Gamma(\sum_k \alpha_k)}{\sum_k \Gamma(\alpha_k)} \right)^\eta \prod_{u=1}^\eta \frac{\prod_k \Gamma(n_k + \alpha_k)}{\Gamma(n_u + \sum_k \alpha_k)}, \quad (10.24)$$

where n_u is the total number of y_i in node u. Now, from (10.21) and (10.22), predictions can again be made, subject to using the appropriate MCMC–MH algorithm to approximate the posterior.

There is an alternative to Bayesian trees for classification as just presented. The idea is to add nodes and form splitting rules at them using Bayes tests; see Pittman *et al.* (2004a) for the methodology and Pittman *et al.* (2004b) for an example of its use. (In principle, a hypothesis-testing approach could also be made with frequentist testing but this seems not to have been done.) If Y is binary then, given a threshold c_j for the explanatory variable x_j, a 2×2 table can be formed at a candidate node, starting with the root node. The number of outcomes for $Y = 0, 1$ are fixed for $x_j \leq c_j$ and for $x_j > c_j$. So, the node represents two binomial variables, conditionally on the number of outcomes $Y = 0$ and $Y = 1$ for $x_j \leq c_j$ and separately for $x_j > c_j$. Now, for fixed c_j, a Bayes test can be done to determine whether the parameters in the two binomials are the same. If they are the same then there is no point in splitting at c_j. If there is a difference, it might be desirable to split at c_j. The methodology proceeds by searching over pairs j and c_j to find the optimal splits according to the corresponding Bayes factors.

To conclude the Bayesian treatment of trees, we look at Bayesian additive regression trees (BARTs). These foreshadow the Bayes model-average (BMA) predictor to be seen in Sec. 11.1. While BART and BMA are both sums of terms, each of which is a tree, in BART the terms are regularized so that no single term is overly influential whereas in BMA the trees are weighted by posterior model probabilities. So, even though BART is closer to a Bayes linear model structure, which permits nonlinear terms, than it is to an average of models, it is an intermediate step between additive models and BMAs.

The basic structure of a BART regression model originates in Chipman *et al.* (2007) and was more fully developed in Chipman *et al.* (2010). The idea is to replace the regression function in

$$Y = T(x) + \epsilon,$$

where $T(\cdot)$ is the function of x defined by the tree T, by the regression function in

$$Y = \sum_{j=1}^m T_j(x) + \epsilon. \quad (10.25)$$

In the one-tree model $E(Y|x)$ is the parameter value of the terminal node of which x is a member; this assumes that the terminal-node functions are constants. In (10.25), x is in the terminal node of each of the m trees and so $E(Y|x)$ is the sum over the m trees of the constants in all the terminal nodes containing x. The prior selection is similar to that described above and is detailed in Chipman *et al.* (1998). The key difference is that the T_j are mostly 'stumps', i.e., trees of depth 1, and this is reinforced by choosing α and β in (10.19) to be relatively large (say, 0.95 and 2, respectively). In fact, the prior selection in

BART is data dependent. This makes BART technically incoherent, but the incoherence is very limited and BART provides good performance.

The output of BART is a posterior distribution given by

$$p(T_1, \ldots, T_m, \sigma \mid Y),$$

where m can be large (say 200), requiring a 'backfitting' MCMC algorithm to approximate. This can be overcome to generate posteriors of the form $p(T_j \mid \sigma)$ as if $Y = T_j(x)$ were the true model. These can be combined to give an overall posterior that can lead to a predictive distribution. The estimation of $f(x_{new})$ or the prediction of $Y(x_{new})$ can then be done in a variety of ways; see Chipman *et al.* (2010, Sec. 3.2).

The BART procedure can also be adapted to classification problems using a probit model. To see this, write

$$p(x) = P(Y = 1 \mid x) = \Phi(G(x)) \quad \text{where} \quad G(x) = \sum_{j=1}^{m} T_j(x).$$

Then, following Chipman *et al.* (2010, Sec. 4), appropriate prior selection leads to draws that give an approximation to $p(x)$ and hence generate predictions.

In BART, it is important that the individual trees be 'weak learners', that is, none may contribute too much to the overall results. If one tree becomes too large, and hence more able to 'explain' Y, it can dominate the regression function, paradoxically degrading performance. The paradox is resolved by realizing that if one tree is so good that it explains Y very well, the other stumps lose their descriptive power because they were weak learners and end up contributing only variability.

Finally, Bayes trees formed by any method in this section can be examined using modifications of the same sorts of evaluation as those for the frequentist trees presented in Sec. 10.2.2. The OOB error can be calculated in much the same way if a bootstrap sample is used to form the Bayes tree. The VIF can be calculated similarly: the OOB should increase when variables that are important are permuted (or otherwise left out). The SIMEX approach to sensitivity analysis can be applied more or less as before. In all these cases, the question would be whether the effect on the final PI as a result of the additional error is large enough to be a problem. An extra measure of the sensitivity of Bayes trees arises from the choice of priors and from the estimates of hyperparameters. To be satisfied with a Bayes predictor, one would want the right amount of stability of the PIs, as reflected in the stability of the parameter estimates and the tree architecture.

10.3 Neural Nets

Like trees, neural nets (NNs)[3] are a very rich, often unstable, class of input–output relations. Both trees and NNs are able to approximate any reasonable function, e.g., one that is continuous or that has only jump discontinuities. Both trees and NNs have nodes, and the functions at the nodes have to be chosen. They both have a type of conditional structure: cutting a tree or NN at a node effectively conditions on the previous nodes or other node functions. Both trees and NNs can be used for regression and classification. Both have variability due to their

[3] Note that here the abbreviation refers to neural nets although earlier it referred to nearest neighbors.

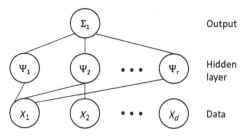

Figure 10.4 A connection graph of a simple neural net. The bottom layer of nodes consists of distinct, possibly vector, inputs. All are inputs to the intermediate 'hidden' layer, which represents a parametrized function that must be specified. (Only the connections for X_1 are shown.) The outputs of the hidden layer are combined into a final output by summation.

architecture and due to their parametrization. In fact, the diagrams typically drawn for trees and NNs sometimes even look similar. However, that is where the similarities end. Trees and NNs differ in almost all their details.

Figure 10.4 shows arguably the simplest useful NN that is possible. The only way in which this NN could be simpler would be if $r = 1$. Neural nets are normally drawn from left to right, but this one is drawn from top to bottom to emphasize the similarity with trees. The idea, however, is that d inputs, denoted X_j, $j = 1, \ldots, d$, form the first (known) layer. In a fully connected NN, such as that indicated here, each X_j is an input to the second layer. The second layer is hidden because it is strictly internal to the NN and not visible to the inputs or outputs. The third layer has one output node denoted Σ_1 even though the output is usually denoted $\hat{Y}$. The input vectors and functions in the nodes of the hidden layer are not indicated. More explicitly, whether for regression or classification, a single-hidden-layer NN is meant to implement a model such as

$$Y = Y(x_1, \ldots, x_d) = \sum_{j=1}^{r} \beta_{2,j} \Psi_j (\beta_{1,j,0} x_0 + \cdots + \beta_{1,j,d} x_d) + \epsilon \tag{10.26}$$

where $x_0 = 1$ and Ψ_j is a transfer function at node j that 'transfers' the value of the linear combination $\beta_{1,0} x_0 + \cdots + \beta_{1,r,d} x_d$ from the hidden layer to the output layer. Note that at each level there are coefficients β, distinguished by their first subscript: β_1, for the one (and only) hidden layer and β_2, for the output layer. The transfer functions Ψ_r do not all have to be the same, but they usually are. A common choice is a sigmoid function, often denoted $\sigma(t)$. A sigmoid function looks like a stretched out letter s; a popular choice is $\sigma(t) = e^t/(1+e^t)$, so that $\sigma(t) \to 0, 1$ as $t \to \mp\infty$.

Neural nets are a very general class of functions. There may be any number of hidden layers, making the expression (10.26) more complicated: each hidden layer gives an extra layer of node functions that have to be weighted and summed. An example NN with two hidden layers can be written as the 'composition'

$$Y = Y(x_1, \ldots, x_d) = \sum_{u=1}^{S} \beta_{3,u} \Phi_u \left(\sum_{j_u=1}^{r_u} \beta_{2,j_u} \Psi_{1,j_u} (\beta_{1,j_u,0} x_0 + \cdots + \beta_{1,j_u,d} x_d) \right) + \epsilon,$$

$$\tag{10.27}$$

in which the notation embedded in the parentheses is an obvious extension of (10.26). Note that S is the number of nodes in the second hidden layer and the r_u do not have to be the same, so different layers may have different numbers of nodes.

It is easy to imagine NNs with more hidden layers even if it is not easy to give explicit mathematical expressions for them. However, if $S = 1$ then the r_1 outputs from the first hidden layer would feed into the one node in the second hidden layer. This would be the same as feeding Σ_1 in Fig. 10.4 into another transfer function, say Φ_1. The resulting network would have two hidden layers even though the second hidden layer would have only one node. In many cases, extra constants are used to express the mathematical form of an NN. Here, however, for compactness of the expressions, X_0 is taken as a constant, unity, so the $\beta_{k,0}$ from layer k represent the constant term in a linear regression (provided that Ψ_{r_u} and one of the Φ_u is forced to be constant as well). Single-hidden-layer NNs can approximate almost any function – as can NNs with multiple hidden layers. This is a feature of NNs that make them hard to use: a complex NN and a simple NN may mathematically be very different yet numerically be very close.

Neural networks originally received their name because it was hoped that they would model how neurons interact. It is now known that they are not very useful for modeling neurons, but the mathematical structure (a form of nonlinear regression) is very useful for many other purposes.

In fact NNs frequently outperform linear regression because they provide such a flexible family of curves. There is no theoretical limit on the number of layers, nodes, or coefficients that they can have. Thus there is uncertainty with respect to both the architecture of an NN and the estimate of the coefficients in it. As with trees, this can be explored and evaluated. Indeed, it is common to penalize the number and size of the parameters in NNs. This 'regularization' is motivated by stability arguments and parallels CC pruning in trees. A strength of NNs is that the Ψ's can be chosen so as to localize a NN in much the same way as a term in an NW estimator can be localized to the region of a particular data point; see Fig. 10.5. This is different from the way in which a tree localizes to a region – the tree tries to capture the interaction between variables while the NN tries to mimic the local input–output relation directly – but all three localize.

All the NNs used here are feedforward, meaning that when their connection graphs are written down there are no cycles. That is, the output from one layer connects only to later, downstream, layers. By contrast, there is an organized theory for NNs that permits feedback; often they are called recurrent because they have loops. These NNs permit 'signals' – outputs from nodes – to travel forward and backward. Such networks often exhibit extremely complicated behavior because the feedback gives them a sort of memory. This makes them dynamic in the sense that they cycle until they reach an equilibrium. New inputs, of course, start the process again and may result in different equilibrium values.

10.3.1 *'Fitting' a Good NN*

Like trees, identifying good NNs requires both an architecture (how many nodes are there and how are they connected) and parameter estimates. If the possible architectures can be narrowed down sufficiently, e.g., to a fully connected single-hidden-layer network with r nodes, then CV, or some variant of it such as mis-classification error, can be used to select r. Doing this assumes that the parameter estimates are accurate enough that the comparison of

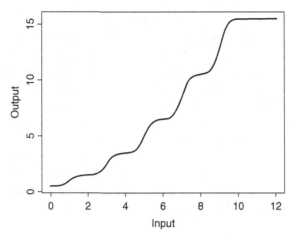

Figure 10.5 The graph of an NN with one unidimensional input fed into five hidden nodes, which then feed into a single output. The mathematical form is $Y = 0.5 + 1/(1 + \exp(-5(x-1))) + 2/(1 + \exp(-5(x-3))) + 3/(1 + \exp(-5(x-5))) + 4/(1 + \exp(-5(x-7))) + 5/(1 + \exp(-5(x-9)))$. There are five terms and each corresponds to a bump on the graph. Innumerable shapes can be created by adding more terms. Indeed, a good approximation to $Y = x$ can be generated by using enough small nodes, indicating that NNs that are structurally very different can be very close numerically.

the architectures is valid; this is always a question with predictive methods that require both a model-selection step and a parameter-estimation step.

Since model selection is typically less efficient than parameter estimation, it makes sense to start by considering a fixed architecture for an NN. This defines the parameters; then a number of techniques can be employed to obtain good estimates of them. The most popular technique for parameter estimation is called backwards propagation ('backprop') because the idea is to define an error criterion for the overall NN and then minimize it by sequentially adjusting the parameters, moving backwards through the net from final outputs to original inputs. This makes most sense when the architecture of the class of NNs is rich enough that model misspecification can be ignored while the SEs of the parameter estimates remain reasonably small.

Let $\text{Net}(x, w)$ denote the function represented by an NN. For instance, $\text{Net}(x, w)$ might be the regression function on the right-hand side of (10.26), with $x = (x_1, \ldots, x_d)$ and a vector of weights $w = (\beta_{2,1}, \ldots, \beta_{2,r}; \beta_{1,1,0}, \ldots, \beta_{1,1,d}, \ldots, \beta_{1,r,0}, \ldots, \beta_{1,r,d})$. As a second instance, $\text{Net}(x, w)$ might represent the regression function on the right-hand side of (10.27), which has three layers of β's; obviously, $\dim(w)$ increases rapidly with the number of connections between nodes, and the NN is only specified once all the values of the β's have been specified. The β's are often denoted by w's and called weights, because they weight the components of x or the nodes that feed into another node. The basic idea for finding values for the β's is to minimize the least squares error or some variant on it. The simplest case is

$$\text{SSE}(w) = \sum_{i=1}^{n} (y_i - \text{Net}(x_i, w))^2 \tag{10.28}$$

where (y_i, x_i) for $i = 1, \ldots, n$ are the available data. As a generality, setting the derivative of SSE(w) equal to zero does not give explicit solutions for w. However, (10.28) can be recognized as the error typically minimized in nonlinear regression problems because NNs have derivatives of all orders and, indeed, NNs without feedback can be regarded simply as a nonlinear regression function (once the architecture has been specified).

Backprop is a gradient descent technique that uses the derivatives of (10.28) to generate moves iteratively for a candidate value $\hat{w}$ of w in whichever direction reduces the error. The amount by which a coordinate in $\hat{w}$ is moved is a product of the derivative of SSE(w) and a 'learning rate' $\eta > 0$. As such it is like the Newton–Raphson updates from a one-term Taylor expansion. To see how this works, consider (10.26). There are two layers of β's, namely, the $\beta_{1,j,k}$ and the $\beta_{2,j}$. The derivatives of SSE(w) with respect to these two classes of β's are different because derivatives with respect to the β's in the hidden layer require an extra use of the chain rule.

The results for the output layer for any $j = 1, \ldots, r$ are

$$\frac{\partial}{\partial \beta_{2,j}} \text{SSE}(w) = 2 \sum_{i=1}^{n} (y_i - \text{Net}(x_i, w))$$

and the results for the hidden layer for any $j = 1, \ldots, r$ and $k = 1, \ldots, d$ are

$$\frac{\partial}{\partial \beta_{1,j,k}} \text{SSE}(w) = 2 \sum_{u=1}^{r} \beta_{2,u} \frac{\partial}{\partial \beta_{1,j,k}} \Psi_u(\beta_{1,u,0}x_0 + \cdots + \beta_{1,u,d}x_d) \sum_{i=1}^{n} (y_i - \text{Net}(x_i, w)).$$

Given a starting value w_0, a new value $\hat{w}_1$ is generated by

$$\hat{w}_i = \hat{w}_{i-1} + \eta_i \nabla \frac{1}{2} \text{SSE}(\hat{w}_{i-1}),$$

where ∇SSE is the gradient of the SSE evaluated at $\hat{w}_{i-1}$. That is, the entries in ∇SSE are the derivatives with respect to the $\beta_{2,j}$ and $\beta_{1,j,k}$. In this case, White (1989) showed that, for appropriate choices of η_i, the estimates $\hat{w}$ are consistent and satisfy an asymptotic normality criterion but that they need not be efficient.

Apart from that for fully connected single-hidden-layer NNs, e.g., the NN described by (10.27), there seems to be little theory regarding $\hat{w}$. However, an obvious analog of the gradient descent method is often advocated. The key difference between the two is that there are more uses of the chain rule and the notation is heavier. At the risk of oversimplifying, the basic idea is to update each $\hat{w}_{\ell,j,k}$ in w by

$$\hat{w}_{\ell,j,k} = \hat{w}_{\ell,j,k} + \eta \nabla \frac{1}{2} E(\ell, j, k),$$

where ℓ indexes a layer of the NN, j and k are assumed to be parameters at the node, $E(\ell, j, k)$ is an error function representing all the nodes affected by $w_{\ell,j,k}$, and the gradient differentiates only with respect to the parameters in $E(\ell, j, k)$. It is tacitly assumed that this sort of gradient descent procedure will ultimately result in a good estimate of w.

One problem with this assumption is that the fitness surface for NNs, i.e., the SSE as a function of the parameters of the NN, is rough: it can have many local maxima, minima, and inflection points. In fact, the SSE should be written as SSE$(\mathcal{A}, w)$ to emphasize the additional dependence on the architecture $\mathcal{A}$ of an NN. This means that the fitness surface for NNs as $\mathcal{A}$ varies along with the parameter values is even rougher, so that it is easy for the iterative procedure to be trapped at an NN with fixed w that is a local maximum or at some other optimality point that gives a poor choice of $\mathcal{A}$ and w. Thus, it is easy to overfit NNs in the sense that the variance is high relative to the bias.

There are two ways around this problem. First, and more commonly, a penalty term can be added to the SSE. Essentially this turns the frequentist problem into a sort of Bayes problem. For instance, the simplest choice is the ridge regression penalty

$$\text{SSE} + \frac{\lambda}{2} w^T w,$$

where λ is a hyperparameter that must be estimated. This expression corresponds to using a mean-zero normal prior on the parameters and a normal error term. This prior has invariance problems in that if an affine transformation is applied to the input variables then the weights do not transform appropriately. An alternative penalty that does not have this invariance problem is that for the case of a single-hidden-layer network:

$$\frac{\lambda_1}{2} \sum_{j=1}^{2} \beta_{2,j}^2 + \frac{\lambda_2}{2} \sum_{j,k=1}^{d,r} \beta_{1,j,k}^2,$$

where λ_1 and λ_2 are hyperparameters. This choice has two hyperparameters but does have some desirable invariance properties; see Srihari (2014). Extensions to more complex NNs are, perforce, more complicated.

A more sophisticated way to apply backprop is to use only a portion of the data for training, reserving a second portion for testing. The hope is that as the iteration of the $\hat{w}_i$ continues the training error will decrease to a constant and, at the same time, the testing error will decrease to a minimum value (after which it may increase) – and the values of $\hat{w}$ that give the minimum testing error are used to specify the NN.

These same two procedures can be used when an NN is being developed for classification purposes where the y_i are, say, zeros and ones. This means that the output of an NN at a value of x, say Net(x, w), must be converted to zeros and ones to match the values of Y. (More generally, Net(x, w) must be converted to a C-ary output if there are C classes.) In fact, one can continue to use the SSE as defined above even in classification problems; it is reasonable to regard a classifier as a regression function and use regression methods to find it. Indeed, the main difference between NNs for regression and for classification is in the interpretation (classes versus a continuous outcome), not in the mathematical representations. An exception is logistic classification, which uses an NN as a regression function, i.e.,

$$\text{logit } P(Y = 1) = \text{Net}(x, w),$$

in which case nonlinear techniques from logistic regression can be used. In this case, it is often assumed that Net(x, w) is in $[0, 1]$ and so can be readily dichotomized. When Net$(x, w) \in [0, 1]$, the log-likelihood function for binary classification comes from a binomial$(n, \text{Net}(x, w))$, i.e.,

$$\sum_{i=1}^{n} \left(y_i \ln \text{Net}(x_i, w) + (1 - y_i) \ln(1 - \text{Net}(x_i, w)) \right). \tag{10.29}$$

For C classes, the corresponding log-likelihood comes from a multinomial with C classes. Backpropagation (backprop) or SA (described below) can be used to find the weights.

At the end of the analysis, the result is a function $\hat{Y}(\cdot) = \text{Net}(\cdot, \hat{w})$. Thus, for a new input, x_{new}, the point prediction for either regression or classification is

$$\hat{Y}(x_{\text{new}}) = \text{Net}(x_{\text{new}}, \hat{w}).$$

Prediction intervals can then be derived using the bootstrapping approach described at the end of Sec. 10.1. Alternatively, for some cases, e.g., single-hidden-layer NNs, an overall SE for the parameter estimates can be derived; see White (1989). Obtaining a PI is then the same as in earlier cases. Write

$$\text{Var}(Y(x_{\text{new}}) - \hat{Y}(x_{\text{new}})) = \text{Var}(Y(x_{\text{new}}) - EY(x_{\text{new}})) + \text{Var}(EY(x_{\text{new}}) - \hat{Y}(x_{\text{new}}))$$
$$= \sigma^2 + \text{SE}^2, \tag{10.30}$$

estimate σ^2 by an MSE and SE^2 by bootstrapping, and take the square root. So, even though an expression for SE is usually hard to find, if one is found it can be used to generate PIs in the $\mathcal{M}$-closed case, with the usual modification for bias in the $\mathcal{M}$-complete case (see (10.17)). In the $\mathcal{M}$-open case, of course, PIs are merely intervals that one might use; a probabilistic interpretation is not valid.

Before turning to choosing an architecture for an NN, it is important to observe that simulated annealing (SA) is a strong competitor for backprop for parameter estimation. As it was discussed in Sec. 9.5, SA is a technique to optimize a functional J on a discrete set S, and it was assumed that this would be used for model selection. However, estimating continuous weights, even in a fixed-architecture NN, can be regarded as the selection of a model. Simulated annealing can be used in this context provided that analogs to the concept of neighbors can be specified. One way to do this is as follows; see Goffe *et al.* (1994).

Adopt SSE (for instance) as an error criterion to be minimized and start with a parameter in an NN, say w_1. Let r be an outcome from Unif$[-1, 1]$. Update w_1 to $w_1' = w_1 + rv_1$, where $v_1 > 0$ is the chosen step length to be associated with w_1. Apply the SA procedure to see whether the move from $(w_1, \ldots, w_k)$ to $(w_1', \ldots, w_k)$ is accepted; here k is the number of parameters in the fixed-architecture NN. Repeat this for w_2 using an independent outcome r and a $v_2 > 0$ appropriate for w_2. Continue in this fashion for $k - 2$ iterations so that all k parameters have had the chance to be adjusted. Repeat the cycling through the k parameters until the values of the error criterion indicate that a minimum has been found. Note that the v_j can be increased to permit larger jumps if the changes in the error seem too small to be important or the convergence seems unduly slow for the size of the problem. Likewise, the v_j can be decreased if it is believed that a small region containing a global minimum has been found. Some authors, e.g., Sexton, *et al.* (1999) have argued that backprop and its variants frequently result in inconsistent and unpredictable performance despite the many modifications that have been attempted. In the words of Sexton, *et al.* (1999): 'such modifications are unnecessary if [there is] a sufficiently complex initial architecture and an appropriate global search algorithm is used for network training. We further show that the simulated annealing algorithm is an appropriate global search technique that consistently outperforms backpropagation for NN training.'

10.3.2 Choosing an Architecture for an NN

Simulated annealing was introduced in Sec. 9.5 as a stochastic search technique that is effective for large problems. Of course, AIC, BIC, and even cross-validation can be used with NNs but the amount of data required for reliable results can be unrealistically large. More recent contributions extending the basic convergence theorem for SA given in Sec. 9.5 can be found in Andrieu *et al.* (2001) and Pelletier (1998). The principle remains the same: fix a collection of architectures, meaning nodes and their connections. Start with an initial choice of architecture. Do a local search of nearby architectures by adding or removing nodes or connections in a variety of ways. If possible, move to a new architecture with a smaller error and repeat to search for a local minimum of the SSE as a function of the class of NNs. At each stage, with some probability jump to a new architecture. This may result in a higher value of the error but may move the NN to a region in which a lower local mode can be found.

More formally, Boese *et al.* (1997) gave a convenient 'pseudo-code' version of SA for NNs. Let the error criterion and cooling schedule be denoted J and T_i, respectively (as in Sec. 9.5), and suppose that neighborhoods $N(u)$ around each NN architecture u have been defined. For M iterations the procedure is as follows.

- □ Start with a random solution $u(0)$.
- □ For $i = 0, \ldots, M-1$:
  - □ Choose $u'(i+1) \in N(u(i))$ (the neighborhood of $u(i)$)
  - □ If $J(u'(i+1)) < J(u(i))$, set $u(i+1) = u'(i+1)$.
  - □ Else set $u(i+1) = u'(i+1)$ with probability $\exp\left([J(u'(i+1)) - J(u(i))]/T_{i+1}\right)$.
- □ Return $u(M)$ or the $u(i)$ with the best $J(u(i))$ (so far).

As in the theorem in Sec. 9.5, if the T_i shrink slowly enough, $u(M)$ converges in probability to a globally optimal solution.

A specific instantiation of SA for NNs that was among the first to be given was due to Tenorio and Shi (1990). The basic idea is to use concepts from the minimum description length (MDL), an information-theoretic concept that is similar in the parametric case to the BIC (see Secs.9.3 and 9.8.3 for a brief summary). Concisely, the self-organizing NN (SONN) procedure uses MDL as its error criterion for SA as well as to choose the functions at the nodes. The nodes and their connections are re-evaluated during each iteration to search a large collection of NNs, and the best of them are pruned by MDL. The procedure is quite complicated but the results (the number of weights, the complexity of the final structure, and the fit) compare favorably to backprop approaches in terms of parsimony, accuracy, efficiency, and computing resources.

However the network architecture is found, it is essential to give PIs that include variability from the uncertainty in the architecture. After all, as has been noted (and is not too hard to prove) multiple NNs may give nearly the same values but actually be mathematically quite different, and at the same time the fitness surface can be very rough (even when regularization is included).

With these caveats in mind, one way to form a PI is to follow what was done generically, at the end of Sec. 10.1, for trees in (10.15) and (10.17), and done partially for NNs in (10.30). First, use all the data and the techniques discussed to generate an estimate $\widehat{\text{Net}}(\cdot, \hat{w})$ of $\text{Net}(\cdot, w)$, where $\widehat{\text{Net}}$ means the estimated architecture of Net and $\hat{w}$ are the estimates of the weights w associated with the architecture $\widehat{\text{Net}}$. Now, for either regression or classification, the obvious point predictor for $Y(x_{\text{new}})$ at a new value x_{new} is $\widehat{\text{Net}}(x_{\text{new}}, \hat{w})$.

For classification, the natural error assessment is the SSE, namely

$$\text{SSE} = \frac{1}{n-1} \sum_{i=1}^{n} (y_i - \widehat{\text{Net}}(x_i, \hat{w}))^2.$$

An analogous quantity serves as an estimate of $\hat{\sigma}^2$ in other regression problems.

Now, suppose that the true function (if it exists) is $G(\cdot)$ and that the NN closest to it in the collection of NNs searched is $\text{Net}_0(\cdot, w_0)$. Then, for x_{new}, write

$$Y(x_{\text{new}}) - \widehat{\text{Net}}(x_{\text{new}}, \hat{w}) = (Y(x_{\text{new}}) - G(x_{\text{new}})) + (G(x_{\text{new}}) - \text{Net}_0(x_{\text{new}}, w_0))$$
$$+ \left(\text{Net}_0(x_{\text{new}}, w_0) - \widehat{\text{Net}}(x_{\text{new}}, \hat{w})\right). \quad (10.31)$$

The first and third terms on the right are independent and the middle term is the bias, a constant. The variance of the first term on the right is σ^2, and an estimate for the variance of the third term, i.e., the variance of estimating $\text{Net}_0(\cdot, w_0)$ by $\widehat{\text{Net}}(\cdot, \hat{w})$, can be found as in (10.15) by bootstrapping.

As described at the end of Sec. 10.1 there are two forms of bootstrapping that can be used with smoothly parametrized nonlinear regression techniques such as NNs. The straightforward approach of drawing bootstrapping samples from the data was explained in Sec. 10.2.1 for trees and would apply for NNs as well. However, since it is usually safe to assume that the class of NNs is so rich that bias is small, here it is just as well to bootstrap over residuals. Thus, (10.31) and (10.17) are conceptually the same in that the variability due to both parameters and architectures must be quantified, but they differ in practice because trees are not smooth and are more likely to exhibit bias. Practically, the difference is between using SA and/or backprop in place of growing and pruning trees.

In more detail, find the n residuals $r_1, \ldots, r_n$, from fitting the $y(x_i)$ with the $\widehat{\text{Net}}(x_i, \hat{w})$. Then, draw B bootstrap samples of size n from the residuals, say $r_{b,i}^*$ for $b = 1, \ldots, B$, $i = 1, \ldots, n$. Using these, form new y_i^b for use in the analog of (10.12) where SA is used to find B architectures and either SA or backprop is used to find B sets of parameter estimates. Thus, B functions $\widehat{\text{Net}}_b(\cdot, \hat{w})$ for $b = 1, \ldots, B$ are found (instead of the parameter values $\hat{\theta}_b$ in (10.12)). Now, the analog to (10.10) can be found, to give an $\text{SE}(x_{\text{new}})$ for any x_{new}, using $\widehat{\text{Net}}_b(x_{\text{new}}, \hat{w})$ in place of $f(x_{\text{new}}, \hat{\theta}^b)$. Ignoring the middle term in (10.31) and assuming asymptotic normality, a $1 - \alpha$ PI for $Y(x_{\text{new}})$ is given by

$$\widehat{\text{Net}}(x_{\text{new}}, \hat{w}) \pm z_{\alpha/2} \hat{\sigma} \sqrt{1 + \text{SE}(x_{\text{new}})}. \quad (10.32)$$

Finally, although the middle term on the right in (10.31) is usually negligible given the richness of most reasonable classes of NNs, it should be noted that $G(x_{\text{new}}) - \text{Net}_0(x_{\text{new}}, w_0)$ can be estimated. It is analogous to $G_0(x_{\text{new}}) - T_0(x_{\text{new}})$ in (10.17) and corrects the prediction intervals (10.32) in the same way as (10.17) does for trees. This is only important when the estimate of $G(x_{\text{new}}) - \text{Net}_0(x_{\text{new}}, w_0)$ is large relative to $\hat{\sigma}$.

10.3.3 Bayes NNs

The Bayes analysis of an NN proceeds in much the same way as the Bayes analysis of any problem once a prior and likelihood have been specified. At least mathematically, the predictive distribution is defined, although sophisticated computational procedures may be necessary to generate the PIs. The extra complexity with NNs, like trees, is that a prior for the architecture and parameters in internal nodes must be specified. Sometimes this is called a prior over models to emphasize that the set of architectures is discrete. A good review of NNs can be found in Titterington (2004); the authoritative text remains Lee (2004). See also Lee (2001, 2005). These focus on classification more than regression, but the analytical problems with regression are similar.

For ease of exposition let $\mathcal{A}$ be the architecture of an NN and let $w = w_{\mathcal{A}}$ be its parameters. Set $\pi(w|\mathcal{A})$ to be the prior on the parameters w required in the architecture $\mathcal{A}$, $\pi(\mathcal{A})$ to be the prior over architectures, and the joint prior to be $\pi(w, \mathcal{A})$. The support of the entries in w is usually $\mathbb{R}$ and the support of $\mathcal{A}$ is whatever collection of NNs is chosen by the user, i.e., the user believes it to include NNs of sufficient complexity as to be adequate for the DG. While the prior $\pi(\mathcal{A}|w)$ could be defined, e.g., as the probabilities that architecture $\mathcal{A}$ has parameters w, it is rarely encountered. However, hyperparameters often occur because they frequently play an important role in architecture selection.

The most typical Bayes NN model for regression has a likelihood of the form

$$Y_i(x_i) \sim N\left(\text{Net}(x_i; w, \mathcal{A}), \sigma^2\right) \tag{10.33}$$

and a conditional prior

$$\pi(w|\mathcal{A}) \propto \exp\left(-\frac{1}{2}\sum_{w_u \in w} \lambda_u w_u^2\right), \tag{10.34}$$

where the notation $w_u \in w$ means that the sum is over all entries in the parameter set w associated with $\mathcal{A}$. The corresponding λ_u are hyperparameters. This means that there are a large number of hyperparameters but in practice empirical estimates of λ_u, say, are often large, meaning that the prior concentrates around zero in the uth dimension. In cases like this, the uth weight can be 'pruned' and the NN simplified. In fact, a procedure called automatic relevance determination uses the same hyperparameter for all weights corresponding to connections from each individual input variable, so that pruning eliminates the explanatory variables that are irrelevant to predicting Y. Another way to simplify the hyperparameters is to equate all the hyperparameters on the weights in a node. In that way pruning might eliminate a node.

Let Y^n and X^n denote the vectors of values of independent data $(Y_i(x_i), x_i)$ for $i = 1, \ldots, n$. For a fixed $\mathcal{A}$, the posterior is the normalized product of the conditional prior and likelihood, thus

$$\pi(w|Y^n, X^n) \propto \frac{\pi(w|\mathcal{A}) \exp\left(-\frac{1}{2\sigma^2}\sum_{i=1}^n (y_i - \text{Net}(x_i; w, \mathcal{A}))^2\right)}{\int \pi(w|\mathcal{A}) \exp\left(-\frac{1}{2\sigma^2}\sum_{i=1}^n (y_i - \text{Net}(x_i; w, \mathcal{A}))^2\right) dw}, \tag{10.35}$$

in which any hyperparameters have been suppressed in the notation. The ratio between σ^2 and any λ_u determines the magnitude of the effect of the penalties on $|w_u|$.

Now, formally, the predictive density for Y_{new} at x_{new} is

$$p(y_{\text{new}}(x_{\text{new}})|Y^n, X^n) = \int \pi(w|Y^n, X^n) p(y_{\text{new}}(x_{\text{new}})|w, \mathcal{A}) dw. \tag{10.36}$$

While (10.36) is valid and in principle generates PIs for any desired α, it hides a lot. First, it can be very hard to compute. Second, repeated analyses (and possibly Bayes hypothesis tests) are usually required to get satisfactory behavior of the hyperparameters. It is tempting to use a fully Bayes approach and integrate over the hyperparameters so that they are included in the prior density; however, this seems to be done only infrequently owing to the complexity of the models. Third, use of this predictive density assumes that exactly one architecture $\mathcal{A}$ is used. Some argue this is not bad: simply choose as large an architecture $\mathcal{A}$ as the data can be assumed to permit and prune it down as much as is reasonable. Moreover, others argue that taking the architecture into account de facto gives a Bayes model average (see Sec. 11.1) that will improve prediction, possibly not in the sense of giving narrower PIs but at least in the sense of giving more accurate PIs.

To see this explicitly in a simple case suppose that K architectures $\mathcal{A}_1, \ldots, \mathcal{A}_K$ are under consideration, i.e., the problem is $\mathcal{M}$-closed. Assume that an across-model prior $\pi(k)$ for $k = 1, \ldots, K$ has been assigned on some rational basis, e.g., so that NN models that are too simple or too complex are downweighted. Then, (10.33) remains the same but (10.34) becomes

$$\pi(w|\mathcal{A}_k)\pi(k) \propto \pi(k) \exp\left(-\frac{1}{2}\sum_{w_u \in w} \lambda_u w_u^2\right), \qquad (10.37)$$

where the notation $w_u \in w$ means that the sum is over all entries in the parameter set w associated with $\mathcal{A}_k$, (10.35) becomes

$$\pi(w, k|Y^n, X^n) \propto \frac{\pi(w|\mathcal{A}_k)\pi(k) \exp\left(-\frac{1}{2\sigma^2}\sum_{i=1}^n (y_i - \mathrm{Net}(x_i; w, \mathcal{A}_k))^2\right)}{\sum_{k=1}^K \pi(k) \int \pi(w|\mathcal{A}_k) \exp\left(-\frac{1}{2\sigma^2}\sum_{i=1}^n (y_i - \mathrm{Net}(x_i; w, \mathcal{A}_k))^2\right) dw}; \qquad (10.38)$$

and (10.36) becomes

$$p(y_{\mathrm{new}}(x_{\mathrm{new}})|Y^n, X^n) = \sum_{k=1}^K \int \pi(w, k|Y^n, X^n) p(y_{\mathrm{new}}(x_{\mathrm{new}})|w, \mathcal{A}_k) dw. \qquad (10.39)$$

Expression (10.39) can be used to generate PIs but the discreteness of k may lead to PIs that are disjoint and hence, in some cases, unnatural. One way around this is to use Bayes testing. It is conceptually easy, but possibly computationally very hard, to find and use

$$p(Y^n, X^n|\mathcal{A}_k) = \int p(Y^n, X^n|w, \mathcal{A}_k)\pi(w|\mathcal{A}_k) dw$$

in order to seek a maximum marginal likelihood architecture $\mathcal{A}_k$, i.e., to do model selection over the architectures. It is actually better conceptually (but usually even harder computationally) to find the posterior probabilities of the $\mathcal{A}_k$,

$$\pi(\mathcal{A}_k|Y^n, X^n) = \frac{\pi(k)p(Y^n, X^n|\mathcal{A}_k)}{\sum_{k=1}^K \pi(k)p(Y^n, X^n|\mathcal{A}_k)},$$

and use actual Bayes tests. Other criteria may also be used to eliminate architectures $\mathcal{A}_k$ that have low individual posterior probability or to restrict the collection of architectures to those that are relatively similar under some criterion. Collectively, these are called Occam's window approaches and while such approaches may be effective in terms of getting reasonable looking PIs, there is the danger that some other region of the possible architectures actually has a higher probability that is unfortunately split so finely amongst competing NNs that, taken together, they have a low posterior probability. This is another example of dilution (seen in Sec. 9.2) and it is of concern for models such as NNs (and trees) that are often dense in the collection of functions being searched.

Bayes classification with NNs in the binary case starts with (10.29). If the output NN is transformed in such a way that

$$y(x) = \frac{1}{1 + \exp^{\left(\text{Net}(x; w, \mathcal{A})\right)}},$$

and hence is in $[0, 1]$, then $y(x)$ is the conditional probability of the class 1 given x, i.e., $y(x) = P(Y = 0|x)$. When $y(x) < 0.5$, class 1 is chosen and when $y(x) > 0.5$ class 2 is chosen. Again, the squared error can be used, but it is often argued that the binomial is a better formulation in the sense of giving better predictions. In the frequentist case, the likelihood was maximized to find w (and $\mathcal{A}_k$). Here, however,

$$P(Y_{\text{new}}(x_{\text{new}}) = 0|Y^n, X^n, x_{\text{new}}, \mathcal{A}_k) = \int y(x; w, \mathcal{A}_k)\pi(w|Y^n, X^n, \mathcal{A}_k)dw$$

and more generally

$$P(Y_{\text{new}}(x_{\text{new}}) = 0|Y^n, X^n, x_{\text{new}}) = \sum_{k=1}^{K} \pi(k) \int y(x; w, \mathcal{A}_k)\pi(w|Y^n, X^n, \mathcal{A}_k)dw.$$

10.3.4 NN Heuristics

Despite our discussion in Sec. 10.3.3, methodological development and intuition for NNs remains mostly frequentist. However, whether the frequentist or Bayes view is adopted, the main difficulties with using NNs in practice are largely computational or arise from the fact that NNs are uninterpretable for most practical purposes – but see Sec. 10.3.5 for further comment. The uninterpretability here is even stronger than that for trees since the topology of a collection of NNs is much less intuitive than the topology of a collection of trees. After all, trees partition an input space into well defined regions on which relatively simple models are assigned, while NNs undulate in complex ways.

One major plus of NNs is that they evade the curse of dimensionality – at least to a limited extent. Prior to 1991 most researchers believed that if x were d-dimensional then, as a rule, $E \int (f(x) - \hat{f}(x))^2 dx$ would grow superlinearly in d where $\hat{f}$ is any estimate, in particular an NN estimate, of a function f. The undesirable property that the mean integrated squared error (MISE) grows faster than d is one formal way of stating the 'curse of dimensionality', from which many methods, e.g., Nadaraya–Watson, splines, etc. suffer. That is, many established techniques would not scale up to the high-dimensional problems that were becoming increasingly common. So, the usual wisdom was that NNs would not do so either.

However, Barron (1991) showed that

$$\int (f(x) - f_r(x))^2 \mu(dx) \leq \mathcal{O}\left(\frac{1}{r}\right),$$

where f_r is a sum of r sigmoidal functions, i.e., nodes. So, estimating the parameters in f_r gives MISE $= \mathcal{O}(1)$. This is an argument for why NNs do well with high-dimensional data. To be sure, there are limitations to Barron's result in that the set of functions to which it applies is defined in terms of Fourier transforms. Nevertheless, the point remains: NNs are in principle one of relatively few methods that escape the curse of dimensionality in some sense. (Projection pursuit also escapes this curse; this is discussed in Sec. 10.7.1.)

A second major plus of NNs is that they perform well with multitype data, at least compared with other methods. This was one of the arguments given in the NN literature in engineering in the 1980s; see Benediktsson *et al.* (1992) and the references therein for one early complex example. This was first observed for classification data, but seems to hold for regression data, too. Besaw and Rizzo (2007) stated '... the algorithm can assimilate significant amounts of disparate data types with both continuous and categorical responses without the burden of computing covariance matrices ... [it is] especially good for spatially autocorrelated variables.'

Other methods employ models that partially solve the multitype data problem by treating different types of data differently. This is typically true for semiparametric models (which usually suffer the curse). Another approach to the multitype data problem when, say, one data source has low dimension and the other data source has a very high dimension, is to limit the number of variables from the high-dimensional data source that enter the model so they will not overwhelm the low-dimensional data source. Again, this treats the data types differently. When NNs are used, their resistance to the curse may enable them to treat the data types more symmetrically, thereby letting the data sort out what is important rather than relying on 'modeling' and the problems attendant with its overuse.

Neural nets also have some notable minuses. As noted, they have a very rough fitness landscape especially when multilayer NNs are used. This is similar to the stability problems that are common with high-dimensional techniques, but worse because stabilizing NNs, e.g., by Bayesian techniques will not smooth the fitness surface very much. Even regularization does not help very much. So, the danger of overfitting is high. One can argue that fit per se is irrelevant to NNs since it is their predictive performance that reflects their shortcomings. Otherwise stated, NNs that do fit well typically fit too well, so, paradoxically, simply seeking a NN that fits well is actually not a particularly good strategy. After all, two networks that have nearly the same graph can be mathematically very different, meaning that a network that has a smooth, simple looking, graph may actually be very complicated.

There have been many efforts to assess the stability of NNs. Questions like 'How much effect can be attributed to a given variable, node, or collection of nodes?' have been asked and partial answers are available via hyperparameters, for instance. A direct analysis of how much an NN changes in response to its inputs is given in Feraud and Clerot (2002) but this does not seem to be commonly used, perhaps because of the computational demands. Likewise, the SIMEX approach may give some indication of the sensitivity of an NN to the the the fine details of the data. It has been empirically observed that regions of multilayer NNs seem to sort out which data from the different types can be usefully related to each other. Given that NNs are a form of nonlinear regression – and fit the paradigm of Sec. 10.1 apart from their richness – it may ultimately be possible to recognize regions of nodes as 'modules' that combine certain data in some useful way.

However, efforts to relate the structure of NNs to anything physical or otherwise interpretable are infrequent and incomplete; some of the most recent thinking is discussed in Sec. 10.3.5. Some references on the interpretability of NNs are Intrator and Intrator (2001), Cechin and Battistella (2007), and Kamimura (2009). Much more has been done, but it is mostly in subject-matter journals where the authors are concerned about the interpretation of an NN in the context of a specific example such as the optical recognition of handwritten digits or images. Hence, many of the findings of these authors exploit idiosyncratic features of their problems. From the purely methodological standpoint, the stability of predictive performance is likely to give more insight into an NN than any effort to formulate generic interpretations, at least for non-small-scale problems. Overall, at present NNs are best regarded purely as a blackbox method for prediction.

In summary, linear models and traditional (smooth) nonlinear models are like donkeys. Treat them right and they'll carry you, grudgingly, as far as they can. They will bray when they are unhappy but you will know how to feed and water them. Unfortunately, they can't travel very far. Neural networks are like a big pile of snakes. Some are poisonous though most are not. They differ enormously in size and shape even if they seem similar. You can grab one that looks right but since you can't see the whole thing it may coil around in unexpected ways. However, if you get the right one it will be strikingly effective for you. Trees are like foxes. They're very clever and run around all over the place. They don't bite; but it is often hard to catch the ones that will solve your problem. However, once you've caught one and trained it, it will generally be well behaved for you.

10.3.5 Deep Learning, Convolutional NNs, and All That

The concept of deep learning rests on the observation that the deeper the NN is, the better the results it can give for classification and regression. On an intuitive level, this is obvious: given an increasing sequence of collections of NNs, the larger the collection, the smaller the error. Noting that the minimum error over a smaller set is larger than the minimum error over a larger set involves the same principle. However, in the NN function-approximation context more can be said. Loosely, there are many functions that require exponentially many hidden nodes in a single-hidden-layer NN to achieve good performance but only require polynomially many hidden nodes to achieve the same performance if multiple hidden layers are permitted; this was observed originally by Hastad (1986, 1991) but only used much later, for instance in Bengio *et al.* (2007) and Stathakis (2009) amongst other publications. Consequently, deep learning generally refers to the use of NNs with at least two hidden layers. Sometimes the gains from extra hidden layers can be negligible; however, often they are substantial, especially when the network architecture is restricted or altered from the fully connected feedforward networks presented here so far.

It should be noted that our presentation did not bound or specify the number of layers but, as a matter of practicality, the focus was on NNs with relatively few layers having relatively few hidden nodes. There are four problems that arise when trying to implement larger NNs.

The first problem is the obvious one – that the computational demands for large NNs can be very high. Second, the number of parameters in a fully connected NN is $\mathcal{O}(dh)$ where h is the number of nodes. So, the use of many hidden neurons may require unrealistic values of n. This means that the estimation of parameters even for single-hidden-layer networks may be difficult; the problem becomes more and more serious as the number of hidden layers increases. Third, the larger NNs are, in terms of the number and sizes of the hidden layers, the more instability can be expected. Indeed, there is some suggestion that as NNs increase in size, their fitness landscape, i.e., a graph of the MSE as a function of the parametrized architectures of the NNs, becomes less rough but has more saddle points. That is, the fitness landscape changes character, relatively speaking, from having many local maxima and minima to having more points that are a maximum and a minimum at the same time depending on the direction from which they are approached; see Choromanska *et al.* (2015) for a useful discussion on the effect of this. Counterintuitively, they argued that such an instability may not make much difference to the performance of the NNs. Fourth, as discussed in Neilson (2012, Chap. 5), parameter estimates in different layers of a multilayer NN may converge at different rates.[4] In particular, parameters in the early layers may converge rapidly compared with parameters in the later layers, which can become 'stuck' near one (suboptimal) value. This may explain in part why multilayer NNs often do not perform much better than single-hidden-layer NNs. Sometimes this is called the vanishing gradient problem, the gradients in question being the derivatives of $SSE(w)$ with respect to the entries in w; it was originally discussed in Hochreiter (1991) and rests on the fact that in gradient descent the gradient may be poorly behaved – usually getting too small but perhaps demonstrating instability and getting too large as well.

Deep learning really only became feasible when the first and fourth problems were solved. That is, advances in computing technology made the running times for algorithms reasonable, and LeCun *et al.* (1998) gave an analysis of why and when backprop works. They showed that most classical second-order methods are impractical for large or deep neural networks and proposed methods that do not have such limitations. Later, Hinton *et al.* (2006) offered a Bayesian algorithm for the estimation of network architecture and weights that resolved the difficulty of differential learning rates across hidden layers. As a result of these two contributions, many subfields of the quantitative sciences have been exploring large-NN approaches to see how far deep learning can be pushed in terms of: multiple 'feature maps' at a layer; depth; and alternative functions linking earlier layers or nodes to later layers or nodes (or the reverse).

The second and third problems remain as such, but getting large enough sample sizes for good inference from complex models is a perennial challenge, and, while it adds computational burden, dealing with the vagaries of fitness landscapes by repeated runs (or other techniques) is now feasible for large NNs when it is necessary (as discussed in Choromanska *et al.* (2015), for instance).

Earlier, see Sec. 10.3.1, the roughness of the fitness landscape was reduced by adding a penalty term or by using a hold-out set, but these can also be regarded as attempts to achieve sparsity because these two techniques would allow parameters or possibly whole nodes to be dropped from the overall NN. Here, these same techniques can be used to achieve sparsity as well, although merging sparsity techniques with procedures for estimating parameters in multilayered NNs seems not to have been directly investigated.

---

[4] This problem can also occur in ensemble methods that involve stacking; see Sec. 11.3. Although not discussed in this book, it is obvious that the predictors being stacked can converge much faster than the stacking weights.

What has been investigated at length is classes of NNs that are smaller than deep, fully connected (feedforward) NNs. That is, the size or number of the hidden layers is restricted in some way. In addition, the connecting functions from one layer to the next are generalized, so that they are not always sigmoids and they do not necessarily take inputs from all nodes in the previous layer. Aside from helping with the second problem, the restrictions and layer-to-layer functions are intended to make the NN more physically interpretable because 'clusters' of nodes may correspond to regions of the input space. Thus, relationships among the inputs may be captured by the NN structure. If in addition nodes, or the parameters at a node, are eliminated when their contribution to the overall fit or predictions generated by the NN are too small, more sparsity and a better reflection of the input space may be achieved.

One of the most popular of these classes is called convolutional NNs (CNNs). They are routinely used for text, image, and speech recognition and are NNs that combine three types of layers: feature map layers, pooling layers, and fully connected output-combining layers in addition to the overall input layer and final output layer. Thus, the smallest CNN has three hidden layers of nodes between the overall input layer and the final output layer, a total of five layers. The input layers feed into the feature-map layer, typically giving several feature maps in this layer. Each map uses a collection of subsets of nodes and common parameters. The idea is that each feature-map in the feature map layer should correspond to some region in the input space. The regions are meant to represent relationships among the input variables. For instance, for a two-dimensional array of inputs representing an image, an individual feature map might represent an edge or a solid mass. The pooling layer often contains the maximum output from a collection of sets of nodes from a single feature map in the previous layer, although in many applications the pooling is done in an L^2 sense rather than using a maximum. There are as many pooled attributes in the pooling layer as there are feature maps in the feature-map layer; the pooling layer is meant to summarize the feature-map layer. The attributes in the pooling layer are then fully connected to a hidden layer that gives the overall output of the NN. Any of the three types of hidden layer may be repeated in any order, so CNNs may have more than five layers.

To be more formal, suppose that there are overall inputs $x = (x_0, \ldots, x_d)$, where $x_0 = 1$, and, to keep the intuition simple, suppose that $d = ab$, where a and b are integers, so the nonconstant inputs form an $a \times b$ matrix. This is analogous to image data, in which each entry of the matrix is a pixel value. However, CNNs do not require a two-way array but only a neighborhood structure on the input nodes. In Fig. 10.6 one neighborhood, labeled R_1, is shown. In practice the neighborhoods are usually square blocks of input variables, often 3×3, that start in the upper left corner and are moved one node a time to the right until they hit the boundary. Then, the process is repeated but the blocks are lowered by one row. In this way an $a \times b$ input matrix with $k \times k$ neighborhoods generates a hidden two-way array with $a'b' = (a-k)(b-k)$ nodes at the second layer. The neighborhoods are called local receptive fields and the function on the neighborhoods of the input is (with some abuse of notation) of the form

$$\sigma_{R_1} = \sigma\left(w_0 + \sum_{\alpha,\beta \in R_1} w_{\alpha,\beta} x_{\alpha,\beta}\right). \tag{10.40}$$

The expression $\alpha, \beta \in R_1$ indicates the set of nodes in the input layer corresponding to R_1; other local receptive fields, say $R_2, \ldots$, will have similar expressions. The difference

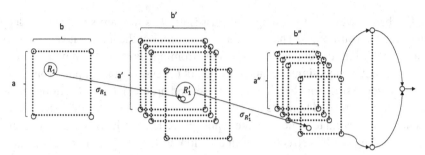

Figure 10.6 A generic CNN showing the input layer, feature map layer, pooling layer, hidden combining layer (indicated by the vertical dots inside the oval), and final output layer (indicated by the arrow on the far right).

between (10.40) and expressions such as (10.26) and (10.27) is the status of the parameters. In principle, w_0 and the $w_{\alpha,\beta}$ in (10.40) are chosen by the user, independently of the data, in an effort to search for features in the array, whereas in the usual NN approach the weights in w are estimated from the data. In practice, of course, the $w_{\alpha,\beta}$ in a CNN are motivated by informal analysis of the input data and what the analyst is seeking, e.g., recognizing handwritten letters or numerals.

Each 'slice' of the $a' \times b'$ array in Fig. 10.6 shows one complete set of outputs for a collection of local receptive fields and for a fixed choice of the parameters w_0 and $w_{\alpha,\beta}$. The array is called a feature map because the constant parameters are assumed to correspond to a specific feature that is being sought over the receptive fields. In principle, there will be as many feature maps as there are important features in the input array and each set of parameters yields a different feature map. Thus the use of many feature maps in the second layer of a CNN is an effort to use the existence of collections of features to help with classification or regression and provide some partial interpretability in what would otherwise be a purely blackbox technique. The second layer in Fig. 10.6 has four parallel feature maps, but any number is possible. As a matter of terminology, the collection of parameters used to define a specific feature map is sometimes called a filter or kernel and the function in (10.40) is sometimes called a convolution. This is so because there is a (loose) parallel between the summation and parameters and the integral and function defining the convolution in a convolution integral.

Once all the feature maps have been formed, a generic CNN has a pooling layer that has one array of nodes for each feature map. Again, a collection of sets of nodes is defined, say $R'_1, \ldots$ on the given feature map and each set of nodes gives one node in the pooling layer; see Fig. 10.6. The formal difference between an array in the pooling layer and an array in the feature map layer is that there are no parameters in the pooling layers. In addition, the pooling layer summarizes the collection of outputs from the feature map layer, so usually $a'' \ll a'$ and $b'' \ll b'$. For instance, a generic function $\sigma_{R'_1}$ from the R'_1 neighborhood in the feature map layer to a node in the pooling layer may have the form of a maximum, i.e.,

$$\sigma_{R'_1} = \max_{\nu \in R'_1} \nu \tag{10.41}$$

where the ν's are nodes in R'_1 and the values of the nodes are their outputs. Again, if $R'_1, R'_2, \ldots$ are shifts of one column of nodes at a time to the right or one column of nodes at a time down then one pooling array is filled out for each feature map array. There are alternative pooling functions to (10.41), such as the square root of the sum of absolute values of the outputs of the nodes in R'_1, but these are less common than a maximum function, possibly because they do not provide as much contrast between the nodes in the pooling layer.

Contrast among the nodes in the pooling layer is important because the third and final, hidden, layer combines all the outputs. It is the fully connected column of nodes indicated in Fig. 10.6. The full connectivity is indicated by the use of curved arrows. To be explicit, fully connected here means there is the usual NN connection from all the nodes over all the pooling arrays to each node in the final hidden layer. The final hidden layer is then fully connected to an output note; again this is indicated by curved arrows.

Figure 10.6 is a generic CNN showing one layer of each possible type. Usually there is a pooling layer after each feature map layer and this can be repeated. Likewise, extra hidden layers may be included at virtually any location. The complexity of the CNN depends on the application. However, in all cases the use of local receptive fields and summarization is intended to build a recognition of aspects of the input space into the structure of the CNN. This is obviously a complex process, and Zeiler and Fergus (2013) is a good place for the interested reader to start. It is important to remember that, as with NNs, the origins of CNNs go back to the 1970s even though most of the early work was of the character explained in the previous subsections. The seminal paper for CNNs is probably LeCun *et al.* (1998) and the reader wishing to use CNNs may be advised to read that paper (and many of the other papers in the volume that contains it).

The interested reader should also be advised that contemporary CNNs often have dozens of layers and use numerous techniques that are more sophisticated than those presented here. However, more layers for the sake of more layers is probably not a good approach to NNs whatever class of NN is being proposed. The goal is good prediction in regression and classification problems, and it must be admitted that the presentation here has by and large been over-positive, in the sense that finding and using NNs of any class is much harder and fraught with many more false starts than has been indicated here. Some problems with deep learning are discussed in Nguyen *et al.* (2015) (images that look like white noise may be classified into a single class with high confidence) and Szegedy *et al.* (2014) (NNs may be confused by the perturbation of images, indicating instability of performance).

The key difference between 'standard' NNs and CNNs is the pooling layer. Otherwise, CNNs are a specific subset of standard NNs. However, pooling layers do not introduce any extra variability (unless different regions of nodes are considered for pooling). So, even though there does not yet seem to be an extension of the main consistency and asymptotic-normality results of White (1989) for the parameters in an NN to CNNs, it would be surprising if analogous results could not be proved; thus reasonable parameter estimates could be obtained by some form of backprop that avoided the vanishing gradient problem. Likewise, extensions of simulated annealing or other model-selection techniques are applicable to CNNs but, because of the depth of the network, may be more difficult. Apart from these caveats, the prediction intervals derived in Secs. 10.3.1–10.3.3 continue to hold for CNNs with only slight modifications. Indeed, the CNN literature is already heavily predictive since the main performance index for CNNs seems to be the generalization error in a classification setting.

To conclude this section we will give a few remarks on other classes of NNs. One popular class is recurrent NNs (RNNs). The idea is that the input is time dependent, so that if $x = (x_1, \ldots, x_d)$ is a fixed input vector then the input is, in the simplest case, a one-dimensional time series $x_1, \ldots, x_t, \ldots$ and the RNN gives an output for each x_i. However, just as the outcomes in a time series are dependent, the RNN is really a sequence of very simple NNs, one for each time step, and the NN for time $i - 1$ feeds the output of at least one node into at least one node of the NN for time i. Many more connections are possible and typically are used. Each component NN is the same, since each is intended to give an output for an input from the same DG, so the parameter estimates and architecture are constant over time. In some settings the periodic updating of parameters and architectures is done, but that is a more complicated case. The point is that the individual NNs for each time are connected one to the next, so that some memory of previous NN activity can be used in forming the prediction for the next input. Backprop, the use of penalty terms, and other standard NN techniques have been adapted to RNNs, and RNNs are most useful for DGs that change over time; they are particularly useful in speech recognition, for instance. The PIs for these cases do not seem to have been given but may be derived in principle if the dependence structure from time step to time step is known in enough detail. Because of the dependence structure it is not clear how well bootstrap-derived PIs will perform.

Another popular class of NNs is called deep belief networks (DBNs). These are focused more on obtaining distributions than point predictors directly. Deep belief networks arise in the following way. First, a Boltzmann machine is a network of binary nodes with an overall 'energy' defined for the entire network. Some nodes receive inputs from the environment; others are internal. Loosely, Boltzmann machines run by repeatedly setting the state of a node and resetting the other nodes in the network. Over time, the state of the network follows a Boltzmann distribution that depends only on the energy. In many respects this is analogous to simulated annealing. Boltzmann machines have not been found to be very effective for learning because often they are just too complex. Consequently, it is more common to study restricted Boltzmann machines (RBMs). The restriction is that the nodes must form a bipartite graph, i.e., the nodes that take inputs from the environment are connected to internal nodes but there are no connections within either class of nodes. (Connections within a class of nodes are permitted in general Boltzmann machines.) Restricting the class of networks has permitted the development of efficient training algorithms for RBMs and this remains an active area of study.

Deep belief networks can be regarded as a generalization of RBMs in the sense that there are more classes of hidden nodes. There is the observed visible layer that gives or receives values from the environment and one or, usually, more sets of hidden nodes that are connected layer to layer but not within layers. For unsupervised learning, a DBN can learn to reconstruct its inputs probabilistically or generate more inputs like those it has seen. This is the sense in which DBNs are generative. The layers of a DBN can then be used to detect features. These networks can also be used as a technique for data compression; see Hinton and Salakhutdinov (2006). A specific sort of DBN is the generative adversarial NN. These, like many DBNs, represent efforts to produce outcomes so representative of a given distribution that they are indistinguishable from actual outcomes from the distribution. Training for DBNs was discussed in Hinton *et al.* (2006) and the techniques are quite different from those presented in earlier sections here. Finally, with supervision a DBN can be further trained to perform classification and hence prediction. While empirically DBNs perform well, there seems to be little theory characterizing their performance predictively.

10.4 Kernel Methods

Recall that the basic problem addressed by ridge regression, in a model of the form $Y_i = x_i^T \beta + \epsilon_i$ with independent data (y_i, x_i) for $i = 1, \ldots, n$ is the near singularity of the design matrix. The proposed solution is to add λ times the identity matrix for some small $\lambda > 0$. This solution can be obtained by backforming the optimization criterion to the minimization of the penalized squared error

$$\sum_{i=1}^{n}(y_i - x_i^T \beta)^2 + \lambda \|\beta\|^2.$$

Otherwise stated, taking derivatives, setting them to zero, and solving gives

$$\hat{\beta} = \left(X^T X + \lambda I\right)^{-1} XY,$$

where $Y = (y_1, \ldots, y_n)^T$, X is the matrix $[x_1, \ldots, x_n]$, and $XY = \sum_{i=1}^{n} y_i x_i$. As written, $\hat{\beta}$ depends on products of the entries in the x_i (and on λ) and also on a linear combination of the x_i using the y_i as coefficients, i.e., $\hat{\beta}$ is in the column space of the design matrix X.

Transform this problem by replacing x_i by $\Phi_i = \Phi(x_i)$, where $\Phi: \mathbb{R}^d \to \mathbb{R}^m$ and it is tacitly assumed that $m > d$ even though $\dim(\Phi) = d$ is possible and, in fact, Φ could be the identity function. Without much confusion, also denote the design matrix of the transformed explanatory variables by Φ. Thus: $\Phi = (\Phi(x_1), \ldots, \Phi(x_n))^T$. Then, using the matrix identity

$$(P^{-1} + B^T B)^{-1} = PB^T (BPB^T + I)^{-1},$$

with $P^{-1} = \lambda I$ and $\Phi = B^T$, the solution becomes

$$\hat{\beta} = (\Phi^T \Phi + \lambda I)^{-1} \Phi Y = (\lambda I)^{-1} \Phi (\Phi^T (\lambda I)^{-1} \Phi + I)^{-1} Y = \Phi (\Phi^T \Phi + \lambda I)^{-1} Y$$

$$= \sum_{i=1}^{n} \alpha_i \Phi(x_i) = \alpha^T \Phi,$$

where $\alpha = (\alpha_1, \ldots, \alpha_n)^T = (\Phi^T \Phi + \lambda I)^{-1} Y$. Again, this means that $\hat{\beta}$ must lie in the column space of the design matrix.

The implication is that a good point predictor for Y_{new} is

$$y_{\text{new}}(x_{\text{new}}) = \hat{\beta}^T \Phi(x_{\text{new}}) = Y^T (\Phi^T \Phi + \lambda I)^{-1} \Phi^T \Phi(x_{\text{new}}) = Y(K + \lambda I)^{-1} \kappa(x_{\text{new}}),$$

where K is the $n \times n$ matrix with (i, j)th entry $k(x_i, x_j) = \Phi(x_i)^T \Phi(x_j)$ and $\kappa(x_{\text{new}})$ is the vector of length n with entries $\kappa_i(x_{\text{new}}) = k(x_i, x_{\text{new}}) = \Phi^T(x_i)\Phi(x_{\text{new}})$. If the transformation Φ does not have a constant term in it (as is commonly the case) it is easy to set $\Phi_0 = 1$, so that there will be a β_0 and hence $\hat{\beta}_0$ to represent a constant bias, meaning that $y_{\text{new}}(x_{\text{new}}) = \hat{\beta}_0 + \hat{\beta}^T \Phi(x_{\text{new}})$.

A similar property is seen if Gaussian process priors are used to form a random effects model; see Sec. 8.4.2. Write $Y(x) = \Phi^T \beta$ but assume that $\beta \sim N(0, (1/\alpha)I)$. The vector of responses is now $Y = \Phi^T \beta$, where Φ is the $m \times n$ design matrix having elements

$\Phi_{ij} = \phi_j(x_i)$ for $i = 1, \ldots, n$ and $j = 1, \ldots, m$, ϕ_j is the jth component of Φ. Since Y is normal, it is enough to find its mean and variance. Thus, $E(Y(x)) = \Phi(x)E(\beta) = 0$ and

$$\mathrm{Var}(Y) = E(YY^T) = \Phi E(\beta\beta^T)\Phi = \frac{1}{\alpha}\Phi^T\Phi.$$

The right-hand side can be written as an $n \times n$ matrix K with (i, j)th element $k(x_i, x_j) = (1/\alpha)\phi(x_i)^T\phi(x_j)$, where again k is a kernel function as in the ridge regression case.

Arguably, the use of kernels is even more important in classification than in regression. The principle is the same: converting methods from their initial formulation by replacing explanatory x's with their transformed values $\Phi(x)$, defining $k(u, v) = \langle\Phi(u), \Phi(v)\rangle$ and making different choices for $k(u, v)$ greatly expands the flexibility of methods. One classical approach starts with $y_i(x_i) = \pm 1$ for explanatory variables x_i and defines two means:

$$m_1 = \frac{1}{\#\{i \mid y_i(x_i) = 1\}} \sum_{i \mid y_i(x_i)=1} x_i \quad \text{and} \quad m_2 = \frac{1}{\#\{i \mid y_i(x_i) = -1\}} \sum_{i \mid y_i(x_i)=-1} x_i.$$

Let $m = m_1 - m_2$. Provided that the denominators of m_1 and m_2 are nonzero, Y_{new} at x_{new} is predicted by looking at whether the angle between the vector $x_{\mathrm{new}} - m$ and the vector $m_1 - m_2$ is smaller than $\pi/2$. That is,

$$Y_{\mathrm{new}}(x_{\mathrm{new}}) = \mathrm{sign}\langle x_{\mathrm{new}} - m, m_1 - m_2\rangle$$

$$= \mathrm{sign}\left(\langle x_{\mathrm{new}}, m_1\rangle - \langle x_{\mathrm{new}}, m_2\rangle + \left((1/2)(\|m_1\|^2 - \|m_2\|^2)\right)\right).$$

$$= \mathrm{sign}\left(\frac{1}{\#\{i \mid y_i(x_i) = 1\}} \sum_{i \mid y_i(x_i)=1} \langle x_{\mathrm{new}}, x_i\rangle - \frac{1}{\#\{i \mid y_i(x_i) = 1\}} \sum_{i \mid y_i(x_i)=-1} \langle x_{\mathrm{new}}, x_i\rangle\right.$$

$$\left. + \left((1/2)(\|m_1\|^2 - \|m_2\|^2)\right)\right).$$

Replacing $\langle\cdot, \cdot\rangle$ by $k(\cdot, \cdot)$ (leaving Φ implicit) gives for the last expression

$$Y_{\mathrm{new}}(x_{\mathrm{new}}) = \mathrm{sign}\left(\frac{1}{\#\{i \mid y_i(x_i) = 1\}} \sum_{i \mid y_i(x_i)=1} k(x_{\mathrm{new}}, x_i) - \frac{1}{\#\{i \mid y_i(x_i) = 1\}}\right.$$

$$\times \sum_{i \mid y_i(x_i)=-1} k(x_{\mathrm{new}}, x_i) + \frac{1}{2}\left(\frac{1}{\#\{i \mid y_i = -1\}} \sum_{(i,j) \mid y_i=y_j=-1} k(x_i, x_j)\right.$$

$$\left.\left. - \frac{1}{\#\{i \mid y_i = 1\}} \sum_{(i,j) \mid y_i=y_j=-1} k(x_i, x_j)\right)\right),$$

in which all that needs to be specified is k.

As a further example, consider the Nadaraya–Watson (NW) kernel regression estimator from Sec. 8.2. Expression (8.48) can be written as

$$\hat{Y}(x_{\mathrm{new}}) = \sum_{i=1}^{n} k_h(x_{\mathrm{new}}, x_i)y_n$$

by setting $k_h(x_i, x) = K(|x - x_i|/h)/\sum_{j=1}^{n} K(|x - x_j|/h)$, using the notation of Sec. 8.2.3. The form of the solution is not changed if a transformation Φ is applied to the x's. Indeed, the k resulting from transforming in an NW solution simply looks more like the earlier solutions from ridge regression and a random effect model that included a transformation.

The point of these examples is to show that often all the information for a point predictor is in a function $k(\cdot, \cdot)$ called a kernel. That is, regardless of $\dim(x)$, the solution to a problem often depends only on an $n \times n$ (symmetric) positive definite matrix defined by k and a vector of length n with evaluations of k at design points and x_{new}. This means that transformations Φ can be easily used in many settings if Φ is used to define a kernel by $k(u, v) = \langle \Phi(u), \Phi(v) \rangle$, as seen in ridge regression, random effect models, and a simple classifier. (In the NW case, an inner product is not needed; it is enough to specify Φ and be able to evaluate it at select points.) The idea is that, when $m > d$, problems may become easier because the larger dimension permits more flexibility and the dependence is on n not $\dim(x)$ or $\dim(\Phi)$. In fact, Φ does not even need to be explicitly specified; it is equivalent to k so it is enough to specify k, i.e., the inner product between the images of points under Φ, explicitly and to do so without reference to whatever the underlying transformation Φ represents. Once k has been fixed, it defines a Hilbert space of functions.

Saying that 'problems become easier' means that, in classification problems, lifting x to a higher-dimensional space may make the classes easier to separate, and in regression problems the kernel gives a natural collection of functions to use as explanatory variables. The strength of this approach is that it is easy to replace inner products with kernels in many procedures, so that manipulations can be done in the kernel-induced Hilbert space rather than the original space of the x's. Since there are so many possible kernels, a large range of techniques can be generated. Of course, this may also be seen as a weakness because visualizing what the inner product k means in terms of x is usually very difficult.

It is a mathematical fact that a kernel as defined above, i.e., as an inner product $k(u, v) = \langle \Phi(u), \Phi(v) \rangle$, is symmetric and positive definite and hence is 'reproducing' on the Hilbert space $\mathcal{H}$ of functions (on a domain) that it defines by using k as an inner product. 'Reproducing' is an abstract property meaning that the value at x of any $f \in \mathcal{H}$ can be represented in terms of its inner product with a fixed function $k_x(\cdot)$.[5] To see this, use k to define an inner product for $\mathcal{H}$, choose $f \in \mathcal{H}$, and choose x in the domain of the functions in $\mathcal{H}$. Let $T_x: \mathcal{H} \to \mathbb{R}$ be defined by $\forall f \in \mathcal{H}, T_x(f) = f(x)$, i.e., T_x is the evaluation functional on $\mathcal{H}$ at x. Since T_x is linear, the Reisz representation theorem gives that for each x there is a $k_x(\cdot) \in \mathcal{H}$ such that

$$f(x) = \langle f, k_x \rangle.$$

Since $k_x(\cdot) \in \mathcal{H}$, the Reisz representation theorem gives for any y that

$$k_x(y) = \langle k_x, k_y \rangle \quad \text{and} \quad k(x, y) = \langle k_x, k_y \rangle,$$

where the last equality is the reproducing property. In particular, if the domain is $\mathcal{X}$ then $k: \mathcal{X}, \mathcal{X} \to \mathbb{R}$ is a reproducing kernel if and only if it is symmetric and positive definite, i.e.,

$$\forall n, \quad \forall x_1, \ldots, x_n \in \mathcal{X}, \quad \forall c_1, \ldots, c_n \in \mathbb{R}, \quad \sum_{i,j=1}^{n} c_i c_j k(x_i, x_j) \geq 0.$$

The Hilbert space defined by a kernel is unique.

---

[5] Technically, a reproducing kernel is defined by the property that it makes every evaluation functional on $\mathcal{H}$ continuous. This definition is hard to motivate even though it is equivalent to the one used here, which makes the reproducing of functions via the kernel explicit.

What is the point of using specially chosen kernels or inner products based on transformations Φ that are (usually) not explicitly specified or used? Concisely, the point is that the representer theorem – and variations on it that include extra constraints on the set over which the minimization is done. Informally, this result states that when optimizing a penalized risk over functions in a reproducing kernel Hilbert space (RKHS), the form of the optimum is a linear combination of evaluations of the kernel. The original form of the theorem is due to Kimmeldorf and Wahba (1971). A more general result is due to Scholkopf *et al.* (2001). Let $L \colon \mathsf{Y} \times \mathbb{R} \to \mathbb{R}^+$ be a loss function and let $f \in \mathcal{H}$, i.e., $f \colon \mathcal{X} \to \mathbb{R}$, where $\mathcal{H}$ is a reproducing kernel Hilbert space (RKHS) with kernel k. Then the risk of f is $EL(Y, f(X))$ where the expectation is over both Y and X. The empirical risk for f from data (y_i, x_i) for $i = 1, \ldots n$ is $(1/n) \sum_{i=1}^{n} L(y_i, f(x_i))$. If the inner product on $\mathcal{H}$ is $\langle \cdot, \cdot \rangle$, write the norm as $\| \cdot \|$. The following two theorems give the key formalizations of kernel-based solutions in the absence of extra constraints.

Theorem 10.1 (Scholkopf *et al.* 2001)   *Let $\Omega \colon \mathbb{R}^+ \to \mathbb{R}$ be a nondecreasing function and suppose that L is an empirical risk, possibly dependent on the x_i as well as on the y_i and $f(x_i)$, but on nothing else. Then, if the minimization problem*

$$\min_{f \in \mathcal{H}} \left(L((x_1, y_1, f(x_1)), \ldots, (x_n, y_n, f(x_n))) + \Omega(\| f \|) \right)$$

has a solution, it is of the form

$$f(x) = \sum_{i=1}^{n} \alpha_i K(x_i, x) \quad where \quad \forall i, \, \alpha_i \in \mathbb{R}. \tag{10.42}$$

Theorem 10.2 (Scholkopf *et al.* 2001)   *Suppose that in addition to the assumptions of Theorem 10.1 there is a set of M real-valued functions ϕ_ℓ, $\ell = 1, \ldots, M$, such that the $n \times M$ matrix $(g_\ell(x_i))_{i,\ell}$ has rank M. Let $f = f_1 + h$ with $f_1 \in \mathcal{H}$ and $h \in \mathrm{Span}(g_1, \ldots, g_M)$. Then, if the minimization problem*

$$\min_{f \in \mathcal{H}} \left(L((x_1, y_1, f(x_1)), \ldots, (x_n, y_n, f(x_n))) + \Omega(\| f_1 \|) \right)$$

has a solution, it is of the form

$$f(x) = \sum_{i=1}^{n} \alpha_i K(x_i, x) + \sum_{\ell=1}^{M} \beta_\ell g_\ell(x) \quad where \quad \forall i, \, \alpha_i \in \mathbb{R}, \beta_\ell \in \mathbb{R}. \tag{10.43}$$

The representer-theorem solutions (10.42) and (10.43) are point predictors in the absence of the usual sorts of modeling information that are built into the formulation of a problem; the major inputs to the representer-theorem predictors are k, L, and Ω, and possibly some g_ℓ. So (10.42) and (10.43) are justified purely by their optimality. Clearly, ridge regression, random effect regression with a GPP and the NW estimator are examples of (10.42) while the offset constant in the classification example makes it an example of (10.43). In classification cases, Y usually takes the values ± 1 and the sign operator must be used to convert the representer-theorem solution to ± 1 values.

The more interesting examples arise when the forms (10.42) and (10.43) are used explicitly to find classifiers or regression functions. These cases obviously depend on the kernel and can be analyzed in either a Bayesian or frequentist way. It is certain that the solutions also depend on the loss function; however, little work seems to have been done on how the loss function affects the coefficients in (10.42). An obvious concern with using, say, (10.42) directly is that n coefficients must be estimated from n data points. That is, no matter how much data is collected there is no guarantee that (10.42) converges to a limit that is independent of the data. This means that if a variance–bias decomposition is done on (10.42) the variance will be elevated. So, in practice, some sort of sparsity is usually imposed. That is, if terms in (10.42) that contribute negligibly are omitted, the predictive error should decrease because the variance will be reduced substantially with only a small increase in bias. The same applies to (10.43).

One of the most important features of (10.42) and (10.43) is that they do not give models in any classical sense. Even if the α_i were known, the solution still uses the x_i. This means that representer-theorem solutions are 'instance-based learners' – instead of learning (estimating) some fixed set of parameters they retain in their mathematical form the data that was used to find them. Thus, representer-theorem solutions are often good predictors because they are more flexible than classical models and they effectively use more of the information in the data. Indeed, they are tied so closely to the specific data set that different data sets can lead to different versions of (10.42) or (10.43) and the solutions do not have to be similar to each other.

Transforming familiar quantities such as ridge regression or hyperplane classification to 'kernelized' forms by recognizing inner products has led to a large number of variations on old techniques, including but not limited to kernelized PCA (Scholkopf *et al.* 1998), kernelized partial least squares (Rosipal and Trejo 2001), kernelized canonical correlation analysis (Fyfe and Lai 2000), and kernelized discriminant analysis (Mika *et al.* 1999). While these are useful, the present section will focus on predictors that are kernel based. Thus, they will be kernelized and will represent either an instance of (10.42) or (10.43) or an instance of the minimizations in Theorems 10.1 and 10.2 under further constraints. Indeed, it will be seen that, for good prediction, there are often constraints that can be imposed to give better kernel solutions that are usually of the same form as (10.42) and (10.43). These generally require Lagrange multipliers and the Karush–Kuhn–Tucker conditions. So, the derivation is more involved, sometimes taking advantage of the details of the formulation of a problem, but the results can be extremely good.

10.4.1 Bayes Kernel Predictors

The simplest kernel methods use (10.42) or (10.43) directly for regression or classification. One formulation of this is due to Tipping (2000), who included a constant term in (10.43) by setting $p = 1$ and $\psi_1 \equiv 1$ and then used a Bayesian analysis to achieve sparsity. The assumption was that n is large enough that there are many small terms in (10.43). Hence, term selection is essential to reduce the variance of the point predictor $\hat{f}$ used to approximate f. The analysis used n hyperparameters τ_i, one for each α_i. When $K(x_i, \cdot)$ is negligible, τ_i should be large and so make the posterior density for α_i concentrate at zero. The resulting regression function is called a relevance vector machine (RVM).

More formally, for regression, choose

$$Y_i \sim N(X_i^T \beta, \sigma^2),$$

where $\beta = (\beta_0, \ldots, \beta_n)^T$ and $X_i^T = (1, K(x_i, x_1), \ldots, K(x_i, x_n)))$. Note that in this specific setting the β that appears in the usual linear regression corresponds to the α that appears in the representer-theorem solution. Thus, for n outcomes,

$$\begin{pmatrix} Y_1 \\ \vdots \\ Y_n \end{pmatrix} = \begin{pmatrix} 1 & K(x_1, x_1) & \ldots & K(x_1, x_n) \\ \vdots & \vdots & \vdots & \vdots \\ 1 & K(x_n, x_1), & \ldots & K(x_n, x_n) \end{pmatrix} \begin{pmatrix} \beta_0 \\ \vdots \\ \beta_n \end{pmatrix} + \begin{pmatrix} \epsilon_1 \\ \vdots \\ \epsilon_n \end{pmatrix}, \quad (10.44)$$

where the $\epsilon_i \sim N(0, \sigma^2)$ are IID and the $(n+1) \times n$ design matrix is denoted $X = [X_1^T, \ldots, X_n^T]$. The MLEs or LSEs for β usually lead to overfitting since $\dim(\beta) = n$.

The simplest priors to use on the β_j are independent $N(0, 1/\tau_j)$ distributions. Letting $\tau = (1/\tau_0, \ldots, 1/\tau_{n-1})$ this leads to the conjugate normal posterior density

$$w(\beta | Y, X, \sigma^2, \tau) = \frac{1}{(\sqrt{2\pi})^{n+1}} (\det \Sigma)^{-1/2} \exp\left(-\frac{1}{2}(\beta - \mu)^T \Sigma^{-1} (\beta - \mu)\right), \quad (10.45)$$

where

$$\Sigma = (X^T B X + T)^{-1} \quad \text{and} \quad \mu = \Sigma X^T B Y, \quad (10.46)$$

in which $T = \mathrm{diag}(\tau_0, \ldots, \tau_n)$ and $B = \sigma^{-2} I_n$. So, using the likelihood from (10.44) and (10.45), integrating out the β_i gives

$$p(Y | \tau, \sigma^2) = \int_{\mathbb{R}^{n+1}} p(Y | X, \beta, \sigma^2) w(\beta | Y, X, \tau, \sigma^2) d\beta \quad (10.47)$$

$$= \frac{1}{(\sqrt{2\pi})^{n/2}} \det |B^{-1} + X \Sigma^{-1} X^T| \exp\left(-\frac{1}{2} Y^T (B^{-1} + X \Sigma^{-1} X^T) Y\right). \quad (10.48)$$

In a fully Bayesian treatment, a prior would be assigned to σ^2 and hyperpriors would be assigned to σ^2 and the τ_i, as well as to any parameters in the kernel. This is discussed in Tipping (2001). However, finding optimal values of σ and τ that, for instance, maximize (10.48) or its fully Bayes counterpart formed by integrating out σ^2 and τ can only be done numerically. In addition, a fully Bayes treatment of this, including an MCMC–MH technique for estimating the posterior (which could be used to approximate the predictive density, cf. (10.47)) is given in Sec. 4.3 of Chakraborty *et al.* (2013).

A pragmatic approach is given in Tipping (2000) and in more detail in Tipping (2001). To summarize, differentiate (10.48) with respect to each τ_i, equate the derivatives to zero, and rearrange to get

$$\tau_i = \frac{\gamma_i}{\mu_i^2} \quad (10.49)$$

where $\gamma_i = (1 - \tau_i) \Sigma_{ii}$, Σ_{ii} being the ith diagonal element in Σ and μ_i the ith element in μ. Analogously, differentiating (10.48) with respect to σ and setting the derivative to zero results in

$$\sigma_i^2 = \frac{\|\beta - X\mu\|}{n - \sum_{i=1}^n \gamma_i}. \quad (10.50)$$

From (10.49) and (10.50) a recursion can be set up. Choose a starting value to get values $\tau_{i,\text{new}}$ for all i from the left-hand side of (10.49). Then, use the $\tau_{i,\text{new}}$ to get a value for the right-hand side of (10.50), update Σ and μ from (10.46), and repeat until the estimates seem to converge. In practice, many of the τ_i become very large, making the corresponding $w(\beta_i | X, Y, \tau, \sigma^2)$ notably spiked at zero, suggesting that the β_i should be pruned.

Once (10.44) has been made sparse (by, for example, pruning), RVM Bayes prediction proceeds as usual. Let the final values of τ and σ be $\hat{\tau}$ and $\hat{\sigma}$. The predictive density is

$$p(y_{\text{new}}(x_{\text{new}}) | X, Y, \hat{\tau}, \hat{\sigma}) = \int p(y_{\text{new}} | x_{\text{new}}, \beta, \hat{\sigma}^2) w(\beta | Y, X, \hat{\tau}, \hat{\sigma}) d\beta, \qquad (10.51)$$

in which both factors in the integrand are normal densities. The first factor is $Y_i(x_{\text{new}}) \sim N((1, K(x_1, x_{\text{new}}), \ldots, K(x_n, x_{\text{new}}))\beta, \hat{\sigma}^2)$ and the second factor is from (10.45) but using $\hat{\tau}$ and $\hat{\sigma}$. So, the integral over β can be evaluated in closed form. Letting $v(x_{\text{new}}) = (1, K(x_1, x_{\text{new}}), \ldots, K(x_n, x_{\text{new}}))^T$, the result is that $p(y_{\text{new}}(x_{\text{new}}) | X, Y, \hat{\tau}, \hat{\sigma})$ is the density of a distribution

$$N\left(\mu^T v(x_{\text{new}}), \hat{\sigma}^2 + v(x_{\text{new}})^T \Sigma v(x_{\text{new}})\right), \qquad (10.52)$$

where the two terms in the normal variance arise in the same way as the two terms in earlier variance expressions (see (10.7), (10.16), and (10.30)). The point predictor from the RVM is $\hat{Y}(x_{\text{new}}) = \mu^T v(x_{\text{new}})$ and it is obvious how to get $1 - \alpha$ PIs from the RVM for $Y_{\text{new}}(x_{\text{new}})$ using (10.52).

The approach advocated in Tipping (2001) is meaningfully different from the fully Bayes approach of Chakraborty *et al.* (2013). The former is aimed at achieving sparsity and so seeks to eliminate the α_j (i.e., the β's in (10.44)), whose posterior is too tightly concentrated around zero. The latter approach, being fully Bayes, retains the probabilistic description of all the α_j, and their concentration near zero, if present, will be reflected in the posterior and hence in the predictive densities. Hence, the seeking of sparsity is not necessary with the fully Bayes approach, although sparsity, if desired, may be achieved. In the Bayes paradigm, at a minimum the posterior is found and inference, predictions in particular, proceeds as usual. One benefit of the fully Bayes approach is that the usual properties of Bayes solutions, e.g., consistency, hold automatically.

Relevance vector machines can also be used for classification but are more abstract because the likelihood is more difficult than that derived from the normal distribution. Also, RVMs for classification do not follow from the representer theorem. Despite their different character, it is reasonable to put them in the same class as RVMs for regression because they use the Bayes formulation to achieve sparsity. Specifically, the likelihood for classification is

$$P(Y = y | \beta) = \prod_{i=1}^{n} \sigma(y_i(x_i) | \beta)^{y_i} (1 - \sigma(y_i(x_i) | \beta)))^{1 - y_i},$$

where $Y = (Y_1(x_1), \ldots, Y_n(x_n))$, $y = (y_1, \ldots, y_n)^T$, and each $Y_i(x_i) = \pm 1$. As with RVMs for regression, let $w(\beta | \tau) = \prod_{i=1}^{n} w(\beta_i | \tau_i)$, where $w(\beta_i | \tau_i)$ is the density of an $N(0, 1/\tau_i)$ random variable, and let $\sigma(y) = 1/(1 + e^{-y})$. Now,

$$w(\beta | \tau, Y) \propto P(Y = y | \beta) w(\beta | \tau);$$

so, maximizing the posterior $w(\beta|\tau, Y)$ over β is equivalent to maximizing

$$\ln P(Y = y|\beta)w(\beta|\tau) = \sum_{i=1}^{n} \left(y_i \ln y_i(x_i) + (1 - y_i)\ln(1 - y_i(x_i))\right) - \frac{1}{2}\beta^T A\beta$$

(10.53)

over β, where $A = \text{diag}(\tau_0, \ldots, \tau_n)$; (10.53) is a penalized logistic log-likelihood so maximization over β is a well-studied problem. Using a Laplace method approximation to find the mode of (10.53), Tipping (2001) obtained

$$\hat{\beta} = \Sigma X^T BY,$$

where

$$B = \text{diag}\left(\sigma(y(x_1))(1-\sigma(y(x_1))), \ldots, \sigma(y(x_i))(1-\sigma(y(x_i))), \ldots, \sigma(y(x_n))(1-\sigma(y(x_n)))\right).$$

The matrix Σ arises from the second derivative of (10.53) (at $\hat{\beta}$) and is

$$\Sigma = (X^T BX + A)^{-1},$$

where A is as in (10.53). Predictions for the probability of $Y(x_{\text{new}}) = 1$ (or -1) can be made as in (10.51) (except that there is no σ) using the Laplace approximation to approximate the integration over β. Thus,

$$P(Y_{\text{new}}(x_{\text{new}}) = 1|X, Y, \hat{\tau}) = \int p(y_{\text{new}}|x_{\text{new}}, \beta,)w(\beta|Y, X, \hat{\tau},)d\beta$$

$$\approx P(Y_{new}(x_{\text{new}}) = 1|x_{\text{new}}, \hat{\beta}).$$

(10.54)

This is a probabilistic prediction and the point prediction would be whichever of 1 or -1 had a higher value, as approximated using (10.54). Of course, sparsity is enforced as in the regression case: those β_i that have τ_i values that are too large are set to zero on the grounds that their posterior density is concentrating at zero, meaning that they have negligible influence and in principle, in the limit of large n, would increase to ∞. Tipping (2000) gave a variety of examples in which RVM regression and classification outperform other sparse methods for complex problems.

Although Gaussian process regression was discussed in Sec. 8.4, it bears mention here because it is kernel based – a point not made earlier. This is not an instance of the representer theorem; as set up here, it is an instance of Bayes' rule. So, suppose that $Y \sim N(c, K)$, where $Y = (Y(x_1), \ldots, Y(x_n))^T$, $c = (c(x_1), \ldots, c(x_n))^T$, and $K = (K(x_i, x_j))_{i,j=1,\ldots n}$. If a new data point, x_{new}, is received and the goal is to predict the value of $Y(x_{\text{new}})$ then without loss of generality write $p(Y(x_{\text{new}})|f, x_{\text{new}}, \sigma^2) \sim N(c(x_{\text{new}}), \sigma^2)$, in which the form of σ^2 will emerge from the derivation. By well-known normal-distribution manipulations, the predictive density is

$$p(y(x_{\text{new}})|x_{\text{new}}, K, Y) = N(c(x_{\text{new}}), \sigma^2(x_{\text{new}})),$$

where $c(x_{\text{new}}) = \kappa^T(K + \sigma^2 I_n)^{-1}Y$ and $\sigma^2(x_x) = K(x_{\text{new}}, x_{\text{new}}) - \kappa^T(K + \sigma^2 I_n)^{-1}\kappa$, in which $\kappa = (K(x_1, x_{\text{new}}), \ldots, K(x_n, x_{\text{new}}))^T$.

Now, the connection between the representer theorem and Bayes' rule for finitely many outcomes or, equivalently, finite-dimensional RKHSs, can be seen. More precisely, the objective function used in the representer theorem is the exponent of a penalized sum of squares that results from using a Gaussian prior and a normal likelihood. To see this, assume that the ϕ_j are orthonormal in the inner product of the RKHS, and recall that when the RKHS is finite dimensional it is identified by a M-dimensional feature map, i.e., $\Phi(x) = (\phi_1(x), \ldots, \phi_M(x))^T$, by setting

$$K(x, y) = \langle \Phi, \Phi \rangle = \sum_{j=1}^{M} \phi_j(x) \phi_j(y).$$

For $\gamma = (\gamma_1, \ldots, \gamma_M)$, functions in the RKHS can then be written as

$$f_\gamma = f(x|\gamma) = \sum_{j=1}^{M} \gamma_j \phi_j(x) = \langle \gamma, \Phi(x) \rangle,$$

leading to $\|f_\gamma\| = \|\gamma\|$. (Analogous properties can be obtained more generally by regarding K as the kernel of an integral operator. The result, Mercer's theorem, expresses K using the eigenvectors of a Hilbert–Schmidt operator.)

Now, using $Y_i(x) = f(x) + \epsilon_i$ with $\epsilon_i \sim N(0, \sigma^2)$, $i = 1, \ldots, n$, the representer-theorem solution can be written as

$$f_R(x) = \alpha_0 + \sum_{i=1}^{n} \alpha_i K(x_i, x) \quad \text{where} \quad K(x_i, x) = \langle \Phi(x_i), \Phi(x) \rangle,$$

and this gives $\|f_R(\cdot|\alpha)\| = \|\alpha\|$ for $\alpha = (\alpha_0, \alpha_1, \ldots, \alpha_n)^T$ if $g_0(x) = 1$. So, the operational part of the likelihood is

$$p(Y|X, f) = N(Y(X), \sigma^2 I_n) \propto \exp\left(-\frac{1}{2\sigma^2} \|Y(X) - f_\gamma(X)\|^2\right).$$

Since the posterior is proportional to the prior times the likelihood, if γ has a normal mean-zero variance-σ^2 prior then the posterior for $f(x_1), \ldots, f(x_n)$ is

$$w(f|X, Y) \propto \exp\left(-\frac{1}{2\sigma^2} \|Y(X) - f_\gamma(X)\|^2 - \|\gamma\|^2\right). \tag{10.55}$$

Thus, minimizing the exponent in (10.55) is equivalent to a special case of the minimization in the representer theorem.

To demonstrate the flexibility of kernel-based methods, we will consider the Bayes version of support vector machines (SVMs). The idea is that, in a two-class classification problem, there should be a boundary between the two classes in the feature space and this boundary should be chosen to maximize the distance (also called the margin) between the points with $Y = 1$ and $Y = -1$. This is impossible in low dimensions if the points are too interspersed, but the goal is to choose Φ so that they are not, as far as is reasonably possible. In other words, the selection of the kernel is intended to make this feasible. Support vector machines involve a different sort of loss function than in that RVMs and, depending on the formulation of the optimization, may include a set of constraints that does not appear in the representer-theorem optimization even though the two formulations lead to similar optima, which depend on a kernel.

The most straightforward formulation is a simplification of the problem studied in Chakraborty (2011). Suppose that (x_i, y_i) are given for $i = 1, \ldots, n$; then the optimization problem can be given as

$$\min_{\mathcal{H}} \left(\sum_{i=1}^{n} (1 - y_i f(x_i))_+ + \lambda \Omega(f) \right),$$

where Ω is a penalty functional and $\lambda > 0$ is a smoothing parameter. The notation $(1 - y_i f(x_i))_+$ denotes the hinge loss, generally written as $\ell(y) = \max(0, 1 - ry)$ where y is the approximate class label (± 1) and r is the correct class label. Thus, when r and y have the same sign the loss is zero, otherwise the loss is 2. Treating the cumulative loss as the negative of a log-likelihood and defining $Y = (y_1, \ldots, y_n)^T$ and $X = (x_1, \ldots, x_n)^T$ gives the likelihood

$$f(Y, X) \propto \exp\left(-\sum_{i=1}^{n} (1 - y_i f(x_i))_+\right). \tag{10.56}$$

Minimizing over a Hilbert space $\mathcal{H}$ and taking $\Omega(f) = \|f\|_{\mathcal{H}}$ gives

$$f(x) = \alpha_0 + \sum_{i=1}^{n} \alpha_i K(x_i, x | \theta),$$

where θ is an index that runs over the kernels that might be used.

One way around finding a normalizing factor for (10.56) is to introduce latent variables. Write

$$f(Y, X | Z) \propto \exp\left(-\sum_{i=1}^{n} (1 - y_i z_i)_+\right),$$

where $Z = (z_1, \ldots, z_n)^T$,

$$z_i = \alpha_0 + \sum_{j=1}^{n} \alpha_i K(x_j, x_i) + \epsilon_i$$

$$= \alpha_0 + K_i \alpha + \epsilon_i \quad \text{and} \quad \epsilon_i \sim N(0, \sigma^2), \tag{10.57}$$

in which the ϵ_i are IID, $K_i = (K(x_i, x_1 | \theta), \ldots, K(x_i, x_n | \theta))^T$, and $\alpha = (\alpha_1, \ldots, \alpha_n)^T$. Furthermore, Chakraborty (2011) suggested a radial basis function kernel, essentially a normal density $K(x, x') = \exp\left(-\|x - x'\|^2 / \theta\right)$, on the grounds that as $\theta \to 0$ the kernel spikes at zero, i.e., localizes quickly.

It remains to specify priors for α_0, α, and σ^2. Chakraborty (2011) proposed $w(\alpha_0, \sigma^2) \propto 1/\sigma^2$ and $(\alpha | \Delta, \sigma^2) \sim N(0, \sigma^2 \Delta^{-1})$, where $\Delta = \text{diag}(\delta_1, \ldots, \delta_n)$. The hyperparameters can then be assigned as $\theta \sim \text{Unif}(u_1, u_2)$, $\delta_i \sim \text{Gamma}(c, d)$, where u_1, u_2, c, and d are user specified. Given this and letting $Z = (z_1, \ldots, z_n)^T$, the full joint posterior $w(Z, \sigma^2, \alpha_0, \alpha, \Delta, \theta | Y, X)$ can be written down (see Chakraborty (2011) expression 16, and omit the even factors). As in any Bayesian procedure, the posterior can be marginalized to give a density on the random variables (parameters) of interest, so the predictive density can be found and used to generate predictions. It is of the form

$$P(Y_{\text{new}} = 1 | x_{\text{new}}, X, Y)$$

$$= \int P(Y_{\text{new}} = 1 | x_{\text{new}}, \sigma^2, \alpha_0, \alpha, \Delta, \theta) w(\sigma^2, \alpha_0, \alpha, \Delta, \theta | Y, X) d(\sigma^2, \alpha_0, \alpha, \Delta, \theta),$$

in which the class (1 or -1) with the higher posterior probability at x_{new} is the prediction $\hat{Y}_{\text{new}}(x_{\text{new}})$.

A Bayesian analysis of SVMs for regression was lucidly presented in Chakraborty *et al.* (2013). They began with Vapnik's ϵ-insensitive loss

$$L(y, f(x)) = |y - f(x)|_\epsilon = \begin{cases} 0 & \text{if } |y - f(x)| \le \epsilon, \\ |y - f(x)| - \epsilon & \text{otherwise.} \end{cases} \tag{10.58}$$

This loss function ignores errors that are less than ϵ. Indeed, if $|y - f(x)| = \epsilon/2$ then the loss is zero and if, say, $|y - f(x)| = 3\epsilon/2$ then the loss is $\epsilon/2$. As with other procedures such as robust cross-validation, the goal is to make inferences less dependent on outliers, partially by ϵ-insensitivity and partially by linearity outside $[-\epsilon, \epsilon]$.

Again consider n latent variables $Z_1, \ldots, Z_n$ such that the Y_i are conditionally independent given the Z_i. Now, the likelihood of y_i, conditionally on z_i, is

$$p(y_i | z_i) \propto \exp(-\rho |y_i - z_i|_\epsilon), \tag{10.59}$$

for $i = 1, \ldots, n$. The Z_i are related to the $f(x_i)$ by

$$Z_i = f(x_i) + \eta_i,$$

where the η_i, as noted in Chakraborty *et al.* (2013), are 'the residual random effects that account for any unexplained source of variation not included in the model'. Assuming that $f(\cdot)$ is of the form of a representer-theorem solution, we have

$$z_i = K_i^T \alpha + \eta_i$$

where $\eta_i \sim N(0, \sigma^2)$ are IID, $K_i = (1, K(x_i, x_1 | \theta), \ldots, K(x_i, x_n | \theta))^T$, and α includes α_0 as its first entry. It remains to assign priors to the parameters.

One selection of priors, implicitly advocated in Chakraborty *et al.* (2013), is

$$
\begin{aligned}
Z_i | \alpha, \theta, \sigma^2, \theta &\sim N(K_i^T \alpha, \sigma^2), \\
\alpha | \sigma, \Delta &\sim N(0, \sigma^2 \Delta^{-1}), \\
\sigma^2 \sim \text{IG}(\gamma_1, \gamma_2), \quad \theta &\sim \text{Unif}(a, b), \\
\rho \sim \text{Unif}(u, v), \quad \delta_i &\sim \text{Gamma}(c, d).
\end{aligned} \tag{10.60}
$$

The remaining parameter is ϵ, which defines the degree of insensitivity. One choice is $\epsilon \sim \text{Beta}(r, s)$; often, however, it is assumed that the hyperparameters are user specified. So, using the normalized likelihood from (10.59) and the priors in (10.60), an expression for the posterior $w(\alpha, \Delta, z_1^n, \sigma^2, \theta, \rho, \epsilon | Y)$ can be given; see Chakraborty *et al.* (2013, expression (4.25)). The consequence is that an MCMC–MH procedure can be devised to give an approximation to this posterior and hence an approximation to the predictive density:

$$
\begin{aligned}
p(y_{\text{new}} | x_{\text{new}}, X, Y) = \int & p(y_{\text{new}} | x_{\text{new}}, \alpha, \Delta, z_{\text{new}}, \sigma^2, \theta, \rho, \epsilon) \\
& \times w^*(\alpha, \Delta, z_1^{n+1}, \sigma^2, \theta, \rho, \epsilon | Y) d(\alpha, \Delta, z_1^{n+1}, \sigma^2, \theta, \rho, \epsilon),
\end{aligned}
$$

where $z_1^{n+1} = (z_1, \ldots, z_n, z_{\text{new}})$. In this expression w^* is the posterior from Y multiplied by the density for z_{n+1}, associated with y_{n+1}, under the prior selection (10.60). The integration is therefore over all parameters including z_{new}. As usual, $1 - \alpha$ PIs for $Y_{\text{new}}(x_{\text{new}})$ can be obtained from $p(y_{\text{new}} | x_{\text{new}}, X, Y)$.

Overall, the presentation in this subsection has focused on the RVM and (Bayes) SVM, including their relationships with Gaussian processes and sparsity. The differences between fully Bayes versus approximately Bayes analyses have also been noted. As will be seen, this is quite different from the frequentist situation.

10.4.2 Frequentist Kernel Predictors

The representer-theorem solution can be regarded as a generalization of linear models in the same way as can ridge regression; see the beginning of Sec. 10.4. Indeed, a predictor can be derived directly. Write

$$f(x) = \alpha_0 + \sum_{i=1}^{n} \alpha_i K(x_i, x) = k^T \alpha, \tag{10.61}$$

where $\alpha = (\alpha_0, \ldots, \alpha_n)^T$ and $k = (1, k(x_1, x), \ldots, k(x_n, x))^T$, and observe that the squared norm of f in $\mathcal{H}$, the RKHS of functions defined by K, is

$$\|f\|_{\mathcal{H}}^2 = \sum_{i=0}^{n} \sum_{j=0}^{n} \alpha_i \alpha_j \langle K(x_i, \cdot), K(x_j, \cdot) \rangle = \sum_{i=0}^{n} \sum_{j=0}^{n} \alpha_i \alpha_j K(x_i, x_j) = \alpha^T K \alpha,$$

where it is understood that $K(x_0, \cdot) = 1$ and K is of the same form as the design matrix in (10.44). The penalized risk to be optimized can now be written as

$$\frac{1}{n} \|Y - K\alpha\|^2 + \lambda \alpha^T K \alpha, \tag{10.62}$$

in which $\lambda > 0$ expresses the tradeoff between the norm of f and how well f fits the data. Expression (10.62) can be minimized as in standard linear models (or ridge regression) by differentiating with respect to each α_i, setting the derivatives to zero and solving. The result is the kernel analog of the normal equations as in linear regression. Explicitly,

$$\frac{1}{n} K(Y - K\alpha) + \lambda K \alpha = 0,$$

which can be rearranged to give

$$\alpha = (K + \lambda n I_{n+1})^{-1} Y.$$

The decay parameter λ can be chosen in a variety of ways. Obviously, cross-validation can be used. However, it is more common to regard λ as a 'term-inclusion parameter' because, as λ increases, the coefficients α_i shrink and it is common to threshold-out coefficients that are too small. Given estimates $\hat{\lambda}$ and $\hat{\alpha}$, they can be substituted into (10.61) to generate point predictions for new values of x. Moreover, the bootstrapping techniques described at the end of Sec. 10.1 can be used to generate PIs with any desired level of confidence. This is reasonable because the representer theorem is only an optimization; it does not include an error structure. So, while point estimates are straightforward, in order to get PIs an error structure must be either assumed to be of a certain form or treated nonparametrically. This parallels the fact that least squares estimators in a linear model do not depend on the error term.

Separately from the direct treatment (10.61), it is possible to modify the optimization in the representer theorem by adding constraints, thereby obtaining different solutions. One important example of this is the frequentist version of SVMs for finding classification and regression functions.

To see this, fix an inner product space $\mathcal{H}$ and suppose that the goal is to make a binary classification ($Y = \pm 1$) for x's using a hyperplane classifier. That is, given x, the task is to find $w \in \mathcal{H}$ and $b \in \mathbb{R}$ such that

$$\hat{Y}(x) = \text{sign}(\langle w, x \rangle + b) \tag{10.63}$$

is an optimal classifier within the set of generalized hyperplanes in $\mathcal{H}$. This is different from the hyperplane classification example at the beginning of this section because, rather than simply choosing the difference between the sample means for the two classes, (10.63) is optimized over w and b. The sense of optimality is given in terms of a 'margin'. The idea is that if the two classes can be perfectly separated then there is a strip in $\mathcal{H}$ that separates them. The strip is called a margin; the maximal width over all margins is the separation. So, the goal can be restated as finding the w and b that make (10.63) a maximal separation (or maximal margin) classifier. If the two classes cannot be perfectly separated by a hyperplane, an extension of this reasoning leads to a classifier that is as close as possible to a maximal margin classifier.

Given a data set (y_i, x_i) for $i = 1, \ldots, n$, and assuming the data are separable, means that there are two parallel hyperplanes that separate the two classes and can be generically written as

$$\langle w, x \rangle + b = 1 \quad \text{and} \quad \langle w, x \rangle + b = -1.$$

The normal vector w and the offset b are common to the two hyperplanes. The distance between the two hyperplanes is reflected in the class labels, ± 1. It is a geometric fact that the distance between these two hyperplanes is $2/\|w\|$. So, maximizing the distance between the two hyperplanes, i.e., maximizing the margin, is equivalent to finding the minimal w for which separating hyperplanes exist.

Maximizing the margin must be done subject to preventing any (y_i, x_i) from landing in the margin. That is, the constraints

$$\langle w, x_i \rangle + b \geq \begin{cases} \geq 1 & y_i = 1, \\ \leq -1 & y_i = -1 \end{cases}$$

must be satisfied. These constraints can be concisely expressed as $y_i(\langle w, x_i \rangle + b) \geq 1$, for all i because y_i must have the same sign as $\langle w, x_i \rangle + b$, $|y_i| = 1$, and $|\langle w, x_i \rangle + b| \geq 1$. The problem to be solved now is

$$\max_{w} \quad \|w\|$$
$$\text{subject to } \forall i \ y_i(\langle w, x_i \rangle + b) \geq 1. \tag{10.64}$$

For convenience, it is just as well to maximize $\|w\|^2/2$, since it is usually differentiable and has a convenient derivative. Then, since a minimum exists, Lagrange multipliers can be used to find it. The Lagrangian is

$$L(w, b, \lambda) = \frac{\|w\|^2}{2} - \sum_{i=1}^{n} \lambda_i (y_i (\langle w, x_i \rangle + b) - 1) + \sum_{i=1}^{n} \lambda_i,$$

where the last term comes from the constraint $\lambda_i \geq 0$ for all i. Taking the derivatives with respect to b and the entries in w leads to the equations

$$\sum_{i=1}^{n} \lambda_i y_i = 0 \quad \text{and} \quad w = \sum_{i=1}^{n} \lambda_i y_i x_i. \tag{10.65}$$

The x_i for which $\lambda_i > 0$ are called support vectors; not all x_i, indeed typically few relative to the number n, are support vectors. Derivatives with respect to the λ_i give

$$y_i (\langle w, x_i \rangle + b) - 1 = 0,$$

i.e.,

$$y_i (\langle w, x_i \rangle + b) = 1 \quad \Rightarrow \quad \langle w, x_i \rangle + b = \frac{1}{y_i} = y_i \quad \Rightarrow \quad b = y_i - \langle w, x_i \rangle.$$

While the last expression is mathematically true, it is more common to estimate b by the average

$$\hat{b} = \frac{1}{n_s} \sum_{i=1}^{n_s} (y_i - \langle w, x_i \rangle)$$

where the x_i have been relabeled so that the first $n_s < n$ of them are the support vectors. So, it remains to reformulate L to take account of the conditions (10.65). The result, cf. Burges (1998, expression (16)), is

$$L_{\text{ref}} = \sum_{i=1}^{n} \lambda_i - \frac{1}{2} \sum_{i=1}^{n} \lambda_i \lambda_j y_i y_j \langle x_i, x_j \rangle. \tag{10.66}$$

For geometric reasons, the maximization in (10.64) becomes a minimization in L_{ref} subject to $\lambda_i \geq 0$ for all i and the constraints in (10.65). Minimizing L_{ref} is a quadratic programming problem that has been amply discussed in a variety of places, e.g., Bottou and Lin (2007). In practice there is a variety of programs that give solutions of reasonable quality, e.g., kernlab in R (R Core Team 2016).

So, given that a solution for w and b has been found, (10.63) can be used to generate predictions in the case where the data are separable by a hyperplane in the inner product space $\mathcal{H}$. The boundary between the two classes is the centerline of the maximal margin and in fact only the support vectors matter. If the rest of the data were omitted from the computations, (10.63) would not be changed.

Support vector classification has two generalizations that are essential. First, the data might not be separable at all. Second, the boundary between two classes might be better expressed if the domain of the x's were transformed, for instance by Φ, so that better class separation could be achieved. Rather than discussing these two cases in any detail, here it is enough to record the problem and solution for each.

First, when the data from the two classes overlap, the optimization problem is harder. However, it can be formulated similarly to (10.64). The difference is that 'slack' variables

$\xi_i \geq 0$ for $i = 1, \ldots, n$ have to be introduced to minimize the distance between a point x_i inside the margin to the edge of the margin associated with its class. (The concept of slack variables is ubiquitous in optimization theory. They arise as a technique to convert an inequality constraint into an equality constraint and a nonnegativity constraint.) The margin boundaries now correspond to the inequalities

$$\langle x_i, w \rangle + b \geq 1 - \xi_i \quad \text{when} \quad y_i = 1,$$
$$\langle x_i, w \rangle + b \leq -1 + \xi_i \quad \text{when} \quad y_i = -1.$$

As derived in Burges (1998), the optimization problem can be expressed as

$$\max_{\lambda} \left(\sum_{i=1}^{n} \lambda_i - \frac{1}{2} \sum_{i=1}^{n} \lambda_i \lambda_j y_i y_j \langle x_i, x_j \rangle \right) \tag{10.67}$$

subject to

$$\sum_{i=1}^{n} \lambda_i y_i = 0 \quad \text{and} \quad \forall i \; 0 \leq \lambda_i \leq C.$$

This is a quadratic optimization problem (with a more complicated Lagrangian requiring the Karush–Kuhn–Tucker conditions for its solution.). The result is that, as in the separable case, the optimal hyperplane is defined by $w = \sum_{i=1}^{n_s} \lambda_i y_i x_i$ and b is estimated by $\hat{b}$. So, given $\hat{w}$ and $\hat{b}$, $\hat{Y}(x_{\text{new}})$ in (10.63) can be found and used as an optimal classifier.

The most powerful setting for frequentist SVM classification occurs when the optimal boundary between the two classes is not linear in the inner product of $\mathcal{H}$. This difficulty is overcome using the 'kernel trick' – the inner product in $\mathcal{H}$ is replaced by choosing a transformation $\Phi \colon \mathbb{R}^d \to \mathcal{H}$, letting $k(x_i, x_j) = \langle \Phi(x_i), \Phi(x_j) \rangle$, and using k in place of $\langle \cdot, \cdot \rangle$ in the solution. In fact, by specifying $k(x_i, x_j)$ the transformation Φ does not even have to be identified. The way in which this is implemented is by looking at (10.66) and (10.67), replacing $\langle x_i, x_j \rangle$ with $k(x_i, x_j)$, and repeating the optimization procedure. After some work, (10.63) becomes

$$\hat{Y}(x) = \text{sign}\left(\sum_{i=1}^{n_s} \lambda_i y_i k(x_i, x) + b \right),$$

where the λ_i are found by quadratic optimization, b is estimated as before, and the first n_s of the x_i are the support vectors corresponding to the λ_i that are strictly positive. Much is known about SVMs for classification – they can be extended to multiclass settings, their complexity can be assessed by the Vapnik–Chervonenkis dimension, they can be computed effectively, and they often give some sparsity because $n_s < n$. They also have limitations in that it is unclear how to choose the 'right' kernel' (although this applies to all kernel methods), and typically they do not work very well with discrete data. For present purposes, however, it is enough to use the expressions they give as point predictors for class labels.

Frequentist SVMs, like their Bayes counterparts, also have a regression form. In parallel to SVMs for classification, there are three cases, the first being to find a hyperplane in an inner product space that represents an unknown function well, that is, to find w and b such that

$$\hat{Y}(x) = \langle w, x \rangle + b$$

provides a good description of the data (y_i, x_i) for $i = 1, \ldots, n$. By reasoning similar to (10.64), the optimization problem is

$$\max_{w} \quad \frac{1}{2}\|w\|^2 \tag{10.68}$$

subject to $\forall i \ y_i - \langle w, x_i \rangle - b \leq \epsilon$

and $\quad \forall i \ \langle w, x_i \rangle + b - y_i \leq \epsilon.$

The constraints in the optimization reflect the fact that the optimization is done using an ϵ-insensitive loss (see (10.58)) written as $L(y, f(x, w)) = \max |y - f(x, w)| - \epsilon, 0)$.

The problem with (10.68) is that it assumes that a function of the form $\langle w, x_i \rangle + b$ approximates all pairs (y_i, x_i) to within ϵ. If ϵ is chosen small relative to $\mathrm{Var}(Y)$ then this will be unlikely, and it will not be possible to find a solution to (10.68). This happens often enough that it is better to focus on a generalization of (10.68) that permits data to lie outside the ϵ-half-width margin; see Smola and Scholkopf (2004, Sec. 1.2). This problem can be solved by introducing slack variables, say ξ_i and ξ_i^*, two for each data point because a data point can be either above or below the regression plane. Doing this, (10.68) becomes

$$\max_{w} \left(\frac{1}{2}\|w\|^2 + C \sum_{i=1}^{n} (\xi_i + \xi_i^*) \right) \tag{10.69}$$

subject to $\quad \forall i \ y_i - \langle w, x_i \rangle - b \leq \epsilon + \xi_i$

and $\quad \forall i \ \langle w, x_i \rangle + b - y_i \leq \epsilon + \xi_i^*$

and $\quad \xi_i, \xi_i^* \geq 0.$

The constant C determines the size of deviations from the regression function and the true function that will be tolerated. Again, the effect of the ϵ-insensitive loss is seen in the constraints.

To solve the optimization problem (10.69) we begin by writing down the Lagrangian:

$$L = \frac{1}{2}\|w\|^2 + C\sum_{i=1}^{n}(\xi_i + \xi_i^*) - \sum_{i=1}^{n}(\eta_i \xi_i + \eta_i^* \xi_i) \tag{10.70}$$

$$- \sum_{i=1}^{n} \lambda_i (\epsilon + \xi_i - y_i + \langle w, x_i \rangle + b)$$

$$- \sum_{i=1}^{n} \lambda_i^* (\epsilon + \xi_i^* + y_i - \langle w, x_i \rangle - b),$$

in which the η_i, η_i^*, λ_i, and λ_i^* are Lagrange multipliers and are required to be nonnegative; see Smola and Scholkopf (2004, Sec. 1.3).

At a point of optimality, the derivatives of L must be zero, to wit:

$$\frac{\partial L}{\partial b} = \sum_{i=1}^{n}(\lambda_i^* - \lambda_i) = 0$$

$$\frac{\partial L}{\partial w} = w - \sum_{i=1}^{n}(\lambda_0 - \lambda_i^*)x_i = 0$$

$$\frac{\partial L}{\partial \xi_i} = C - \lambda_i - \eta_i = 0$$

$$\frac{\partial L}{\partial \xi_i^*} = \lambda_i^* - \eta_i^* = 0.$$

(The derivative with respect to w is actually d-dimensional but this is suppressed in the notation.) Substituting these four equations into the Lagrangian gives

$$\max_{\lambda, \lambda^*} \left(-\frac{1}{2} \sum_{i,j}^{n} (\lambda - \lambda_i^*)(\lambda_j - \lambda_j^*) \langle x_i, x_j \rangle - \epsilon \sum_{i=1}^{n} (\lambda_i + \lambda_i^*) + \sum_{i=1}^{n} y_i (\lambda_i - \lambda_i^*) \right) \quad (10.71)$$

$$\text{subject to} \quad \sum_{i=1}^{n} (\lambda_i - \lambda_i^*) = 0 \text{ and } \forall i \; \lambda_i, \lambda_i^* \in [0, C],$$

where $\lambda = (\lambda_1, \ldots, \lambda_n)$ and $\lambda^* = (\lambda_1^*, \ldots, \lambda_n^*)$. This optimization can be solved computationally; see Smola and Scholkopf (2004).

For the present, suppose that a solution exists. Then, from $\partial/\partial w(L) = 0$, we have $w = \sum_i (\lambda_i - \lambda_i^*)$ and hence the optimum is of the form

$$Y(x_{\text{new}}) = \sum_{i=1}^{n} (\lambda_i - \lambda_i^*) \langle x_i, x_{\text{new}} \rangle + b. \quad (10.72)$$

The offset b can also be estimated, but this requires a careful examination of the constraints; see Smola and Scholkopf (2004). Thus, if estimates of λ_i, λ_i^*, and b are found, the regression function in (10.72) can be specified and used to give point predictors. One feature of this predictor is that it has some built-in sparsity. Only when $|f(x_i) - y_i| \geq \epsilon$ will the corresponding Lagrange multipliers be nonzero. The x_i with $|f(x_i) - y_i| < \epsilon$ do not contribute to the regression function. That is, like SVM classification, SVM regression depends on a small number of points that may not be very stable although on average (in some sense) they should give a good performance.

Nothing is changed in the optimization if the kernel trick (replacing the inner product with $\langle \Phi(x), \Phi(x) \rangle$ for some $\Phi \colon \mathcal{X} \to \mathcal{H}$) is used. Then, (10.72) becomes

$$Y(x_{\text{new}}) = \sum_{i=1}^{n} (\lambda_i - \lambda_i^*) \langle \Phi(x_i), \Phi(x_{\text{new}}) \rangle + b = \sum_{i=1}^{n} (\lambda_i - \lambda_i^*) k(x_i, x_{\text{new}}) + b, \quad (10.73)$$

with the appropriate parameter estimates. This gives SVM regression a great deal of flexibility and in practice can lift the d-dimensional data points into a higher-dimensional space, making them easier to fit. There may be a cost due to instability in higher dimensions, but the optimization is valid. So, again (10.73) gives point estimates. In order to get $1 - \alpha$ PIs from (10.72) or (10.73), an error structure must be assumed. Since it is difficult to do so explicitly with error terms, the nonparametric approach to generating PIs described at the end of Sec. 10.1 may be more effective.

In general, the contrast between RVMs and SVMs (for regression or classification) is hard to make reliably. However, there are some patterns. First, RVMs tend to rely on fewer support vectors than SVMs. The reason is partly that the Bayes formulation has some inbuilt sparsity and partly because it is more common to impose sparsity conditions on RVMs than on SVMs. Both RVMs and SVMs can be unstable even though they are optimal. This is so because they rely heavily on a small number of points that may be outliers or inliers. The idea behind SVMs is to map out a boundary between classes – so SVM support vectors behave like boundary points. The idea behind RVMs is to find points that are prototypical for an exhaustive collection of disjoint subsets of the data, so that RVM support vectors tend to behave somewhat like cluster centers, summarizing locations where points accumulate. Both SVMs and RVMs are optimal, and in principle give good solutions – at the possible cost of stability in the sense that a different data set from the same DG might give a different SVM or RVM. In Chapter 11, techniques to stabilize good predictors will be discussed, thereby partly remedying this problem.

To conclude this section, we continue the animal analogy at the end of Sec. 10.3. If linear models are donkeys, NNs are snakes, and trees are foxes, then kernel methods are goats. They can handle whatever you throw at them; they tend to have an independent streak but are often robust across settings. They can be hard to train because default settings are often ineffective, and even when trained they can be unpredictable. However, they are determined creatures: they will find optima whether you like them or not.

10.5 Penalized Methods

Penalized methods can be regarded as a variation on kernel methods because the objective functions for the two settings are so similar. However, that is where the similarity ends, because penalized methods do not invoke an RKHS and do not permit x's to appear in the loss function (outside the model). In addition, the penalized expression need not be related to a norm of any sort but always has a decay parameter λ in it to control the tradeoff between the empirical risk and the complexity of the model. One common form for a penalized method is

$$\min_{\beta} \left(\sum_{i=1}^{n} L(y_i, x_i^T \beta) - \lambda \Omega(\beta) \right), \tag{10.74}$$

where L is a loss (almost always taken to be the squared error), (x_i, y_i) for $i = 1, \ldots, n$ are assumed independent, x is d-dimensional, $\beta \in \mathbb{R}^d$, $\lambda > 0$, and Ω is the penalty term. In this sense, (10.74) is similar to the MSPs of Sec. 9.2. However, these penalized methods are different from those in Sec. 9.2 because the penalty is intended to represent complexity and the tradeoff between complexity and empirical risk is most important when $d \gg n$, i.e., the model-selection problem includes so many candidate models that the problem is regarded as $\mathcal{M}$-open rather than $\mathcal{M}$-closed and hence model uncertainty is likely to be the dominant form of uncertainty.

Penalized methods are often called shrinkage methods because the penalty term tends to force parameter estimates to be closer to zero than they would otherwise be. That is, these methods tend to shrink the coefficients toward zero. Thus, shrinkage as defined here is qualitatively similar to James–Stein shrinkage even though the mathematics is quite different. An exception to this is the intercept parameter β_0 in β. Although suppressed in the notation, Y is almost always centered, so that the intercept parameter in β can be taken as zero. If this is not done then the penalization of β_0 would make the solution depend on the origin of Y – an undesirable property.

One of the earliest examples of a penalized method is ridge regression, described at the beginning of Sec. 10.4. It arises if Ω in (10.74) is taken as $\|\beta\|^2$. This can be regarded as a penalty on models that have too many $\beta_j \neq 0$ and is merely one of many formalizations of the concept of complexity. To use ridge regression (or any other penalized method), λ must be chosen and there are usually many ways to do this. One obvious choice is to find the value of λ with minimal cross-validation error. A different way to choose λ is to define the degrees of freedom df(λ) associated with β. This is possible because ridge regression is linear (in the sense used in Chapter 8). As a function of λ,

$$\mathrm{df}(\lambda) = \mathrm{tr}(X(X^T X + \lambda I_d)X^T) = \sum_{j=1}^{d} \frac{\tau_j^2}{\tau_j^2 + \lambda},$$

where the τ_j are the singular values of X. Since there is a one-to-one relationship between λ and df, it is possible to choose the number of degrees of freedom that seems appropriate to a given problem and solve for λ.

Write $\beta = \beta_\lambda$ to indicate the dependence of β on λ. Then, if β and λ are estimated by $\hat\beta$ and $\hat\lambda$, the obvious point predictor at x_{new} is

$$\hat{Y}(x_{\mathrm{new}}) = x_{\mathrm{new}}^T \hat\beta_{\hat\lambda}. \tag{10.75}$$

To get a PI at x_{new}, suppose that L is the squared error and observe that, for each fixed λ, $\mathrm{Var}(\hat\beta_\lambda) = \sigma^2 (X^T X + \lambda I_d)^{-1} X^T X (X^T X + \lambda I_d)^{-1}$. So, neglecting the variability due to $\hat\lambda$, a naive $(1 - \alpha)$ PI for $Y(x_{\mathrm{new}})$ is

$$\hat{Y}(x_{\mathrm{new}}) \pm z_{1-\alpha/2} \hat\sigma \sqrt{1 + \mathrm{Var}(\hat\beta_{\hat\lambda})},$$

where $\hat\sigma$ is the root residual sum of squares. Essentially, this reflects the assumption that Y and the regression function, in this case $x^T \beta$, differ by an $N(0, \sigma^2)$ error term. If L is not the squared error, no convenient expression for $\mathrm{Var}(\hat\beta)$ exists. So, either the variance would have to be approximated numerically or a PI would have to be found by bootstrapping, as described at the end of Sec. 10.1.

More generally, expression (10.74) is simultaneously Bayes and frequentist. Indeed, (10.74) is proportional to the log of the density:

$$p(Y|\beta, \lambda) \propto e^{\lambda \Omega(\beta)} \prod_{i=1}^{n} e^{-L(y_i, f(x_i|\beta))}, \tag{10.76}$$

in which the penalty term corresponds to a prior on β, λ is a hyperparameter, and $f(x|\beta)$ is the regression function for a given β. So, for fixed λ, minimizing (10.74) to get $\hat\beta$ is equivalent to maximizing (10.76) over β, which in turn is the same as finding the posterior mode of $w(\beta|\mathcal{D}, \lambda)$, i.e., the Bayes estimator under classification loss, where $\mathcal{D}$ is the data. For any L, this would give the same point predictor as (10.75), assuming λ were estimated in the same way in both cases.

The empirical Bayes $1 - \alpha$ PI would be derived from the predictive density

$$\int p(y_{\mathrm{new}}|x_{\mathrm{new}}, \beta, \hat\lambda) w(\beta|\mathcal{D}, \hat\lambda) d\beta,$$

possibly by choosing the highest posterior density region. A fully Bayes treatment would put a prior on λ as well. Thus, in principle, the posterior $w(\beta, \lambda | \mathcal{D})$ could be found and a $1 - \alpha$ PI would be derived from

$$\int p(y_{\text{new}} | x_{\text{new}}, \beta, \lambda) w(\beta, \lambda | \mathcal{D}) d\beta d\lambda.$$

As long as n is relatively large, the two Bayes PIs and the frequentist PI should be very close for each α when L is the squared error. Indeed, on general principles, all three should asymptotically be the same for fixed d when $n \to \infty$, for all well-behaved loss functions and penalties or priors.

One form of the optimal behavior of penalized or shrinkage methods with increasing n is summarized in the oracle property; see Fan and Li (2001) and Zou (2006). Suppose that the model $y_i = x_i^T \beta + \epsilon_i$ is to be fitted, where the ϵ_i are IID mean zero and variance σ^2 and that the explanatory variables in x belonging to the model have indices in $\mathcal{A} = \{j | \beta_j \neq 0\}$, i.e., have nonzero coefficients, with $\#(\mathcal{A}) = r < d$. Let $\eta > 0$ and let $\hat{\beta}$ denote an estimator of β. Denote $\hat{\mathcal{A}}_\eta = \{j | |\hat{\beta}_j| > \eta\}$. Then $\hat{\beta}$ has the oracle property if and only if it satisfies

1. $\hat{\mathcal{A}}_\eta \to \mathcal{A}$ in probability for some $\eta > 0$, and
2. $\sqrt{n}(\hat{\beta}_\mathcal{A} - \beta_\mathcal{A}) \to N(0, \Sigma)$ in distribution, where Σ is the variance matrix of the correct subset model and the subscript $\mathcal{A}$ indicates which entries are selected.

Both items hold as $n \to \infty$ and, in the second item, the subscript used is the correct $\mathcal{A}$, not its estimate $\hat{\mathcal{A}}$, although by item 1 these are asymptotically equivalent.

The oracle property is a sparsity condition – relatively few of the d explanatory variables appear in the true model and they can be found, at least approximately, by using $\hat{\beta}$. In practice, only the large entries of β are retained and a common choice is to discard $\hat{\beta}_j$ that have absolute values less than a cutoff such as $z_{1-\alpha/2}\hat{\sigma}/\sqrt{n}$. Of course, this amounts to doing d hypothesis tests of the form $\mathcal{H}_0$: $\beta_j = 0$, so when $d \gg n$ the usefulness of the method is open to question. In fact, rather than estimating λ or using a cutoff, many analysts use decreasing λ values to order the x_j for inclusion in the model. Since $\lambda = \infty$ usually means that no x_j are included, the first x_j to be allowed in the model as λ decreases is often taken as the most important. Note that as λ is allowed to decrease more x_j are included, so that λ can be chosen to allow the x_j into the model that seem most reasonable.

Most importantly, in the oracle property d and r are fixed; only n increases. The problem is that, as a generality, in cases where sparsity is important $d \gg n$, so asymptotics in d are more important; see Fan and Lv (2008) and the discussion at the end of Sec. 9.1. Thus, the oracle property is best understood as the sense in which an $\mathcal{M}$-complete problem is approximated by an $\mathcal{M}$-closed problem with d variables, and the $\mathcal{M}$-complete problem has an $\mathcal{M}$-closed solution with r variables.

There are many criteria of the form (10.74) that satisfy the oracle property. Two of the best known use $\Omega(\beta) = \sum_{j=1}^d w_j |\beta_j|$, i.e., the penalty corresponding to using a Laplace prior called the adaptive LASSO (ALASSO) and the Ω corresponding to the smoothly clipped absolute deviation (SCAD), the penalty that starts out as linear and then becomes flat. Smaller penalties Ω distort the parameter estimates less than larger penalties do, so SCAD often gives results that perform well. Both the ALASSO and SCAD satisfy the oracle property; the hypotheses are important to note because they provide intuition about how the methods will behave.

First, the ALASSO is based on the least absolute shrinkage and selection operator (LASSO), see Tibshirani (1996), which corresponds to using a double Laplace prior on the β_j or equivalently taking Ω to be an L^1 penalty on the vector β. Unlike the ALASSO, LASSO does not satisfy the oracle property and can be inconsistent for variable selection. To define the ALASSO, start with an estimator $\hat{\beta}$ for which there is a sequence $\langle a_n \rangle$ with $a_n \to \infty$ such that $a_n(\hat{\beta} - \beta) = \mathcal{O}_P(1)$. Then, choose $\gamma > 0$ and let $\hat{w} = (\hat{w}_1, \ldots, \hat{w}_d) = (1/|\hat{\beta}_1|^\gamma, \ldots, 1/|\hat{\beta}_d|^\gamma)$. The ALASSO estimator is

$$\hat{\beta}_\lambda = \arg\inf_\beta \left(\|Y - X\beta\|^2 + \lambda \sum_{j=1}^d \hat{w}_j |\beta_j| \right)$$

and the natural estimator of $\mathcal{A}$ is $\hat{\mathcal{A}} = \{j \mid |\hat{\beta}_j| > \eta\}$, where η is smaller than any of the true β_j. In some cases η is set to zero, so $\hat{\mathcal{A}} = \{j \mid |\hat{\beta}_j| \neq 0\}$. The key theorem about ALASSO is the following.

Theorem 10.3 (Zou 2006) *Assume that the $x_j = (x_{j1}, \ldots, x_{jn})$ are linearly independent explanatory variables and that $Y_i = x_i^T \beta + \epsilon_i$, where $x_i = (x_{1i}, \ldots, x_{di})$ and the ϵ_i are IID mean zero with variance unity for $i = 1, \ldots, n$. Write $X = [x_1, \ldots, x_n]^T$ as the matrix of explanatory variables and assume that $(1/n)X^T X \to C$, where C is a positive definite $d \times d$ matrix. Then, if $\lambda = \lambda_n$, $\lambda_n/\sqrt{n} \to 0$ and $a_n^\gamma \lambda_n/\sqrt{n} \to \infty$, the ALASSO parameter estimators satisfy the oracle property. In this case $\Sigma = C_{11}^{-1}$, the upper left-hand $r \times r$ block of C.*

In addition to the oracle property, ALASSO, like LASSO, SCAD (see below), and many other shrinkage methods, permits the derivation of a closed-form expression for estimates $\hat{\beta}_j$ of the β_j. At some stage in an analysis, e.g., when checking stability, the values of the $\hat{\beta}_j$ may be important, but here the focus is on prediction and $\hat{\beta}_j$ is simply the change in the predictor $\hat{Y}$ for a unit change in x_j, assuming that the effect of the penalty on the $\hat{\beta}_j$ is tolerable.

The trick that makes the ALASSO work, aside from the asymptotics in λ_n, is that the weights $\hat{w}_j$ increase to infinity when $\beta_j = 0$ but tend to a finite limit when $\beta_j \neq 0$. The result is that, in principle, the zero and nonzero β_j can be separated. The pseudo-estimator $\hat{\beta}$ is a technical device needed for the proof. Obviously, the usual least squares estimator can be used if it exists, but shrinkage methods are intended for the case $d \gg n$ where least squares estimators are ineffective. Guidance on choosing $\hat{\beta}$ and γ is scant. Nevertheless, the obvious point predictor, whether ALASSO is regarded as a frequentist penalized likelihood procedure or as a Bayes procedure, is $\hat{Y}(x_{\text{new}}) = x_{\text{new}}^T \hat{\beta}_\lambda^*$, where $\hat{\beta}_\lambda^*$ is the vector of estimates $\hat{\beta}_j$ that are not too small and hence set to zero.

To get a frequentist $1 - \alpha$ PI, one naive way is to form intervals using $\hat{\beta}_\lambda^* \sim N(0, \sigma^2 C_{11}^{-1})$, where σ^2 is estimated by $\hat{\sigma} = \sqrt{\text{MSE}}$. Thus, an approximately $1 - \alpha$ PI is found by writing

$$Y(x_{\text{new}}) - \hat{Y}(x_{\text{new}}) = (Y(x_{\text{new}}) - x_{\text{new}}^T \beta) + (x_{\text{new}}^T \beta - \hat{Y}(x_{\text{new}}))$$

and therefore obtaining

$$\hat{Y}(x_{\text{new}}) \pm z_{1-\alpha/2}\hat{\sigma}\sqrt{1 + x_{\text{new}}^T \hat{C}_{11}^{-1} x_{\text{new}}},$$

because $\hat{Y}(x_{\text{new}}) \sim N(x_{\text{new}}^T \hat{\beta}_{\hat{\lambda}}, \hat{\sigma}^2 \hat{C}_{11})$ where $\hat{C}_{11}$ is the approximation to C_{11} using the available data. A more sophisticated estimator for the SE of the β_j is described in Zou (2006, Sec. 3.6) and leads to PIs with a different form; it is not clear how close they are numerically to the naive approach. The Bayesian approximately $1 - \alpha$ PIs follow from forming the predictive density for $Y(x_{\text{new}})$ using the analog of (10.76).

To get a Bayes $1 - \alpha$ PI, one obvious, but possibly naive, way is to form highest predictive density intervals from

$$\int \phi(y(x_{\text{new}})|\beta)w(\beta|\mathcal{D}, \hat{\lambda})\mathrm{d}\beta,$$

where $\phi(x|\beta)$ is an $N(x^T \beta, \sigma^2)$ density and $\hat{\lambda}$ is chosen by some minimal-error criterion such as cross-validation. More pragmatically, β can be replaced by β^* and the integral approximated by a plug-in procedure. A fully Bayes procedure, as opposed to an empirical Bayes procedure, would put a prior on λ.

Although SCAD came before ALASSO, it was intended as an alternative to LASSO. The motivation for SCAD, see Fan and Li (2001), was to achieve three goals: (i) the near unbiasedness of the estimators for β, i.e., that the penalty should be small; (ii) sparsity, in the sense that there is a natural thresholding rule that sets small coefficients to zero, i.e., the penalty should have 'corners' on the axes of the parameter space; and (iii) continuity, in the sense that the model selection and estimation of β should not be unstable as a function of the data.

Given a log-likelihood function ℓ, typically obtained from the product of normals and hence equivalent to the squared error loss, SCAD minimizes

$$Q(\beta) = \sum_{i=1}^{n} \ell(y_i, x_i^T \beta) - n \sum_{j=1}^{d} p_\lambda(|\beta_j|), \tag{10.77}$$

where each term in the penalty is of the form

$$p_\lambda(|\beta_j|) = \begin{cases} \lambda|\beta_j| & \text{if } |\beta_j| \le \lambda, \\ -\dfrac{(|\beta_j|^2 - 2a\lambda|\beta_j| + \lambda^2)}{2(a-1)} & \text{if } \lambda < |\beta_j| \le a\lambda, \\ \dfrac{(a+1)\lambda^2}{2} & \text{if } |\beta_j| > a\lambda. \end{cases}$$

That is, if a parameter β_j gets large then it is truncated at $a\lambda$, but near zero the penalty is linear. The intermediate case smoothly interpolates between the linear part of the penalty and the flat part of the penalty. There is no requirement that the hyperparameters a and λ be the same for all j although, for simplicity, here it will be assumed that a and λ are independent of j. In practice, if there were reasons to believe that some variables were more important to retain in the model than others, adjusting a and λ to ensure this might be reasonable.

Establishing the predictively important properties of SCAD requires two steps. Recall that $p(y|x, \beta)$ is the density of Y for design point x and parameter β. Assume that the third derivatives of $p(\cdot|x, \beta)$ with respect to entries in β are locally (in β) dominated by an integrable function (in Y). Write the true β as $\beta_0 = (\beta_{10}, \ldots, \beta_{d0})^T = (\beta_{10}^T, \beta_{20}^T)^T$, where all the β_j in β_{20} are zero and all the β_j in β_{10} are nonzero. Then, Fan and Li (2001) established that the $\hat{\beta}_j$ exist and satisfy the oracle property. More formally:

Theorem 10.4 (Fan and Li (2001))

(I) *Suppose that*

$$\max\{p''_{\lambda_n}(|\beta_{j0}|)|\beta_{j0} \neq 0\} \to 0$$

as $n \to \infty$; then there exists a local maximum $\hat{\beta}$ of $Q(\cdot)$ with the property that

$$\|\hat{\beta} - \beta_0\| = \mathcal{O}_P\left(\frac{1}{\sqrt{n}} + a_n\right),$$

where $a_n = \max\{p'_{\lambda_n}(|\beta_{j0}|)|\beta_{j0} \neq 0\}$.

(II) *Assume that*

$$\liminf_{n\to\infty} \liminf_{\eta\to0^+} \frac{p'_{\lambda_n}(\eta)}{\lambda_n} > 0.$$

Then, if $\lambda_n \to 0$ and $\sqrt{n}\lambda_n \to \infty$ as $n \to \infty$, the $\hat{\beta}$ from clause I satisfies

(i) *Sparsity: $\hat{\beta}_{02} = 0$, and*

(ii) *Asymptotic normality and efficiency:*

$$\sqrt{n}(I_1(\beta_{10}) + \Sigma)\left(\hat{\beta}_{10} - \beta_{10} + (I_1(\beta_{10}) + \Sigma)^{-1}b\right) \to N(0, I_1(\beta_{10}))$$

in distribution, where $I_1(\beta_{10}) = I(\beta_{10}, 0)$, the Fisher information for β when $\beta_{20} = 0$, s is the number of nonzero β_j,

$$\Sigma = \mathrm{diag}(p''_{\lambda_n}(|\beta_{10}|), \ldots, p''_{\lambda_n}(|\beta_{s0}|))$$

and

$$b = (\mathrm{sign}(\beta_{10})p'_{\lambda_n}(|\beta_{10}|), \ldots, \mathrm{sign}(\beta_{s0})p'_{\lambda_n}(|\beta_{s0}|)).$$

Given this, the natural frequentist point predictor is $\hat{Y}(x_{\mathrm{new}}) = x_{\mathrm{0new}}^T\hat{\beta}_{01}$ if appropriate values $\hat{\lambda}$ and $\hat{a}$ are chosen. This is essentially the same as the Bayes point predictor under classification loss, the mode of the posterior formed by exponentiating (10.77), i.e., regarding $\prod_{j=1}^d \exp p_\lambda(|\beta_j|)$ as the prior.

As noted in Fan and Li (2001), under appropriate thresholding, if $\lambda_n \to 0$, i.e., $\hat{\lambda}$ is chosen to be small, and a is chosen well, clause II of the theorem gives that the asymptotic variance of $\hat{\beta}_{10}$ is approximately $I_1^{-1}(\beta_{10})/n$, which can be estimated by $I_1^{-1}(\hat{\beta}_{10})$. Thus, an approximately $1 - \alpha$ frequentist PI is given by

$$x_{\mathrm{0new}}^T\hat{\beta}_{01} \pm z_{1-\alpha/2}\hat{\sigma}\sqrt{1 + I_1^{-1}(\hat{\beta}_{10})/n}.$$

Bayesian PIs follow in a way similar to those found for ALASSO. That is, naively,

$$\int p(y_{\mathrm{new}}|x_{\mathrm{new}}, \beta)w(\beta|\mathcal{D}, \hat{a}, \hat{\lambda})d\beta$$

can be used to generate an empirical Bayes highest predictive density region where the hyperparameters a and λ have been estimated by $\hat{a}$ and $\hat{\lambda}$ (or chosen by the user) and $p(\cdot|x_{\text{new}}, \beta)$ is the density for the next outcome. Indeed, $\hat{\beta}$ could be used in the density for y_{new} to give an even simpler plug-in method. However, in a fully Bayes analysis both a and λ would be assigned a hyperprior and then integrated out. As with the ALASSO, $w(\beta|\mathcal{D}, \hat{a}, \hat{\lambda})$ is the posterior formed from treating the exponential of the penalized log likelihood of (10.77) as the product of a likelihood (the ℓ's) and a prior (the p_λ).

Some general discussion of penalized or shrinkage methods is important. First, they are a way to achieve sparsity, but that is mostly because of the prior selection implicit in the choice of λ and Ω in (10.74). That is, the asymptotics are as $n \to \infty$ for fixed d. Like other methods they can probably be extended to permit $d = d_n$ to be slowly increasing with n, but that is not the point: Bayes methods in general and posterior densities in particular are consistent with increasing n under fairly mild conditions. So, even for poorly chosen λ's and Ω's the posterior would concentrate at the true value of β. The benefit given by penalized methods is to speed convergence to the true model by effective prior selection (Ω), effective hyperparameter selection (λ), and thresholding out β_j that are close to zero.

Second, penalized methods are parametric and in many cases can be used to good effect even when $n > p$. This means only that their solutions converge quickly in the sense that they approach their oracle-function limiting behavior even when n is small relative to p. This is the main reason why such methods are called sparse. In the absence of careful choices for Ω and λ, penalized methods would simply amount to standard Bayes.

Third, the quantity being penalized is a cumulative loss or a log-likelihood. For some parametric families the log-likelihood is a loss function. This is true for the normal, which gives the squared error, and for the Laplace (double exponential), which gives the absolute error. In either case, the parameters are multiplicative factors on the x_j because the models are additive, indeed, linear. However, this is not necessary. For instance, NNs are also typically penalized merely by assuming a normal-based prior on their β_j. The formal difference between Sec. 10.3 and the present section is that in an NN model class there are multiplicative parameters on the nodes as well as the x_j. In practice, there is a different motivation: the penalization of NNs is above all to achieve stability of model or predictor selection, whereas the penalization of a cumulative loss or likelihood is primarily to achieve *sparsity*. Hence the penalty term (or prior) and λ are chosen much more carefully for shrinkage methods. It is likely that, to some extent, the penalty in penalized methods does stabilize the solution, but this stability is not as important as with NNs because the additive model classes used with penalized methods are much smaller than those associated with NNs. Indeed, penalization is most typically applied to linear models that can suffer instability due to collinearity, a problem that is worsened by having many explanatory variables x_j. Hence, orthogonalization of the explanatory variables is often more important than the choice of penalty (or even λ, at least within a range) to decrease instability in penalized methods.

Fourth, some have argued that the good performance of penalized methods, at least in inference, is simply super-efficiency. This is implicit in Fan and Li (2001) but developed in detail in Leeb and Pötscher (2005) and Leeb and Pötscher (2008). The basic idea is that 'whatever the loss function, the maximal risk of a sparse estimator – in large samples – is as bad as it can possibly be.' While this is true, neither the oracle property nor its possible poor risk behavior addresses whether sparse estimators give poor predictions. Indeed, there is some evidence that the best penalized methods are comparable in predictive performance to some of the better model-averaging techniques, designed for prediction; these will be presented in Chapter 11. The issue is that sparse methods will give a good performance predictively when an $\mathcal{M}$-complete or large $\mathcal{M}$-closed problem can be well approximated by the leading terms of a (smaller) $\mathcal{M}$-closed problem.

Finally, amongst the more recent contributions to the general area of penalized methods is Ročková and George (2016), which explores the effect of hyperparameters in the 'spike' part of a spike-and-slab prior and includes cases where the hyperparameter and spike cannot be separated into two different penalty terms, one of which is familiar from LASSO. Thus, even though there is no magic in penalized methods, further work may find that in many cases they give generally good solutions for prediction.

10.6 Computed Examples

Although the techniques in this chapter are described as blackbox, it is possible to develop an intuition about how they work. The first computed examples in this section will use the techniques, described in the previous sections, for curve fitting with IID data generated from a known curve with known noise. The Doppler function is a good choice for this because it includes oscillations of various frequencies and amplitudes. It should be noted that, in practice, it is rare to use a single tree even though using single NNs, SVMs, RVMs and penalized methods is common. Here, see Eq. (10.78), a single tree will be used for comparability with the other methods. Commonly, however, ensemble methods, to be seen in Chapter 11, are used: They pool the results from many components and on average perform better than merely using any one component. In Chapter 11 multiple trees (often called random forests), NNs, SVMs, RVMs, and penalized methods will be combined in an effort to get improved prediction.

Second, an example with a much more complicated data structure will be shown. The goal is pragmatic: find the best predictor for a vegetative greenness index, as an assessment of vegetative growth at a given location. The greenness index is measured at various locations over a span of years, so the predictor will depend on data from other similar locations during the same and earlier years, on data that are independent of location but not year, and on data that depend on year but not location. The predictors developed are different from those in earlier examples, because they are adaptive in the sense that they are nontrivially updated each year.

10.6.1 Doppler Function Example

Consider the general Doppler function given by

$$g(x) = \sqrt{(x(1-x))} \sin((2\pi(1+e))/(x+e)), \tag{10.78}$$

where $x \in (0, 1)$ and the offset is denoted e. To generate data that would be representative of a Doppler signal, set $e = 0.05$ and assume that $Y(x) = g(x) + \epsilon$, where $\epsilon \sim N(0, 0.1^2)$. Choose 100 IID values of X, using a Unif[0,1] distribution, and 100 values of ϵ using an $N(0, 0.1^2)$. Now, the correct model for (Y_i, X_i) for $i = 1, \ldots, 100$ is $Y(x_i) = g(x_i) + \epsilon_i$.

Strictly speaking, since the X_i are random, the following analysis is conditional on $X^{100} = x^{100}$ since it treats the x_i as fixed design points.

For explanatory variables, choose the first ten Legendre polynomials (LPs), not counting the zeroth-order LP, which is a constant. The LPs are orthonormal on $(-1,1)$ under the L^2 norm. They are orthogonal on $(0,1)$ as well but only approximately normalized. This would correspond (roughly) to sphering the explanatory variables, i.e., transforming them so that the new covariance matrix is the identity matrix, rather than just Studentizing them. This means that the analysis should be relatively easy for the various methods, and their properties will be seen more easily. .

For illustration, consider fitting a single tree in R using rpart with constants in each leaf given by the average of the data points. The most common splitting criterion for this kind of data is to minimize SST − (SSL − SSR), where SST is the total sum of squares, treating the y_i as 100 outcomes of a single random variable and ignoring the explanatory variables; SSL and SSR are analogous sums of squares (also from the y_i) but for the y_i in the left and right nodes, respectively, in a binary split of one explanatory variable. This is equivalent to finding the binary split of an explanatory variable that maximizes the between-groups sum of squares, in the sense of a variance analysis. The stopping-splitting rule is the default which is a grow-and-prune strategy, as discussed in Sec. 10.2. The rpart procedure also can be used to generate a variable importance plot (seen in Sec. 1.2.3).

The default output of rpart is shown in Fig. 10.7. The left-hand panel shows the function (10.78) together with the data (joined by dashed lines) and the fitted curve (dotted lines). The tree diagram is shown in the right-hand panel. In this right-hand panel the fitted curve is nearly zero on $[0, 0.2]$. This is seen in the right-hand leaf on the left-hand side of the first split, i.e., the seventh LP is at least 0.493 and the tenth LP is at least 0.558. The data from the Doppler function in the left-hand panel is chaotic yet symmetric around zero, so that a tree that treats it as zero is a good approximation. This kind of damping to zero is seen with the larger oscillations as well. For instance, all three maxima starting from the right are underestimated by the tree; these are the values 0.358 and 0.314, the former holding for the two intervals where the tenth LP is bigger than 0.078. The analogous overestimation is

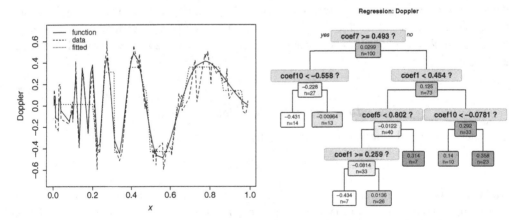

Figure 10.7 Doppler function with a tree. Left: Data, fitted curve, and actual function. Right: Diagram of the tree structure of the fitted curve.

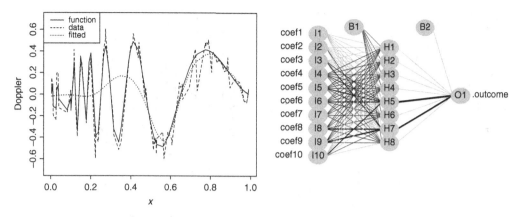

Figure 10.8 Doppler with NN. Left: Data, fitted curve, and actual function. Right:
Diagram of the node structure of the fitted curve.

true of the first two minima from the right (the tenth LP is less than -0.558 and the first
LP is greater than 0.259), but not the earlier minima, where there is slight underestimation.
Aside from the largest oscillation (on the right) the tree formed by using ten LPs is mostly
unable to match the regions of increase or decrease, essentially squaring them off into step
functions. A larger sample size would reduce (but not eliminate) this problem. The largest
residuals (in absolute value) are near local optima or in regions of high oscillations (in, say,
$[0, 0.2]$), where the noise dominates.

The same data can be used with NNs, and the result is somewhat different. As can be seen
in the left-hand panel of Fig. 10.8, the larger oscillations have better fits and it is only the
rightmost oscillation that is fitted well. As with trees, the NN is essentially flat on $[0,0.2]$.
Overall, the residuals are large where the oscillations are either too small and rapid (at the
left) or simply too rapid ($[0.2, 0.5]$). A larger sample size would enlarge the region where the
fitted curve fits well, but the problem near zero would remain. In this version of NNs, nnet
was used and it defaults to using an $N(0, 1)$ prior on the coefficients. This smooths them
and makes the result, formally, an empirical Bayes solution because the decay parameter
λ is estimated by $\hat{\lambda} = 0.1$, invoking a minimum root MSE (RMSE) criterion. Without the
penalty term, the fitted curve would be much rougher. Several NN architectures were tested.
Two had five nodes in the first hidden layer and either two or four nodes in the second hidden
layer. The third simply had eight nodes in the one and only hidden layer. It turned out that
the last was the best of the three (see Fig. 10.10.) It is plotted in the right-hand panel of Fig.
10.8. The thickness of the lines indicates the size of the coefficients and the color of the lines
(grey or black) indicates the sign (negative or positive) of the coefficients. Loosely speaking,
the first few nodes corresponding to the first three or four LPs do not contribute much to the
final output. The nodes B1 and B2 indicate bias – constants that were optimal to add to the
various nodes for better fit. They are analogous to the intercept terms in a linear regression
model.

The same data can be analyzed using kernel methods and RVM and SVM regression.
The left-hand panel in Fig. 10.9 shows the output of the RVM procedure in kernlab using
the radial basis function kernel, essentially a normal density with the value of σ estimated
internally to the program. Two relevance vectors are found, observations $x_{(76)}$ and $x_{(88)}$

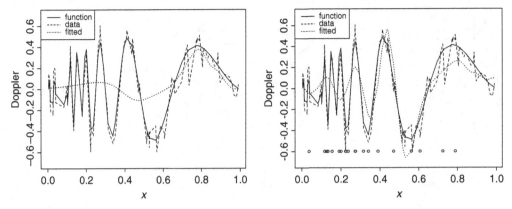

Figure 10.9 Doppler with radial basis function kernels. Left: RVM regression. Right: SVM regression under ϵ-insensitive loss; the support vectors are indicated along the bottom of the panel.

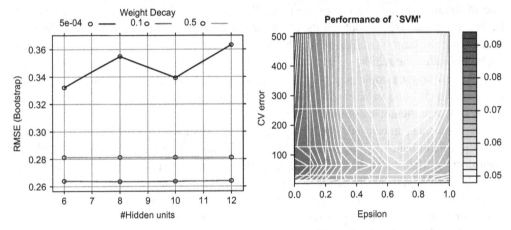

Figure 10.10 Left: Tuning to find the NN in Fig. 10.8 by selecting the number of nodes and decay parameter. Right: Tuning to find the SVM regression function in the right-hand panel of Fig. 10.9.

from the Unif[0,1], with values approximately 0.77 and 0.88, respectively. The fit is seen to be quite poor, even for the largest oscillation, because the Doppler is just not sparse as a function of LPs. The fitted curve essentially misses all features of the Doppler function until about 0.65, being relatively flat and insensitive to the data on $[0, 0.65]$. By contrast, the SVM regression fit under ϵ-insensitive loss with $\epsilon = 0.8$ has 22 support vectors and is comparatively quite good. The fit only really captures the three or four largest oscillations, but this is better than the other methods. Part of the reason for this is that the Doppler function is not sparse and SVMs impose nowhere near the sparsity of RVMs. Indeed, in the present example, SVM regression imposes less sparsity than the other four methods considered in this subsection.

It was noted in Secs. 10.3 and 10.4 that the implementation of NN and RVM procedures requires auxiliary quantities to be estimated. The left-hand panel of Fig. 10.10 shows a graph of the RMSE found using bootstrap samples for different numbers of hidden nodes and different values of λ. It is seen that the best value for λ is $\hat{\lambda} \approx 0.1$, since smaller and larger

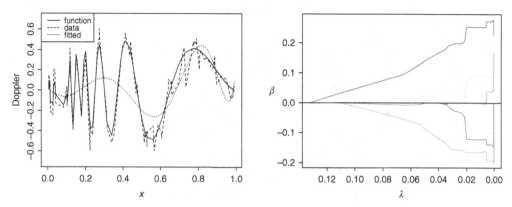

Figure 10.11 Doppler with SCAD. Left: Data, fitted curve, and actual function. Right: The β–λ plot showing how the coefficients β vary with λ.

values give higher RMSEs. Loosely speaking, any number of nodes between eight and 12 is reasonable, although this tuning procedure was only performed for a single-hidden-layer architecture. Since there is essentially no variance–bias tradeoff in terms of the number of nodes, for parsimony eight nodes were used in Fig. 10.8. The right-hand panel of Fig. 10.10 shows a graph of the ten-fold cross-validation error for different values of ϵ in the ϵ-insensitive loss and of C in (10.67) or (10.69). This coefficient C is denoted 'cost' on the vertical axis. The darker regions correspond to lower values of the error; $C \approx 512$ and $\epsilon \approx 0.8$ were used to generate the right-hand panel in Fig. 10.9.

As a fifth and final analysis of the Doppler simulated data with the first ten LPs as explanatory variables, consider using one of the more successful shrinkage methods, namely, smoothly clipped absolute deviation (SCAD). The left-hand panel in Fig. 10.11 shows the true function, the data, and the fitted curve. Unsurprisingly, SCAD does not do very well because although it is a successful shrinkage technique that imposes sparsity, the Doppler function is not sparse in the LP basis. The SCAD fit misses essentially all the structure of the Doppler, including the largest oscillation. The right-hand panel is a λ–β plot. That is, it shows the value of each β_j as λ increases. The first β that is not zero comes off at around 0.14; other β's come off for smaller values of λ, i.e., as the SCAD penalty gets less weight. In this example, the value of the decay parameter was found to be $\hat{\lambda} = 0.00013$ by minimizing the penalized sum of the squared errors. The final model has a nonzero intercept and all β's nonzero except for the coefficient for the ninth LP.

Visually, the best fit comes from SVM and the second best comes from trees. The RVM and NN fits come next and provide roughly the same quality of fit even though NN is not a sparse method. The SCAD method gives the weakest performance, possibly because it imposes more sparsity than RVMs in this example. The reader is invited to redo this analysis with the function $(\sin x)/x$ on $[-10, 10]$ and in addition to compare the results with those using nonregularized NNs, which will generally give a rougher fit than that seen in Fig. 10.8.

10.6.2 Predicting a Vegetation Greenness Index

To see how predictive ideas apply in a realistic setting, involving missing data and data for which Y depends differently on the different explanatory variables, consider the Vegout data

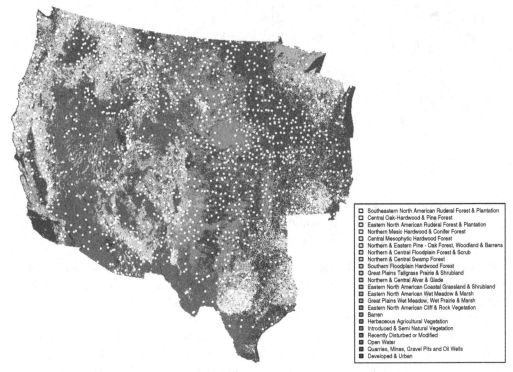

Figure 10.12 The white dots on the map indicate the locations of 1402 weather stations, but not all stations reported all their data. The exact type of ecological terrain is indicated in the legend to the right of the map.

set presented and analyzed in Harms *et al.* (2009). The variable to be predicted, Y, is the satellite-observed seasonal greenness (SSG6) at six weeks from the 13th biweek, in early July, of the calendar year for the 22 most western states of the USA. This region is shown in Fig. 10.12, shaded according to the ecological terrain; weather stations are indicated as dots. There are, in fact, many more weather stations than the map shows, as certain stations were eliminated from the analysis for various reasons. Some had not been in operation long enough, some had data that were so incomplete that it would be pointless to include them, some were no longer in operation, and some were too close to an urban area or body of water to provide a representative assessment of vegetation conditions. The SSG6 was standardized relative to the satellite-observed seasonal greenness at time zero (SSG0), measured during the 13th biweek of the calendar year. The data were gathered for $n = 18$ years, 1989–2006, so for each year SSG0 and SSG6 were available (although obviously SSG6 was not available at the time when the corresponding SSG0 was measured). Hence SSG0 can be used as an explanatory variable for predicting SSG6. In this example, the predictions serve as early warnings about future vegetation conditions. Both SSG0 and SSG6 depend on year and location.

The other explanatory variables are in three categories. The first category consists of eight biophysical variables that depend on location but not time: latitude (Lat), longitude (Long), elevation (Elev), multiresolution land characteristics (MRLCs), i.e., land-cover

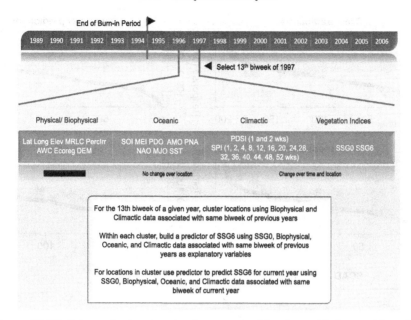

Figure 10.13 Schematic showing the data structure for the prediction problem. The first five years were used as burn-in for the predictors. The explanatory variables for SSG6 were the eight physical, eight oceanic, and 17 climactic variables, along with SSG0.

class, percentage of irrigated land (PerrIrr), available soil water capacity (AWC), ecological region (Ecoreg), and digital elevation model (DEM). See Harms *et al.* (2009) and `gapanalysis.usgs.gov` for further details.

The second category consists of eight oceanic variables that depend on year but not location: southern oscillation index (SOI), multivariable El Niño and SOI (MEI), Pacific decadal oscillation (PDO), Atlantic multi-decadal oscillation (AMO), northern Atlantic oscillation index (NAO), Madden–Julian oscillation (MJO), and sea surface temperature anomalies (SSTs).

The third category consists of 17 climactic variables that, like SSG0 and SSG6, depend on both year and location. They are the Palmer drought severity index (PDSI), measured one week and two weeks prior to SSG0, and the standardized precipitation index (SPI), measured at 1, 2, 4, 8, 12, 16, 20, 24, 28, 32, 36, 40, 44, 48, and 52 weeks prior to the 13th biweek. The 15 SPI measurements at a given location therefore summarize that location's precipitation over the previous year.

The strategy presented here to find good predictors started by summarizing the data by K-means clustering. The first five years of data (1989–2003) were used as burn-in and to choose a reasonable number of clusters; this turned out to be six. The clustering of all weather stations into six clusters was carried out for each year from 1995 and for all the data from previous years that was used. These clusters were based on all the relevant explanatory variables – biophysical, climactic, and SSG0. The oceanic variables were omitted since they did not depend on location and, of course, SSG6 was omitted since it was the response. Within each cluster, for a given year and location, a predictor for SSG6 was found based on all the data in the cluster up to and including the 13 biweek of the year prior to the given

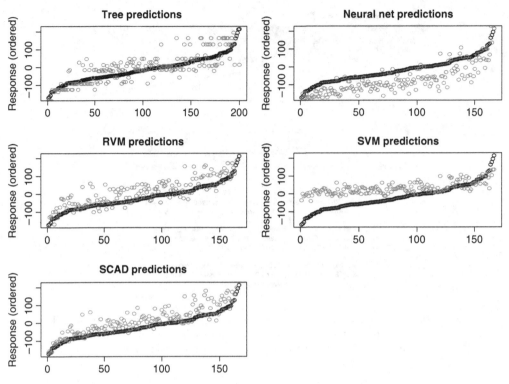

Figure 10.14 Predicted versus actual results for the five predictors, for the first cluster in the first year after the burn-in. From left to right and row to row: trees, NNs, RVM, SVM, and SCAD.

year. This predictor was used to predict the SSG6 for the given year at each location in the cluster, using the given year's values for the explanatory variables (i.e., the 'new' x). Thus, for each year, six predictors were found, one for each cluster. The data were reclustered every year so as to improve accuracy, and the clusters changed from year to year, adding an extra layer of prequentiality to the analysis. This was done for each of the five predictors in Sec. 10.6.1 using essentially the same procedures.

To see what the results look like consider the first cluster, in 1994, the first year after the burn-in. This is shown in Figure 10.14. The observations from the weather stations are in increasing order and the black circles indicate their values. The lighter circles indicate the predicted values obtained from using each method. In this specific case, trees and RVMs appear to do best, with trees giving slightly better results in terms of location at the cost of higher spread. Both tend to overpredict somewhat. The NN results underpredict the observations and SCAD overpredicts them, while SVM does not seem to capture the shape of the ordered observations adequately.

A few limitations of this analysis should be noted. First, the clusters do not have any obvious physical interpretation. Here, clustering is used merely to partition the data into more homogeneous groups as a way to get improved prediction. One of the clusters, 2, is quite small and it is not clear how much it helps the analysis. Second, many tuning parameters (or architectures) had to be chosen to generate predictions; this was done for each method

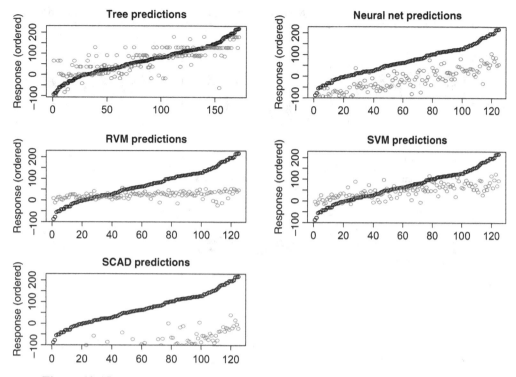

Figure 10.15 Predicted versus actual results for the five predictors, for the first cluster in the 12th year after the burn-in. From left to right and row to row: trees, NNs, RVM, SVM, and SCAD.

as described in Sec. 10.6.1. Third, while it is clear physically that the results at the different weather stations are mutually dependent this fact is not built into the analysis, which treats the data points as independent. Fourth, while not apparent in Fig. 10.14, there are many cases where so many data were missing that it was impossible to get predictions from some methods. These were dropped from the analysis.

For contrast with Fig. 10.14, consider Fig. 10.15. It shows the errors for cluster #1 in 2005, i.e., year 12. Since the data are reclustered every year, the clusters change from year to year; however, the changes are not large. Nevertheless, it is easy to see the differences between Fig. 10.15 and Fig. 10.14. First, although SVM is giving a slightly better performance, for all practical purposes both SVM and RVM miss the shape of the actual data. The degree of underprediction by NNs has increased and SCAD has changed from slight overprediction to severe underprediction. Only trees have been relatively stable in terms of performance. However, as will be seen shortly, no method seems to be stable in terms of converging to a true model. This high degree of model uncertainty is not unusual when analyzing $\mathcal{M}$-open data.

A more comprehensive way of looking at the performance of the five methods is seen in Fig. 10.16. In this figure, the predictive errors for each method over all six clusters and 13 years of predictions are shown. Note that some methods have 13 years of results while others have 12 years of results; this is due to missing explanatory data in year 13. Trees are the most stable in performance in that the 'predictuals' are symmetrically distributed tightly around

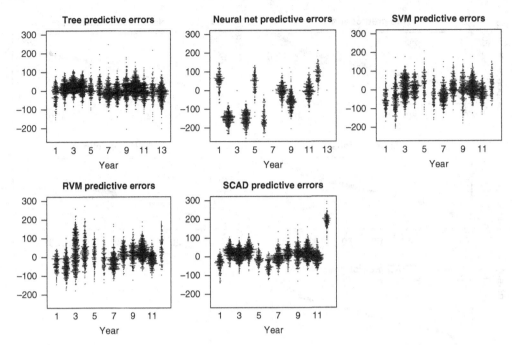

Figure 10.16 Predictuals for the five predictors over the 13 years, pooled over the six clusters. The vertical scales on all panels are the same.

zero, at least compared with the other four methods. Neural nets are the least stable in the sense that the centroids of the shaded regions vary more than for other methods. Although it is not apparent from the figure, there were cases where the NN procedure did not converge properly. The SVM and RVM results are roughly comparable while, SCAD is either very good (years two through eleven) or very bad (years one and twelve).

A more detailed summary is given in Figs. 10.17–10.21. Clearly, Fig. 10.17 is consistent with the performance of trees as seen in Figs. 10.14–10.16, i.e., the boxplots are relatively stable and the central 50% is under 50 for all clusters, with only a few exceptions.

By contrast, Fig. 10.18 looks poor. The effect of missing data is too severe for NNs to handle well. This, too, is consistent with Figs. 10.14–10.16 because the NNs are seen to be unstable and to underpredict the response, sometimes only a little and sometimes quite severely. This is seen in Fig. 10.18 because the 'boxes' show that many predictuals are high – well over 50, for instance.

Figures 10.19 and 10.20 are very similar, just as the SVM and RVM panels in Fig. 10.16 are very similar. If one were forced to choose, SVMs are slightly better than RVMs because the spread of the RVM errors is a little larger than the spread of the SVM errors. The slight difference can be seen in Fig. 10.16 and by comparing Fig. 10.19 with Fig. 10.20; the RVM absolute errors are typically a little larger. It should be noted that both SVMs and RVMs perform poorly compared with trees and that, even though they perform better than NNs, this does not mean that either SVMs or RVMs are capturing the behavior of the data well; see the relevant panels in Figs. 10.14 and 10.15.

The corresponding plot for SCAD in Fig. 10.21 shows that the SCAD results in Fig. 10.16 are not particularly representative of how SCAD performs. Arguably, the SCAD results in

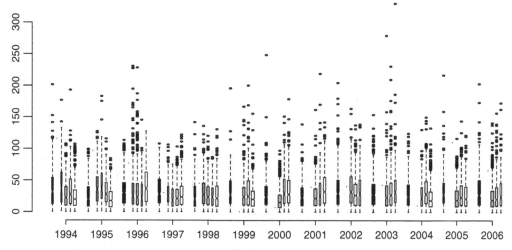

Figure 10.17 Boxplots of the absolute errors for the six clusters over the 13 years using trees. Clusters #2 and #3 (after 1998) had too much data missing to be included.

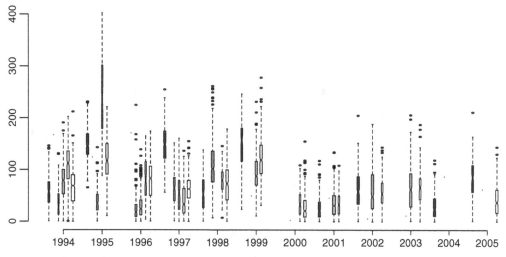

Figure 10.18 Boxplots of the absolute errors for the six clusters over the 12 years using NNs. Clusters often had too much missing data and had to be dropped. Neural nets performed very poorly for 1995.

Fig. 10.15 are more representative in that the SCAD predictor captures some of the correct shape, albeit poorly. Indeed, SCADs overall performance is worse than that of the other methods. For instance, in the later years, say after 2000, the absolute errors are generally increasing. The paradox is that SCADs overall performance is worse than that of the SVM and RVM even though SVMs and RVMs do not necessarily capture any features of the actual data; see the middle row of Fig. 10.15.

As a diagnostic for the stability of the five predictors, Fig. 10.22 shows how the predictor from each class varies with cluster and year. All methods are seen to be relatively unstable, possibly because the data is so complex. The range for trees was 29 to 102 leaves, although

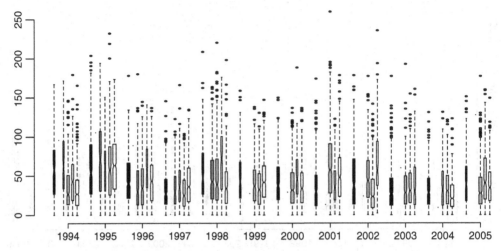

Figure 10.19 Boxplots of the absolute errors for the six clusters over 12 years using SVMs. Clusters 2 and 3 (after 1998) had too much missing data to be included.

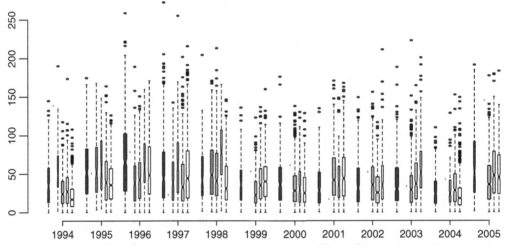

Figure 10.20 Boxplots of the absolute errors for the six clusters over 12 years using RVMs. Clusters #2 and #3 (after 1998) had too much missing data to be included.

numbers in the 30s and 40s were typical. The most typical architecture for NNs was (3, 2) – i.e., three nodes in the first hidden layer and two in the second. The NN architectures permitted in this analysis were no larger than (4, 3). The range for support vectors was 2 through 183. While there is no range that is obviously typical, the number of support vectors generally increased with year. The range for relevance vectors was 2 through 140. As with SVMs there was no typical range for the number of relevance vectors. However, the number of relevance vectors also tended to increase with year as well as to be lower than the number of support vectors. This is consistent with RVMs' enforcing sparsity more than SVMs.

As a final overall summary of the relative performance of the five methods, Fig. 10.23 gives boxplots showing the year on year absolute errors pooled over clusters. The location of the 'box' is an indicator of how well the method performs, but the 'whiskers' are so long

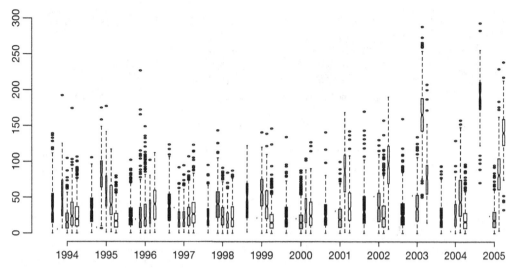

Figure 10.21 Boxplots of the absolute errors for the six clusters over the 12 years using SCAD. Clusters #2 and #3 (after 1998) had too much missing data to be included.

that in some cases the effect of the extreme values shifts the mean to a position substantially above the median. Nevertheless, comparing the median absolute errors indicates that trees give the best performance and NNs give the worst performance; RVMs and SVMs gave performances that were roughly equivalent and slightly better than SCAD, which seemed to perform worse in later years than earlier years. It must be emphasized that boxplots by themselves do not indicate whether a method captures important features of the data. If a method gives errors that are small enough then one may be led to believe that it captures essential features of the data, but the boxplots by themselves do not show this. For instance, Fig. 10.23 shows that NNs perform worst in the sense of errors but nevertheless Figs. 10.14 and 10.15 show that NNs capture some essential features of the data. By contrast, the same figures show that RVMs and SVMs do not necessarily capture any features of the data but still have smaller errors than NNs.

To conclude this example, recall that RVMs and SCAD are sparsity methods, i.e., they seek parsimonious models, so it is no surprise that they do not perform particularly well when the data are as complex as they are here. This is accentuated by the fact that there are really only 13 years of data. The fact of missing data automatically means that NNs will often not give results at all. Even when NNs do give predictions, since the NNs here are not penalized, parameter estimates will tend to be unstable. Moreover, there may be convergence problems with the parameter estimates. Indeed, even though (3, 2) is the most commonly occurring architecture, Fig. 10.22 shows that it is obviously appropriate only for cluster #2 (which is small) and cluster #3. Support vector machines impose some sparsity but not as much as RVMs or SCAD, so they tend to do better. The usual interpretation of support vectors in classification is that they roughly outline the boundary between regions. It is not clear how this interpretation carries over to a regression context, but the support vectors will remain as points indexing the kernel in such a way as to give as good an approximation as possible in (10.73). By contrast, relevance vectors loosely resemble prototypical points. If

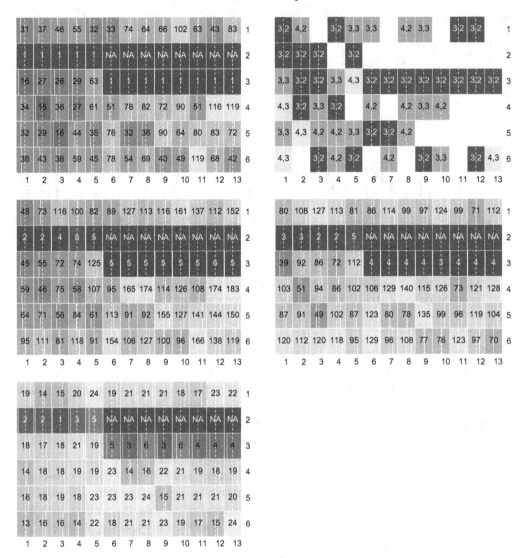

Figure 10.22 Stability of the five predictors. Top, left: For each year and cluster the corresponding box shows the number of leaves in the tree used to make the next prediction; NA means that a predictor was unable to be found. Top, right: For each year and cluster the pair of integers in the corresponding box indicates the numbers of nodes in the first and second hidden layers respectively. Middle: The numbers of (on the left) support vectors in the SVM and (on the right) relevance vectors in the SVM used to make the next prediction. Bottom: The number of terms, chosen by the defaults in the R package, used to identify a SCAD model in order to make the next prediction.

there are no prototypical points, because the data are so complex for instance, then RVMs will typically not perform very well. This may be why SVMs provide a better predictor, though not by much, than RVMs.

Trees do not impose parsimony (unless they are designed to), so it is no surprise that they outperform SCAD, RVMs, and SVMs. In addition, while trees are not particularly

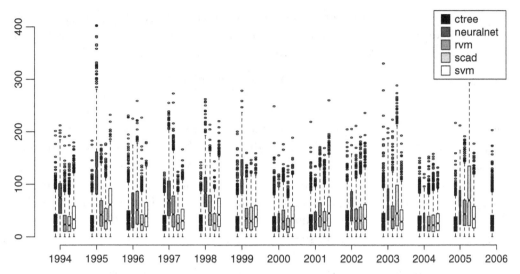

Figure 10.23 Year on year boxplots of the absolute errors for the five methods. This omits the cases where a given method was unable to give a prediction. Judging by the central 50% of absolute errors, NNs gave the worst performance and trees gave the best performance. The single method for 2006 was trees.

stable, they are not as unstable as NNs because they are nonparametric and resistant to small perturbations. Hence, in this example trees gave the best performance among the five methods.

10.7 Endnotes

There are many blackbox techniques that have not been discussed in this chapter. Even though several more are presented in these endnotes (projection pursuit, logic trees, hidden Markov models, and errors-in-variables models), still omitted are, e.g., MARS (due to Friedman (1991)), wavelets (see Mallat (1989) amongst others), and, more generally, frames (see Christensen and Jensen (1999) for an accessible treatment in general settings).

10.7.1 Projection Pursuit

Suppose that data (X, Y) are random, with $X \in \mathbb{R}^d$ distributed as P while $Y \in \mathbb{R}$. Then, it is well known that $f(x) = E(Y|X = x)$ is the best predictor of Y in a squared error sense. The projection pursuit regression (PPR) model is stated as

$$f(x) \approx \hat{f}_m = \sum_{j=1}^{m} g_j(x^T \beta_j),\tag{10.79}$$

where $\beta_j \in \mathbb{R}^d$ and, for all j, $\|\beta\| = 1$. The g_j, β_j, and m are assumed unknown. Then, following Friedman and Steutzle (1981), a representation for $f(\cdot)$ is found recursively. Initialize by setting $m = 0$, so that the residuals from fitting the zeroth model are $r_i = y_i$ for $i = 1, \ldots, n$. To find the first nontrivial term, fix a value of β_1 and obtain the values $x_i^T \beta_1$ in

$\mathbb{R}$. Now, any unidimensional nonparametric regression method (see Chapter 8) can be used to find a regression function $\hat{g}_1$. Since $\hat{g}_1$ is a function of the one-dimensional argument $x^T \beta_1$, it remains to specify $\beta_1 \in \mathbb{R}^d$. Given $\hat{g}_1$, the function

$$D(\beta_1) = 1 - \frac{\sum_{i=1}^n (r_i - \hat{g}_1(x^T \beta_1))^2}{\sum_{i=1}^n r_i^2}$$

summarizes how well $\hat{g}_1(x^T \beta_1)$ as a function of β_1 matches the residuals. Now, for a given $\hat{g}_1$, $\hat{\beta}_1 = \arg\sup_{\beta_1} D$ is the best choice for β_1. To find $\hat{\beta}_2$ the process is repeated using residuals $r_i = y_i - \hat{g}_1(x_i^T \hat{\beta}_1)$. For each $\beta_2 \in \mathbb{R}$ a unidimensional regression function $\hat{g}_2$ can be found. This defines a new objective function from which $\hat{\beta}_2$ can be determined. Continuing this procedure gives a sum having the form of the right-hand side of (10.79). Some sort of stopping rule must be imposed on m, e.g., one continues adding terms $\hat{g}_j$ with projections $\hat{\beta}_j$ until $D(\hat{\beta}_m) \geq c$ for some preassigned $c > 0$ that is not so high that it leads to overfit.

Obviously, this procedure is open to misuse because of overfitting, especially when finding the $\hat{g}_j$. However, used judiciously, PPR has many desirable properties, such as evading the curse of dimensionality, like NNs; it does this by limiting each stage to a univariate regression that nevertheless can approximate a nonparametric collection of functions, i.e., functions that are not parametrizable by a finite-dimensional parameter. Various versions of PPR (using different ways to find the $\hat{g}_j$, for instance) have also been shown to be consistent in various senses; see Huber (1985) and Jones (1987).

Projection pursuit regression gives the obvious point predictor

$$\hat{Y}(x_{\text{new}}) = \sum_{i=1}^m \hat{g}_j(x_{\text{new}}^T \hat{\beta}_j),$$

which may perform well in general settings if ϵ is small enough and n is large enough that the generalization error is not too great.

The only obvious way to get an interval predictor is via bootstrapping and using the residual sum of squares to estimate the variance, assuming it exists, i.e., by resampling with replacement from multiple data sets indexed by $b = 1, \ldots, B$. Each data set can be used to give a value of $\hat{Y}_n(x_{\text{new}})$, say $\hat{Y}_{n,b}(x_{\text{new}})$. Then, the SE at x_{new} can be estimated from

$$\text{SE}^2(x_{\text{new}}) = \frac{1}{B-1} \sum_{b=1}^B \left(\hat{Y}_{n,b}(x_{\text{new}}) - \frac{1}{B} \sum_{b=1}^B \hat{Y}_{n,b}(x_{\text{new}}) \right)^2 \qquad (10.80)$$

leading to $1 - \alpha$ PIs of the form

$$\hat{Y}(x_{\text{new}}) \pm z_{\alpha/2} \hat{\sigma} \sqrt{1 + \text{SE}^2(x_{\text{new}})}$$

where $\hat{\sigma}^2 = (1/n) \sum_{i=1}^n (y_i - \hat{f}_m(x_i))^2$, cf. (10.10) and (10.11). There is some redundancy between the use of $D(\beta_j)$ to find β_j and the use of the residual sum of squares to estimate σ^2. However, the procedure for generating $\hat{f}_m$ is greedy in that at each step it performs an optimization. Over several iterations this may or may not give a value close to $\hat{\sigma}^2$. However, this issue does not seem to have been investigated.

10.7.2 Logic Trees

Logic trees are due to Ruczinski *et al.* (2003). The key idea is that trees as presented in Sec. 10.2 are based on using an 'exclusive or' at each node as a function on a specified explanatory variable. A logic tree is more general in that it permits 'and', as well as being more restricted in that essentially it applies only to binary data, e.g., the presence or absence of a trait that may be linked to an outcome Y that need not be binary. Of course, trees (or treed models – trees that permit a general regression expression in each leaf rather than a constant) can accommodate 'and' under suitable redefinition of the nodes, but for the most part this is rarely done since trees are already very flexible.

The basic model expression for a logic tree uses d binary explanatory variables, X_j, say, and can be written as

$$g(EY) = \beta_0 + \sum_{k=1}^{K} \beta_k L_k,$$

where g is user specified (usually g is the identity or logit function) and L_k is a boolean combination of any of the d explanatory variables. That is, L_k assumes the values one or zero and is a function of variables that also assume the values one or zero, in this case the X_j. In the expression for $g(EY)$ the notational similarity between logic trees and generalized linear models can be seen. The goal in logic regression is feature selection, i.e., to choose the L_k at the same time as parameter estimation, i.e., finding values $\hat{\beta}_j$.

Usually, logic tree models are found by simulated annealing. There are four steps: (i) choose an optimality criterion (the quantity J in Sec. 9.5); (ii) choose a finite set of logic trees to be considered (the set S in Sec. 9.5); (iii) choose the maximum number K of leaves to be allowed in the tree; and (iv) carry out the simulated annealing procedure. Usually, the initialization sets all have $L_k = 0$. Moves, i.e., a change to an L_k, are then proposed. For each move find the $\hat{\beta}_k$ that provide a least squares fit to Y. The move is accepted or not according to the optimality criterion and cooling schedule. The usual optimality criterion is the residual sum of squares for $\hat{\beta}_0 + \sum_{k=1}^{K} \hat{\beta}_k L_k$. (This procedure is the obvious extension of using simulated annealing to find a single logic tree.) See Ruczinski (2003) for a useful summary. Over many independent runs, the SA output gives a histogram over the possible logic trees from which one can be chosen. Thus, if g is the identity, the resulting fitted model gives predictions

$$\hat{Y}(X_{\text{new}}) = \hat{\beta}_0 + \sum_{k=1}^{K} \hat{\beta}_k \hat{L}_k(X_{\text{new}}),$$

where X_{new} is a new value of the vector of explanatory variables. If g is the logit function, then the expression above must be transformed to get a point predictor. Similarly, if g is some other choice then point predictions are obtained essentially by inverting or otherwise manipulating g. If Y is discrete then obviously one chooses the permissible value of Y closest to the value that logic regression gives.

If Y is continuous then the bootstrapping approach in Sec. 10.7.1 or (10.10) and (10.11) can be used. The result is that SE is estimated from B bootstrap samples as in (10.80) and a $(1 - \alpha)100\%$ approximate PI is given by

$$\hat{Y}(x_{\text{new}}) \pm z_{\alpha/2}\hat{\sigma}\sqrt{1 + \text{SE}^2(x_{\text{new}})},$$

where $\hat{\sigma}$ is the root residual sum of squares of the logic tree model (found by SA on the original data) at which the PI is located. Logic regression has been found to work reasonably well in classification settings, and to be comparable with other good techniques, but so far its purely predictive properties have not been sufficiently explored.

10.7.3 Hidden Markov Models

Stated succinctly, a first-order finite-state discrete-time Markov chain is a sequence of random variables X_i for $i = 1, \ldots$ with the property that

$$P(X_{i+1} = k_{i+1}|X_i = k_i, \ldots, X_0 = k_0) = P(X_{i+1} = k|X_i = k_i),$$

where $k = 1, \ldots, K$ are the states and 'first-order' means that conditioning on the full past is equivalent to conditioning on the most recent random variable. The $K \times K$ transition matrix from time step i to time step $i + 1$ is $T_{i,i+1} = (P(X_{i+1} = k|X_i = j))_{k,j}$, where k indicates the column and j indicates the row. Let $v_i = (w_{i1}, \ldots, w_{iK})^T$ be a probability vector giving the marginal probabilities of X_i. Then the marginal distribution of X_{i+1} is $T_{i,i+1}v_i$, where the typical entry is $\sum_{j=1}^{K} w_{ij}P(X_{i+1} = k|X_i = j)$. Now, given an initial marginal probability vector v_0 for, say, X_0, the sequence of marginal probability vectors is $v_1 = T_{0,1}v_0$, $v_2 = T_{1,2}T_{0,1}v_1 = T^2v_0$, and so on. This can be simplified if the Markov chain is stationary, i.e., there is a single matrix $T = T_{i,i+1}$, where T depends only on the fact that time step i and $i + 1$ are one unit apart, not on the specific value of i. If the values of the X_i are observable, the only questions remaining are (i) estimating the elements of T and (ii) choosing an initial starting probability vector. The elements of T can be estimated empirically since they are conditional probabilities. If v_0 is not known, usually the stationary distribution of the Markov chain is employed, i.e., the vector v that satisfies $Tv = v$. The specific choice of v_0 is often not very important because its effect decreases with n. The probability vector at time $n + 1$, v_{n+1}, gives the probabilities of each state at time $n + 1$ and without further assumptions it is unclear what a good point predictor would be.

More interestingly, the Markov process may be partially hidden so that, instead of observing $X_i = x_i$, only a function of X_i is observed. That is, there is some f such that the observations are $Z_0 = f(X_0)$, $Z_1 = f(X_1)$, $Z_2 = f(X_2)$, $\ldots$ It is usually assumed that f does not change from time step to time step or, more precisely, that the way in which an unseen observation X_i from the underlying Markov chain is converted into an observation is independent of the time step. A typical sequence from a hidden Markov model (HMM) is $z^n = (z_1, \ldots, z_n)$, where the z_i are of the form $z_1 = f(x_1), z_2 = f(x_2), z_3 = f(x_3), \ldots$ If f is the identity function then, for all n, $Z_n = X_n$ and the HMM is just a Markov chain.

While there are many questions that can be asked within the HMM framework, two that are of concern here are (i) how to predict X_{n+1} given the previous unseen values of the Z_i, i.e., given $z^n = (x_1, \ldots, x_n)$, and (ii) how to predict Z_{n+1} given X_{n+1}. Clearly, (i) and (ii) are linked. Since prediction with HMMs is complicated, an example will highlight the reasoning. Suppose that K is known, $\langle X_n \rangle$ is stationary, and Z_n assumes one of finitely many values generically denoted $1, 2, \ldots, M$. Then there is a collection of *KM* probabilities of the form $P(Z_n = z_n | X_n = x_n)$.

To accomplish (i) and (ii) it is necessary to estimate T and the *KM* conditional probabilities. This is done by the Baum–Welch algorithm, which is a type of EM algorithm. Given this, one natural pair of predictors for (i) and (ii) is

$$\hat{X}_{n+1} = \arg \max_{u=1,\ldots,K} P(X_{n+1} = u | z^n)$$

and

$$\hat{Z}_{n+1} = \arg \max_{v=1,\ldots,M} P(Z_{n+1} = v | \hat{x}_{n+1}),$$

respectively; see Ye *et al.* (2013). So, if $\hat{X}_{n+1}$ is found then the estimates of $P(Z_{n+1} = v | \hat{x}_{n+1})$ from the Baum–Welch algorithm automatically give $\hat{Z}_{n+1}$. To obtain $\hat{Z}_{n+1}$ write

$$P(X_{n+1} = u | z^n) = \sum_{k=1}^{K} T_{k,u} P(X_n = k | z^n),$$

where (with a small abuse of notation) $T_{k,u}$ is the (k, u)th element of T. The Baum–Welch algorithm estimates the $T_{k,u}$, and the $P(X_n = k | z^n)$ can be found from the appropriate factorization; see Ye *et al.* (2013). In principle, $\hat{Z}_{n+1}$ can now be found.

Other prediction problems are possible within the context of HMMs, e.g., the prediction of Z_{n+1} given z^n or x^n. (The latter problem requires estimating the outcomes x_i, and can be accomplished by the Viterbi algorithm, for instance.) Moreover, K is not usually known and HMMs are not unique, because two or more distinct HMMs, with different K's, can be consistent with an observed sequence. In these cases, a Bayes information criterion is often used to choose K. Treatments of HMMs have also been given in the Bayesian context; see, for instance, Gassiat and Rousseau (2014).

10.7.4 Errors-in-Variables Models

Errors-in-variables models are not blackbox methods. Indeed, they have a more elaborate interpretation than fixed effect linear models because they include measurement error on the design points. They are worth commenting on because they show that, even with a viable interpretation, modeling is much more difficult than is commonly realized.

To give a simple example, suppose that $Y_i = \beta_0 + \beta_1 X_i + \epsilon_i$, where the ϵ_i are independent $N(0, \sigma^2)$ outcomes for $i = 1, \ldots, n$. The idea is that X_i is the outcome of a random variable $X_i = X_i^* + \eta_i$, where η_i is the measurement error, taken to be, say, IID $N(0, \sigma_x^2)$, and X_i^* is the correct value. If the usual least squares approach is taken with this regression problem then the results are the same as in simple linear regression since the estimates $\hat{\beta}_0$ and $\hat{\beta}_1$ are independent of the error structure. However, $\hat{\beta}_1$ converges to $\beta_1 \sigma^2 / (\sigma^2 + \sigma_x^2)$ in probability. Even though $\hat{\beta}_1$ is inconsistent for β_1, it is consistent for the parameter that gives the best linear predictor of $(Y | X)$. The classic reference for these models remains Fuller (1987); a more modern treatment focusing on the nonlinear case can be found in Carroll *et al.* (2006, Chap. 3 in particular).

It should be noted that errors-in-variables models are probably more suited to regression problems than conventional fixed effect models but they are not used as widely as they could be, partly because they are more difficult to implement and partly because less is known about them. For instance, although not discussed here, the class of errors-in-variables models is large, with many particular subclasses that require different analyses. It may not be clear which subclass is appropriate and there are subclasses for which little is known; this is somewhat like the situation for nonlinear mixed effects (fixed and random) models.

11

Ensemble Methods

> In the predictive modeling disciplines an ensemble is a group of algorithms that is used to solve a common problem ...Each modeling algorithm has specific strengths and weaknesses and each provides a different mathematical perspective on the relationships modeled, just like each instrument in a musical ensemble provides a different voice in the composition. Predictive modeling ensembles use several algorithms to contribute their perspectives on the prediction problem and then combine them together in some way. Usually ensembles will provide more accurate models than individual algorithms which are also more general in their ability to work well on different data sets ...the approach has proven to yield the best results in many situations.
>
> Miner *et al.* (2012, p. 906)

Ensemble methods comprise a class of statistical techniques in which the components, elements of the ensemble, are combined in some way to produce a single composite object. It is hoped that the composite will routinely perform an assigned task (whatever that might be) better than any of its individual components. Model averaging is a type of ensemble method where the results from several distinct models are combined, usually to achieve better prediction. The collection of models is often called a model list rather than an ensemble. Again, the hope is that the composite object will routinely perform better than any of its components. Ensemble methods as a class are more general than model averaging because they include the ensembling of other things, such as clusterings, ranked sets of variables (as opposed to the models they might form), and optimal actions from decision theory problems. These other ensemble methods often have modeling implications e.g., the ensembling of clusterings often leads to a better clustering and this may be important for predicting class membership. Here, however, the focus is on model averaging or, more typically, averaging the predictions that models generate.

Model averages (MAs), as a generality, are uninterpretable. In fact, predictions from models that result from incompatible assumptions may be averaged to good effect. This may seem odd, but the intuition is that pooling the predictions from different models resting on different assumptions takes model uncertainty (or model misspecification) into account. Taking model uncertainty into account means that the models – or their predictions – are weighted according to their 'worth', i.e., their usefulness in making good predictions. Hence, if the models in an ensemble or model list are chosen reasonably well, using the weighted average of their predictions will usually be better than using the predictor from any individual model. Taking model uncertainty into account can be likened to regarding the models as independently generated by a process that is unbiased for the true model, but possibly highly variable. It does not matter whether the models used to generate the model average rest on compatible assumptions because the errors cancel out.

Even better for the purpose of point prediction, the models themselves do not have to be known – it is enough to know their predictions. This was observed as early as the start of the twentieth century by Galton (1907), who took the median of a collection of predictions from experts as the actual prediction and found that it was extremely close to the actual outcome. Taking the median was reasonable because the empirical distribution of the guesses was not normal. Implicitly, the experts were well informed, had every incentive to make their best guess using whatever information they found to be relevant, and made their guesses independently. This corresponds to having a model list that is well chosen, is centered (however weakly) at the true model, and can be regarded as having independently generated members, i.e., the models on the model list are regarded as the independent outcomes of a 'model generating' process.

A simple experiment will illustrate the principle. Suppose that someone has a jar of jelly beans. The jar is clear glass so simply looking at it gives partial information about how many jelly beans are in it. If the goal is to guess the number of jelly beans in the jar then one can do modeling, i.e., one can estimate the volume of the jar and the volume of an individual jelly bean. This gives an empirical upper bound on the number of jelly beans. Then one can reduce this optimal number to account for the irregular shape of the jelly beans and any tapering of the jar. Essentially, this amounts to formulating the problem as estimating a population size (the number of jelly beans), and hence it would fall under the survey sampling model for variability.

It is just as well to regard this problem as merely embedded in a sequence of similar jelly-bean-number guessing problems and therefore regard it purely as a prediction problem. In this formulation, an MA approach makes sense. Choose 100 people at random and ask them how many jelly beans there are in the jar. Let them give you answers obtained however they want. Then drop the ten largest and ten smallest answers and take the average of the middle 80. This is what you predict as the number of jelly beans in the jar. The fact that this procedure usually gives good answers is sometimes called the 'wisdom of crowds'. As with the example in Galton (1907), it rests on assuming that the individuals in the crowd are giving you their best guesses independently, i.e., their guesses are given without reference to each other, possibly using different sources of information, and can be combined to a single value that has a relatively small bias with a variability that is not impossibly high. The question is how to choose the people to give guesses and how to combine their guesses into one value.

A more sophisticated version of this predictive procedure is to plot the guesses and use the best location for the histogram. This may mean transforming the histogram so that it looks approximately normal, taking the mean, and then transforming the mean back to the original scale. When the appropriate transformation is the logarithm, this leads to using the geometric mean (for which a response is as likely to be twice the size of the correct answer as it is to be half the size of the correct answer) as a guess for the outcome. There are many variations on this procedure and several youtube.com videos are available that demonstrate them; do a search on 'wisdom of crowds', for instance. All of them conclude that using an MA predictor will give better results than selecting a single model and using it to make predictions. Indeed, in the econometrics literature, arguments for this use of MAs began with Gordon (1924); see Clemen (1989) for a historical perspective.

In addition to empirical results showing that MAs are better for prediction than model selection, there are theorems that state the same thing especially in a Bayes setting; see, e.g., Raftery and Zheng (2003). It is well known in many examples that in the presence of model uncertainty the best predictors come from averages of models whereas prediction after selecting a single model is usually suboptimal. At root, the problems with the latter approach stem from model misspecification, Draper (1995), and abandoning interpretability permits optimization over a larger class of models and hence a stronger optimum. Thus, instead of finding a model, understanding it, and then checking its performance, MAs are a way to get optimal predictive performance directly. Then, the MA can be examined to see whether it has any physical meaning. An uncertainty principle between interpretability and prediction can be restated as follows: outside $\mathcal{M}$-closed problems, the better you think you know the true model, the worse your predictive performance is likely to be, and the converse also holds, especially if it is expressed as a contrapositive. Because interpretability depends on whether there is anything stable enough in the DG to model, for complex DGs MAs give better predictive performance since they do not rely as heavily on the stability of the DG. Loosely, the only situations in which one can get better prediction than by using an MA are those where the DG is simple or extremely well approximated by a simple model; in practice this is a rare case.

In the examples above, the people giving guesses, often called 'experts', correspond to models that are not explicitly known and the values they guess are actions they would take under loss functions of their own choosing. This is true despite the fact that the loss functions, like the models, are not explicitly specified. It is enough that the loss functions ensure the guesses are close to the actual value. Moreover, no assumption is made that any of the unknown models is correct or even that a correct model exists. So, this sort of procedure applies to $\mathcal{M}$-open problems. In the special case where the models are explicitly specified, they need not be assumed true in any sense; it is enough to regard them as giving actions (predictions) that can be usefully combined. The procedure still can be used in $\mathcal{M}$-open problems although the predictions will depend on the models chosen and how the actions they generate are combined. Such procedures are purely predictive since the 'modeling' involved is mostly in choosing the model list. That is, even when models are used, their status is not as candidates for the DG. They are merely mathematical devices to help construct good predictors. A special case of this in which the models being averaged are assumed known is treated in Sec. 11.7.1. The idea is to track the performance of the best model, i.e., 'expert', without regard to whether the model represents anything true about the DG.

Obviously, these techniques can be applied to $\mathcal{M}$-complete and $\mathcal{M}$-closed problems. However, as $\mathcal{M}$-closed problems have relatively distinguishable models, it is possible to identify a true model and in the easier $\mathcal{M}$-complete problems it is possible to identify a good approximation to the DG; the degree of approximation decreases as n increases. So, in this chapter the $\mathcal{M}$-closed case is mostly neglected. With $\mathcal{M}$-complete problems, the question is how well they can be approximated by $\mathcal{M}$-closed problems. If they can be approximated well then, again, little needs to be said beyond the $\mathcal{M}$-closed case. However, when $\mathcal{M}$-complete problems cannot be well approximated by $\mathcal{M}$-closed problems, in practice they are often similar to $\mathcal{M}$-open problems so the techniques of this chapter may be relevant.

Conceptually, MA can be regarded as an intermediate methodology between model selection (Chapter 9) and nonparametrics (Chapter 8). Since the results of many models are being combined one can imagine the net result being well represented by a nonparametric model, although the sample sizes required for the convergence of nonparametric models are usually unattainable outside low-dimensional problems. At a minimum, the model space for nonparametric methods contains the convex hull of the model list, i.e., its MAs.

However, the success of model selection depends on the quality of the model list. Model selection really only gives good results when at least one model is close to the true model (assuming it exists), and the performance of an MA rests on how well the true model can be represented by the model list, understanding that 'representing' is only to be understood in a predictive sense, not necessarily in any approximation sense. Of course, when exactly one models on the model list is the closest to the true model then an MA will typically put more and more weight on it as $n \to \infty$. In some cases this follows from the main theorem in Berk (1966), which states, informally, that the posterior distribution over models on a model list converges to unit mass at the wrong model that is closest to the true model in relative entropy. When this holds, taking the model in the MA with the highest posterior probability is essentially a model selection principle (MSP), which is consistent when the true model is on the model list and otherwise gives the best choice from the list.

Informally, MSPs are nearly equivalent to MAs. If an MSP assigns a nonnegative worth to each model on a finite model list then these can be normalized and taken as the weights $\hat{w}_i$. Conversely, if the MA has positive weights that sum to unity, then the model having the largest weight is the natural model to choose, thereby giving an MSP. In this intuition, it is implicitly assumed that the data are IID or at least approximately independent ($AR(p)$, α-mixing, etc.) without too much nonidenticality, so that, as $n \to \infty$, asking for the consistency of an MSP is reasonable. If the true distribution is a mixture of IID parametric families, for instance, then this intuition may break down. Moreover, in many prediction settings n cannot be assumed large. Consequently, it is more difficult to formulate intuition, let alone prove theorems. Indeed, asymptotically optimal predictors can often be outperformed in finite samples; see Wong and Clarke (2004) and Clarke *et al.* (2013). In settings where the data are far from independent, few results are available that characterize the performance of MA methods even though the basic principle that MAs outperform individual models holds.

A technical point is that there is a difference between an MA and the prediction it gives. The MA is literally an average of models and hence itself a model. The predictions are actions, under a loss function, that one might take given a model list as one input to the action space. Thus, different loss functions can lead to different actions for the same model list, i.e., different MA predictors. In practice, there is little lost in regarding an MA as equivalent to the predictions it gives, as long as it is clear how the models on the list are to be combined to give a prediction.

It is interesting to compare the use of MAs with the use of individual models. As a generality, MAs are better for prediction than individual models, but there are some caveats about how the model list is chosen. Consider an $\mathcal{M}$-complete problem and assume the data are generated by a single model, say p. If p is outside the model list, how does model averaging compare with other single-model-based predictors? The answer is not simple. First, assume that $p^* \neq p$ is another model, which is not on the model list from which a prediction might be generated. If p^* is closer to p in relative entropy than any element of the model list then typically p^* will lead to better predictors. However, this only happens when p^* is an a priori plausible model that was not included in the list, i.e., the list was constructed poorly. On the other hand, if one or more models on the list are closer in relative entropy to p than p^*, the model average will typically perform better. Typically, if p is in the middle of the cloud of models on the list, the MA will perform better than p. If p is equidistant from several elements of the model list or includes a weighted average of some or all of them, the reasoning is more complicated. The size and spread of the model list relative to p strongly affects how an MA will perform compared with p.

The foregoing discussion emphasizes the importance of the selection of the model list. This has long been recognized but rarely formalized. For instance, usually the more different the models in the MA are, the better. This was first observed in Breiman (1994). The idea is that this makes the models on the list seem closer to independently generated, given that they are all meant to be close to a true, or otherwise best, model. Second, there is a variance–bias tradeoff that operates on the level of model lists: Richer model lists should have lower bias but an MSP will require a larger n for convergence, while smaller model lists may have higher bias but the MSPs may converge faster. Hence, just as it is possible for model uncertainty to dominate the variability of a predictor, it is possible for model list uncertainty to dominate the variability of an MA predictor. Third, a model list may lead to dilution, discussed near the end of Sec. 9.2. Dilution can be regarded as the effect of model-space redundancy; see Chipman *et al.* (2001) and George (2010). Heuristically, suppose that there are models $p_k(y_{new}|\hat{\theta}, x_{new})$ for $k = 1, \ldots, K$ giving predictors $\hat{Y}_k(x_{new})$ and that there is an MA predictor that combines them into $\hat{Y}_{ma}(x_{new}) = \sum_{k=1}^{K} \hat{w}_k \hat{Y}_k(x_{new})$, where $\hat{w}_k$ is the weight that gives the worth of model k. Now, it is easy to see that if the model list, the p_k, is chosen badly – with many models having roughly equal worth and giving predictors with $|\hat{Y}_k(x_{new})| \leq B$ for some bound $B > 0$ – and the $\hat{w}_k \geq 0$ sum to one, then $\hat{w}_k \propto 1/m$ so that $\hat{Y}_{ma}(x_{new}) \to 0$ as $m \to \infty$ for fixed n. This can occur in Bayesian MAs as well as in other MAs. The performance of MSPs and MAs depends on the model list. With MAs, there is an uncertainty from the model list and in the models themselves given the list.

One way to formalize these comments, at least partly, is to write $\mathcal{F} = \{f_1, \ldots, f_K\}$ as the model list, f_T as the true model, and f^* as another candidate model and assume that a signal-plus-noise model holds for $Y(x)$ with $\text{Var}(Y) = \sigma^2$. For each model assume that $f(x) = f(x|\beta)$ and write $\hat{f}(x) = f(x|\hat{\beta})$ where $\hat{\beta}$ is an estimate of β. If $\hat{Y}_k(x)$ is the predictor from $\hat{f}_k$ then, by judicious addition and subtraction, the mean squared error of the MA, MSE(MA), is equal to

$$E\left(Y_{n+1}(x_{n+1}) - \sum_{k=1}^{K} \hat{w}_k \hat{Y}_{k,n+1}(x_{n+1})\right)^2$$

$$= \sigma^2 + E\left(EY_{n+1}(x_{n+1}) - \sum_{k=1}^{K} E(\hat{w}_k \hat{Y}_{k,n+1}(x_{n+1}))\right)^2$$

$$+ E\left(\sum_{k=1}^{K} E(\hat{w}_k \hat{Y}_{n+1}(x_{n+1})) - \sum_{k=1}^{K} \hat{w}_k \hat{Y}_{k,n+1}(x_{n+1})\right)^2,$$

where the second term is the bias of the model average and the third term is the variance of the model average. Thus, there are two variance terms in the MSE for an MA. The same sort of decomposition can be found for MSE($\hat{f}^*$). The key difference is in the last two terms:

whichever of an MA or a single-model predictor gives the smaller MSE asymptotically is likely to be the better choice. Usually, the bias of an MA is smaller than the bias of a single model and the extra variance in an MA is little more than for a single model, because the single model, to be competitive with a model average, often has to be very complicated and hence more variable.

To conclude this introduction to the intuition behind model-average methods we note three points. First, it is often best to ignore correlations between the predictors being averaged in estimating weights. Sometimes this is called 'idiot's Bayes' but it's effectiveness is borne out in classification problems; see Hand and Yu (2001) and Bickel and Levina (2004). The same property is likely to carry over to typical regression problems. Second, combining simple forecasting methods tends to outperform complex models. That is, using an MA is not the same, at least in practice, as finding a hugely complicated model and forecasting from it – even though an MA can be regarded as a hugely complicated model. In the Bayes context, this was observed in Minka (2002). Monteith *et al.* (2011) tried to reconcile model averaging with model combination, arguing that the fact that MAs come from a larger class of models is the main reason why they perform better than individual models. However, this can neglect the effect of the complexity of the single model representing the model average. For further elaboration in the classification context see Kim and Ghahramani (2012). Third, as a general principle, robust combinations of robust components, i.e., elements of the model list or ensemble, tend to be better than nonrobust components and combinations. This is no surprise but it must be qualified by noting that too much robustness can also be a problem, and the principle may not apply to nonrobust prediction goals. For example, usefully predicting the maximum of a random variable, although in this case predicting, say, the outcome of a 99th percentile, may make more sense.

For the rest of this book, DGs are assumed to be $\mathcal{M}$-open or at the difficult end of $\mathcal{M}$-complete problems unless noted otherwise. That is, the identification of a true model for the DG is regarded essentially hopeless or at least as including unavoidable model misspecification that cannot be satisfactorily addressed. Therefore the strategy will not be to get more data (which is usually impossible) but to use the data in different ways, i.e., in predictors from different models that capture different aspects of the data. As noted above, it is as if there were a model-generating process that gives a stream of candidate models, which amounts to a separate data stream that can be used with the actual data to generate better predictions.

In the rest of this chapter, the main MA predictors are presented and their properties discussed. First, Bayes model averaging (BMA) is presented in Sec. 11.1. Then bootstrap aggregation (bagging) is presented in Sec. 11.2, stacking in Sec. 11.3, and median methods in Sec. 11.5. Section 11.4 presents boosting for classification and for regression, even though the latter procedure is less successful. Section 11.6 looks at how these methods perform in practice. Section 11.7 gives notes on 'prediction along a string' – a predictor designed for the $\mathcal{M}$-open case – and a brief summary of the 'no free lunch' theorems.

11.1 Bayes Model Averaging

Bayes model averaging was probably the first organized MA method; see Hoeting *et al.* (1999) for a good review with examples and Clyde and George (2004). The basic idea is to assume that the K models in a model list correspond to densities in $\mathcal{M}_k = \{p_k(\cdot\,|\theta_k, \mathcal{M}_k)|k = 1, \ldots, K\}$ indexed by parameters θ_k with $\dim(\theta_k) = d_k$ and having within-model priors $w(\theta_k|\mathcal{M}_k)$. The conditioning on $\mathcal{M}_k$ indicates that different models p_k may depend on different sets of explanatory variables. Letting $p(\mathcal{M}_k)$ be an across-model prior gives a single trivariate model with joint density

$$p(y, \theta_k, \mathcal{M}_k) = p(y|\theta_k, \mathcal{M}_k)p(\theta_k|\mathcal{M}_k)p(\mathcal{M}_k),$$

thereby satisfying the 'containment principle' of Bayesian statistics (that the entire inference problem is situated in a single probability space). It would be more compact but less explicit to write k in place of $\mathcal{M}_k$; often p_k is denoted p when no confusion about k will result.

Now, the laws of probability can be applied to get expressions for objects of inference. The marginal likelihood for $\mathcal{M}_k$ is

$$p(y|\mathcal{M}_k) = \int p(y|\theta_k, \mathcal{M}_k)p(\theta_k|\mathcal{M}_k)\mathrm{d}\theta_k.$$

The posterior model probabilities are

$$p(\mathcal{M}_k|y) = \frac{p(\mathcal{M}_k)(y|\mathcal{M}_k)}{\sum_{k=1}^{K} p(\mathcal{M}_k)(y|\mathcal{M}_k)} = \frac{p(\mathcal{M}_k)\int p(y|\theta_k, \mathcal{M}_k)p(\theta_k|\mathcal{M}_k)\mathrm{d}\theta_k}{\sum_{k=1}^{K} p(\mathcal{M}_k)\int p(y|\theta_k, \mathcal{M}_k)p(\theta_k|\mathcal{M}_k)\mathrm{d}\theta_k},$$

where the density is with respect to counting measure and hence is a probability mass function. If a single outcome y is replaced by $y^n = (y_1, \ldots, y_n)^T$, the expressions only change cosmetically.

The Bayes MA (BMA) is a weighted sum of posterior quantities, where the weights are the $p(\mathcal{M}_k|y)$. More formally, let Z be a quantity of interest (a future outcome, utility, etc.) that has a posterior density. The BMA for Z is the density

$$p(z|y^n) = \sum_{k=1}^{K} p(z|y^n, \mathcal{M}_k)p(\mathcal{M}_k|y^n).$$

(It should perhaps be called the Bayes averaged model to emphasize that it is a model, but BMA has become standard.) For example, the BMA for a future value $Y_{n+1} = Y_{n+1}(x_{n+1})$ is

$$p(y_{n+1}|y^n) = \sum_{k=1}^{K} p(y_{n+1}|y^n, \mathcal{M}_k)p(\mathcal{M}_k|y^n), \tag{11.1}$$

the usual predictive density but now averaged over models as well as parameters. Explicitly, for each fixed k,

$$p(y_{n+1}|y^n, \mathcal{M}_k) = \int p(y_{n+1}|\theta_k, \mathcal{M}_k)p(\theta_k|\mathcal{M}_k, y^n)\mathrm{d}\theta_k,$$

where

$$p(\theta_k|\mathcal{M}_k, y^n) = \int p(y^n|\theta_k, \mathcal{M}_k)p(\theta_k|\mathcal{M}_k)\mathrm{d}\theta_k.$$

All these expressions can be derived using factorizations on the probability space defined by $(Y^n, Y_{n+1}, \Theta_k, \mathcal{M}_k)$.

Point and interval predictions can be derived from the BMA (11.1). First, a natural point predictor arises from taking the expectation:

$$\hat{Y}_{n+1} = E(Y_{n+1}|y^n) = \sum_{k=1}^{K} E(Y_{n+1}|y^n, \mathcal{M}_k) p(\mathcal{M}_k|y^n), \tag{11.2}$$

in which the expectation in the summation is the posterior mean with respect to the distribution of Y_{n+1} that is conditional on y^n, so that the θ_k and $\mathcal{M}_k$ are integrated out. This is optimal under squared error loss. Although not used very often, another natural choice for a point predictor would be the median of the BMA for Y_{n+1},

$$\hat{Y}_{n+1} = \text{med}\left(p(y_{n+1}|y^n)\right), \tag{11.3}$$

optimal under absolute value loss. Thus, a single BMA leads to multiple point predictors depending on the loss function.

A special case of (11.2) occurs when the p_k correspond to linear models. Suppose that $Y^n = X\beta + \epsilon^n$, where the design matrix is $n \times p$ and formed from $(x_1^T, \ldots, x_n^T)$, the errors ϵ_i are independent $N(0, \sigma^2)$, $\beta \in \mathbb{R}^d$, and there is a prior for (β, σ^2). Then the expectation in the summation in (11.2) is $f_k(x_{k,n+1}|\hat{\beta}) = x_{k,n+1}^T E(\beta_k|y^n)$, assuming that the entries in $x_{k,n+1}$ have been selected to correspond to model k. So, $\hat{Y}_{n+1}$ is the weighted average of the predictions from the K models, where the weights are their posterior probabilities. In this case, the coefficients of the K predictions are positive and sum to unity and the posterior probability of $\mathcal{M}_k$ is the 'worth' of model k.

Since posterior probabilities usually can't be found in closed form, they are often found using MCMC. In the special case of posterior model probabilities and large enough n, the $p(\mathcal{M}_k|y^n)$ can also be well approximated using the BIC discussed in Sec. 9.3. From the derivation of the BIC it is easy to see that

$$p(\mathcal{M}_k|y^n) \approx \frac{\exp\{-0.5\text{BIC}_k\}}{\sum_{k=1}^{K} \exp\{-0.5\text{BIC}_k\}}, \tag{11.4}$$

where BIC_k is the BIC value for the model $\mathcal{M}_k$. Essentially, this means that as $n \to \infty$, posterior probabilities become independent of the prior and are determined by likelihood ratios. Thus, in its simplest (linear-model) form the BMA prediction is

$$\hat{Y}_{n+1}(x_{n+1}) = \sum_{k=1}^{K} \hat{w}_k f_k(x_{n+1}|\hat{\beta}) = \sum_{k=1}^{K} \hat{w}_k x_{n+1,k}^T E(\beta|y^n),$$

where the $\hat{w}_k$ are from the approximation (11.4) and the subscript k on x indicates that the variables chosen represent the kth model.

The natural way to obtain PIs is to use (11.1). As long as suitable values for the posterior model probabilities and the predictive distribution for each of the K models on the list can be found, at least approximately, the model determined by the BMA can be approximated. Consequently $1 - \alpha$ PIs for Y_{n+1} can be found in principle. Let $t_{\alpha/2}$ and $t_{1-\alpha/2}$ be the $(\alpha/2)100$th and $(1 - \alpha/2)100$th percentiles of (11.1). Then, $[t_{\alpha/2}, t_{1-\alpha/2}]$ is a $1 - \alpha$ PI for Y_{n+1} because

$$P(Y_{n+1} \in [t_{\alpha/2}, t_{1-\alpha/2}]|y^n) = 1 - \alpha.$$

Another way to obtain $1 - \alpha$ PIs is to use the highest predictive density regions. That is, to choose h_α so that under (11.1)

$$P(Y_{n+1} > h_\alpha | y^n) \geq 1 - \alpha.$$

Both these ways to find PIs have shortcomings. Depending on the model list, they may give PIs that are very large or not connected. Consequently, PIs from BMAs are rarely used – even though a large or disconnected PI may be accurate.

If a PI from a BMA is desired, a better one to use might be the usual 'mean plus or minus two SDs'. The posterior variance of $Y_{n+1}(x_{n+1})$ is

$$\mathrm{Var}(Y_{n+1}|y^n) = \sum_{k=1}^{K} \left(\mathrm{Var}(Y_{n+1}|y^n, \mathcal{M}_k) + \hat{Y}^2_{n+1,k} \right) P(\mathcal{M}_k|y^n) - E(Y_{n+1}|y^n)^2,$$

where $\hat{Y}_{n+1,k}$ is the predictor from model k; see Hoeting *et al.* (1999). In principle, the PI $\hat{Y}_{n+1} \pm z_{1-\alpha/2}\sqrt{\mathrm{Var}(Y_{n+1}|y^n)}$ could be used and regarded as approximately $1 - \alpha$. The problem is that this effectively assumes approximate normality, which does not hold very often, chiefly because the predictions from models and the posterior distributions over models are usually not normal. The consequence is that PIs from BMAs are not often used; it is much more typical to use the BMA to give a point predictor and argue that this is representative of the BMA.

Using a BMA is different from doing Bayes model selection. Recall that, for a Bayesian, the model uncertainty is summarized by the vector $(p(\mathcal{M}_1|y), \ldots, p(\mathcal{M}_K|y))$. So, a Bayesian could choose the modal model $\hat{k} = \arg\max_k p(\mathcal{M}_k|y)$ (optimal under zero–one loss), and this is equivalent to what the BIC and Bayes testing would give. The BMA is a weighted average of the models on the model list, with their worths given by the $p(\mathcal{M}_k|y)$. The Bayes model average and Bayes model selection are only identical when there is one model with $p(\mathcal{M}_k|y) = 1$, i.e., there is no model uncertainty. Since BMAs typically converge to the model on the list closest to the true model (when it exists), Bayes model selection and BMAs are usually asymptotically equivalent. However, when there is no unique closest model, BMAs do not have to converge; Berk (1966) gives an example with a continuous parameter.

To see more explicitly what BMAs look like, consider the linear-model setting and recall the structure developed by Clyde and George (2004), presented in Sec. 9.3. Write $Y = X_\gamma \beta_\gamma + \epsilon$, where γ is a string of 1's and 0's indicating whether or not a given X_k is in the model. (In this notation, either Y can be centered or β_0 can be associated with the explanatory variable $X_0 \equiv 1$.) Most commonly the data are centered, so γ can assume any one of 2^K values. Each within-model prior is on one β_γ that is conditional on γ, and the across-model prior is on the γ's themselves. Sometimes IID Bernoulli priors are put on the individual γ_k. Thus,

$$p(\gamma|w) = w^{K-p_\gamma}(1-w)^{p_\gamma},$$

where p_γ is the number of 1's in the K entries of γ and $w \in [0, 1]$ is a hyperparameter. The 'noninformative' choice of $w = 0.5$ leads to a uniform distribution over models but as a random variable $p_K \sim \mathrm{binomial}(K, 1/2)$, so that $E(p_k) = K/2$. Since there are $C(K, j)$ models with j explanatory variables, and $C(K, j)$ increases with j until $j = K/2$ after which $C(K, j)$ decreases, the uniform distribution over models often is not representative of the prior information about model size. The more general choice of w in $p(\gamma|w)$ means that the user can choose w to reflect the expected number of explanatory variables to be included in the models. This is often more reasonable. However, another problem with Bernoulli's as across-model priors is that they may lead to dilution when there are too many similar models, e.g., in the presence of high collinearity.

One common choice in the linear-models context is Bernoulli across-model priors with Zellner g-priors (see Secs. 4.2 and 9.3) used within models. This means that there are two hyperparameters to choose, g and w. In this case, for $g > 0$,

$$p(\beta_\gamma|\gamma, g) = N(0, g\sigma^2(X_\gamma^T X_\gamma)^{-1}),$$

where the dimension of the normal distribution is p_γ. To complete the prior specification it is common to set

$$p(\beta_0, \sigma^2|\gamma) \propto \frac{1}{\sigma^2}.$$

One benefit of this choice is that it leads to closed-form expressions for $p(y^n|\gamma)$; see Clyde and George (2004, Sec. 3). Thus,

$$p(\gamma|y^n) = \frac{p(y^n|\gamma)p(\gamma)}{\sum_{\gamma \in \{0,1\}^K} p(y^n|\gamma)p(\gamma)}$$

can be found relatively easily. If empirical Bayes estimates of w and g are used then, since γ is merely a reparametrization of $\mathcal{M}_k$, the BMA point predictor in (11.2) becomes

$$\hat{Y}_{n+1} = \sum_{\gamma \in \{0,1\}^K} E(Y_{n+1}(x_{n+1})|y^n, X_\gamma, \gamma, x_{n+1})p(\gamma|y^n), \tag{11.5}$$

in which the expectation on the right-hand side can be derived from (9.12). Note that large g's give Bayes factors (BFs) typically favoring the null. Moreover, a small w and large g often puts high posterior weights on parsimonious models with large coefficients, while a large w and small g often puts high posterior weights on large models (many coefficients) with small coefficients. Fully Bayes approaches tend to lead to closed-form marginals with Cauchy-like tails; see Clyde and George (2004) for details and references.

The generic form

$$Y = X_\gamma^T \beta_\gamma + \epsilon \quad \text{for} \quad \gamma \in \{0, 1\}^K \tag{11.6}$$

is not as restrictive as it seems. Indeed, Clyde and George (2004) gave three examples that demonstrate its flexibility. First, any basis expansion for a unidimensional Y can be used to generate a linear model. Hence, a multiresolution wavelet basis generates a linear model if one writes

$$E(Y|x) = \beta_0 + \sum_{j=1}^{J-1} \sum_{i=1}^{m2^{j-J}} \beta_{ji}\phi_{ij}(x), \tag{11.7}$$

where $\phi_{ji}(x) = 2^{-j/2}\psi(2^{-j}x - i)$ are the translations and scalings of a mother wavelet $\psi(\cdot)$ for fixed J and m. That is, treating the wavelets as explanatory variables reduces wavelet regression to (11.6). In this setting it may make sense to put more mass on the γ's representing models with lower resolution terms than on those with higher-resolution terms, so that the data can determine which high-resolution terms are important. In this case, closed-form expressions for the coefficients may not exist, but MCMC–MH can provide estimates and the posterior can be used to get model weights. Thus, expression (11.5) can be adapted to wavelets.

Second, univariate spline models are also of the form (11.6). Write

$$E(Y|x) = \sum_{i=1}^{q} \alpha_i x^i + \sum_{j=1}^{J} \beta_j (x - t_j)_+^q, \tag{11.8}$$

where q is the order of the spline, $(\cdot)_+$ indicates the positive part of a function, and the t_j are the knots for $j = 1, \ldots, J$. Removing one $(x - t_j)_+^q$ is equivalent to removing the knot at t_j, so uncertainty about knot locations directly corresponds to the uncertainty in variable selection. So, as with wavelet bases, expression (11.5) can be adapted to splines. There are analogs to (11.8) and (11.7) (and hence (11.5)) for any linear regression technique; see Chapter 8 for further examples.

Third, the linear model structure underlies generalized linear models (GLMs), which can be defined by writing

$$g(E(Y|X_\gamma)) = X_\gamma \beta_\gamma,$$

using the same γ-notation, but now $g(\cdot)$ is a link function giving the best predictor for the parametric family used in the conditional expectation. This parametric family is always of exponential form and in addition to β_γ has a scale parameter ϕ. Ibrahim and Laud (1991) provided a Bayes analysis of GLMs based on using the Jeffreys prior within each candidate model. This can, in principle, be generalized to arbitrary priors by using MCMC–MH. Thus, an expression for $p(\beta_\gamma | X_\gamma, y^n, \phi)$ can be given. Assigning a prior to γ means that the posterior model probabilities can be found, again by MCMC–MH, and so, even though the model is not of the form (11.6), the posterior model weights and the predictors within each model can be combined into an expression like (11.5).

The BMA procedure can be applied to many other model classes such as NNs and trees. Recalling the Bayes analysis of individual NNs in Sec. 10.3.3, a joint prior $\pi(w_k, \mathcal{A}_k) = \pi(k)\pi(w_k | \mathcal{A}_k)$ over architectures $\mathcal{A}_k$ with parameter vectors w_k, respectively, must be specified for $k = 1, \ldots, K$. Using a normal likelihood for the error of the NN model, the posterior distribution over k and w_k is given by (10.38):

$$\pi(w_k, k | Y^n, X^n) \propto \frac{\pi(w_k, \mathcal{A}_k) \exp\left(-\frac{1}{2\sigma^2} \sum_{i=1}^{n} (y_i - \text{Net}(x_i; w_k, \mathcal{A}_k))^2\right)}{\sum_{k=1}^{K} \pi(k) \int \pi(w_k | \mathcal{A}_k) \exp\left(-\frac{1}{2\sigma^2} \sum_{i=1}^{n} (y_i - \text{Net}(x_i; w_k, \mathcal{A}_k))^2\right) dw_k}, \tag{11.9}$$

in which $\mathcal{A}_k$ has been abbreviated to k where no confusion will result. From (11.9) the posterior for the kth architecture $\mathcal{A}_k$ can be obtained by integration,

$$\pi(k | Y^n, X^n) \propto \frac{\pi(k) \int \pi(w_k | \mathcal{A}_k) \exp\left(-\frac{1}{2\sigma^2} \sum_{i=1}^{n} (y_i - \text{Net}(x_i; w_k, \mathcal{A}_k))^2\right) dw_k}{\sum_{k=1}^{K} \pi(k) \int \pi(w_k | \mathcal{A}_k) \exp\left(-\frac{1}{2\sigma^2} \sum_{i=1}^{n} (y_i - \text{Net}(x_i; w_k, \mathcal{A}_k))^2\right) dw_k},$$

and used in (11.2). As derived in Sec. 10.3.3,

$$p(y_{\text{new}}(x_{\text{new}})|Y^n, X^n, k) = \int \pi(w_k|Y^n, X^n, \mathcal{A}_k) p(y_{\text{new}}(x_{\text{new}})|w_k, \mathcal{A}_k) dw_k.$$

can be used to give the expectation in (11.2). The computations may be difficult, since MCMC–MH must be used to find various posterior quantities, but in principle the usual BMA predictor under L^2 for a class of neural nets can be found.

Bayes model averages using NNs for classification purposes follow from much the same procedure as above. See Sec. 10.3.3 for some details that, combined with the expression for $\pi(k|X^n, Y^n)$ above, permit a derivation of the resulting classifier.

Just this sort of procedure has been used in many examples. For instance, Liang (2005) uses a BMA of NNs in a nonlinear time series context to provide one-step-ahead forecasts. A normal prior is used on the weights, an inverse gamma prior is used on σ, and a truncated Poisson(λ) is used as the across-model prior on the connections. The computing is done using MCMC–MH, and the BMA of NNs compares favorably with several other standard methods in complex settings. Ahmad *et al.* (2010) used a more involved and ad hoc technique. Splitting the data into test, train, and validation subsets, and bootstrapping from the training data, gives a collection of 20 single-hidden-layer NNs. These are equipped with a uniform across-model prior, updating it much as in (11.4). Rather than assigning priors to the within-model weights and hence using Bayes estimates of them, Ahmad *et al.* (2010) used Levenburg–Marquart optimization (with regularization and early stopping), a procedure based on nonlinear least squares. Estimates of σ were found by assuming a normal likelihood and finding the MSE. In two complex prediction settings, Ahmad *et al.* (2010) found that BMA on NNs outperformed several other predictors and weighting schemes for model averages. Hassan *et al.* (2013) also partitioned the data into train, test, and validation subsets but used five classes of single-hidden-layer NNs (with 5, 7, 10, 12, and 15 nodes). A normal likelihood and (11.4) were used to get the weights; σ was estimated by an auxiliary procedure. Empirically, the BMA gave a better prediction than any of its components. Finally, in a more complicated approach to prediction, Vlachopoulou *et al.* (2015) used (11.4) as posterior weights based on a normal likelihood in order to compare empirically a BMA formed from several model classes – NNs, autoregression, support vector regression, random forests (see Sec. 11.2), and sparse regression (see Sec. 10.5) – with individual components from those classes. The BMA over the five classes outperformed each individual class. This probably means that, in some complex data settings, representatives from different classes can capture different features of the data.

A procedure for BMA over regression trees is similar in structure. The elements that must be found for (11.1) are the model class, the within-model prior, and the across-model prior. One way to specify the across-model prior is via the Chipman *et al.* (1998) construction given in Sec. 10.2.3. The resulting posterior probability for a tree T, $w(T|X, y^n)$, is given in (10.21). The point predictor from a tree T can be obtained by substituting estimates $\hat{\theta}_k$ for the θ_k in (10.18). These may be Bayes estimators using the priors on θ and σ in (10.20). Putting all these pieces together is routine for BMAs using regression trees.

A variation of BMAs over regression trees is provided by BMAs over classification trees. The same across-model and within-model priors can be used but with the likelihood (10.23). The θ in the likelihood must be estimated to get a predicted class from a given tree T.

Posterior model probabilities can be derived from (10.24) using (10.21). Combining the predicted classes with the posterior probabilities of the trees that gave these classes gives the BMA.

As described at the end of Sec. 10.3.3, BART can also be regarded as a BMA using stumps (trees with a small number of splits) for both classification and regression. Each stump has a posterior probability and gives a predicted class, so these can be combined to give a BMA.

In principle, kernel methods such as SVMs and RVMs can also be combined to give BMAs. The natural analog of a model is given by the kernel function, so an across-kernel prior would have to be combined with within-kernel priors. For instance, SVMs for regression could be regarded as the location for a normal likelihood with variance σ^2. Then, the BMA would combine the predictions from the various kernels, using posterior weights for those kernels derived from the likelihood and priors. This possibility does not seem to have been explored extensively in the literature on BMAs but see Seeger (2000) and Sollich (2002) for Bayes analyses that could lead naturally to BMAs formed from SVMs (or RVMs).

Separately from how BMAs are formed from a model list, the effective model list can be modified. Two ways in which this can be done are by using Occam's window (see Raftery *et al.* (1993) for linear models and Madigan and Raftery (1994) for graphical models) and by using 'local' Bayes factors (see Yu *et al.* (2013)). The core idea behind Occam's window is that a BMA such as (11.1) overstates the model uncertainty because it may include models that the data discredit, especially if a poor model list is chosen, that is, e.g., far from the true model if there is one or otherwise too big and/or spread out. That is, the posterior probabilities of inadequate models may strongly influence BMA predictions. (This is the reverse of dilution, where many good models split the posterior probabilities too finely.) Thus, to seek a reasonable prediction, some, perhaps many, models on a list of length K should be omitted on a rational basis. Given a model list, within-model priors, and an across-model prior, two sorts of model are excluded from the BMA, essentially on Bayes testing criteria. The first set of models to be excluded have low posterior probability. Thus, for properly chosen $c > 0$, one excludes the models in

$$\mathcal{A}_1 = \left\{ \mathcal{M}_k \middle| \frac{W(\mathcal{M}_k|y^n)}{\max_k W(\mathcal{M}_k|y^n)} \leq c \right\}.$$

By analogy with p-value cutoffs, Madigan and Raftery (1994) suggested $c = 20$. The second set of models to be excluded are those that are not parsimonious, i.e., models in $\mathcal{A}_1^c$ that have a submodel with higher posterior probability. Thus, one excludes

$$\mathcal{A}_2 = \left\{ \mathcal{M}_k \middle| \exists \mathcal{M}_\ell \in \mathcal{A}_1^c \text{ such that } \mathcal{M}_\ell \subset \mathcal{M}_k \text{ and } \frac{W(\mathcal{M}_\ell|y^n)}{W(\mathcal{M}_k|y^n)} > 1 \right\}.$$

The reduced set of models is now

$$\mathcal{A} = \{ \mathcal{M}_k | \mathcal{M}_k \notin \mathcal{A}_1 \cup \mathcal{A}_2 \} = \mathcal{A}_1^c \cap \mathcal{A}_2^c.$$

Letting $K' = \#(\mathcal{A})$ and replacing K by K' in the foregoing gives the Occam's window version of BMA prediction. Just as BMA can be done 'dynamically' – rechoosing the weights used for the prediction from each model at each time step – Occam's window BMA prediction can be done dynamically; see Onorante and Raftery (2014).

Another Bayes testing approach rests on the observation that when a true model exists it will often not be on the model list and so Bayes procedures will not converge to it. In fact, the posterior probability will converge to the model on the list closest to the true model but different models on the list will often provide better approximations of the true model on different regions of the covariate space. Global weights such as are used in BMA are unable to take this into account.

Following Yu *et al.* (2013), to explain the basics of the technique, it is enough to consider two models, i.e., $K = 2$, since the extension to more models is straightforward. Since the weights in BMA are posterior model probabilities, taken together they have the same information as Bayes factors (BFs). Thus, the BF in favor of $\mathcal{M}_1$ over $\mathcal{M}_2$ is B_{12} defined from the $W(\mathcal{M}_k|y^n)$ by writing

$$\frac{W(\mathcal{M}_1|y^n)}{W(\mathcal{M}_2|y^n)} = \frac{W(\mathcal{M}_1)p(y^n|\mathcal{M}_1)}{W(\mathcal{M}_2)p(y^n|\mathcal{M}_2)} \equiv B_{12}\frac{W(\mathcal{M}_1)}{W(\mathcal{M}_2)},$$

where the last fraction is the prior odds ratio. Given a partition of the covariate space $\mathcal{X}$ into R disjoint regions $\mathcal{X}_1, \ldots, \mathcal{X}_R$, Yu *et al.* (2013) defined local BFs by restricting the covariate space to $\mathcal{X}_r$:

$$\mathrm{BF}_{12r} = \frac{p(y_r^n|\mathcal{M}_1)}{p(y_r^n|\mathcal{M}_2)} \quad \text{where} \quad y_r^n = \{x_i \in \mathcal{X}_r\}.$$

Omitting precise definitions of the likelihoods, $\ln B_{12} = \sum_{r=1}^{R} \ln B_{12r}$. Simplifying the remainder of the steps, the method then proceeds by arbitrarily splitting the data into two equal subsets Z_1 and Z_2, finding the posterior $w(\theta_k|Z_1)$ for $k = 1, 2$, using this to find the marginal distribution M_{ik} of a data point (y_i, x_i) in Z_2 under model k ($k = 1, 2$), finding $z_i = \log \mathrm{BF}_i = \log(M_{1i}/M_{2i})$ for the (y_i, x_i) in Z_2, and taking the z_i with explanatory vectors x_i as new data to input to a regression tree procedure that determines the $\mathcal{X}_r$. Predictions for new x's follow from using the local weights and local models for the region containing x. Effectively, Yu *et al.* (2013) partitioned the covariate space (using trees) while Occam's window partitions the model list. It is natural to ask whether combining the two techniques would improve prediction further.

11.2 Bagging

Bagging is a contraction of 'bootstrap aggregation', a strategy to improve the predictive accuracy of a procedure. Like BMA, it is another way to generate and combine predictors, possibly based on a model, in the hope that the resulting predictor will encapsulate whatever is stable in the DG and therefore give improved predictions. The basic procedure is simple: given a sample (y_i, x_i) for $i = 1, \ldots, n$, define a base model or base predictor $\hat{f}(x)$ using n data points and consider predicting the response for a new x_{new}. To do this, draw B bootstrap samples from the training data; i.e., draw B samples of size n, with replacement, from the original sample. Each bootstrap sample is used to find another $\hat{f}$, say $\hat{f}_b$, for $b = 1, \ldots, B$. The bagged point predictor for Y_{new} is

$$\hat{f}_{\mathrm{bag}}(x_{\mathrm{new}}) = \frac{1}{B}\sum_{b=1}^{B} \hat{f}_b(x_{\mathrm{new}}). \tag{11.10}$$

Although rarely used, bagging also leads to PIs. Suppose that $Y(x) = f(x) + \epsilon$ with the usual assumptions. Setting $\hat{Y}(x) = \hat{f}_{\text{bag}}(x)$, a simple identity gives

$$\text{Var}(\hat{Y}(x_{\text{new}})) = \text{Var}(Y(x_{\text{new}})) + \text{Var}(\hat{Y}_{\text{bag}}(x_{\text{new}}) - Y(x_{\text{new}})) \qquad (11.11)$$

The first term is σ^2, which can be estimated by the MSE of $\hat{Y}$. The second term can be estimated by using a repeated bootstrap procedure to get, say, B' values $\hat{Y}_{\text{bag},b'}(x_{\text{new}})$. The average of these B' values can be used in place of Y, so that

$$\text{Var}(\hat{Y}_{\text{bag}}(x_{\text{new}}) - Y(x_{\text{new}})) \approx \sum_{b'=1}^{B'} \left(\hat{Y}_{\text{bag},b'}(x_{\text{new}}) - \frac{1}{B'} \hat{Y}_{\text{bag},b'}(x_{\text{new}}) \right)^2.$$

So, a $1 - \alpha$ PI for $Y(x_{\text{new}})$ is

$$\hat{f}_{\text{bag}}(x_{\text{new}}) \pm z_{\alpha/2} \sqrt{\hat{\sigma}^2 + \text{Var}(\hat{Y}(x_{\text{new}}))}.$$

If desired, a variance–bias decomposition of the right-hand term in (11.11) can be used and gives a slightly different PI.

A set of bootstrap samples from a sample of size n is taken to be of size n unless noted otherwise. Since the data points are chosen with replacement according to a discrete uniform distribution on n points, the probability that a fixed x_i is not chosen for a given bootstrap sample is $(1 - 1/n)^n \to 1/e \approx 0.37$ as $n \to \infty$. Hence, a given x_i has probability 0.63 of being in a given bootstrap sample and the expected number of distinct points in a bootstrap sample of size n is $0.63n$. If a bootstrap sample is drawn from a bootstrap sample then it can be verified computationally that the expected number of distinct points in it is $0.47n$.

Assuming that the base predictor $\hat{f}$ is 'good', $\hat{f}_{\text{bag}}$ typically improves it. What does 'good' mean and why should bagging a predictor improve it? In the present context, a procedure is 'good' if it is drawn from a large enough class of procedures that it can, in principle, find one (or more) that has small error. This often means that the collection of predictors over which $\hat{f}$ may vary is large, and hence there is a great deal of predictor uncertainty relative to n. Otherwise stated, the selection of the predictor is unstable – not robust to repeated selection – but predictor selection is nevertheless consistent or asymptotically unbiased (assuming that there is a model). Bagging can improve these good but unstable predictors, making them closer to optimal. It does this by weighting the terms in the average equally so that the resampling ensures that the predictors in the average are more representative of an optimal predictor (again assuming that one exists). Essentially, the resampling is a proxy for having more independent data sets that can be used to provide more averaging thereby moving $\hat{f}$ closer to its mean.

For instance, well-chosen trees are usually good but unstable predictors. The class of all trees is a collection of step functions of K explanatory variables and so is dense in the set of measurable functions of those K variables. Therefore, if limitations due to sample size are ignored, trees can approximate any of these functions. In fact, there will usually be many trees, with very different structures, that are equivalent in terms of function approximation. Hence there will be high model uncertainty or instability even though trees can provide excellent approximations. Accordingly, one expects that bagging trees will give better predictions than those from using a single tree. This is borne out by a popular technique called random forests. The same argument applies to neural nets, subset selection in linear regression, and most kernel methods because these involve large model classes in which stability is usually more important than bias. However, as a generality, bagging does not provide much improvement of good but stable predictors such as nearest neighbor methods, even though the model class is nonparametric.

One of the original arguments for bagging is due to Breiman (1994). The argument exploits the possibility that a procedure may predict well on average, e.g., in the limit of large n the predictor may correspond to the true model, but the procedure may have poor generalization error for finite n. To see this, let $\hat{\phi}(x, D)$ be a predictor for Y when $X = x$, where the data are $D = \{(y_1, x_1), \ldots, (y_n, x_n)\}$. Then, under squared error loss, the expected average prediction error (APE) for $\hat{\phi}(x, D)$ is

$$\text{APE}(\hat{\phi}) = E_D E_{Y,X}(Y - \hat{\phi}(X, D))^2 = E_{Y,X} E_D(Y - \hat{\phi}(X, D))^2, \tag{11.12}$$

where E_D is the expectation with respect to the n data points regarded as random variables and $E_{Y,X}$ is the expectation with respect to a new, $(n+1)$th data point.

By contrast, the expectation over the data of $\hat{\phi}(x)$ for fixed x is $\phi_A(x) = E_D \hat{\phi}(x, D)$ and the APE of $\phi_A(\cdot)$ is

$$\text{APE}(\phi_A) = E_{Y,X}(Y - \phi_A(X))^2 = E_{Y,X}(Y - E_D \hat{\phi}(X, D))^2. \tag{11.13}$$

Jensen's inequality gives that, pointwise in $X = x$, $(E_D \hat{\phi}(x, D))^2 \leq E_D \hat{\phi}^2(x, D)$ almost everywhere. So, the relationship between (11.12) and (11.13) is

$$\text{APE}(\hat{\phi}) = E_Y Y^2 - 2E_{Y,X,D} Y \phi_A + E_{Y,X} E_D \hat{\phi}^2(X, D) \geq E_{Y,X}(Y - \phi_A)^2 = \text{APE}(\phi_A), \tag{11.14}$$

with equality if and only if $(E_D \hat{\phi}(X, D))^2 = E_D \hat{\phi}^2(X, D)$ almost everywhere in X.

The reasoning behind inequality (11.14) highlights the role of variability. If the effect of D on $\hat{\phi}(x, D)$ is small, i.e., $\hat{\phi}(x, D)$ is already stable, then $\text{APE}(\hat{\phi}) - \text{APE}(\phi_A)$ should be small. On the other hand, if averaging over D has a big effect, the two sides of (11.14) will be quite different, $\text{APE}(\phi_A)$ being much smaller. Since the bagged estimate $\hat{\phi}_B(x)$ of $\hat{\phi}$ is a proxy for ϕ_A, bagging improves $\hat{\phi}(x, D)$ most when it is much less stable than ϕ_A. In practice, $\hat{\phi}_B(x)$ can actually do worse than $\phi_A(x)$. So, there is a crossover point in terms of stability. Above a certain level of stability, bagging is little or no help, possibly even worsening the predictor. Otherwise stated, a good, stable, procedure $\hat{\phi}$ should vary tightly around an optimal predictor ϕ_{opt} so that $\hat{\phi} \approx \phi_A \approx \phi_{\text{opt}}$. Below a certain level of stability, however, bagging may be of great use in stabilizing a predictor to achieve $\hat{\phi}_B \approx \phi_A \approx \phi_{\text{opt}}$.

The same sort of argument applies in the classification context, but the loss function is zero–one rather than the squared error. Following Breiman (1994), suppose that $\phi(x, D)$ predicts a class label $j \in \{1, \ldots, J\}$ at x given data D. If (X, Y) are drawn from P independently of D then

$$P_{X,Y}(Y = \phi(X, D)) = \sum_{j=1}^{J} P_{X,Y}(j = \phi(X, D)|Y = j) P_Y(Y = j).$$

Let $Q(j|x) = P_D(\phi(x, D) = j)$, where P_D is the probability of the data D, so that the averaged classifier is $\phi_A(x) = \arg\max_j Q(j|x)$. The probability of correct classification by ϕ_A is

$$\sum_{j=1}^{J} E(Q(j|X)|Y=j) P_Y(Y=j) = \sum_{j=1}^{J} Q(j|x) P_Y(Y=j|x) P_X(dx),$$

where P_X is the marginal for X, assumed to exist. The 'risk' of ϕ_A is

$$R_A = \sum_{j=1}^{J} \int \chi_{\{\arg\max_j Q(j|x)=j\}} P(j|x) P_X(dx);$$

i.e. it is the expected value of the loss function.

The optimal classifier is $\phi_{\text{opt}}(x) = \arg\max_j P(j|x)$. So, ϕ_{opt} and ϕ_A agree on the set

$$C = \{x | \phi_{\text{opt}}(x) = \phi_A(x)\}$$

and a larger C indicates that ϕ_A is closer to ϕ_{opt}. Now, R_A can be written as

$$R_A = \int_C \max_j P(j|x) P_X(dx) + \int_{C^c} \sum_{j=1}^{J} \chi_{\{\phi_A(x)=j\}} P(j|x) P_X(dx).$$

The risk of the optimal classifier is

$$R_{\text{opt}} = \int \max_j P(j|x) P_X(dx),$$

which is similar to the first term in R_A. When $C^c = \varnothing$, $R_A = R_{\text{opt}}$ so that $\phi_A = \phi_{\text{opt}}$, i.e., ϕ_A is optimal. However, for fixed $x \in C$, it is possible that

$$\sum_{j=1}^{J} Q(j|x) P(j|x) < \max_{j=1,\dots,J} P(j|x).$$

Thus, if ϕ is unstable and $P_X(C) \approx 1$ then ϕ may be very different from ϕ_A even though ϕ_A is close to ϕ_{opt}. Otherwise put, when C is large and ϕ is good but unstable, ϕ_A is a nearly optimal improvement on ϕ. If ϕ is relatively stable, the improvement may be small and ϕ_A may even be worse than ϕ. Since the bagged version $\hat{\phi}_B$ of ϕ is a proxy for ϕ_A, these conclusions apply to $\hat{\phi}_B$. Hence, again, bagging improves good but unstable classifiers primarily by stabilizing them.

The formula (11.10) is a mean – the terms in the sum have equal weight. The differences from term to term in the sum lie in the values of the data set that are resampled and hence in the function estimators given by the resampled data. Hence, it is equivalent to regard the function estimators as generated randomly by the resampling. This is a sense in which taking the weighted mean of the models in the model average is a parallel to taking an average of observations. Just as using a sample mean is more informative for prediction than using an individual outcome, so using a model average is more informative than using a single model. 'Informative' here simply means more efficient or less biased. The same argument applies when the terms in the sum have unequal weights, as long as the weights represent the plausibility or probability of their respective terms, as in BMA or other model averages. This is the sense in which the models being averaged can be regarded as the independent outcomes of an auxiliary process representing a collection of experts (or good predictors), as discussed at the beginning of this chapter.

Bagged classification trees are regarded as one of the most successful classifiers. Indeed, classification is the context in which bagging is most developed and this procedure is often called 'random forests'. The idea is that each tree in the collection of trees, or forest, is a classifier: each of its leaves assigns a response Y to a class, these are usually denoted 0 and 1 in the case of binary classification. To form a random forest, draw a collection of B bootstrap samples of size n. From each of these bootstrap samples generate a classification tree using one of the techniques described in Sec. 10.2, but do not prune. For each tree, a new value x_{new} assigns a class $T_b(x_{new})$ for $b = 1, \ldots, B$. The majority voting technique then assigns x_{new} to whichever class gets the majority of votes, 0 or 1, from the $T_b(x_{new})$.

There are three extra procedures that are often used with random forests (RFs). First, usually the splits are done on a random selection of variables, often taken to be $\sqrt{K}$ where $K = \dim(x)$. This ensures that the trees do not just keep the best of several strongly associated explanatory variables. Different trees in the forest will be able, on average, to capture the benefits of each variable. Second, there is a natural way to evaluate the overall performance of RFs. Since approximately 63% of the original n data points occur in each bootstrap sample, the tree formed from each bootstrap sample can be evaluated on the 37% of data points not used. Thus, each tree has an 'out-of-bag' (OOB) classification error, so called because bootstrapping is like choosing a 'bag' of data from n data points with replacement. The proportion of times that each tree misclassifies on its OOB set can be averaged over the trees to give an assessment of how well the overall forest is performing. It has been suggested that this average gives a consistent value for the misclassification of the forest. Third, there is also a natural way to assess the importance of each variable going into the RFs procedure. Simply, use the usual RFs ensemble method and then pick one of the X_k. Randomly permute the n values of X_k. Given the misclassification proportions from the trees in the forest, evaluate the misclassification rate of each tree on the n data points with the permuted values of X_k. The average of these differences in misclassification rates over the trees in the forest gives the importance of the variable X_k. Notice that each of these terms is phrased in terms of classification but has a natural analog for a forest of regression trees if one simply replaces the misclassification error with the average of the residuals. Thus, RFs amount to an improved version of bagging.

A natural question is how a simple interpretable model-based method for classification such as logistic regression compares with trees and RFs for classification. The answer is usually that the main effect terms in the logistic regression are different from the most important variables in RFs. The same holds true for linear regression and RFs for regression. The reason is that the main effect terms for logistic and linear regression are found using all the data i.e., the modeling is not conditional on using a subset of the data. By contrast, each split in a tree is conditional on the previous splits and hence on the region of outcomes defined at the node being split. So, with the possible exception of the initial split in a tree, there are no unconditional main effects. This holds for RFs as well – the splits in the trees after the initial split are conditional, and when the classifications for the trees in the forest are combined they do not become unconditional. The exact relationship between an RF and either a logistic or linear regression is harder to characterize because the conditioning may vary from tree to tree, but it is rare for an average over conditional structures to duplicate main effects.

Breiman (2001b) established an important property of RFs for classification. Imagine bagging J binary tree classifiers taking a majority vote to make a class assignment. The average number of correct classifications is $\mathrm{AV}(Y) = \sum_{j=1}^{J,*} \mathbf{1}_{h_j(x)=y}$ and the average number of misclassifications for an incorrect class k is $\mathrm{AV}(k) = \sum_{j=1}^{J,*} \mathbf{1}_{h_j(x)=k}$. The notation $\sum^*$ indicates that the sum is divided by its number of nonzero summands (which is random). Let $\mathrm{DA}(X,Y) = \mathrm{AV}(Y) - \max_{Y \neq k} \mathrm{AV}(k)$, the biggest difference between the average number of votes for the correct class and the average number of votes for the most probable incorrect class.

Larger values for DA reflect the lower error of the majority vote classifier formed from $h_1, \ldots, h_J$. The predictive error of this classifier is

$$\mathrm{PE} = P(\mathrm{DA}(X,Y) < 0),$$

in which X and Y are treated as random. In RFs, the h_j are random and can be represented as $h_j(X) = h(X, \Theta_j)$, in which Θ is a random variable indexing the tree chosen for a given data set. This leads to the following.

Theorem 11.1 (Breiman 2001b) *Almost surely, for all sequences $\Theta, \ldots, \Theta_J$,*

$$\mathrm{PE} \to P\left(P_\Theta(h(X, \Theta) = Y) - \max_{Y \neq j} P_\Theta(h(X, \Theta) = j) \right)$$

Proof See Breiman (2001b, Appendix I). □

This theorem means that as individual tree classifiers are included in the majority vote classifier, the probability of misclassification converges to the generalization error. So, using more and more individual trees leads to a limiting value of the PE, not to overfitting the classification function.

The intuition behind this theorem leads to the conjecture that RFs for regression trees will have an analogous property. Briefly, every regression problem can be converted to a limit of classification problems, where the limit is found as the number of classes increases. Assume that the dependent variable Y is monotonic on a compact region of the covariate space. Partition its range into D sets of equal length, say $C_1, \ldots, C_D$, and let Y_D be the discretized version of Y,

$$Y_D = Y_{\mathrm{mid},d} \quad \text{when} \quad Y \in C_d, \quad d = 1, \ldots, D$$

where $Y_{\mathrm{mid},d}$ is the midpoint of C_d. Breiman's theorem applies to Y_D for each fixed D as the number of individual tree classifiers increases. So, D can be allowed to increase slowly with n until Y_D satisfies $\max_d \left(\max |Y_D - Y| \chi_{C_d} \right) < \epsilon$ for any pre-assigned $\epsilon > 0$. That is, as D and n increase at suitable rates, Y_D converges to Y and Y_D does not lead to overfitting as $J \to \infty$.

Recall that any regression or classification problem has explanatory variables that can be treated as either deterministic, i.e., as design points, or random. In BMA this is blurred because conditioning on the data means that they are treated as fixed even though they might have been random. (In fact, here they are treated as fixed because no likelihood for X is used.) For RFs, the X's are usually treated as random and one consequence is that the aggregation over trees effectively reduces the variance. To be precise, let $T_b(\cdot)$ for $b = 1, \ldots, B$

be the trees formed from B bootstrap samples of size n. The bagged classifier or regression function is given by (11.10) and for fixed x_{new} has variance

$$\text{Var}\left(\frac{1}{B}\sum_{b=1}^{B}T_b(x_{\text{new}})\right) = \frac{1}{B^2}\sum_{b=1}^{B}\left(\sum_{b'\neq b}\text{Cov}(T_b(x_{\text{new}}), T_{b'}(x_{\text{new}})) + \text{Var}(T_b(x_{\text{new}}))\right)$$

Letting $\sigma^2 = \sigma^2(x_{\text{new}}) = \text{Var}(T_b(x_{\text{new}}))$ and $\rho = \rho(x_{\text{new}}) = \text{Corr}(T_b(x_{\text{new}}), T_{b'}(x_{\text{new}}))$ independently of b and b', since all B samples are independent and identical,

$$\text{Var}\left(\frac{1}{B}\sum_{b=1}^{B}T_b(x_{\text{new}})\right) = \frac{1}{B^2}\sum_{b=1}^{B}((B-1)\rho\sigma^2 + \sigma^2) = \rho\sigma^2 + \sigma^2\frac{1-\rho}{B}.$$

Thus, as B increases, the second term becomes small and if a smaller number of randomly selected explanatory variables than $\sqrt{K}$ is selected then the first term decreases. Hence, if $\dim(x)$ increases, the trees in the bagging predictor become less and less correlated. Thus, an RF predictor should scale up well to high-dimensional X's because its variance can be made small. The same argument applies to bagging any predictor, so the reduction in correlation is a property of bagging rather than a property of the predictors being bagged. For a more thorough presentation of the theory behind bagging and RFs see Louppe (2012), especially Chaps. 4 and 6.

Obviously, bagging can be applied to any procedure that uses the n data points (and automatically specifies any other inputs to the bagging procedure). Thus the Bayes procedures in Sec. 10.2.3 can be bagged – although the resulting procedure (Bayes bagging) might be regarded as outside the Bayes paradigm. In this context it is worth recalling BART. The idea is to combine many 'weak learners' (stumps or piecewise-constant tree functions taking a small number of values, often only two or three) in the hope that combining them will give a better predictive performance than RFs. Very loosely, BART can be a regarded as a BMA of stumps formed by bagging, i.e., a modification of what might be termed Bayes bagging. The posterior weights of the stumps and the explanatory power of each stump must be managed so that no one stump dominates. Thinking of stumps as parallel to empirically constructed (by bagging) basis elements, BART finds a basis expansion for Y in terms of an empirical basis of stumps. In this sense, BART is a hybrid of two techniques, namely, a BMA over stumps and bagging. There is some evidence that extra layers of averaging tend to give a better performance for more complex data, and this may partly account for the strong performance of BART. Indeed, BART, BMA, and the other nonbagging model-averaging techniques in this chapter can be bagged as well, regardless of the components in the average. This does not seem to have been systematically explored.

Recent work on RFs, and on bagging more generally, includes that of Biau (2012), who argued that the convergence rate of RFs depends more on how may explanatory variables dominate in the predictor function than on the number of explanatory variables that may be available. In this sense, RFs are a sparsity method in the same sense as penalized methods that have the oracle property; see Sec. 10.5. Denil *et al.* (2014) compared several variants on the overall RF procedure in an effort to assess the reasonableness of the different simplifications made in theoretical work on RFs. Ren *et al.* (2015) provided another variant on RFs in which pruning and growing are done using a global rather than a local criterion. These are merely some of many efforts to modify RFs in order to get better results. Rather than analyzing versions of RFs, Jones and Linder (2015) looked at what features RFs (and trees more generally) can model well. Their goal was to help develop intuition about how RFs typically behave, e.g., how variable importance can be assessed, especially when the explanatory variables are mutually associated, how missing values can be overcome, and how trees can capture local interactions between explanatory variables better than linear or logistic regression.

Even though bagging can be applied generally, it has, for the most part, been thoroughly investigated theoretically only for trees.

For illustration, consider the case of NNs. They are good for nonlinear regression or classification in the sense of having low bias, but they are apt to be unstable. This is partly because the error surface they define is rough, so it's easy for the iterative fitting procedure to get caught in local minima. The instability also arises because two NNs may be close numerically yet far apart in terms of their parameter values. The result is that, even with a penalty term, NNs can overfit. It can be argued that bagging NNs is one way to reduce the degree of overfit and hence improve the generalization error. This is a different argument from those used and formalized in Breiman (1994). However, it seems to be valid as a principle. Chaouachi *et al.* (2012), for instance, bagged NNs for the short-term forecasting of solar power. They used up to two hidden layers and permitted autoregressive input data. They compared these bagged NNs with individual NNs, radial basis function NNs (essentially NNs with the logit transfer function replaced by a normal density), and certain types of recurrent NNs.[1] Chaouachi *et al.* (2012) argued that bagged NNs typically have a better performance because they have increased redundancy and hence a decrease in the degree of overfit built into the predictor. Chaouachi *et al.* (2012) also suggested this was not uniformly true because, when power generation was at its highest or lowest, the recurrent NNs consistently gave slightly better predictions. That is, while the use of bagged NNs avoided trapping of the fitting procedure in local minima, when the predictand was very high or very low then averaging NNs was detrimental.

By contrast, NNs were bagged for classification and regression purposes in Islam *et al.* (2008). In their work they found that the variance-based arguments of Breiman (1994) carried over heuristically to both the bagging and boosting of NNs (see Sec. 11.4). The difference may be due to the nature of the data being analyzed, in that Islam *et al.* (2008) probably had an $\mathcal{M}$-complete problem whereas Chaouachi *et al.* (2012) probably had an $\mathcal{M}$-open problem.

Other contributions in the general area of bagging NNs include that of Maclin and Opitz (1997), who confirmed that bagging NNs (or decision trees) for classification essentially never overfits and observed that bagging also handles noisy variables well. They noted that while boosting (see Sec. 11.4) may sometimes give better results, it is prone to overfitting. Zhou *et al.* (2002) offered a refinement to bagging NNs for classification and regression, somewhat like the way in which RF refines bagging for trees. The idea is that the weights on the NNs in the ensemble are found by an auxiliary procedure called a genetic algorithm.

---

[1] Recurrence in NNs was mentioned briefly in Sec. 10.3. The idea is that certain neurons have outputs that feed into layers behind their origin, including the layer of nodes from which they were produced. These networks have different convergence properties because the recurrence can lead to stability or instability of the network, depending on its exact form.

Genetic algorithms provide a search strategy for solving constrained optimization problems by defining a collection of artificial organisms (candidate solutions), letting them reproduce and mutate (change values or evolve), and selecting the best adapted, i.e., those that come closest to optimal. The resulting ensemble consists of only some of the bagged NNs and can be regarded as an analog of Occam's window for bagging. A variant on this was advocated by Martínez-Muñoz and Suárez (2006), who ordered bagging ensembles and pruned them. The pruned 'subensembles' seem to improve the generalization error in classification problems. In a methodological point, Petersen *et al.* (2007) argued that cross-validation should not be used internally to a bootstrap-generated element in a bagging procedure. It is better to use cross-validation only after the bagged predictor has been formed. The problem seems to be that regular cross-validation may lead to overfit within the summands of a bagged predictor that is not corrected by the averaging. It is likely that this principle applies to other optimizations performed within a bootstrap step. However, it is unclear whether the principle holds when the bootstrap sample is itself divided into independent subsamples for different purposes; see Sec. 9.4.

Robustness and stability have always been issues with bagging in general and bagged NNs in particular. Cunningham *et al.* (1999) argued that K-fold cross-validation is a good way to choose the NNs that emerge from bagging and that one should aggregate the output of several NNs chosen in this way. This is a variant on the usual bagging procedure which effectively tries to choose the better randomly generated NNs. Zhang (1999) tried to stabilize bagged NNs by dividing the data into training and testing sets in a variety of ways, generating a network from each partition, combining the results through a principal-components regression, and giving CIs and PIs by further bootstrapping. Although apparently unused, one way to assess the stability of any method, ensemble or otherwise, is to perturb the data by adding randomly generated noise to it from $N(0, \sigma^2)$ random variables and observing how the stability of predictions or other inferences decreases as σ increases; see the end of Sec. 9.5. The edited volume by Sharkey (1999), although dated, provides a collection of perspectives on aggregating neural nets that remains relevant.

Bagging has also been applied to kernel methods such as support vector machines (SVMs), mostly for classification. Perhaps the earliest contribution in this direction was Kim *et al.* (2003) for SVM classification. Several standard ways to combine the bootstrapped SVMs, e.g., majority voting, outperformed individual SVMs. (This was also shown when SVMs were 'boosted'; see Sec. 11.4.) Essentially this technique was used in Zarasiz *et al.* (2012). The simplest bagged SVMs were found to provide a highly accurate classification of leukemias in terms of the receiver operator curve (ROC) and other assessments. In an effort to refine bagging (again as a parallel to the way in which RFs refine bagging for trees), Valentini and Dietterich (2003) found that bagging 'low-bias SVMs' (as determined by OOB assessments) never increased the generalization error and often decreased it over well-tuned single SVMs and even over bagged individually well-tuned SVMs. In a further refinement, Zaman and Hirose (2009) used 'double bagging' – using one bootstrap sample to construct a classifier and its complement to construct another that can be combined with it. They found that this procedure compared favorably in terms of the generalization error for messy data. In some sense this is not a surprise since it is known that SVMs can overfit, and double bagging ensures that the whole sample is used (albeit the two parts in different ways), thereby correcting for the slight reduction in variance in a bootstrap sample when compared with the original sample. As a generality, bagging has been found to provide high-quality classifiers in a variety of complex settings.

Comparisons of several classification techniques are common, but Liang *et al.* (2011) provided one of the first few relatively comprehensive empirical studies of bagging predictors for different base predictor classes. This study used 12 predictor types and 48 benchmark data sets but did not include a systematic simulation study. The focus was not only on accuracy but also on stability and robustness as they defined it. Insisting on these properties helped ensure a better performance for the bagged predictors that they studied than any other single predictor except for K-nearest-neighbors and Bayes. It should be recalled that K-nearest-neighbors, being nonparametric, will be robust but will not scale up to high dimensions very well, while Bayes classifiers are optimal but hard to find in practice. Bagging NNs and trees often gave the best performance.

In a more general context, an extensive comparison of ensemble methods was also provided by Song *et al.* (2013). These authors used benchmark data sets and simulations to compare a wide variety of techniques including a new one they proposed: basically, the bagging of generalized linear models (GLMs). Their new method necessitated the construction of good GLMs for each bootstrap sample and hence various auxiliary techniques such as variable selection. They argued that their method provided as good predictions as any other technique e.g., RFs, outperforming most of them as well as permitting a degree of interpretability which the others lacked. Being based on GLMs their method could be used for both binary and continuous data. In effect it finds a more useful form of sparsity than many other methods, at least when used as individual rather than ensembled predictors.

Finally, bagging can be applied to other individual predictors. Often it helps as noted above but, in many cases, e.g., that of multivariate adaptive regression splines, it does not; see Arleina and Otok (2014). One would not expect bagging to improve classical nonparametrics. An obvious variant on bagging would be to use the median or trimmed mean of the f_b over b; it should be almost as good as bagging and much more stable. The effect of bagging on projection pursuit, penalized methods, relevance vector machines, logic trees and many other 'learners' seems unexplored, at least systematically. Also, combining bagging versions of two or more classes of predictors at the same time does not seem to have been explored. There are many promising variations and settings in which bagging has not yet been tested.

11.3 Stacking

Stacking is a model-averaging technique that weights models by coefficients derived from an optimality criterion based on cross-validation (CV). Stacking was invented by Wolpert (1992a), see also Wolpert (1992b), and studied by Breiman (1996a) amongst many others. The setting is the case where Y can be expressed as a signal plus noise and K distinct models $\mathcal{M}_k$, $k = 1, \ldots, K$, are available, represented as functions of an explanatory variable x of dimension J, and indexed by a parameter θ_k with $\dim(\theta_k) = d_k$. Thus, $\mathcal{M}_k$ is of the form $\{f_k(\cdot|\theta_k)|\theta_k \in \Theta_k\}$, where Θ_k is the parameter space for the kth model. The $\mathcal{M}_k$ may be trees, NNs, or other classes of functions. Although this is not common, different $\mathcal{M}_k$ may also be from different function classes.

The task is to combine the f_k into a single predictor. First, suppose that data of the form (y_i, x_i) for $i = 1, \ldots, n$ are available. In the special case of leave-one-out (LOO) CV, there are n distinct subsets of the data formed by leaving out one (y_i, x_i). Let $\hat{f}_k^{-i}(x) = f_k(x|\hat{\theta}_d)$, where it is understood that $\hat{\theta}_d$ is an estimate of θ_d using all but the ith data point. Thus, for each k, there are n estimates $f_k^{-i}(\cdot)$ and hence n values $f_k^{-i}(x_i)$. These values represent the best efforts of f_k to provide a point predictor for $Y(x_i)$. Hence, empirical weights $\hat{w}_k$ for the $\hat{f}_k$ can be found from the training data. The key condition defining the vector $\hat{w} = (\hat{w}_1, \ldots, \hat{w}_K)$ is

$$\hat{w} = \arg\min_w \sum_{i=1}^{n} \left(y_i - \sum_{k=1}^{K} w_k \hat{f}_k^{(-i)}(x_i) \right)^2. \tag{11.15}$$

Typically the set over which $w = (w_1, \ldots, w_K)$ varies in (11.15) is a subset of $\mathbb{R}^K$. Common choices are (i) all w_k nonnegative, (ii) $\sum_{k+1}^{K} w_k = 1$, or (iii) both (i) and (ii). Less common, but sometimes more effective, is to allow the w_k to be any real numbers. Whatever set is chosen, the $\hat{w}_k$ that arise from (11.15) tend to be close to zero for f_k that have poor LOOCV accuracy in the training sample.

Now, the point predictor at x_{new} from a stacking model average (SMA) is

$$\hat{f}_{\text{stack}}(x_{\text{new}}) = \sum_{k=1}^{K} \hat{w}_k \hat{f}_k(x_{\text{new}}), \tag{11.16}$$

where $\hat{f}(\cdot) = f(\cdot|\hat{\theta}_k)$, in which it is understood that $\hat{\theta}_k$ is found using all the data. Since the deficiencies of LOOCV are well known, the stacking weights used in (11.15) and (11.16) are more typically found by k-fold CV. For instance, a 'fifths' approach is common: partition the data set at random into five subsets as close to equisized as possible. Use four of the subsets to find the $\hat{f}_k^{-u}$, where $u = 1, \ldots, 5$ indicates the subset not used. Then find the sum of the residuals on the uth subset, cycle through the u's, and find $\hat{w}$ by minimizing the sum of the residuals over the values of u. Since the squared error is used the optimizations can usually be done by quadratic optimization procedures. If other loss functions are used, the computing may be more demanding.

To get a $1 - \alpha$ PI from an SMA consider the following empirical procedure. As in earlier cases, write

$$\text{Var}(Y(x_{\text{new}}) - \hat{f}_{\text{stack}}(x_{\text{new}})) = \sigma^2 + \text{bias}^2(E\hat{f}_{\text{stack}}(x_{\text{new}})) + \text{Var}(\hat{f}_{\text{stack}}(x_{\text{new}})).$$

The obvious estimator for σ^2 is

$$\hat{\sigma}^2 = \frac{1}{n - K} \sum_{i=1}^{n} (y_i - \hat{f}_{\text{stack}}(x_i))^2,$$

where K is the number of models. (A logical alternative would use $K = \dim(\theta_1) + \cdots + \dim(\theta_K)$ to account for all degrees of freedom.) The bias term is $f_T(x_{\text{new}}) - E\hat{f}_{\text{stack}}(x_{\text{new}})$, assuming that a true model exists. The expectation $E\hat{f}_{\text{stack}}(x_{\text{new}})$ can be found by bootstrapping: use B bootstrap samples to form B estimates $\hat{f}_{\text{stack},b}(\cdot)$, $b = 1, \ldots, B$, so that $E\hat{f}_{\text{stack}}(x_{\text{new}}) \approx (1/B)\sum_{b=1}^{B} \hat{f}_{\text{stack},b}(x_{\text{new}})$. Then use another bootstrap average of non-parametric estimators of f_T, say, $\sum_{b=1}^{B} \hat{f}_{\text{np},b}(x_{\text{new}})$. The difference gives an estimate of the bias term. Finally, again take B bootstrap samples to form $\hat{f}_{\text{stack},b}(\cdot)$ for $b = 1, \ldots, B$ and hence $\bar{f}(x_{\text{new}}) = \sum_{b=1}^{B} \hat{f}_{\text{stack},b}(x_{\text{new}})$. Then, the variance term can be estimated by $(1/B) \sum_{b=1}^{B} (\hat{f}_{\text{stack},b}(x_{\text{new}}) - \bar{f}(x_{\text{new}}))^2$. Thus, subject to several regularity conditions, an approximate $1 - \alpha$ PI for $Y(x_{\text{new}})$ is

$$\hat{f}_{\text{stack}}(x_{\text{new}}) - \text{bias}(E \widehat{\hat{f}_{\text{stack}}(x_{\text{new}})}) \pm z_{1-\alpha/2} \sqrt{\hat{\sigma}^2 + \frac{1}{B} \sum_{b=1}^{B} (\hat{f}_{\text{stack},b}(x_{\text{new}}) - \bar{f}(x_{\text{new}}))^2}.$$

As presented here, stacking combines models, and hence the result is a model average. Stacking can also be used to combine densities (see Smyth and Wolpert (1998, 1999)) and classifiers (see Džeroski and Ženko (2004) and Rokach (2010)). In all cases, similar questions arise: what set of w's and f_k should be used? These same questions arise in BMA because (i) the analog of w is the set of posterior model weights as derived from an across-model prior that must be chosen, and (ii) the analog of the $\hat{f}_k$ is the set of densities denoted $p(y|\theta_k, \mathcal{M}_k)$ in Sec. 11.1 (which includes within-model priors). Bagging is different because the weights on the $\hat{f}_b$ are uniform, $1/B$ for each of the B bootstrap samples, and only one model f is chosen. So, unless the f in bagging represents a particularly large class, the span of BMAs or SMAs will be larger than that of bagging MAs. Overall this means that the most important design problem in MA is deciding which models to include. This is similar to the selection of the model list to be used with an MSP.

Even if the same models are used in two different MAs, the behavior of the MAs is not in general the same. This arises from the way in which the coefficients on the models are derived. For instance, thinking of Berk (1966), it is likely that the BMA converges to the model on the list closest to the true model (assuming it exists). The convergence of bagging is examined in Biau *et al.* (2008a). Random forests are consistent for the true function, whatever it may be, in many cases, in particular for the 1-nearest-neighbor case. However, it may also be inconsistent in some cases (see Proposition 8 in Biau *et al.*). Loosely, as long as the class of f_b is large enough, e.g., nonparametric, bagging or RFs is quite effective. Like bagging or RFs, the limit of an SMA does not have to be a model on the list. An SMA may converge to a model on the list, i.e., be consistent for one of the f_k, but more generally the consistency of SMAs depends on the behavior of the CV; see Sec. 9.4. As a generality, SMA has the linear span of the set of $\hat{f}_k$ and this may depend on the choice of $\hat{\theta}_k$. So, even though the span of an SMA is finite dimensional, for each selection of estimators the span of an SMA may be overall quite large.

A strategy for finding a good BMA is to choose models that surround the true model (if it exists) and then to correct for dilution if necessary. That is, an optimal choice for a model list for BMA is one for which the true model is near its geometric center. Hence, if nothing is known about where the true model lies, the natural collection of models is one for which any model not on the list is close to at least one model on the list. Here, 'close' means in the sense of the distance used to define the BMA predictor. For SMA, the intuition is often the exact opposite: Breiman (1996a) argued heuristically that the f_k should be chosen to be as different from each other, and hence as far apart, as reasonably possible. The effect is to maximize the span. In principle, an SMA can converge to any element of the span of the models in it. It does not in general converge to the model in the average closest to the true model.

Turning to the w's, the stacking coefficients downweight models that have a large MSE, i.e., models that are ill-fitting or are too complex, and depend only on the data. In BMA, the coefficients represent model plausibility and are functions of the likelihoods of the models as well as of the data. Minimizing the lack of fit and complexity is heuristically similar to maximizing plausibility, but the two concepts are different. This is seen in the different asymptotics of the stacking and the BMA coefficients. As a consequence of the dependence of BMA coefficients on the likelihoods, the BMA coefficients represent considerably more information than the stacking coefficients and hence converge to a narrow range of limits.

In contrast with stacking and BMA, which explicitly put weights on models, bagging pools over repeated evaluations. It can be argued that if the class of functions being evaluated in a bagging average is large enough, bagging amounts to implicitly finding the plausible models empirically and weighting them according to how often they are found. Thus, loosely, bagging is an empirical version of BMA or stacking. In fact, it can be proved that bagging reconstructs a BMA (see Le and Clarke (2016)) and that stacking is asymptotically a Bayes action; see Clyde (2012) and Le and Clarke (2017).

In practice, stacking and Bayes are often equivalent when the model list chosen is good. If the model list is perfect (i.e., there is a j such that $f_j = f_T$) then BMA usually does better because the BMA coefficients depend on the likelihood and so converge to zero or one faster than the stacking coefficients. However, as model list misspecification increases, stacking does better in predictive performance relative to BMA because stacking does not depend as much on the likelihood and also the span of the models being stacked is larger and so often is more representative of the data generator. Exactly how far a model on the model list from the true model must be for stacking to outperform BMA and still give reasonably good predictions is unclear. Usually the difference amounts to two to four terms.

Like other MAs stacking can be applied to many objects. In addition to densities and classifiers, SMAs may be formed by taking a number of bagging predictors and stacking them, or by bagging a single stacking predictor. Bayes model averages can be stacked and then bagged, or bagged and then stacked. The models used in the stacking average may be found by using part of the data to generate a collection of, say, Nadaraya–Watson estimators. The bottom line is that more elaborate model averaging tends to work better with more complex data. However, the cost is in the analysis of the components of the variability. With both pure BMA and pure stacking, i.e., given a fixed model list, the variability of the selection of the f_k to combine is zero. Otherwise stated, all SEs are conditional on the model list, and variability in the choice of model list is neglected. This neglect of higher-order variability is familiar from the model selection literature.

Of particular interest are the methodologies for combining stacking and bagging in Wolpert and Macready (1996). The basic idea is to form bootstrap averages using different base functions that correspond to the f_k in stacking. Then the bagged versions are stacked. As noted by Wolpert and Macready (1996): 'bagging is designed to improve an algorithm's variance and stacking can be viewed as also (or perhaps primarily) concerned with correcting the bias.' Thus, loosely, stacking helps overcome 'model mismatch', the difference between a model's f_k and f_T. When mismatch is low, stacking after bagging will not provide much gain in accuracy. When mismatch is high, there is a better chance that stacking after bagging will provide gains especially when bagging reduces variability as well; cf. Clarke (2003).

Stacking was brought into a Bayes $\mathcal{M}$-open problem setting by Clyde and Iversen (2012). The essence of the idea is that the models that are used to form an average should be regarded as actions and the coefficients on them regarded as weights to be found, so that the action space is the collection of linear combinations of the models. So, even though there is no prior on the models (since it is a priori assumed that none of them is true), the coefficients on the models that would correspond to a prior in an $\mathcal{M}$-closed setting simply define the full action space. This space is searched for the action with the maximal posterior expected utility (assuming that a true model exists) or searched empirically for the action that gives the best prediction.

Recall that the collection of w's over which the stacking weights are optimized must be specified. If the w's are in the K-dimensional simplex then their entries are nonnegative and sum to one. This is the same set as that containing all the posterior probabilities of a BMA, although the weights cannot in general be regarded as posterior probabilities. In this context, Clyde (2012) obtained $w_k \propto 1/\text{MSE}(\text{model } k)$, where the MSE is derived from the CV error. However, Clyde (2012) also argued that the sum-to-one constraint on the w_k should be released. She considered the case $K = 2$, with two linear models having orthogonal design matrices as models for the same response Y. Releasing the sum-to-one constraint means that the best predictive model (a sum of the two models each weighted by 1 rather than by 1/2) can be found and would satisfy the positivity constraint. Furthermore, relaxing the positivity of the w_k makes sense because the span of the SMA is increased, giving stronger optima, so a larger class of problems can be solved using stacking. If the w_k are allowed to be negative, there is a loss of interpretability regarding the influence of each component f_k in an overall stacking average. However, the primary goal is good prediction and allowing negative weights is a generalization that often helps in regression problems. Indeed, this aspect of interpretability may be more important in classification than regression. On the other hand, it has been argued that, in some classification settings, positivity constraints have little effect on a model's predictive performance; see Ting and Witten (1997).

In the classification context SMA can overfit, for instance with logistic-regression based classifiers, especially when the component models are well tuned and therefore likely to be correlated. One solution to this problem, discussed in detail in Reid and Grudic (2009), is regularization of the classifiers, e.g., by ridge regression or LASSO; see Sec. 10.5. Regularization usually gives sparsity (in terms of the number of explanatory variables) and reduces variance at the cost of only slightly increased bias. Since the resulting classifiers do not overfit, and combining them via stacking should not increase their bias, the overall bias and variance may be reduced.

As with BMA and bagging, any models can be used in SMA and many have been. In fact, one strength of the SMA is that several distinct varieties of models can be used in the same SMA. Sometimes these are called 'heterogeneous' models; see Sesmero *et al.* (2015). As noted there, a diversity of model types often helps increase accuracy in classification because the different classifiers effectively partition the space of explanatory variables into regions where each is most effective. This is a refinement of the observation of Breiman (1996b) that the components in stacking should be as different from each other as possible.

A finding that indicates the limits of this intuition was presented in Bourel and Ghattas (2013), in the context of using an MA of density estimators. They compared stacking histograms, bagging histograms, an average of histograms formed by randomly perturbing the break points, and several other MA methods. They did this for 12 densities and assessed the performance of the MAs by the mean integrated squared error. Concisely, stacking and stacking-related methods did best overall when n and K, the number of histograms being stacked, were not too large. As n and K increased, the performance of stacking deteriorated relative to other methods and, in fact, the best method in their large-scale examples was based on the random perturbation of histogram break points. (An intermediate method was based on boosting; see Sec. 11.4.) Thus, heuristically, in small problems, stacking tends to outperform other methods perhaps because its coefficients, not depending on likelihoods, converge relatively slowly. In other words, the coefficients reduce bias so much that their variability is not harmful and the SMA retains useful flexibility compared with other MAs. However, when n and K are large enough, other methods give a more accurate approximation of the optimal predictor because they have less bias (if K is large) and small variability (if n is large). Essentially, with large K and sufficiently large n for the chosen value of K, the search over candidate f_k determines the relative effectiveness of stacking compared with other methods. Since bias decreases as K and the spread of the f_k increase, other methods, e.g., the random generation of densities from a large class of histograms, can outperform stacking in larger problems. By the same logic, when K is small and the f_k can be chosen to be suitably dispersed, stacking performs well especially in the presence of bias due to the greater variability in the coefficients.

While many statistical objects can be stacked, the two most commonly occurring SMAs, aside from linear models, involve NN and SVM classifiers. It may be that these are cases where stacking is most effective, but it is more likely that they are simply the cases that investigators have studied.

For instance, Ghorbani and Owrangh (2001) examined stacking neural networks for classification on 'statistically neural problems'. These are problems in which randomly selecting from the population has a uniform probability across classes. For instance, consider the parity problem. Let C_N be the collection of strings $a = (a_1, \ldots, a_N)$ of length N of zeros and ones. Define the class of a string a to be $\sum a_k \bmod 2 = 0, 1$. Then, a priori, any representative sample will assign a marginal probability 0.5 to each of the two classes, i.e., the probability is over the two classes. Statistically, neural problems are generally believed to be amongst the hardest classification problems. In a series of these problems Ghorbani and Owrangh (2001) showed that stacking neural networks with the node function taken to be the composition of sine with the usual sigmoid function outperforms any individual neural network used in the SMA.

Mohammadi and Das (2015) had a more complicated structure involving two layers of averaging. First, they used a collection of NNs (and subunits of them) as features to be fed into an SVM. Doing this several times gives a collection of SVMs that can be ensembled into a single output. In this case, the NNs are being used for feature selection rather than directly for stacking. The SVMs are then stacked. The resulting classifier gives evidence of being able to solve several classification problems simultaneously. This is evidence that increasing the layers of ensembling improves the generalization error of classification (or regression, probably), not necessarily by lowering it for a specific data generator (DG) but by making the generalization error small over a larger class of DGs.

The reverse of this procedure has also been developed. Abdullah *et al.* (2009) and Chen *et al.* (2009) stacked SVMs, essentially using SVMs as a way to build feature extraction into the formation of classifers that can then be combined. The general conclusions of these methods include the idea that (i) stacking is generally better than majority-vote classification schemes, (ii) including too many f_k in an SMA can reduce its performance because of overfitting, and (iii) the use of various methods helps identify which features are most important to include in a SMA. That is, the recurrence of variables or functions of variables among the f_k to be stacked indicates which features may be most informative. The same principles apply to stacking NNs.

Most recently, stacking has been applied to classifiers that are themselves qualitatively different. Dinaker *et al.* (2014) stacked SVMs with linear and radial basis function kernels and another classifier called gradient-boosted decision trees. Their goal was purely predictive as they implicitly recognized that the problem they were addressing (assessing teenage distress using the online data platform *A Thin Line*) was $\mathcal{M}$-open. These authors echoed (in sociological parlance) a common finding: the components of an ensemble method can exploit the presence of co-occurring patterns of explanatory variables. In another text-mining application, involving medical records, Kim and Riloff (2015) ensembled four qualitatively different ternary classification methods using stacking and majority voting. They found that stacking outperformed majority voting not in terms of accuracy but in terms of ease of use. Adding new classifiers to a stacking ensemble is easier than adding new classifiers to a majority voting scheme because in their application the thresholds for majority voting do not change. That is, it is more difficult to assess the effect of adding new models to a majority voting scheme. Using a simple majority vote is coarse, they note, because 'small changes can sometimes produce dramatic effects', i.e., such a scheme is not robust. By contrast, stacking automates the weighting of the components in an ensemble, so that adding new components only requires retraining of the classifier. Kim and Riloff (2015) also provided a comprehensive list of references.

One final methodology that rests on stacking is the 'super learner' concept described in van der Laan *et al.* (2007), which is used in targeted learning; see van der Laan and Rose (2011). The intuitive idea is that CV for parameter estimation is a 'super learner' because it satisfies an oracle inequality. Fundamentally, an oracle inequality is a finite-n statement about parameter estimation saying that the 'risk' of the estimator is bounded from above by the minimal risk plus a term of the form $\mathcal{O}(1/n)$. This is more general than requiring the asymptotic normality of an estimator, because no asymptotic distribution need exist, but more restrictive because the mode of convergence is much stronger than in distribution. Pragmatically, any estimator $\hat{\theta}$ that is consistent for θ and for which $\sqrt{n}(\hat{\theta} - \theta)$ has an asymptotic distribution is a good candidate for an oracle inequality if $\sqrt{n}(\hat{\theta} - \theta)$ is uniformly integrable. This is heuristically valid even if $\hat{\theta}$ is discrete, e.g., when θ is the number of explanatory variables in the true model or best predictor.

In the simplest case, the sort of oracle inequality that is useful here is of the following form. Let $L(Y - \theta(X))$ be a loss function giving the loss of using θ when $(Y, X) \sim P$. Here, θ is a parameter indexing functions that transform X into a good approximation for Y. Let $\Delta(\theta, \theta_{\text{opt}}) = L(Y - \theta(X)) - L(Y - \theta_{\text{opt}}(X))$ be the difference between the loss of using θ versus the loss of using the optimal value $\theta_{\text{opt}} = \arg\min_\theta \int L(Y - \theta(X)) \mathrm{d}P(x, y)$. Suppose that finitely many models of the form $f_k(x_k|\theta_k)$, $k = 1, \ldots, K$ are available and that there is an estimation scheme based on CV (where the CV error is found using L in place of the squared error) that gives estimates $\hat{\theta}$, i.e., minimizes the empirical risk over the parameters in the kth model and over all K models. Let $\hat{\theta}^*$ be the parameter value achieving the smallest CV error over all f_k and their parameter spaces under L, when P is the correct joint distribution for independent copies of (Y, X). Thus, $\hat{\theta}^*$ is not an estimator because it requires knowledge of P. However, an all-knowing 'oracle' would know P and so could find $\hat{\theta}^*$, whence the term oracle inequality. Combining results from van der Laan and Dudoit (2003), van der Laan *et al.* (2004), van der Vaart *et al.* (2006), and van der Laan *et al.* (2007) gives the following.

Theorem 11.2 (van der Laan *et al.*) *Under various regularity conditions, $\exists C_1, C_2 > 0$ such that the difference in loss from using the empirical CV parameter estimate versus the optimal value θ_{opt} is bounded by the difference in risk from using the oracle CV parameter value versus the optimal value θ_{opt} plus an $\mathcal{O}(1/n)$ term, i.e.,*

$$E\Delta(\hat{\theta}, \theta_{\text{opt}}) \le C_1 E\Delta(\hat{\theta}^*, \theta_{\text{opt}}) + C_2 \frac{\ln K}{nd}, \tag{11.17}$$

where d is the correct number of parameters in the best model among the f_k.

Otherwise stated, using an empirical estimator such as $\hat{\theta}$ does only a little worse than the value $\hat{\theta}^*$ that an oracle (which knew P) would use. Asymptotically, the empirical estimator and the oracle's value are equivalent in terms of the reduction of loss since the difference is $\mathcal{O}(1/n)$ which goes to zero. Even though there are oracle inequalities that predate (11.17), it is likely that this is the first use of an oracle inequality with a general loss L to justify CV-based parameter estimates.

The estimator $\hat{\theta}$ is a single L-based CV value that gives a single parameter identifying a single model. However, it has been repeatedly argued that MAs outperform individual models. So, it remains to convert this result to an MA scheme that can be used in prediction. To this end, Polley and van der Laan (2010) defined

$$\theta_k(X) = \arg\min_{\theta} EL(Y, \theta_k(X)), \tag{11.18}$$

where θ_k is the generic parameter for a model $f_k(X|\theta_k)$ for $k = 1, \ldots, K$. Then, using the empirical forms of (11.18) as estimates of the predictors $f_k(\cdot|\hat{\theta}_k)$ gives K models to stack. The stacking coefficients can be found using L in place of the squared error in (11.15) to give an L-version of (11.16). The improvement is that now the overall stacking predictor, as stated in Polley and van der Laan (2010), 'performs asymptotically as well as the best possible weighted combination'. That is, the stacking predictor constructed in this way satisfies an oracle inequality. This result was not stated and proved explicitly but seems to follow from the results given in the references prior to the theorem.

The limitation of this reasoning, as in other cases, is the choice of the K models. There is going to be a tradeoff between n and the model list because the oracle inequality guarantees only that the stacking predictor is the best combination of the K models. A stacking predictor using a smaller n but based on a better collection of models can easily give better predictions than a stacking predictor using a larger n but a worse collection of models. Thus, if the real problem is $\mathcal{M}$-open or $\mathcal{M}$-complete and complex then the spirited defense of the intensely parametric inference methodology developed in van der Laan and Rose (2011) does not directly address the prediction of future outcomes. It relies heavily on modeling, even as it admits the inadequacies of modeling such as bias and the possibility that no model exists. Again, this is the reverse of what is advocated here, namely, that good prediction should be achieved first and the resulting predictor studied for whatever modeling inferences it can give about the DG. While one motivation for the framework in van der Laan and Rose (2011) was causal interpretations, cf. Pearl (2000), causal interpretations are not entirely rejected in the predictive approach: causality is accepted pragmatically when all reasonable 'noncausal' predictors have failed to give adequate prediction, 'reasonable' and 'adequate' being context dependent. Otherwise, a model is just a simplification of a predictor.

The notion of a super learner is central to the concept of targeted learning in van der Laan and Rose (2011, see the diagram on p. 38 and also Chap. 3). Stripped to its basics, the paradigm statistical problem is formulated as one of estimation in $\mathcal{M}$-complete settings that are not too close to the $\mathcal{M}$-complete–$\mathcal{M}$-open boundary. The idea is to express the parameter to be estimated in the form $\theta = \theta(P)$, where P is the unknown probability of the DG. Then P, or the important part of it, say Q, is estimated by a nonparametric or semiparametric method that has an oracle property e.g., stacking. Figure 3.3 in van der Laan and Rose (2011) shows the effectiveness of stacking in several cases. The parameters in the first-stage estimation problem using Q are then refined for interpretation purposes possibly by finding an alternative parametrization. Within this refined context, an estimate achieving an optimal variance–bias tradeoff is found. The resulting estimate can be interpreted either as a 'statistical' parameter or as a 'causal' parameter, according to nontestable assumptions of the modeling procedure.

There are several conceptual problems with this 'targeted learning' approach. First, it is narrow: $\mathcal{M}$-open problems are left out entirely and it is an open question how well complex $\mathcal{M}$-complete problems would fit into a scenario that requires an oracle inequality using the notion of a true model as a fundamental ingredient. Second, there are many oracle inequalities, refined parametrizations, and nontestable modeling assumptions that might be invoked. Third, other methods (BMA, see Sec. 11.1, bagging, see Sec. 11.2, boosting, see Sec. 11.4, median methods, see Sec. 11.5, Shtarkov solutions, see Sec. 11.7.1, etc.) have other optimality properties besides oracle inequalities and it is not a priori clear which optimality is most desirable.

None of these criticisms applies to the predictive approach. Predictive optimality can be invoked regardless of problem class. It does not depend particularly on extraneous concepts such as oracle inequalities or parametrizations; pragmatically, whatever works best should be used. Finally, the multiplicity of optimality properties are bypassed simply by regarding them as methods for constructing predictors that must be evaluated by how well they predict. Good prediction is the common yardstick for performance because it is empirical and does not include artificial constructs apart from a loss function, needed so that predictions can be compared with realized values. Other properties such as stability may require the introduction of other constructs e.g., random noise, but they are the consequence of wanting to assess predictors beyond their purely predictive performance.

As a final comment, inference problems that start with a formulation such as $\theta = \theta(P)$, i.e., a representation of a parameter as a functional of a probability to be estimated, have been studied extensively over many years. See Wasserman (2010, Chap. 8 for a introductory treatment of statistical functionals and Chap. 19 for a treatment of causal inference).

11.4 Boosting

Boosting comprises a collection of procedures that originated in classification. The underlying idea is that a 'weak learner' can be improved to become a 'strong learner', whence the term boosting. In this context, a learner is a classifier, i.e., a function $f(x)$ assuming values in, say, $\{1, \ldots, C\}$, where x is an explanatory variable. Often $x \in \mathbb{R}^d$, but x may also include discrete components. A weak classifier is one that does better than chance, but not by much. A strong classifier is one with low error, i.e., $P(Y = f(X))$ is close to one. The term 'learner' originates in computer science and means a procedure by which a good classifier may be identified. Also, in computer science a classifier is often called a 'concept'; a collection of classifiers is often called a 'concept class'. Thus, to a computer scientist, an analyst finding a good classifier is finding a learner that will choose the best concept from a specified concept class. In computer science there are also the notions of a hypothesis and a hypothesis class. These are essentially the empirical versions of a concept and a concept class. To avoid the reinvention of established terminology, statistical terminology will be used here.[2]

The central idea of boosting was first expressed in Schapire (1990). Start with a classifier, say $\hat{f}$, obtained by some procedure using independent data (y_i, x_i) for $i = 1, \ldots, n$ where the y_i are $\{1, \ldots, C\}$. Try to boost the classifier, i.e., lower its misclassification rate, by looking at the points x_i that it misclassified, i.e., for which $\hat{f}(x_i) \neq y_i$ and finding a new classifier for which the penalty of repeated misclassification on those x_i is higher. This forces the new empirical classifier to be more accurate for the values of x_i that its predecessor $\hat{f}$ misclassified. This procedure is done successively, possibly many times, and the sequence of classifiers is usually averaged or its majority vote taken. That is, a classifier that is not very good can be iteratively improved to one that is much better, often with a nontrivial improvement occurring at each iteration up to a specific number of iterations. The question is when to stop iterating, but it is much easier to answer this than to design a method that improves a classifier. A further issue is determining the sample size required to ensure that the final empirical classifier achieves a given misclassification rate.

Sometimes a boosting 'principle' is stated to the effect that weak learners, i.e., poor empirical classifiers, can be combined to give a strong learner, i.e., a good empirical classifier. In fact this is possible if there are enough weak learners that the regions of the x-space where they are good overcome the regions where they are poor. Bayesian additive regression tress (BARTs), Sec. 10.2.3, form a key example of this. Random forests can be regarded as satisfying this principle too, if not as completely as BARTs, because many trees in the forest can be regarded as weak learners relative to the whole forest. By contrast, BMA is not an example of combining weak learners to get a strong learner: the model weights typically go to zero or one, hence identifying a specific 'best' model from the model list (regardless of how good it is when compared with models that have not been considered). Stacking is different again: the goal is not to combine weak learners per se but merely to combine predictors that are different enough that they are worth combining.

The 'magic' of boosting is that the collection of classifiers has to be flexible enough that, when a high penalty is put on the points that have been misclassified, the new classifier does not simply become good on regions of the x-space where the initial classifier was poor

---

[2] Computer science terms include statements about running times that have been ignored here.

and poor on the regions of the x-space where the initial classifier was good. As a heuristic, usually weak learners don't overfit (which would give a low variance) even though they can have high bias and thus not be effective for hard classification problems. Strong learners tend to have the opposite properties. So, selecting a good classifier can be regarded as minimizing a variance plus bias squared criterion.

Freund (1995) offered two more boosting algorithms. One used Schapire's central idea and combined it with bootstrapping as a way to generate classifiers empirically. Again, a classifier is improved iteration by iteration but not all the iterates qualify as 'weak learners'. (One can argue that the classifiers found on subsamples of the data are weak, but the output of each iteration is intended to be better than the previous output.) The other boosting algorithm used a variant of acceptance–rejection sampling and required the generation of further data at each stage. The analysis of these algorithms amounted mostly to an evaluation of their efficiency, i.e., how much data they required to achieve certain levels of performance on average. In computer science parlance, efficiency is often called 'sample size complexity'. These two boosting algorithms were more efficient than the procedure in Schapire (1990).

While there are numerous other boosting algorithms – see Freund and Schapire (1995), Freund and Schapire (1997), Schapire and Singer (1999), Freund (2001), and Kalai and Servedio (2005) amongst many others – the most popular member of this collection of algorithms (and most thoroughly studied) is Adaboost, introduced in Freund and Schapire (1996). Moreover, while there have been efforts to adapt boosting to regression, they have been less successful overall than methods that boost classifiers. Hence, the focus in this section will be on Adaboost for prediction in classification problems.

11.4.1 Boosting Classifiers

An accessible exposition of boosting can be found in Rojas (2009); see Bühlmann and Hothorn (2007) and Hothorn *et al.* (2010) for a more general treatment. Given the data, form a classifier, say, $f_0 = \hat{f}(x) = h_0(x)$. Suppose that f_0 is not very good and so the goal is to improve it by reducing an empirical version of its misclassification rate as a proxy for its actual misclassification rate. Let h_1 denote the improved version of f_0 and write

$$f_1(x) = \text{sign}(C_1(x)) = \text{sign}\,(a_0 h_0(x) + a_1 h_1(x)),$$

where it is now assumed that the classifier f_1 takes values in $\{\pm 1\}$ and $a_0, a_1 > 0$. The notation can be iterated to give a sequence of classifiers of the form

$$f_m(x) = \text{sign}(C_m(x)) = \text{sign}\,(a_0 h_0(x) + \cdots + a_m h_m(x)),$$

where C_m is the weighed sum of the h_j, $j = 0, \ldots, m$, in which the a_j are the (strictly positive) weights and the h_j correspond to the growing number of classifiers in the average. Loosely, the idea is that, as m increases, there are more and more classifiers that do well on hard-to-classify regions, thereby improving f_m over the earlier f_j.

Consider improving C_{m-1} to C_m. Write $C_m(x) = C_{m-1}(x) + a_m h_m(x)$ and try to find a useful choice of a_m and h_m. It does not matter much what h_0 is, so, for simplicity, assume that $h \equiv 0$. No generality is lost by this choice. Suppose that there will be T iterations, i.e., $m = 0, \ldots, T$. For the mth stage the cumulative empirical error (CEE) is

$$\text{CEE}(m) = \sum_{i=1}^{n} \exp\left(-y_i(C_{m-1}(x_i) + a_m h_m(x_i))\right)$$

$$= \sum_{i=1}^{n} \exp\left(-y_i C_{m-1}(x_i)\right) \exp\left(-y_i a_m h_m(x_i)\right)$$

$$= \sum_{i=1}^{n} w_i^m \exp\left(-y_i a_m h_m(x_i)\right),$$

in which $w_i^m = \exp\left(-y_i C_{m-1}(x_i)\right)$, for $i = 1, \ldots, n$. For $m = 0$, $h_0 = 0$, the initial values are $w_i^0 = 1$, independently of i. As m increases, w_i^m represents the weight of the pair (x_i, y_i) among the n data points. Write the weight vector at stage m as $w_m = (w_1^m, \ldots, w_n^m)$. When a w_i^m value is large, this implies that classifying x_i correctly is important, as it has a relatively large contribution to $\text{CEE}(n)$ especially when it is misclassified. Because of the use of exponentials, even classifying x_i correctly contributes to the error. The issue is the size of the contribution relative to the other data points.

Next, partition $\text{CEE}(m)$ into two summations, one where h_m is correct and one where it is not. Since $y_i h_m(x_i) = \pm 1$ accordingly as h_m is correct or incorrect,

$$\text{CEE}(m) = \sum_{\{i \mid y_i = h_m(x_i)\}} w_i^m e^{-a_m} + \sum_{\{i \mid y_i \neq h_m(x_i)\}} w_i^m e^{a_m} = e^{-a_m} W_c + e^{a_m} W_{err} \qquad (11.19)$$

because each term has a common factor, e^{-a_m} or e^{a_m}. To isolate the exact quantity to optimize to find h_m write

$$e^{a_m} \text{CEE}(m) = W_c + e^{2a_m} W_{err} = W_c + W_{err} + e^{2a_m} W_{err} - W_{err} = W_c + W_{err} + (e^{2a_m} - 1) W_{err}.$$

At this iteration, therefore, it is enough to choose h_m so as to minimize $E_{err} = E_{err}(m)$ since $a_m > 0$ does not affect where the minimum occurs. While each correct classification increases $E_c = E_c(m)$, it does so by less than the amount by which correct classification reduces W_{err}. Thus,

$$h_m = \arg\min_{h_m} W_{err}$$

is the natural choice for h_m. However, while there is an explicit requirement that h_m be good where C_{m-1} was poor – this becomes more obvious on examining (11.20) below – there is no explicit requirement that h_m be overall better than C_{m-1} or any of the h_k in C_{m-1}. Despite this, iterations typically improve f_m. Otherwise stated, the h_m may be weak learners but this is not necessarily or even typically so.

It remains to find the weight a_m. So, differentiate (11.19) with respect to a_m, set the derivative equal to zero, and solve. The result is

$$a_m = \frac{1}{2} \log \frac{W_c}{W_{err}} = \frac{1}{2} \log \frac{W - W_{err}}{W_{err}} = \frac{1}{2} \log \frac{1 - \epsilon_m}{\epsilon_m},$$

where $W = W_c + W_{err}$ and $\epsilon_m = W_{err}/W$, the weighted average of the errors as a ratio of the weighted average over all the data. Given this choice of a_m and the previous choice of h_m, new weights w_i^{m+1} can be found and hence new values a_{m+1} and h_{m+1}, so the process can be iterated to generate f_{m+1}, f_{m+2} through to f_T. If h_m does not perform better than random guessing then $a_m = 0$, i.e., it is not included in f_m and the iterations terminate. However, if $W_{err} = W$, i.e., h_m is always wrong, then $a_m = -\infty$ and h_m would be included as its reverse.

It is more standard, though also more mysterious, to present the boosting procedure by taking the weights a_i^m as leading to a distribution and not partitioning CEE(m) into two summations. In this version of the derivation, write the distribution in which to evaluate the misclassification error for h_m as the probability vector $D_m = (D_m(1), \ldots, D_m(n))$ of length n initialized with $D(0) = (1/n, \ldots, 1/n)$. Starting with h_0 at $m = 0$, define iterates h_m for $m = 1, \ldots, T$ by setting the mth stage misclassification error to be

$$\epsilon_m = P_{D_m}(h_m(X_i) \neq Y_i) = \sum_{i\,:\,h_m(x_i) \neq y_i} D_m(i),$$

so that ϵ_m is the probability of misclassification under D_m of h_m on the x_i.

To find h_{m+1} given h_m and D_m, set

$$a_m = \frac{1}{2} \log \frac{1 - \epsilon_m}{\epsilon_m}$$

and update D_m to D_{m+1} by

$$D_{m+1}(i) = \frac{D_m(i) \exp(-a_m y_i h_m(x_i))}{C_m},$$

in which C_m is a normalization factor to ensure that D_{m+1} is a probability vector. As before, the exponential upweights the cost of misclassifications and downweights the cost of correct classifications. Set $h_{m+1}(x)$ to be as follows:

$$h_{m+1}(x) = \arg\ \min_{g \in G} \sum_{i=1}^{n} D_t(i) \mathbf{1}_{\{y_i \neq g(x_i)\}}, \tag{11.20}$$

where G is a large collection of permissible classifiers and, with each iteration, add h_{m+1} to the growing sum. That is, the updated weighted-majority-vote classifier is

$$h_{m+1}(x) = \text{sign}\left(\sum_{j=0}^{m+1} \alpha_j h_j(x) \right)$$

and the final classifier, h_T, is the boosted version of h_0.

The obvious predictor for a new value of x is $\hat{Y} = h_T(x_{\text{new}})$. There is essentially no obvious assessment of variability as there would be in, say, Bayes classification, because the output of boosting is not probabilistic. In principle, the a_m and h_m could lead to some sort of assessment of the variability of the point predictor but it is not clear how to scale them to reflect uncertainty. At root, boosting is a way to find a sequence of classifiers that drives down the empirical error, possibly to zero, even though this is not the same as achieving good prediction.

Boosting is a very popular procedure and a lot is known about it. However, as will be seen, little of what is known about boosting pertains directly to its predictive performance. If the probability of error is $P(Y_i \neq h_T(X_i))$ then Freund and Schapire (1996) showed that

$$P(Y_i \neq h_T(X_i)) \leq \hat{P}(Y_i \neq h_T(X_i)) + \mathcal{O}\left(\sqrt{\frac{Td}{n}}\right)$$

with high probability, where d is the Vapnik–Chervonenkis dimension of the collection of classifiers, G being searched.[3] Freund and Schapire (1999) also established bounds on the training error. If the error of h_m is $\epsilon_m = 1/2 - \gamma_m$ then

$$\prod_{m=1}^{T} \left(2\sqrt{\epsilon_m(1 - \epsilon_m)}\right) = \prod_{m=1}^{T} \sqrt{1 - 4\gamma_m^2} \leq \exp\left(-2\sum_{m=1}^{T} \gamma_m^2\right).$$

More recently, the book by Freund and Schapire (2012) gave a relatively complete compendium of results on boosting.

There have also been numerous empirical comparisons of boosting with other methods; see Bauer and Kohavi (1999) for one of the earliest large comparisons. They found that, typically, bagging reduced the variance while boosting reduced both the variance and the bias. However, boosting and bagging increased the variance of naive Bayes, known to be optimal under classification loss and very stable. As a generality, under Adaboost larger trees tended to give smaller errors than smaller trees, up to the point where the variance became large. They also found that voting methods tended to reduce MSE more than nonvoting methods and that boosting could be sensitive to outliers as well as to 'hard' regions. Their comparisons covered variants on these methods, including Arc-x4 (see Breiman (1998)), not treated here.

More recently, Caruana and Mizil (2006) undertook a large-scale comparison using ten methods, eight performance measures, and 11 data sets. They found that, overall, calibrated boosted trees gave the best classification method. They noted that RFs came a close second followed by uncalibrated bagged trees, calibrated SVMs, and uncalibrated NNs. They also noted that average behaviors masked sharp differences between the classifiers: a classifier that was good on average could perform unexpectedly poorly and a classifier that was poor on average could perform unexpectedly well, depending on the data set and performance measure. Given the key idea in Sec. 11.7.2 this is not surprising. However, these authors refrained from trying to characterize when one method should perform better than another.

An even more extensive comparison of methods was provided by Bibimoune *et al.* (2013). They found that the rotation forest family of classifiers (variants on RFs) generally gave the best performance with or without calibration. They state: '...the success of this approach is closely tied to its ability to simultaneously encourage diversity and individual accuracy [of trees] via rotating the feature space and keeping all principal components.' They also found that ensembles of trees using randomization as a search strategy was very competitive with other methods in terms of accuracy, even for small sample sizes.

We have presented here the use of boosting for binary classification. However, boosting has also been extended to the multiclass setting, and computations in that setting were given in Sabarian and Vasconcelos (2011) (and the references therein) and in Mukherjee and

---

[3] The Vapnik–Chervonenkis dimension is a measure of the complexity of a class of functions. It is the maximal number of points that the functions in the class are able to partition in all possible ways and hence is well suited to classification problems.

Schapire (2013). Separately, Freund and Schapire (2012 Chap. 10) also discussed multiclass classification, but in a more abstract way. The treatment of this topic is beyond our present scope.

In addition to the foregoing contributions, there are two more that bear discussion in some detail here, namely, the logistic regression interpretation of Friedman *et al.* (2000) and the counterexamples of Mease and Wyner (2008).

The key observation of Friedman *et al.* (2000) was that boosting is not based on the zero–one loss typically used in classification. Rather, boosting is based on an exponential loss: in CEE(m) correct classifications are penalized by $e^{-a_m} > 0$ and incorrect classifications are penalized at the higher rate of e^{a_m}, since $a_m > 0$. It is hard to motivate this choice of loss function apart from the fact that it leads to an effective binary classification procedure. Some reflection leads one to recollect that logistic regression is a form of probabilistic classification in which the probabilities are typically the exponential of a regression-like expression. Hence, action of the boosting algorithm can be regarded as fitting an additive logistic regression model via Newton-like updates by minimizing the functional

$$J(f) = E(e^{-Yf(X)}).$$

More formally, Friedman *et al.* (2000) established the following. Essentially they provided a representation for the optimal choice of a_m in a population sense rather than an iterative sense.

Proposition 11.1 (Friedman *et al.* 2000) *The functional J is minimized by*

$$f(x) = \frac{1}{2} \log \frac{P(y = 1|x)}{P(Y = -1|x)}.$$

This leads to the logistic regression classifier

$$P(Y = \pm 1|x) = \frac{e^{\pm f(x)}}{e^{f(x)} + e^{-f(x)}},$$

in which it remains to find f. As in boosting f is found iteratively, as the central results from Friedman *et al.* (2000), stated informally here, established.

Theorem 11.3 (Friedman *et al.* 2000) *The boosting algorithm fits an additive logistic regression model via Newton-like updates for minimizing $J(f)$. Equivalently, the boosting algorithm fits an additive logistic regression model by using the stagewise and approximate optimization of $J(f)$.*

Within the proofs of these results the sequence of h_m to be used in the logistic regression function for classification was derived, including the sequential reweighting and re-optimizing that is the most striking feature of the boosting procedure. The import of these results is to render boosting no longer mysterious: boosting is simply a variant on logistic regression in which the regression function is constructed nonparametrically, cf. Sec. 4.5. A signature data-mining procedure from computer science is simply a nonstandard formulation of a problem in a long-established branch of statistics. This is not to say that boosting does not work well, often exceedingly well. Indeed, many subject-matter researchers use logistic regression because it is relatively interpretable and is regarded by them as 'stable', even if it is not quite clear what is meant by that.

However, like any technique it is to be expected that there will be classes of problems on which boosting (and the other techniques presented here) will perform poorly. Indeed, Mease and Wyner (2008) reviewed a variety of boosting algorithms, many of them variants on Adaboost, in an effort to highlight how little is actually known about their performance. As they noted, a wide variety of tacit assumptions are made about boosting, e.g., that it rarely if ever overfits, no matter what the size of T, or that if it overfits then it is enough not to let T become too large. Other issues touched on include whether a variant on Adaboost called Logitboost (see Friedman *et al.* (2000)) performs better for binary classification with noisy data (see also Long and Servedio (2010)), whether smaller trees should be used in the T-iterations when the Bayes error is larger, and whether regularization should be applied to avoid overfitting. They even question whether boosting can be consistent without regularization. There may be sensible, heuristic, answers to many of these questions, but the conditions under which the heuristic answers are valid seem largely unexplored. For instance, one may argue that the resistance to overfitting seen in boosting is the result of the fact that successive h_m may chiefly affect the regions where misclassification occurs, so that, as T increases, h_m affects smaller and smaller regions; hence overfitting in the limit of large T is rare. This would be consistent with the Newton-like updates noted in Friedman *et al.* (2000), which typically shrink as the iterations converge. However, this heuristic seems not to have been explored.

There have been numerous comments on the paper of Mease and Wyner (2008), some critical, and the rejoinder section of the paper answers them well. Taken together, however, the paper and discussion lay out a wide variety of aspects of boosting that, despite representing commonly held views, do not appear to hold as generally as is often assumed. Where a neutral third party reader will stand on these issues seems to depend on how reasonable the Mease and Wyner (2008) examples are thought to be. The more reasonable they are, the more serious the limitations on boosting-type algorithms. It should be noted that many of these examples are considered to be hard problems – those either at the complex end of $\mathcal{M}$-complete or genuinely $\mathcal{M}$-open. One natural interpretation is that, among very hard problems, there are some classes where boosting or its variants perform well but other classes where they do not. There are also classes of difficult problems where boosting performs comparably with other 'good' methods. For a large class of simple to moderate problems, boosting is often better than other methods, but again such classes have not been clearly identified despite much work.

Overall, the debate about when each good classification technique is best remains unresolved. At present, roughly speaking, in the two main camps are the advocates of RF methods and the advocates of boosting-based methods. That there are classes of problems in which a specific technique – whether trees, boosting, or some other method – is effectively best is beyond doubt. The issue is characterizing those classes usefully so that one can choose which classifier to use in order to get the best predictions.

11.4.2 Boosting and Regression

Heuristically, if Y is continuous then it can be discretized to Y^D and each discrete value – say, the midpoint of an interval – may be taken as a class, so that $(Y - Y^D)$ is small when the intervals are short. Thus, a regression problem is equivalent to a classification problem with an increasing number of classes. This means that an extension of boosting for C-class classification converges to a form of boosting for regression as $C \to \infty$. While various versions of this approach have been developed, how to carry out boosting for regression most effectively must still be regarded as an open problem. Nevertheless, several techniques have been proposed.

The earliest approach was due to Drucker (1997). It was largely an adaptation of Adaboost to a regression setting. At iteration m assign a weight, say w_i^m, to each pair (x_i, y_i), taking all $w_i^0 = 1$ for simplicity. Assign a probability $w_i^m / \sum_{i=1}^n w_i^m$ to (x_i, y_i) and, under this distribution, take a bootstrap sample of size n from the n data points. Using the bootstrap sample construct a regression function, say h_m, and find $h_m(x_i)$ for each of the n values x_i. Let $D = \sup_i |h_m(x_i) - y_i|$ and take the weighted residual error of h_m on the sample as

$$\bar{L} = \bar{L}_m = \sum_{i=1}^n w_i^m L_i, \quad \text{where} \quad L_i = \frac{|h_m(x_i) - y_i|}{D^2}.$$

Other choices for L are possible. Let $\beta = \bar{L}/(1 - \bar{L})$; it can be regarded as a measure of the quality of h_m, the lower the better. Now the w_i^m can be updated by setting $w_i^{m+1} = w_i^m \beta^{1-L_i}$. A small L_i indicates a large reduction in w_i^m, making it less likely that this data point will be selected in the next bootstrap sample. Running this procedure T times gives $h_0, \ldots, h_T$. The point prediction is the median

$$f_T(x) = \inf \left\{ y \,\middle|\, \sum_{m:\, h_m(x) \le y} \log \frac{1}{\beta_m} \ge \frac{1}{2} \sum_{m=1}^T \log \frac{1}{\beta_m} \right\}$$

taken in the discrete distribution generated by $\log(1/\beta_m)$, where β_m is the value of β for the mth iteration. An SE for this prediction can be found only by bootstrapping, as discussed in Sec. 11.3 or at the end of Sec. 10.1.

In discussing the performance of this adaptation of boosting to regression, Drucker (1997) showed that his approach gave better results for trees than for bagging on Friedman's #1, #2, and #3 functions. Using a median will tend to stabilize predictions and so is probably a better choice for regression since regression is a more complex problem than classification. Whether the weights are optimal in any sense remains an open question. Indeed, one can argue that the extensive use of bootstrap samples means that Drucker (1997) did not actually boost regression so much as possibly improve bagging by weighting the predictions from the leaves of the trees. As a generality, neither this method nor any others have been much used in practice, at least partly because they need to be tested more extensively especially on hard predictive problems with real data.

Another adaptation of boosting to regression was due to Ridgeway *et al.* (1999). Heuristically, it is an attempt to implement the intuition at the beginning of this subsection. However, rather than treating a continuous function as a single multiclass classification problem, they treated it as a series of binary classification problems. For simplicity, suppose that $Y \in [0, 1]$ is the variable to be predicted using, say, two explanatory variables X_1 and X_2. As in Ridgeway *et al.* (1999) suppose that there are two data points, say $(X_1, X_2, Y) = (0.6, 0.4, 0.3), (0.8, 0.5, 0.9)$. Then, fix a vector $S = (s_1, \ldots, s_K)$ of K equally spaced points in $[0, 1]$ and create a set of auxiliary binary variables $Y_k^* = \chi_{\{s_k \ge Y\}}$. The point

$(0.6, 0.4, 0.3)$ now becomes K points of the form $(0.6, 0.4, Y_k^*)$, where the Y_k^* are zero for those s_k that are strictly less than 0.3 and one for all later s_k. At this point, the methodology in Ridgeway *et al.* (1999) was to develop a classifier on the augmented data set, i.e., the data set including the Y_k^*, and use this to obtain predictions about whether Y is above or below each given element s_k, effectively approximating Y. The remaining task was to find a good classifier for the augmented data set. Ridgeway *et al.* (1999) proposed a boosted naive Bayes classifier. Finally, this was used to develop a boosted naive Bayes regression algorithm. It has steps that are formally reminiscent of other boosting procedures, e.g., the updating of weights w_i^m that now depend on the s_k as well, adding one logit term to a growing sum at each iteration, and taking a weighted median similar to that of Drucker (1997). Developing this in detail is beyond our present scope. It suffices to note that in practice the specific formulation of boosted regression given in Ridgeway *et al.* (1999) gives reasonable fits in some cases but in a variety of settings does not outperform other established techniques. As a predictor, aside from generating a point predictor, its SE would have to be found by bootstrapping.

A third attempt to extend boosting from classification to regression was due to Friedman (2001). It was concisely described in Ridgeway (1999) as follows. Suppose that the goal of a function estimation problem is to find a regression function $\hat{F}$ that minimizes the expectation of a loss function $L(y, F)$, as in

$$\hat{F}(\cdot) = \arg\min_F E_{Y,X} L(Y, F(X)) = \arg\min_F E_X E_{Y|X} L(Y, F(X))|X).$$

This is not the same as finding a good predictor, but it may generate predictors that can be validated. One way to generate a function estimate, akin to a boosting procedure, is to minimize the empirical risk

$$\hat{J}(F) = \sum_{i=1}^{n} L(y_i, F(x_i)).$$

The negative gradient (with respect to F) of the functional $\hat{J}$ gives the direction in function space that locally decreases $\hat{F}$ the fastest. Given that this direction, denoted $\nabla \hat{J}(F)$ can be found, an estimate $\hat{F}$ would be updated to $\hat{F} - r\nabla\hat{J}(F)$, where $r > 0$ is the size of step taken in the direction of steepest descent. Obviously this can be iterated; Ridgeway (1999) applied it to several examples. While this technique is called gradient boosting and is an effort to boost performance by using the gradient, it is essentially Newton–Raphson in function space and bears only an abstract resemblance to other boosting algorithms. However, it gives a point predictor even though how to generate a PI or SE is unclear.

Although boosting for regression has not really proved itself a viable technique, there are packages that implement various versions of it; see Schonlau (2005), Ridgeway (2006), and Mayr *et al.* (2014). These permit a more extensive evaluation of its performance relative to other techniques for generating predictors. A relatively rare comparison of trees and boosting in the regression context was due to Barutçuŏglu and Alpaydin (2003). As they noted, different boosting algorithms seem to require different complexities of base models and at their best outperform bagging. However, they recommend bagging to cover simple and complex base models simultaneously.

To conclude this section, note that every complex function has regions that are hard to estimate, because the behavior of the function is complex (rapidly oscillating, for instance), or actually impossible to estimate because there's no data. This problem worsens as the dimension increases. If there are too many such regions and too wide a range of Y values, boosting can't fix them. Boosting seems to generate good predictors only in some cases in which Y takes finitely many, ideally two, values. This does not preclude taking a limit of boosted estimators (which are themselves a limit) having more and more classes, but this does not seem to have been tested.

11.5 Median and Related Methods

As has been seen in the previous sections, ensemble predictors follow a general pattern. Once the data has been obtained, it is partitioned in various ways to generate training and test sets. Given each training and test set, predictors that attempt to reproduce the relationship between Y_{n+1} and x_{n+1} can be formed. Sometimes the training and test data are used only once; sometimes they are used many times. Either way, an ensemble of predictors is created. The final step is to combine the predictors in the ensemble to give a single prediction. A taxonomy of the main techniques is given in Rokach (2009).

Depending on the application, there is a variety of ways to combine the predictions from the elements of the ensemble. One of the most popular ways to ensemble predictors is to take a convex combination. This is seen in BMA, boosting, and bagging. Majority voting in classification can also be regarded as taking a convex combination. More generally, non-convex combinations can be used, e.g., in stacking (Sec. 11.3) since the weights need not be positive or sum to one.

An alternative way to approach ensembling is to use a different summary statistic, i.e., not just a weighted mean. The obvious candidates can be regarded as medians, modes, or trimmed means, treating the individual predictions from the elements of the ensemble as if they were data points. However, any combination rule for predictors is permitted, including varying the distribution in which a summary statistic is evaluated. In this section various of these alternative ways to combine predictors are covered, together with a description of various components that might be used to generate predictions to be combined. The section concludes with a general discussion of the heuristics of choosing an ensemble method appropriate for a given problem.

11.5.1 Different Sorts of 'Median'

One median point predictor has already been seen in Sec. 11.1: expression (11.3) is the median $\hat{Y}_{\text{med}}(x_{n+1})$ of the BMA that is conditional on the data. It is the point predictor from BMA under L^1 loss; other loss functions give other point predictors. (For instance the loss function $y = ax$ for $x > 0$ and $y = bx$ for $x < 0$ with $a > 0$ and $b < 0$ gives the (a/b)th percentile as the point predictor.) Deriving a $1 - \alpha$ PI would require obtaining the posterior distribution of $\hat{Y}_{\text{med}}(x_{n+1})$ and identifying $(\alpha/2)100$th and $(1 - \alpha/2)100$th percentiles for it. This can be done (in practice) by finding the posterior model probabilities $p(\mathcal{M}_k|y^n)$ by MCMC and computing the usual predictive distribution, namely, the mixture (11.1). Apart from special cases, all model probabilities converge to zero except for the model closest to the true model, which converges to one; cf. Berk (1966). So, for large n, it is enough to obtain PIs from the component in the BMA with the largest posterior model probability. However, pre-asymptotically, it may make sense to include all $\mathcal{M}_k$ that are located relatively close to the median or that have large posterior probabilities. These do not have to be the same distributions; which to include depends on the location of the main mass of each $\mathcal{M}_k$ and the weight assigned to it. This is an Occam's window approach; it depends on the spread and locations of the models in the average.

Just as a median can be obtained from a BMA, the median or a trimmed mean can be used in place of the mean in both stacking and bagging. This would tend to stabilize these predictors at the possible cost of efficiency. Indeed, instead of finding the model that minimizes the sum of squares in standard CV in stacking, there are cases where the distribution of the errors is sufficiently skewed that minimizing the median of the squared errors gives better behavior. The reason is that squared-error CV tends to induce sparsity by failing to capture small terms. Robust forms of CV, e.g., those based on the truncated squared error, are less prone to losing small terms than regular CV, but median methods retain small terms most effectively. However, in prediction, as opposed to model identification, there may be a benefit in dropping small terms because then the variance decreases.

A different approach to using the median for prediction is based on the 'median model' introduced in Barbieri and Berger (2004) and recently studied in Ghosh (2015). Roughly, a Bayesian wants to use the highest-probability model. Especially in large model spaces with dependent regressors, this model may be problematic because it can have an unrealistically low probability owing to dilution, for instance. That is, the highest-probability model and the best prediction model do not in general coincide and can be very different. One way around this is to seek a model that gives optimal predictions directly by pooling over the various ways in which the posterior probability can be split among good models. The median model does this by finding the posterior inclusion probabilities of the explanatory variables.

At root, the median model is constructed from those variables that have posterior inclusion probability equal to at least 1/2. In the linear-model context, let model $\mathcal{M}_\gamma$ for $\gamma \in \Gamma$ be $Y = X_\gamma^T \beta_\gamma + \epsilon$ with $X^T = (1, X_1, \ldots, X_K)$ and $\beta^T = (\beta_0, \beta_1, \ldots, \beta_K)$. The model index $\gamma \in \{0, 1\}^{K+1}$ indicates which of the $K + 1$ variables are present: for $k = 0, \ldots, K$, $\gamma_k = 1$ means X_k is present; $\gamma_k = 0$ means X_k is not present; and X_γ, β_γ are X and β with the $\gamma_k = 0$ entries omitted. As usual, the error ϵ is assumed IID mean-zero with finite variance σ^2.

Assume that a likelihood for ϵ is available and depends on σ^2. Also, assume that priors $w(\beta_\gamma, \sigma^2 | \gamma)$ have been assigned to the parameters within the $\mathcal{M}_\gamma$ and that a prior $w(\gamma)$ has been assigned across the $\mathcal{M}_\gamma$. Now, any posterior probability can be found. In particular, the inclusion probability for X_k is

$$p_k = \sum_{\gamma | \gamma_k = 1} W(\mathcal{M}_\gamma | y^n) \qquad (11.21)$$

where $y^n = (y_1, \ldots, y_n)^T$ are n independent outcomes and the x_i have been suppressed in the notation. Given this, the median model is the model for Y constructed by including only those X_k for which $p_k \geq 1/2$. Here, the median model is well defined only if it has a nonzero prior probability, so that the across-model posterior can converge to it. That is, to be well defined the median model must be an element of Γ. This property of Γ is called a 'graphical model structure'.

Often the median model and the modal model will be the same, for instance, if the modal model has posterior probability that is over 1/2. Another case where the median and modal model coincide occurs when $W(\mathcal{M}_\gamma | y^n) = \prod_{k=1}^{K} p_k^{\gamma_k} (1 - p_k^{\gamma_k})$. In general, however, the median and modal models will not be the same and it is the median model that is predictively optimal.

To see this, consider linear models under squared error loss and assume that $\mathcal{M}_\gamma$ is true for some fixed γ. Then the optimal predictor for a future value $Y(x_{n+1})$ is

$$\hat{Y}_\gamma(x_{n+1}) = x_{n+1} H_\gamma \tilde{\beta}_\gamma,$$

in which $\tilde{\beta}_\gamma$ is the posterior mean of β_γ using the within-model prior $w(\beta_\gamma, \sigma^2 | \gamma)$. The matrix H_γ is $K \times K_\gamma$, where K_γ is the dimension of β_γ and the (i, j)th entry of H_γ is

$$h_{ij} = \begin{cases} 1 & \gamma_i = 1 \text{ and } j = \sum_{r=1}^{i} \gamma_r, \\ 0 & \text{otherwise.} \end{cases}$$

More simply, $x_{n+1} H_\gamma$ is the subvector of X corresponding to the nonzero entries in γ. As seen in Sec. 11.1, under squared error the optimal point predictor over all the $\mathcal{M}_\gamma$ is the BMA

$$\hat{Y}(x_{n+1}) = x_{n+1} \left(\sum_\gamma W(\mathcal{M}_\gamma | y^n) H_\gamma \tilde{\beta}_\gamma \right) \equiv x_{n+1} \bar{\beta}.$$

This leads to the following.

Theorem 11.4 (Barbieri and Berger 2004) *Assume that $Q = E_P(X^T X)$ exists and is positive definite and that the X_i are drawn IID from P. Then, under squared error loss, the optimal model for predicting Y_{n+1} is the $\mathcal{M}_\gamma$ that achieves*

$$\min_\gamma R(\gamma) = \min_\gamma \left((H_\gamma \tilde{\beta}_\gamma - \bar{\beta})^T Q (H_\gamma \tilde{\beta}_\gamma - \bar{\beta}) \right).$$

If, in addition, Q is diagonal with entries $q_k > 0$ and $\tilde{\beta}_\gamma = H_\gamma^T \tilde{\beta}$, i.e., the posterior mean of any β_γ can be extracted from the posterior mean of γ, then

$$R(\gamma) = \sum_{k=1}^{K} \tilde{\beta}_k^2 q_k (1 - p_k),$$

where p_k is as in (11.21) and $\tilde{\beta}_k$ is the posterior mean of β_k. Finally, if Γ has a graphical model structure, the median model is the best predictive model in the sense that it achieves the minimum of $R(\gamma)$ over γ.

The median model, i.e., the model containing explanatory variables with inclusion probabilities greater than 1/2, can be defined in general Bayes settings, e.g., for trees and neural nets, but the optimality only holds for linear models. In fact, optimality may extend to model classes that are effectively approximated by linear models, but this has not been explored.

The median model point predictor at a new value $x_{n+1} = (x_{1,n+1}, \ldots, x_{K,n+1})^T$ is

$$\hat{Y}_{\text{med}}(x_{n+1}) = \sum_{k=1}^{K} \chi_{\{p_k \geq 0.5\}} x_{k,n+1} \tilde{\beta}_k. \tag{11.22}$$

Expression (11.22) corresponds to the model average

$$p(y_{n+1}|y^n) = \frac{1}{\Gamma^*} \sum_{\gamma^* \in \Gamma^*} p(y_{n+1}|y^n, \mathcal{M}_{\gamma^*}), \tag{11.23}$$

in which Γ^* contains a set of γ's whose union of nonzero entries is exactly $\{k | p_k \geq 0.5\}$, with no duplication. Now, an approximate $1 - \alpha$ PI can be obtained computationally from (11.23) for (11.22) by simply taking a highest posterior density set or, if this is not connected, by choosing the smallest connected set with posterior probability $1 - \alpha$. In general, however, this will not be unique because Γ^* is not unique. However, the difference from one choice of Γ^* to another should be small and should depend on #(Γ) more than the elements of Γ^*.

Another way to form a predictive interval for $\hat{Y}_{\text{med}}$ is to write

$$\hat{Y}_{\text{med}} - Y_{n+1} = \sum_{k=1}^{K} x_{k,n+1} \left(\chi_{\{p_k \geq 0.5\}} \tilde{\beta}_k - \beta_{k,T} \right) + \epsilon,$$

in which $\beta_{k,T}$ is the true value of β_k. Ignoring the variability in $\chi_{\{p_k \geq 0.5\}}$, adding and subtracting $E(\tilde{\beta}_k)$, squaring, taking expectations, and recognizing that the cross-terms are zero gives the standard approximate $1 - \alpha$ PI for Y_{n+1} of the form

$$\hat{Y}_{\text{med}}(x_{n+1}) \pm z_{1-\alpha/2} \sqrt{\sigma^2 + \sum_{k=1}^{K} (E(\tilde{\beta}_k) - \beta_{k,T})^2 + \sum_{k=1}^{K} x_{k,n+1}^2 \text{Var}(\tilde{\beta}_k)}. \tag{11.24}$$

Typically, $n \text{Var}(\tilde{\beta}_k) \to \sigma^2$ so the PI is asymptotically of the form $\sigma \sqrt{(1 + c/n)}$, apart from the bias term, which, in simple settings, is $o(1/n)$. The parameter σ^2 can be estimated by an MSE or by its posterior mean.

It is interesting to compare $\hat{Y}_{\text{med}}$ to predictors obtained from an Occam's window approach, discussed briefly in Secs. 10.3.3 and 11.1. An Occam's window predictor can be written as

$$\hat{Y}_{\text{Occam}}(x_{n+1}) = \sum_{\gamma} W(\gamma|y^n) \chi_{\{W(\gamma|y^n) \geq t\}} x_{n+1,\gamma}^T \tilde{\beta}_\gamma \tag{11.25}$$

and corresponds to the model average

$$p(y_{n+1}|y^n) = \sum_{\gamma} p(y_{n+1}|y^n, \mathcal{M}_\gamma) \chi_{\{W(\gamma|y^n) \geq t\}} W(\mathcal{M}_\gamma|y^n). \tag{11.26}$$

Expression (11.26) is not a BMA in general because it omits some model terms. It is an Occam's window approach in the sense that the window is defined by a threshold t for posterior model probabilities rather than by a neighborhood in model space.

Ignoring the indicator function, it is clear that the expectation of (11.26) gives the point predictor (11.25). Likewise, if the posterior model probabilities are found (for instance by MCMC), (11.26) can be used to give an approximately $(1 - \alpha)100\%$ PI for Y_{n+1}. The approximation arises because in general (11.26) does not integrate to one. It can be renormalized, and this probably gives a good approximation when $\sum_\gamma \chi_{\{W(\gamma|y^n)\geq 0.5\}} W(\mathcal{M}_\gamma|y^n)$ is large enough.

Of course, the variance of (11.25) or (11.26) can also be found. Write

$$Y_{n+1} - \hat{Y}_{\text{Occam}}(x_{n+1}) = (Y_{n+1} - EY_{n+1}) + (EY_{n+1} - E\hat{Y}_{\text{Occam}}) + (E\hat{Y}_{\text{Occam}} - \hat{Y}_{\text{Occam}}(x_{n+1})).$$

Fixing σ, squaring both sides, taking the expectation, and recognizing that the cross-terms are zero gives

$$\text{MSE}(Y_{n+1} - \hat{Y}_{\text{Occam}}(x_{n+1})) = \sigma^2 + (EY_{n+1} - E\hat{Y}_{\text{Occam}})^2 + \text{Var}(\hat{Y}_{\text{Occam}}),$$

in which closed-form (but somewhat messy) expressions for the two rightmost terms can be found using (11.25). Hence, a $(1-\alpha)100\%$ PI for $Y(x_{n+1})$ is given by

$$\hat{Y}(x_{n+1}) \pm z_{1-\alpha/2}\sqrt{\sigma^2 + (EY_{n+1} - E\hat{Y}_{\text{Occam}})^2 + \text{Var}(\hat{Y}_{\text{Occam}})}, \qquad (11.27)$$

where the last term in the square root is $\mathcal{O}(1/n)$ and σ can be estimated by an MSE or by its posterior mean.

An alternative to the median model predictor is the posterior weighted median (PWM) predictor; see Clarke *et al.* (2013). Again, consider a class of models $\mathcal{M}_\gamma$ for $\gamma \in \Gamma$. The linear-model case is easiest but the methodology extends to any Bayes model-selection problem. Assuming that model $\mathcal{M}_\gamma$ is equipped with a prior $w(\theta_\gamma|\gamma)$, where θ_γ is the parameter in $\mathcal{M}_\gamma$, and that an across-model prior $W(\gamma)$ is assigned then in principle $W(\gamma|y^n)$ can be found. Also, each model $\mathcal{M}_\gamma$ can be used to generate a prediction $\hat{Y}_\gamma(x_{n+1})$ for Y_{n+1}, perhaps by estimating θ_γ by its posterior mean. Now, the predictions can be used to generate an order statistic over models. Write this as

$$\left(\hat{Y}_{(\gamma_1)}, \ldots, \hat{Y}_{(\gamma_K)}\right),$$

where the subscript parentheses indicate the order statistics for the model indices they enclose. In parallel with this write the probability vector for the ordered posterior probabilities of the models giving the predictions $\hat{Y}_{(\gamma_k)}$ as

$$\left(W((\gamma_1)|y^n), \ldots, W((\gamma_K)|y^n)\right).$$

This means that the prediction $\hat{Y}_{(\gamma_k)}$ is drawn with posterior probability $W((\gamma_k)|y^n)$. Otherwise stated, the K predictions can be regarded as the outcomes of a random variable with distribution given by the posterior model probabilities. Now, for the problem of choosing an L^1 optimal predictor for the outcome of this distribution it is natural to choose the median of the $\hat{Y}_{(\gamma_k)}$ in the posterior distribution across models, i.e., to use the probabilities $W((\gamma_k)|y^n)$.

The optimality of this method is in L^1, but is conditional on the data. Formally,

$$\hat{Y}_{PWM}(x_{n+1}) = \arg\min_u \sum_{\gamma \in \Gamma} |u - \hat{Y}_\gamma(x_{n+1})| W(\gamma|y^n).$$

Conceptually this is different from the predictions from the median model. First, a graphical model structure is not required of Γ and $\hat{Y}_{PWM}$. Second, the PWM minimizes the effect of dilution, not by summing over model probabilities but by effectively putting the predictions from the good models in the middle of the vector of predictions, so that one of them will be chosen as the PWM prediction. It is not clear how to obtain a PI for this point predictor mathematically. Although messy, analogs to (11.24) and (11.27) can be written down. However, bootstrapping can be used to generate an approximate histogram for the PWM predictor, and this can be used to obtain approximate $1 - \alpha$ PIs. In practice, the PWM predictor seems to work as well or slightly better than the median model predictor even though it has less formal justification.

The methodology behind the PWM can be altered to use a trimmed mean or (approximate) mode as well. A trimmed mean could be slightly more efficient than a trimmed mode or median. However, in the cases where a mode would be most reasonable, it would give essentially the same point predictor as the median. Moreover, the median of the predictions need not be taken using the posterior model probabilities. Any scheme that assigns a finite 'worth' to each model can be normalized to give a distribution. For instance, the AIC weights or $1/CVE$, where CVE is the cross-validation error, could be used and would give different results in general than using the posterior.

A third sort of median predictor arises from Bayes model selection. The idea is to use Bayes testing based on a median intrinsic Bayes factor (see Berger and Pericchi (1998) for a formal definition) so as to rule out many models on a list. The final model (or models) can be used to generate predictions. While the predictor itself is not the direct result of taking a median of Bayes factors, it is the result of a procedure based on taking a median and as such may give more robust inferences than conventional Bayes testing.

As a generality, median methods for prediction have been underused for a variety of reasons, one being that in many cases sums such as occur in bagging, stacking, or boosting are sufficiently stable. However, many aspects of combining the outputs of a collection of models without using a mean should be explored, because in sufficiently complex settings means often do not give good prediction. That is, in many of these settings there is so little stability in the data stream that means are effectively ruled out. For instance, for predicting $\mathcal{M}$-complete and especially $\mathcal{M}$-open DGs there may be effective predictors, e.g., Shtarkov's solution given below in Sec. 11.7.1, but they are qualitatively different from those treated in the main sections of this chapter. Indeed, many of the predictors described here, i.e., predictors derived from conventional models using means, will typically underperform relative to combinations of predictors that do not correspond to conventional models, e.g., ensembles of kernel methods or Shtarkov's solution. Combinations of point predictors that go beyond the use of means and models are the main hope for finding good predictors in the most complex settings.

11.5.2 *Median and Other Components*

When using median methods to combine predictors, it often makes sense to use components (i.e., single predictors) that are also based on medians. Then both the components and the way in which they are ensembled will be more stable. This is most important for DGs that are $\mathcal{M}$-open but may be effective for complex $\mathcal{M}$-complete DGs also. Moreover, there is some evidence that increasing the stability of predictors by forming them as an ensemble of predictors that are *themselves* the result of ensembling may give better prediction for the most difficult DGs.

For the present, it is worth listing a few (nonensemble) predictors that are not based on taking means. These can then be ensembled and hopefully improved for $\mathcal{M}$-complete and $\mathcal{M}$-open problems.

In the simplest cases, the mean in an optimality property can be replaced by a median or the squared error can be replaced by an absolute value. For instance, instead of standard least squares in regression, the least median of squares can be used; see Rousseeuw (1984) for the original description and Kim and Pollard (1990) for its asymptotic properties. Although it has not been tested extensively, an intermediate between using the mean and the median for regression is the trimmed mean. As shown in Cizek (2004, 2005) this optimality criterion gives results intermediate between using the mean and median. Separately, the L^1 error can be used in place of the L^2 error in regression; see Eakambaram and Elangovan (2015) for a recent treatment. These approaches are mostly useful when the quantity to be predicted is a 'central' rather than an 'extreme' value, because it is in the central case that stability is most useful. An exception is the possibility of a quantile-based decision theory, discussed at the end of Sec. 9.3, which was originally intended for high quantiles, say 99.5%, representing unlikely but very costly events – even though it can be used for percentiles such as the median as well.

In the case of extreme values, the predictive situation is more complex. It is well known that if data $Y_1, \ldots, Y_n$ are drawn from a distribution F then the distribution of the sample maximum $Y_{(n)}$ depends on whether F is in the domain of attraction of a Frechet, Weibull, or Gumbel distribution, as these are the only nondegenerate limiting distributions that $Y_{(n)}$ can have. If multiple independent copies of $Y_{(n)}$ are available and n is large enough then the parameters in the limiting distribution can be estimated and the resulting distribution can be used to generate PIs. These predictions can also be ensembled, and this is most useful when the different predictors use different sets of explanatory variables. See Wang and Dey (2010) for an example of this sort of regression. If multiple copies of $Y_{(n)}$ are not available, see Reiss (1989) for a thorough treatment.

11.5.3 Heuristics

To conclude this section we will compare the various ensemble methods in terms of their anticipated predictive performance, at least tentatively.

First, in the case of regression, BMA will typically converge quickly to one of the models on the list. So, in the presence of bias, BMA often does not perform very well. It is really only suited to $\mathcal{M}$-closed problems. However, for truly $\mathcal{M}$-closed problems and a list of parametric models, BMA is likely to be nearly equivalent to using any good model-selection technique and making a prediction from the selected model. The predictions will usually be more stable when the BMA predictor is found using L^1 rather than L^2, but the speed of convergence of the posterior model weights should make the differences small. By contrast, the stacking weights do not converge as fast as BMA weights, and stacking is better at accommodating bias. So, in $\mathcal{M}$-closed problems stacking is usually not as good as BMA and, outside $\mathcal{M}$-closed problems, stacking tends to outperform BMA. When the bias is high enough, neither stacking nor BMA will perform well because both are limited by the span of the model list (even if the ways in which they are thus limited are different). In $\mathcal{M}$-closed problems, methods such as bagging, the median model, and PWM often perform well, but they are really intended for more complex settings.

Far enough away from $\mathcal{M}$-closed problems, neither BMA nor stacking perform very well unless the components being averaged are from a flexible family, e.g., SVMs, Nadaraya–Watson kernel estimators, etc. This flexibility effectively removes the limitation of the model list. In these cases, stacking usually outperforms bagging and the relative performance of BMA does not seem to have been explored. In $\mathcal{M}$-complete settings, both the median model and PWM perform relatively well and are comparable even though PWM is defined more generally. Bagging and stacking also typically perform very well provided that the components in the stack or the bagging average are sufficiently flexible, e.g., they are trees. In the case of $\mathcal{M}$-open problems and flexible components, usually stacking outperforms bagging and the other methods (median model, PWM, BMA) do not seem to have been explored yet. As noted earlier, boosting for regression does not seem to be competitive.

Second, in the case of classification, Bayes classifiers such as BMA are formally optimal in $\mathcal{M}$-closed problems. Other methods typically perform nearly as well, e.g., bagging and boosting. Stacking often does not combine classifiers very effectively. Outside $\mathcal{M}$-closed problems, boosting and bagging trees (random forests) tend to perform well over a wide range of problems. Bagging is probably more generally applicable than boosting because (i) boosting requires that incremental improvement by gradient descent be effective, and (ii) bagging can usually identify more complex regions in the space of explanatory variables relative to boosting. The effectiveness of BMA or median methods for classification does not seem to have been explored. Conditionally on sufficient flexibility of the classifiers being averaged, median methods are likely to perform well, if not quite as well as bagging or boosting. However, at this time, if one had to choose one method for a wide range of very complex problems, bagging trees would probably be it.

A concept that underlies much model averaging, whether in regression or classification, is the concept of the 'weak learner'. The idea is that weak learners can be combined to give a 'strong learner'. This philosophy makes good sense provided that the selection of weak learners is large enough and broad enough for there to be enough weak learners in every region of the space of explanatory variables that majority-vote or some other averaging method will combine them effectively. Implementing this, e.g., in BART, requires careful control of the weak learners, and such control amounts to having extra information not embedded automatically in the weak learners themselves. In fact, while this philosophy has motivated the development of boosting, boosting really works because of incremental improvement in regions where prediction is poor; this is not quite the same intuition as combining weak learners.

The weak learner philosophy is in contrast to a 'greedy' philosophy in which each component tries to be as strong a learner as it can be. In fact, this is also consistent with boosting because at each stage in the procedure one is seeking the strongest learner. There is good evidence, e.g., from random forests, that the greedy philosophy is effective as well and that it may be effective more generally than the weak learner philosophy. The cost will be the tendency to get trapped in local rather than global optima. Detecting this and correcting for it also requires information not automatically embedded in the learners themselves. It seems that, whatever philosophy is adopted, making effective predictions requires that knowledge, presumably derived from the data, be applied judiciously. That is, there is no automatic method that will routinely give good performance without tuning. Hence, no philosophy seems to lead to overall better methods in general.

11.6 Model Average Prediction in Practice

Having discussed a variety of ensembling techniques, it is time to see how they work in practice for prediction. The first subsection gives a simulation study in which five ensembling techniques (and one set of nonensembled techniques) will be seen in a simple setting – a univariate explanatory variable in two relatively narrow function classes. Many other function classes could have been used, possibly giving different results. However, the point is to understand ensembling methods, not function classes or complex prediction problems. The second subsection presents a re-analysis of the Vegout data used in Sec. 10.6.2, using ensemble techniques so that the results can be compared with the earlier nonensembled blackbox techniques. The third subsection gives a general discussion of the strengths and weakness of various model-averaging approaches, including some not presented here.

11.6.1 Simulation Study

For the sake of demonstrating the use of ensemble or model-averaging methods in a regression context, consider two classes of functions. The first is called 'treed' because they are a generalization of one-dimensional trees (which look like a sequence of blocks). They are generalized in two senses: the cutpoints or block endpoints are randomly chosen and the leaves have random polynomials in them. They are of the form

$$f_{\ell,a,c_0,c_1,c_2,d}(x) = \sum_{j=1}^{d+1} \chi_{a_{(j-1)},a_{(j)}}(x) \sum_{u=0}^{\ell_j} c_{j,u} x^u, \quad x \in [0,1],$$

where $d \sim \text{DUnif}[2,10]$ is the number of cutpoints, $\ell_j \sim \text{DUnif}[0,2]$, $j = 1,\ldots,d+1$, is the order of the polyomial (here limited to a maximum of 2), $a_j \sim \text{Unif}[0,1]$, $j = 1,\ldots,d$ (with $a_0 = 0$ and $a_{d+1} = 1$), are the actual cutpoints, and all $c_{j,u} \sim \text{Unif}[-5,5]$ are the coefficients on the terms in the polynomials in the leaves. The $a_{(j)}$ are order statistics and $\chi_{a_{(j-1)},a_{(j)}}(\cdot)$ is the indicator for the region between $a_{(j-1)}$ and $a_{(j)}$.

The second function class is Doppler. This class of functions is of the form

$$f_{a,b}(x) = \sqrt{x(1-x)} \sin \frac{b\pi(1+a)}{x+a}, \quad x \in [0,1],$$

where $a \sim \text{Unif}[0,0.5]$ and $b \sim \text{Unif}[-10,10]$. These functions look like stretched-out sinusoids with the stretching increasing with x.

Since the domain of both function classes was $[0,1]$, the values of x were chosen according to a Unif[0,1]. The sample size was set at $n = 200$ and $N(0,1)$ noise was added to each function value at each x to generate the corresponding y. Data for 30 replications were generated, i.e., 30 data sets from each function class, and the first 50 points of each data set were used as a burn-in for all model average methods. Thus, for each replication, 150 one-step predictions were made and compared with the actual value of Y generated. Representative examples of the functions from the classes chosen and the data with noise are given in Fig. 11.1.

In contrast with the data generators are the methods used to predict them. In all cases the components of the model averages were trees or neural nets (NNs), but the implementation used to generate the trees or NNs for the model averages depended on the ensembling method. This was necessary because in many cases established functions or packages were

Table 11.1 *Ensembling methods and functions or packages used for each component function class and model-averaging method.*

Model average	Component class	R function/package
Non-ensembled	trees	rpart
	NNs	nnet
BMA	trees	BART
	NNs	nnet
Bagging	trees	treebag
	NNs	avNNet
Stacking	trees	rpart and ctree in party
	NNs	nnet
Boosting	trees	blackboost in caret
	NNs	nnet
PWM	trees	rpart
	NNs	nnet

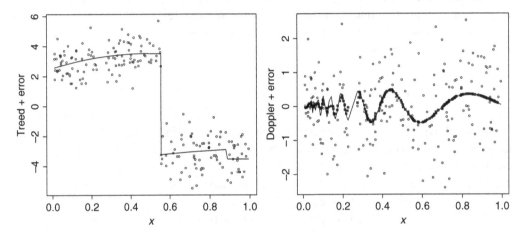

Figure 11.1 Examples of treed functions (left) and Doppler functions (right). The solid line indicates the function value and the dots indicate the data. On the left there are three leaves and, on the right, the Doppler function shape is tracked closely by the five-point moving average. Outside time series or other settings with specialized knowledge, e.g., that the function was a Doppler, one would not know that the data should be represented by a moving average.

either specific to one method or not available; some original coding of methods was required to obtain the results presented. See Brownlee (2016) for a good background on the coding of ensemble methods in R. Table 11.1 shows the functions or packages that were used, subject to the following comments.

First, in the non-ensembled cases, only one tree or NN was generated. The function rpart generates trees using the means of the points in the leaves as the regression function; this is less general than treed functions permit. This function allows binary splits and chooses them on the basis of a minimized mean squared error criterion (see Sec. 10.2.1). Also, nnet uses the usual sigmoid function and optimizes over single-hidden-layer NNs by the Broyden–Fletcher–Goldfarb–Shanno procedure, which can be regarded as a variant on backprop.

Model selection is done over the number of components in the hidden layer, usually by cross-validation. (In principle, more elaborate ensembling could be done over NNs with multiple hidden layers, but the focus here is on the ensembling, not on the components of the ensembling.)

Second, the coding for the BMA of trees was taken from the Bayes additive regression trees (BART) package; see Chipman *et al.* (2010). Essentially, the BMA is built into BART; 100 trees were generated and their coefficients were computed internally to the program. The coding for the BMA of NNs, however, was written especially for the purpose, following the procedure described in Chitsazan *et al.* (2015). Nine NNs were generated by varying the decay parameter and number of nodes in the hidden layer. There were three possible values for the decay parameters and three possible values for the number of hidden nodes.

Third, bagging merely combined the output of the R functions treebag or avnnet. For trees, the default of 10 bootstrap samples was used. For NNs, the default of five bootstrap samples was used.

Fourth, stacking was coded to purpose but some delicacy was required to ensure that a diversity of trees and NNs were used in their respective ensembles. This is important for avoiding multicollinearity when determining the stacking weights or coefficients. It necessitated changing control parameters such as parameter settings, cutpoints, the number of hidden nodes, etc. Five trees and nine NNs were used in these predictors.

Fifth, for trees, the software for boosting was available; however, for NNs the software had to be coded to purpose following the procedure described in Cao *et al.* (2010); see Charpentier (2015) for useful background. The sequential fitting is not greedy; a decay parameter is used to avoid overfitting residuals. For boosting trees, 100 iterations were used; for boosting NNs ten were used.

Finally, PWM was the only median method that could be readily used since the Barbieri–Berger method requires a graphical model structure. Intrinsically, neither the trees nor the NNs were Bayesian, so a posterior had to be constructed for them to generate predictions. For trees and NNs, the technique here was similar to that in Chitsazan *et al.* (2015). Details can be found in Chi (2010). Loosely, a BIC value for each tree is found using the MSE and $\log n$ and taking the number of leaves in the tree as the number of parameters. For NNs the number of parameters is already well defined. For PWM using trees, 36 trees were generated. In many cases, several trees turned out to be the same. These replicate models were counted with multiplicity in forming the posterior. This problem did not occur with NNs, partly because only nine were used to find the PWM and partly because NNs, being a richer class than trees, have a greater tendency to be unique.

There are six methods, two model classes, and two component classes, giving a total of 24 performances to be compared, each over 30 replications. Proceeding in sequence, the results begin with the no-ensembling case (as in Chapter 10), using trees and NNs, respectively. The first plots, shown in Fig. 11.2, show for comparison the average L^1 errors using predictions generated from single trees and NNs. The bold solid lines give the average for trees and the bold dashed lines give the average for NNs; ±SD regions are shaded gray.

The region of darkest gray indicates where the ±SD error regions overlap. In some places a light or medium gray can be seen. The light gray (bounded by a dotted line) indicates those places where the tree had a larger error than the NN and the medium gray (bounded by a dashed line) indicates those places where the NN had a larger error than the tree.

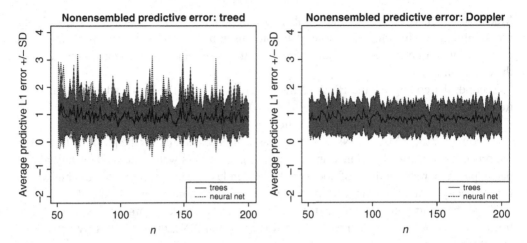

Figure 11.2 Left: L^1 prediction errors averaged over 30 replications for 30 randomly chosen treed functions using trees or neural nets. Right: Similar, but for a Doppler function using trees or neural nets.

Figure 11.2 shows that trees and NNs are roughly equally good at predicting a treed or a Doppler function, because the region of darkest gray is by far the largest of the gray regions. The region does not shrink as the predictive iteration increases, suggesting that, within a given replication, single trees or NNs are unable to 'find' the true function, possibly because the true function is neither an NN nor a Doppler. This is typical, in that the function class containing the true function is almost always unknown. The error for predicting the Doppler function is slightly smaller than the error for predicting the treed function, independently of whether a tree or NN was used, and the sample size makes no difference. It is tempting to see a slight improvement in the treed case because the dark gray band seems to narrow slightly towards the right. However, any improvement is very small indeed.

The stability of the four methods can be assessed by looking at boxplots of the predictuals over the 30 replications, as shown in Fig. 11.3. In the upper row, a treed function is the target and in the lower row a Doppler function is the target. The scales on the two rows are different, showing that the error for predicting a treed function tends to be more variable (using either a tree or an NN) than that for a Doppler function. It is harder for the ensembles to predict a treed function because of its sudden jumps. However, the continuous and simple regions of a treed function may permit the overall error to decrease (slowly) with n, as these regions are found to have small errors. As a result, the overall L^1 error is comparable with the average error for the Doppler functions, which are smooth but complicated.

Turning to the first of the ensemble methods, BMA, the average L^1 errors are shown in Fig. 11.4 for all four cases. Again, most of the region is dark gray, indicating that the two methods are roughly equivalent. However, the predictuals for the treed function are around three times the size of the predictuals for the Doppler. This may be so because treed functions have a wider range on [0, 1]. As in the nonensembling case, the average errors do not decrease with the number of iterations. It can seen that BMA with trees or NNs gives much larger errors than nonensembling for treed functions. With Doppler there seems to be a slight improvement by using BMA over a single tree or NN.

Looking at individual replications of length 200 gives the boxplots in Fig. 11.5. Since the location of the 'box' varies widely in the top row, BMA with trees or NNs leads to much

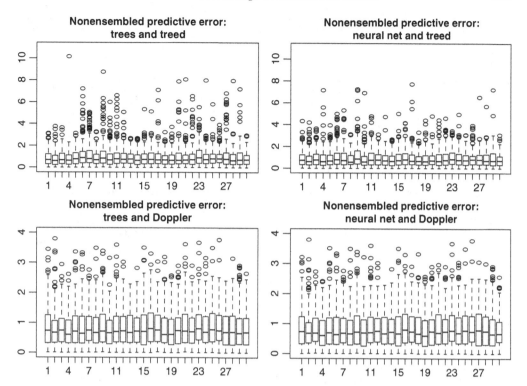

Figure 11.3 Boxplots of the predictuals from the 30 replications without ensembling. There are four cases. Upper left: A treed function predicted by a tree. Upper right: A treed function predicted by an NN. Lower left: A Doppler function predicted by a tree. Lower right: A Doppler function predicted by an NN.

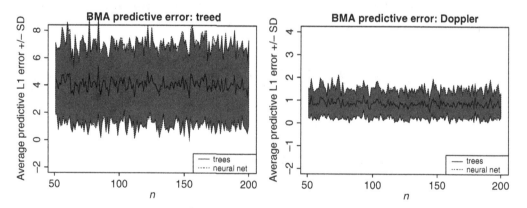

Figure 11.4 Left: L^1 prediction errors averaged over 30 replications for 30 randomly chosen treed functions using a BMA of trees or neural nets. Right: Similar, but for a Doppler function using trees or neural nets.

greater variability for treed functions than for the Doppler function. This is consistent with the fact that the dark band on the left-hand panel of Fig. 11.4 is wider than the dark band on the right. Indeed, BMA often performs poorly owing to instability.

Comparing Fig. 11.6 and Fig. 11.4 shows that the performance of bagging trees or of NNs for treed functions is clearly better than for BMA, in terms of the average L^1. The

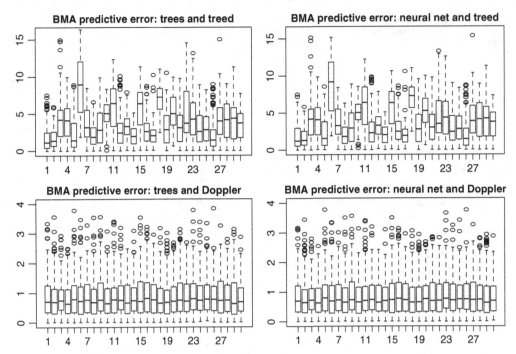

Figure 11.5 Boxplots of the predictuals from the 30 replications using BMA.
There are four cases. Upper left: Treed functions predicted by a BMA of trees.
Upper right: Treed functions predicted by a BMA of NNs. Lower left: Doppler
functions predicted by a BMA of trees. Lower right: Doppler functions predicted by
a BMA of NNs.

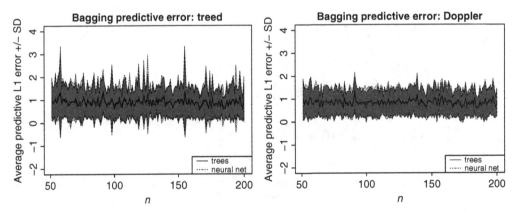

Figure 11.6 Left: L^1 prediction errors averaged over 30 replications for 30
randomly chosen treed functions using bagged trees or neural nets. Right: Similar,
but for Doppler functions using bagged trees or neural nets.

performances of bagging and of BMA for Doppler functions seem comparable. The components in the model averages do not seem to matter much because the bands of dark gray are
nearly the same for both target function classes. Also, the dark gray bands do not narrow as
the number of iterations increases, so little if any improvement over iterations is indicated.

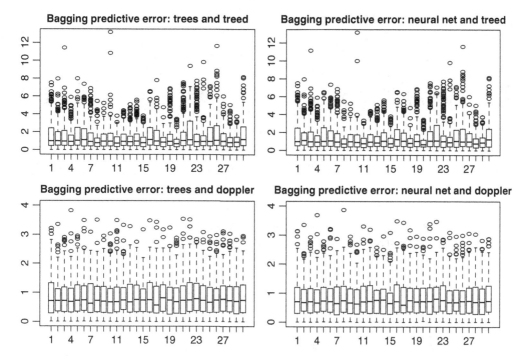

Figure 11.7 Boxplots of the predictuals from the 30 replications using bagging. There are four cases. Upper left: Treed functions predicted by bagging trees. Upper right: Treed functions predicted by bagging NNs. Lower left: Doppler functions predicted by bagging trees. Lower right: Doppler functions predicted by bagging NNs.

In terms of the overall L^1 error, bagging performs a little better than the nonensembled case for treed functions but is fairly comparable for NNs; cf. Fig. 11.2 and Fig. 11.6.

Boxplots of the predictuals from individual replications using bagging can be seen in Fig. 11.7. The average errors for treed functions have a larger band than for Doppler functions, as seen in Fig. 11.6, and the boxplots of the predictuals in Fig. 11.7 show that bagging for the treed class is much more unstable than for the Doppler class. Thus, when bagged predictors are used for treed functions they will be close to the true value quite often but will be far wrong sometimes, while the bagged predictors will be more reliably close, regardless of the components. The very high level of instability seen with BMA in Fig. 11.5 is not present here.

The average of L^1 errors for stacking trees or NNs, for treed and Doppler functions, is shown in Fig. 11.8. The overall averages are comparable although possibly slightly larger than those for bagging or for nonensembling, Figs. 11.6 and 11.2, and stacking tends to be more variable. In contrast with BMA, stacking performs nearly as well for Doppler functions and much better for treed functions.

The increased variability in performance over bagging or nonensembling is borne out in Fig. 11.9 – the top row for treed functions is much more variable than the bottom row for Doppler functions, but in both cases the scales are larger than for other ensembling methods.

The performance of boosting is generally worse on average for treed functions but better on average for Doppler functions than the earlier methods; see Fig. 11.10. Moreover, in the

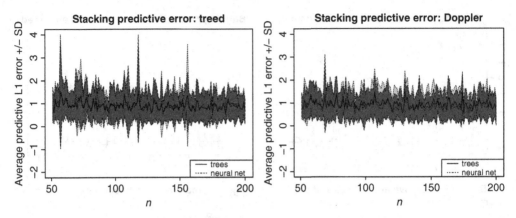

Figure 11.8 Left: L^1 prediction errors averaged over 30 replications for 30 randomly chosen treed functions using a stack of trees or neural nets. Right: Similar, but for Doppler functions using trees or neural nets.

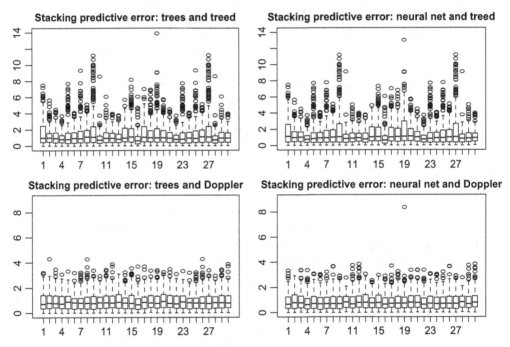

Figure 11.9 Boxplots of the predictuals from the 30 replications using stacking. There are four cases. Upper left: A treed function predicted by stacking trees. Upper right: A treed function predicted by stacking NNs. Lower left: A Doppler function predicted by stacking trees. Lower right: A Doppler function predicted by stacking NNs.

left-hand panel of Fig. 11.10 the difference in the components being boosted is enhanced. There are many more light gray regions than medium or dark gray regions, meaning that trees gave a worse performance than NNs. However, for the Doppler class the components were roughly equivalent (shown as a single narrow dark gray region) and the error region

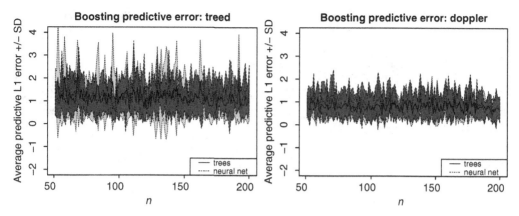

Figure 11.10 Left: L^1 prediction errors averaged over 30 replications for 30 randomly chosen treed functions using using boosted trees or neural nets. Right: Similar, but for Doppler functions using trees or neural nets.

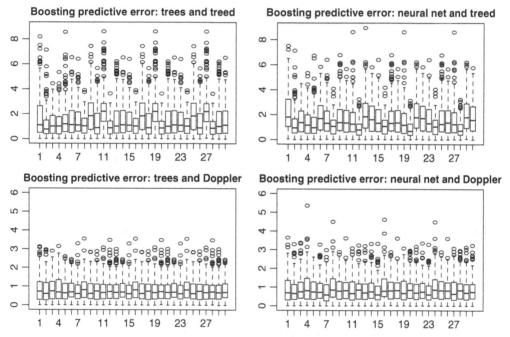

Figure 11.11 Boxplots of the predictuals from the 30 replications using boosting. There are four cases. Upper left: A treed function predicted by boosting trees. Upper right: A treed function predicted by boosting NNs. Lower left: A Doppler function predicted by boosting trees. Lower right: A Doppler function predicted by boosting NNs.

tends to narrow as the number of iterations increases – perhaps not by much, but enough to notice. That is, the boosting procedure is 'learning' the unknown function, at least on average.

In terms of the individual replications, Fig. 11.11 shows that, for treed functions, boosting trees or NNs is more stable than any previous method (smaller scale) and, for Doppler

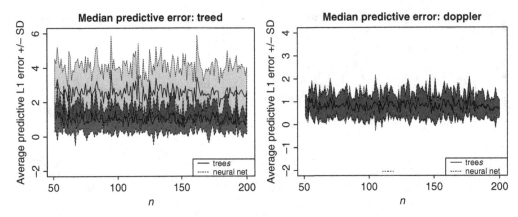

Figure 11.12 Left: L^1 prediction errors averaged over 30 replications for 30 randomly chosen treed functions using PWM with trees or neural nets. Right: Similar, but for Doppler using trees or neural nets.

functions, the boxplots tend to be comparable or a little smaller than for the other methods. Thus, so far, boosting is the best method for these prediction problems, although not by a wide margin, and much of the improvement may be attributable to a reduction in variability rather than in bias.

Finally, Fig. 11.12 shows the averaged L^1 errors for the posterior weighted median (PWM). As seen in the right-hand panel for the Doppler functions, PWM tends to give the best predictions, slightly better even than boosting. Most of the region is dark gray, meaning that the different components did not have much impact; however, the dark gray region narrows noticeably as n increases, suggesting that the components in PWM are 'learning' the unknown function. This is consistent with the very stable boxplots for PWM with Doppler in Fig. 11.13 for both trees and NNs. Thus, of all the methods for the Doppler class, PWM appears to give the best performance.

The situation for treed functions and PWM is different. The right-hand panel of Fig. 11.12 shows that, as with boosting, the nature of the components make a difference, with NNs giving a better performance than trees (i.e., there are more light gray regions than medium gray regions). As seen in the upper row of boxplots in Fig. 11.13, trees tend to be more unstable than NNs. So, PWM with trees on treed functions actually performs noticeably worse than the other methods (except for BMA on treed functions). Meanwhile, PWM with NNs on treed functions performs better than the other methods, sometimes noticeably, and sometimes only marginally.

If the results presented here are taken as being representative of a wider class of problems (a very strong hypothesis) then one is led, as a generality, to infer that PWM with NNs outperforms boosting NNs and these two methods tend to outperform the other methods, including those using individual trees or NNs. Moreover, BMA on treed functions was the worst case, while stacking tended to be poor. Bagging performed a little better than not ensembling at all. In short, the best of the ensembling methods were better than not ensembling, and the worst of the ensembling methods were worse than not ensembling. As a generality, the differences among methods seem to arise from their variability. Of course, the above hypothesis is usually false so this kind of generalization should not be taken too seriously.

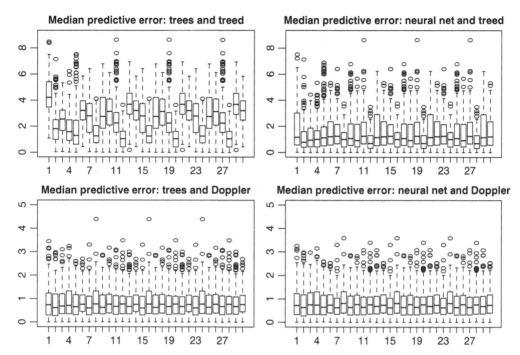

Figure 11.13 Boxplots of the predictuals from the 30 replications using PWM.
There are four cases. Upper left: A treed function predicted by a PWM of trees.
Upper right: A treed function predicted by a PWM of NNs. Lower left: A Doppler
function predicted by a PWM of trees. Lower right: A Doppler function predicted
by a PWM of NNs.

11.6.2 Reanalyzing the Vegout Data

In this subsection, the same five model-averaging techniques (BMA, stacking, bagging,
boosting, median) as exemplified in Sec. 11.6.1 are used to reanalyze the greenness index
in the Vegout data described in detail in Sec. 10.6.2. These five model-averaging predictors
are compared with baseline predictors formed without using any ensembling. In all cases
the ensembles are the same as those used in Sec. 11.6.1, which were formed by combining
trees and NNs. The non-ensembled methods are individual trees and NNs.

Figure 11.14 shows the predicted and actual values for six methods for the first cluster and
first year after burn-in (first time step). The rows correspond to model averages: the top row
shows no averaging, the middle row is for BMA, and the bottom row is for bagging. Trees
were used to generate the left-hand column and NNs were used to generate the right-hand
column. Figure 11.15 is a continuation of Fig. 11.14 in that it has the same structure but is
for three further model averages, namely, stacking, boosting, and medians (using the same
components, i.e., trees and NNs).

Before comparing the predicted values with the actual values, we give some details on
how the predicted values were found. First, the single trees and NNs were found as in Table
11.1. Second, BART was used for the BMA of trees, using the default of 50 trees, and neuralnet was used for the BMA of neural nets with a range of one to five hidden nodes. Third, for
bagging, the caret package was used with trees and ten bootstrap samples; for NN, avNNet
within caret was used with five NNs with one to five hidden nodes. Fourth, for stacking, five

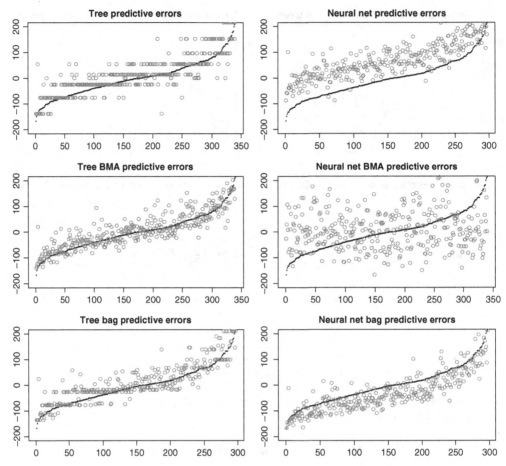

Figure 11.14 Predicted versus actual values for the six predictors for the first cluster in the first year after the burn-in. The left-hand column uses trees; the right-hand column uses NNs. Top row: An individual tree and NN. Middle row: A BMA of trees and NNs. Bottom row: Bagged trees and NNs.

trees from rpart models were ensembled by varying the splitting rule and stopping criterion; nine NNs were ensembled by varying the decay parameter and the number of hidden nodes (two, four, six). Fifth, for boosting trees the caret package (blackboost) was used with 50 iterations while boosting for NN was self-coded for two nodes using ten iterations. Sixth, and finally, the median method for trees was based on 36 trees that varied according to splitting rule and stopping criterion while, for NNs, nine were found by varying the decay parameter and the number of hidden nodes (two, four, six). The running time for boosting was generally the longest, followed by that for NN methods. This is why sometimes fewer NNs were used than trees. Although not a completely fair comparison, the results indicate the relative merits of the methods for the Vegout data.

It is seen that, among the 12 techniques featured in Figs. 11.14 and 11.15, BMA with trees gives the best predictions in terms of their being relatively tightly clustered around the actual measurements. Second best is boosting with trees; it has a low variance but the predicted values are biased upwards. Bagging trees appears to have nearly zero bias but

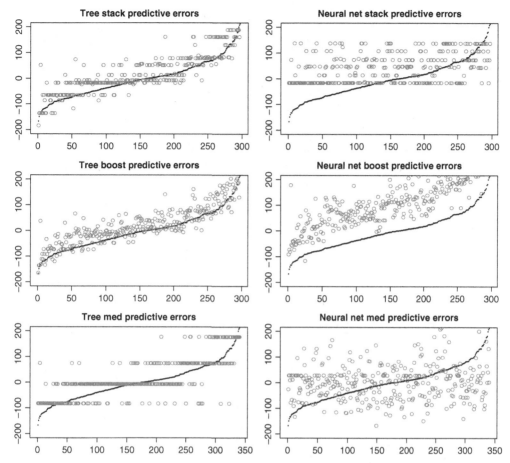

Figure 11.15 Similar to Fig. 11.14, but for six more methods. Predicted versus actual values for the five predictors for the first cluster in the first year after the burn-in. The left-hand column uses trees; the right-hand column uses NNs. Top row: Stacks of trees and NNs. Middle row: Boosted trees and NNs. Bottom row: Median trees and NNs.

a larger variance. An individual tree has an even larger variance (but remains essentially unbiased). Median with trees also has a large variance but is essentially unbiased. Predictions from model averages of NNs tend to be biased upwards (individual, boosting, and even more so, stacking) or downwards (bagging and median). By contrast, BMA with NNs and median methods with NNs give the worst performance in terms of the MSE; NNs with boosting or BMA are obviously biased upwards while the other NN averages are roughly equivalent in the sense that the combination of their variance and bias yields MSEs that seem about the same. Both Figs. 11.14 and 11.15 show that trees may give horizontal lines as predictors. This may indicate that the trees had too few leaves, e.g., the effect of important variables was missed.

Next, we consider the same figures for time step 12, cf. Fig. 10.15. Figures 11.16 and 11.17 show the 12 collections of predictions (six methods, two sorts of components) and actual values for the first cluster at the 11th iteration (predictions for the 12th year after

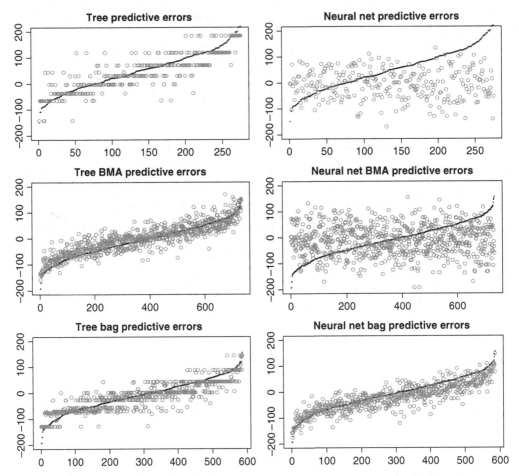

Figure 11.16 Similar to Fig. 11.14. Predicted versus actual values for the first six predictors for the first cluster but in the 12th year after the burn-in.

burn-in). Simple visual inspection shows that the best two methods in terms of the MSE are BMA and boosting with trees; they are essentially equivalent. However, bagging and median methods with NNs were the worst in the same sense, namely, that of their MSE. The BMA method and boosting for NNs were clearly biased upwards, and have a slightly larger variability than the best methods, but the bias is the serious problem. The other methods are generally not very biased but have greater variability and hence greater MSE than the best methods while being better than the worst methods. As with Figs. 11.14 and 11.15, Figs. 11.16 and 11.17 sometimes gave horizontal lines for predicted values, suggesting that the trees had too few leaves, i.e., omitted important variables.

In comparing the first time step predictions with the 11th time step predictions, it is seen that in both cases BMA and boosting with trees gave the best results in terms of MSE. Likewise, bagging and the median method gave the worst results in both cases, in terms of MSE. Other methods were often biased and had higher variability than the best methods. Contrasting Figs. 11.14–11.17 with Figs. 10.14 and 10.15 shows that, as a generality, ensemble methods give better performance than individual components, and the best of the ensemble

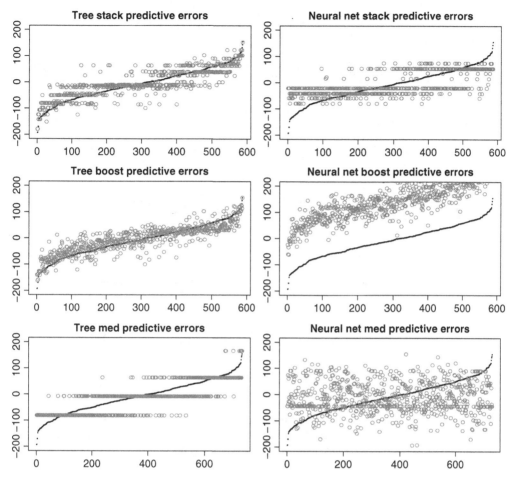

Figure 11.17 Similar to Fig. 11.16. Predicted versus actual values for the second six predictors for the first cluster but in the 12th year after the burn-in.

methods give a performance that is far better than individual components in terms of both bias and variance.

To make a cumulative comparison of the six methods, figures analogous to Fig. 10.16 were generated. These are seen in Figs. 11.18 and 11.19. Taken together there are 12 panels and the methods are presented in the same order as the panels for the predicted and actual in Figs. 11.16 and 11.17. Starting with trees, the predictuals from boosting trees appear to have the tightest distributions around zero with the least variability from year to year. Stacking and BMA with trees are nearly equivalent and arguably the best methods. The median method and bagging with trees are equally tightly distributed around zero for some years but show more variation from year to year. An individual tree is also more variable year to year than the tree-based ensembles. By contrast, using NNs, bagging methods give the best results – relatively tight around zero and moderate year to year variability. Moreover, it is immediately seen that the predictuals from the middle right panel (BMA with NNs) in Fig. 11.18 are overall the biggest. However, they are nearly symmetric about zero and are stable in the sense that they are all at about the same level. However, the year on year

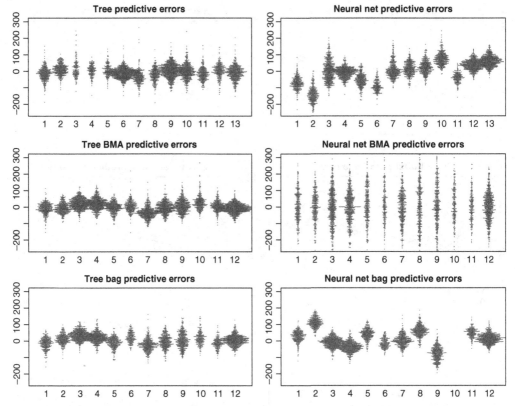

Figure 11.18 Similar to Fig. 10.16. Predictuals for the first six predictors over the 13 years pooled over the six clusters. The vertical scales on all panels are the same. The order of the panels is the same as in Fig. 11.15. No beeswarm was possible for year 13 due to missing data.

variability for individual NNs, boosted NNs, and median NNs is quite high. Only boosted NNs come close to the bagged NN predictor. Indeed, all the ensembles of NNs gave a worse performance than all the ensembles of trees.

A problem with NNs was that the existing software does not accommodate missing values very well. Accordingly, the 'beeswarm' plots for bagging in year 10 in Fig. 11.18 and for boosting in year six in Fig. 11.19 were not found.

As a third comparison of the performance of the point predictors side by side, boxplots can be generated for the 12 years for the six clusters. These are shown in Fig. 11.20 for boosting with trees (the best method, upper panel) and for the median method with NNs (the worst method with NNs, lower panel). The key qualitative difference between the two methods is that NNs will give a continuous predictor as a function of its arguments while trees will not. However, the value of continuity is in question because the predictuals from the best of the 12 ensemble methods (boosting trees) are visibly smaller than for the worst of the ensemble methods using only NNs (the median method) – even without taking the difference in scales into account. With the exception of the individual component predictors, i.e., a single tree or NN, these panels (and the corresponding panels for other ensemble methods, not shown) give boxplots that are generally shorter than the nonensemble methods

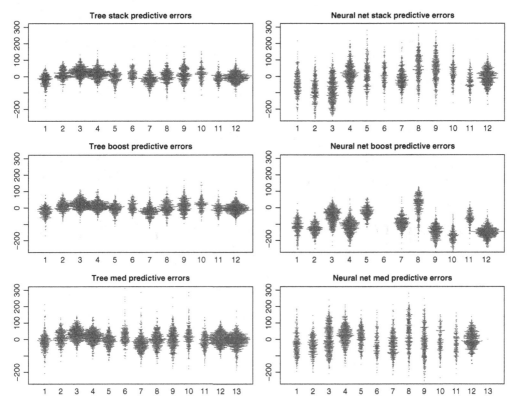

Figure 11.19 Similar to Fig. 11.18, but for six other methods. Predictuals for the second six predictors over the 13 years pooled over the six clusters. The vertical scales on all panels are the same. The order of the panels is the same as in Fig. 11.17.

shown in Fig. 10.23. (Figure 10.23 also differs from Fig. 11.20 in that the former gives a comparison of all methods over all clusters while the latter compares single methods over individual clusters.)

Finally, having seen a comparison of the best predictor overall (bagging or boosting with trees) and the worst predictor using only NNs (bagging or median) in terms of predictuals, i.e., residuals from the predictions, we will compare them in terms of stability. The usefulness of this comparison is that it establishes the range of stability for these methods. Another natural comparison would be of the stabilities of the best of the tree-based ensembles with the best of the NN-based ensembles. However, this is less interesting because, while it is obvious that the best tree-based ensembles (BMA or boosting) are more stable than the best NN-based ensembles (bagging), it is not clear what scale to use to assess this difference. Otherwise stated, comparing 'best to worst' gives a range that in principle can be used for assessing other differences in stability. Moreover, this approach may suggest a relationship between predictive performance and stability, as will be discussed briefly at the end of this subsection. Note that the stability analysis here is different from the stability analysis at the end of Sec. 10.6.2, because here it is the stability of the predictions themselves that will be examined rather than the stability of the predictors.

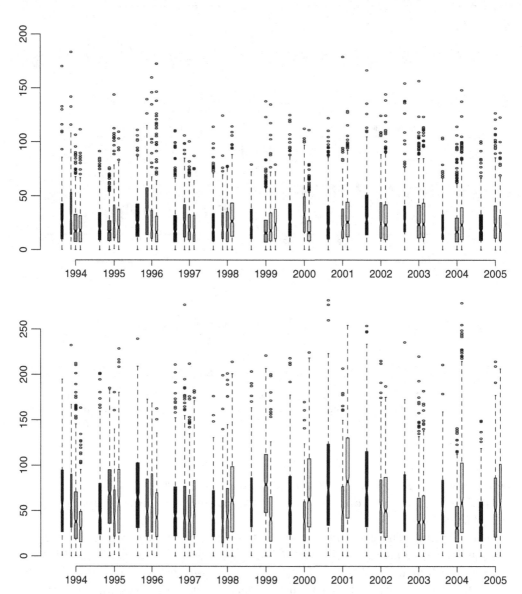

Figure 11.20 Top: Boxplots of the absolute errors for the six clusters over the 12 years using boosted trees. Bottom: Analogous boxplots using the median method with NNs. In some cases, clusters were too small (trees, NNs) or had too much missing data (NNs) and had to be dropped.

To assess the stability of, for instance, the boosted tree predictor versus the median NN predictor, the data was perturbed by adding independent $N(0, \gamma \sigma_p^2)$, $\gamma \geq 0$, noise to each real-valued data point. Here, σ_p was estimated by $\hat{\sigma}_p = 90$, the SD of the five years of burn-in data for cluster 1. This procedure was done 100 times for boosted trees, but only 50 times (for running time reasons) for NNs.

Under these circumstances, an example of predicted versus actual values was generated for one perturbed data set and the results are given in Fig. 11.21. This was done for cluster #1 only, using $\gamma = 0$, 0.2, 0.6, and 0.8, but the panels are qualitatively representative of the other runs. As before, the data points in cluster #1 are along the horizontal axis in no particular order (although the order is constant over all figures). Similar values of the SD were found for other large clusters but some clusters were too small or had too much missing data for predictors such as used here to be easily calculated. Likewise, the corresponding SD using all the data was close to 90, but it seemed more reasonable to use only the data for cluster #1. Note that the first and fourth panels in the first column essentially reproduce the corresponding panels in Fig. 11.15.

The upper three rows are for the first three years of predictions for boosting with trees (1994, 1995, 1996); the lower three rows are for the first three years of predictions for the median with NNs. Looking at the stability of predictions three years into the future seems, in context, to be about as far as makes sense physically. For theoretical purposes the analysis could have been extended as far as data were available. Each column corresponds to values of $\gamma = 0$, 0.2, 0.6, 0.8. As γ increases, the variability among the predicted values versus the actual values increases – there are more and more dark regions indicating where predicted and 'actual but perturbed' values do not match, for both methods. This is no surprise.

For $\gamma = 0.8$, the median NN method follows the data a little better than boosting trees; this is surprising given the method's relatively poor performance as seen in Fig. 11.19. However, median NNs seem to have more variability although boosting gives better predictions overall. That is, median may have a lower bias and higher variance compared with boosting, which may have a higher bias but lower variance. However, for $\gamma = 0.2$, 0.6 there are years in which the exact reverse also occurs in some cases: the median of NNs has higher bias and lower variability, and boosting with trees tracks the data and its variability better; compare the panels with $\gamma = 0.2$ in 1994 and 1995 or $\gamma = 0.6$ in 1995. The occurrence of this sort of reversal may indicate the limit of stability analysis.

The implication of Fig. 11.21 is that, despite occasional reversals, boosted trees are substantially more stable than median NNs in the sense of having generally smaller biases and variances. It should be noted that the trees were only allowed to have five levels of splits, and boosting was taken out to 50 iterations, while for NNs only one to five hidden nodes were allowed and the decay parameter was set at 0.01 in the package neuralnet. Since boosting often converges very quickly while NNs usually take longer and since NNs can approximate any continuous function arbitrarily well, a more extensive stability analysis might have yielded the reverse conclusion. Still, the main point stands: often, as here, under realistic conditions boosted trees give more stable predictive performance than median NNs. As noted, Fig. 11.21 shows the results of only one of a 100 boosted trees or 50 median NNs. However, the other runs gave qualitatively the same results.

Next we look at a summary of the effect of the perturbations. Recall that the perturbation of each data point by independent $N(0, \sigma^2)$ noise terms gives a perturbed actual value and the method gives a prediction. So, each prediction gives an absolute error. There are four values of γ and three prediction years. Figure 11.22 gives a qualitatively representative collection of side-by-side beeswarm plots of the L^1 errors over cluster #1 for each of the 12 settings. Obviously, even at this aggregated level, increasing γ gives taller plots within each year. For boosted trees (left) the beeswarms thicken in the middle (even if their heights

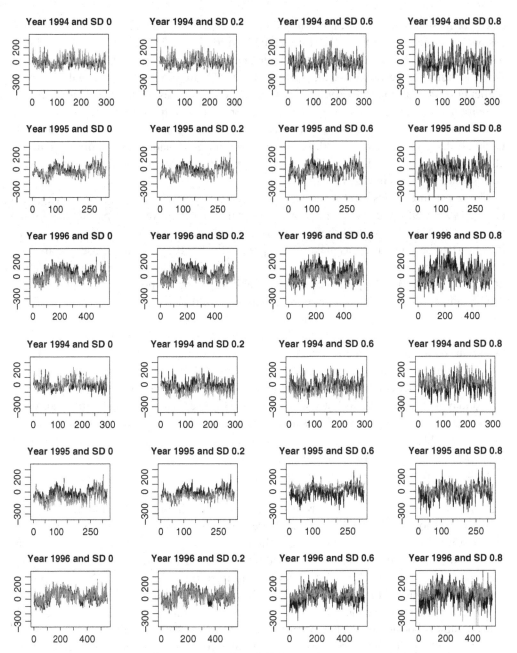

Figure 11.21 Upper three rows: Predicted (light gray) versus actual (dark gray) values for one, two, and three years after burn-in, for boosted trees, for the first cluster. Lower three rows: Similar, but for the median method with NNs. The four columns correspond to increasing random perturbations of the data to be predicted.

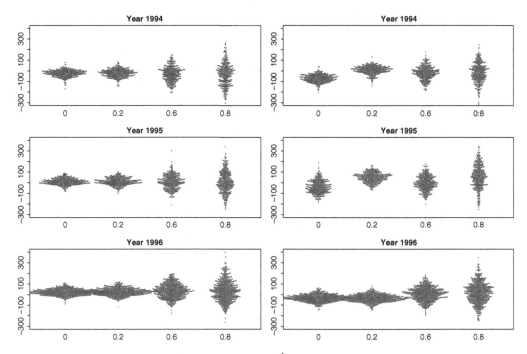

Figure 11.22 Left: Beeswarm plots of the L^1 errors for boosted trees, 100 iterations, three years. Right: Same but for median NNs and 50 iterations.

do not change much), indicating decreased bias and some decrease in variance – although, curiously, a greater improvement is seen from 1995 to 1996 than from 1994 to 1995. For median NNs, the different beeswarms are at different levels for different values of γ within each year, although they smooth out by 1996. The variability with γ in 1994 and 1995 is higher than for boosted trees and, in general, the swarms are at least as tall as for boosted trees. Taken together this suggests slightly higher bias and variance. By 1996, boosted trees and median NN are more similar than in 1994 because median NNs improve substantially from 1995 to 1996 – more than boosted trees do. This may be an artifact of the computational methods since NNs are smooth. However, again, from the aggregate perspective, boosted trees are more stable than median NNs. This is consistent with the implications of Fig. 11.21.

As a final look at the stability of the two methods, consider Fig. 11.23. Rather than showing predicted versus actual plus perturbation for a single run, as in the previous figure, this figure shows the means of errors (not absolute) of the predictions over the three years and four values of γ, for the 100 (resp. 50) perturbations of the original data for boosted trees (resp. median NNs). The vertical scales are different on the left and right but all three beeswarms on the left have the same (smaller) scale and all three beeswarms on the right have the same (larger) scale. Clearly, for trees, the mean error deteriorates with γ for 1994 and 1996 even though they go in different directions. For 1995, the mean error is relatively stable and better than for 1994 and 1996. The flat lines may be an artifact of the restriction on the trees to six leaves, which greatly limits their variability.

On the right, the corresponding means for median NN predictors are shown for the four values of γ. The beeswarms lengthen a little from year to year and, despite the different

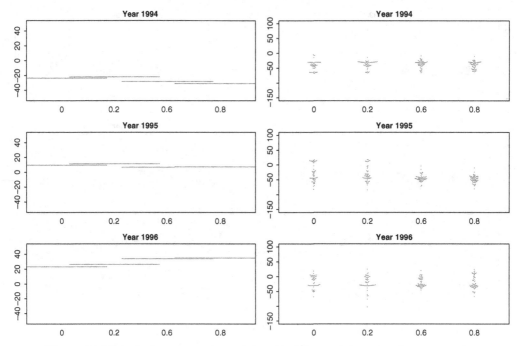

Figure 11.23 Left: Beeswarm plots of the 100 differences between the average predicted values and the average perturbed actual values for the first three years of prediction for boosted trees with $\gamma = 0, 0.2, 0.6, 0.8$. Right: Same but for 50 differences using median NNs. Note that the scales on the left and right are different but from row to row are the same.

scale from that for the boosted trees, the mean errors are larger and more variable. These two properties may follow because NNs are generally more sensitive than trees. Overall, little to no improvement is seen from year to year with the NNs. Indeed, it looks as though NNs often deteriorate year over year for each value of γ. Thus, Fig. 11.23 is an aggregate assessment of the stability of the two methods and is broadly in conformity with the earlier figures: boosted trees are more stable than median NNs as well as being better in a point predictor sense (at least for this problem). Indeed, it is tempting to conjecture that methods that are more stable under perturbation will be better under a point prediction error criterion, since such errors often increase more rapidly for unstable predictors than for stable predictors.

11.6.3 Mixing It Up

Clearly, model average predictors may consist of a variety of components. Here, trees and NNs have been discussed because they are common and are interesting from a variety of applied, methodological, and theoretical standpoints. However, there is evidence that ensembling kernel techniques or penalized methods may work very well, too. For instance, stacking RVMs or bagging penalized methods may give good results in some settings. It should be noted that one point of kernel or penalized methods is to obtain sparsity, so ensembling them is contrary to this aim. However, one can argue that sparse methods are precisely those that should be ensembled since they are (hopefully) adept at pulling out the important terms efficiently. For instance, forming a model average by bagging RVMs or predictors found from penalized methods may permit different variables to appear (because of the bootstrapping) that would otherwise fail to be caught by a single model.

There is no rule against ensembling kernel methods over a variety of kernels or combining them with penalized methods using a variety of penalties. Indeed, there is no rule against combining the predictors arising from different ensemble methods, and there is some evidence that, as the complexity of the DG increases, more complex methods such as these give better results. Some have even speculated that there may be a complexity-matching principle – the complexity of a good predictor will 'match' to the complexity of a DG even though the concepts of complexity for DGs and predictors are different. Moreover, the complexity of the components of a good predictor may reveal aspects of the processes implicit in a DG when it is $\mathcal{M}$-closed or $\mathcal{M}$-complete.

On the contrary side, recall that there is also evidence that bagging a good predictor may worsen it by increasing its variability. For instance, usually bagging a BMA worsens it, although, to be fair, BMAs often do not perform well in small-to-moderate sample settings, especially those with complex DGs, even though they are asymptotically optimal. As a generality, model-average methods have not been tested extensively, and the testing that must be done should be done predictively since cross-validation (CV) is often not effective. Indeed, unless used very carefully, CV is internal to the data set so a law of large numbers will often control the CV error. This makes the CV error less effective than genuine prediction, which is similar to unguided extrapolation. (Extrapolation simply extends the range of the explanatory variables in a prescribed way; prediction allows the elements of the range to be chosen randomly by the DG.)

It should be remembered that, if there really is reliable modeling information, this can be helpful, especially for ruling out techniques that are unlikely to give good prediction. For instance, in $\mathcal{M}$-closed or $\mathcal{M}$-complete problems, if the true function is a step function then there is little to be gained (and perhaps much to be lost) by combining trees and NNs rather then just using trees. If a true function is known to be both smooth and highly oscillatory then NNs will be likely to outperform trees, except in the limit of very large sample sizes where the lack of smoothness of the trees is slight compared with the scale and sample size of the data. However, because there are so many predictor classes, modeling information will probably be more useful for ruling out candidate predictor classes than for selecting a predictor class.

11.7 Endnotes

There are many aspects of complex prediction that have not been covered here. One that bears comment is the effort to use variable selection to reduce the dimension of the explanatory variables in prediction. The importance of this is well known in various contexts. For instance, in clustering, when the dimension of the data points gets too high then all clustering procedures break down because all data points are equidistant from each other; see Beyer *et al.* (1999) for one formulation or Koepke and Clarke (2011) for a simpler one. This is a property of an ultrametric topology; see Murtagh (2004). The same problem occurs in classification; see Fan and Fan (2008, Theorem 1).

As a solution, some argue that functions of the explanatory variables carefully based on physically meaningful 'state variables' should be chosen. There is some truth in this –

see the music example of Sec. 1.2.3. However, in many settings it is not feasible because the required information is unavailable. Indeed, in clustering, the common wisdom is that variable reduction with high-dimensional data is mostly ineffective when it's not actually misleading. One problem is that in many settings responses are the result of many small influences that defy condensation into a small number of variables, however well chosen. Hence, dimension reduction in complex data is only treated minimally here, e.g., in Sec. 10.5. On the other hand, Fan and Lv (2008) argued that a simple t-test will be sufficient for high-dimensional variable selection. The hypotheses are discussed at the end of Sec. 9.1.

Another aspect of complex prediction that bears comment is the definition of complexity itself. Here, complexity has been treated only heuristically, with the implicit assumption that the DG consists of a large number of different but connected parts or is simply hard to understand and its outcomes hard to predict well. While this is intuitively reasonable, it should be noted that there are numerous formal definitions of complexity, including the Vapnik–Chervonenkis dimension, the description length, the Kolmogorov complexity (a variant on the running time complexity), and the sample size complexity (essentially the efficiency). These have not been treated here because (i) they require a lengthy exposition to be understood well and (ii) outside specific examples, it is not yet clear how to use them to help choose or evaluate predictors. Admittedly, the MSE (variance plus bias squared) can be regarded as a complexity that can be used to guide predictor selection. However, it is easier to regard the MSE as a distance.

To conclude this chapter, two important topics are addressed. First we consider the main technique devised for prediction in $\mathcal{M}$-open problems where no assumptions are made about the DG. Second, we discuss the no free lunch theorems, which effectively say that no technique can be universally optimal.

11.7.1 Prediction along a String

For the sake of completeness, it is important to consider an $\mathcal{M}$-open predictive setting in which average properties are not available. The most important is the Shtarkov (1988) solution. Suppose that a game is played between Nature and a Forecaster conducted by an referee. The point of the game is for the Forecaster to come up with the best (probabilistic) predictions of the outcomes Nature emits. The better the Forecaster does, the smaller the loss incurred. The order of play is that first the Forecaster announces a prediction density. Second, Nature issues an outcome. Third, the referee looks at the prediction density and the outcome and makes the two players settle up in whatever coin they have agreed to use. More formally, suppose the Forecaster's goal is to predict a univariate outcome x emitted by Nature using the $\mathcal{M}$-open DG X. There is no information except that (i) success is to be measured by a scoring rule, in this case using a log, and (ii) the Forecaster has access to the predictions of a collection of 'experts'. Scoring rules are defined and discussed in Sec. 5.5. For convenience, index the experts by θ and denote the prediction of expert θ by the density $p(x|\cdot)$. Write the density that the Forecaster announces as $q(x)$. One way for the Forecaster to find an optimal q is to reason as follows. In terms of the logarithm, the difference between the Forecaster's prediction and expert θ's prediction is

$$\log \frac{1}{q(x)} - \log \frac{1}{p(x|\theta)} = \log \frac{p(x|\theta)}{q(x)}. \tag{11.28}$$

The largest difference between the Forecaster and the collection of experts is

$$\sup_{x} \sup_{\theta} \log \frac{p(x|\theta)}{q(x)}, \tag{11.29}$$

which is minimized when Forecaster chooses q to be

$$\arg\min_{q} \sup_{x} \sup_{\theta} \log \frac{p(x|\theta)}{q(x)} = q_{\text{opt}}(x) = \frac{p(x|\hat{\theta})}{\int p(x|\hat{\theta})}, \tag{11.30}$$

assuming the integral exists. This q_{opt} was first derived in Shtarkov (1988). Sufficient conditions for the Shtarkov solution to exist are given in Rissanen (1996) and computational details are given in Kontkanen and Myllymaki (2007); see also Roos (2008). (More recently, see Barron *et al.* (2014).) Once the DG issues the value x, the referee makes sure that the Forecaster pays $\log q_{\text{opt}}(x)$ to Nature and the game is over. If Nature wanted to maximize the Forecaster's loss and knew q_{opt} then Nature would choose $\arg\max_{x} q_{\text{opt}}(x)$. Otherwise, Nature would simply generate x by whatever means the DG represented.

In practice q_{opt} can be used to generate point predictors, e.g., a mean, median, or mode, with an associated standard deviation. Alternatively, a $1 - \alpha$ PI can be found from q_{opt} by finding regions on which it is highest. Clearly, the effectiveness of these predictions depends heavily on the collection of 'experts', i.e., the parametric family $p(\cdot|\theta)$, about which nothing has been assumed. In principle, the experts could even be purposefully misleading.

Expression (11.28) is called the 'regret' because it represents how much worse q is predicting than the θth expert. Analogously, (11.29) is the maximum regret and (11.30) is the minimax regret. The idea is that the Forecaster has given up predicting the outcomes of the DG directly and is simply trying to track the performance of the best expert. Thus, no distribution is associated with X; it is simply an agent emitting outcomes, i.e., this is an $\mathcal{M}$-open problem.

This game can be played sequentially and the data from earlier rounds may be used by the 'experts' to help them predict Nature's next outcome. That is, at the $(n + 1)$th round of the game, the Shtarkov solution may be written as

$$\arg\min_{q} \sup_{x_{n+1}} \sup_{\theta} \log \frac{p(x_{n+1}|\theta, x^n)}{q(x_{n+1})} = q_{\text{opt}}(x_{n+1}) = \frac{p(x_{n+1}|\hat{\theta}, x^n)}{\int p(x_{n+1}|\hat{\theta}, x^n)}.$$

That is, if the Forecaster always adopts a strategy achieving the minimax regret, each round concludes with the Forecaster paying Nature $\log q_{\text{opt}}(x_{n+1})$, in which q_{opt} is found using the first n outcomes. However, the sequence $(x_1, x_2, \ldots)$ is only a string of outcomes of the DG and no effort is made to generate a model for it.

There are numerous variations on this sequential game. For example, Xie and Barron (2000) and Cesa-Bianchi and Lugosi (2006) considered the inclusion of side information. The latter publication provides the best single location for theory related to the Shtarkov solution as well as other sequential predictors, albeit in an $\mathcal{M}$-complete context. A broader theoretical treatment of sequential prediction, still from the $\mathcal{M}$-complete perspective, is Rakhlin and Sridharan (2014). The relation between the Shtarkov and Bayes solutions is studied in Clarke (2007) and the actual performance of the Shtarkov solution is studied in Le and Clarke (2016).

11.7.2 No Free Lunch

The central point of no free lunch theorems is that, for any procedure, 'any elevated performance over one class of problems is offset by performance on another class', as stated in Wolpert and Macready (1997). In the context of optimization problems they noted that this means 'if some algorithm a_1 has superior performance relative to that of another algorithm a_2 over some set of optimization problems, then the reverse must be true over the set of all other optimization problems.' There is a similar principle in the context of supervised learning problems: 'for any two learning algorithms, there are just as many situations (appropriately weighted) in which algorithm one is superior to algorithm two as vice versa'; see Wolpert (2001).

To state a no free lunch theorem, an extended Bayes formalism must be introduced; see Wolpert (2001). The idea is that extending the Bayes formulation by distinguishing between the estimand and the estimator allows one to correct for the inability to assess bias within the Bayesian paradigm. It is left to the reader to decide the degree of success of this extension.

Let $\mathcal{X}$ and $\mathcal{Y}$ be countable sets representing input and output spaces with elements x and y respectively. This is sufficient for data collected by finite precision equipment and manipulated by finite precision computers. A data set of size n consists of n pairs from $\mathcal{X} \times \mathcal{Y}$, $\mathcal{D}_n = \{(x_1, y_1), \ldots, (x_n, y_n)\}$. Assume that there is a collection of functions $\mathcal{F} = \{f : \mathcal{X} \to \mathcal{Y}\}$ and that the true function $f_T \in \mathcal{F}$. Estimates $\hat{f}$ of f are also in $\mathcal{F}$. Thus, $\hat{f}(\cdot) \in \mathcal{F}$ is a function on $\mathcal{X}$ determined by $\mathcal{D} = \mathcal{D}_n$. Assume that there is a prior w on $\mathcal{F}$ taking values $w(f)$ and that there is a likelihood $p(\mathcal{D}|f)$, so the posterior $w(f|\mathcal{D})$ is well defined. Let $L(f(x), \hat{f}(x))$ be the loss associated with announcing $\hat{f}(x)$ when $f(x)$ is correct. As a function of $x \in \mathcal{X}$, $L(Y_f(x), Y_{\hat{f}}(x))$ can be regarded as a random variable $C = C(x)$. Now, for instance, the misclassification error can be written as

$$E(C) = Err(f, \hat{f}, \mathcal{D}) = \sum_x \pi(x)(1 - \delta(f(x), \hat{f}(x)))$$

where δ is the Kronecker delta and an off-training-set (OTS) error can be written as

$$E_{\text{OTS}}(C) = \frac{\sum_{x \notin \mathcal{D}} \pi(x)(1 - \delta(f(x), \hat{f}(x)))}{\sum_{x \notin \mathcal{D}} \pi(x)},$$

where π is some weighting scheme on $\mathcal{X}$. Since C is a random variable, its conditional expectation given $\mathcal{D}$ for misclassification error is

$$E(C|\mathcal{D}) = \sum_{h, f} Err(h, f, \mathcal{D}) W(f|\mathcal{D}) P(h|\mathcal{D}), \qquad (11.31)$$

with corresponding expressions for other losses and the OTS risk. Note that in (11.31) the sum over f is with respect to the posterior for f and the probability $P(h|\mathcal{D})$ is of unit mass on $\hat{f}$ if $\hat{f}$ is uniquely determined by $\mathcal{D}$ and permits more general randomized procedures. This leads to multiple no free lunch theorems.

Theorem 11.5 (Wolpert 2001) *Consider the off-training-set risk. Let $W_1(h|\mathcal{D})$ and $W_2(h|\mathcal{D})$ be the posteriors for two learning algorithms and denote the expectations with respect to those posterior by E_k for $k = 1, 2$. Then:*

1. *uniformly averaged over all f, $E_1(C|f) = E_2(C|f)$;*
2. *uniformly averaged over all f, $E_1(C|f, \mathcal{D}) = E_2(C|f, \mathcal{D})$;*
3. *averaged with respect to $P(f)$, $E_1(C) = E_2(C)$; and,*
4. *averaged with respect to $P(f)$, $E_1(C|\mathcal{D}) = E_2(C|\mathcal{D})$.*

That is, in these senses of risk, *all algorithms are equivalent on average.* There are other no free lunch theorems for supervised problems (Wolpert 1996). See Clarke *et al.* (2009) for a more mathematically explicit treatment of one form of a no free lunch theorem. Analogous results hold for search procedures; see Wolpert and Macready (1996).

There are two gaps to be filled to ensure that no free lunch theorems apply in prediction in the way they appear intuitively to do so. First, an extension to continuous problems with noise as opposed to purely discrete problems must be established. This is discussed in Sec. 2 of Wolpert (1992c) and Sec. 3 of Wolpert (1996). Details are not given, but the argument that the discrete case converges to the continuous case is plausible. Second, technically, no free lunch theorems apply to procedures that seek something fixed, whether an optimal value or a true function. Hence, they are designed for the $\mathcal{M}$-complete case (but could be tested in the $\mathcal{M}$-open case with suitable caveats). Accordingly, they do not apply directly to predictors. However, they do provide a way to construct predictors, since the search for a good predictor can be done decision-theoretically as a matter of either inference or optimization. In the $\mathcal{M}$-complete context, no free lunch theorems say that there are just as many predictive settings (appropriately weighted) in which one predictor can be expected to outperform another as vice versa. Hence the need for the empirical validation of a predictor; it is often difficult to determine a priori whether a given predictor is superior to another in a given predictive setting.

Because so many methods have been presented in this chapter (and indeed the whole book), it is worth concluding with a quote from Wolpert (1996): 'Consider any of the heuristics that people have come up with for supervised learning: avoid overfitting, prefer simpler to more complex models, boost your algorithm, bag it, etc. The no free lunch theorems say that all such heuristics fail as often (appropriately weighted) as they succeed. This is true despite formal arguments some have offered trying to prove the validity of some of these heuristics.'

12

The Future of Prediction

... the definitive property of good theory is predictiveness. Those theories endure that are precise in the predictions they make across many phenomena and whose predictions are easiest to test by observation and experiment.

E. O. Wilson (1998) – noting the centrality of prediction
even in the modeling of $\mathcal{M}$-closed or $\mathcal{M}$-complete DGs.

To conclude this treatment of prediction in statistics, in this chapter we survey a variety of recent developments. We also discuss developments that, although not particularly recent, remain important and have not been covered elsewhere in this book. No single book (that can can be carried by a single person as opposed to, say, a forklift truck) can do justice to an ubiquitous subject such as prediction and the present text is no exception. As can be seen from the earlier chapters, topics have been chosen to provide an overview of predictive ideas and to see how these ideas play out in several traditional fields and how they are embedded to a greater or lesser extent in fields that have emerged recently.

As a conclusion, the present chapter provides an orientation to several fields in which researchers are currently very active and in which prediction is a central feature. This includes recently emerged data types and also data types that have been around for many years but have only recently become feasible to analyze, perhaps because costs of data collection have decreased or computing power has increased.

Before turning to these topics, there are two points that should be borne in mind. First, two of the biggest trends in the history of data collection are its decreasing cost and the development of data types that are large in the sense of requiring massive storage. The latter is different from data that is ever more complex. For instance, the data collected by CERN involves colliding particles – and collisions do not happen very often. So, even though petabytes of data are collected, they are extremely sparse and not complex. Another example is next-generation sequencing (NGS) data. Even though collecting this is a complex procedure, the cost has decreased rapidly. (Relative to Sanger sequencing the reduction in cost is even more considerable.) However, viewed 'as is', NGS data is simple: strings of length around 120 of the nucleotides A, T, C, and G from a specific genome or metagenome. One thing that makes NGS data difficult is the storage and manipulation of the data sets. By contrast, the Vegout data (analyzed in Secs. 10.6.2 and 11.6.2) is 'small' in terms of storage but is very complex. Functional data and signal analysis data, amongst other data types, are also often complex but small. In these cases, the process of generating the data may also be very complex, but that is a separate issue. The data itself is complex and even though it may not require much storage, the mathematical object it represents is a curve, plane, or hyperplane, a much more complex object than a string of nucleotides viewed without reference to the three-dimensional configuration of DNA. Thus, it is helpful to distinguish between data on two axes: complex versus simple and large versus small. All four categories are possible and, informally, when people use the term 'big data' they often mean high-dimensional data that is both complex and hard to store because of its sheer volume.

Second, as data become larger or more complex, new problems must be resolved. Model misspecification usually becomes more detectable, and hence model uncertainty is a greater problem. One way to think about this is that typically data become less representative of the DG or, more precisely, both biased sampling and heightened variability become unavoidable because the range of possible data sets that could be produced by the DG is so large. To a substantial extent this may be a factor underlying the irreproducibility of reported results in scientific studies that has been noted of late, even where there is no evidence of malfeasance or incompetence. This is on top of the fact that many DGs are $\mathcal{M}$-open, so that modeling is not as useful as before.

A key feature that has achieved new importance recently is validation, or more precisely, the evaluation of how good a predictor is. This is a cornerstone of the predictive approach. Above all, elevating the prediction of future outcomes to be the sine qua non of scientific veracity is an effort to minimize the irreproducibility of inferences and outcomes. This effort is limited conceptually by the facts that (i) big data often have counterintuitive properties such as nonidentifiability in variable-selection contexts (see the end of Sec. 9.1), (ii) all data points are roughly equally far from each other in the limit of high dimensions (Beyer *et al.* 1999), (iii) classification breaks down (Fan and Fan 2008), (iv) clustering breaks down (Koepke and Clarke 2011), and (v) many methods e.g., penalized or kernel, are unstable or at best merely approximate (with little ability to evaluate the degree of approximation). For further properties of high-dimensional data in general, see Murtagh (2009), Murtagh and Contreras (2011), Murtagh (2004), and Hall *et al.* (2005). The basic lesson of these studies is that, loosely speaking, in high dimensions all data points are equally distant from each other and hence form an ultrametric space.[1] The topology of an ultrametric space is strikingly different from the topologies of lower-dimensional spaces, and this means that many standard low-dimensional techniques become completely ineffective in sufficiently high dimensions. The problems with complex low-dimensional data ('small data') are qualitatively different – bias and variance as opposed to instability. (Here, instability essentially means that any solution often seems as good as any other solution, so there is less basis for comparison.)

The topics discussed in this chapter start with recommender systems in Sec. 12.1 – ways used by many companies to present a purchaser with those products he or she is likely to buy. Section 12.2 discusses streaming data. This includes sensor data, a key source of streaming data. This section also introduces questions such as online feature extraction and decision making. Sections 12.3–12.5 present summaries of the current predictive thinking on spatio-temporal data (this includes remote sensing data), network models (arising from actual networks under study), and multitype data, i.e., the extraction of information by combining two or more qualitatively different data types. Following these formalities. In

---

[1] In an ultrametric space the metric d satisfies $d(x, z) \leq \max(d(x, y), d(y, z))$ for any x, y, z in the space; this is a much stronger criterion than the triangle inequality.

Sec. 12.7 a discussion is given on aspects of predictors that have not been covered in the detail they deserve. Section 12.6 gives a partial listing of other topics that should have been in this book in but are not, owing to lack of time and space. Finally, a discussion of the future role of prediction is presented in Sec. 12.8. This chapter is discursive; technical details are kept to a minimum.

12.1 Recommender Systems

Recommender systems (RSs) were of interest in predictive statistics starting in the early 2000s but only burst onto the general statistical consciousness after the 2006 Netflix contest. The first conference on RSs occurred in 2007. Since then research on RSs has been intense because devising better RSs means that companies can make more money, product creators can reach a wider audience, and buyers can be introduced to products that they may want. The data available to recommender systems are often nonstandard: they can be big data, as discussed in the introduction; they can be online data, meaning that a list of recommendations must be made sequentially for users, and the 'inferences' from it can rarely be evaluated by familiar techniques.

The core idea of a recommender system is the following. Imagine a population of potential buyers and a list of products. The task is to present a new buyer with products he or she might want to buy, ranking them in terms of the buyer's preferences as best this can be determined. Obviously, this is an $\mathcal{M}$-open prediction problem: the buyer in principle does not know what he or she is seeking (and neither does anyone else) and hence there cannot be a model for what to buy. The RS merely outputs items for the buyer's consideration in the absence of any model necessarily being true. (In this way, RSs are like the Shtarkov solution; see Sec. 11.7.1.) Recommender systems are typically not 'pure' prediction problems because the usual output is only a list of products ranked in order of the buyer's preferences. So, the buyer may not buy anything and the preference order may not correspond to a unique probability distribution. In some cases, RSs may correspond to a pure prediction problem if the preferences can be expressed as a random variable, i.e., are numerical and can be combined into a probability – including the probability that the buyer buys nothing.

There are several classes of RSs, including collaborative filtering (CF), content-based (CB), and knowledge-based (KB). Hybrid forms of these are common, too. An important aspect of RSs is their evaluation. However, this will be discussed only briefly. These topics amongst others are presented in the text by Jannach *et al.* (2011). More recent developments are in Ricci *et al.* (2011). All these methods can be regarded as input–output systems, the four possible inputs being (i) information about the new buyer, (ii) information about other buyers, (iii) the properties of the products, and (iv) any extra knowledge that might be available. The output is a recommendation list, i.e., a list of products that the new buyer might want with a score for each indicating how strongly he or she might be inclined to buy it.

12.1.1 Collaborative Filtering Recommender Systems

Start with a set of users $\{u_1, \ldots, u_n\}$ and a set of products $\{p_1, \ldots, p_m\}$ and assume that the purchase history of the users for the products is known. Furthermore, let the $n \times m$ rank matrix $R = (r_{ij})_{i=1,\ldots,n; j=1,\ldots,m}$ correspond to rank user i being assigned product j. Now consider a user, say u_{n+1}, who has a purchase history, who may have ranked various products, and who is now coming back as a de facto new user. One task is to choose a product, say p_{m+1}, that the user has not seen and assign a value to the unknown rank $r_{n+1,m+1}$. There are five basic ways to use R to obtain recommendations: (i) nearest neighbors using the rows of R, (ii) an item-based procedure using the columns of R, (iii) a matrix factorization of R, e.g., a singular value decomposition (SVD), (iv) association rules based on the concepts of support and confidence[2] derived from making the entries in R binary, say zeros and ones, and (v) extending this to develop classifiers based on a discretized version of R. Obviously, (iv) and (v) are similar in spirit; the difference is that the latter defines a classification model.

Before giving some mathematical details, it should be noted that these techniques have arisen from a real physical problem in commerce. Moreover, R is generally sparse – few users will have bought more than a few products and many products will have attributes that can make them 'similar' to each other from a user's perspective. The entries of R are often taken to be Likert, meaning that they are on an integer scale from 1 (extremely dislike) to 5 (extremely like). Scales with seven, ten, or even continuous values have also been used and there is some evidence that in practice continuous scales are preferred by users and are more effective. In discrete cases, the scales are ordinal and not numerically accurate. i.e., a rank of 5 does not mean that the product is ranked 1.25 'better' than one ranked 4.

The simplest type of RS is a collaborative filter (CF) that uses the rows $r_1, \ldots, r_n$ of R to define a neighborhood of previous users around the vector $r_{n+1} = (r_{n+1,1}, \ldots, r_{n+1,m})$. One way to do this is to find the (Pearson) correlation between r_{n+1} and each r_i using the first m entries in each row. Since the different rows of R typically have missing values, these correlations can be found only using the columns for which all the r_{ij} are pairwise complete. Another constraint is that all the r_i for $i \leq n$ that are used must have a value in column $m+1$. In this way a set of n correlations $\rho_i = \text{Corr}(r_{n+1}, r_i)$ is found and the r_i with the highest ρ_i are chosen, thereby providing a 'neighborhood' of users around u_{n+1}. From this collection of rows, the corresponding $r_{i,m+1}$ can be found and combined into a value for $r_{n,m+1}$, for instance, by taking an average. Doing this for a variety of new products ranks them, and the new products with ranks above a prespecified threshold for user $n+1$ would be presented to a prospective buyer, in order of their rankings.

Obviously this is too simple for most applications. One better approach may be to weight the chosen r_i by their similarity to r_{n+1} on the products (columns) with nonvoid entries that they have in common. One way to do this is by correlation. In this case, writing $\text{Corr}(u_{n+1}, u)$ for $\text{Corr}(r_{n+1}, r)$ to emphasize the fact that the user is new, the prediction is

$$\hat{Y}(u_{n+1}(p_{m+1})) = \bar{r}_{n+1} + \frac{\sum_{u \in \mathcal{N}} \text{Corr}(u_{n+1}, u)(r_{u,p_{m+1}} - \bar{r}_u)}{\sum_{u \in \mathcal{N}} \text{Corr}(u_{n+1}, u)},$$

where $\bar{r}_i$ is the average of the ranks of user i and $\mathcal{N}$ is the neighborhood of users around u_{n+1}. Thus, instead of a simple average, an average weighted by the correlations with u_{n+1} is used to generate the predicted rank. Other assessments of similarity besides the Pearson correlation have been tested, e.g., Spearman correlation, L^2 distance, the application of transformations to the ranks so that the ranks of more important products will contribute

---

[2] Here, the term 'confidence' means a specific conditional probability that arises in RSs; it does not have its established statistical meaning.

more to the overall sum, etc. Various ways of defining the 'neighborhood' around u_{n+1} have also been studied; however, the default for many purposes is simply to use the nearest neighbor under Pearson correlation.

A slightly more complicated CF technique is item based; it is somewhat like a nearest neighbor approach but uses the columns of R augmented by r_{n+1}. That is, examine the first m column vectors of $R^* = [R^T, r_{n+1}^T]^T$ to find those that are similar to the column vector for product $m + 1$, excluding $r_{n+1,m+1}$ since that is the value to be assigned. Then, using the column vectors that are most similar to that for product $m + 1$, set $\hat{Y}(u_{n+1}(p_{m+1}))$ to be the average or, better, the weighted average, of the ratings that user $n + 1$ assigned to these similar items. In parallel with the Pearson correlation, it is common to use the cosine similarity for products p_j and p_k, namely

$$\text{sim}(p_j, p_k) = \frac{c_j \cdot c_k}{\|c_j\| \, \|c_k\|},$$

where c_j and c_k are distinct columns of R. Now, the assigned rating to product $m + 1$ for user u_{n+1} is

$$\hat{Y}(u_{n+1}(p_{m+1})) = \frac{\sum_{p \in \mathcal{N}(u_{n+1})} \text{sim}(p, p_{m+1}) r_{n+1,p}}{\sum_{p \in \mathcal{N}(u_{n+1})} \text{sim}(p, p_{m+1})},$$

where $\mathcal{N}(u_{n+1})$ is the set of items with ratings vectors c_j in the neighborhood of c_{m+1}, defined, for instance, in terms of the cosine similarity.

Matrix factorization methods are the most mathematically sophisticated. The simplest of these is a singular value decomposition; it generalizes the eigenvector decomposition for nonnegative-definite square matrices. Write

$$R^T = V \Sigma U^T.$$

Since R^T is $m \times n$, V can be chosen to be an $m \times m$ unitary matrix, U^T is an $m \times n$ unitary matrix, and Σ is an $m \times m$ diagonal matrix with nonnegative real numbers on the main diagonal. Loosely, V corresponds to products and U corresponds to users. Since the rows and columns of unitary matrices form orthonormal bases of their vector spaces, V and U^T are not simply rotations; they also preserve lengths and angles between vectors, i.e., for any vectors w and z and a unitary matrix such as V (or U^T), $\langle Vw, Vz \rangle = \langle w, z \rangle$. Thus the largest, say, k values in Σ provide an approximation to R^T consisting of the first k columns, say V_k, of V and the first k rows, say U_k^T, of U^T. That is, $R^T \approx V_k \Sigma_k U_k^T = R_k^T$ where Σ_k is the upper left $k \times k$ block of Σ.

Effectively, the kth singular value approximation to R^T projects the rating vectors for the ith user or the jth product onto $\mathbb{R}^k$, i.e., the rows and columns of R_k can be regarded as points in $\mathbb{R}^k$. Likewise, the rating vector from u_{n+1} for the first m products can be projected into $\mathbb{R}^k$; that is,

$$r_{n+1}(k) = r_{n+1} V_k \Sigma_k^{-1} \in \mathbb{R}^k.$$

(The projection can be done for the n ratings of the $m + 1$ product analogously.) Now, in k dimensions there are $n + 1$ points and $r_{n+1,m+1}$ can be assigned a value by looking at the nearest neighbors of $r_{n+1}(k)$ or by using on item-based approach, as has already been discussed. An extension of this general approach that uses content information can be found in Nguyen and Zhu (2013).

Association rules are another way to generate recommendations of products to users. First, instead of obtaining a matrix of product rankings by users, look only at purchases. Since several purchases may be made by a user at the same time, define a transaction T to be a 'pair' in the data of the form $(u, p_1, \ldots, p_\ell)$, where u is a user and $p_1, \ldots, p_\ell$ are the products that the user purchased at one time, and let $\mathcal{T}$ be the collection of transactions. The idea is to find sets S_1 and S_2 of products with the property that if a user bought S_1 then the same user is very likely to buy S_2; these statements are called *association rules*. Often this is written as if it were an implication, $S_1 \Rightarrow S_2$. The support s of an association rule such as $S_1 \Rightarrow S_2$ is the empirical probability that $S_1 \cup S_2$ was bought by the users in their transactions, i.e.,

$$s = \frac{\#\{T \in \mathcal{T} \mid S_1 \cup S_2 \in T\}}{\#(\mathcal{T})},$$

with mild abuse of notation because $S_1 \cup S_2$ can be only a subset of the product entries in $T \in \mathcal{T}$. Likewise, the confidence c of an association rule is the conditional probability of S_2 given S_1, i.e.,

$$c = \frac{\#\{T \in \mathcal{T} \mid S_1 \cup S_2 \in T\}}{\#\{T \in \mathcal{T} \mid S_1 \in T\}}.$$

The goal is to find association rules with a high c and, if possible, high s also. This search can be done separately from the online process of making recommendations to u_{n+1} and is not discussed here.

The rank matrix R can be made binary by replacing all entries above a threshold with ones and all other entries with zeroes. This assumes that there are no void entries. Although unrealistic, this assumption can be accommodated in practice by using only the relevant submatrices of R, i.e., those for which all entries are nonvoid. This requires more detailed notation than is used here and would needlessly clutter the exposition since the goal here is only to outline the main ideas. Moreover, it is henceforth assumed that a set of association rules have been found that meet minimum s and c criteria. The details of how to conduct such searches are omitted since they are usually highly dependent on the specific context of the users and products.

Now it is straightforward to generate recommendations in simple cases. Let the set of association rules of the form $S_1 \Rightarrow S_2$ that are relevant to user $n + 1$ be those for which S_1 is a subset of the products u_{n+1} that has bought and liked. Given this set of association rules, let $S_{2,n+1}$ be the union of the products in the subsets S_2 that have not been bought by u_{n+1}. Order the products in $S_{2,n+1}$ in terms of decreasing values of c for the association rule that generated them. The top k of these products (with their respective c's) are the ordered recommendations.

A related way to generate recommendations is to let $\mathcal{L}(j)$ be the collection of association rules that recommend product p_j and to assign a value $v(j)$ to p_j by the product and sum rules familiar from introductory probability theory:

$$v(j) = \sum_{l \in \mathcal{L}} s_l c_l,$$

in which s_l and c_l are the support and confidence of rule l.

As suggested by the use of association rules, generating recommendations can be regarded as a classification problem when the r_{ij} are discrete, and here assume only finitely many values. One example of this, which is oversimplified but shows some of the key ideas, is the following. The task is to assign to each product the rank on the Likert scale that a given new user would assign to it, i.e., to identify the rank value with the highest probability given the past rankings of a new user. This is a Bayes classifier of the form $\arg\max_\ell P(L = \ell | r_{n+1,1}, \ldots r_{n+1,m})$, where L is a random variable assuming values ℓ that are the possible ranks of p_{m+1} for u_{n+1}. Now,

$$P(L = \ell | r_{n+1,1}, \ldots r_{n+1,m}) \approx \frac{P(L = \ell) \prod_{j=1}^{m} P(r_{n+1,j} | L = \ell)}{P(r_{n+1,1}, \ldots r_{n+1,m})}$$

if the ratings are approximately conditionally independent. The factor $P(L = \ell)$ can be estimated by the fraction of earlier users, $u_1, \ldots, u_n$, who gave product p_{m+1} the rank ℓ. This corresponds to the prior in a Bayesian formulation. The factors $P(r_{n+1,j} | L = \ell) = P(r_{n+1,j}, L = \ell) / P(L = \ell)$ can be estimated in the usual way in which conditional probabilities are estimated, namely, by the ratio of the number of times p_{m+1} received rank l and p_j received rank $r_{n+1,j}$, amongst the earlier users, and the number of times p_{m+1} received rank l amongst the earlier users. This corresponds to the likelihood in a Bayesian formulation. Finally, since $P(r_{n+1,1}, \ldots, r_{n+1,m})$ is only a normalizing constant, it can be omitted. Now, taking the $\arg\max_\ell$ of the $P(L = \ell | r_{n+1,1}, \cdots r_{n+1,m})$ gives a prediction (and a score – the probability) of how much u_{n+1} will like p_{m+1}. If the probability is high enough then p_{m+1} is recommended for u_{n+1}.

There are many more CF RSs that include clustering, demographic information about the user and the user base, and extra domain knowledge amongst numerous other complications. A relatively recent exposition of many of these aspects of RSs in general contexts can be found in Meyer (2012).

12.1.2 Content-Based (CB) Recommender Systems

A separate class of RSs are the content-based (CB) systems. By the 'content'is meant the products' characteristics, so these systems are qualitatively different from CF RSs, which use only the rankings from previous purchases. A product's characteristics are expressed as the evaluation of a variable on the product, e.g., its cost, size, color, manufacturer, and so forth. There are three main ways to use content for RSs. They are (i) the term frequency–inverse document frequency (TF–IDF) method, (ii) similarity methods, and (iii) naive Bayes methods. The fundamental point is to try to match user preferences to product characteristics. The user preferences are expressed by what the user liked in the past or seems to be seeking in the present. Although CB systems can be applied to any sort of product, they are most commonly applied to products that are all in the same class; the paradigm example that led to the development of CB RSs is a collection of text-based objects. Thus, the product characteristics are often taken as keywords within documents. The usual terminology of CB RSs is used here, and bears this out: most work on CB RSs is written as if the products were texts, i.e., the characteristics are zero–one variables indicating the absence or presence of various keywords. However, these techniques can be applied to any collection of products (songs, power tools, images, cars, etc.) provided that the characteristics chosen, sometimes called the meta-information, are appropriate for characterizing the products.

First, observe that with text recommendation an exhaustive list of keywords can be identified at least in principle – one simply takes the union of all the documents in the collection to be searched, orders this set, and summarizes a document by a string of zeros and ones indicating which keywords are present in it. If a profile of a user is represented as a string of zeros and ones indicating interest in each keyword then recommending documents to the user is simple: find the documents that have the greatest number of ones in common. In fact this does not work well because keywords are usually not equally important and the procedure will tend to favor longer documents. A variation on this would be to find those documents with minimal Hamming distance from their string to the user's string. Again, keywords are rarely equally important to a user. Also, since the strings often have only a few ones, the Hamming distance can be made small by simply choosing the void document whose string is all zeros. So, this can unduly favor short documents.

The TF–IDF technique is as follows. Regard each document as a point in a vector space where the entries are determined by the term frequencies (TFs) and the inverse document frequencies (IDFs). The TF for keyword i is the normalized number of times it appears in document j. (Normalization is important to avoid favoring long documents.) Let $F(i, j)$ be the raw frequency of keyword i in document j and, for fixed i, let $O(i, j)$ be the set of keywords, other than keyword i, that appear in document j. Then $F(i, j)$ can be normalized by using

$$O_{\max}(i, j) = \max\{F(w, j)|w \in O(i, j)\},$$

the maximum number of keywords other than i that appear in j. This gives

$$\text{TF}(i, j) = \frac{F(i, j)}{O_{\max}(i, j)}$$

as a reasonable assessment of how often a term occurs in a document.

The IDF for a keyword i is its relative importance among all the other keywords. Essentially, it reduces the importance of keywords that appear in too many documents on the premise that they provide less discrimination across documents than rarer keywords do. Suppose that $i = 1, \ldots, M$, where M is the number of documents being searched, and let $M(i)$ be the number of documents in which keyword i appears. The IDF for keyword i is

$$\text{IDF}(i) = \log \frac{M}{M(i)}.$$

So, the greater $M(i)$ is, the smaller $\text{IDF}(i)$ is; the argument of the logarithm is always positive and the $\text{IDF}(i)$ values are not too large because $\log x \ll x$ for large x.

Now the TD–IDF weights for given i and j are

$$(\text{TF} - \text{IDF})(i, j) = \text{TF}(i, j)\,\text{IDF}(i).$$

That is, document j is represented as the vector $(\text{TF} - \text{IDF}(1, j), \ldots, \text{TF} - \text{IDF}(n_K, j))$, where n_K is the number of keywords, rather than a vector of zeroes and ones. So, given a user profile, the documents closest to it can be identified and presented in order of proximity as recommendations, with the distance serving as a score. The question of getting a suitable user profile remains unaddressed so far, but seeking feedback directly from users is often a good way to determine the weights that the user implicitly assigns to different keywords.

More complex versions of the TF–IDF vector space approach are discussed in Jannach *et al.* (2011) and Ricci *et al.* (2011): TF–IDF can be improved by a variety of techniques even though some limitations will remain. Difficulties include the fact that implementing TF–IDF can be complicated and TF–IDF is frequently highly sensitive to the specific context of users and products.

Similarity methods are likewise broad, especially when implementation in specific settings is considered. These RSs typically work by evaluating the similarity of a new item for a user to items that the user has liked in the past. Again, consider a very simple approach: if a user likes documents of a specific genre (say) then the RS could simply find all documents of that genre amongst the N documents in the collection. Similarly if a user has given several preferences, the RS could again search all N documents to find those that satisfied these preferences. This would roughly correspond to the problematic zero–one approach discussed earlier in this subsection.

Another option is to measure the similarity of the overlap in keywords between the user's preferences and the documents. Let $K(p_j)$ be the set of keywords in document p_j and define the Dice coefficient between two documents as

$$D(j,k) = \frac{2\#(K(p_j) \cup K(p_k))}{\#(K(p_j)) + \#(K(p_k))}.$$

Under this similarity, documents would be presented in order of decreasing size of the Dice coefficient between the user's preferences and candidate documents – essentially a nearest neighbor technique on the preference and keyword vectors. This similarity is simple to implement and can adapt easily to changes in users. Also, in practice, a relatively small number of preferences is enough to get reasonable predictions. However, it has a key limitation in that it simply might not accurately reflect the distances between the user's preferences and the available documents. That is, other techniques not based on nearest neighbors often give better results.

One of these methods, called Rocchio's algorithm, see Rocchio (1971), goes back to the late 1960s but continues to be used and studied. It refines the user's preferences, now regarded as 'queries' to a database, sequentially to help the user find the most relevant documents. As before, the term 'documents' can be generalized to any class of products. Rocchio's algorithm assumes that there is a sufficient purchase history and set of ratings for each given user, that some method for generating recommendations from a given query is available, and that even though the user is unable to formulate his or her preferences accurately, he or she will be able to identify how closely the presented documents are to what he or she wants. Thus, over several iterations the RS will present, hopefully, those texts that the user prefers.

The method is deceptively simple. Let $Q_{u,v}$ be a preference vector for user u at step v that is of the same form as the vectors expressing the characteristics of the documents. Treat $Q_{u,v}$ as a query in the sense that a recommender system, e.g., TF–IDF, can be used to generate a collection of positive recommendations R^+ that try to match it and a collection of negative recommendations R^- that try to be as far from it as possible. Rocchio's algorithm is based on the formula

$$Q_{u,v+1} = \alpha Q_{u,v} + \beta \frac{1}{\#(R^+)} \sum_{r \in R^+} r - \gamma \frac{1}{\#(R^-)} \sum_{r \in R^-} r.$$

That is, the updated query is related to $Q_{u,v}$ by the coefficient α, increased by a coefficient β times the centroid of the positive recommendations, and decreased by a coefficient γ times the centroid of negative recommendations. The latter two terms represent the way in which the updating rule improves $Q_{u,v}$. This formula has only an empirical justification – no theory seems to lead to it as the most appropriate choice of updating rule. Empirically, the parameter values (α, β, γ) are often in proportion to $(2, 4, 1)$, so positive recommendations are worth more than negative recommendations and, beyond a small number of iterations, improvement can be quite small.

As with CF, CB RSs can be developed from formal classification problems. The simplest case occurs where each document has an associated vector of zeroes and ones indicating whether or not the document has a given keyword. However, it is not the keyword per se that is of interest so much as the predicted response of the user, like or dislike. Thus, there are two classes, like and dislike, and the task is to ascertain which has a higher conditional probability given the data from other users on the documents and the past responses to keywords of the current user, say u. The construction of the naive Bayes classifier is much the same as in the CF case.

Suppose that there are N documents, that $N - 1$ of them have been rated by u on $j = 1, \ldots, J$ keywords, and that u has also rated the Nth document for the first $J - 1$ keywords. What is the predicted rating of u for the Jth keyword in the Nth document? If $C = c$ denotes the classes, $c = 1$ for like and $c = 0$ for dislike, then

$$P(C = c | r_{N,1}, \ldots, r_{N,J-1}) \approx \frac{P(C = c) \prod_{j=1}^{J-1} P(r_{N,j} | C = c)}{P(r_{N,1}, \ldots, r_{N,J-1})},$$

where $r_{N,j}$ is one or zero accordingly as u likes or dislikes keyword j and the individual ratings of like or dislike are approximately conditionally independent. As in the CF case, the factors in the likelihood are estimated by the usual estimators for conditional probabilities. For instance, here $\widehat{P}(r_{N,j} = 1 | C = 1)$ is the number of times that the first $N - 1$ documents were ranked one for keyword 1 divided by the number of times the first $N - 1$ documents were ranked one for the Jth keyword. This corresponds to the likelihood portion of the posterior. The prior part of the likelihood is estimated by $\widehat{P}(C = 1)$, the number of times keyword J was ranked one amongst the first $N - 1$ documents. Finally, the denominator can be ignored since it does not depend on C. By maximizing the posterior over c, a value one or zero can be assigned to u's Jth keyword and Nth document. Thus, how interesting a document is to u can be assigned on the basis of its properties and user preferences rather than rankings of previous purchases. There are many extensions of the methods presented here; see Manning *et al.* (2008), for example, for a more sophisticated classification method that also takes into account the number of occurrences of keywords.

12.1.3 Other Methods

Two further types of RS are worth discussing. The first is knowledge-based RSs and the second is hybrid systems. Knowledge-based (KB) RSs are designed for settings in which user preferences evolve, purchases are infrequent, and users want more interaction with the recommender to ensure that their requirements – which may not be well formulated – are met. Loosely, KB RSs are those that use data sources different from CF and CB, i.e., such sources as ratings of past purchases and features of the various products, explicit knowledge of the customer, and fitness of the products for the customer's purpose. By contrast, hybrid methods are those that combine the sources of information used in any of the CF, CB, or KB approaches, possibly including other sources of information such as the community to which a user belongs.

Knowledge-based RSs fall roughly into two categories: constraint-based and case-based. In both categories, the user must specify requirements that a product must satisfy and the RS outputs the products meeting those requirements. If no product satisfies the requirements then they must be modified or the user must give up. So, KB systems rely more on detailed knowledge about products than CB RSs. The difference between the two categories is that constraint-based RSs search a product list for products that satisfy a set of constraints, while case-based RSs retrieve products on the basis of their similarity to product requirements. The two systems and are similar in that they both focus on the properties of the products; however, they differ substantially in that CB RSs are based on user ratings of product properties rather than the product properties themselves.

Constraint-based RSs correspond to a constraint satisfaction problem. A set of variables, $\mathcal{V}$, has to be identified, along with ranges for each variable and a collection of constraints $\mathcal{C}$ that solutions must satisfy. Thus, if $\#(\mathcal{V}) = d$ then a solution is a point in the d-dimensional space of values that the variables in $\mathcal{V}$ can assume while satisfying the constraints. Typically, there are two types of variable, one for the customer (the price, the use to which a product will be put, etc.), say $\mathcal{V}_C$, and one for the properties of the product, say $\mathcal{V}_P$ (its weight, manufacturer, etc.). Obviously there may be some overlap between $\mathcal{V}_C$ and $\mathcal{V}_P$, but for simplicity they are often made disjoint. Typically, there are three types of constraint set. One, say $\mathcal{C}_C$, is the set of compatibility constraints. For instance, a variable in $\mathcal{V}_1$ may be the price of a product and a variable in $\mathcal{V}_2$ might be some measure of its quality. If a customer is willing to spend only a given amount, some products might be of such high quality that they are ruled out. A second set, say $\mathcal{C}_F$, is the set of 'filter' constraints. These constraints may be determined purely by the products: a camera that can generate sufficiently large photos clearly may require a certain minimal resolution. A third set of constraints, say $\mathcal{C}_P$, is the list of available products.

Now, the customer's requirements may be expressed in terms of bounds on the variables in $\mathcal{V}_C \cup \mathcal{V}_P$. Combining these with the constraints in $\mathcal{C} = \mathcal{C}_C \cup \mathcal{C}_F \cup \mathcal{C}_P$ defines the search space from which solutions must be found. Thus, recommendations can be generated without using a notion of similarity; however, they cannot then be ranked. However, if the constraints are strong enough that the result is only a few recommendations, it may not be important to rank them.

To present case-based KB RSs, it helps to define explicitly the customer's requirements as a set, say $\mathcal{R}$. For each $r \in \mathcal{R}$ suppose that there is a function defined on the product list $\phi_r : \{p\} \to \mathbb{R}$. Then, abstractly, the similarity Sim between a product p and $\mathcal{R}$ can be written as the weighted sum

$$\mathrm{Sim}(p, \mathcal{R}) = \frac{\sum_{r \in \mathcal{R}} w_r \mathrm{sim}(p, r)}{\sum_{r \in \mathcal{R}} w_r},$$

where the weights are $w_r \in \mathbb{R}^+$ and $\mathrm{sim}(p, r)$ is the extent to which r matches p, as expressed in terms of $\phi_r(p)$. Loosely, $\phi_r(p)$ is the desired value (or range) of a product feature (price, quality, etc.), so the specification of reasonable ϕ_r values is tied closely to the exact details of the users and products.

Assuming that the ϕ_r can be specified, there are three natural forms for $\mathrm{sim}(p, r)$ corresponding to cases where a customer's goal is to minimize a feature (e.g., price, maintenance costs, etc.), maximize a feature (e.g., durability, performance, etc.), or obtain a value as close as possible to a desired value (e.g., there may be an ideal size or weight of a product). These can be expressed in terms of similarities such as the following:

$$\mathrm{sim}(p, r) = \frac{\phi_r(p) - \min_{r \in \mathcal{R}}(r)}{\max_{r \in \mathcal{R}}(r) - \min_{r \in \mathcal{R}}(r)},$$

$$\mathrm{sim}(p, r) = \frac{\max_{r \in \mathcal{R}}(r) - \phi_r(p)}{\max_{r \in \mathcal{R}}(r) - \min_{r \in \mathcal{R}}(r)},$$

$$\mathrm{sim}(p, r) = \frac{|\phi_r(p) - r|}{\max_{r \in \mathcal{R}}(r) - \min_{r \in \mathcal{R}}(r)},$$

amongst other possibilities. See Jannach *et al.* (2011, Chap. 4) for more details, including (i) the various ways in which similarity functions such as these can be used to rank products in terms of how well they meet user-specified criteria, and (ii) the use of utility functions, which are an important form of extra knowledge about the user that might be available but not used in CF or CB RSs.

A final class of RSs that is important comprises hybrid RSs. As noted above these combine multiple data sources to make recommendations. Equally importantly, they combine two or more algorithms in order to make use of the data sources. Loosely, hybrid RSs can be regarded as falling into one of three classes; see Jannach *et al.* (2011) for details and some formalities. First, there are monolithic hybrid RSs, which combine the features of two or more RSs in one implementation. One example of this occurs when one uses a nearest neighbors approach to the combined features of a CF RS and a CB RS in order to generate a single RS. Second, there are parallelized hybrid RSs. These involve running two or more RSs from different classes separately and combining the results, creating an ensemble of qualitatively distinct RSs. Third, there are pipelined hybrid RSs, in which the output of one RS serves as the input to another RS. The downstream RSs may also use data initially fed into one of the upstream RSs. A good but early survey of these techniques can be found in Burke (2002). More recently, Ricci *et al.* (2011) has chapters covering many topics discussed here and going beyond them. The ongoing development of RSs remains substantial; for instance, the Association for Computing Machinery holds an annual conference (ACM RecSys) on RSs. The most recent proceedings volume is Cremonisi and Ricci (2017); a conference and a proceedings volume are planned for 2018 (and presumably beyond).

It can be seen overall that, in terms of forming recommendations, KB systems (not unlike CF and CB RSs) are nearer computer science than statistics in the sense that they rest on database searching and optimization over the results of the search. So, statistical properties are not necessarily strongly brought to bear on the formulation of the candidate solutions – except possibly in the association rule or naive Bayes formulation, where probabilistic expressions are used regardless of their empirical validity. However, all these RSs are simply ways to generate predictors that must be evaluated – and the evaluation is statistical.

12.1.4 Evaluation

Because the basic problem that RSs are intended to solve is $\mathcal{M}$-open, in general the performance of an RS cannot be evaluated accurately by looking at expectations, probabilities, or other model-derived quantities. Even when model-based approaches such as treating a CF RS as a classification problem are used, the probabilistic assumptions on the data are generally not met. That being said, it is possible to evaluate RSs via simulation, and this has the usual advantages that the performance of an RS under specific conditions such as the distribution of user properties, products, or rating sparsity can be evaluated.[3] The problem is that these conditions are usually unknown and it is impossible to know how realistic the chosen conditions would be in a given real-world setting. Indeed, there is a substantial risk that the simulation conditions may unfairly disfavor RSs that in practice would be extremely good. So, simulation techniques are mostly useful for verifying that RSs do not have glaring qualitative flaws or for assessing running times, not for any reliable evaluation of performance. Overall, it is more common to compare RSs on historical data sets or real users even though the results of these evaluations cannot be directly compared.

Consider using a historical data set to evaluate an RS and suppose that the goal is to predict the rating that a user would have for a given product. If the user ratings are simplified to 1 or 0 for like or dislike then, if an RS predicts '1', i.e., that the user will like the product, and the user does like it, i.e., rates it a '1', the result is a true positive. If the RS prediction does not match, the result is a false negative (the RS gives 0 and the user reported 1) or a false positive (the RS gives 1 and the user reported 0). If both RS and user gave 0 then that is a true negative. The problem is that this can be done only for products for which the user has given a rating; no assumption can be made about unrated products. If unrated products are treated as 0's, the RS will tend to give unreliable false positives and, if they are treated as 1's, the RS will tend to give unreliable false negatives.

This is less of a problem when dealing with real users, who can be asked whether they liked a given product (assuming that they answer). So, false negatives and false positives can be managed (at least to some extent) in real settings. The problem with missing data arises from the inability to assess how users would react to products that have not been presented to them. Aside from missing-data problems, the familiar techniques from classification – receiver-operating characteristic curves, stability, sensitivity and specificity, etc. – can be used to compare RSs on binary ratings.

In addition to treating binary historical data using classification techniques, cross-validation (CV) can be used. Briefly, the users and their ratings are split into collections of testing and training sets, each training set generates a RS that is used on a test set of responses, and a CV error for predicting the ratings can be calculated; this is often the mean absolute error based on L^1.

Finally, RSs can be evaluated in laboratory studies and field studies. In the first type of study, a group of users is selected expressly for the purpose of comparing RSs and in the second the RS or collection of RSs is implemented and allowed to run in real time in a preselected real-world environment. On the one hand laboratory studies allow a greater control of subjects but have the familiar problems of self-selected samples and the failure of samples

---

[3] Often the sparsity S is defined to be $S = 1 - \#(R)/(mn)$, where $\#(R)$ is the number of nonvoid entries in R, m is the number of products, and n is the number of users.

to be representative of a population (assuming that the statistical notion of a population is even meaningful). Ensuring that the selected participants behave as they normally would is a tall order. On the other hand, field studies, which can often be more accurate because they are done in the real world may have problems due to the lack of controls, e.g., the desire of the company in the real-world setting to be profitable.

In all these settings the cumulative predictive error (CPE) can also be used to assess predictive performance. First, a predicted rating can be compared with an actual rating and this can be done via a sum of squares or similar function. This compares RSs with other RSs. Second, predicted ratings can be compared with actual purchases. For instance, if a rating is above a threshold then the RS may predict that the user will buy the product. Then, the RS has a loss equal to 0 if the user buys the product and a loss equal to 1 otherwise. In this case, the performance of an RS is effectively assessed by profitability. In both cases, the assessment follows the prequential principle and the predictors can be updated in response to errors and evaluated for stability, complexity, and so forth, as has been seen for other predictors in earlier chapters.

Note that the goals of these evaluations are varied and may be, amongst others, to increase the profits from the sale of products, to improve the products or RSs, or to achieve greater user satisfaction with the RSs.

As a final point in this subsection, RSs have applications well beyond sales or document finding. For instance, Lum and Isaac (2016) discussed how police forces in various jurisdictions are using quantitative techniques that are essentially RSs to predict crimes and assign duties to police officers. This domain for RSs brings in even more interactivity because the predictions are sequential and the population can respond to what the police officers do. Also, RSs have been employed in medical contexts; see Weissner and Pfeiffer (2014) for an introduction. Moreover, RSs can be regarded as a general class of solutions for many prediction problems. For a general survey of the more common uses of RSs, see Lu *et al.* (2015). As noted at the end of the last subsection, research on RSs, including how to evaluate them effectively, is ongoing.

12.2 Streaming Data

The concept of streaming data, sometimes called online data, has been around for decades and has always been important especially when decisions have to be made while new data are being received. In the general case, the data have no particular properties except that they were received in a particular order. They do not have to represent any population let alone any model. They do not have to be bounded, received at regular intervals, or even be the same data type. Even so, there are many questions that can be asked of such data – and answered under extremely mild assumptions. For instance, it may be important to keep a running tally of how many distinct elements are in the stream or data sequence, or what are the approximate median, distribution, or moments of the stream. Of course, there are many other questions that may be asked (and answered), so only a few of the basic problems can be discussed here. More comprehensive and up-to-date treatments on the analysis of streaming data include Ellis (2014), McGregor (2017), and Gupta and Saxena (2016). In addition, lucid surveys of the basic material can be found in Muthukrishnan (2009) and Chakrabarti (2014), at `http://cs.au.dk/~gerth/stream11/`

and `http://stellar.mit.edu/S/course/6/fa07/6.895/`, although there are many other sources for the basics including overview lectures from conferences.

Section 12.1 on recommender systems can be regarded as an example of streaming data since information is gathered on users. However, limitations on data storage were not considered there but will be considered here. Like recommender systems, it will be seen that in the general case streaming data is $\mathcal{M}$-open, arguably the paradigm case of $\mathcal{M}$-open data. Even if populations and reasonable samples from it could be identified (which is not the typical case), often there is nothing stable enough in the data stream to permit conventional inference. The usual notions of variance and bias do not apply, so common ways of thinking, e.g., quantifying the sources of error and trying to reduce the largest, cannot be used. There are streaming data applications that can be regarded as $\mathcal{M}$-complete, and some are treated at the end of this section. They can be regarded as $\mathcal{M}$-complete mostly because they can be fitted, however uncomfortably, into the conventional inferential framework of random variables on measure spaces and are stable enough that conventional inference yields useful conclusions.

In general, the task with streaming data is to estimate some aspect of a very long, possibly unbounded, vector or string of outcomes, say $\sigma = (y_1, \ldots, y_n, \ldots)$. With little loss of generality (in practice), assume that the y_i are drawn from a set $\{1, \ldots, T\}$. One goal is to process the vector of the y_i using limited working memory, e.g., s bits. Shannon source-coding theory gives that the best possible s would satisfy $s = \mathcal{O}(\ln n + \ln T)$ on average. It may be more natural to use logarithms to the base 2 to represent bits, but using logarithms to the base e is more common. Sometimes σ is called a stream of data. Whether called a stream, vector, or string, the point is that the methods to be used only see the data once, i.e., in one pass or in one sequence (of a given order). This makes streaming data different from time series or longitudinal data but is equivalent to saying that, from the analysis standpoint, one y_i is revealed and processed at a time. There are methods that permit two or more passes over the same string, but one pass is preferred for computational efficiency. The methods here will satisfy this preference.

12.2.1 Key Examples of Procedures for Streaming Data

Many problems have been well studied in the streaming-data literature, leading to a plethora of procedures. Here, four of these problems, arguably the most common, will be presented and discussed. They are (i) estimating the number of distinct elements in a stream σ; (ii) estimating the probabilities of the distinct elements in the stream σ; (iii) estimating moments and medians from the stream σ; and (iv) finding a representative sample from the stream σ. Of course, none of these problems can be solved perfectly; the point is to find procedures with provably good performance under certain $\mathcal{M}$-complete conditions. Other procedures have been developed for other problems, such as finding correlations, linear regressions, or increasing subsequences; these are treated in the references given at the start of this secton. Clearly, getting useful answers to the first two problems is the bare minimum needed for prediction. Problem (iii) is a way to find predictors such as a moment or median value presuming an L^2 or L^1 loss, respectively. Obtaining a good solution to problem (iv) is the natural way to permit the streaming usage of any prespecified predictor.

Two important concepts in streaming data are hash functions and the performance assessment of procedures. Put simply, a *hash function* is a function on a set taking values in another set that is typically smaller. Often both the domain and range are finite and the hash function is not invertible. In some cases, the domain consists of arbitrary-length arguments that the hash function maps into a set of fixed-length arguments. A hash function h is 'collision free' if and only if it is hard to find two arguments, say y_1 and y_2, for which $h(y_1) = h(y_2)$. That is, the hash function does not compress the data too much. This statement is left deliberately vague since various notions of 'hard to find' are used in practice.

Now, write a set $\mathcal{H}$ of hash functions as $\mathcal{H} = \{h | h : \mathcal{T} \rightarrow \mathcal{N}\}$, where $\mathcal{T}$ and $\mathcal{N}$ are sets of the first T and first N nonnegative integers, respectively. A family $\mathcal{H}$ of hash functions is *independent* if and only if (i) $\forall x \in \mathcal{T}$, $h(x)$ is uniformly distributed on $\mathcal{N}$ when $H = h$ is drawn from Unif$(\mathcal{H})$, and (ii) $\forall x_1, x_2 \in \mathcal{T}$, with $x_1 \neq x_2$ and any pair h_1, h_2 chosen independently from Unif$(\mathcal{H})$, $h_1(x_1)$ and $h_2(x_2)$ are independent. Sometimes such $\mathcal{H}$'s are called '2-universal'. The independence of hash functions is a technical condition that makes performance evaluations such as (12.1) and (12.2), see below, possible to prove. Essentially, a hash function converts variability in the y_i to variability in H, so that a probability P can be meaningfully defined. Remembering that it is $h \in \mathcal{H}$ that is random, an independent family satisfies the property that

$$\forall x_1, x_2 \in \mathcal{T}, \ \ \forall y_1, y_2 \in \mathcal{N}, \ \ x_1 \neq x_2 \Rightarrow P(H(x_1) = y_1 \text{ and } H(x_2) = y_2) = 1/N^2.$$

The paradigm example of an independent family of hash functions is, for a prime number p, the set

$$\mathcal{H} = \{h_{a,b}(x) = ax + b \bmod p | 0 \leq a, b \leq p - 1\}.$$

Assessing the performance of procedures is obviously a good way to compare them, bearing in mind that the assessment rests on a collection of assumptions, e.g., that the problem is $\mathcal{M}$-complete, that may not be true. Let $\mathcal{A} = \mathcal{A}(\sigma)$ be the output of a procedure on the data stream σ and let $\phi(\sigma)$ denote the function that $\mathcal{A}$ is meant to approximate. Then, $\mathcal{A}$ is a multiplicative (ϵ, δ) approximation of ϕ if and only if

$$P\left(\left| \frac{\mathcal{A}(\sigma)}{\phi(\sigma)} - 1 \right| \geq \epsilon \right) \leq \delta. \tag{12.1}$$

The additive version of this condition is

$$P\left(\left| \mathcal{A}(\sigma) - \phi(\sigma) \right| \geq \epsilon \right) \leq \delta. \tag{12.2}$$

In both cases, the assumption that there is a probability measure that can be used may amount to a very strong condition. It is quite possible that the population and the distribution on the population, even if they exist, may change dramatically from element to element in the stream σ, making any solution unstable. Of course, this is not as difficult as a truly $\mathcal{M}$-open problem, but it does show the limitation of model-based assessments: a true model must exist for them to be meaningful.

In fact, in the $\mathcal{M}$-open case, only empirical measures are available and the analogs to the concept of an (ϵ, δ) approximation would be expressed on substrings of σ that would, unfortunately, require more than one pass to be recalculated. That is, one empirical multiplicative assessment would be

$$\frac{1}{m} \sum_{j=1}^{m} \left| \frac{\mathcal{A}(\sigma_j)}{\phi(\sigma_j)} - 1 \right| \tag{12.3}$$

where the σ_j are substrings of σ and would be assessed in terms of a cutoff using the empirical distribution of the summands in (12.3), roughly as in permutation testing. The additive case would be similar. Another alternative to (12.1) is its empirical form i.e., estimating the right-hand side probability from the available data by breaking σ into m substreams and checking whether the frequency of the left-hand event was less than δ for a given ϵ. Note that neither of these measures are CPEs. In the present case, the multiplicative CPE would be $\sum_{j=1}^{m} |\mathcal{A}_j(\sigma_j)/\phi(\sigma_j) - 1|$, giving a direct comparison of predictions with actual values. This measure may be conceptually preferable to other empirical measures that are closer in spirit to (12.1).

There are various procedures for estimating the number of distinct elements in a stream σ. As a set of elements, σ is counted with multiplicity (and hence is sometimes called a multiset). Each element $y_i \in \sigma$ is in $\mathcal{T}$. So, denote the qth power of the q-norm of σ by

$$F_q(\sigma) = \|\sigma\|_q^q = \sum_{k=1}^{T} |n_k|^q, \tag{12.4}$$

where n_k is the number of occurrences of k in σ. Clearly, for $q = 0$, $F_0(\sigma)$ is the number of distinct elements amongst the y_i in σ. Now the task can be redefined as finding a streaming algorithm that estimates $F_0(\sigma)$ for given σ, with the added problem of quantifying its performance. The main way to do this has been via (ϵ, δ) approximations regardless of how appropriate this criterion is for a given stream. A more statistical way to think of the number of distinct elements is to write a probability vector $f = (f_1, \dots, f_M)$ and set $d = \#(\{k | f_k > 0\})$. Clearly, $d = F_0(\sigma)$; it just counts the number of distinct y_i in σ without multiplicity.

One algorithm for determining d is due to Bar-Yossef et al. (2004); see Chakrabarti (2014). To present this, let $Z(r)$ be the number of zeroes on the right of the binary expansion of any integer $r \geq 1$. That is, let $Z(p) = \max\{k | 2^k \text{ divides } p\}$. Also, let $\epsilon > 0$. Choose a random hash function $h : \mathcal{T} \to \mathcal{T}$ from an independent set and another random hash function $g : \mathcal{T} \to \mathcal{Q}$ from another independent set, where $\mathcal{Q}$ is the set of nonnegative integers not exceeding $(b \log_2 T)/\epsilon^4$. (Here, $\log_2$ denotes the logarithm to the base 2 and b is a user-chosen constant that influences algorithm performance.) Let z_i be an integer initialized by $z_0 = 0$ and let B_i be a set initialized by $B_0 = \varnothing$, the void set. For fixed $c > 0$, in pseudo-code the procedure is as follows:

- Let $\sigma = (y_1, \dots, y_n)$ be a stream and proceed as follows for $i = 1, \dots, n$:
- If $Z(h(y_i)) \geq z$ then replace B_{i-1} by $B_i = B_{i-1} \cup \{(g(y_i), Z(h(y_i)))\}$.
- While $\#(B) \geq c/\epsilon^2$ update z_{i-1} to $z_i = z_{i-1} + 1$ and shrink B by removing all (α, β) with $\beta < z_i$ (where (α, β) denotes a random element of B).
- Output $\#(B_n) 2^{z_n}$.

This procedure does more than simply taking the maximum value of $Z(y_i)$ in the stream. (If it did, the output would be essentially 2^z.) The extra stage in the procedure is to count the number of y_i with $Z(h(y_i)) \geq z$. If there are d distinct values in the stream then $d/2^z$ should be in B_n. Thus, $d \approx \#(B_n) 2^{z_n}$. Note that $1/\epsilon^2 = o(n)$, i.e., the algorithm offers no savings over simply looking for distinct elements in a simple one-pass way.

There is a space complexity analysis that can be given to show that the storage requirements increase as $\mathcal{O}(\log n + (1/\epsilon^2) \log(1/\epsilon) + \log \log T)$. Likewise, there is an analysis of the quality of the output as an estimate for d that shows that the Bar-Yossef *et al.* (2004) procedure above is an $(\epsilon, 1/3)$ approximation for d (in the sense of (12.1)). This can be improved to an (ϵ, δ) approximation (for $\delta \in (0, 1/3)$) at a cost of a $\mathcal{O}(1/\delta)$ factor of increase in space usage. The improvement is due to the 'median trick', i.e., running k copies of this procedure in parallel, using mutually independent hash functions, and outputting the median of the k answers; see Chakrabarti (2014) for the details of the analysis. Essentially, this is an ensemble method similar to the posterior weighted median with a uniform posterior, see Sec. 11.5.1. As noted above, the theoretical properties of this procedure are valid only under the assumption that σ is generated in a stable probabilistic way. A more complex (and arguably more successful) algorithm for determining the number of distinct elements in a stream is due to Indyk (2006); see Mehlhorn and Sun (2014, Sec. 3.3.3) for a succinct description.

Assuming that the number of distinct elements in σ has been adequately determined, the next logical question is to ask how often each element occurs. This can be represented as an empirical probability vector for a string σ by $\hat{f} = (\hat{f}_1, \dots, \hat{f}_T)$, in which $\hat{f}_k = \#(\{i \mid y_i \in \sigma \text{ and } y_i = k\})$. Thus, $\hat{f}_k = n_k/\#(\sigma) = n_k/n$, as usual. Given that the length n of the stream σ is known, it is only the frequencies n_k that must be found to obtain $\hat{f}$; having a way to estimate the n_k will enable prediction. (The use of the word 'estimate' here corresponds to its usage in common English rather than its specific meaning in statistical terminology.)

To estimate the n_k, consider a slightly more general framework for streaming data than before. Suppose that a stream σ is initialized as the void set $\sigma_0 = \varnothing$ and then the stream of data modifies σ_0 in such a way that σ_n is treated as an unordered set of data, i.e., it is simply counted with multiplicity. In this formulation, at each time step, any of three operations may be performed on a given σ_i, yielding a sequence of σ's. The three operations are (i) adding y, i.e., $\sigma_{i+1} = \text{ADD}(\sigma_i, y) = \sigma_i \cup \{y\}$; (ii) subtracting y, i.e., $\sigma_{i+1} = \text{SUBTRACT}(\sigma_i, y) = \sigma_i \setminus \{y\}$; and (iii) to obtain the frequency of y in σ_i, finding $\text{COUNT}(\sigma_i, y) = \#\{i \mid y_i = y\}$. Sometimes this formulation of streaming data is called the 'turnstile' model. It is understood that the y's used in these operations were found from the first problem, which aims to determine the number of distinct elements that could occur in σ. Of course, these are idealized techniques and in reality the data that is streaming may change – a given y may stop occurring or a new y may start occurring. Such complexities are ignored here for ease of exposition.

Within the turnstile model there are several techniques that can be applied to obtain values for the n_k. Among these will be sufficient to present one, called the 'Count-Min' sketch. Another, called the 'Count' sketch, provides a better (ϵ, δ) approximation guarantee but is more complicated (and uses more storage space). A procedure is a *sketch* if it is possible to combine the results of the procedure, as applied to two streams σ and σ', in a way that is efficient at each stage, to be equivalent to the result of applying the procedure to the concatenated stream (σ, σ'). It is important to be aware of this definition in practice, but the discussion here will not use it. Also, even though the presentation of the Count-Min sketch is given in the context of the turnstile model, it obviously simplifies to the case in which the operation SUBTRACT never occurs. That is, it reduces to the earlier model for streaming data used to estimate the number of distinct elements in a stream.

Let (ϵ, δ) be given, $\epsilon, \delta > 0$, $d = \lceil \ln(1/\delta) \rceil$, and $w = \lceil 2/\epsilon \rceil$. Let C be a $d \times w$ matrix, initialized to have all entries zero, and write the (u, v)th element of C as $C(u, v)$. In the Count-Min sketch procedure for estimating (or, better, approximating) n_k it is assumed that operations ADD, SUBTRACT, and COUNT arrive one at a time in a long sequence. The procedure itself was originally proposed in Cormode and Muthukrishnana (2012). The following pseudo-code has been modified from Mehlhorn and Sun (2014).

- Choose t independent hash functions $h_\ell : \mathcal{T} \to \mathcal{W}$ for $\ell = 1, \dots, d$.
- If ADD(σ_i, y_{i+1}) arrives for some $y_{i+1} \in \mathcal{T}$ then update C_i to C_{i+1} by setting $C_{i+1}(\ell, h_\ell(y_{i+1})) = C_i(\ell, h_\ell(y_{i+1})) + 1$, for $\ell = 1, \dots, d$.
- If SUBTRACT(σ_i, y_{i+1}) arrives for some $y_{i+1} \in \mathcal{T}$ then update C_i to C_{i+1} by setting $C_{i+1}(\ell, h_\ell(y_{i+1})) = C_i(\ell, h_\ell(y_{i+1})) - 1$, for $\ell = 1, \dots, d$.
- If COUNT(σ_i, y) arrives for some $k \in \mathcal{T}$ then output $\hat{n}_k = \min_{1 \le \ell \le d} C_i(\ell, h_\ell(k))$.

An accessible analysis of the Count-Min sketch can be found in Mehlhorn and Sun (2014) or Chakrabarti (2014). In these analyses the simplifying assumption is made that the elements of C_i are never negative, even though the procedure permits this. (Other procedures, such as the Count sketch, see Charikar *et al.* (2004), do not require this extra assumption to be analyzed.) The key result is that the Count-Min procedure produces $\hat{n}_k$ for stream σ_i with two properties: (i) for any $k = 1, \dots, T$, $\hat{n}_k \ge n_k$ because $h_\ell(k) \ge n_k$, and (ii) if the first moment of σ_i is denoted F_1 then $P(\hat{n}k \le n_k + \epsilon F_1) \ge 1 - \delta$. In this expression, the probability derives from the probability on hash functions, not from the probability of elements in the stream. A small comparison of four procedures for estimating frequencies is given in Chakrabarti (2014). The Count-Min sketch procedure has a smaller storage requirement than the Count sketch procedure. However, the Count procedure permits estimated 'frequencies' to be less than their true value (assuming that term has a well-defined meaning in a streaming context).

At this point, in principle, using the procedures in sequence (identifying the distinct values in a stream and providing estimates of their frequencies) means that crude predictions can be made. Obviously, if the distinct values are taken as known real numbers and their distribution is taken as known, it is a simple matter to regard the elements in the stream as draws from a multinomial (with dependence). So, the modal cell may be a good predictor for the next outcome; the empirical expectation or median of the possible future values may also be a good predictor, as may any other predictor that does not rely on knowing a dependence structure. Of course, one could in principle include dependence structures by first seeking distinct pairs of elements in the stream along with their frequencies and then doing the same for individual elements. This would allow for the computation of analogs of conditional probabilities and the generation of predictors that include one-step-back dependence.

One problem with these procedures is that the randomness in the analysis is based on hash functions, not on the outcomes y_i themselves. Thus, 'estimating' or 'approximating' quantities such as the number of distinct elements or the normalized frequency is formally undefined unless σ is generated under a stable probability – a very strong assumption. Indeed, a true value for the number of distinct elements, let alone the limiting values of the n_k/n, may not exist outside special cases so trying to determine such values by procedures that rest on probabilistic properties may be doomed to failure. This is the problem with $\mathcal{M}$-open DGs: the data are generated by some mechanism but the usual notion of random variables on a probability space may not be a good way to express this formally.

Nevertheless, since it is important to provide answers, it is worthwhile turning to problem (iii), that of estimating frequency moments and medians from a data stream σ. Problem (i), estimating the number of distinct elements in a stream, can be regarded as the zeroth case in (12.4) under the convention $0^0 = 0$. The first moment of the frequency vector f is trivial to compute exactly: it is $\|\sigma\|_1 = \hat{f}_1 + \cdots + \hat{f}_T = n$ and the first moment of the stream is $\sum_{j=1}^{T} j\hat{f}_j/n$. Indeed, finding the qth moment of a stream is easy – simply raise j to the qth power and take the same sum. The harder part is to estimate the q-norms of the frequency vector from σ, i.e., $F_q(\sigma)$ as in (12.4). It can be imagined that, even though this is not a prediction problem in and of itself, it may be an ingredient used in a solution to a prediction problem. For instance, the q-norm of the frequency vector may be used to generate the q-norm of a probability vector, which is a natural quantity in some solutions to prediction problems depending on a loss function. Chakrabarti (2014) motivated this problem, at least for $q = 2$, by looking at self-joins in a relational database.

There are several procedures for doing this; perhaps the earliest was given by Alon *et al.* (1999), who offered two algorithms. One was intended to give good performance in an (ϵ, δ)-probability sense for $\|\sigma\|_2$. This algorithm is called a tug-of-war procedure (in the turnstile model) because it updates iterates by pulling them to the right when $h(y_i)$, for a randomly chosen hash function h, is positive and to the left otherwise. This procedure produces a sketch that requires only logarithmic storage. Alon *et al.* (1999) also gave a procedure for estimating q-norms that requires polynomial storage and does not produce a sketch but is applicable for any choice of q. It is this second procedure that will be discussed here for the general case.

To present this procedure recall that $\sigma = (y_1, \ldots, y_n)$ and fix a value $k \in \mathcal{T}$. Let $r_{k,\ell} = \#\{y_i \mid y_i = k, i \geq \ell\}$ for some ℓ chosen according to a Unif($\mathcal{N}$). Regard $n(r_{k,n}^q - (r_{k,n}-1)^q)$ as the output for a given n; in fact it is an estimator for the empirical moment of the frequency vector. Following Chakrabarti (2014), the procedure can be presented as follows. Initialize the vector (n, r, k) at $(0, 0, 0)$ and assume that one new value in the stream is received at each time step.

- For each $i = 1, \ldots, n$ in turn, draw an outcome $z_i \in \{0, 1\}$ from a Bernoulli (p) distribution with $p = 1/i$.
- If $z_i = 0$ then do not update the value of $r_{k,i}$.
- If $z = 1$, suppose that $y_i = k$.
- For this k update $r_{k,i}$ to $r_{k,i+1} = r_{k,i} + 1$.
- For each $k = 1, \ldots, m$, output $n(r^q - (r-1)^q)$, where $r = r_{k,n}$.

The interpretation of the first step, in which a Bernoulli outcome is obtained, is that the updating of $r_{k,i}$ occurs only forward from the occasions where $z_i = 1$. The randomization means that the value k of Y is chosen uniformly at random from $\mathcal{T}$ without knowledge of the length of the stream σ. Note that the iteration over the values $k = 1, \ldots, m$ assumed by the y_i is a consequence of processing the y_i. From simple coding theory, it is easy to see that the storage required for this procedure is $\mathcal{O}(\ln n + \ln m)$ bits (or nats). Also, although this procedure is expressed in terms of the simple streaming data model, it extends to the turnstile model.

Although it is not obvious that this procedure should estimate $\|\sigma\|_q$, in fact it is unbiased for $\|\sigma\|_q^q$. This can be seen by the following clever argument. Think of the procedure as picking a random value from σ by a two-stage procedure: first pick a value of k according to the empirical probability vector $(f_1/n, \ldots, f_m/n)$ on T. Then search σ for the locations where $y_i = k$ occurs and pick one at random, again using a discrete uniform distribution. Since randomness is built into the algorithm, the output is an outcome from a random variable. In addition, the values of k and r are from random variables, say, K and R. Denoting the random output by $\widehat{M}_q$,

$$E(M_q) = \sum_{k=1}^{K} P(K = k) E(M_q | K = k) = \sum_{k=1}^{K} \frac{f_k}{n} E(n(R^q - (R-1)^q) | K = k)$$

$$= \sum_{k=1}^{K} \frac{f_k}{n} \sum_{j=1}^{f_k} \frac{1}{f_k} n(j^q - (j-1)^q)) = \sum_{k=1}^{K} \frac{f_k}{n} \frac{1}{f_k} n(f_k^q - 0^q)$$

$$= \sum_{k=1}^{K} f_k^q = F_q(\sigma). \tag{12.5}$$

The expectation does not use the distribution of the data. Indeed, the value of the expectation in (12.5) is a closed-form function of the data that corresponds to the data being held fixed. Omitting the details, a similar line of reasoning leads to an upper bound on the variance. The result is

$$\mathrm{Var}(M_q) \leq E(M_q^2) \leq q\, F_1(\sigma) = F_{2q-1}(\sigma),$$

which can be bounded in turn by $q M^{1-1/q} F_{2q}(\sigma)$.

It is hard to assess the quality of the result of the procedure. However, applying one more step – taking the median of an average – gives a general result. Chakrabarti (2014) established the following, which implicitly defines a distribution, assuming that σ is available.

Theorem 12.1 *Let C have a finite second moment and be unbiased for any real $Q > 0$. Suppose that $\{C_{uv}\}$ is a set of IID copies of C for $u = 1, \ldots, U$ and $v = 1, \ldots, V$. Then, given $\delta, \epsilon > 0$ and setting $V = 3\lceil \mathrm{Var}(C)/(\epsilon^2 E(C)^2) \rceil$ and $Z = \mathrm{med}_u (1/V) \sum_{v=1}^{V} C_{uv}$, there is a universal constant c such that, for $U = \lceil c \ln(1/\delta) \rceil$,*

$$P(\|Z - Q\| \geq \epsilon Q) \leq \delta.$$

Applying this result to IID outputs of the procedure provides an empirical assessment of the performance of the resulting median of means without assuming any particular properties of σ. So this result holds even in $\mathcal{M}$-open cases, de facto treating σ as a population. Again it is seen that a variant on the posterior weighted median, a model-averaging method (see Sec. 11.5.1), gives a demonstrably good result. This may be so because the median stabilizes the means, which are themselves good estimators.

Recall that the second part of problem (iii) was to obtain running values of the median. If the elements of $\mathcal{M}$ are ordered and numerical, the median can be regarded as a point predictor for the next outcome. This has been seen in simpler cases than that of streaming data. For instance, in finite-sample independent data, the median is a nonparametric predictor (see Sec. 4.1), is the predictor under L^1 loss (see the discussion after (4.7)), and arises in survival analysis (Sec. 7.1.2) and model averaging (Sec. 11.5). More generally, in the streaming data context, there are many procedures for obtaining medians but they do not seem to have been studied in an (ϵ, δ) performance sense, only in a running time or storage sense. These are important but do not address how good the medians are, even in $\mathcal{M}$-complete problems.

Despite this, there are in fact two procedures that are representative of the various methodologies. One traces its origin back to Munro and Paterson (1980)[4] and is now often called the 'median' heap. The Munro–Paterson algorithm uses a sequence of intervals whose endpoints bracket the median and converge to a point, while the median heap separates the large and small values in the stream, effectively creating an interval between them that should shrink. The other technique is called the 'median of medians' because that is its central feature. Neither is easy to put into pseudo-code, but the following descriptions should suffice.

The idea behind the median heap is to partition the stream of data into two 'heaps' or groups of size differing by at most one. Assume a starting value 'med' for the median; the max-heap has all the values below med and the min-heap has all the values above med. Thus, as the two heaps grow the median will always be a function of the maximum of the heap of smaller values or the minimum of the heap of larger values. Suppose that a new y_i arrives. It is put in the max-heap if it is smaller than the current median and in the min-heap if it is larger than the current median. If the cardinality of two heaps differs by two, move the largest element of the max-heap into the min-heap, or the reverse, so that the difference in cardinality is as small as possible. In this procedure, the two heaps will never differ by more than one element. Then, take the median using the largest element of the max-heap and the smallest value of the min-heap as before. Iterating this gives a sequence of medians. The problem is that with this algorithm, all data must be stored. So, even though the running time is small, the storage is a problem. There are various ways to decrease the storage requirements, e.g., randomly remove data points from the min-heap and max-heap that are away from those used to find the median. The more data retained, the better the expected answer but the higher the storage requirements. The original procedure from Munro and Paterson (1980) controlled the storage requirements by using bounds on the ranks of subsets of σ so as to zero-in on a value for the median.

The idea behind the median of medians is to partition the stream σ into substreams of odd size, e.g., five or seven, and then take the median again. With small loss of generality, assume that at each stage the median can be chosen uniquely as the middle value of a set of numbers. If not, appropriate changes must be made in the procedure described next. So suppose σ is partitioned into σ_j for $j = 1, \ldots, n/5$, so that $\#(\sigma_j) = 5$. Within each σ_j find the median as the center value and denote it m_j. Now, find the median m_1 of the m_j – the median of the medians. If the time required to find the median of, say, m elements is $T(m)$ then the time taken to find m_1 is $T(n/5)T(5)$, in which $T(5)$ can be treated as a small factor. Now, use m_1 to partition the elements in σ into two subsets $\sigma_{1,L}$ and $\sigma_{1,R}$ consisting of the elements of σ that are strictly below or above m_1 and return the index ℓ_1 of the y_i that gives m_1, i.e., $y_{\ell_1} = \mathsf{m}_1$. If $\sigma_{1,L}$ and $\sigma_{1,R}$ have the same cardinality then m_1 is the overall median and the procedure stops. Otherwise, $\#(\sigma_{1,L}) \neq \#(\sigma_{1,R})$. If so, then choose whichever is larger. Without loss of generality, suppose that this is $\#(\sigma_{1,R})$, so the median is in $\#(\sigma)_{1,R}$ and the task is to find the $((n-1)/2 - \ell_1)$th smallest number in it.

---

[4] This is commonly regarded as the first refereed paper on streaming data in the contemporary sense, i.e., not in the sense of sequential data. A preliminary version appeared in 1978 in the *Proceedings of the 19th IEEE Symposium on Foundations of Computer Science*, pp. 253–258.

Essentially this requires a generalization of the whole procedure to be used again, this time to find the kth smallest element of a set. That is, $\sigma_{1,R}$ is split into $\sigma_{2,L}$ and $\sigma_{2,R}$ and a new $y_{\ell_2} = m_2$ is found, with m_2 closer to the actual median than m_1. This procedure is a modification of the iteration just described that is a combination of two well-known procedures, Quickselect and Partition. These are procedures that have a good average-case performance but can have a poor worst-case performance; often they have running time $\mathcal{O}(n)$. The Munro–Paterson algorithm can have a smaller running time at the cost of more passes over the data. See Cormen *et al.* (2009, Chap. 2), for a more detailed exposition.

Obviously there are many ad hoc ways to obtain PIs from a sequence of medians, and a simpler (but less adaptive) technique is to obtain the $\hat{f}_k/n$ and use them as an empirical distribution from which to extract a median. As the $\hat{f}_k/n$ are updated with more data, the median they generate will be updated and an estimated SE can be found. More adaptively, the medians found at various stages of the iterations can be used to find an SE, but it is unclear how effective this is since defaulting to a normal or multinomial may give PIs that are not accurate. This is so because a probability may not be available and the knowledge that there is a true but unknowable probability (the $\mathcal{M}$-complete case) may not be enough to give good results in worst-case scenarios. In $\mathcal{M}$-open cases, this sort of analysis is inapplicable, so storage and running time arguments may be all that is available outside specific cases.

Problem (iv), finding a representative sample of σ, is the most general of the four problems presented here. Many techniques used to address this come down to online clustering, in which the cluster centers, ensured to be elements of $\mathcal{T}$, are taken as the representatives of σ. For instance, assuming the y_i in σ are from a metric space $(\mathcal{T}, \delta)$, where $\delta = \delta(x, y)$ for $x, y \in \mathcal{T}$ is a metric, the problem can be rephrased as partitioning σ into $K \geq 1$ clusters (or subsets) and summarizing each cluster by one representative from it that is also an element of σ. Call the set of representatives $R \subset \sigma$ and suppose that we require $\#(R) \leq K$. For any set S and $x \in \mathcal{T}$, extend δ to $\Delta(x, S) = \min_{y \in S} \delta(x, y)$. Natural choices for Δ in order to find R include the familiar $\Delta_c(\sigma, R) = \max_{x \in \sigma} \Delta(x, R)$ (complete linkage) and $\Delta_s(\sigma, R) = \min_{x \in \sigma} \Delta(x, R)$ (single linkage). However, any dissimilarity can be used, including the L^1 and L^2 errors, of the form $\Delta_1 = \sum_{x \in \sigma} \Delta(x, R)$ and $\Delta_2 = \sum_{x \in \sigma} \Delta(x, R)^2$, respectively. Similarly to the case of clustering, the goal is to produce an R that minimizes $\Delta(\sigma, R)$. To do this, the main procedure presented here (the doubling algorithm, originally presented in Charikar *et al.* (1997)) focuses on complete linkage and assumes that the clusterings are convex, that is, for any $y \in \sigma \setminus R$, y will be assigned the representative $\arg\min_{r \in R} \Delta_c(y, r)$. Related techniques are presented in Dasgupta and Paturi (2013) and Ding *et al.* (2015).

To present the doubling algorithm, start with the following initialization. Let S_{K+1} be the first $K + 1$ elements of $\sigma = \{y_1, \ldots\}$ and let

$$(x, y) = \arg\min_{u, v \in S_{K+1}} \delta(u, v).$$

Now set $\tau_{K+1} = d(x, y)$ and $R_{K+1} = S_{K+1} \setminus \{x\}$. The algorithm proceeds by considering $y_{K+2}, y_{K+3}, \ldots, y_n$ in sequence. Following Chakrabarti (2014), an iteration is of the following form.

- For each $i = K + 2, \ldots, n$, if $\min_{r \in R_i} d(y_i, r) > 2\tau_i - 1$, set $R'_i = R_{i-1} \cup \{y_i\}$.
- If $\#(R'_i) > K$, set $\tau_i = 2\tau_{i-1}$.

- Let $R_i = \max_{Q \subseteq R_i'} \{Q | \forall r, s \; s \neq s, \delta(r, s) \geq \tau_i \}$
- If $\#(R_i') \leq K$, proceed to $i + 1$.

There are three facts that make this procedure give valid answers. They are

1. Let $y_1, \ldots, y_{K+1} \in \sigma \cap \mathcal{M}$ satisfy $\delta(y_i, y_j) \geq \tau$ for $i \neq j$. Then, for any set $R \subset M$, $\Delta_c(\sigma, R) \geq \tau/2$.
2. $\forall r, s \in R_i$, for any i, $r \neq s \implies \delta(r, s) \geq \tau$, and,
3. for any i, $\Delta_c(\sigma, R_i) \leq 2$.

Taken together these facts guarantee that R_n is a collection of points that represents σ roughly in the same sense as cluster centers represent a data set. Of course, this only represents the spread of the data, not the cluster size. Guha's algorithm, see Guha *et al.* (2003), is considered a substantial improvement over the original doubling algorithm described here. It uses multiple copies of the doubling algorithm with increasing initial τ's in an effort to get a lower 'cost' in terms of Δ_c.

Although the output of these procedures as stated do not give clusterings that reflect the size of clusters, it is possible that running them multiple times will do so – possibly requiring smaller τ and larger K values. Indeed, there are streaming clustering techniques that are more sophisticated than these. See, for instance, Ailon *et al.* (2009), which gives a streaming version of K-means, and Dasgupta and Paturi (2013) for a more general discussion. As a generality, the size and spread of representative sets for σ can be found by using multiple runs of these procedures, with different initial conditions and well-chosen (usually large) K's, which can then be clustered in the usual, nonstreaming, way. In all these cases, it is important to observe that the set R is representative of σ, not necessarily representative of the DG of σ. The usual interpretations of clusters and clusterings may not apply, especially in $\mathcal{M}$-open cases. There are many other procedures that can be used to analyze and predict aspects of streaming data. Some of these have associated relevant and valid statistical analyses, at least in the $\mathcal{M}$-complete case. Although only a small number of the most important procedures have been described here, the references at the beginning of this section provide further procedures and details.

12.2.2 Sensor Data

A sensor is a device that detects a physical quantity input and responds to it, usually outputting a digital signal. As a generality, sensors are often treated as always 'on' because, once set up, they are able to gather large amounts of data at low cost. The data are most often streaming in the sense that they are rarely treated as a fixed data set. Indeed, sensor data provide a motivating example for Sec. 12.2, in part because of ongoing thinking about the 'internet of things'. Sensors come in a vast array of types. They are used to monitor meteorological patterns, moisture at various levels of soil depth, medical patient's personal statistics, energy usage, engine function in planes and cars, etc. Often these data are used for predictive purposes but they can also be used as a way to decide when to repair or replace equipment optimally. Sensors often have a spatial pattern, i.e., are spread over a range of fixed pre-chosen locations. In many cases, sensors combine inputs from other sensors before transmitting a final output to a monitor. While problems relating to the technical functioning of sensors and sensor networks have been substantially resolved, problems with data collection, storage, computation, and analysis remain and have increased as the numbers of sensors in networks have increased, owing to the increase in the amount of data collected.

An individual sensor can yield query data, event data, or continuous data. Query data are measurements made by the sensor in response to an operator's specific demand. Event data are those that accumulate in response to events. Events may occur at irregular times; however, continuous time is treated as uniformly discretized and the sensor responds in some way each time an event is detected. That is, the sensor sends one of a finite set of values. In some cases, this can be regarded as a Markov chain of order $k \geq 1$ and sometimes this analysis extends to multiple sensors. More generally, event data are often treated as the outcome of a stochastic process which can then be used to give predictions; see Sec. 5.5. By contrast, continuous data essentially generate a time series that can be analyzed using techniques from Chapter 5. Three inferential problems are typically associated with sensor data: predicting future values; summarizing the data, for instance, by online clustering (see the end of Sec. 12.2); and anomaly detection, i.e., determining whether a given data point is consistent with normal performance or indicates an aberrancy that requires some action. Event sensor data and continuous sensor data are examples of streaming data; query data will not be considered further in this section, since it is often not streaming. Anomaly detection will be discussed in Sec. 12.2.3.

Commonly, in statistical problems, there are three stages to each inference made from sensor data or streaming data. First, the data must be converted to a form that is analysis-ready. This often requires data cleaning – checking that each point is valid and, if it is not, deciding whether to keep it, remove it, or replace it by some technique for dealing with missing data or data processing. For instance, in some cases, points are regarded as outliers if they are outside a confidence band around, say, a spline estimator for the function that the data seem to represent or if they must be transformed intelligently so that an analysis will be feasible. Second, an actual analysis is done to generate an inference, whether the goal is prediction, summarization, or anomaly detection. Finally, some assessment is made of the quality of the inference. This includes assessments of the validity of a prediction or clustering, possibly combining the results with other information for decision making, and presenting the output in some form that is readily intelligible to users.

All this is more complicated when many sensors are responding, because there may be spatial, temporal, or other dependence effects. Consequently, the focus in the rest of this section will be on networks of sensors. One way to organize the output from a collection of sensors is via the 'data snapshot' framework; see Armenakis (1992) or Appice *et al.* (2014, Chap. 1) for a more general discussion. Let S be a finite set of sensors. The data snapshot framework assumes that a data point is of the form $(K_t, Z_t(\cdot))$ for some $t \in \mathcal{T}$, where $K_t \subset S$ is the subset of sensors that are sending values $Z_t(s)$ of a function Z at time t for $s \in K_t$. Putting data snapshots together in sequence gives a data stream of the form $(K_t, Z_t(\cdot))$ for $t = t_1, \ldots, t_n, \ldots$

Sensor data streams are usually so large that as such they cannot be effectively stored. However, often they do not have to be – the data can be summarized and only the summary stored. This may lead to a tradeoff between how complete the summary is and how wide a range of questions can be answered later. Windowing is one common way to reduce the volume of data stored. The simplest is to take disjoint time intervals of snapshots. That is, if T is an interval containing equispaced values u_j for $j = 0, \ldots, m$, the data is partitioned into m sequences of snapshots for $u_j \le t_i < u_{j+1}$. Next the chosen summaries are found and saved for each subsequence of snapshots. Then the raw data are discarded. A variation on this is to choose a window size w and use data snapshots of length w but allow the windows to overlap. Thus, the first w windows are summarized, the results are stored, and the same procedure is used for, say, the snapshots from $w/2$ to $3w/2$, from $3w/2$ to $5w/2$, and so forth (thereby allowing an $w/2$ overlap). Again, the summaries are saved and the raw data is discarded. The overlap links the summaries for adjacent windows. In both cases, only finitely many windows are summarized and stored.

What sort of summaries are common? Obviously, it is possible to take a random sample from each snapshot or window. Depending on the structure of the network of sensors this may or may not be a good summary. If it is not then stratified sampling over groups of sensors may be better. Alternatively, if the data are numeric, correlations between sensors within a window can be calculated and only the data from sensors with high enough or low enough correlation, as appropriate, are kept. If K_t does not vary much with t then it may be sufficient to store means and variances from the snapshots and analyze them as a time series to make predictions. More generally, a meta-analysis approach to the summaries may be taken; conversely, a proposed technique for meta-analysis may motivate data summarization. This is an open question, like so many others, in the area of sensor data analysis. Histograms may also be found, for each snapshot or window, and saved. Storing prespecified quantiles for each snapshot or window would be a more severe summary. If the sensor data is effectively a collection of time series then the values can be discretized into bins. Streaming cluster analysis can also be used, over snapshots or windows, and the centroids (for instance) of the clusters saved. A more recent development is the concept of a trend cluster; see Ciampi *et al.* (2010). The basic idea is to cluster the sensors by the sequence of values they emit, take a summary at each time step (or window) of their values (say a mean or median), and then store these as a trend. That is, the summarization is the trendline for a cluster of sensors with similar trendlines. If the sensors have a spatial pattern, the expectation is that the clusters will match it and this may also be helpful for dealing with missing data and outliers.

A special case of interpolating a missing data point or replacing a data point deemed to be incorrect in some way (because it is an outlier or corrupted, etc.) assumes that the sensor network has both a time and spatial structure and that each sensor generates a time series. Obviously, if K_t does not change too much over time, Kriging, inverse distance weighting, or radial basis functions can be used spatially and time series methods can be used temporally to fill in missing or incorrect data points. Roughly the same techniques can be used for prediction. The two types of methods may or may not agree. If they do not, the problem is then which value to choose. Methods which combine the two may give better results. For instance, if the neighborhood around a sensor that gave a missing or incorrect value does not change with t then spatial interpolation can be done first and this can be followed by a temporal interpolation method such as splines or a time series model. More sophisticated techniques combining spatial and temporal methods have started to be developed; see Appice *et al.* (2010) for one of them. Chapter 3 of Appice *et al.* (2014) has a good discussion of these techniques, including spatio-temporal Kriging.

In addition, prediction in a sensor data context can be done in a variety of ways. The simplest is to predict the next value of a single sensor, assuming that all its previous values (or summaries of its values) are available, fit a time series model, and then obtain predictions from it. An alternative approach is to apply more or less the same procedure but to the cluster trend, providing the sensors in the cluster do not change much over time. In this case, the same prediction would be made for all the sensors in the cluster. If trend clusters are used for prediction, it is important to verify that the clustering is stable, so that the predictions will be stable and hopefully more reliable. Standard perturbation methods based on the adjusted RAND or Jaccard indices often suffice. A third technique is to find the sensors that are highly correlated with a given sensor and use them to generate a prediction for the next value of the given sensor.

Event data is typically discrete and may be numeric or categorical. An event may be a *primitive* in the sense that it occurs at a specific time point or does not occur at all. Sometimes it is called a *point event*. A *composite event* is higher level; it is a function of a collection of primitive events and may sometimes be called an *interval event*. Event processing – detecting whether events of interest have occurred – is usually the first stage of analyzing event data, whether it involves reading, transforming, or editing events. The second stage of analyzing event data is pattern detection, discovery, or actual prediction. The categories of techniques are sometimes described as statistical, topographical, and edge detection algorithms (although all three are actually statistical). Typical examples of these include: estimating a distribution by a nonparametric density estimation technique; collecting a time series of data maps and identifying complex events through their similarity to pre-identified spatio-temporal patterns; and using image processing or classification techniques to identify sensors that are near the boundary of a region. A survey of this is given in Wang *et al.* (2013). It should be noted that **a pattern detection or discovery problem is de facto a prediction problem** because announcing the presence of a pattern effectively means that certain decisions are better than others, and a decision problem amounts to predicting which action is going to be optimal. This differs from the usual prediction problem, in which the predictand is a future value of a random variable. That is, detection is little other than a summary statistic that can be used to help predict the best action. This is essentially the same argument as was used to regard hypothesis testing as prediction in Sec. 1.2.2, but framed as an input to a prediction problem.

As a generality, there are few techniques that give good predictions over the range of possible sensor data because this range is so large. After all, sensor networks may generate both event and time series data, with a huge variety of dependence structures. An overview of networks more generally is given in Section 12.4. Given that the sensors generating data may change over time, may change their locations over time, and even when functioning well may have a high degree of noise, the majority of analyses of sensor data are particular to specific applications. This is seen in the various compendia of contributions (e.g., Aggarwal (2007) and Aggarwal (2013a)). Chapter 4 in the latter publication discusses event data and Chapter 5 presents a further collection of predictive techniques for sensor data, mostly based on time series. Little work seems to have been done to combine event and time series data or to develop principles for experimental design with sensor data. However, an example of the latter is Wu *et al.* (2012). Up to date treatments of the range of sensor data problems can be found in the *Journal of Sensors*, the *International Journal of Distributed Sensor Networks* and in a wide variety of engineering and computer science journals and proceedings volumes.

12.2.3 Streaming Decisions

One reason why it is important to analyze streaming data in a single pass is that it is frequently online and requires a decision to be made before the next data point arrives. There are two cases: the decision made at time i does or does not affect later data points $i + 1$, $i + 2$, and so on, in the stream. If later data is affected then the problem is more complicated than if it does not. For ease of exposition, the focus here is on the easier case. Although there are many problems that can be posed, only two will be discussed here. They are (i) online anomaly detection (which comes in two forms, supervised and unsupervised) and (ii) streaming decision theory using loss functions to generate optimal actions, which can be assessed by CPE, for instance. Item (i) is closely related to outlier and change-point detection, but the present discussion will not include these since they have a long-established literature in statistics. Moreover, given the range of streaming decision problems, only a small number will be described here; the references cover many more, often in more detail. Taken together, Chandola *et al.* (2009, 2012), Zimek *et al.* (2012), Aggarwal (2015), De Rosa (2016), and the references therein provide a good overview of many techniques.

Anomaly detection refers to the problem of finding patterns in a data set that do not conform to expected behavior. An anomaly may be an outlier, an observation that is not valid, or simply one or more observations that are aberrant or indicate an aberrancy in the DG. Somewhat problematically, the notion of an aberrancy is little more than that of finding some pattern of interest. So, anomaly detection effectively includes novelty detection. In any case, finding an anomaly usually means that some corrective action must be taken in response.

As with analyzing sensor data, the exact way in which one goes about detecting anomalies is highly particular to each given application. Nevertheless, there are some techniques that can be used in, or adapted to, many settings. Supervised methods include classification methods (of which there are many), and a variety of other inference methods, parametric and nonparametric. Unsupervised methods include clustering, nearest neighbor methods, and, in the case of high-dimensional data, subspace methods.

One popular classification technique uses trees, whether Bayesian or frequentist; see Sec. 10.2. These can be extended to streaming data simply by using the fourth (last) procedure in Sec. 12.2.1 to find a sample that is representative of the stream or by simply using a sliding window of past data – say, the last n data points – to construct the tree. The hope is that n can be chosen to be large enough that it would approximate a single correct tree, if one existed. In this case, the data is a stream of values $Y_i = 0, 1$ for given explanatory variables x_i, where $Y_i = 1$ means that a prespecified anomaly has been detected at the ith step. Updating the tree regularly by using a sliding window means that, even if the DG drifts, the ability to detect anomalies and predict optimal actions in response is maintained (at least somewhat); this is not true if a representative sample of the entire stream is used. In either case, if x_{i+1} is received then Y_{i+1} can be predicted and if (y_{i+1}, x_{i+1}) are received together then the appropriate action for $y_{i+1} = 1$ can be taken. As instances, Bayes model averaging (Sec. 11.1) and bagging (Sec. 11.2) can be used in an effort to get a better detector and predictor in the Bayes or frequentist context, respectively.

Other classifiers are also often used with either representative samples of the stream or a sliding window from the stream. Nearest neighbor classification, like nearest neighbor regression, has the benefit of robustness but may not scale up well to high-dimensional x's; see Secs. 4.7 and 8.3. However, in either case, a representative sample may be maintained or a sliding window used. Another variant on nearest neighbor is to use streaming clustering to generate a large number of small clusters, the centroids of which can be taken as virtual data points. Then, k-nearest-neighbors classification can be applied on the level of the centroids rather than on the level of the data points. The gain from using this method is mostly in running time, because fewer data points need be considered in the classification step. Like trees, nearest neighbor methods require that the type of anomaly be specified prior to constructing the detector. Logistic, neural net, kernel-based (with a support or relevance vector), and boosting classifiers can be treated similarly. Indeed, regardless of the choice of classifier, the problem is how to modify its computation so that only one pass over the stream is required.

In some cases of streaming decisions, only the y_i are available, i.e., there are no explanatory variables. Often the y_i are discrete values. The paradigm example would be a single sensor measuring temperature, sending 0 at each time step for which the temperature is below a threshold and 1 when the temperature is above the threshold. A pattern that had, say, five or more 1's in sequence might be an indicator of a problem. In this case there are many 'normal' sequences, a collection of sequences that indicate a problem, and sequences that may indicate that a problem is arriving (say three 1's in a row). In this setting, an anomaly score can be assigned to each sequence of y's. Simply count the number of times that a sequence of five 1's in a row is seen, possibly adding a small amount for time steps where a sequence of four or fewer 1's is found. A high anomaly score indicates an anomaly that may require action; see Chandola *et al.* (2012) for details.

A variant on this is to compare the frequencies of strings of outcomes to their expected frequencies, assuming that the latter are available. This procedure determines which outcomes are observed and their frequencies (using the procedures from Sec. 12.2.1). Comparing the estimated frequencies with their expected (or desired) values can be done using a variety of statistical procedures; perhaps the chi-squared goodness-of-fit test is the most elementary and will be satisfactory provided that its power is not too low. Otherwise a Kolmogorov–Smirnov, Bayes or other test may be used.

This kind of setting can also sometimes be modeled using Markovity. If there are finitely many possible values of Y (and no explanatory variables) then a kth-order Markov chain model can be found provided that a suitable value of k exists and can be found as well. The entries in a transition matrix are of the form $P(Y_{i+1} = k | Y_i = j)$ and are independent of i if the chain is stationary. Thus, the transition matrix can be estimated and used to generate predictions. If it is accurate, the predictions can lead to a decision being taken before the next time step. In the case where there are explanatory variables, a Markov chain can be found using regions of the space of explanatory variables.

A more sophisticated version of this is a hidden Markov model (HMM), see Sec. 10.7.3 for a brief overview, which posits that the observed data arises from an unseen underlying state that evolves Markovianly without drift. The central idea is to find an HMM with the correct number of hidden states, say K, that has a long enough subsequence from σ. This uses the Baum–Welch algorithm. Now, for any sequence of values of Y_i, the optimal sequence of states in the underlying Markov chain can be found using the Viterbi algorithm. Doing this for any candidate anomalous subsequences gives a sequence of underlying states to which any other anomaly-detection procedure can be applied. Hidden Markov Model predictions can also be obtained for future values of Y, thereby permitting actions to be taken before the next time step.

There is a variety of other statistical methods, many of them elementary, that are worth noting as well. Most of these are based on 'typicality', i.e., an anomalous outcome is atypical in the sense that it is in a low probability region of a stochastic model. This only makes sense in an $\mathcal{M}$-closed or $\mathcal{M}$-complete setting. More generally, for anomalous observations, there is often reason to suspect that they were not output by normal functioning of the DG. In this category, standard quality-assurance techniques such as control charts and hypothesis tests are the most elementary, in the absence of explanatory variables. Data that is more than three SDs away from a distribution mean may be anomalous, assuming the data is normal. The Chebyshev inequality gives a conservative version of the same principle: a data point k SDs away from the mean occurs with probability bounded by a decreasing function of k. A similar assessment comes from the boxplot rule – data above the third quartile plus half the interquartile region (IQR) or below the first quartile minus half the IQR is regarded as anomalous. Likewise, tests for the maximum deviation of a data point from the mean of the data – essentially asking whether a residual is so large as to suggest that the data point that generated it is invalid – are available. Student's t-test and its multivariate generalization Hotelling's test are the most used examples of this.

In the presence of explanatory variables, the various techniques for outlier detection from the standard theory of linear, generalized linear, and nonlinear models are available in many standard texts and often extend to include random effects. In this $\mathcal{M}$-complete context, the models are often chosen by AIC, BIC, or other criteria. Many of these techniques for outlier detection can also be extended to time series contexts, whether AR, MA, ARMA, or, most generally, SARIMA.

Outliers can also be assessed by where they land relative to a histogram of the full data or of a representative sample of the stream. The further out in the tails the candidate outlier is, the more likely it is to be an outlier. This is simply an alternative way to use the boxplot test and can be applied to the histogram of residuals from a model-fitting technique. Indeed, the residuals from nonparametric techniques such as the Nadaraya–Watson estimator (Sec. 8.2.3 or smoothing splines (Sec. 8.6.1) can also be used in this way.

Unsupervised anomaly detection often uses clustering. The idea is that a data point or pattern that is anomalous is far from any cluster, while data points that are near at least one cluster are representative of the normal or desired function of the DG. Since the data is streaming, all the clustering techniques must be streaming as well. One way to achieve this is to update a representative sample of the stream continually as explained in the last procedure in Sec. 12.2.1, and cluster it by any clustering technique. Another way is to modify how clustering is done in such a way that it can be updated time step by time step as data in the stream arrive. A simple version of this is to regard the output of the representative sample as a collection of cluster centers. More generally, there are streaming versions of K-means and K-medians clustering, as referenced at the end of Sec. 12.2.1 – though K must be chosen sensibly, possibly by stability assessments.

The limitation of this approach is the assumption that anomalies are outliers. In fact, anomalies may be in clusters in their own right. Thus, a more general form of these techniques arises when it is assumed that typical data from a DG lie in large densely filled clusters while atypical data lies in small or sparse clusters. Again, a formal way to assess whether a given data point or pattern belongs to a given cluster and a way to decide whether that cluster represents an anomaly must be specified; this is highly application dependent.

The principle of nearest neighbors can also be applied in an unsupervised setting. The operative principle is that typical data occur in dense neighborhoods with many close nearest neighbors while the nearest neighbors for anomalous data are comparatively distant. One simple way to measure this is, for each data point, say x_i, to find $d_i = d(x_i, x_{i,(k)})$ where $x_{i,(k)}$ is the data point in $x_1, \ldots, x_n$ that is kth closest to x_i and d is a distance measure on the space from which the x_is are drawn. If d_i is bigger than a threshold, x_i is deemed to be anomalous; otherwise it is not. The same principle can be extended to sequences of x_i, to assess whether their kth nearest neighbor sequence is sufficiently close. Another way to formalize the operative principle is to fix a distance d_0 and find the density at x_i, $\#(\{x_j \mid \|x_i - x_j\|/n < d_0\})/n$. If the density is too small then x_i is an anomaly, otherwise it is not. A small dense cluster that is far from the other clusters will tend to have a low density for a large enough value of d_0 and hence be regarded as an anomaly. There are numerous variants on these two approaches, e.g.: using a 'local outlier factor' (LOF) by taking the ratio of the density at x_i formed by using a sphere containing the k nearest neighbors of x_i to the average of the densities at all the x_i; clustering the data, grouping the distances to the kth nearest neighbor for each cluster, identifying clusters that cannot be anomalies, removing these data, and then searching the remaining data for anomalies, amongst many others.

In the case of unsupervised high-dimensional data, the concept of anomaly breaks down because all points are equally distant with norm equal to their expected value in the limit of increasing dimension; see Beyer *et al.* (1999). Zimek *et al.* (2012) showed that this problem can be serious for dimension as small as ten but that LOFs have the potential to overcome it because, being ratios, they permit outliers to be detected ever more clearly as the dimension increases. This is called the concentration effect. However, it relies on assuming that all the entries in the high-dimensional vector are relevant. For the purpose of outlier detection, if there are too many irrelevant dimensions, the outlier can remain 'hidden in the noise', so reducing the noise by looking within subspaces of much smaller dimension may be helpful. That is, the 'outlyingness' of a given x_i can be magnified by looking at the right subset of attributes. Often the subset of attributes depends on x_i, and various techniques are used to select subspaces of attributes for the purpose of outlier, and more generally anomaly, detection; see Zimek *et al.* (2012) for a review.

Decision theory for streaming data can be developed as a parallel to the standard Wald (or Rostek) formulation of decision theory. Recall that $L(\delta, y)$ is the loss in making a decision $\delta \in \Delta$ when $Y = y$ is the correct value. Assuming that a model is available, the risk is $E(L(\mathcal{D}, Y_{n+1}(X_{n+1}))$ where E is taken in the joint distribution of, say, $n + 1$ IID copies of (X, Y) and $\mathcal{D}$ refers to the first n copies. Denote the joint distribution of (X, Y) by its density $p(x, y) \in \mathcal{P}$ and assume a prior probability W over $\mathcal{P}$. Then the Bayes risk can be written $E_W E_P(L(\mathcal{D}, Y_{n+1}(X_{n+1})))$, where the subscripts on the expectations indicate the distributions they use. Now, the posterior risk is

$$E_{Y_{n+1}|\mathcal{D}, x_{n+1}} L(\delta_{\mathcal{D}}(x_{n+1}), Y_{n+1}(x_{n+1})). \tag{12.6}$$

At this point, the risk of any δ as a predictor for $Y_{n+1}(x_{n+1})$ can be evaluated in a streaming data context provided that a good estimate of the distribution $(Y_{n+1}|\mathcal{D}, x_{n+1})$ can be found. In discrete cases, or discrete approximations to continuous cases, the events with nonzero probability can be identified and their frequency estimated using the techniques of Sec. 12.2.1. If the data are not IID then the problem is more complicated, but some dependence structures, e.g., kth order Markovity, can be handled. Indeed, in some cases, it is likely that a combination of techniques from HMMs and Sec. 12.2.1 can be used to approximate risks or posterior risks and find good, if not optimal, δ's for an even more general class of distributions. Exactly how to do this remains an open problem. Getting results for general stochastic processes by approximating them by their corresponding kth-order Markov processes, i.e., for a general process P approximating $P(x_{n-m}, \ldots, x_n | x_1, \ldots, x_m)$ by $P_k(x_{n-m}, \ldots, x_n | x_{m-k}, \ldots, x_m)$ for $k < m$, so that P_k converges to P as k increases, may also be possible.

The real problem is, however, more general and reminiscent of the formulation in Wong and Clarke (2004). It is one thing to condition on all the data or a set of representative data points as data accumulate, but it is another matter to allow conditioning on arbitrary statistics. Since it is well known that conditioning on a statistic is equivalent to conditioning on the σ-field that the statistic generates, (12.6) can be generalized to

$$E_{Y_{n+1}|\mathcal{F}_n} L(\delta_{\mathcal{D}}(x_{n+1}), Y_{n+1}(x_{n+1})), \tag{12.7}$$

where $\mathcal{F}_n$ is a sub-σ-field of $\sigma(\mathcal{D})$, possibly either generated by a set of summary statistics or the result of feature selection. Then the optimal choice for δ depends not just on the class of distributions $\mathcal{P}$ and W, but also on the conditional distributions of $\mathcal{P}$. Clearly, a better δ can be found if an extra optimization over $\mathcal{F}$ is built into the overall procedure; it amounts to finding the data or functions of the data most germane to the prediction problem. There is scant work on this topic, but it is promising if only as a more general formulation of the online prediction problem. In particular, it would be enough to optimize over possible streaming estimates of the statistics used to generate $\mathcal{F}$ in (12.7), find streaming estimates of their distributions, and then optimize over δ, cf. Vehtavi and Ojauen (2012).

To a greater or lesser extent the vast majority of methods used for streaming or online decisions[5] require some sort of automated feature extraction or, more precisely, data summarization, whether via clustering, extraction of a representative sample, or computing statistics. The coarsest version of this is simply using a sliding window of data points, and this would be appropriate if the DG drifts, i.e., there is nothing stable enough over its operation to be identified. The next coarsest is maintaining a representative sample of the data of fixed size (assuming no DG drift). More sophisticated techniques include online clustering, online extraction of principal components (if appropriate for the setting), or other online ways to choose statistics and update them. Even techniques like self-organizing maps, Chernoff faces, and multidimensional scaling can be regarded as de facto feature selection provided that they are stable enough. As noted before, a sufficiently detailed knowledge of the DG may lead to the identification of a reasonable set of summary statistics. However, typically neither time nor knowledge is available to permit a cogent derivation

---

[5] The technical difference is that online data permits only one pass because it is an infinite sequence that we see, one data point at a time. Streaming data, in principle, is finite and so permits multiple passes. Here this distinction is ignored.

of the key features of a data stream. So, automatic feature selection remains either an application-dependent question or an optimization (over $\mathcal{F}_n$ in (12.7)) problem.

12.3 Spatio-Temporal Data

Spatio-temporal (ST) data represent a data type with a challenging, complex structure because such data combine the difficulties of time series with the difficulties of spatial processes. The commonality is that observations that are close (in space or time) are expected to be similar even when the dependence structure amongst the observations is complex. In contrast with IID data, dependent data usually represent less information per observation. A simple version of this can be seen by supposing that $Y = (Y_1, \ldots, Y_n)^T \sim N(0, \Sigma)$, where the (i, j)th entry of Σ is $\mathrm{Cov}(Y_i, Y_j) = \rho\sigma^2$ for some $\sigma > 0$ and $\rho \in [-1, 1]$. It is easy to show that $\mathrm{Var}(\bar{Y}) = (1 + (n - 1)\rho)\sigma^2/n$. When $\rho = 0$ this variance is minimized; conversely, it increases for $\rho \geq 0$. Cases with $\rho < 0$ are frequently unstable and rarely occur outside conditional quantities, e.g., $\mathrm{Cov}(Y_i, Y_j | \bar{Y})$ is often negative because, for fixed $\bar{y}$, Y_i can only increase when Y_j decreases and vice versa.

There are three major classes of ST data. The first comprises point data, meaning that an observation is at a point and in principle an observation at any point (within a region) can be obtained. In practice this may not be feasible; for instance, although remote sensing data is typically point data, in remote sensing the resolutions of the devices making the observations may be on different scales. In some cases this is resolved by 'upscaling' or 'downscaling', a class of techniques discussed below in Sec. 12.3.2; see Schmidli *et al.* (2006) and Berrocal *et al.* (2012), respectively, for accessible examples that highlight the main issues. Sensor data (Sec. 12.2.2) is also often an example of ST point data when the measurements of the sensors over time at different locations are related by a dependence structure.

The second class comprises point process data. In this case, one stochastic process specifies the locations where events happen and another stochastic process specifies which events happens at each location. The locations themselves are like random zero–one data: the locations specified are the 'ones' and all the other locations are the 'zeroes'; these evolve over time. The process governing the events that occur at the ones also evolves over time. The key feature here is that the process has an intensity indicating how likely it is that a point from a given region will be chosen at each time. This was seen in Sec. 5.5 for the Poisson and compound Poisson processes. Since the intensity of a Poisson process is constant, it is often taken as the paradigm example of a point process, but stochastic processes are much more general. This structure is reminiscent of functional magnetic resonance imaging (fMRI) data, and other temporal imaging modalities, where images of an object are made over time and the pixel intensities may be thresholded to indicate regions of high activity.

The third class comprises area data. Usually this is a regional count such as the number of cases of a disease in a county. The point is that the area in question has some meaning beyond simply being a region that contains the point at which an observation was made. So, it is usually not natural to regard area data as point data by shrinking a county, or parts of it, to a point or points, or to try to define something like a density on the county, because the data are meant to be an aggregate already.

Even though time may be treated as discrete or continuous, the biggest qualitative differences among the three classes of data reflect their spatial aspects. Indeed, point data can often be regarded as a multivariate continuous time series and area data can often be regarded as a multivariate discrete time series. From an analysis standpoint, it is not usually helpful to think of ST data in this way, but it is an aid to understanding its structure. Individual time series were treated in some detail in Chapter 5. However, spatial processes have not been treated, so a concise summary for the purposes of the next four subsections may be helpful.

The simplest and most important case is, arguably, point data and the most important spatial prediction technique is called Kriging. Kriging originated in Krige (1951) and was developed in Matheron (1962). It uses linear regression on a collection of points $(y_1, x(s_1)), \ldots, (y_n, x(s_n))$ that are assumed to have a nondiagonal covariance matrix predicting Y_{n+1} at s_{n+1} using the $x(s_i)$. The $s_1, \ldots, s_n$ are points in a region that, for present purposes, can be imagined to be in a well-behaved open set (convex, differentiable boundary, etc.) in $\mathbb{R}^2$. At each s_i there is a measurement $y_i = y_i(s_i)$ with error $\epsilon_i = \epsilon(s_i)$, and covariates $x_i = x_i(s_i)$, assumed to be the same for all s_i. If the x_i are univariate, the model can be written as

$$Y = X\beta + \epsilon, \tag{12.8}$$

using familiar matrix notation in which $E(\epsilon) = 0$ and $\text{Var}(\epsilon|X) = \Sigma$, a positive definite $n \times n$ matrix with entries σ_{ij}. (If the x_i are multivariate, obvious modifications can be used.) It is well known that the generalized least squares estimator for β is $\hat{\beta}_{GLS} = (X^T \Sigma^{-1} X)^{-1} X^T \Sigma^{-1} Y$, assuming that Σ is known. In practice, Σ is rarely known. Hence, it is parametrized and the parameter is often estimated by assuming normality and maximizing the likelihood; this aspect will be ignored here but see Schabenberger and Gotway (2005, Chap. 4) for some examples.

To derive the Kriging predictor for $Y_{n+1}(s_{n+1})$, observe that the best linear unbiased predictor $\hat{Y}_{n+1}$ under squared error satisfies

$$\min_{w \in \mathbb{R}^n} E(Y_{n+1} - w^T Y)^2 \quad \text{subject to} \quad w^T X = X_{n+1}^T, \tag{12.9}$$

where $\hat{Y}_{n+1} = w_{n+1}^T X_{n+1}$ and $E(\hat{Y}_{n+1}) = EY_{n+1}$. Expression (12.9) follows because unbiasedness implies $w^T X\beta = X_{n+1}^T \beta$ for all β, which in turn implies $w^T X = X_{n+1}^T$. Adding and subtracting $EY_{n+1} = X_{n+1}^T \beta$ and $E(w^T Y) = w^T X\beta$ inside the expectation of (12.9) and using (12.8) converts (12.9) to

$$\min_{w \in \mathbb{R}^n} E(\epsilon_{n+1} - w^T \epsilon)^2 \quad \text{subject to} \quad w^T X = X_{n+1}^T. \tag{12.10}$$

Note that w is used to form $\hat{Y}$ (and depends on β) while Y_{n+1} depends on β directly.

Minimization of (12.10) leads to the Lagrangian

$$E(\epsilon_{n+1} - w^T \epsilon)^2 + (w^T X - X_{n+1}^T)\lambda, \tag{12.11}$$

where $\lambda \in \mathbb{R}$.

In the special case where the ϵ_i are uncorrelated, (12.11) becomes

$$E\epsilon_{n+1}^2 + E(w^T \epsilon)^2 + (w^T X - X_{n+1})\lambda = \sigma^2(1 + w^T w) + (w^T X - X_{n+1}^T)\lambda.$$

Setting the derivatives with respect to w and λ equal to zero and solving gives $2\sigma^2 w = X\lambda$ and $X^T w = X_{n+1}$ and leads to $\lambda = 2\sigma^2(X^T X)^{-1} X_{n+1}$, which gives $w = X(X^T X)^{-1} X_{n+1}$, and the predictor reduces to $\hat{Y}_{n+1} = w^T Y = X_{n+1}^T (X^T X)^{-1} X^T Y$, which is just $X_{n+1} \hat{\beta}_{OLS}$ from ordinary least squares, cf. Sec. 4.2.

When the ϵ_i are correlated, minimizing (12.11) is more complicated because the different variances must be accommodated. Write $\Sigma_0 = \text{Var}(\epsilon)$, $\Sigma_{01} = \text{Cov}(\epsilon, \epsilon_{n+1})$, and $\Sigma_1 = \text{Var}(\epsilon_{n+1})$. Expression (12.11) remains the Lagrangian; now, setting the derivatives with respect to w and λ equal to zero gives $2\Sigma_0 w - X\lambda = 2\Sigma_{01}$ and $X^T w = X_{n+1}$. Routine manipulations give

$$\lambda = 2(X^T \Sigma_0^{-1} X)^{-1} (X_{n+1} - X^T \Sigma_0^{-1} \Sigma_{01})$$

and

$$w = \Sigma_0^{-1} \left(\Sigma_{01} + X (X^T \Sigma_0^{-1} X)^{-1} (X_{n+1} - X^T \Sigma_0^{-1} \Sigma_{01}) \right).$$

Finally, $\hat{Y}_{n+1} = w^T Y = X_{n+1} \hat{\beta}_{GLS} + \Sigma_{01} \Sigma_0^{-1} (Y - X\hat{\beta}_{GLS})$ is the Kriging point predictor, sometimes called 'universal' on the grounds that X and X_{n+1} are arbitrary. The derivation here was modified from Sun (2015); a more sophisticated derivation using the variogram (a function of the covariance between Y's at nearby points) and giving SEs for frequentist point prediction is found in Lichtenstern (2013, Chap. 8).

A Bayesian approach to Kriging has been developed also. It's a reformulation of the Kriging problem into a Bayesian structure rather than a variant on frequentist Kriging. Because it generates predictors for the same setting, it's called Bayesian Kriging even though the procedure does not much resemble frequentist Kriging. In addition to having the usual theorems on consistency and asymptotic normality, it tends to extend more readily to a wider class of functions of the explanatory variables and statistical distributions. The idea can be compactly expressed as follows. Suppose that $\theta \sim N(0, \tau^2)$, $\epsilon \sim N(\theta, \Sigma(\theta))$, and $Y|\epsilon, \Theta \sim N(X\beta + \epsilon, \tau^2)$. Of course, τ and β are unknown, so a prior must be put on each of them. A benefit of hierarchical Bayes is that the priors on the higher levels tend to matter less and less. One effect of the priors is that distinct Y_i are positively correlated: $\text{Cov}(Y_i, Y_j | \beta) = \text{Cov}(X_i^T \beta + \epsilon_i, X_j^T \beta + \epsilon_j) = \tau^2$, for $i \neq j$. It is easy to verify that the correlation is positive even without the conditioning on β. Now, the posterior $w(\beta, \theta, \tau^2 | Y, X)$ can be found and the usual Bayes predictor under squared error loss, the predictive distribution, can be produced for any new location s_{n+1}. In this context it is seen that identifying a form for the covariance is a central problem in spatial statistics; it is even more important in spatio-temporal statistics since the covariance is even more complicated (and important). Discussions are given in Cressie and Wikle (2011), Wikle (2015), and Fuentes (2011); see also Faye *et al.* (2015) for a detailed presentation.

Kriging in the spatial setting can be extended to the ST setting just as point process data in the spatial setting can be extended to the ST setting. The difference remains, however, in that the use of point data assumes the locations are fixed, essentially design points, while the use of point process data assumes the locations are random. Area data are qualitatively different again, because usually there is no micro-scale structure to exploit. That is, the regions are the regions (i.e. there are no subregions) and the time scale is coarse. This makes the techniques qualitatively different from those used for point data or point process data. Moreover, there are many qualitatively different predictors for each class of ST data. This field is growing and evolving rapidly at present. Only a small fraction of the simplest predictors can be presented here and the focus will be on point data and point process data.

12.3.1 Spatio-Temporal Point Data

Spatio-temporal point data are of the form $Y(s_i, t_i)$, for $i = 1, \ldots, n$, where the s_i are points or locations in a region and the t_i are times. So, $Y(s, t)$ can be regarded equivalently as $Y_s(t)$, a spatially varying time series model or as $Y_t(s)$, a temporally varying spatial model. Although the number of points and times is the same, n neglects the multiplicity. Data sets range from few locations and many times at each location to many locations and few times at each location. These problems are qualitatively different: the first is better for predictions for future times and the second is better for predictions for future locations. The DGs are usually complex so, often, whatever data are available is insufficient and modeling assumptions must be invoked to get results. Whether they are useful depends critically on whether the DG is $\mathcal{M}$-complete. If it is then there is a good chance that generating predictions from a model will often be somewhat successful in the short run. However, when the DG is $\mathcal{M}$-open, or nearly so, it may be impossible to assess effectively what, if any, aspects of the DG can be successfully encapsulated by modeling. That is, there may be essentially no features of the DG that are stable enough to improve prediction above, say, linear extrapolation, nearest-neighbor extrapolation, the Shtarkov predictor (see Sec. 11.7.1), or other predictors that do not require credible modeling assumptions.

To begin, consider adapting standard regression modeling to the ST context. This means writing $Y(s, t)$ as a sum of two terms, one representing a mean or first-order effect, the other representing a 'residual' or second-order effect. Thus, $Y(s, t) = \mu(s, t) + \epsilon(s, t)$. Both the mean function and the residual term can be elaborated to

$$Y(s, t) = [x(s, t)\beta(s, t)] + [W(s, t) + \varepsilon(s, t)] \tag{12.12}$$

in which the $x(s, t)$ are covariates (regarded as deterministic), the entries in $\beta(s, t)$ are ST-varying parameters, $W(s, t)$ is a mean zero stochastic process, and ε is a mean-zero Gaussian process with ST-varying covariance matrix. The first stage of (12.12) is $(Y|\mu, W) \sim N(\mu + W, \varepsilon)$. This structure can be extended to other distributions such as GLMMs using a link function g and writing $g(E(Y(s, t))) = \mu(s, t) + W(s, t)$; see Secs. 6.3 and 6.4. A difficulty with (12.12) is that $W(s, t)$ and ε depend on (s, t), so often both must be simplified. For instance, $W(s, t)$ may be written as $\alpha_t + W(s)$, i.e., as a sum of temporal and spatial components, in which α is usually $N(0, \sigma_\alpha^2 \Sigma(\phi))$ for some variance matrix parametrized by ϕ, or as $W_{t-1}(s) + \eta_t(s)$ in which the $\eta_t(s)$ are independent spatial process increments. Also, the $\varepsilon(s, t)$ are often taken as independent $N(0, \tau_t^2)$ distributions.

One benefit of these structures is that they are hierarchical, so that, for given s_{n+1}, t_{n+1} and data Y, the likelihood $f(Y|\theta)$ is normal for appropriately chosen stochastic processes and parameter vector θ. Therefore the conditional density $f(Y(s_{n+1}, t_{n+1})|Y, \theta)$ can be found and, if θ is assigned a prior density $w(\theta)$, $w(\theta|Y)$ can be obtained. Natural point and interval predictors for $Y(s_{n+1}, t_{n+1})$ now follow from finding the predictive density

$$f(Y(s_{n+1}, t_{n+1})|Y) = \int f(Y(s_{n+1}, t_{n+1})|Y, \theta)w(\theta|Y)d\theta. \tag{12.13}$$

The parallel frequentist approach would be to use a point estimator such as the MLE for θ and thereby generate point predictions. An SE for $\hat{Y}(s_{n+1}, t_{n+1})$ does not seem to have been derived although this is, in principle, feasible.

A more purely frequentist predictor arises from using empirical orthogonal functions as a basis expansion. For convenience, let the data be regarded as n-times series, i.e., as a matrix $Y = (Y_{t_j}(s_i))_{ij}$ for $i = 1, \ldots, n$, $j = 1, \ldots, T$. When $n > T$ a singular value decomposition (SVD) (see Sec. 12.1.1) can be applied to Y. Write it as

$$Y = V \Sigma U^T = \sum_{\ell=1}^{T} d_\ell v_\ell u_\ell^T = \sum_{\ell=1}^{T} d_\ell (v_\ell(1), \ldots v_\ell(T))(u_\ell(1), \ldots, u_\ell(n))^T,$$

where V is an $n \times n$ orthogonal matrix with columns $v_\ell = (v_\ell(s_1), \ldots, v_\ell(s_n))$, U is a $T \times T$ orthogonal matrix with columns $u_\ell = (u_\ell(t_1), \ldots, u_\ell(t_T))$, and Σ is an $n \times T$ diagonal matrix with elements d_j arranged in decreasing order. (Thus Σ has $n - T$ rows of zeroes at the bottom.) This means that

$$Y(s_i, t_j) = \sum_{\ell=1}^{T} d_\ell v_\ell(s_i) u_\ell(t_j). \tag{12.14}$$

If estimates of the functions $v_\ell(s)$ and $u_\ell(t)$, possibly nonparametric, see Chapter 8, can be given then they generate an obvious point predictor for $Y(s_{n+1}, t_{n+1})$.

Another approach to prediction is ST Kriging. Essentially, the time t is taken as another spatial dimension and the spatial Kriging predictor from the last subsection is extended by including an extra layer of variability. To see this, treat the data as a vector $Z = (Z(s_1, t_1), \ldots, Z(s_m, t_m))$, where $m = nT$ in the previous notation. The observed Z is now regarded as a noisy outcome of the response $Y(s, t)$, i.e., $Z = Y + \epsilon$ where ϵ has a mean-zero finite-second-moment distribution, often taken as $N(0, \sigma^2)$ and independent of Y for each s and t.

Spatio-temporal Kriging requires second-order stationarity. Parallel to the concept in time series, the covariance function for Y is

$$\text{Cov}(Y(s, t), Y(s', t')) = E\big(Y(s, t) - E(Y(s, t))\big) E\big(Y(s', t') - E(Y(s', t'))\big).$$

Second-order stationarity means that $\text{Cov}(Y(s, t), Y(s', t'))$ is a function of $s - s'$ and $t - t'$. That is, there is a function C such that $\text{Cov}(Y(s, t), Y(s', t')) = C(s - s', t - t')$. This condition in s was also required for purely spatial Kriging but was automatically satisfied by the use of the variances $\Sigma_0, \Sigma_{01}, \Sigma_1$.

In the present case, universal Kriging is not needed because there are no explanatory variables. It is enough to predict $Y(s_{n+1}, t_{n+1})$ optimally using a linear predictor of the form $w^T Z + C$, where $w \in \mathbb{R}^n$ and C is a constant. With some loss of generality, assume that $E(Y(s_{n+1}, t_{n+1})) = 0$, so that $C = 0$; Lichtenstern (2013, Chap. 6) called this the simple Kriging case. To regain the generality, a model for the mean of $Y(s, t)$ would have to be obtained and the unbiasedness condition explicitly carried through the minimization over w; see Agarwal (2011, Chap. 3) for a version of this. Here, unbiasedness is satisfied because $E(Y(s_{n+1}, t_{n+1})) = 0$.

As in Wikle (2015) the simple Kriging predictor is $\hat{Y}(s_{n+1}, t_{n+1}) = w^T Z$, where

$$w = \arg \inf_{w \in \mathbb{R}^m} E\left(Y(s_{n+1}, t_{n+1}) - w^T Z\right)^2. \tag{12.15}$$

Routine manipulations give $\hat{Y}(s_{n+1}, t_{n+1}) = c(s_{n+1}, t_{n+1})^T \Sigma_Z^{-1} Z$, in which $\Sigma_Z = \text{Cov}(Z)$ and $c(s_{n+1}, t_{n+1})^T = \text{Cov}(Y(s_{n+1}, t_{n+1}), Z) = \text{Cov}(Y(s_{n+1}, t_{n+1}), Y)$. The SE of the simple Kriging estimator in this case is

$$\sigma(s_{n+1}, t_{n+1}) = \sqrt{\text{Var}(Y(s_{n+1}, t_{n+1})) - c(s_{n+1}, t_{n+1})^T \Sigma_Z^{-1} c(s_{n+1}, t_{n+1})}.$$

A Bayes version of ST Kriging is given in Faye *et al.* (2015); see Xu *et al.* (2015) for a variation on Bayes Kriging and You *et al.* (2016) for a complex application combining Bayesian and frequentist methods. A comprehensive treatment is found in of Cressie and Wikle (2011, Chap. 6).

To make this operational, a form of the covariance must be specified; this is analogous to choosing a form of $\Sigma = \Sigma(\theta)$ in the purely spatial setting. Often this is done in both the spatial and ST settings by proposing specific parametric forms of the covariance, although it is more common to use a variogram, a quantity that is slightly more general than the covariance. A variogram is the function

$$\gamma(u, v) = C(0, 0) - C(u, v) = \text{Var}(Y(s + u, t + v) - Y(s, t)),$$

assuming, for all (s, t), that $Y(s, t)$ is second-order stationary with stationary covariance function $C(s, t)$. The slight increase in generality is that technically the variogram (often called a semivariogram) does not require the second or product moments of Y to be finite. Sometimes it is easier either to specify a parametric form for γ rather than for C or to represent Kriging solutions in terms of γ.

Taking the Fourier transform of $C(s, t)$ or regarding $C(s, t)$ as the Fourier transform of some other function has also been proposed for ST data. These Fourier transforms are called spectral measures or spectral densities, respectively. Moreover, it is a useful simplification if C is a product of two functions, $C(s, t) = C(s)C(t)$ or $\gamma(u, v) = \gamma(u)\gamma(v)$, a property termed separability. This property often does not hold in reality, but when it does the spatial and temporal parts of the analysis can be separated. Spectral function methods are important primarily because they are ways to motivate and identify useful covariance functions; see Fuentes (2011). Technically, most ST processes occur in continuous time even though most measurements are made in discrete time. Here, this distinction is ignored; whichever treatment of time gives the more useful solution is assumed although it is clear that some models lend themselves to one formalism over the other.

The foregoing modeling can be extended to GLMs. For instance, let $Y(s, t)$ be Poisson, i.e., $Y(s, t)|\lambda_t(s) \sim \text{Poisson}(\lambda_s(t))$. Following Wikle (2015), consider the 'process' model

$$\log(\lambda_t(s)) = \mu_t + h_{st}^T u_t + \eta_t(s),$$
$$\mu_t = \mu_t + \epsilon_t,$$
$$u_t = M(\delta, \alpha)u_{t-1} + \gamma_t, \qquad (12.14)$$

in which the error terms η_t, ϵ_t, and γ_t (sometimes called 'innovations') are IID $N(0, \sigma_\eta)$, $N(0, \sigma_\epsilon)$, and $N(0, \Sigma(\theta_\gamma))$, respectively. That is, the mean structure evolves over time with a one-step lag governed by M and the spatial effects from point to point are stochastically the same. To complete the model, priors must be assigned. For instance, $\delta|\beta, \sigma_\delta^2, R_\delta \sim N(\Phi\beta, \sigma_\delta^2 R_\delta)$ and $\alpha \sim N(\alpha_0, \sigma_\alpha^2)$. The initial condition u_0 requires a prior, as does the vector of parameters β and possibly the variances. Given this specification, data $(s_1, t_1), \ldots, (s_n, t_n)$

and routine (but complex) computations can generate an estimate for $\lambda_t(s)$ and therefore generate predictors for $Z(s, t)$. The same structure can be used for normal-based models e.g., a model of the form (12.12) for $Y(s, t)$ can be proposed. In these cases, the parts of the model in which the data appear are called the measurement equations (although there is usually only one) and the other equations for how the terms in the measurement equation evolve are called the transformation equations (but the structure is similar to (12.14)). See Cressie and Wikle (2011, Chaps. 7 and 8) for further details, in particular the computations necessary to generate predictions.

Arguably the most general model-based predictive structure in ST statistics is a dynamic ST model (DSTM); see Wikle (2015). It has three parts: a data or measurement model, a process model, and parameter models. The form of a DSTM starts with a data model

$$Y(s, t) = \mathcal{H}(Z(s, t), \theta_h, \epsilon(s, t)), \tag{12.15}$$

in which $Y(s, t)$ is observed, $\mathcal{H}$ is the function relating the observations to an underlying process $Z(s, t)$ that is not directly observed, θ_h is a parametrization for the hidden functions $\mathcal{H}$, and $\epsilon(s, t)$ is an error term that perturbs Z to give Y. The process model for Z, the unseen entity giving rise to Y, can be taken to be

$$Z(s, t) = \mathcal{M}((Z(s, t - 1), Z(s, t - 2), \ldots), \theta_m, \eta(s, t)), \tag{12.16}$$

in which θ_m parametrizes $\mathcal{M}$ and $\eta(s, t)$ is the process error. Often (12.16) is assumed to be Markov of some order. Finally, parameter models must be specified for θ_h, θ_m, ϵ, and η, in a Bayesian analysis. Explanatory variables, when they exist, are usually built into the parameter distributions. Clearly, (12.14) is an example of a DSTM and the Kalman filter is a special case of a DSTM when there is no spatial dependence.

As an instance of the generality of a DSTM, note that it can incorporate basis function representations. For instance, set

$$Y(s, t) = \mu(s, t) + \Phi \alpha(s, t) + \Psi \beta(s, t),$$

where Φ is the matrix of evaluations of basis functions at the (s, t)'s with parameter expansion $\alpha = M_\alpha \alpha_{t-1} + \eta_{\alpha,t}$; cf. (12.14). Linear DSTMs arise from choosing $\mathcal{H}$ in (12.15) and $\mathcal{M}$ in (12.16) to be linear. In some cases, partial differential equations serve as process models. Moreover, nonlinear DSTMs are currently studied in a variety of settings but are beyond our present scope, as are the analyses that give rise to predictors in these general settings, but see Cressie and Wikle (2011, Chaps. 7 and 8) and Wikle (2015) for guidance.

12.3.2 Remote Sensing Data

Remote sensing data refers to data that is gathered remotely from the subject being observed. Typically, this data is geospatial or geo-ST point data, of the form discussed in the last subsection, gathered by satellites or drones recording the electromagnetic energy reflected or emitted by surfaces usually on the Earth such as soil, water, vegetation, or other landforms. One problem associated with this class of data is the fact that there are often multiple scales of measurements that must be considered. Either different scales have to be merged somehow or the scale on which data of one type were collected differs from the scale on which predictions and other inferences from the data are relevant. Technically, reading a traffic sign and other image segmentation problems are also examples of remote sensing data but these are usually amenable to image analysis and outside the scope of our present treatment. (There are any number of excellent image analysis, image processing, and image segmentation textbooks, as a web search will reveal.) The focus of this subsection will therefore be on upscaling and downscaling for the purposes of geostatistical prediction. Introductions to remote sensing can be found in NRC Tutorial (2016), Tempfli *et al.* (2009), and Aggarwal, S. (2013), amongst others.

For the present, it is enough to know that the amount of radiation from an object or region is called its radiance and is influenced by both the properties of the object or region and the radiation hitting it – whether naturally (e.g., the passive radiation from the sun) or intentionally (e.g., the active radiation from equipment). Within the visible spectrum of electromagnetic radiation (EMR) humans can, in principle, analyze and interpret the variation in color and intensity of remote sensing data. Outside the visible spectrum (gamma rays, X-rays, ultraviolet, infrared, microwave, etc.) various types of equipment and techniques are used to make electromagnetic radiation visible, analyzable, and interpretable. Solar, and other types of energy, are often modified as they pass through the atmosphere, by scattering, absorption, and refraction. The usual sequence of steps (much simplified) for obtaining remote sensing data is:

1. EMR is emitted.
2. The EMR travels from the source to some target e.g., the Earth's surface, and may be affected by passing through a medium, e.g., the atmosphere.
3. The EMR that hits the target interacts with the target.
4. The EMR then travels from the target to the remote sensor, again through a medium that may affect it.
5. The remote sensor detects the received EMR and generates output data from it.

Thus, there are many steps that must occur in sequence for the received data to be generated. Moreover, it is often desirable to combine the data from multiple sources in the same region to improve prediction.

If the sources of data are on the same scale e.g., they amount to repeated measurements at points, this problem is difficult but essentially falls within the category of multivariate analysis or relatively standard ST modeling. However, often two sets of data are on different scales, e.g., they have different resolutions or different time scales, and cannot simply be combined in the form in which they are received. Thus, data are *downscaled* when they are coarse and have to be adapted to a finer level. Data are *upscaled* when they are at a fine level and have to be aggregated to a coarser level. The problem of putting two sets of data onto the same scale has been around for decades but has received new urgency, and hence new attention, owing to the vastly increased quantity of remotely sensed data now readily available. The discussion below is based on recent work in spatial data but can be extended to ST data, bearing in mind the computational difficulty.

Numerous downscaling methods have been developed; see Atkinson (2013), Hoy and Quaife (2015), and Ha *et al.* (2013). Essentially, these are methods to predict the value of the Earth's surface at a point given the coarse-level data. Overall, the simplest approach is to take the coarse-level data for regions and assign its value to all points individually. The limitation of this is that the result is not, in general, a continuous surface over the entire region and the magnitude of the approximation error is therefore unclear. So, usually some smoothing is applied. Sometimes a simple mathematical interpolation approach is taken; at other times extra knowledge may be available about what the surface should look like. This latter is sometimes called regularization, since the point is to use extra information, possibly an equation believed to describe the surface at a fine scale, to adjust the edges of each region.

Other more involved downscaling methods have also been used. For instance, regression methods (linear and nonlinear) take the coarse data on regions and the finer-scale data to form a regression function that predicts locations at the finer scale that were not measured. Sometimes this is called 'area to point' prediction, and a variation on Kriging can be applied. A third approach is to bypass the downscaling of numerical data and try to go directly to a classifier for each point to identify the surface type, e.g., in terms of the type of landcover. Sometimes this is called a super-resolution mapping. The strategy is to develop a coarse-level classifier that can be refined. The refinement requires that multiple coarse-level measurements are available, so that interpolation is possible. Numerous interpolation methods have been proposed including neural networks and genetic algorithms amongst others. Sometimes these methods rely on an empirical variogram to express the dependence between points; see Wu and Li (2009, Sec. 4.4) for this and other less commonly used but traditional methods.

Upscaling can also be regarded as a prediction problem. The task is to take fine-grained measurements and use them to predict an aggregated measurement; this is analogous to taking many temperature measurements at different locations and using them to assign an overall temperature to a region. Obviously, there is a variety of simple methods. For instance, simply averaging the measurements in a region can be taken as the value for a region. As with the simplest downscaling technique, this does not, in general, lead to a continuous surface and so may have to be modified. Another approach is to threshold the fine-scale values and treat the result as a fraction, e.g., of pixels that are black or white, thereby coarsening the fine-scale data to regions. These fractions can then be combined over patches to aggregate more smoothly (or less smoothly if desired); see Hufkins *et al.* (2008). A more involved method for averaging is used in Stoy *et al.* (2009). Based on Kolaczyk *et al.* (2012) the idea is to represent the fine-scale data by a tree structure, i.e., a partition on the space that might be induced by a dyadic tree. Then, finite-mixture models are fitted to various regions and the best one chosen through maximum likelihood. This is then used to segment the region (or data) to the desired scale.

Although an earlier publication, Hay *et al.* (1997) provides several methods that appear not to overanalyze the data as much as the finite-mixture approach. The idea is to use multiple-sized windows of the fine-scale data to identify shapes prior to upscaling. Then, the identified shapes are retained in the upscaled data. Shapes are defined essentially by a form of autocorrelation, and a separation between two shapes can be identified by a sudden change in variance. Effectively, a variance threshold defines the shapes. The shapes then have an associated mean, variance, and region. Within each area, a kernel method with fixed bandwidth (or another nonparametric method) is used to smooth the mean data on regions, using the area per point as the assessment to achieve the desired scale.

Most recently, complex computationally intensive techniques have been used for upscaling and downscaling. For instance, an empirical hierarchical Bayes model for downscaling was proposed in Hobbs *et al.* (2016). The idea is that Y is observed and the main task is to identify the state X that gave rise to it. The core equation in this approach is $Y = F(X, \beta) + \epsilon$, where F is either a known model or to be estimated. Kang *et al.* (2016) proposed merging fine-scale computations from a physical model with coarse-scale observed data as a way to downscale with greater accuracy, via a model similar in spirit to (12.12) . (This assumes that the physical model 'data' is valid.) Marchetti *et al.* (2016) proposed upscaling by using a hierarchical clustering on high-resolution data in a way that preserves the covariance structure. Finally, Johannesson *et al.* (2016) used a mix of up- and downscaling to interpolate field image data between two different times, in the process of correcting for cloud and cloud shadows.

It should be recalled that in all these methods questions of the location of the true model (if it exists) in the collection of models being used to generate predictions will determine the accuracy of predictions outside small ST neighborhoods. This will generally only be successful for DGs that are $\mathcal{M}$-complete and readily approximable. Predictive accuracy for continuous $\mathcal{M}$-open problems may also be locally valid.

12.3.3 Spatio-Temporal Point Process Data

An introduction to spatio-temporal point process data was given in Sec. 5.5 for the purpose of comparing this method with time series methods. There, the concepts of a point process, Poisson process, and compound process were presented, in which the evaluation of predictors is based on scoring functions. Inferences as to which predictors are best under scoring function criteria are often unstable, compared with using CPE, because scoring functions often rest on probabilities or densities rather than on simply comparing predictions with observations. Even if scoring function criteria satisfy some form of the Prequential Principle, they are more distant from observables than the CPE. Here, the task will be to see some of the rudiments of how the very general class of ST point processes (STPPs) can be used to generate predictors. Only a small fraction of all the STPPs that are possible have been studied in any detail, so there is hope that in the future STPPs may be found that will provide better prediction for complex DGs than at present. For a recent comprehensive review, see González *et al.* (2016).

In general, an STPP is a stochastic process, say $Y = Y(\omega)$, taking a set of values generically denoted by s and ranging over a region R that may depend on t, for $t \in [0, T]$. Strictly, Y generates a random finite or countable subset of $R \times [0, T]$. If Ω denotes the underlying measure space then $\omega \in \Omega$ is a reminder that the (s, t) generated are the outcomes of a random process. Often an STPP is defined such that $s \in \mathcal{D}(t)$, i.e., the index set for s may be selected at random from the domain of t. In short, an STPP combines a time series with a spatial process and STPPs are point processes because they take values in $\mathbb{Z}^+$ for each t. The outcomes of a point process are usually assumed to be indistinguishable except by their locations and times. In this sense, as noted at the beginning of Sec. 12.3, STPPs are zero–one data indicating locations where an event happens at a particular time but not by themselves describing the event. *Marked STPPs* are a more complex formulation that permits other information to be associated with a location–point pair but are beyond our present scope, which is to present predictors in the simplest cases.

When necessary, $Y(S, \tau)$ will indicate the random number of points landing in the spatio-temporal region $S \times \tau$, where S is a spatial region and τ is a time interval. If $S_1 \times \tau_1$ and $S_2 \times \tau_2$ are disjoint then $Y(S_1, \tau_1)$ and $Y(S_2, \tau_2)$ are independent. Well-behaved STPPs also have a density. It amounts to a theorem that, for a given k and $g \geq 0$, there is a density h_k such that

$$E\left(\sum_{\xi_1,\ldots,\xi_k \in R \times [0,T]} g(\xi_1,\ldots,\xi_k)\right) = \int_{R \times [0,T]} \cdots \int_{R \times [0,T]} g(\xi_1,\ldots,\xi_k) h_k(\xi_1,\ldots,\xi_k) \prod_{i=1}^{k} d\xi,$$

where $d\xi = ds\,dt$ indicates a small region around $\xi = (s,t)$ with volume $|ds\,dt|$. The intensity function of an STPP, which is essentially equivalent to h_k, assumes that $S \to s$ and $\tau \to t$ and is given by

$$\lambda(s,t) = \lim_{\text{Vol}(S \times \tau) \to 0} \frac{EY(S \times \tau)}{\text{Vol}(S \times \tau)},$$

where it is assumed that $\text{Vol}(S \times \tau) = \text{Vol}(S)\text{Vol}(\tau)$ and both factors go to zero. Now write

$$\mu(S \times \tau) = \int_S \int_\tau \lambda(s,t)ds\,dt.$$

That is, the integral of the intensity function λ gives the intensity measure μ. The marginal spatial and temporal intensities are

$$\lambda(s) = \int_T \lambda(s,t)dt \quad \text{and} \quad \lambda(t) = \int_R \lambda(s,t)ds,$$

while the conditional spatial intensities are defined by the relation

$$\lambda(s,t|\mathcal{H}_t)ds\,dt = E(Y(ds\,dt)|\mathcal{H}_t),$$

where $\mathcal{H}_t$ is the history of the STPP up to time t. More symmetrically, the conditional intensities are

$$\lambda(s|t) = \lim_{|ds| \to 0} \frac{Y(ds,t)}{|ds|} \quad \text{and} \quad \lambda(t|s) = \lim_{|dt| \to 0} \frac{Y(s,dt)}{|dt|}.$$

To generate predictions from Y requires an estimator $\hat{\lambda}$ of $\lambda : R \times [0,T] \to \mathbb{R}$. If $\lambda(s,t) = \lambda(s)\lambda(t)$, i.e., λ is separable, and data $(s_1,t_1),\ldots,(s_n,t_n)$ are available then kernel estimators with kernel K and bandwidths h and h' can be used. Let

$$\hat{\lambda}(s) = \frac{1}{n}\sum_{i=1}^{n} K_h(s - s_i) \quad \text{and} \quad \hat{\lambda}(t) = \frac{1}{n}\sum_{i=1}^{n} K_{h'}(t - t_i)$$

and write $\hat{\lambda}(s,t) = \hat{\lambda}(s)\hat{\lambda}(t)$. Obviously, other estimators – including function estimators that do not require separability – could be used. Now, since μ exists, one natural point predictor of the number of points in a region $S \times \tau$ is

$$\hat{N}(S \times \tau) = \widehat{EY}(S \times \tau) = \int_{S \times \tau} \hat{\lambda}(s,t)ds\,dt. \tag{12.17}$$

To obtain an SE for this predictor requires the concept of a second-order intensity. This is the straightforward generalization of the (first-order) intensity:

$$\lambda_2(s_1,t_1;s_2,t_2) = \lim_{|ds_1 dt_1|,|ds_2 dt_2| \to 0} \frac{E(Y(ds_1 dt_1)Y(ds_2 dt_2))}{|ds_1 dt_1||ds_2 dt_2|}.$$

Marginal and conditional second-order intensities are defined similarly. So, the natural way to obtain an SE for $\hat{N}(S \times \tau)$ is to find

$$\widehat{N^2}(S \times \tau) = \widehat{EY^2}(S \times \tau) = \int_{S \times \tau} \hat{\lambda}_2(s, t; s, t) \, ds \, dt, \qquad (12.18)$$

where $\hat{\lambda}_2(s_1, t_1; s_2, t_2)$ is an estimator of $\lambda_2(s_1, t_1; s_2, t_2)$, possibly Nadaraya–Watson, but more likely one that has a better performance in moderate dimensions. From (12.17) and (12.18) an SE for $\hat{N}(S, \tau)$ can be found.

An STPP is characterized by its intensity and the intensity is often assumed to be a member of a parametric family, so nonparametric estimators, which often require a lot of data to be effective, do not have to be used. In the special case where the STPP has a constant intensity, i.e., it is derived from a Poisson process, it is natural to estimate λ by $\hat{\lambda} = n/\text{Vol}(R \times [0, T])$, and the estimated intensity for finding the SE can be taken as $\hat{\lambda}^2$; see Dorai-Raj (2001, pp. 83 and 78).

An inhomogeneous Poisson ST process is one that has a nonconstant intensity. One simple form of this uses an intensity function λ that is piecewise continuous. Now, the likelihood as a function of λ is

$$L(\lambda) \propto \prod_{i=1}^{n} \exp\left(-\int_R \lambda(s, t_i) \, ds\right) \prod_{i=1}^{n} \lambda(s_i, t_i)$$

in which it is assumed that each t_i is in a different 'piece' of the intensity. Then, $\hat{\lambda}$ can be obtained and used in (12.17) and (12.18) to generate predictions. In general, the dependence of Y on parameters and covariates (parametrically or nonparametrically) can be built into λ in a variety of ways. A more complex example of an inhomogeneous ST Poisson process was used in Liang *et al.* (2010).

More generally, the log-likelihood for λ given data is

$$L(\lambda) = \sum_{i=1}^{n} \ln \lambda(s_i, t_i) - \int_R \int_{[0,T]} \lambda(s, t) \, ds \, dt;$$

see González *et al.* (2016, p. 24). So, as noted there, it is particularly useful if λ can be treated by regression methods. For instance, set

$$\ln \lambda(s, t) = \sum_{j=1}^{p} \beta_j X_j(s, t),$$

in which the X_j are covariates that depend on location and time with coefficients β_j. Diggle (2014, Chap. 8) gives an example of this. Again, (12.17) and (12.18) can be used to give a PI for the number of occurrences in a region over a time span. Of course, in these cases, marginal (or conditional) intensities can be used to predict the number of locations or times (or conditional locations or times), using straightforward modifications of (12.17) and (12.18). There are numerous other variations on inhomogeneous processes; see Diggle (2014) and González *et al.* (2016).

Another way to identify an intensity function (either parametrically or not) is via the K-function; see Wang (2013, Chap. 1), for instance. Let $N_r(s, t)$ be the number of further events within a specified range r of (s, t). Then, the idea is to set $K_r(s, t) = E(N_r(s, t))/\lambda(s, t)$.

There is a variety of standard forms that can be used to obtain least squares estimates for the parameters in λ and hence K_r. Indeed, there are numerous variants on inhomogeneous Poisson STPPs, numerous other ways to generate predictors for them, and many processes that have not been discussed (see González *et al.* (2016) and Diggle (2014)) or perhaps not even been discovered that may better encapsulate aspects of DGs in those $\mathcal{M}$-complete applications for which this form of modeling can be effective.

12.3.4 Areal Data

Suppose that a fixed number n of regions – counties, postal codes, etc. – with fixed boundaries are of concern and data is available from each of them at certain times, say $t_1, \ldots, t_T$. Additional information in the form of covariates for each region, also over time, are also typically available. This may be count data from the regions, averages over outcomes in a region, and so forth. The basic model for this situation remains

$$Y(i, t_j) = \mu_{ij} + \epsilon(i, j) = W(i, t_j) + \varepsilon(i, j),$$

in parallel to (12.12), where the $W(i, t_j)$ are spatio-temporal random effects and $\varepsilon(i, j)$ is an error term; in some cases it is necessary to transform the raw data at (i, t_j) to the form given for $Y(i, t_j)$. In this expression for the model, μ_{ij} denotes fixed effects.

The simplest case is linear: an intercept and covariates with coefficients of the form $x_{it_j}\beta_f$, the intercept having been absorbed into the x_{it_j}. Usually, $W(i, t_j)$ is assumed to be a sum of two terms, one for random effects that is often taken as linear, i.e., of the form $z_{it_j}\beta_r$ for covariates z_{it_j} and random coefficients β_r, and one that is a sum of a conditionally autoregressive (CAR) term, say, θ_{it_j}. The error term is often taken to be $\varepsilon(i, j) \sim N(0, \tau_j^2)$ for some values $\tau_j > 0$ or even simply $N(0, \tau^2)$ so that it is independent of both i and j. One of the most important subject matter domains for this class of models is epidemiology; see Höhle *et al.* (2007) and Meyer *et al.* (2015).

Here, CAR is in an ST sense. First, let $A = (a_{k,\ell})_{k,\ell=1,\ldots,n}$ be the adjacency matrix of the overall region. That is, for $k \neq \ell$,

$$a_{k,\ell} = \begin{cases} 1 & \text{regions } k \text{ and } \ell \text{ have a common boundary,} \\ 0 & \text{otherwise,} \end{cases} \tag{12.19}$$

for $k = 1, \ldots, n$, $a_{k,k} = 0$. Other choices for A that have (positive) entries decreasing as the regions they represent become further apart are also possible. Since A is not time dependent, write $\theta_j = (\theta_{1t_j}, \ldots, \theta_{nt_j})$ as an outcome of $\Theta_j \sim \text{CAR}(A, \sigma_j)$, which are IID over j. In a Bayesian setting, $\sigma_j^2 \sim \text{IG}(a, b)$ for some $a, b > 0$; see Guha and Ryan (2006). The notation $\text{CAR}(A, \sigma_j)$ means that the distribution of Θ_j conditional on $\hat{\Theta}_j = \hat{\theta}_j = (\theta_1, \ldots, \theta_{j-1}, \theta_{j+1}, \ldots, \theta_T)$ depends only on the neighbors of θ_j in $\hat{\theta}_j$ as defined by A in an autoregressive sense, e.g., for some weights $\tilde{\alpha}_{jk} = \alpha_{jk}/\sum_k \alpha_{jk}$,

$$\Theta_j | \hat{\Theta}_j = \hat{\theta}_j \sim N\left(\sum_{k=1}^{T} \tilde{\alpha}_{jk}\theta_k, \sigma^2 \Big/ \sum_k \alpha_{jk}\right),$$

in which $\sum_k \alpha_{jk}$ is the number of neighbors of region j. Guha and Ryan (2006) cited a result giving a closed-form expression for $(\Theta | A)$. The net effect of this is that a MCMC procedure can be developed, and hence the posterior predictive for future outcomes can be found, so that point and interval predictions can be given. A related analysis using spatio-temporal CAR distributions is given in Mariella and Tarantino (2010); they used an explicit neighborhood structure and regarded the outcomes at a given time as random fields that are sequentially assembled. As noted in Fuentes (2011), areal unit data are often nonnormal, e.g., sparse counts, that might be better modeled by a Poisson. In this case, the analysis of Guha and Ryan (2006) could in principle be extended, although this does not seem to have been done yet.

There are numerous other contributions to the general topic of ST areal data analysis via models, much of it from the Bayes perspective. One that exemplifies perhaps the most successful methodology to date because of its relative simplicity and computational effectiveness is due to Vivar and Ferreira (2009). Begin with a lattice $\{1, \ldots, S\}$; this is analogous to the regions earlier indexed by j. Suppose that the lattice has a neighborhood structure $\{N_s | s = 1, \ldots, S\}$, where N_s is the set of neighbors of s. For $t = 1, \ldots, T$, the variable Y_{ts} is observed. For fixed times t these form a random field and for fixed positions s they form a time series. Write $Y_t = (Y_{t1}, \ldots, Y_{tS})^T$ and assume that, conditionally on a latent random field process $X_1, \ldots, X_T$, the Y_t are independent. Suppose that the observational equation

$$Y_t = F_t^T X_t + \epsilon_t, \quad \epsilon_t \sim \text{PGMRF}(0_S, V_t^{-1})$$

where 0_S is a vector of zeroes of length S and the $\epsilon_t = (\epsilon_{t1}, \ldots, \epsilon_{tS})^T$ are independent. In the expression for Y_t, F_t is a matrix connecting the observations to the latent process and it is assumed to be either known or unknown up to a small number of parameters. Typically, F_t is sparse because of the neighborhood structure. The $S \times S$ matrices V_t represent the covariance structure of the errors and PGMRF stands for a proper Gaussian Markov random field. A normal distribution is attached to each s and these distributions are Markov-conditional on the neighborhood structure. The adjective 'proper' means that none of the variance matrices for any s is singular. Coupled with the observation equation is the evolution equation

$$X_t - \mu_x = G_t(X_{t-1} - \mu_X) + \omega_t, \quad \omega_t \sim \text{PGMRF}(0_S, W_t^{-1}),$$

where the error terms ω_t are independent and μ_X is the mean. Now, Vivar and Ferreira (2009, equation (5.1)), provides an expression for $p(y_{t+1} | \mathcal{D}_t)$, the predictive density for Y_{t+1} using all the data up to and including time t. This expression is dependent on the ST model assumed; however, the dependence can be removed by model selection via Bayes factors or, in principle, it can be averaged over by a BMA.

To conclude this section, we observe that spatio-temporal models are highly varied and almost always highly simplified compared with the DGs that they attempt to describe. This is true even of the work in Wikle (2015), which incorporates extra information via differential equations (partial and ordinary, regular and stochastic). This is not to say they are ineffective. Rather, it means that they generally have uncorrected, possibly uncorrectable, bias that may simply have been folded into the error terms. In fortunate cases, the error is small enough that the model makes good predictions for a short period of time or a short distance in space. However, even if the model is 'good' and the problem is $\mathcal{M}$-complete, the error structure generally makes the model ineffective for new data points that are too far (in time or space) from those in the data set. It is reasonable that model predictions may diverge owing to the accumulation of error terms: they become somewhat like a random walk if, in fact, error terms are accumulating in ways that do not cancel each other out. However, the inability to predict very far out may reflect only the limitations of the modeling approach. After all, for many complex DGs, after data is observed it is common to hear people say things like 'I should have seen that coming' – and this reaction represents the presence of bias, not increasing randomness. That is, the result was foreseeable although poorly foreseen. The analog of this may occur in $\mathcal{M}$-open problems also, when there is some feature stable enough to permit useful predictions.

12.4 Network Models

There are numerous classes of network models and this section treats only one of them, networks that are presumed to exist in some physical sense and describe binary relationships among a class of physical entities. The task is to understand their structure so that predictions can be made about the binary relationships or the evolution of communities of the entities. Other types of networks, e.g., Bayesian networks (which describe the dependence among a set of random variables), percolation networks, input–output relation networks, agent-based modeling, etc. are exceedingly important in many applications. However, they have been extensively studied elsewhere and will not be discussed here owing to space limitations. Descriptions of these other classes of network problems and references that present them can be found in Goldenberg *et al.* (2009, Sec. 1.2).

Let $\mathcal{G}$ be a graph, i.e., $\mathcal{G}$ is defined by a set $\mathcal{V}$ of vertices and a set $\mathcal{E}$ of edges. It is understood that the edges lie only between two elements of $\mathcal{V}$. The edges may be directed or not; here, they will be assumed undirected unless indicated otherwise. Edges are sometimes allowed to be between a given vertex and itself; this will generally not be allowed here, again unless indicated. Edges may have weights associated to them and some graphs permit multiple edges between a given pair of vertices. A distinction is often made between graphs as just defined and networks: a network is often taken as the collection of relationships amongst entities in the real world, modeled respectively as the edges and vertices in a graph. Thus a collection of graphs constitutes a collection of models for a given network, and the best graph (however defined) within a collection of graphs constitutes the best estimate for a given network using the given collection of graphs.

Graphs are equally well defined by their adjacency matrices, $A = (a_{k\ell})_{k,\ell=1,\dots,n}$, where n is the number of vertices and $a_{k\ell}$ is defined as in (12.19) with 'common boundary' replaced by 'edge'. Thus, $a_{k\ell} = 1$ if and only if there is an edge between vertex k and vertex ℓ and is zero otherwise. If no vertices have self-edges then all $a_{kk} = 0$. If the graph is undirected then $a_{k\ell} = a_{\ell k}$. The ith row of A indicates all the links of vertex i to other vertices and the number of elements along the ith row gives the degree of vertex i, the number of vertices to which it is linked. Similar statements hold for the columns of A. A path between two vertices v_i and v_j, if it exists, is a sequence of edges to be traversed in order to travel from v_i and v_j with no repeated vertices. A walk from v_i to v_j is more general in that it allows repeated vertices, which are disallowed for paths. The distance between vertices v_i and v_j is the length of the shortest path between them. The paths that achieve this need not be unique but the value of the distance is.

A subgraph $\mathcal{G}'$ of $\mathcal{G}$ is a graph for which $\mathcal{V}' \subset \mathcal{V}$ and $\mathcal{E}' \subset \mathcal{E}$, i.e., the edges in $\mathcal{E}'$ are between vertices in $\mathcal{V}'$. Two graphs, say $\mathcal{G}$ and $\mathcal{H}$, are isomorphic if and only if there is a function $\phi : \mathcal{V}_\mathcal{G} \to \mathcal{V}_\mathcal{H}$ such that, for any $v_i, v_j \in \mathcal{G}$, $(v_i, v_j) \in \mathcal{E}_\mathcal{G} \Leftrightarrow (\phi(v_i), \phi(v_j)) \in \mathcal{E}_\mathcal{H}$. It is possible for a subgraph of a graph to be isomorphic to another graph; sometimes this is called subisomorphism and the subgraph is called a motif.

Two important examples of motifs are cliques and hub-and-spoke graphs. A clique is a collection of vertices for which all possible edges are included, i.e., each vertex is joined by an edge to each other vertex. A clique of size c in a graph is maximal if it is impossible to add a vertex to create a clique of size $c+1$. In a hub-and-spoke motif there is one vertex to which many other vertices are joined. It has been observed that airlines design their flight routes based on hub-and-spoke motifs: certain airports have many more flights into them and out of them than other airports do. This leads to the concept of the degree sequence for a graph. This is the vector of degrees of the vertices in the graph, usually sorted to be decreasing. The empirical probability on the nodes formed by using the degrees is often called the degree distribution. (In directed graphs, i.e., those for which edges have a direction assigned, it is important to distinguish between the numbers of edges leading to and from a vertex. This is reflected in A and the parallel statements for directed graphs can be readily derived.) In practice, degree distributions are usually strongly right-skewed with heavy tails, indicating that the degree does not drop off rapidly. Graphs are bipartite (resp. tripartite, etc.) if their vertex sets can be partitioned into two (resp. three or more) disjoint subsets. This is important when trying to identify communities of vertices. Note that although vertices are equivalent in the mathematical formulation of graphs, they are not in general equivalent in networks. Loosely, graph models are usually $\mathcal{M}$-complete, if not $\mathcal{M}$-closed, while real intervals are usually $\mathcal{M}$-complete if not $\mathcal{M}$-open.

Graphs as models for networks have another limitation, which can be severe: graphs represent only binary links (i.e., between two vertices) well. If the relationship between two vertices involves a third (or more) vertex then it may be difficult for a graph to encapsulate it. For instance, if vertex i buys from vertex j only when vertex k is not selling then it is not clear what graph should be used. Boolean networks are often better for more complex settings such as this. A Boolean network has a collection of Boolean functions at each vertex, i.e., each function accepts a zero–one argument from the connected vertices and gives a single zero–one output for the vertex. In many settings, e.g., biochemical networks, this is more realistic than a graph. See Atias *et al.* (2014) for but one of many examples. However, binary graphs are a smaller model class and hence more convenient. For instance, $\sum_{k,\ell} a_{k,\ell}$ is the number of directed edges in a graph, $\sum_{k \neq \ell} a_{k,\ell}$ is the number of nonreflexive edges in a directed graph, and $\sum_{k > \ell} a_{k,\ell}$ is the number of nonreflexive edges in an undirected graph. Also, the maximum number of edges for n vertices is $\binom{n}{2}$. In short, it is easier to assign distributions to graphs than to Boolean networks and calculations can often be carried out effectively, if not in closed form at least numerically.

The most important work on graphs or networks for our present purposes falls into two categories: static and dynamic. As the name suggests, the task in static graphs is to identify a single optimal, or at least good, graph to represent a network. A considerable amount of work has been done on this from the standpoint of the sampling and generation of graphs; only a fraction of it can be surveyed here, although we note that often the goal is either to predict whether an edge exists or to predict the whole graph. In some cases, the graph can be used to make predictions about the structure of the network but that is beyond our present scope. Dynamic graphs are more complicated because they allow that, just as networks evolve over time, so too must a sequence of graphs intended to represent the stream of networks. Again, only a small fraction of this topic can be surveyed, partly because so much work has already been done but also because it is an active area of research. Again, one of the main tasks is prediction. For instance, the identification of communities is an intellectually interesting question in its own right for both static and dynamic graphs, but obviously this is not of much use if it has no predictive power.

12.4.1 Static Networks

For simplicity, begin with an $\mathcal{M}$-closed problem. In the present setting this means that there is a finite list of graphs and one is known to be true. The task is to identify it. To help achieve this task, data is collected. Obviously there are many ways to do this badly – simply ask questions that will not elicit the information required. For example, suppose that the goal is to identify the network of friendships for a group of individuals. One question could be to ask each person (vertex) who his or her three best friends are. Automatically, three becomes an upper bound on the out-degree of the vertex whether or not it's correct. Then there are the other usual problems with getting accurate answers on survey questions, such as wanting to look good to the interviewer, the phrasing of questions, etc.

More pragmatically, it is often possible to start with a simple random sample (SRS) of vertices $\mathcal{V}^*$ or edges $\mathcal{E}^*$ from $\mathcal{V}$ or $\mathcal{E}$, respectively. Sampling from vertices and putting only those edges (i, j) in $\mathcal{E}^*$ for which $i, j \in \mathcal{V}^*$ and $(i, j) \in \mathcal{E}$ always leads to a graph $\mathcal{G}^*$ with vertex degrees less than or equal to the vertex degrees of $\mathcal{G}$. This follows because the sampling may miss edges that are in $\mathcal{G}$ but were not included in $\mathcal{G}^*$. Taking an SRS of edges and then adding the vertices at the ends of those edges to form $\mathcal{G}^* = (\mathcal{V}^*, \mathcal{E}^*)$ can be readily made unbiased in many situations, because the edge is the appropriate sampling unit. For instance, in a financial transaction the appropriate sampling unit is not two vertices (say, banks) but a transaction between them. Likewise, with protein binding, it is more natural to regard the sampling unit as the bond, not the entities bonding. In the case of an SRS over edges, one can argue that bias can still be a problem because the sampling distributions over the collections of graphs generated by SRS over vertices and over edges can be very different. Although not discussed here, one advantage of SRS over edges rather than vertices is that its results can in principle be scaled up to give consistent estimators of the expected number of edges or vertices.

There are other methods for sampling graphs as a way to estimate them, and these other methods may be useful in specific circumstances. For instance, if sampling weights are known (which is rarely the case) then the Horvitz–Thompson estimator may be used to estimate means. This estimator is consistent and even has a closed form for its variance. Consequently, it can also be used as a predictor. Thus, if a property of a network can be expressed as an average of units (understanding that units may be vertices, edges, or motifs) then modifications of classical survey sampling (see Secs. 2.1 and 2.3.3 for a brief review and see Lohr (2010) for more details) can often be used to improve an estimator to make it consistent and thereby regard it as a predictor for a future outcome. The SRS-of-edges estimator and other estimators are studied in more detail in Kolaczyk (2009, Chaps. 4 and 5).

A different approach is taken via random graph models. The idea is that there is a finite collection of graphs and each can be the output of a graph-valued random variable. In the simplest case a collection of undirected graphs has n vertices, and the probability of an edge between any two vertices is $p \in (0, 1)$. It is assumed there are $\#(E)$ edges chosen randomly from the $\binom{n}{2}$ possible edges, so that each graph with $\#(E)$ edges has the same probability of being generated. Since each edge is a Bernoulli(p) outcome, this leads to a binomial likelihood, i.e.,

$$\ell(\mathcal{G}(n, p)) \text{ has } \#(E) \text{ edges} | p) = p^{\#(E)} (1 - p)^{\binom{n}{2} - \#(E)}. \tag{12.20}$$

More compactly, in terms of A, this is

$$\ell(A|p) = \prod_{i<j} p^{a_{ij}} (1 - p)^{1 - a_{ij}}.$$

The data are $A = a$, i.e, the collection of edges $\mathcal{E}$ indicated by the nonzero a's with $\mathcal{V}$ taken as the vertices those edges require. Obviously, if $p = 0$ then the vertices form n disjoint connected components and, if $p = 1$, the whole graph is a clique of size n. In between, more can be said: as noted in Goldenberg *et al.* (2009), there are three asymptotic forms for this random graph DG, and they depend on $\lambda = pn$. They are as follows.

1. $\lambda < 1$: In the limit of large n, the random graphs output by the probability will have no connected components bigger than $\mathcal{O}(\ln n)$.
2. $\lambda = 1$: In the limit of large n, there will be a subgraph of size $\mathcal{O}(n^{2/3})$.
3. $\lambda \to c > 1$: In the limit of large n, one connected component of the output graph will have a unique largest part (containing a positive fraction of the vertices) and no other connected component will have more than $\mathcal{O}(\ln n)$ vertices.

Since all edges are equiprobable and each of the n vertices can be linked to the other $n - 1$ vertices, the degree of any vertex is Binomial($n - 1, p$). Thus the mean degree of any vertex is $(n - 1)p$, which goes to ∞ for fixed p. Hence, $\lambda' = \lambda - p$, so that λ and λ' are nearly the same. However, the p-parametrization and the λ'-parametrization exhibit different limiting properties. Specifically, if λ' is not held constant while n increases, item #3 gives a dense graph sequence in the limit, whereas if λ' is held constant while n increases, p decreases so the graph gets sparser and sparser. That is, the distribution of the mean degree of a vertex (a binomial) converges in distribution to a Poisson(λ'). Thus, Poisson random graphs can be regarded as a limit of binomial random graphs; this is the case described under item 1 of the above list. This sort of property is familiar from normal approximation.

From (12.20), the log-likelihood can be simplified to

$$\ell(p) = \left(\sum_{i<j} \ln(1 - p) + a_{ij} \ln \frac{p}{1 - p} \right), \quad \text{i.e.,} \quad \hat{p} = \sum_{i<j} \frac{a_{ij}}{\binom{n}{2}}, \tag{12.21}$$

the MLE for p. In this case, the likelihood is maximized at the observed graph and, since the likelihood is from an exponential family, $\hat{p}$ is sufficient, consistent, asymptotically normal, and efficient.

These random graph models are mathematically tractable but are rarely seen in practice because they require all vertices to be symmetric and all edges to be equally important and independent, neither of which hold outside toy models. To construct a larger collection of graphs consider the following procedure; see Wasserman (2006). (i) Regard the absence or presence of each edge as a zero–one random variable. (ii) Specify a dependence structure on the edges. (iii) Generate a specific graph using (i) and (ii). (iv) Simplify any parameters by imposing constraints. (v) Estimate the parameters and assess the fit of the graph to the underlying network.

As a step toward implementing this strategy, regard A as a random matrix. Then the Clifford–Hammersley theorem gives the following statement:

$$P(A = a) = \frac{1}{\kappa} \exp \left(\sum_{E \in \mathcal{N}_{\mathcal{D}}} \lambda_E \prod_{\substack{(i,j) \in E}} a_{ij} \right), \tag{12.22}$$

where $\prod_{(i,j) \in E} a_{ij}$ is the suffucent statistic for λ_E, E ranges over the sets of vertices $\mathcal{N}_{\mathcal{D}}$ of the dependence graph $\mathcal{D}$ for A, i.e., the elements of $\mathcal{E}$ for the random variables in A whose edges represent pairs of the random variables assumed to be conditionally independent (given the values of all other random variables), and κ is a normalizing constant. In (12.22), $\lambda_E = 0$ when the subgraph induced by the vertices of E is not a clique of $\mathcal{D}$. The content of this is that the probability of a graph depends on the maximal cliques of $\mathcal{D}$.

The binomial graph or any graph with independently generated edges is trivially an example of (12.22). A simple but nontrivial example is the block model. The idea is that the vertices of $\mathcal{G}$ are divided into subsets (the blocks) $L_1, \ldots, L_k$ and that the edges still individually follow a Bernoullli(p). That is,

$$P(A_{ij} = 1 | Z_i = r, Z_j = s) = b_{rs}, \tag{12.23}$$

where $B = (b_{rs})_{r,s=1,\ldots,k}$ is called the affinity matrix. The event $Z_i = r$ means that vertex i is in block r. This requires that for, $n_r = \#(L_r)$, $n = \sum_{r=1}^{k} n_r$. The content of (12.23) is that whether or not the edge (i, j) is present depends on the blocks that contain its vertices. See Shalizi (2016) for more details. Exchangeable graphs are also possible; see Goldenberg *et al.* (2009, Sec. 3.3).

Given these examples and the fact that it is unlikely that they represent networks well, outside highly symmetric cases, it is important to be able to take one of these graphs, possibly found through maximum likelihood, and increase or decrease its edges or vertices to get a better answer. Given that the graph has been constructed, all four of these procedures can be regarded as predictions about whether a vertex or edge belongs in the graph. This can be regarded as a fit, a residual, or even an imputation; however, given that the graph is random and the task is to predict the presence or absence of a random quantity (an edge or vertex), it is more properly regarded as predictive. One likelihood-based procedure is given in Wasserman (2006).

Consider (12.21). It is the special case of a logistic regression but exemplifies the properties of many other distributions that might be used. The same applies to (12.23). Another important example is

$$p(a|\theta) \propto \exp \left(\theta_1 e(a) + \theta_2 t(a) \right), \tag{12.24}$$

where $e(a)$ is the number of edges in $\mathcal{G}$ and $t(a)$ is the number of triangles (although any motif could be used). In all these, and many other cases, likelihood methods can be used as the basis for prediction. For instance, let a_{+ij} and a_{-ij} be instances of A in which the edge joining vertex i to vertex j is present or absent, respectively. Now, generalizing

(12.20), (12.22), (12.23), or (12.24), write the likelihood as an exponential family, cf. Shalizi (2016):

$$p(a_{+ij}|\theta) = \frac{\exp(T(a_{+ij}) \cdot \theta)}{C(\theta)}, \qquad (12.25)$$

where θ is the natural parameter, T is the natural sufficient statistic, and $C(\theta)$ is the normalizing constant. This easily leads to

$$\ln \frac{p(a_{+ij}|\theta)}{p(a_{-ij}|\theta)} = \Delta_{ij} \cdot \theta,$$

where $\Delta_{ij} = T(a_{+ij}) - T(a_{-ij})$, essentially a logistic regression. Clearly, adding or deleting edges (and their vertices) can be used to increase the log-likelihood in order to give a prediction for whether an edge should be included.

A more sophisticated version of this procedure is to be found in Wasserman (2006). Recall that by definition $\mathcal{D}$ has vertices $\mathcal{V}_D = \{(i,j) \in \mathcal{V}, i \neq j\}$ and edges

$$\mathcal{E}_D = \{(i,j), (k,\ell) | A_{ij} \text{ and } A_{k\ell} \text{ are conditionally independent}\}.$$

Now assume that the bias in the observation $A = a$ leads to incorrectly omitting edges or vertices rather than incorrectly including them. Then

$$\mathcal{V}_c = \mathcal{V}_o \cup \mathcal{V}_m \quad \text{and} \quad \mathcal{E}_c = \mathcal{E}_o \cup \mathcal{E}_m,$$

i.e., the complete vertex or edge set is the (disjoint) union of the observed vertex or edge sets and the missing vertex or edge sets.

Suppose that the random matrix A is in fact a function of statistics of the unknown graph that are approximately normal and independent. This leads to a likelihood of the form (12.25) or of the form (12.22) (assuming that it is parametrized by θ, say), both exponential families. The Wasserman (2006) procedure is to use $\mathcal{D}$ to posit a distribution for A and then find the maximized likelihood

$$\ell(\hat{\theta}, A = a) = \arg\max_a \ell(\theta, a)$$

based on $\mathcal{V}_o$ and $\mathcal{E}_o$. Letting $U = \#(\mathcal{E}_c)$ and $W = \#(\mathcal{E}_o)$, the maximum number of possible edges is $\binom{U}{2}$, so the possible number of omitted edges is in the set $T = \{1, \ldots, \binom{U}{2} - W\}$. For $i \in T$, i.e., a possible edge in $\mathcal{E}_m$, let $a(i)$ be the modification of $A = a$ that includes it. Re-estimate θ, obtaining say $\hat{\theta}_i$, and find $L(\hat{\theta}_i, a(i))$. Now set $L_{\text{new}} = \max_i L(\hat{\theta}_i, a(i))$ and consider updating L_{new} by the addition of vertices from $\mathcal{V}_m$. Sequentially going through the addition of each vertex in $\mathcal{V}_m$ leads to $L_{\text{max}} = \max_{i \in \mathcal{V}_m} L_{\text{new}}$. This is the maximum likelihood over the addition of any vertex from $\mathcal{V}_m$ and any edge from $\mathcal{E}_m$. If $L_{\text{max}} > \ell(\hat{\theta}, A = a)$, update a to reflect the maximized likelihood; otherwise, default back to $\ell(\hat{\theta}, A = a)$. Essentially this method incrementally tests whether adding one vertex or edge improves the likelihood. It may be improved, especially for large graphs, by considering the addition of multiple edges or vertices at the same time.

Again, it is important to note that this is prediction in what is usually an $\mathcal{M}$-open context. It is possible to imagine settings in which the problem is simple enough e.g., basic metabolic networks, where some version of this sort of procedure will be $\mathcal{M}$-closed in the sense of uncovering a unique correct network when it exists. This can occur only in the presence of enough information – that a uniquely correct network exists and its constituent reactions can be listed along with the essential information about them, such as the enzymes, reducing power, ATP consumption, DNA coding capacity, etc., that they require to operate. However, even in these cases, it is known that enzymes are not perfectly specific to compounds, i.e., to an extent one enzyme from a given family of enzymes can catalyze multiple reactions with varying degrees of success, so the list of enzymes and compounds, roughly corresponding to edges and vertices, may not be as cleanly defined as it initially seemed to be. Thus, once the full complexity of the real problem is taken into account, it may turn out to be $\mathcal{M}$-complete, or effectively $\mathcal{M}$-open rather than $\mathcal{M}$-complete.

Directed graphs introduce more complications and make the limitations of using undirected graphs more obvious. Much of the work in this area comes out of social network theory, namely, the 'p_1' and 'p_2' models. They are also random-edge models, but the generation of the edges is dependent on the vertices. Following Goldenberg *et al.* (2009), consider vertices i and j and let $P_{ij}(u, v)$ be the probability of linkages between i and j. Specifically, no link between i and j corresponds to $u = v = 0$, a link from i to j corresponds to $u = 1$ and $v = 0$, a link from j to i corresponds to $u = 0$ and $v = 1$, and a bidirectional link from i to j corresponds to $u = v = 1$. The full p_1 model sets are

$$\ln P_{ij}(0,0) = \lambda_{ij},$$
$$\ln P_{ij}(1,0) = \lambda_{ij} + \alpha_i + \beta_j + \theta,$$
$$\ln P_{ij}(0,1) = \lambda_{ij} + \alpha_j + \beta_i + \theta,$$
$$\ln P_{ij}(1,1) = \lambda_{ij} + \alpha_i + \alpha_j + \beta_i + \beta_j + 2\theta + \rho_{ij}. \tag{12.23}$$

In (12.23), the λ_{ij} are normalizing constants for the four terms that can be associated with each pair (i, j), θ is a base rate for edge propogation, α_i is the 'expansiveness', expressing the effect of an outgoing edge from i, β_j is the 'popularity', expressing the effect of an incoming edge for j, and ρ_{ij} is the 'reciprocity', expressing a sort of interaction between the incoming and outgoing edges for a pair (i, j). The effect of ρ_{ij} is to increase or decrease the probability of an edge from i to j or an edge from j to i by a factor of ρ_{ij} over the probability of edges being generated independently.

As written, the model (12.23) is unidentifiable, in addition to being an obvious oversimplification; for instance, the probability of an edge between i and j may be unrelated to the probability of other edges. Correcting the first of these problems requires imposing constraints on the parameters. Several are used routinely. The most extreme is to set all α_i, all β_j and all ρ_{ij} to zero. If only $\theta \neq 0$ then each directed edge has the same probability of appearance and the model reduces to the directed edge version of (12.20). If all $\rho_{ij} = 0$ then the random generation of edges depends primarily on the in-degree and out-degree of the n vertices, essentially a variant on (12.24). Two other constraints that do not obviously reduce to the random generation of graphs as already presented are the cases where (i) all $\rho_{ij} = \rho$, a common value, and (ii) $\rho_{ij} = \rho + \rho_i + \rho_j$. These are harder to interpret in physical terms, but (ii) can be regarded as a form of edge-dependent reciprocation or as simply folding the reciprocity factor into the other parameters. The MLEs for (i) and (ii) were derived in Holland and Leinhardt (1976) and Fienberg and Wasserman (1981). The difference is that the change in the probability of reciprocal edges does not depend on the vertices in (i) but may do so in (ii).

In (i), the case of constant reciprocation, the p_1-likelihood for (12.23) can be written in exponential form. Specifically, if $A = a$ is the $n \times n$ adjacency matrix,

$$\ln P(a) \propto a_{++}\theta + \sum_{i=1}^{n} a_{i+}\alpha_i + \sum_{j=1}^{n} a_{+j}\beta_j + \sum_{i,j=1}^{n} a_{ij}a_{ji}\rho,$$

where $+$ denotes summation of the corresponding elements in a. Even though Holland and Leinhardt (1976) gave an iterative algorithm for finding the MLE (with some added constraints) in this context, a key issue is that standard asymptotics do not apply to these problems. Even when point estimates can be obtained, it is unclear how to generate SEs for future edges or nodes as in the Wasserman (2006) procedure – apart perhaps from some properly designed bootstrap procedure. The problem is that the numbers of parameters α_i and β_j increase with n, the number of vertices. To date, however, it seems that these theoretical problems have not been overcome, even in a pragmatic sense.

There is a Bayesian analysis of the p_1-model. Sometimes this Bayesian analysis is called a p_2-model because assigning priors to the parameters in (12.23) amounts to turning them into random effects; see Thiemichen *et al.* (2016) and the references therein for a recent treatment of this formulation that includes explanatory variables. The basic model is logistic, i.e., it proposes a functional form for $\ln P(A_{ij} = 1|\beta, \phi)$ given explanatory variables x, where ϕ is a collection of parameters corresponding to the α's and β's in (12.23), which are usually assigned an $N(0, cI_n)$ prior for some $c > 0$. (The other parameters are set to zero.) The estimation of the parameters follows from MCMC–MH; see Sec. 9.6. So, given the posterior for the parameters and the model for a new vertex, in principle the mode of the predictive distribution over possible edge and vertex additions and deletions can be found and used as a prediction for a possibly new graph modeling a given network summarized by $A = a$. Again, this line of inquiry seems not to have been pursued yet.

The concept of a 'power law' arises in some aspects of graph analysis for networks. The basic idea is to back away from identifying the architecture of a graph and merely try to model one feature of it, namely, its degree distribution. Recall that the degree of a vertex in an undirected graph is the number of edges meeting it. Digraphs have an in-degree and an out-degree, but this will not be considered here, even though there is an analog to the present reasoning. The distribution of the degrees of a graph, therefore, is worth studying since in the models so far the graphs have been randomly generated. The next result is that the distribution for the degrees of the graph can be used to predict what fraction of the n vertices have degrees within a given range. This does not identify the graph fully but can be a useful quantity for prediction in various settings, e.g., to answer questions such as: are there a few regions of the graph that are highly connected, like cliques, amid a sea of low connections (think of hub-and-spoke type graphs) or are the vertices roughly equally well connected to each other, so that the graph is relatively uniform in its connections?

Graphs that satisfy a power law tends to be those with degree distributions that are strongly right-skewed but do have a finite second moment. Unfortunately, in practice, power law distributions do not fit all that well – other heavy-tailed distributions such as log-normals or negative gammas often fit as well. That is, if the 'true' distribution is P and d is a distance on distributions then any reasonable neighborhood such as $\{Q|d(P, Q) < \epsilon\}$ will contain many equally good models, which may give different predictions. This high model uncertainty remains to be resolved – either decreased or regarded as a feature of the DG that must be accommodated somehow.

To be more formal, a power law is just the subject-matter term for a Pareto distribution. In the vertex-degree context, it corresponds to the assumption that vertices with more edges are likely to attract more edges than vertices with fewer edges. This follows from the derivation of the Pareto distribution using the simple differential equation and solution $y'(t) = \alpha y(t)$ and $y = y_0 e^{\alpha t}$, for some $\alpha > 0$, where $y_0 = y(0)$ is the initial condition. If objects, in this case edges, appear at a uniform rate λ then the arrival times are Poisson(λ) and it is well known that the waiting time is $T \sim$ exponential(λ), which has the memoryless property. Now, a straightforward calculation (see Shalizi (2016)) verifies that, for $y_0 \leq k \in \mathbb{Z}^{\geq 0}$,

$$P(Y \geq k) = \left(\frac{k}{y_0}\right)^{\lambda/\alpha},$$

a Pareto, notable for its heavy tails. If F is the DF for a Pareto then $1 - F(k)$ is maximal at y_0 and falls quickly, tailing off slowly as k increases. A log–log plot of a Pareto is a straight line with slope $-\lambda/\alpha$.

This derivation implicitly assumes continuous time and that n, the number of vertices, is so large relative to the number of edges that, even though the limit as the number of edges goes to infinity is formally undefined, the limiting behavior becomes apparent. Thus, in these simple cases, it is not unreasonable to estimate λ and α and extract predictions about the degree distribution of the vertices from the Pareto. (The SE of the Pareto can also be used to get approximate PIs.)

To be more formal still, suppose that Y, the degree for a vertex in a randomly generated graph, is Pareto(α), i.e., $P(y) \propto y^{-\alpha}$. Then, it is not hard to show that, for some $C > 0$, $P(Y \geq y | Y \geq z) = Cy^{-\alpha+1}$. This is the scale-free (or self-similar) property. Scale-free does not mean the same as memoryless; the exponential is memoryless but not scale-free. Pre-2000 or so, Pareto distribution parameters were often estimated by the simple approach of binning the data, writing $P(b)$ for the proportion of data in bin b, and regressing $P(b)$ on $C - \alpha \ln b$ to get an estimate for α. The deficiencies of this approach are obvious: regression assumptions do not hold and, even when $\hat{\alpha}$ is consistent, the approach is inefficient.

Contemporary treatments assume that y_0 is known and are likelihood based. The log-likelihood and MLE are

$$\ln L(\alpha; D) = \prod_{i=1}^{n} c(\alpha, y_0) y_i^{-\alpha} \quad \text{and} \quad \hat{\alpha} = 1 + \frac{n}{\sum_{i=1}^{n} \ln y_i},$$

where the data $D = \{y_1, \ldots, y_n\}$ comprise the set of degrees of the n vertices, y_0 is the smallest observed degree, and $c(\alpha, y_0)$ is a normalizing constant that depends on α and y_0. If the DF were known (or conjectured) then the Kolmogorov–Smirnov test could be used to compare it with the $\hat{F}_n$ from the MLE procedure. However, F is rarely known, so a bootstrap approach is often used. That is, a candidate F can be found by bootstrapping from $A = a$ to form multiple $\hat{F}_n$ from the MLE procedure that can be averaged to give a candidate F. Alternatively, the EDF F_n can be used in place of F. A good check on this procedure is to compare the final model with other heavy-tailed distributions using the BIC or other assessments of model fit. It is also worthwhile verifying that the point predictors and the PIs from the numerous variations on this sort of method are not unacceptably different – or seeking explanations if they are. At the present time, it is difficult to be more precise about which methodologies will prove effective for specific classes of applications. This is an obvious direction for future work, notwithstanding contributions such as Fosdick *et al.* (2016), Crane and Dempsey (2016), Lovász (2012), Snijders (2011), and Easley and Kleinberg (2010) amongst others.

Another prediction problem associated with networks is to predict the graph that represents them. This is a type of model-selection problem except that the model (the graph) is random. Given such a graph predictor, the Wasserman (2006) procedure described earlier can be used to refine a graph as well. It remains to identify a satisfactory graph predictor. One class of predictors that has been proposed and studied comprises adaptions of cross-validation (CV) to the graph context. As earlier, the vertices are assumed fixed and known while the edges are random. Three forms of CV are given in Hoff (2008), Chen and Lei (2015), and Dabbs and Junker (2016); they differ in the exact way in which the 'folds' in K-fold CV are defined. Their basic structure has three steps, cf. Sec. 9.4. First, partition edges $A_{ij} = a_{ij} \neq 0$ into K disjoint folds $V_1, \ldots, V_K$. This leads to a fold-assignment matrix $T = (t_{ij})_{i,j=1,\ldots,n}$ defined by $t_{ij} = k \Leftrightarrow a_{ij} \in V_k$. Second, for each fold k define the training set to be $A_{(k)} = \{a_{ij} | t_{ij} \neq k\}$ and, for each a_{ij} in fold k ($t_{ij} = k$), estimate p_{ij} by $\hat{p}_{ij} = \hat{p}_{ij}(A_{(k)})$, for instance, using one of the likelihood methods already discussed. Third, compare $\hat{p}_{ij}$ with a_{ij} within each of the K folds, for example by a predictive risk of the form

$$\hat{R}_{CV}(A, (\hat{p}_{ij})_{i,j=1,\ldots,n}) = \frac{1}{n(n-1)} \sum_{k=1}^{K} \sum_{a_{ij}|t_{ij}=k} L(a_{ij}, \hat{p}_{ij}), \qquad (12.24)$$

where L now denotes a loss function. The graph that minimizes (12.24) over the collection of graphs under consideration is the CV-based predictor.

To specify fully the minimum requires L, K, and a method for assigning edges to the folds. One natural choice for L is the squared error, in which case (12.24) reduces to

$$L_{MSE}(A, (\hat{p}_{ij})_{i,j=1,\ldots,n}) = \frac{1}{n(n-1)} \sum_{i \neq j} (a_{ij} - \hat{p}_{ij})^2, \qquad (12.25)$$

which can be regarded as a prediction error. The L_{MSE} has a well-known variance–bias decomposition (to see this, add and subtract p_{ij} and $E(a_{ij})$). If the L_{MSE} is large, a variance–bias may help identify why a particular collection of estimates $\hat{p}_{ij}$ is poor. The number of folds K lies between $K = 1$ (leave-one-out CV), which is known to be poor, and $n/2$. In conventional model-selection settings, Shao (1993) suggested using as small a number of folds as possible, so that as much data as possible can be used for testing, although in practice the choice of K often makes little difference within a 'reasonable range'. Expressions (12.24) and (12.25) are direct generalizations of CV error; meaningful skewness in the histograms of the terms in either might make robust CV or median CV more applicable; see Yu and Clarke (2015).

Unsurprisingly, the way in which edges are assigned to folds can have a large effect on (12.24). At least three ways that have been proposed. One, called 'random CV' (Hoff

2008), obtains a fold assignment matrix by choosing uniformly at random among all possible fold assignment matrices with an equal number of edges in each fold. This ensures that the matrices have the same number of edges but does not ensure that the vertices of the folds are comparable. A second way to perform network CV (NCV) was proposed by Chen and Lei (2015). They used an extra fold ($k = 0$) to contain edges that could not be unambiguously assigned to one of the $k = 1, \ldots, K$ folds. That is, suppose that the vertices are assigned to folds randomly. Then, if two vertices v_i and v_j are in the same fold, a_{ij} is assigned to that fold; otherwise, a_{ij} is assigned to fold $k = 0$. Thus, $t_{ij} = k$ when $v_i, v_j \in V_k$ and $t_{ij} = 0$ otherwise. When $t_{ij} = 0$, the edge between v_i and v_j is assigned to every training set and never used in (12.24) to evaluate the performance of a graph. The benefit of this procedure is that each training set contains all the edges between the vertices in it; the cost is that the test sets are typically smaller than they would otherwise be.

A third way called 'Latin CV' was proposed by Dabbs and Junker (2016). The idea behind Latin CV is to overcome the fact that fold $k = 0$ in NCV reduces the amount of data available to the CV. Recall that in experimental design a Latin square is a two-factor design in which each level of each factor is assigned to exactly one treatment condition. If the levels of the two factors are taken as the rows and columns of a matrix and the entries in the matrix are taken as the treatment assignment, the Latin square design is any matrix in which each row and column has exactly one occurrence of each treatment. If the folds are regarded as treatments then the vertices are the levels. Assuming that there are more vertices than folds, the collection of Latin square matrices consists of all matrices with equal occurrences of each fold assignment along each row and along each column of the matrix. Thus, to form a Latin CV, begin with a fixed fold-assignment matrix T where each row and column of T has an equal number of occurences of each fold. Permuting the rows and columns of T independently in all possible ways under a discrete uniform distribution leads to a random Latin CV fold-assignment matrix.

Dabbs and Junker (2016) provided a comparison of these three CV methods in the block model context. Since block models are parametric, Dabbs and Junker (2016) also included AIC and BIC (see Sec. 9.2) in their comparisons. Their major finding about CV was that methods that use all possible edges for testing, namely, random CV and Latin CV, outperform methods such as NCV that do not. They also argued that AIC, BIC, and other methods ('infomap' and 'common community detection') often lead to overfit, and the overfitting becomes worse as the graph size increases. They also noted that random CV and Latin CV extend to other graph-theoretic models besides the block model, and it is likely that the principles they identified generalize to give good predictors for these more complex settings.

As can be seen, these methods do not, strictly speaking, require the network problem to be $\mathcal{M}$-closed. In fact, they can generally be used with $\mathcal{M}$-complete and $\mathcal{M}$-open problems, in the sense that they give answers even if the quality of those answers cannot be assessed very well, if at all. Indeed, despite the results in Chen and Lei (2015), there seems to be little theory to guide prediction in network models even for $\mathcal{M}$-closed problems. More difficult problem classes can be expected to be even more recondite. Even for edge prediction the value of p might be small, in which case it is unclear how important it is to include a particular a edge from a practical prediction standpoint. This adds an extra layer of difficulty to $\mathcal{M}$-complete problems on top of the inexpressibility of the true model. $\mathcal{M}$-open problems are correspondingly more difficult.

12.4.2 Dynamic Networks

A probabilistic model for dynamic networks is a stochastic process, say $Y(t)$, where $Y(t)$ is the value of the process at time t. The values of the process are graphs; the time can be continuous or discrete. Virtually all graph models for dynamic networks are based on Markovity, either directly in that Y is a Markov process or indirectly in that Y is a hidden Markov process, i.e., a function of a Markov process. Often a distinction is made between networks of states and networks of events; see Snijders (2011). A network of states represents a state as being between two vertices (connected by an edge or not) whereas in a network of events a vertex is an event that may lead to other events/vertices over time. That is, the vertex represents an occurrence rather than an agent. Often, networks of states can be represented by Markov or hidden Markov processes; networks of events are usually better represented by hidden Markov processes. In either case, over time, the edges or vertices may be dependent on each other and some edges or vertices may be added or dropped. Therefore the network data is longitudinal and the models must be, too. Thus $\mathcal{G}_t = (\mathcal{V}_t, \mathcal{E}_t)$ and the task is to determine which properties of $\mathcal{G}_t$ are useful for predicting structures in $\mathcal{G}_s$ for $s > t$.

Most work to date has focused on static networks rather than on dynamic networks–because of the added complexity and dearth of data sets to analyze; see Goldenberg *et al.* (2009). This situation is in the process of changing and some would say it already has done so; see Zhang *et al.* (2016) for an example. However, dynamic networks have tended to focus on states rather than events, overall, and many are straightforward generalizations of static graphs to a dynamic context.

For instance, the basic random graph with $\mathcal{V}_t = \{v_1, \ldots, v_n\} = \mathcal{V}$, where the v_i are independent of t, generates a dynamic process: start with $\mathcal{V}$, let t be discrete (for instance, integer valued). At each time step add a new edge to $\mathcal{G}_t$ i.e., to $\mathcal{E}_t$, with probability $p = \#(\mathcal{E})/\binom{n}{2}$, where $\mathcal{E}$ is a fixed number of edges from which to choose. This process can be extended to allow the random inclusion of more vertices; the key condition is to ensure that an edge is not added at time t unless both its vertices are already in $\mathcal{G}_t$. This model is so simple that it does not allow for the removal of edges or vertices, although again it is obvious how it can be extended to permit this. (For instance, any edge can be removed in a time step, with a fixed probability, but a vertex can be removed only when all edges connecting to it have been removed.) Durret (2006) provided a discussion of these cases and their links to random walks and branching processes. One criticism of this approach is that the graphs generated are not scale-free, i.e., the vertex-degree distribution does not satisfy a power law, whereas most networks encountered in real problems are in fact scale free. So while this approach may lead to edge or vertex prediction (of addition or deletion) it is not likely to be a very useful predictor outside special cases.

One variant on this procedure that does generate scale-free graphs is called the dynamic preferential attachment (DPA) model. The key idea is to start with n_0 unconnected vertices at time $t = 0$ and add a new vertex with $m < n_0$ edges joining it to the other vertices. The edges are not added using a discrete uniform distribution but are added using a discrete uniform distribution at time t. That is, an edge is added between the new vertex and an existing vertex on the basis of the degree of the existing vertex. This is a multinomial distribution with, say, n_t urns (vertices), each urn having probability $p_i = \deg(v_i)/\sum_{j=1}^{n_t} \deg(v_j)$, where $\deg(v_j)$ is the degree of v_j. This is preferential because vertices that already have high degrees are favored for more edges than vertices with low degrees. This can be used to identify the most probable graph to occur at the next time step and in principle can be allowed to run over many time steps to give an overall prediction of the best graph.

Block models, see (12.23), have also been generalized to give dynamic graphs. As with the earlier two models, $V_t = V$ continues to be taken as time independent. Following Zhang *et al.* (2016), partition n vertices into k blocks and assign each block a parameter θ_i. It is assumed that the parameters are constrained for the sake of identifiability by

$$\sum_{j=1}^{n} \theta_{L(v_j)} \delta_{L(v_j),r} = 1,$$

where $L(v_j)$ is the block to which v_j belongs, $\delta_{L(v_j)r} = 1 \Leftrightarrow L(v_j) = r$, and r is an index over blocks. There are k such constraints. Generate an initial state drawn from a static block model and then generate a trajectory of graphs over time by adding edges between each pair of distinct vertices (v_i, v_j) at a Poisson rate $\lambda_{ij} = \mu_{rs}\omega_{rs}\theta_i\theta_j$ and remove existing edges independently at a rate μ_{rs} where $r = L(v_i)$ and $s = L(v_j)$. In this formulation μ_{rs} is related to the mean of the Poisson and ω_{rs} controls the density of edges between and within blocks. Zhang *et al.* (2016) provided an extensive discussion of how to use and interpret this class of dynamic graphs.

The basic random-graph model can also be extended in several ways to discrete-time Markov chains with probabilities in exponential form somewhat like (12.24) and (12.25). Following Goldenberg *et al.* (2009), write

$$P(A(t) \mid A(t-1)) = \frac{1}{c} \exp\left(\sum_{k-1}^{K} \beta_k s_k(A(t), A(t-1))\right), \qquad (12.25)$$

where $A(t)$ is the adjacency matrix at time t, the s_k are sufficient statistics, the β_k are the natural parameters, and $c = c(\beta_1, \ldots, \beta_K)$ is the normalizing constant. The s_k can be of various forms, for instance, the density of edges over the n vertices at time t so that $s_1(A(t), A(t-1)) = (1/(n-1))\sum_{ij} a_{ij}(t)$ (independently of $t-1$). Other choices for s could indicate the stability of the graph over time (e.g., how much it changes), reciprocity (e.g., how the edges from vertex i to vertex j compare with those from vertex j to vertex i), and transitivity (i.e., how likely it is that an edge from vertex i to vertex k is followed by an edge from vertex k to vertex j); see Goldenberg *et al.* (2009). As before, this can be used for edge prediction and over time this amounts to predicting the entire graph, assuming that the vertex set does not change. It is likely that these models can be generalized to accommodate additions and deletions to V_t, but this does not seem to have been investigated.

An extension of the random-graph model to continuous-time Markov chains is also possible although less tractable. The key point is to write the transition probabilities in a useful form. Suppose that the transition probabilities between any pair of times s and t with, say, $s < t$, depend only on $t - s$, so that $P(A(t-s)) = P(A(t) = a(t)|A(s) = a(s))$ for realized adjacency matrices $a(t)$ and $a(s)$ representing the edge configurations of a network with n vertices. Then it can be proved that $P(t) = e^{tQ}$ for some fixed Q called the *intensity* matrix. This can be written as $Q(A, A') = (q(A, A'))$ and the elements q can be regarded as derivatives of the conditional probabilities, in the sense that

$$P(A(t+\epsilon) = A^* \mid A(t) = A^{\dagger}) \approx \epsilon q(A^{\dagger}, A^*),$$

where ϵ now denotes a small positive increment in time. This leads to a series of models roughly parallel to those in (12.23). The natural predictor at a given time adds the edge that maximally increases the probability on the left-hand side. Doing this repeatedly should lead to stationary graphs that would be good overall predictors, but this approach does not seem to have been pursued.

In addition, a feature of continuous-time Markov processes is that it is possible to distinguish between edge- and vertex-oriented dynamics; see Snijders (2006). In both cases the intensity matrix can be factored: one factor represents the possible changes and the other factor represents the likelihood of a change. That is, the single continuous-time Markov process is regarded as two subprocesses. The first process remains in continuous time, determining when a change occurs, and the other process determines the probability of a specific discrete event occurring. This is much like a point process with finitely many possible jumps. In all these cases, edge prediction and overall graph prediction are possible but do not seem to have been developed. Nevertheless, a brief description of the setting is helpful.

Edge-oriented dynamics are similar in appearance to exponential models. Let $y(i, j, z)$ be a function of the graph that indicates the presence or absence of an edge between vertices v_i and v_j by $z = 1$ or 0, respectively. Then write $q_{ij}(y) = \rho p_{ij}(y)$, where

$$p_{ij}(y) = \frac{\exp\left(f(y(i, j, 1 - a_{ij}))\right)}{\exp\left(f(y(i, j, 0))\right) + \exp(f(y(i, j, 1)))}$$

This means that in edge-oriented dynamics each edge follows an independent Poisson process, so that the waiting time until an event occurs has an exponential distribution with parameter ρ. In this version of the model, the graphs are assumed directed and the event of an edge from v_i to v_j is that the direction is reversed with probability $p_{ij}(y)$. The function f appearing in the definition of p_{ij} is called a *potential* function but in fact is usually taken as a linear function of the network statistics of the form

$$f(y) = \sum_{k=1}^{K} \beta_k s_k(y),$$

for K statistics s_k. Goldenberg *et al.* (2009) listed several common choices for s_k and discussed some aspects of using these models.

Vertex-oriented dynamics define an intensity matrix by $q_{ij}(y) = \rho_i p_{ij}(y)$, where the ρ_i are parameters for independent Poisson processes determining the edge-change opportunity for the vertices (indexed by i). The probability function is

$$p_{ij}(y) = \frac{\exp(f_i(y(i, j, 1 - a_{ij})))}{\sum_{\ell \neq i} \exp(f_i(y(i, j, 1 - a_{i\ell})))}.$$

Since the edge changes now depend on the vertices, each vertex has its own potential function, of the same form as before apart from an extra index. Thus,

$$f_i(A(t)) = \sum_{k=1}^{K} \beta_k s_{ik}(A(t)).$$

The difference between this and the function f above is that with vertex-oriented dynamics only the edges coming into or going out of v_i affect the potential. Thus, at vertex i, the edge change that leads to the greatest-increase in potential is probabilistically favored over the other edge changes. (Edge- and vertex-oriented dynamics were explored in Snijders (2006) but are difficult to work with and thus are omitted here.)

A large number of analyses can be performed on dynamic networks for nonpredictive inferences; see, for instance, Aggarwal and Subbian (2014) for a review. However, here it will be enough to present the idea of how a predictive analysis should proceed. Recall the basic structure of a linear dynamical system: $x(t)$ is a vector of length n and

$$\exists \text{ an } n \times n \text{ matrix } B : \frac{\mathrm{d}}{\mathrm{d}t} x(t) = B x(t).$$

In discrete time, the analogous equation is $x(t+1) - x(t) = Bx(t)$, which gives $x(t+1) = Cx(t)$ where $C = B + I_n$. When C is amenable, say, it has n orthogonal eigenvectors w_i with eigenvalues λ_i, then for $t = 1, 2, \ldots,$

$$x(0) = \sum_{i=1}^{n} \langle x(0), w_i \rangle w_i \quad \text{and} \quad x(t) = C^t x(0),$$

which, over t time steps, leads to

$$x(t) = \sum_{i=1}^{n} \lambda_i^t \langle x(0), w_i \rangle w_i.$$

Now regard the n dimensions of x as the vertices of a graph and consider the linear dynamical system defined by

$$\frac{\mathrm{d}}{\mathrm{d}t} x_i(t) = r \sum_{j \neq i} a_{ij}(x_j - x_i), \tag{12.26}$$

where $x_i(t)$ is the time evolution of a quantity at vertex i. The interpretation of this dynamical system is that each x_i looks at the corresponding quantities of its neighbors, the x_j with $a_{ij} = 1$, at time t and if $x_j > x_i$ then x_i is incremented at a rate $r > 0$. Otherwise it is decremented at the rate r. Over enough iterations, the x_i at vertices i should even out so that the limit, $x_i(\infty)$, is constant as a function of i.

Expression (12.26) can be written equivalently as

$$\frac{\mathrm{d}}{\mathrm{d}t} x(t) = -rLx(t),$$

where $L = D - A$ with $D = \mathrm{diag}(\deg(x_1(0)), \ldots, \deg(x_n(0)))$, and the $n \times n$ matrix L is called the graph Laplacian. For undirected graphs, the graph Laplacian is symmetric, so that all eigenvalues exist and are real numbers. One eigenvalue is zero since $L\mathbf{1} = 0$ and $D_{ii} = \sum_j a_{ij}$. The eigenspace of the eigenvalue zero will have dimension equal to the number of connected components of the graph.

The graph Laplacian can be used to partition the graph into subsets of vertices that are like modules, in the sense that there should be more connections within the modules than between the modules. As noted in Shalizi (2016), suppose that a graph has one connected component. Then, $\mathbf{1}_n$ is the eigenvector with eigenvalue zero and the next largest eigenvalue must be $\lambda_2 < 0$ with eigenvector w orthogonal to $\mathbf{1}_n$. That is, $0 = \langle w, \mathbf{1}_n \rangle = \sum_i w_i$. Hence w has both positive and negative entries and provides a division of the graph into the two parts having the least flow between them in the sense of (12.26). Further eigenvectors continue to partition the graph into smaller subsets with minimal pairwise flow between them; this is the same principle as is used in spectral clustering.

If L is used to define the time evolution of the graph then $x(t) \to \alpha \mathbf{1}_n$ as the dominant eigenvector, and the values at the vertices will become equal. However, the longer a component of the partition of the graph by the eigenvectors lasts, the more reasonable it is to regard the component as a real module, i.e., a collection of vertices that somehow belong together because they have high and persistent connectivity amongst themselves and lower connectivity to other vertices. Thus, this process can be used to predict which vertices are related to which other vertices and therefore to predict which components will persist, i.e., taken together, they give a prediction for the graph at the next time step. Note that the graph Laplacian is not used as a model for the generation of graphs; it is only used to identify components that can be assembled to provide a prediction for the graph at the next time step. Just as spectral clustering is effective only for simple clustering problems, so it is expected that this approach to predicting graphs will be successful only for simple graph trajectories (although this does not seem to have been explored extensively). However, for simple $\mathcal{M}$-closed problems this procedure may be adequate. It also gives answers outside $\mathcal{M}$-closed problems but here, too, the usefulness of these answers as yet seems unexplored.

12.5 Multitype Data

The multitype or multisource data problem (or 'integrative data analysis', as it is now sometimes imprecisely called) dates in its contemporary form to the late 1980s. The reason is that in the 1970s neural nets were recognized as a poor model for real neurons but often as a very good technique for nonlinear regression, particularly in fields where multitype data were common; see Gong (1996) and the references therein for one example. The thinking was that there were enough parameters even in moderate sized NNs that the interrelationships among different data types could be captured effectively. There was some success with this, but fundamentally even NNs were generally not the right structure within which to analyze multitype data in order to make predictions. For about two decades the multitype data problem was a niche interest in terms of research even though its fundamental importance was admitted across various application domains.

The early thinking about multitype data as encapsulated in NNs, trees, or kernel methods was simple: ignore the specific features of each data type and throw all the data together into one model big enough to accommodate them. This left unaddressed the problem of how to combine supervised problems such as classification or regression with unsupervised problems such as clustering. In any event, treating all data types in the same way in a model ignored the specific features of the data types and meant that a lot of information was not available for the analyses, making them informationally weaker than they might have been.

One obvious alternative to this was to analyze each data type separately and combine the analyses in some way, possibly by an ensemble method. While this approach would enable the analyst to derive all the information within each data type, for the most part it did not make full use of the dependences between data types and hence sacrificed inferential power. Again, this ignored how to combine supervised and unsupervised data and these analyses were informationally weaker than they could be.

More recent approaches will be discussed in Sec. 12.5.2.

The goal in this class of problems is to include all the information in each data type as well as the dependences among data types so as to get the best analysis possible. Thus the problem as discussed here is restricted to combining data over a common set of subjects. There are multiple measures on each subject and the measurements segregate into naturally disjoint classes. The paradigm example of this comes from biostatistics: physicians will generally have demographic and health history data on subjects prior to gathering any other data. The new data gathered could be image data, diagnostic test data, and any of a large class of genomic data types. It is often known, or at least believed, that some demographic or test data variables are important; however, if all the data is combined as in early methods, model-selection techniques can easily eliminate the demographic and test data in favor of the genomic data simply because the dimension of the genomic data is so large. On the other hand, analyzing each data type separately is known to be inadequate because there are relationships between, say, demographic data and test data, e.g., the interpretation of a medical test may depend on the age or sex of a human subject.

The multitype data problem differs from the meta-analysis problem of combining different data types (or the same data type) over disjoint sets of subjects. Meta-analysis has been studied extensively, especially in biomedical statistics, and is much easier; see Brockwell and Gordon (2001) and Higgins and Thompson (2002) as examples. In these contexts, in principle the likelihoods from the various experiments can be combined and the analysis redone. However, meta-analysis is often carried out by looking only at the p-values or other summary outputs from a set of experiments. In some cases, meta-analysis is regarded as a type of integrated data analysis, but here the focus will be on multitype data analysis (which in principle could be fed into a meta-analysis if the subjects were not all the same), as dealing with the multiple types of data, in general, remains an unsolved problem.

One reason why the multitype data problem has seen new activity comes from the increase in new data types, especially in genomics and remote sensing. Having discussed sensor data in two contexts (Secs. 12.2.2 and 12.3.2), it is important to consider some typical genomics (or simply 'omics) data briefly, along with the usual single-data-type analyses that are often done to provide some context, before discussing methods that might yield improved prediction using multitype data.

12.5.1 'Omics Data

A distinctive feature about 'omics data is that the data are often collected to study the entire genome of an organism as opposed to a single gene or other region of a chromosome or piece of genetic material. The genome of an organism may include coding and noncoding DNA from the nucleus, mitochondria, or other substructures of a cell or other organism such as a virus. It also includes messenger RNA. In addition to being the first stage of transcription of DNA in eukaryotes – and hence a natural way to study the genome – RNA is also the genome of some viruses.

Loosely the various types of genomic data can be divided into three groups: (i) those that provide direct measurements on the genome; (ii) those that are defined by the type of compound they represent; and (iii) those that are defined by their function in a cell or other organism. Data types in the first group include single-nucleotide polymorphisms, methylation data, and other sequencing data from DNA whether Sanger or 'next generation'. The second group includes proteomics, metabolomics and lipidomics, the name indicating the type of compounds represented. The third group includes the transcriptome, translatome, and secretome (proteins that are secreted). This is of course only a partial list and it can be safely assumed that the technology required to gather high-quality genomic data may be strikingly different from type to type.

Pooling the information in these data types is one way to try to understand the overall function of the genome and hence is one way to answer important questions such as: how does the proteome (the proteins made by the genetic material) interact with the regulome (the regulatory structures that are also often made by the genetic material) to generate the trans-latome (the collection of compounds that the genetic material finally produces)? These sorts of question are often more varied and focused for diseases that have a genetic component such as cancer or depression.

One example of a data type from group (i) is single nucleotide polymorphisms (SNPs, pronounced 'snips'). Consider a population of individuals so genetically similar that their DNA strands line up respectively from individual to individual. Oversimplifying, suppose that, at a specific number of nucleotides from one end of one DNA molecule, the majority of individuals have an A–T pair of nucleotides. If a small fraction of people (say 10%) in the population have a C–G pair of nucleotides, it is called a SNP because it is regarded as a change in the consensus base pair of the DNA. This is detected by knowing in advance what the consensus sequence of base pairs is for the population – sequencing is not 'error free' so enough individuals have to be sequenced a sufficient number of times that a consensus sequence is available. This presumes that the DNA being searched for a SNP has been sequenced often enough that the SNP will be accurately detected – again because sequencing is not error free. Since human DNA is in chromosomes that are paired, an individual may have a given SNP in zero, one, or two chromosomes in a given pair.

The typical question asked with SNP data is whether a SNP or collection of SNPs has any biological effect. In some cases this can be addressed by some kind of regression analysis using the presence of the SNP as a dummy variable. Examples of this include determining how a SNP or a group of SNPs affect a response such as yield in plants or susceptibility to disease in humans. Analyses of this sort can be directly evaluated in a predictive sense. An alternative way to address this question is to separate the population into two groups on the basis of some other characteristic, so that each group has a 1×3 contingency table for the number of zeroes, ones, and twos of a given SNP among the members of the group. Together this gives a 2×3 contingency table for each SNP. Now a χ-squared test, Fisher's exact test, or other contingency table test can be used to determine whether the SNP profiles of the two subpopulations are different. As with micro-array data, this leads to multiple comparison problems when two or more SNPs are involved. These can be evaluated predictively but it is not as straightforward as with regression techniques. A more detailed overview of SNPs and their analyses can be found in Kirk *et al.* (2002).

A second type of group (i) genomic data is methylation data. Like SNP and microarray data it is biochemically complicated to generate, so only a very simple description will be given here; see Bock (2012) for a more detailed review. Methylation data rests on the idea that replacing an H on a cytosine (C) in the DNA by a methyl group (CH_3) makes a segment of DNA hard if not impossible to transcribe. That is, the methylation of C inhibits transcription. In normally functioning cells, some C's, for instance 5-MeCs, are methylated (by DNA methyltransferase) as part of the regulation of their function while other C's are not. The question is how to distinguish between these two forms of C. The process for answering this question is to treat DNA with bisulfite. Bisulfite de-aminates the unmethylated C's, converting them to uracil (U) nucleotides, but does not affect the 5-MeC's. Thus, it is as if there are five nucleotides rather than four because some C's have become U's. Then, just as it is possible to sequence DNA or RNA, it is possible to sequence the bisulfite-treated DNA. The result is a long string of A's, T's, 5-MeC's, U's, and G's.

Usually it is not the sequence that is analyzed directly but fractions of the form $\rho = \#(U)/(\#(U)+\#(mC))$, where $\rho \in [0, 1]$, for various segments of the DNA. When ρ is small for some segment, it suggests that transcription of the segment is not very active. When ρ is large, near unity, it suggests the segment is actively transcribed. Often ρ is calculated for a variety of segments of the DNA, usually disjoint, and possibly the entire DNA molecule.

It is often difficult if not impossible to go into a living organism to add or remove methyl groups to or from the C's to see what happens, so the most common sort of experiment is to obtain methylation data on a subject, apply a treatment to the subject, and then obtain a second set of methylation data for the subject. The result is a collection of paired fractions (before and after) for each subject. The simplest analysis that can be done on this sort of data is to look at the difference of the paired fractions marginally (on a fixed segment of DNA) for all n subjects and use a test for the difference of proportions of U's in the region. This can be done for many segments to determine which parts of the DNA have been activated or deactivated by the treatment. Again there are multiple comparison problems, so often the false discovery rate (FDR) is used instead of regular p-values. Predictive evaluations for inferences with methylation data are straightforward: use the data to form a classifier or other predictor and evaluate it for new subjects or, invoking the Prequential Principle, evaluate it on permutations of the existing subjects.

A common way to generate group (i) genomic data is next-generation sequencing (NGS), seen briefly in Sec. 1.2.2. The classical Sanger sequencing of DNA takes a long time because it proceeds methodically and sequentially, identifying one nucleotide at a time. The idea behind NGS is to parallelize sequencing. Thus, the genetic material is cleaved into relatively short segments called reads; sometimes these are as short as 35–50 base pairs although longer lengths are often used, too. When the amount of genetic material (DNA) available is insufficient, it is often increased by polymerase chain reaction (PCR) techniques that cause more of it to be made. (When NGS is used on RNA, a modification of PCR, reverse transcriptase PCR, is necessary; the reverse transcriptase turns the RNA into DNA to which PCR can be applied.) Then, through another procedure, 'adapters' are attached to each end of the reads. One adapter attaches the read to the surface of a slide; the other adapter stops the generation of a complementary segment of nucleotides. The slide is then flooded with nucleotides and DNA polymerase, so that many copies of the reads are generated. This happens one base at a time, so that each time a nucleotide is added it can be identified. Loosely, the nucleotides are fluorescent, each nucleotide being given a different color. So, each time a nucleotide is added the process stops, a picture is taken, and the just-added nucleotide is identified. Then the process is repeated for each subsequent nucleotide until a complete complementary read corresponding to each original read is generated. The complementary reads often overlap, so it is possible to use the overlaps to reconstruct the whole DNA molecule, a process called assembly. If a reference genome is available then reads can be 'mapped' to the reference, a process called alignment. This gives a very simple overview of the generation of NGS data; see Pabinger *et al.* (2014) or Vierstraete (2012) for a detailed exposition.

Analyses using NGS data are numerous and varied, since almost anything that can be asked about a genome can be answered, possibly with a lot of work, by analyzing NGS data. The cost is that often NGS data have hundreds of thousands, if not millions, of short reads that have to be included. So, managing and manipulating the data require relatively sophisticated analysis pipelines. Nevertheless, a common application is to variant detection. Variant detection asks what the differences are between a genome from an individual and the consensus or reference genome for the population from which the individual was drawn. In general, this sort of analysis does not require that one know the population, only that a collection of consensus sequences be available. This collection of consensus sequences permits the short reads to be mapped back to their likely origin. This allows the detection of insertions, deletions, reversals, etc. Thus, the data generated from NGS for DNA comprise a collection of read counts from different regions of one or more genomes. It is possible that the same read can be mapped back (or 'aligned') to two or more genomes, reducing the information about the identity of the individual's genetic material. In some cases, reads do not align back to any region; these may or may not be errors. Although the reads are statistically dependent, in practice the dependence is often small enough to be ignored to a first approximation. So, read counts that align to regions have been modeled as a binomial, Poisson, gamma, negative binomial, multinomial, etc., in the hope that the models are not far wrong and can answer questions of interest such as the differential expression of genes. Often, however, the dispersion is higher than these models permit, so modifications of them are required. A good recent tutorial on the basics of these methods is found in Sec. 10 of González (2014). There are numerous important and difficult problems that NGS data can be used to address; however, at this time there is no single 'standard' analysis using NGS data. For example, see *Statistics and Its Interface*, vol. 8, no. 4. for a collection of methodological papers using NGS data in wide range of analyses, many of which permit predictive validation. Moreover, as it is an active field, there are many more examples of NGS data analyses throughout the published literature.

As an example of the 'omics data in group (ii), consider proteomics. The proteome is the entire collection of proteins produced by an organism or cellular system. Proteomics is the comprehensive study of a specific proteome including the abundances of each protein, their interactions, and their variations, in an effort to understand biochemical processes in cells. Proteomics is difficult because proteins are continually undergoing changes due to reactions, binding, and degrading, as well as changes in the organism in response to the environment. The most common way to generate proteomic data is via mass spectrometry. The mass spectrum is a plot of the intensity as a function of the mass-to-charge ratio (MCR) of a collection of molecules and is the typical form for proteomic data. An intensity peak at a value of the MCR suggests a high prevalence of molecules with that MCR. Again oversimplifying, the process for generating the data is as follows. Proteins are dissolved into a solution that contains crystalized molecules and is then spread on a metal plate. The solvent containing the proteins and crystals vaporizes, leaving the proteins embedded in the crystals. The protein–crystal matrix is then put inside a mass spectrometer, where a laser is then shot at it. This causes the proteins to become ionized, i.e., acquire a charge. When an electric field is applied, the charged protein molecules detach from the plate and travel a certain distance depending on their MCRs. Smaller, higher charged particles travel further than heavy particles with less charge. This is repeated until a mass spectrum graph is generated satisfactorily, e.g., it has recognizable peaks. The peaks themselves may occur at peptides rather than whole proteins because the proteins are fragile and may break into two or more pieces during the process. Thus, several peaks may correspond to a single protein. See Clark *et al.* (2013) for a more detailed exposition.

The main point of proteomic analyses is to identify what proteins of interest are present or absent by looking at local maxima. Before this can be done, the data usually have to be 'cleaned'; that is, the mass spectrum plot may be trimmed to boundaries that are believed to be physically meaningful, smoothed, and interpolated so that baseline noise can be removed, thereby identifying contaminants. Then the analysis amounts to bump hunting. For instance, a mixture model of normal densities can be fitted to the plot using some technique such as the BIC to choose the number of components and another technique such as the E-M algorithm to estimate the coefficients and parameters. From the molecular weights, the proteins present in a single population may be identified, in principle, given sufficient domain knowledge. If there are mass spectrum plots for a number of individuals from two populations, it may be of interest to look for differences between the two populations. One way to do this is to develop a classifier that is a function of a mass spectrum plot and predicts the population that generated the plot. This is a large-p small-n problem because there are typically many more peaks on the mass spectrum plot than there are classes in an experiment. Support vector machine classifiers, see Sec. 10.4, are often used because they frequently give the right degree of term-wise sparsity and can give 'double sparsity' if the dimension of the argument of the kernel is also reduced, perhaps by using only selected functions of the mass spectrum plot; see Plechawska-Wojcik (2012) for details on this sort of analysis. Loosely, given this, the molecular weights of the proteins that make the two populations different can be obtained and to a degree the proteins can be identified, given sufficient domain knowledge. In both the one- and two-population settings, a predictive validation of the proteins that have been identified can be achieved readily.

As a final example, transcriptomics is one of the most important of the group (iii) type of 'omics data. Transcriptomics is basically the study of the RNAs produced by DNA and one way to do this is to generate microarray data, sometimes misleadingly called DNA microarray data. Loosely speaking, since these biochemical processes are highly complex, the portion of genetic material to be studied is put in a solution. Enzymes are added to cut the genetic material at known locations. Through another process, the resulting lengths of nucleotides are 'tagged' with other chemicals that make them fluorescent under various conditions, e.g., temperature. Separately, a rectangular chip with a regular grid of indentations in it (called wells) is prepared. Each well is designed to form hydrogen bonds with one type of string of the tagged lengths of nucleotides. The solution is spread over the chip so that, hopefully, the fluorescent strands will bond noncovalently with their complementary nucleotide sequence. The chip is then washed and the fluorescence is photographed. Experimental conditions vary enough from one region of a chip to another region that the direct readings from each chip must be standardized over the regions of the chip and with respect to other chips in the same batch. There are several techniques for doing this; descriptions of them are important but are omitted here. An accessible overview of the entire microarray-data-generating process can be found in Buck and Lieb (2003), though much has been learnt since its publication.

The output of this procedure, which is what a statistician sees most often, is a list of subjects and a vector of standardized fluorescences – a string of real numbers that is usually thousands of entries long. In this case, the usual concept of a real dimension is relevant except that there are thousands of them and it is often not known whether a given dimension corresponds to a gene or only a part of a genetically meaningful nucleotide sequence. Usually many chips are prepared for subjects from two tissue classes (say cancerous and not cancerous) and the goal is to look for differences in gene expression. There are many analyses of DNA microarray data, some more convincing than others; the most basic is a *t*-test between corresponding entries in the vector of standardized fluorescences, and this is the basic question that has reinvigorated the study of multiple comparison problems. Here, the importance of knowing how the data is generated and used gives a limit to what a researcher can expect to learn from analyzing DNA microarray data. That is, if a classifier (based on a selection of nucleotide strands) can be found that predicts sufficiently accurately which subjects have, say, a certain type of cancer, then one may be led to infer that the nucleotide strands whose fluorescences figure in the classifier relate to portions of the genome that play a role in the cancer. The variability inherent in the experimental procedure makes any conclusions tentative at best, not least because model uncertainty is very high; however, these types of analyses remain near the state of the art because DNA microarray technology adapts to other compounds.

Next-generation sequencing technology can be used with RNA in place of DNA, and microarray technology can be used for DNA rather than RNA. However, there is a trade-off between microarray data and sequencing data. With microarrays, the chips are designed in advance so the experimenter knows which compounds should bind to the probes on the microarray. Microarrays are cheaper and generally provide less information than sequencing through NGS. However, while NGS for RNA (called RNA-seq) or NGS for chromatin immunoprecipitation (called ChIP-seq) is more expensive and the reads show more variability, they provide more information even though the RNA reads have to be turned back into DNA reads. This is not 'hard' but there is uncertainty due to the error rates of sequencing, for instance. One way in which this latter procedure is more informative is that it provides information on the abundances of 'unknown' RNAs. (Some RNAs may be genuinely unknown, others may be unknown simply because they are not on the microarray chip.) Similar tradeoffs apply to other genomic data because there are also chip (microarray) methods and sequencing procedures for obtaining, for example, SNP data.

There are other important genomic data types that are omitted here. See `https://en.wikipedia.org/wiki/Omics` for a more complete listing with many follow-on limbs.

For some types of genomic data, e.g., microarrays, the concept of a real dimension is meaningful. In some cases, such as that of SNPs or proteomics, a real dimension is not directly meaningful; one can in principle calculate statistics from the output data i.e., do feature selection, but it is unclear how well this will summarize the data. When the dimension concept is not helpful, the size of the data, i.e., big data where the size of the data is measured by storage requirements rather than dimension, may be more meaningful. The complexity of the data is also an issue but to date has been largely unexplored. For example, proteomics data may be more complex than NGS data even though they generally require less storage per subject. Obviously, the interplay of dimension, complexity, feature selection, and storage along with the tradeoff among the informativity of the data types, amongst many other questions, merits more study in this general area.

To conclude this subsection, note that the data types and typical problems discussed can be regarded as $\mathcal{M}$-closed, on the grounds that there are finitely many nucleotides, proteins, RNAs, etc. However, there are often so many that the problem is virtually $\mathcal{M}$-complete, and it is almost certainly $\mathcal{M}$-complete (if not genuinely $\mathcal{M}$-open) when chemical dynamics and changes in the environment over time are taken into consideration. So, in fact, even though much reliable modeling information is available, it amounts to only a small fraction of what is not known and, as the problem enlarges from single reactions to networks of reactions and to entire cells or organisms, the fraction of modeling information decreases to the point that it constitutes minimal constraints on the model. A visual might help appreciate the scale: it is one thing to imagine strands of DNA as being like so many pieces of spaghetti. It is another thing to regard strands of DNA as so many pieces of spaghetti that have lengths comparable to the circumference of the earth.

12.5.2 Combining Data Types

In addition to the general benefits of combining the information in multiple types of data, as discussed at the beginning of Sec. 12.5, combining data can compensate for missing or unreliable data points in any single data type. Importantly, combining different data types may be necessary to make predictions because each includes information from different aspects of the DG. The use of multiple data types that lead to the same, or at least compatible, predictors also ensures a robustness of the predictors to the selection of explanatory variables and possible modeling strategy.

The two most basic approaches to addressing multitype data, discussed in Sec. 12.5, are the batch approach (throw all the data into one big model) and the hierarchical approach (analyze each data type individually and then combine the results by some kind of ensembling). These two approaches can be applied to either supervised or unsupervised data separately. However, the merging of supervised and unsupervised data largely remains an open problem. So, the first subsection here will focus on prediction using multitype supervised data and the second subsection on prediction using multitype unsupervised data.

Without further comment, it will be assumed here that any problems with data quality and standardization have been resolved and the data are analysis-ready. This is a highly nontrivial assumption because quality control can generally be done only in aggregate, for instance, via summary statistics, normalization (so that all the data are comparable, at least within a type), and any applicable data-type specific quality-assurance techniques. In particular, variables suspected to have been badly measured may have to be discarded or replaced. Data may have to be transformed prior to analysis so that the information they contain is more accessible.

Supervised Data

Currently, neither the batch approach nor the hierarchical approach is particularly common, at least not without extra steps. The two most popular analyses are probably multi-staged analysis and meta-dimensional analysis; see Ritchie *et al.* (2015). For a contrasting taxonomy of multitype data approaches, see Bersanelli *et al.* (2014), in which existing methods are partioned into four categories (network-based versus not-network-based and Bayes versus frequentist) and examples from each category are described. These later categories are based on the formal statistical techniques being used rather than on the properties of the data types collected.

The idea behind a multi-staged analysis is that the data types are brought into the analysis sequentially, possibly including domain-specific knowledge at some steps. Ritchie *et al.* (2015) gave a three-step example that starts with SNP data. First, the SNP data is examined for associations with phenotypes of the subjects, where the threshold for saying that a given SNP is associated with a given phenotype is based on a genome-wide significance level. Second, the SNPs found to be 'significant' are tested with another sort of 'omics data such as gene expression, methylation, metabolomic, or proteomic data. This can be done for one or more types of 'omics data in sequence. Third, the SNPs that survive from the second step are tested for correlation (or other measure of association) with the phenotype data and again reduced to those with the strongest relation to the phenotypes. Obviously, the order in which the data are used can be permuted and the more permutations that lead to a given SNP, the more robust is the inclusion of that SNP, and hence the more likely that the particular SNP is not spurious. Finally, the SNPs that survive most frequently can be combined into a classifier that predicts which phenotype a given pattern of SNPs is most likely to represent.

Step one in this example can be accomplished using many different statistical techniques – linear regression (perhaps with dummy variables), logistic regression, or any other sort of evaluation of dependence. Moreover, there are numerous variants on how domain-specific knowledge may be brought in; see Ritchie *et al.* (2015) for details on how functional or pathway information may be applied.

The main limitation of this approach is that, because it is sequential, a SNP ruled out at one stage is not brought back at a later stage. This means that it does not allow for phenotypes that are the result of combining the effects from two or more levels of data. For instance, a phenotype that results from the interaction of SNPs, methylation, and proteins (proteomic data) is unlikely to be successfully 'explained'. A strength of this approach is that it helps to ensure that the extra information above the SNP level is used to help identify the right SNPs more accurately.

Meta-dimensional analysis – basically, using all the data in one large analysis rather than a series of analyses – comes in three 'flavors.' The first is the batch analysis already described. Often shrinkage methods such as those discussed in Sec. 10.5 are applied to do feature selection. This means that it is possible to include interactions between terms from different data types. However, putting all these variables and, say, their products on the same scale – usually a prerequisite for shrinkage methods – can be a challenge because SNP data give values in $\{0, 1, 2\}$, methylation data give proportions in $[0, 1]$, and gene expression data (RNA) give positive real values. This is important because it is well known that penalized methods are sensitive to multicollinearity (dependence among explanatory variables), so Studentizing, if not sphering, the data is necessary to get replicable results; the decay factor on the penalty term must also be chosen well. A separate but important problem is that the number of variables in the data types ranges from thousands (for proteomic data, to millions (SNP data), so, ensuring that no one data type dominates spuriously may also be a challenge.

Second, instead of combining the data into one batch, sometimes each data type can be reduced to a common form e.g., a network or other graph model. These are then comparable and features of them may be used to generate predictors. This is a more difficult approach, but it is reasonably successful when it can be implemented well. The problem with this approach is that, in reducing each data type to a common form, features of the data type that interact with other data types may be lost, partly defeating the point of using multitype data. This limitation disappears if it can be ensured that the common form does not lose any useful information about interactions between the data types, but it is difficult to achieve this.

The third 'flavor' is essentially the hierarchical approach: a predictor is developed for each data type, hopefully preserving its properties, and these predictors are then combined into one. As noted, the limitation is that the interactions between data types may be lost unless the predictor from each data type retains all relevant information, including those variables that interact with the observations from the other data types. Again, this is difficult to justify by argument and to ensure.

Regardless of the technique adopted to generate predictors in the six cases just described, the validation problem still remains, perhaps more than in other settings. The data are highly variable, the model classes used are typically very rich, and any feature selection is probably far from unique. Hence, even good predictors are likely to give predictions with large SEs and so validation may be less reliable than desired. Indeed, good predictors may nevertheless validate only weakly and the validation itself may exhibit high variability. It many cases, it may simply be too difficult to obtain enough data for convincing validation and the collection of predictors that are roughly equally good may be quite large.

Undergirding all these techniques is the problem of interpretability. As noted in Wei (2015), machine learning techniques such as boosting, random forests, and SVMs have been used to good effect in analyzing cancer data, at the cost of being generally uninterpretable physically. As a consequence, Wei (2015) focused on the selection of interpretable features via matrix methods – matrix factorization (not unlike factor-analytic, principal-component, or partial least squares models). Wei (2015) also treated the inclusion of multitype data in survival models (see Chap. 7) and how penalized methods (Sec. 10.5) or sure independence screening (Fan and Lv (2008); see Sec. 9.1 of the present text) to achieve sparsity can be used in Cox proportional-hazards models. One of the more novel aspects of multitype data analysis treated in Wei (2015) is the prediction of one type of genomic profile from another, e.g., in transcription factor binding and gene expression. These are cases where even if the variables available to the predictors are only of one type, the variables to be predicted number in the thousands or tens of thousands. Citing the use of single-hidden-layer neural networks, Wei (2015) noted that this opens up a new possibility in multitype analysis that in principle can be used to fill in missing data or ensure that the inferences from different data types will be compatible. After all, if the prediction for a variable in one data type using another data type differs from the measured value, potentially there are compatibility problems between the data types, a problem that seems not to have received the attention it deserves.

Interpretability also motivates an interesting example of matrix factorization methods that tries to achieve double sparsity; see Okimoto *et al.* (2016). Once the data is analysis-ready, the idea is to stack the matrices of data into a single large $p \times n$ matrix $\mathcal{D}$ and find a rank-1 approximation to $\mathcal{D}$ of the form $D = uv^T$, where $u \in \mathbb{R}^p$ and $v \in \mathbb{R}^n$, by minimizing the error criterion

$$E(u, v, \lambda) = \|\mathcal{D} - uv^T\|_F + \lambda \|u\|_{\ell^1},$$

in which λ is a decay parameter, subject to the constraint that $v = \mathcal{D}^T u$. Thus, only one high-dimensional vector u is used and is sparse (it has few nonzero entries) relative to p while $\mathcal{D}$ is approximated by a rank-1 matrix, i.e., there is double sparsity. In the above error criterion the first term uses the Frobenius norm (the L^2 norm on matrices) and the second term uses the ℓ^1 sequence norm. The optimization of $E(u, v, \lambda)$ is effected by an iterative algorithm that finds an optimal λ and ensures that $v = \mathcal{D}^T u = \sum_{m=1}^{M} \mathcal{D}_m u_m$, where there are M data types with data matrices $\mathcal{D}_m$ and $u = (u_1, \ldots, u_M)^T$; u_m corresponds to data type m. Okimoto *et al.* (2016) compared this method with a variety of other methods and found that it works relatively well. The problem that the analysis does not adequately address, and perhaps cannot do so at this time, is the appropriate sense in which sparse-model validation should be done for a system that is unlikely to be sparse.

Another recent way to address the construction of predictors in the multitype data context follows from a series of papers (Hwang *et al.* 2009), Turnbull *et al.* 2013, and Ghosal *et al.* 2016). The central idea in these papers is to adapt forward selection, as commonly used with low-dimensional regression data, to sparse high-dimensional data ignoring the data types (and their dimensions) except in terms of their usefulness in explaining Y. In its most basic form, the setting is linear regression, $Y_i = X_i^T \beta + \epsilon_i$. When $p > n$, penalized methods are often used for variable selection when many of the p explanatory variables are thought to have negligible influence, i.e., the regression is sparse. Hence, the goal of this approach is to construct a shrinkage estimator through the sequential inclusion of variables in the predictor with a one-dimensional absolute value penalty.

In the simplest version, the variables are Studentized, so that $\bar{X}_d = 0$ and estimates for the β_j, $j = 1, \ldots, p$, are formed for $r = 1, \ldots, R$ iterations, starting with $Y_{i,1} = Y_i$ by setting $\hat{\beta}_{j,r} = \sum_{i=1}^{n} X_{ij} Y_{ir} / n\bar{X}_j^2$. The initial value for the prediction vector is $\hat{\mathbf{f}}_1 = 0$. The shrinkage version of $\hat{\beta}_r$ is taken under some penalty, e.g., LASSO, adaptive LASSO, or elastic net, etc. Then, forward selection is used. In the LASSO case this gives

$$j_r^* = \arg\min \sum_{i=1}^{n} (Y_{ir} - X_{ij} \hat{\beta}_{jr}^L)^2. \tag{12.27}$$

Then r is updated to $r+1$, Y_{im} is updated to $Y_{i,m+1}$, and $\mathbf{f}_r$ is updated to $\mathbf{f}_{r+1} = \mathbf{f}_r + \mathbf{X}_{j_r^*} \hat{\beta}_{j_r^*,r}^L$ where $\mathbf{X}_{j_r^*}$ is the j_r^*th row of the data matrix. An analogous treatment can be given for any shrinkage method that corresponds to a closed-form expression for the $\hat{\beta}$'s or, more generally, for which the $\hat{\beta}$'s can be calculated. Usually, R is chosen so that a Cauchy-like convergence criterion is satisfied for a pre-chosen $\epsilon > 0$. Once the variables have been selected, they can be used in a standard least squares regression. This often gives a smaller prediction error than using a shrinkage method for regression as well as for forward variable selection. As a generality, the results can be improved if the variables X_j are Studentized or sphered; the result is orthogonal forward selection. Moreover, the basic technique extends to maximum-margin classifiers such as SVMs. Again, there is a shrinkage penalty that makes the penalized error amenable to forward selection as in (12.27), with a termination rule based on a convergence criterion; see Ghosal *et al.* (2016, Sec. 2).

Finally, a promising approach that is under development is the idea of making the different data types 'informationally disjoint' in the sense of eliminating the overlap of information between two data types. In that way each data set can be analyzed separately and, when the results are put together, no interaction information will be lost. The simplest example of this is to consider developing a classifier Y that assumes one of K values using two data types with matrices W and Z. Suppose that W is $n \times p_1$ and Z is $n \times p_2$, with $p_2 \gg p_1$, and that any preprocessing by, say, Studentization or sphering has been done. Also suppose that some information in W duplicates information in Z.

More precisely, assume that it is known that some of the p_1 variables in W must be in the classifier even though, under a simpler method, they would be omitted spuriously because p_2 is so large. An instance of this comes from agronomy, where W represents phenotypic data and Z represents some high-throughput data type such as SNPs. To avoid Z washing out W, it is natural to try to remove the information in W that is already in Z and then develop a classifier for Y using the residuals from this process and the data in W. That is, do p_1 penalized forward selections of W on Z in a linear regression context, resulting in residuals of the form $\hat{e} = W - Z\hat{\beta}$ where $\hat{\beta}$ is the result of sparse regression. The penalized forward selection can be repeated but on a multinomial logistic regression of Y on $(\hat{e}, Z)$. Essentially, this process corrects W to give residuals that contain only information not already in Z. This makes sense because frequently the variables in W have no less information than those in Z. Thus, the logistic regression classifier can now be generated by penalized forward selection as in Hwang *et al.* (2009), Turnbull *et al.* (2013), or Ghosal *et al.* (2016) over the two sets of variables that have been made informationally disjoint.

Further examples of analyses of multitype data can be found in Xie and Ahn (2013), Gowen and Fong (2013), and Arakawa and Tomita (2013).

Unsupervised Data

When the data are unsupervised there is no response Y and the data set is of the form $\{x_1, \ldots, x_n\}$, in which each x_i is a vector of multitype data of length p. One prediction problem amounts to using some of the data to predict other parts of the data, i.e., the problem is constructed for supervision by selection of a Y. In some cases this is by imputation; more typically, however, the data are clustered, so that the cluster index becomes the Y for a new data point x_{n+1} Otherwise stated, given a new data point x_{n+1}, the problem is to predict its correct cluster. One uses the clusters to define classes and then to find a good classifier. Finding such a classifier is often done simply by choosing a distance measure on the data and assigning x_{n+1} to the class cluster whose center (however defined) is closest. This is a type of adaptive classification since the 'classes' evolve as the clusters change with the accumulation of data. In this subsection the focus will be on generating the clustering in the first place, since the previous subsection treated supervised data; finding a good classifier using multitype data is sometimes similar to the techniques presented there, although more often it rests primarily on the data summarization techniques used.

In the simplest cluster analyses of multitype unsupervised data, each type is analyzed separately and the results are combined. Usually it is the subjects that are clustered rather than the variables measured on the subjects. Thus, as long as a data type can be clustered – it might be difficult for some data types such as graphs or streaming data – a consensus clustering can be found. Since n is usually small this is often done manually and simplified by looking only at a subject's membership in clusters, i.e., not using the data that generated the cluster, although stability techniques are often used as well.

Consensus clustering is often, though not always, based on the similarity between data points in terms of their membership over multiple clusterings. That is, the ratio of the number of clusterings in which two data points are in the same cluster to the total number of clusterings available can be regarded as the similarity between any pair of data points. This is a similarity in the same sense as in hierarchical clustering. Thus, any hierarchical method can use the similarity matrix to generate a consensus clustering. There are numerous other ways to generate a consensus clustering (also sometimes called an ensemble clustering or an aggregate clustering); see Ghosh and Acharya (2014).

One limitation of this approach is that clustering individual data types often requires dimension reduction to avoid degeneracy (Koepke and Clarke 2011), but dimension-reduction techniques do not work well across data types; see Shen *et al.* (2009). That is, effective dimension reduction usually is data-type dependent. A second problem is that the clustering must take account of the within-data-type covariances and the between-data-type covariances separately in order to take into consideration the ways in which some data correspond to the same cluster, e.g., if all subjects have the same disease, while other data distinguish between the clusters. This is important if, as is often the case, subjects have some features in common but other features that create meaningful distinctions between clusters; e.g., there are often subtypes of a given type of cancer.

Clustering over variables rather than subjects would identify groups of variables that were likely to be related in some way; this problem has been addressed in the context of a single data type (using principal component analysis, canonical correlation analysis, etc.) but seems largely unaddressed in the multitype data context. The single or multitype data problem may be interesting from a subject-matter perspective; however, from a strictly predictive perspective it is unclear for what prediction problem it would represent, apart from data summarization, a step towards predictor formation. Given a clustering over variables for a set of subjects, assigning a new variable to a cluster would seem to be artificial because it would implicitly assume that a sequence of variables is measured on the same subjects.

One way to cluster multitype data is to define a similarity that takes into account the inter- and intra-relationships of the data types; see Wang *et al.* (2003). For any two data points x and x' write the similarity as

$$S(x, x') = \alpha s_f(x, x') + \beta s_{\text{intra}}(x, x') + \gamma s_{\text{inter}}(x, x')$$

in which $\alpha, \beta, \gamma \geq 0, \alpha + \beta + \gamma = 1$. The function s_f gives the similarity between the features over the M data types while s_{intra} and s_{inter} are the intra-data, and inter-data similarities. Once S is well defined, see Wang *et al.* (2003) for the details, any hierarchical method can be applied. This methodology can be improved by 'mutual reinforcement'. That is, the overall clustering is updated by using clusterings on the individual data types. This complex procedure is explained in Wang *et al.* (2003).

Another way to cluster multitype data is a variant on factor analysis. So, suppose that there is a small set of latent traits that underlie the clusters. This technique, called integrated clustering or iCluster, originated in Shen *et al.* (2009) and was later used in Shen *et al.* (2013) for several examples and comparisons. Suppose also that the data types are numerical, e.g., methylation and gene expresion data. Write the data matrix from the mth data type, $m = 1, \ldots, M$, as

$$\mathcal{D}_m = W_m Z + E_m,$$

in which $\mathcal{D}_m$ is the mean-centered $p_m \times n$ data matrix, where p_m is the dimension of the mth data type, W_m is the $p_m \times (K-1)$ loading matrix, Z is the $K-1 \times n$ matrix of $K-1$ latent traits for the n subjects, and E_m is the $p_m \times n$ error term. Assuming normal errors,

$$\mathcal{D}_m | Z \sim N(W_m Z, \Phi_m),$$

where $\Phi_m = \text{diag}(\sigma_1^2, \ldots, \sigma_{p_m}^2)$. If $Z \sim N(0, I_n)$ then, marginally, $\mathcal{D}_m \sim N(0, \Sigma_m)$ where $\Sigma_m = W_m W_m^T + \Phi_m$. Now, the EM algorithm can be used to obtain estimates of the W_m and the Φ_m. Also, instead of $K-1$ loading vectors one can select $\ell \leq K-1$ vectors; ℓ can be chosen by model-selection techniques. When the p_m are large, penalized methods can be used to achieve sparsity. Suppressing the extensive details, K-means clustering can finally be applied to Z. Indeed, any clustering method can be applied to Z; however, the K-means method is the most natural because Z has been assumed normal. Shen *et al.* (2013, Sec. 5) briefly considered predicting cluster membership for the purposes of a stability analysis, but did not otherwise evaluate predictive performance.

Overall, this technique is intended to do dimension reduction (to ℓ underlying traits) and seems to perform better than using K-means alone on the full data set or using a consensus version of K-means over the data types separately. It also seems to give better clusterings than a matrix factorization method based on a sparse SVD and a clustering method based on penalized Gaussian mixture models. On the other hand, iCluster has strong assumptions; it only applies when the data are continuous real numbers, and it may suffer the usual problems that factor analytic techniques have, such as nonuniqueness. However, nonuniqueness may not be a problem because the method gives sparsity and the stability of the clusterings seems good. The latter benefit assumes sparsity, whether or not this is actually the case.

Bayesian approaches to clustering multitype data have been proposed by Wang *et al.* (2011) and by Lock and Dunson (2013). These are called Bayes cluster ensembles (BCEs) and Bayes consensus clustering, respectively. The BCE method starts with a base clustering and assumes that it corresponds to a Bayes graphical model. Then, if each x_i has an underlying mixed membership to different consensus clusters, let θ_i denote the latent mixed-membership vector for x_i. If there are K clusters, then Θ_i has a (discrete) distribution over the K clusters. So, it is possible to regard $\Theta_i \sim \text{Dirichlet}(\alpha)$. The clusters in the ensembled clustering also have a latent distribution for the indices of their elements. Thus, if an x truly belongs to the ensembled cluster h, x will follow a distribution that can be proposed in closed form. From this a posterior density can be derived and used and an ensemble clustering identified; see Wang *et al.* (2011) for details of the construction.

Bayes consensus clustering is based on a different approach. Assume that there is an overall clustering using all the data types, with $C(x)$ giving the index of the cluster to which x belongs, and that there are M clusterings, one for each data type, with $L_m(x)$ giving the index of the cluster, in the mth data type, clustering to which x belongs. Now

$$P(L_m(X) = k | C(x)) = \begin{cases} \alpha_m & \text{if } C(X) = k, \\ \dfrac{1 - \alpha_m}{K - 1} & \text{otherwise,} \end{cases} \tag{12.28}$$

where $\alpha_m \in [1/K, 1]$ is the 'adherence' estimated from the data. Now, the conditional probability $P(L_m(X) = k | C(X), \Theta_m, \mathcal{D}_m)$, where Θ is the parameter vector for the mth data type, can be specified and used to find the overall probability $P(C(X) = k | (L_m, \alpha_m)$ by

Bayes' rule. Thus, the posterior can be written down explicitly and found by MCMC–MH, as detailed in Lock and Dunson (2013).

A methodology in the same spirit as Wang *et al.* (2003), called integrating phenotype framework (iPF), is presented in Kim *et al.* (2015). Assuming the data are preprocessed and merged into vectors of multitype data, a dissimilarity matrix between any two features across the data sets can be defined. This is used in multidimensional scaling (MDS), rather than in a clustering procedure, so as to reduce the dimension of the data vectors. The resulting points are then smoothed and clustered as points in a plane, assuming that the MDS compresses the data to two dimensions; see Kim *et al.* (2015) for the details. The results presented suggest that iPF is as robust and accurate as iCluster and Bayesian consensus clustering, as well as being computationally easier and having the ability to accommodate different variable types.

Numerous other methods have been proposed. For instance, Liang *et al.* (2015) suggested a methodology based on graphical models that have a visible and a hidden component and enough constraints to make these models similar to feed-forward neural networks. Arguably, the hidden components are analogous to the latent variables in iCluster. Gwalani *et al.* (2017) treated the problem of multiple data types arising from different levels of granularity – blocks, census tracts, and counties – by putting them on the same scale, somewhat in the same spirit as the downscaling and upscaling as seen in Sec. 12.3.2. Overall, problems involving multitype data began to be important in the 1980s and are likely to remain an active area of research as gathering data becomes generally less expensive.

12.6 Topics that Might Have Been Here ... But Are Not

No single book can comprehensively cover a major field such as statistical prediction. What has been covered is the conceptual and theoretical undergirding of statistical prediction (see Chaps. 1 through 4), the way in which these ideas play out in five different traditional subfields of statistics (Chaps. 5 through 9), and the way in which they appear to be playing out in numerous emerging (or rapidly developing) subfields of statistics (Chaps. 10 through 12). Every effort has been made to choose topics judiciously: not for convenience but rather to ensure a well-structured presentation with as few important topics as possible omitted. That being said, there is much of great importance that has not been covered, in order that this book could be finished within the lifespan of its authors. Apologies to all those whose contributions have been given short shrift or ignored. Here, some of these important omitted topics are listed – partly as an apologia and partially to encourage further research to elucidate the predictive view of statistics thoroughly. So, herewith is a list of topics that might have been covered, or covered in more detail.

First, the most successful predictions usually come from experimental data – and experimental design for predictive purposes has been entirely neglected. The all too common 'dry cleaner's model' of statistical analysis has been implicitly taken. The experimenter brings some data to a statistician. The design of the data collection procedure is problematic and the data themselves have probably been collected badly. The statistician faces a shoe box of measurements that must be cleaned before they can be used. This includes the questions of missing data and imputation, which have not been treated here, e.g., propensity scores. After much work, a statistician with a knack for producing order from chaos finally produces a nice-looking scientific suit from the scraps and the consultee trots off happily.

Second, the discussion of data types and their properties has been superficial. Many very smart people have spent the bulk of their careers developing individual aspects of data-collecting procedures, in the hope that the data quality after their improvements will be noticeably better, if still not entirely satisfactory, for most scientific purposes. Also, the range of contemporary data types presented here has been downright Spartan. In fact there has been a proliferation of data types, on a par with the increase in data volume, that represents an intellectual smorgasbord for a world that hungers for quantitative analyses and the information they yield. Some of these data types are as follows: functional data, image data, manifold learning (from explanatory variables), and many types of network data not covered in Sec. 12.4, e.g., reaction networks.

Third, a chapter on the complexity of data and predictors was not included even though it might seem central to prediction. This would have included the Vapnik–Chervonenkis dimension (VCD), usually for function spaces; the VCD commonly occurs in error bounds for estimation but can apply to prediction as well. So far, it is largely a theoretical construct, but it is ripe for use in more practical senses. Also, concepts of complexity that stem from codelength are important: one can make a whole approach to statistics from an information-theoretic standpoint. See Ebrahimi *et al.* (2010) for one important predictive contribution in this area. There are also numerous concepts of complexity that are on the boundary between statistics and computer science. They have much to teach mainstream statisticians, e.g., algorithm design for efficient running time, sample complexity (basically an extension of the concept of efficiency to settings where the Fisher information does not meaningfully exist), and probably approximately correct (PAC) learning; see Haussler (1990) for an introduction.

Fourth, there are many important analytic techniques that fit comfortably into a predictive view but have not been treated. For instance structural equation modeling, percolation models, agent-based models, the selection of random effects in linear mixed models, generalized linear mixed models, and nonlinear mixed models can all be considered predictively. Kalman filter models, graphical models, and semiparametric models fit into this category too. Arguably, so do frames and wavelets, even though they are essentially only for one explanatory variable and one response variable at a time.

Finally, aspects of analysis that involve dimension reduction (as opposed to feature selection) and their effect on prediction could not be treated as fairly as they deserve. These aspects include the classical techniques of factor analysis, principal components, partial least squares, and latent variables. Sometimes there is a physical motivation; at other times the motivation is purely the desire to get a simple predictor that can be explicitly identified. Meta-analysis, especially as a follow-on to multitype data analysis, often falls into this category as do penalized methods to achieve sparsity, an oracle property, and oracle inequalities more generally. Feature selection from a physical standpoint has barely been touched, on the simplistic grounds that it requires modeling information that is usually only sporadically available and even then only at great cost and often dubious reliability.

Of course, treating all these topics would require an extra volume with the concomitant considerations of $\mathcal{M}$-closed, $\mathcal{M}$-complete, and $\mathcal{M}$-open problems.

12.7 Predictor Properties that Remain to be Studied

Just as there are model classes and aspects of data that have not been discussed adequately, so there are aspects of predictors for which this is true. In many cases it is not that the material is 'out there' in the literature or can be adapted from the literature so much as that many aspects of prediction have not yet been sufficiently explored. As such they are wide open as research questions.

First of these is probably the adaptive reselection of predictors. In many examples in this book predictors were reselected as data accumulated, but they were reselected from within the same, often small, model class. Assuming that the data set is large and rich, it is desirable to let the class of predictors change as data accumulate. The class could simply get larger, so that there are, e.g., more nodes in a neural network, more splits in a tree, or more features in a penalized method, or it could change character completely, e.g., it could move from generalized linear models to kernel methods, according to some criterion for fit or a CPE. Another example is clustering to generate empirical classes for use when one is developing classifiers sequentially; see Sec. 12.5.2.

Second is the evaluation of predictors beyond simply CPE. Other factors are important besides simply making good one-step-ahead prediction. For instance, one should assess the complexity of a predictor relative to the complexity of the data in the hope that the two match or are at least not too different. Also, the importance of variance–bias concerns and robustness, especially in adaptive prediction, is high. A good thing about adaptive prediction is that, because it is adaptive, it can conform itself to the stable aspects of the DG, thereby improving prediction in principle. A bad thing about adaptive prediction is that, because it is adaptive, it can be more highly variable as it tries to encapsulate stable features of the DG. This can worsen prediction by increasing the length of PIs. In some cases, ensemble methods may give better stability (assuming that stability is desired) than predictor selection does.

As an a side, the view taken here is that scoring rules do not give satisfactory performance because they are so flexible that they can give any answer desired. There are, however, other researchers, more eminent than the authors, who would disagree vehemently. For instance, Dawid and Musio (2015) made a conspicuously strong counter-argument because they considered a prequential setting.

Third, just as there are limitations to the modeling approach in statistics, there are limitations to the predictive approach. Some DGs may be impossible to predict, i.e., they may be extremely $\mathcal{M}$-open so perhaps only weak predictions are possible – large PIs with comparatively low confidence. This can occur when there is too much variability in the DG or when small changes in the data from the DG may be unexpectedly amplified, e.g., in streaming data or network data. This means that much more thought has to be given to what is predictable by modeling versus what is effectively only predictable by, say, ensemble methods (that do not generally correspond to a single model), the Shtarkov solution or other predictors yet to be invented. In some cases like this, predictions may be done using propensity scores even if this is really imputation more than pure prediction. Nevertheless, imputation can be done sequentially and this technique may be sufficiently nonparametric and data-driven that it works when model based methods perform poorly.

Finally, however devoted one is to any particular form of inference, each has its limitations. The no free lunch theorems (see Sec. 11.7.2) set limits on the range of optimality of any method. That is, each methodology has a 'catchment area' where it is optimal or nearly so. Often, intuitively, if the optimality is particularly strong then the effectiveness of the methodology falls off more quickly outside its catchment area than if its optimality were not so strong. Boosting is a case in point: it seems so well suited to binary classification that efforts to date to extend it to give effective classification (or regression) more generally have not been very successful. Overall, it remains to characterize the catchment areas where each class of predictors performs optimally, performs generally well, or breaks down.

12.8 Whither Prediction?

This section is an epilogue for this particular book since there can be no epilogue for a topic like statistical prediction. Even though prediction feels familiar to statisticians, it differs enough from other forms of inference that many of its properties are hard to quantify and sometimes unexpected. For instance, there are asymptotically optimal predictors in a variety of settings, but the optimality is only asymptotic and can be beaten in finite samples; this is usually done by making the predictor more adaptive to the data stream. As another instance, feature selection for prediction can be different from feature selection for modeling. The features that express the state of an economy at a particular moment, for example, may not be very good indicators for the state of the same economy in a year. As a generality, in this book we have railed against using modeling information too much because often it has not been satisfactorily validated – hence the antidote is greater validation. This differs from the modeler's approach which seeks to use all available information, often uncritically or on the basis of the authority of a subject-matter specialist (with his or her own biases). It must be conceded that when the modeling information is reliable and germane it can help an analysis tremendously. The problem is that this is the exception, not the rule, and even when reliable modeling information is available it is unclear how to assess its usefulness for the available data, especially as data volume and complexity increase.

One reason why subject-matter specialists do not focus on validation is that their desire to understand the mechanism inside the DG often overtakes a critical evaluation of their findings. The stance of this book is that, until accurate prediction has been achieved, 'understanding' the DG is mostly a fool's errand. Nevertheless, subject-matter specialists put a high value on the interpretability of models and the predictors they generate even though, in our view, they should put a higher value on the validation of such predictors and an even higher value on finding optimal predictors. If nothing else, insisting that a predictor be interpretable severely limits the range of predictors that can be proposed and hence may lead to higher error. Interpretation when it is valid is very powerful. When it is not valid, it is just wrong – often seriously so. For most questions there are many more wrong answers than right ones, so it is no surprise that a wrong answer can be found much more easily than a right one.

This leads to the issue of sources of information. Multitype data has been discussed; modeling information versus the information in the data has been discussed; the prior information that a Bayesian has is also well known (see Clarke (1996) and Lin *et al.* (2007) for one way to 'diagnose' this). Thus, one task that the statistician must address is the proportion of information from the various sources, i.e., data types, models, and priors, including dimension reduction and feature selection are part of the mix. One must not use too much modeling or other pre-experimental information or it may overwhelm the data. One must not overuse the information from one data set or type since it may produce spurious results. Likewise, poor dimension reduction or feature selection may lose much of the information in the data. How to achieve the optimal balance of information from the various sources is not obvious, but it badly needs debate and much work. It is in this context that stability with respect to the sources of information is most important.

Some of this work has started, particularly in a Bayesian setting where it has been recognized as an issue for over 20 years. Recently, Hahn and Carvalho (2015) sparsified a Bayes predictor to get another predictor that was nearly as good but more interpretable than the nonsparsified solution. Effectively they did a Bayesian analysis in two stages – the first gave the optimal predictor and the second made it interpretable. They have extended their technique to variable selection for more complex models, including some nonparametric Bayes models see Hahn *et al.* (2017) and Puelz *et al.* (2017). Of course, one must evaluate the stability and generality of the technique but early indications are positive. Shahbaba and Johnson (2013) wrestled with a similar problem in an applied context.

One effect of recently emerged data types is that sparsity is ever more important. The problem arises when sparsity does not hold – a situation that is plausible for many systems that generate big and complex data sets. One way to accommodate this is to develop a new probability theory. It is not that probability theory is wrong – it may just not scale up effectively to these settings, which are more complex than any that were envisaged when Kolmogorov first proposed his axioms. Indeed, genomic, remote sensing, streaming, network, and image data and the DGs they represent may exhibit more variability than traditional probability measures can effectively encapsulate.

The idea can be visualized by recalling that the cardinality of $[0, 1]$ is $\aleph_1$ and it has a dense subset (the rationals in $[0,1]$) of cardinality $\aleph_0$. The continuum hypothesis is that $\aleph_2$ is the cardinality of the real line and that $\aleph_2 > \aleph_1 > \aleph_0$, with no cardinals in between them. So, if the class of events of interest is large – e.g., it cannot be approximated by a set of events of cardinality less than or equal to $\aleph_1$ – then the range of a probability measure, which is less than or equal to $\aleph_1$, may not represent the events of interest well. In short, probabilities that take values in a set of cardinality $\aleph_2$, e.g., a Banach space, may be necessary.[6]

There already exists an extension of probability theory to Banach-space-valued random variables, since the limits needed for probabilities are precisely those abstracted by the concept of a Banach space. Perhaps it is time to dust off those books from the 1980s and see whether they now have applications that their writers did not imagine. In short, statistics may need a new mathematical undergirding to describe randomness, since our understanding of randomness has expanded. A concomitant of this is that concepts of variability, such as SEs or entropies, may have to be updated in a new mathematics for variability.

On a more tangible level, taking a predictive approach to statistics would force statisticians to change in ways that are at once traditional and unfamiliar. For instance, statisticians would need to be closer to the generation of the data, including the design of data collection and foreseeing what sort of analyses will be feasible post data collection. In this way, statisticians would be better able to see the limitations of the analyses they can perform. They would need to pay more attention to validation as a way to cope with the multiple sources of information they must pool as they try to understand what the data and any other reliable

---

[6] It is easy to see that a probability loosely corresponds to a metrization of the event space – write $W(B(x_0, r))$ where $r \in \mathcal{R}^+$. Then, as r increases, $W(B(x_0, r))$ increases too. However, here the events have cardinality $\aleph_1$. If r ranged over a space of size $\aleph_2$, there would be large regions on which $W(B(x_0, r))$ would have to be constant. So, if a random variable assumes values in a Banach space of cardinality $\aleph_2$, at most only $\aleph_1$ of its events can be assigned a real probability. Hence, it can be seen that the range of a probability measure must be greater than or equal to the cardinality of the collection of events unless there is some simplifying assumption such as measurability.

information have to say. Such validation will necessitate prediction even if the question is one of classical inference, e.g., parameter estimation or hypothesis testing. However, given the heightened variability of modern data types, prediction will assume a new importance – indeed it has been doing so over the last few years – because model uncertainty is recognized as being higher than before and must be accommodated (by ensemble methods), overcome by predictive performance, or managed satisfactorily in some other way. Finally, it is not a stretch to forecast that, given all the intellectual ferment of contemporary statistics, consulting will become another branch of statistics, not unlike visualization, statistics education ('stat ed'), Bayes nonparametrics, or time series. Indeed, consulting may well end up as a leading branch since it interfaces with other fields and helps the rest of us formalize what kind of generic problems are important for us to try to solve.

In short, statistics has been revolutionizing itself every decade or so and there is no evidence that the pace will slacken. Does anyone really think that the way in which we will conceptualize statistical problems in 50 years – or even 20 – will be the same as we do today? Indeed, does anyone believe that we will be thinking about these *same* ideas in 50 years? Predicting the future of statistics, or even just that of statistical prediction, is an $\mathcal{M}$-open problem par excellence – and that is part of the charm of our field.

References

ABDULLAH, A., VELTKAMP, R., AND WIERING, M. (2009) Spatial pyramids and two layer sacking SVM classifiers for image categorization: a comparative study. In: *Proc. Joint International Conf. on Neural Networks*, pp. 9–12. (Cited on p. 476.)

ABRAHAM, C., BIAU, G., AND CADRE, B. (2004) On the asymptotic properties of a simple estimate of the mode. *ESAIM: Probability and Statistics*, **8**, 1–11. (Cited on p. 90.)

AFIFI, A. AND AZEN, S. (1972) *Statistical Analysis: A Computer Oriented Approach*. Academic Press, New York. (Cited on p. 348.)

AGARWAL, G. (2011) A new approach to spatio-temporal Kriging and its applications. MS dissertation, Program in Computer Science and Engineering, Ohio State University. (Cited on p. 560.)

AGGARWAL, C. (2007) *Data Streams: Models and Algorithms*. Springer, New York. (Cited on p. 550.)

AGGARWAL, C. (2013) *Managing and Mining Sensor Data*. Springer, New York. (Cited on p. 550.)

AGGARWAL, C. (2015) A survey of stream classification algorithms. In: Aggarwal, C. (ed.), *Data Classification: Algorithms and Applications*, pp. 245–274, CRC Press, Boca Raton, FL. (Cited on p. 551.)

AGGARWAL, C., AND SUBBIAN, K. (2014) Evolutionary network analysis: a survey. *ACM Comp. Surveys*, **47**, Art. 10. (Cited on p. 584.)

AGGARWAL, S. (2013) Principles of remote sensing. See: http://www.wamis.org/agm/pubs/agm8/Paper-2.pdf. Accessed May 15, 2017. (Cited on p. 563.)

AGRESTI, A. (2002) *Categorical Data Analysis*, 2nd edn. Wiley and Sons, New York. (Cited on p. 166.)

AHMAD, Z., HA, T., AND NOOR, R. (2010) Improving nonlinear process modeling using multiple neural network combination through Bayes model averaging. *IIUM Eng. J.*, **9**, 19–36. (Cited on p. 460.)

AILON, N., JAISWAL, R., AND MONTELEONI, C. (2009) Streaming K-means approximation. In: Bengio, Y., Schuurmans, D., Lafferty, J., Williams, C., and Culotta, A. (eds.), *Advances in Neural Information Processing Systems*, vol. 22. (Cited on p. 547.)

AITCHISON, J. (1975) Goodness of prediction fit. *Biometrika*, **62**, 547–554. (Cited on pp. 54, 55, 141, 310, and 329.)

AITCHISON, J. AND DUNSMORE I. (1975) Statistical Prediction Analysis. Cambridge University Press, Cambridge. (Cited on p. 206.)

AKAIKE, H. (1973) Information theory and an extension of the maximum likelihood principle. In: Petrov, B. and Csaki, F. (eds.), *Proc. 2nd International Symp. on Information Theory*, Tsahkadsor, Armenia, September 2–8, 1971; Budapest: Akademiai Kiado, pp. 267–281. (Cited on p. 321.)

ALBERT, J. (2009) *Bayes Computation with R*. Springer, New York. (Cited on p. 243.)

ALEXANDERSSON, H. (1985) A simple stochastic model of the precipitation process. *J. Clim. Appl. Meteorol.*, **24**, 1285–1295. (Cited on p. 152.)

ALGOET, P. (1992) Universal schemes for prediction, gambling, and portfolio selection. *Ann. Prob.*, **20**, 901–941. (Cited on pp. 285 and 286.)

ALON, N., MATIAS, Y., AND SZEGEDY, M. (1999) The space complexity of approximating the frequency moments. *J. Comp. Syst. Sci.*, **58**, 137–147. (Cited on p. 543.)

ALTMAN, D. AND BLAND, J. (1994) Diagnostic tests. 1: Sensitivity and specificity. *British Medical J.*, **308**, 1552. (Cited on p. 113.)

ANDO, T. (2011) Predictive Bayes model selection. *Amer. J. Math. Management Stud.*, **31**, 13–38. (Cited on p. 354.)

ANDRIEU, C., BREYER, L., AND DOUCET, A. (2001) Convergence of simulated annealing using Foster–Lyapunov criteria. *J. Appl. Prob.*, **38**, 975–994. (Cited on p. 393.)

ANGUITA, D., GHELARDONI, L., GHIO, A., ONETO, L., AND RIDELLA, S. (2012) The K in K-fold cross-validation. In: *Proc. 20th European Symp on Artificial Neural Networks*, Computational Intelligence and Machine Learning, Bruges, Belgium, pp. 441–446. (Cited on p. 337.)

ANTONIAK, C. (1974) Mixtures of Dirichlet processes with application to Bayesian nonparametric problems. *Ann. Stat.*, **2**, 11521–1174. (Cited on p. 93.)

APPICE, A., CIAMPI, A., MALERBA, D. AND GUCCIONE, P. (2010) Using trend clusters for spatiotemporal interpolation of missing data in a sensor network. *J. Spatial Info. Sci.*, **6**, 119–153. (Cited on p. 549.)

APPICE, A., CIAMPI, A., FUMAROLA, F., AND MALERBA, D. (2014) *Data Mining Techniques in Sensor Networks*. Springer, London. (Cited on pp. 548 and 549.)

ARAKAWA, K. AND TOMITA, M. (2013) Merging multiple omics datasets in silico: statistical analyses and data interpretation. In: Alper, H. (ed.), *Systems Metabolic Engineering*, pp. 459–470. Humana Press, Springer, New York. (Cited on p. 596.)

ARLOT, S. AND CELISSE, A. (2010) A survey of cross-validation procedures for model selection. *Stat. Surveys*, **4**, 40–79. (Cited on p. 334.)

ARCONES, M. (1995) Asymptotic normality of multivariate trimmed means. *Stat. Prob. Lett.*, **25**, 43–53. (Cited on p. 90.)

ARLEINA, O. AND OTOK, B. (2014) Bootstrap aggregating multivariate adaptive regression splines (bagging MARS) for poor households Classification in region of Jombang. *J. Sains Dan Seni Pomits*, **3**, D-91– D-96. (In Indonesian. English version: http://papers.ssrn.com/sol3/papers.cfm?abstract_id=2489898. Accessed 15 May, 2017.) (Cited on p. 471.)

ARMENAKIS, C. (1992) Estimation and organization of spatio-temporal data, In: *Proc. Canadian Conf. on GIS92*, pp. 900–911. (Cited on p. 548.)

ATIAS, N., GERSHENZON, M., LABAZIN, K., AND SHARAN, R. (2014) Experimental design schemes for learning Boolean network models. *Bioinformatics*, **30**, 445–452. (Cited on p. 571.)

ATKINSON, P. (2013) Downscaling in remote sensing. *Int. J. App. Earth Obs. Geoinform.*, **22**, 106–114. (Cited on p. 563.)

BARBIERI, M. AND BERGER, J. (2004) Optimal predictive model selection. *Ann. Stat.*, **32**, 870–897. (Cited on pp. 490 and 491.)

BARNETT, G., KOHN, R. AND SHEATHER, S. (1997) Robust Bayesian estimation of autoregressive-moving-average models. *J. Time Series*, **18**, 11–28. (Cited on p. 141.)

BARRON, A. (1991) Universal approximation bounds for superpositions of a sigmoidal function. *IEEE Trans. Inform. Theory*, **39**, 930–945. (Cited on p. 397.)

BARRON, A. AND COVER, T. (1991) Minimum complexity density estimation. *IEEE Trans. Inform. Theory*, **37**, 1034–1054. (Cited on pp. 65 and 355.)

BARRON, A. ROOS, T., AND WATANABE, K. (2014) Bayesian properties of normalized maximum likelihood and its fast computation. See: http://arxiv.org/pdf/1401.7116.pdf. Accessed 15 May, 2017. (Cited on p. 521.)

BARUTÇUŎGLU, Z. AND ALPAYDIN, E. (2003) A comparison of model aggregation methods for regression. In: Kaynak, O., Alpaydin, E., Oja, E., and Xu, L. (eds.), *Artificial Neural Networks and Neural Information Processing ICANN/ICONIP 2003*, Lecture Notes in Computer Science, vol. 2714, pp. 76–83, Springer, Berlin. (Cited on p. 488.)

BAR-YOSSEF, Z., JAYRAM, T., KUMAR, R., SIVAKUMAR, D., AND TREVISAN, L. (2004) Counting distinct elements in a data stream. In: Rolim, J. and Vadhan, S. (eds.), *Randomization and Approximation Techniques in Computer Science, RANDOM 2002*, Lecture Notes in Computer Science, vol. 2483, pp. 1–10, Springer, Berlin. (Cited on pp. 540 and 541.)

BAUER, E. AND KOHAVI, R. (1999) An empirical comparison of voting classification algorithms: bagging, boosting, and variants. *Mach. Learning*, **36**, 105–142. (Cited on p. 484.)

BAYES, T. AND PRICE, R. (1763) An essay towards solving a problem in the doctrine of chances. By the late Rev. Mr. Bayes, communicated by Mr. Price, in a letter to John Canton, M.A. and F.R.S. *Phil. Trans. Roy. Soc. London*, **53**, 370–418. (Cited on p. 34.)

BEDRICK, E., EXUZIDES, A. JOHNSON, W. AND THURMOND, M. (2002) Predictive influence in the accelerated failure time model. *Biostatistics*, **3**, 331–346. (Cited on pp. 245 and 246.)

BENEDETTI, J. (1977) On the nonparametric estimation of regression functions. *J. Roy. Stat. Soc.* Ser. B, **39**, 248–253. (Cited on pp. 268 and 269.)

BENEDIKTSSON, J., SWAIN, P., AND ERSOY, O. (1990) Neural network approaches versus statistical approaches in classification of multisource remote sensing data. *IEEE Trans. Geosci. Rem. Sensing*, **28**, 540–552. (Cited on p. 398.)

BENGIO, Y., LAMBLIN, P., POPOVICI, D., AND LAROCHELLE, H. (2007) Greedy layer-wise training of deep networks. See: `https://papers.nips.cc/paper/3048-greedy-layer-wise-training-of-deep-networks.pdf`. (Cited on p. 399.)

BERGER, J. (1980) *Statistical Decision Theory*. Springer-Verlag, New York. (Cited on p. 49.)

BERGER, J. (1994) An overview of robust Bayesian methods. *TEST*, **3**, 5–58. (Cited on p. 65.)

BERGER, J. AND PERICCHI, L. (1998) Accurate and stable Bayes model selection: the median intrinsic Bayes factor. *Sankhya* Ser. B, **60**, 1–18. (Cited on p. 494.)

BERGER, J., GHOSH, J., AND MUKHOPADHYAY, N. (2003) *J. Stat. Planning and Inference* **112**, 241–258. (Cited on p. 329.)

BERK, R. (1966) Limiting behavior of posterior distributions when the model is incorrect. *Ann. Math. Stat.*, **37**, 51–58. (Cited on pp. 452, 457, 473, and 490.)

BERLINER, L., AND HILL, B. (1988) Bayesian nonparametric survival analysis. *J. Amer. Stat. Assoc.*, **83**, 772–779. (Cited on p. 218.)

BERNARDO, J. AND SMITH, A. (2000) *Bayesian Theory*. John Wiley and Sons, Chichester, UK. (Cited on pp. 36, 51, 68, 69, and 336.)

BERROCAL, V., CRAIGMILE, P., AND GUTTORP, P. (2012) Regional climate model assessment using statistical upscaling and downscaling techniques. *Environmetrics*, **23**, 482–492. (Cited on p. 556.)

BERSANELLI, M., MOSCA, E., REMONDINI, D., GIAMPIERI, E., SALA, C., CASTELLANI, G. *et al.* (2014) Methods for the integration of multi-omics data: mathematical aspects. *BMC Bioinformatics*, **17**, Suppl. 2:15 (Cited on p. 593.)

BERTSIMAS, D. AND TSITSIKLIS, J. (1993) Simulated annealing. *Stat. Sci.*, **8**, 10–15. (Cited on p. 339.)

BESAW, L., AND RIZZO, D. (2007) Stochastic simulation and spatial estimation with multiple data types using artificial neural networks. *Water Resources Res.* **43**, W11409. (Cited on p. 398.)

BETHEL, J., AND SHUMWAY, R. (1988) Asymptotic properties of information theoretic methods of model selection. Technical Report No. 112, Statistics Department, University of California at Davis. (Cited on pp. 320, 321, 322, 323, 324, and 325.)

BEYER, K., GOLDSTEIN, J., RAMAKRISHNAN, R. AND SHAFT, U. (1999) When is 'nearest neighbors' meaningful? In: Beeri, C. and Buneman, P. (eds.), *Database Theory – ICDT99*, Lecture Notes in Computer Science, vol. 1540, pp. 217–235. Springer-Verlag, Berlin. (Cited on pp. 519, 525, and 554.)

BHATIA, N. AND VANDANA, J. (2010) Survey and nearest neighbor techniques. *Int. J. Comp. Sci. Info. Security*, **8**, 302–305. (Cited on p. 116.)

BIAU, G. (2012) Analysis of a random forests model. *J. Mach. Learning Res.*, **13**, 1063–1095. (Cited on p. 468.)

BIAU, G., DEVROYE, L., AND LUGOSI, G. (2008) Consistency of random forests and other averaging classifiers. *J. Mach. Learning Res.*, **9**, 2015–2033. (Cited on p. 473.)

BIAU, G., BLEAKLEY, K., GYORFI, L. AND OTTUCSAK, G. (2008) Nonparametric sequential prediction of time series. See: `http://arxiv.org/pdf/0801.0327.pdf`. Accessed May 15, 2017. (Cited on pp. 285 and 286.)

BIAU, G., CHAZAL, F., COHEN-STEINER, D., AND DEVROYE, L. (2011) A weighted k-nearest neighbor density estimate for geometric inference. *Elec. J. Stat.*, **5**, 294–237. (Cited on p. 280.)

BIBIMOUNE, M., ELGHAZEL, H., AND AUSSEM, A. (2013) An empirical comparison of supervised ensemble learning approaches. In: *Proc. International Workshop on Complex Machine Learning Problems with Ensemble Methods*, pp. 123–138. (Cited on p. 484.)

BICKEL, P. AND LEVINA, L. (2004) Some theory for Fisher's linear discriminant function, naive Bayes, and some alternatives when there are more variables than alternatives. *Bernoulli*, **10**, 989–1010. (Cited on p. 454.)

BILMES, J. (1998) A gentle tutorial of the EM algorithm and its application top parameter estimation for Gaussian mixture and hidden Markov models. Technical Report 97-021, Computer Science Division, Department of Electrical Engineering and Computer Science, University of California at Berkeley.

BJØRNSTAD, J. (2010) Survey sampling: a necessary journey in the prediction world. Discussion Paper 608, Division for Statistical Methods and Standards, Statistics Norway. (Cited on p. 58.)

BLAHA M., BUDOFF M., DEFILIPPIS A., BLANKSTEIN R., RIVERA J., AGATSTON A. *et al.* (2011) Associations between C-reactive protein, coronary artery calcium, and cardiovascular events: implications for the JUPITER population from MESA, a population-based cohort study. *Lancet*, **378**, 684–692. (Cited on p. 8.)

BOCK, C. (2012) Analysing and interpreting DNA methylation data. *Nat. Rev. Genetics*, **13**, 705–719. (Cited on p. 587.)

BOESE, K., FRANKLIN, D., AND KAHNG, A. (1997) Training minimal artificial neural networks architectures for subsoil object detection. In: *Proc. SPIE Aerosense-95: Detection Technologies for Mines and Minelike Targets*, vol. 2496, pp. 900–911. (Cited on p. 393.)

BOOTH, J. AND HOBERT, J. (1998) Standard errors of prediction in GLMM's: *J. Amer. Stat. Assoc.*, **93**, 262–272. (Cited on p. 193.)

BOOTH, J. AND HOBERT, J. (1999) Maximizing generalized linear mixed model likelihoods with an automated Monte Carlo EM algorithm. *J. Roy. Stat. Soc.* Ser. B, **61**, 265–285. (Cited on p. 193.)

BOTTOU, L. AND LIN, C.-J. (2007) Support vector machine solvers. In: Bottou, L., Chapelle, O., DeCoste, D., and West, J. (eds.), *Large Scale Kernel Problems*, pp. 1–28. MIT Press, Cambridge, MA. (Cited on p. 418.)

BOUREL, M. AND GHATTAS, B. (2013) Aggregating density estimators: an empirical study. *Open J. Stat.*, **3**, 344–355. (Cited on p. 475.)

BOVE, D. AND HELD, L. (2011) Hyper *g* priors for generalized linear models. *Bayes Anal.*, **6**, 387–410. (Cited on p. 99.)

BOX, G. (1979) Robustness in the strategy of scientific model building. In: Launer, R. and Wilkinson, G. (eds.), *Proc. Workshop on Robustness in Statistics*, Academic Press, San Diego, CA. (Cited on pp. 73 and 76.)

BOX, G. AND JENKINS, G. (1970) *Time Series Analysis: Forecasting and Control*. Holden-Day, San Francisco. (Cited on pp. 129 and 139.)

BOZDOGAN, H. (1993) Choosing the number of component clusters in the mixture model using a new informational complexity criterion of the inverse Osher information matrix. In: Opitz, O., Lausen, B. and Klar, R. (eds.), *Studies in Classification, Data Analysis, and Knowledge Organization*, pp. 40–54, Springer, Heidelberg. (Cited on p. 321.)

BREIMAN, L. (1994) Bagging predictors. Technical Report 421, Department of Statistics, University of California at Berkeley. (Cited on pp. 17, 453, 464, and 469.)

BREIMAN, L. (1996a) Stacked regressions. *Mach. Learning*, **24**, 49–64. (Cited on pp. 471 and 473.)

BREIMAN, L. (1996b) Bagging predictors. *Mach. Learning*, **24**, 123–140. (Cited on pp. 17, 71, and 475.)

BREIMAN, L. (1998) Arcing classifiers. *Ann. Stat.*, **26** (3), 801–849. (Cited on p. 484.)

BREIMAN, L. (2001a) Statistical modeling: The two cultures (with discussion). *Stat. Sci.*, **16**, 199–231. (Cited on pp. 73 and 123.)

BREIMAN, L. (2001b) Random forests. *Mach. Learning*, **45**, 5-32. (Cited on pp. 466 and 467.)

BREIMAN, L., FRIEDMAN, J., STONE, C., AND OLSHEN, R. (1984) *Classification and Regression Trees*. CRC Press, Boca Raton, FL. (Cited on pp. 28 and 380.)

BRESLOW, N. AND CROWLEY, J. (1974) A large sample study of the life table and product-limit estimates under random censorship. *Ann. Stat.*, **2**, 437–453. (Cited on pp. 210, 211, and 213.)

BROCKWELL, P. AND DAVIS, R. (1987) *Time Series: Theory and Methods*. Springer, New York. (Cited on pp. 129 and 134.)

BROCKWELL, S. AND GORDON, I. (2001) A comparison of statistical methods for meta-analysis. *Stat. Med.*, **20**, 825–840. (Cited on p. 586.)

BROWN, P., VANUCCI, M., AND FEARN, T. (1998) Multivariate Bayes variable selection and prediction. *J. Roy. Stat. Soc.* Ser. B, **60**, 627–641. (Cited on p. 180.)

BROWNLEE, J. (2016) How to build an ensemble of machine learning algorithms in R (ready-to-use boosting, bagging and stacking). See: http://machinelearningmastery.com/machine-learning-ensembles-with-r/. Accessed May 15, 2017. (Cited on p. 498.)

BRYANT, P. AND CORDERO-BRANA, O. (2000) Model selection using the minimum description length principle. *Amer. Statist.*, **54**, 257–268. (Cited on p. 41.)

BUCK, M. AND LIEB, J. (2003) ChIP-chip: considerations for the design, analysis, and application of genome-wide chromatin immunoprecipitation experiments. *Genomics*, **83**, 349–360. (Cited on p. 591.)

BÜHLMANN, P. AND HOTHORN, T. (2007) Boosting algorithms: regularization, prediction, and model fitting. *Stat. Sci.*, **22**, 477–505. (Cited on p. 481.)

BURGES, C. (1998) A tutorial on support vector machines for pattern recognition. *Data Mining and Knowledge Discovery*, **2**, 121–168. (Cited on pp. 418 and 419.)

BURKE, R. (2002) Hybrid recommender systems: survey and experiments. *User Modeling and User-Adapted Interaction*, **12**, 331–370. (Cited on p. 535.)

BURNHAM, K., AND ANDERSON, D. (2002) *Model Selection and Multimodel Inference: A Practical Information-Theoretic Approach.* 2nd edn. Springer-Verlag, New York. (Cited on p. 134.)

BURRIDGE, J. (1981) Empirical Bayes analysis of survival time. *J. Roy. Stat. Soc.* Ser. B, **43**, 65–75. (Cited on p. 234.)

CAMPBELL, G. AND HOLLANDER, M. (1982) Prediction intervals with a Dirichlet process prior distribution. *Can. J. Stat.*, **10**, 103–111. (Cited on p. 92.)

CANADA CENTRE FOR MAPPING AND EARTH OBSERVATION (2016) *Fundamentals of Remote Sensing.* See: http://www.nrcan.gc.ca/node/9309. (Cited on p. 563.)

CAO, D.-S., XU, Q.-S., LIANG, Y.-Z., ZHANG, L.-X., AND LI, H.-D. (2010) The boosting: a new idea of building models. *Chemo. Int. Lab. Syst.*, **100**, 1–11. (Cited on p. 499.)

CARROLL, R., RUPPERT, D., STEFANSKI, L. AND CRAINICEANU, C. (2006) *Measurement Error in Nonlinear Models.* Chapman and Hall, Boca Raton, FL. (Cited on p. 447.)

CARUANA, R. AND MIZIL, A. (2006) An empirical comparison of supervised learning algorithms. In: *Proc. 23rd Internat. Conf. on Machine Learning, ICML'06 .* (Cited on p. 484.)

CECHEN, A. AND BATTISTELLA, E. (2007) Interpretation of feedforward neural networks for secondary structure prediction. *Int. J. Hybrid Intell. Sys.*, **4**, 3–16. (Cited on p. 399.)

CESA-BIANCHI, N. AND LUGOSI, G. (2006) *Prediction, Learning and Games.* Cambridge University Press, Cambridge. (Cited on p. 521.)

CHAKRABARTI, A. (2014) Data stream algorithms lecture notes. CS49, Dartmouth College. See: www.cs.dartmouth.edu/ ac/Teach/CS49-Fall11/Notes/lecnotes.pdf. Accessed May 9, 2017. (Cited on pp. 537, 540, 541, 542, 543, 544, and 546.)

CHAKRABORTY, S. (2011) Bayesian semi-supervised learning with support vector machine. *Stat. Method.*, **8**, 68–82 (Cited on p. 414.)

CHAKRABORTY, S. (2012). Bayesian multiple response kernel regression model for high dimensional data and its practical applications in near infrared spectroscopy. *Comp. Stat. Data Anal.*, **56**, 2742–2755 (Cited on p. 17.)

CHAKRABORTY, S., GHOSH, M., AND MALICK, B. (2012). Bayesian non linear regression for large-p small-n problems. *J. Mult. Anal.*, **108**, 28–40. (Cited on p. 17.)

CHAKRABORTY, S., GHOSH, M., AND MALICK, B. (2013). Bayesian hierarchical kernel machines for nonlinear regression and classification. In: Damien, P., Dellaportas, P., Polson, and N., Stephens D. (eds.), *Bayesian Theory and Applications*, pp. 50–69. (Cited on pp. 410, 411, and 415.)

CHANDLER, R. AND SCOTT, E. (2010) *Statistical Methods for Trend Detection and Analysis in the Environmental Sciences.* John Wiley and Sons, London. (Cited on pp. 129, 139, and 158.)

CHANDLER, R. AND SKOURAS, R. (1998) *Forecasting Class Notes.* Department Statistical Science, University College, London. (Cited on pp. 129, 134, 137, and 138.)

CHANDOLA, V., BANNERJEE, A., AND KUMAR, V. (2009) Anomaly detection: a survey. *ACM Comput. Surv.*, **41**, Art. 15 (Cited on p. 551.)

CHANDOLA, V., BANNERJEE, A., AND KUMAR, V. (2012) Anomaly detection for discrete sequences: a survey. *IEEE Trans. Know. Data. Eng.*, **24**, 823–839. (Cited on pp. 551 and 552.)

CHANG, M. AND SHUSTER, J. (1994) Interim analysis for randomized clinical trials: simulating the log-rank test statistic. *Biometrics*, **50**, 827–833. (Cited on p. 218.)

CHAOUACHI, A. AND NAGASALA, K. (2012) A novel ensemble neural network based short-term wind power generation forecasting in a microgrid. *ISESCO J. Sci. Tech.*, **8**, 2–8. (Cited on p. 469.)

CHARIKAR, M., CHEKURI, C., FEDER, T., AND MOTWANI, R. (1997) Incremental clustering and dynamic information retrieval. In: *Proc. 29th Annual ACM Symp. on Theory of Computing*, pp. 626–635. (Cited on p. 546.)

CHARIKAR, M., CHEN, K., AND FARACH-COLTON, M. (2004) Finding frequent items in data streams. *Theor. Comput. Sci.*, **312**, 3–15. (Cited on p. 542.)

CHARPENTIER, A. (2015) An attempt to understand boosting algorithms. See: `http://www.r-bloggers.com/an-attempt-to-understand-boosting-algorithms/`. Accessed May 15, 2017. (Cited on p. 499.)

CHATFIELD, C. (1995) Model uncertainty, data mining and statistical inference. *J. Roy. Stat. Soc.* Ser. A, **158**, 419–44. (Cited on p. 86.)

CHEN, K. AND LEI, J. (2015) Network cross-validation for determining the number of communities in network data. See: `https://arxiv.org/abs/1411.1715v2`. Accessed May 15, 2017. (Cited on pp. 579 and 580.)

CHEN, K. AND LO, S. (1997) On the rate of uniform convergence of the product limit estimator: weak and strong laws. *Ann. Stat*, **25**, 1050–1087. (Cited on pp. 210, 211, and 212.)

CHEN, J., WANG, C., AND WANG, R. (2009) Using stacked generalization to combine SVMs in magnitude and shape feature spaces for classification of hyperspectral data. *IEE Trans. Geosci. Rem. Sensing*, **47**, 2193–2205. (Cited on p. 476.)

CHI, E. AND REINSEL, G. (1989) Models for longitudinal data with random effects and AR(1) errors. *J. Amer. Statist. Assoc.*, **84**, 452–459. (Cited on pp. 189 and 190.)

CHI, M. (2010) Decision trees and model selection (AIC and BIC). See: `https://www.cs.cmu.edu/~aarti/Class/10701/recitation/decisiontree_modelselection.pdf`. Accessed May 15, 2017. (Cited on p. 499.)

CHIB, S. AND GREENBERG, E. (1994) Bayes inference in regression models with $ARMA(p,q)$ errors. *J. Econometrics*, **64**, 183–206. (Cited on pp. 158 and 159.)

CHIB, S. AND GREENBERG, E. (1995) Understanding the Metropolis–Hastings algorithm. *Amer. Stat.* **49**, 327–325. (Cited on pp. 340, 344, and 345.)

CHIPMAN, H., GEORGE, E., AND MCCULLOCH, R. (1998) Bayesian CART model search. *J. Amer. Stat. Assoc.*, **93**, 935–960. (Cited on pp. 383, 384, 385, and 460.)

CHIPMAN, H., GEORGE, E., AND MCCULLOCH, R. (2001) The practical implementation of Bayes model selection. In: *Model Selection*, IMS Lecture Notes, vol. 28, 67–116. (Cited on p. 453.)

CHIPMAN, H., GEORGE, E., AND MCCULLOCH, R. (2007) Bayesian ensemble learning. In: Schölkopf, B., Platt, J., and Hoffman T., (eds.), *Advances in Neural Information and Processing Systems*, vol. 19, 265–272, MIT Press. (Cited on p. 385.)

CHIPMAN, H., GEORGE, E., AND MCCULLOCH, R. (2010) BART: Bayesian additive regression trees. *Ann. Appl. Stat.*, **4**, 266–298. (Cited on pp. 385, 386, and 499.)

CHITSAZAN, N., NADIRI, A., AND TSAI, F. (2015) Prediction and structural uncertainty analyses of artificial neural networks using hierarchical Bayesian model averaging. *J. Hydrology*, **528**, 52–62. (Cited on p. 499.)

CHOROMANSKA, A., HENAFF, F., MATHIEU, M., AROUS, G., AND LECUN, Y. (2015) The loss surfaces of multilayer networks. See: `https://arxiv.org/abs/1412.0233` (Cited on p. 400.)

CHRISTENSEN, O. AND JENSEN, T. (1999) An introduction to the theory of bases frames and wavelets. See: `http://citeseerx.ist.psu.edu/viewdoc/summary?doi=10.1.1.17.241`. Accessed May 15, 2017. (Cited on p. 443.)

CHRISTENSEN, R., JOHNSON, W., BRANSCUM, A. AND HANSON, T. (2011) *Bayesian Ideas and Data Analysis*. Chapman & Hall/CRC Press. (Cited on p. 110.)

CIAMPI, A., AND THIFFAULT, J. (1988) Recursive partitioning in biostatistics: stability of trees and choice of the most stable classification. In: *Compstat*, International Association for Computing. Physica-Verlag Heidelberg for IASC. (Cited on p. 382.)

CIAMPI, A., APPICE, A., AND MALERBA, D. (2010) Summarization for geographically distributed data streams. In: *Proc. 14th International Conf. on Knowledge-Based and Intelligent Information and Engineering Systems*, Lecture Notes in Computer Science, Vol. 62–78, 339–348. Springer. (Cited on p. 549.)

CIZEK, P. (2004) Asymptotics of least trimmed squares regression. Humbold University of Berlin, School of Business and Economics. Center Discussion Paper No. 2004–72. (Cited on p. 495.)

CIZEK, P. (2005) Least trimmed squares in nonlinear regression under dependence. *J. Stat. Planning and Inference*, **136**, 3967–3988. (Cited on p. 495.)

CLARK, A., KALETA, E., ARORA, A., AND WOLK, D. (2013) Matrix-assisted laser desorption ionization time of flight mass spectrometry: a fundamental shift in the routine practice of clinical microbiology. *Clin. Microbiology Rev.*, **26**, 547–603. (Cited on p. 590.)

CLARKE, B. (1996) Implications of reference priors for prior information and for sample size. *J. Amer. Stat. Assoc.*, **91**, 173–184. (Cited on p. 602.)

CLARKE, B. (2003) Comparing Bayes model averaging and stacking when model approximation error cannot be ignored. *J. Mach. Learing Res.*, **4**, 683–712. (Cited on pp. 100 and 474.)

CLARKE, B. (2007) Information optimality and Bayesian modeling. *J. Econ.*, **138**, 405–429. (Cited on p. 521.)

CLARKE, B. (2010) Desiderata for a predictive theory of statistics, *Bayes Anal.*, **5**, 283–318. (Cited on pp. 65 and 339.)

CLARKE, B. (2013) Comment on HAMPSON, L. AND JENNISON, C. (2013) Group sequential tests for delayed responses (with discussion). *J. Roy. Stat. Soc.* Ser. B, **75**, 45–46. (Cited on p. 251.)

CLARKE, B. AND BARRON, R. (1988) Information-theoretic asymptotics of Bayes methods. Technical Report 26, Deptartment of Statistics, University of Illinois. (Cited on p. 329.)

CLARKE, B. AND FOKOUE, E. (2011) Bias variance tradeoff for prequential model list selection. *Stat. Papers*, **52**, 813–833. (Cited on p. 320.)

CLARKE, B. AND SEVERINSKI, C. (2010) Subordinators, adaptive shrinkage and a prequential comparison of three sparsity methods. In: *Bernardo, J. et al.* (eds.), Bayesian Statistics, vol. 9, pp. 523–528, Oxford University Press. (Cited on p. 10.)

CLARKE, B. AND SUN, D. (1999) Asymptotics of the expected posterior. *Ann. Inst. Stat. Math.*, **51**, 163–185. (Cited on p. 251.)

CLARKE, B., FOKOUE, E., AND ZHANG, H. (2009) *Principles and Theory for Data Mining and Machine Learning*. Springer Series in Statistics, New York. (Cited on pp. 111, 115, 205, 316, and 523.)

CLARKE, B., CLARKE, J. AND YU, C.-W. (2014) Statistical problem classes and their links to information theory. *Econometric Rev.*, **33**, 337–371. (Cited on pp. 6, 302, and 321.)

CLARKE, J. AND CLARKE, B. (2009) Prequential analysis of complex data with adaptive model reselection. *Stat. Anal. and Data Mining*, **2**, 274–290. (Cited on p. 64.)

CLARKE, J., CLARKE, B., AND YU, C.-W. (2013) Prediction in $\mathcal{M}$-complete problems with limited sample size. *Bayes Anal.*, **8**, 647–690. (Cited on pp. 8, 100, 134, 302, 452, and 493.)

CLARKE, B., VALDES, C., DOBRA, A., AND CLARKE, J. (2015) Bayesian approach to metagenomic profiling in bacteria. *Stat. Interface*, **8**, pp.173–185. (Cited on p. 23.)

CLEMEN, R. (1989) Combining forecasts: A review and annotated bibliography. *Int. J. Forecasting*, **5**, 559–583. (Cited on p. 451.)

CLYDE, M. (2012) Bayes perspectives on combining models. Slides from invited talk at ISBA 2012; pdf on request from the author. (Cited on pp. 474 and 475.)

CLYDE, M. AND GEORGE, E. (2004) Model uncertainty. *Stat. Sci.*, **19**, 81–94. (Cited on pp. 332, 454, 457, and 458.)

CLYDE, M. AND IVERSEN, E. (2012) Bayes model averaging in the $\mathcal{M}$-open framework. In: Dellaportas, P., *et al.* (eds.), *Bayesian Theory and Applications*, pp. 484–498, Oxford University Press. (Cited on p. 475.)

COLE, T. AND GREEN, P. (1992) Smoothing reference centile curves: the LMS method and penalized likelihood. *Stat. Med.* **11**, 1305-1319. (Cited on pp. 204 and 205.)

COLLETT, D. (1994) *Modeling Survival Data in Medical Research*. Chapman & Hall, London. (Cited on pp. 213, 216, and 217.)

COOK, J., AND STEFANSKI, L. (1994) Simulation-extrapolation estimation in parametric measurement error models. *J. Am. Stat. Assoc.*, **89**, 1314–1328. (Cited on p. 382.)

COOK, N. (2008) Statistical evaluation of prognostic versus diagnostic models: beyond the ROC curve. *Clin. Chem.* **54**, 17–23. (Cited on p. 221.)

CORMEN, T., LEISERSON, C., RIVEST, R., AND STEIN, C. (2009) *Introduction to Algorithms* (3rd ed.). MIT Press and McGraw-Hill. (Cited on p. 546.)

CORMODE, G. AND MUTHUKRISHNAN, S. (2012) Approximating data with the count-min sketch. *IEEE Software*, **29**, 64–69. (Cited on p. 542.)

CORTEZ, P. AND MORAIS, A. (2007) A data mining approach to predict forest fires using meteorological data. In: Neves, J. *et al.* (eds.), *Proc. 13th EPIA Portuguese Conf. on Artificial Intelligence*, pp. 512–523. (Cited on p. 20.)

COX, D. (1972) Regression models and life tables (with discussion). *J. Roy. Stat. Soc.* Ser. B, **74**, 187–220. (Cited on pp. 227 and 229.)

COX, R. (1961) *Algebra of Probable Inference*. Johns-Hopkins University Press, Baltimore, MD. (Cited on p. 51.)

CRANE, H. AND DEMPSEY, W. (2016) A framework for statistical network modeling. See: https://arxiv.org/abs/1509.08185v4, Accessed May 15, 2017. (Cited on p. 579.)

CREMENISI, P. AND RICCI, F. (eds.) (2017) *Proc. Rec Sys' 17: 11th ACM Conf. on Recommender Systems*. Association for Computing Machinery, New York. (Cited on p. 535.)

CRESSIE, N. AND WIKLE, C. (2011) *Statistics for Spatio-Temporal Data*. Wiley, Hoboken, NJ. (Cited on pp. 558, 561, and 562.)

CUNNINGHAM, P. AND DELANY, S. (2007) k-nearest neighbor classifiers. Technical Report UCD-CSI-2007-4. (Cited on p. 116.)

CUNNINGHAM, P., CARNEY, J., AND JACOB, S. (1999) Stability problems with artificial neural networks and ensemble selection. *Art. Intell. in Med.*, **20**, 217–225. (Cited on p. 470.)

CZADO, C. (2009) Predictive model assessment for count data. *Biometrics*, **65**, 1254–1261. (Cited on pp. 153 and 156.)

CZEPIEL, A. (2002) Maximum likelihood estimation of logistic regression models: theory and implementation. See: http://czep.net/stat/mlelr.pdf. Accessed May 15, 2017. (Cited on p. 109.)

DABBS, B. AND JUNKER, B. (2016) Comparison of cross-validation methods for stochastic block models. See: https://arxiv.org/abs/1605.03000v1. Accessed May 15, 2017. (Cited on pp. 579 and 580.)

DANIELS, M., CHATTERJEE, A., AND WANG, C. (2012) Bayes model selection for incomplete data using the posterior predictive distribution. *Biometrics*, **68**, 1055–1063. (Cited on p. 354.)

DASGUPTA, S. (2008) *Asymptotic Theory of Statistics and Probability*. Springer, New York. (Cited on p. 88.)

DASGUPTA, S. AND PATURI, M. (2013) Online streaming algorithms for clustering. See: https://cseweb.ucsd.edu/ dasgupta/291-geom/. Accessed May 15, 2017. (Cited on pp. 546 and 547.)

DAVIDIAN, M. AND GILTINAN, D. (2003) Nonlinear models for repeated measurements: an overview and update. *J. Agricultural, Biological, and Environmental Stat.*, **8**, 387–419. (Cited on p. 194.)

DAVIS, C. (2002) *Statistical Methods for the Analysis of Repeated Measurements*. Springer, New York. (Cited on p. 166.)

DAVIS, R. AND DUNSMUIS, T. (1997) Least absolute deviation estimation for regression with *ARMA* errors. *J. Theor. Probab.*, **10**, 481–497. (Cited on p. 158.)

DAWID, A. (1982) The well-calibrated Bayesian. *J. Amer. Stat. Assoc.*, **77**, 605–610. (Cited on pp. 59, 150, 153, and 329.)

DAWID, A. (1984) Present position and potential developments: some personal views: statistical theory: the prequential approach. *J. Roy. Stat. Soc.* Ser. B, **147**, 278–292. (Cited on pp. 64, 82, and 153.)

DAWID, A. (1992) Prequential analysis, stochastic complexity and Bayesian inference (with discussion). In: Bernardo, J., Berger, J., Dawid A., and Smith, A., *Bayesian Statistics*, vol. 4, pp. 109–125, Oxford University Press. (Cited on p. 3.)

DAWID, A. AND MUSIO, M. (2015) Bayes model selection based on proper scoring rules. *Bayes Analysis*, **10**, 479–499. (Cited on p. 601.)

DAWID, A. AND VOVK, V. (1999) Prequential probability: Principles and properties. *Bernoulli*, **5**, 125–162. (Cited on p. 82.)

DAWID, A., LAURITZEN, S., AND PARRY, M. (2012) Proper local scoring rules on discrete sample spaces. *Ann. Stat.*, **40**, 593–608. (Cited on pp. 26 and 219.)

DE BOOR, C. (2001) *A Practical Guide to Splines*, revised edn. Springer, New York. (Cited on p. 303.)

DE GOOIJER, J., ABRAHAM, B., GOULDE, A., AND ROBINSON L. (1985). Methods for determining the order of an autoregressive-moving average process: a survey. *Int. Stat. Rev. A*, **53**, 301–329. (Cited on p. 132.)

DE ROSA, R. (2016) Confidence decision trees via online and active learning for streaming (big) data. See: `https://arxiv.org/abs/1604.03278v1`. Accessed May 15, 2017. (Cited on p. 551.)

DENIL, M., MATHESON, D., AND DE FREITAS, N. (2014) Narrowing the gap: random forests in theory and practice. In: *Proc. 31st Int. Conf. on Machine Learning*, Beijing. (Cited on p. 468.)

DENNISON, D., MALLICK, B., AND SMITH, A. (1998) A Bayesian CART algorithm. *Biometrika*, **85**, 363–377. (Cited on p. 383.)

DEVROYE, L. AND GYORFI, L. (1985) *Nonparametric Density Estimation: The L^1 View.* John Wiley and Sons, New York. (Cited on p. 89.)

DEVROYE, L., AND WAGNER, T. (1977) The strong uniform consistency of nearest neighbor estimates. *Ann. Stat.*, **5**, 536–540. (Cited on p. 275.)

DEVROYE, L., KRZYZAK, A., AND LUGOSI, G. (1994) On the strong universal consistency of nearest neighbor regression function estimates. *Ann. Stat.*, **22**, 1371–1385. (Cited on p. 116.)

DIAZ, I., HUBBARD, A., DECKER, A., AND COHEN, M. (2013) Variable importance and prediction methods for longitudinal problems with Missing Variables. University of California at Berkeley Division of Biostatistics Working Paper Series, No. 318. See: `http://biostats.bepress.com/ucbbiostat/paper318`. (Cited on p. 161.)

DIGGLE, P. (2014) *Statistical Analysis of Spatial and Spatio-Temporal Point Patterns*. 3rd edn. Chapman and Hall/CRC Press, Boca Raton. (Cited on pp. 567 and 568.)

DIGGLE, P., LIANG, K., AND ZEGER, S. (1996) *Analysis of Longitudinal Data*, Oxford University Press, Oxford. (Cited on p. 188.)

DIGGLE, P., HEAGERTY, P., LIANG, K., AND ZEGER, S. (2002) *Analysis of Longitudinal Data*, 2nd edn. Oxford University Press, Oxford. (Cited on pp. 175 and 183.)

DINAKER, K., WEINSTEN, E., LIEBERMAN, H., AND SELMAN, R. (2014) Stacked generalization learning to analyze teenage distress. In: Adar, E., Resnick, P., Choudhury, M., Hogan, R., and Oh, A. (eds.), *Proc. 8th Int'l Association for the Advancement of Artificial Intelligence Conf. on Weblogs and Social Media*, The AAAI Press. (Cited on p. 477.)

DING, S., WU, FULIN, QIAN, J., JIA, H., and JIN, F. (2015) Research on data stream clustering algorithms. *Artif. Intell. Rev.*, **43**, 593. (Cited on p. 546.)

DOMINGOS, P. (2000) A unified bias–variance decomposition for zero–one and squared loss. In: *Proc. 17th National Conf. on Artificial Intelligence and 12th Conf. on Innovative Applications of Artificial Intelligence (AAAI/IAAI-00)*, pp. 564–569. (Cited on pp. 65 and 339.)

DOMINGOS, P., AND PAZZANI, M. (1997) On the optimality of the simple Bayesian classifier under zero–one loss. *Mach. Learning*, **29**, 103–130. (Cited on p. 111.)

DORAI-RAJ, S. (2001) First and second order properties of spatio–temporal point processes in the space–time and frequency domains. PhD thesis, Virginia Polytechnic University. (Cited on p. 567.)

DRAPER, D. (1995) Assessment and propagation of model uncertainty. *J. Roy. Stat. Soc.* Ser. B, **57**, 45–97. (Cited on pp. 8, 9, 93, 218, 310, and 451.)

DRAPER, D. (1997) On the relationship between model uncertainty and inferential/predictive uncertainty. Unpublished manuscript. (Cited on p. 8.)

DRAPER, D. (2014a) Bayesian model specification: toward a theory of applied statistics. In preparation. (Cited on pp. 356 and 357.)

DRAPER, D. (2014b) Personal communication. (Cited on p. 333.)

DRUCKER, H. (1997) Improving regressors using boosting techniques. In: *Proc. 14th International Conf. on Machine Learning*, pp. 107–115, Morgan Kaufmann, San Francisco, CA. (Cited on pp. 487 and 488.)

DUDA, R., HART, P., AND STORK, D. (2000) *Pattern Classification*. Wiley, New York. (Cited on p. 115.)

DURRET, D. (2006) *Random Graph Dynamics*. Cambridge University Press. (Cited on p. 581.)

DUTTA, S. AND GHOSH, A. (2016) Some novel transformations of data for nearest neighbor classification in high dimensions. *Mach. Learning*, **102**, p. 5783. (Cited on p. 116.)

DŽEROSKI, S. AND ŽENKO, B. (2004) Is combining classifiers with stacking better than selecting the best one? *Mach. Learning*, **54**, 255–273. (Cited on p. 473.)

EAKAMBARAM, S. AND ELANGOVAN, R. (2015) *Least Absolute Deviation: Theory and Methods*. Lambert Academic Publishing, Germany. (Cited on p. 495.)

EASLEY, D. AND KLEINBERG, J. (2010) *Networks, Crowds, and Markets*. Cambridge University Press, Cambridge. (Cited on p. 579.)

EBRAHIMI, N., SOOFI, E., AND SOYER, R. (2010) On the sample information about parameter and prediction. *Stat. Sci.*, **25**, 348–367. (Cited on p. 600.)

EFRON, B. (2014) Estimation and accuracy after model selection. *J. Amer. Stat. Assoc.*, **109**, 991–1006. (Cited on p. 312.)

EHLERS, R. AND BROOKS, S. (2013) Bayesian analysis of order uncertainty in arima models. See: `http://citeseerx.ist.psu.edu/viewdoc/download?rep=rep1&type=pdf&doi =10.1.1.129.2692`. Accessed May 15, 2017. (Cited on p. 141.)

ELLIS, B. (2014) *Real-Time Analytics: Techniques to Analyze and Visualize Streaming Data*. John Wiley and Sons. (Cited on p. 537.)

FAN, J. AND FAN, Y. (2008) High dimensional classification using features annealed independence rules. *Ann. Stat.*, **36**, 2605–2637. (Cited on pp. 519 and 525.)

FAN, J. AND LI, R. (2001) Variable selection via nonconcave penalized likelihood and its oracle properties. *J. Amer. Stat. Assoc.*, textbf96, 1348=1360. (Cited on pp. 70, 424, 426, 427, and 428.)

FAN, J. AND LV, J. (2008) Sure independence screening for ultrahigh dimensional feature space. *J. Roy. Stat. Soc. Ser B*, **70**, 849–911. (Cited on pp. 319, 363, 424, 520, and 594.)

FAYE, P.-A., DRUILHET, P., AZZAOUI, N., AND YAO, F. (2015) Bayesian spatio-temporal kriging with misspecified black-box. See: `https://hal.archives-ouvertes.fr/hal-01226169`. Accessed May 15, 2017. (Cited on pp. 558 and 561.)

FELLER, W. (1948) On the Kolmogorov–Smirnov limit theorems for empirical distributions. *Ann. Math. Stat.*, **19**, 177–189.

FELLER, W. (1968) *An Introduction to Probability Theory and Its Applications*, vol. 1, 3rd edn. Wiley and Sons, New York. (Cited on p. 152.)

FERAUD, R. AND CLEROT, F. (2002) A methodology to explain neural network classification. *Neural Networks*, **15**, 237–246. (Cited on p. 398.)

FERGUSON, T. (1973) A Bayesian analysis of some nonparametric problems. *Ann. Stat.*, **1**, 209–230. (Cited on p. 91.)

FERGUSON, T. (1974) Prior distributions on spaces of probability measures. *Ann. Stat.*, **2**, 615–629. (Cited on p. 289.)

FIENBERG, S., AND WASSERMAN, S. (1981) An exponential family of probability distributions for directed graphs: comment. *J. Amer. Stat. Assoc.*, **76**, 54–57. (Cited on p. 576.)

FISHER, R.A. (1936) The use of multiple measurements in taxonomic problems. *Ann. Eugenics* **7**, 179–188. (Cited on p. 113.)

FITZMAURICE, G., LAIRD, N., AND WARE, J. (2004) *Applied Longitudinal Analysis*. Wiley Series in Probability and Statistics, New York. (Cited on pp. 162 and 183.)

FITZMAURICE, G., DAVIDIAN, M., VERBEKE, G., AND MOLENBERGHS, G. (2009) *Longitudinal Data Analysis*, edited volume. Chapman and Hall/CRC, Boca Raton, FL. (Cited on p. 183.)

FIX, E., AND HODGES, J. (1951) Discriminatory analysis, nonparametric discrimination: consistency properties. Report No. 4, Project No. 21-49-004, USAF School of Aviation Medicine. (Cited on pp. 116 and 275.)

FLIGNER, M., AND WOLFE, D. (1979) Nonparametric prediction for a future sample median. *J. Amer. Stat. Assoc.*, **74**, 453–456. (Cited on p. 250.)

FOLDES, A., AND REJTO, L. (1981) Strong uniform consistency for nonparametric survival curve estimators from randomly censored data. *Ann. Stat.*, **9**, 122–129. (Cited on pp. 210 and 211.)

FOSDICK, B., LARREMORE, D., NISHIMURA, J., AND UGANDER, J. (2012) Configuring random graph models with fixed degree sequences. See: https://arxiv.org/abs/1608.00607v1. Accessed May 15, 2017. (Cited on p. 579.)

FOSTER, D., AND STINE, R. (2004) Variable selection in data mining: building a model for bankruptcy. *J. Amer. Stat. Assoc.*, **99**, 303–313. (Cited on p. 354.)

FRANSES, P. (2002) Testing for residual autocorrelation in growth curve models. *Tech. Forecasting and Soc. Change*, **69**, 195–204. (Cited on p. 203.)

FRANZ, T. (2007) Ecohydrology of the upper ewaso Ngiro river basin, Kenya. MS thesis, Department of Environmental Engineering, Princeton University. (Cited on p. 194.)

FRANZ, T., CAYLOR, K., NORDBOTTOM, J., RODRIGUEZ-ITURBE, I., AND CELIA, M. (2010) An eco-hydrological approach to predicting regional woody species distribution patterns in dryland ecosystems. *Adv. Water Res.*, **33**, 215–230. (Cited on p. 194.)

FRANZ, T., WANG, T., AVERY, W., FINKENBINER, C., AND BROCCA, L. (2015) Spatio-temporal characterization of soil moisture fields using cosmic ray neutron probes and data fusion. *Geophys. Res. Lett.*, **42**(9), 3389–3396. (Cited on p. 295.)

FREEDMAN, D. AND PURVES, R. (1969) Bayes method for bookies. *Ann. Math. Stat.*, **40**, 1177–1186. (Cited on pp. 51 and 330.)

FRUEND, Y. (1995) Boosting a weak learning algorithm by majority. *Inform. Comp.*, **121**, 256–285. (Cited on p. 481.)

FREUND, Y. (2001) An adaptive version of the boost by majority algorithm. *Mach. Learning*, **43**, 293–318. (Cited on p. 481.)

FREUND, Y., AND SCHAPIRE, R. (1995) A decision-theoretic generalization of online learning and an application of boosting. In: *Proc. 2nd European Conf. on Computational Learning Theory*, pp. 23–37. (Cited on p. 481.)

FREUND, Y., AND SCHAPIRE, R. (1996) Experiments with a new boosting algorithm. In: Saitta, L. (ed.), *Proc. 13th International Conf. on Machine Learning*, pp. 148–156, Morgan-Kauffman. (Cited on pp. 481 and 483.)

FREUND, Y., AND SCHAPIRE, R. (1997) A decision-theoretic generalization of online learning and an application of boosting. *J. Comp. Syst. Sci.*, **55**, 119–139. (Cited on p. 481.)

FREUND, Y., AND SCHAPIRE, R. (1999) A short introduction to boosting. *J. Japanese Soc. Artifical Intell.*, **44**, 771–780. (Cited on p. 484.)

FREUND, Y., AND SCHAPIRE, R. (2012) *Boosting: Foundations and Algorithms.* MIT Press, Cambridge, MA. (Cited on pp. 484 and 485.)

FREY, J. (2013) Data-driven non-parametric prediction intervals. *J. Stat. Planning Inference*, **143**, 1039–1048. (Cited on p. 251.)

FRIEDMAN, J. (1991) Multivariate adaptive regression splines. *Ann. Stat.*, **19**, 1–67. (Cited on p. 443.)

FRIEDMAN, J. (2001) Greedy function approximation: a gradient boosting machine. *Ann. Stat.*, **29**, 1189–1232. (Cited on p. 488.)

FRIEDMAN, J. AND STUETZLE, W. (1981) Projection pursuit regression. *J. Amer. Stat. Assoc.*, **78**, 817–823. (Cited on p. 443.)

FRIEDMAN, J., HASTIE, T. AND TIBSHIRANI, R. (2000) A statistical view of boosting. *Ann. Stat.*, **28**, 337–347. (Cited on pp. 31, 485, and 486.)

FUENTES, M. (2011) Spatio-temporal model. See: http://www.stat.ncsu.edu/people/fuentes/courses/madrid/lectures/spacetime.pdf. Accessed May 15, 2017. (Cited on pp. 558, 561, and 569.)

FULLER, W. (1987) *Measurement Error Models.* Wiley and Sons, New York. (Cited on p. 447.)

FYFE, C. AND LAI, P. (2000) Kernel and nonlinear canonical correlation analysis. *Int. J. Neural Syst.*, **10**, 365–377. (Cited on p. 409.)

GALLANT, R. (1987) *Nonlinear Statistical Models.* John Wiley and Sons, New York. (Cited on p. 365.)

GALTON, F. (1907) Vox populi (The wisdom of crowds). *Nature*, **75**, 450–451. (Cited on p. 450.)

GASSER, T. AND MÜLLER, H. (1984) Estimating regression functions and their derivatives by the kernel method. *Scand. J. Stat.*, **11**, 171–185. (Cited on pp. 267, 268, and 269.)

GASSIAT, E. AND ROUSSEAU, J. (2014) About the posterior distribution in hidden Markov models with unknown number of states. *Bernoulli*, **20**, 2039–2075. (Cited on p. 447.)

GEISSER, S. (1993) *Predictive Inference: An Introduction.* Chapman and Hall, New York. (Cited on pp. 1 and 43.)

GELFAND, A., AND GHOSH, S. (1998) Model choice: a minimum posterior predictive loss approach. *Biometrika*, **85**, 1–11. (Cited on pp. 354 and 355.)

GELMAN, A., STEVENS, M., AND CHAN, V. (2003) Regression modeling and meta-analysis for decision making: a cost benefit analysis of incentives in telephone surveys. *J. Bus. Econ. Stat.*, **21**, 213–225. (Cited on p. 57.)

GELMAN, A., CARLIN, J., STERN, H., AND RUBIN, D. (2004) *Bayesian Data Analysis*, 2nd edn. Chapman and Hall/CRC Press, Boca Raton, FL. (Cited on pp. 110, 163, and 354.)

GELMAN, A., AND SHIRLEY, K. (2011) Inference from simulations and monitoring convergence. In: Crooks, S. *et al.* (eds.), *Handbook of Markov Chain Monte Carlo*, pp. 163–174, Chapman and Hall/CRC, Boca Raton, FL. (Cited on p. 345.)

GEORGE, E. (1999). Sampling considerations for model averaging and model search. Invited discussion of model averaging and model search, by M. Clyde. In: Bernardo, J., Berger, J., Dawid, A. P., and Smith, A. (eds.), *Bayesian Statistics 6*, 175–177. Oxford University Press. (Cited on p. 327.)

GEORGE, E. (2010) Dilution priors: compensating for model space redundancy. In: *Borrowing Strength: Theory Powering Applications. A Festschrift for Lawrence D. Brown*, IMS Collections, vol. 6, 158–165. (Cited on pp. 327 and 453.)

GEORGE, E., AND MCCULLOCH, R. (1993) Variable selection via Gibbs sampling. *J. Amer. Stat. Assoc.*, **88**, 881–889. (Cited on p. 346.)

GEORGE, E., LIANG, F., AND XU, X. (2012) From minimax shrinkage estimation to minimax shrinkage prediction. *Stat. Sci.*, **27**, 82–94. (Cited on p. 310.)

GEYER, C. (2011) Introduction to Markov chain Monte Carlo. In: Crooks *et al.* (eds.), *Handbook of Markov Chain Monte Carlo*, pp. 3–48, Chapman and Hall/CRC, Boca Raton, FL. (Cited on p. 345.)

GHORBANI, A., AND OWRANGH, K. (2001) Stacked generalization in neural networks: generalization on statistically neutral problems. In: *Proc. International Joint Conf. on Neural Networks*, pp. 1715–1720. (Cited on p. 476.)

GHOSH, J. (2015) Bayes model selection using the median probability model. *WIREs Comput. Stat.*, **7**, 185–193. (Cited on p. 490.)

GHOSH, J. AND ACHARYA, A. (2014) Cluster ensembles: theory and applications. In: Aggarwal, C. and Reddy, C. (eds.), *Data Clustering Algorithms and Applications*, pp. 551–570, Chapman and Hall, Boca Raton, FL. (Cited on p. 597.)

GHOSH, S. AND GHOSAL, S. (2006) Semiparametric accelerated failure time models for censored data. In: Upadhyay, S., Singh, U., and Dey, D. (eds.), *Bayesian Statistics and Its Applications*, Chapter 15, Anamaya Publishers, New Delhi. (Cited on p. 245.)

GHOSH, M. AND MEEDEN, G. (1997) *Bayesian Methods for Finite Population Sampling.* Chapman and Hall, Boca Raton, FL. (Cited on p. 57).

GHOSH, J. AND RAMAMOORTHI, R. (2003) BAYESIAN NONPARAMETRICS Springer, New York. (Cited on pp. 91, 220, and 291.)

GHOSH, M., MERGEL, V., AND DATTA, G. (2008) Estimation, prediction, and the Stein phenomenon under divergence loss. *J. Mult. Anal.*, **99**, 1941–1961. (Cited on pp. 309 and 310.)

GHOSHAL, S. (2010) Dirichlet process, related priors, and posterior asymptotics. In: Hjort, N., Holmes, C., Muller, P., and Walker, S. (Eds.), *Bayesian Nonparametrics*, pp. 35–79, Cambridge University Press. (Cited on p. 94.)

GHOSHAL, S., GHOSH, J., AND VAN DER VAART, A. (2000) Convergence rates for posterior distributions. *Ann. Stat.*, **28**, 500–531. (Cited on p. 291.)

GHOSAL, S., TURNBULL, B., ZHANG, H., AND HWANG, W. (2016) Sparse penalized forward selection for support vector classification. *J. Comp. Graph. Stat.*, **25**, 493–514. (Cited on pp. 595 and 596.)

GIESBRECHT, F. AND GUMPERTZ, M. (2004) *Planning, Construction, and Statistical Analysis of Comparative Experiments.* Wiley and Sons, Hoboken, NJ. (Cited on p. 169.)

GILL, R. (1983) Large sample behavior of the product limit estimator on the whole line. *Ann. Stat.*, **11**, 49–58. (Cited on p. 210.)

GIRI, N. (2004) *Multivariate Statistical Analysis: Revised and Expanded.* Dekker, New York. (Cited on p. 165.)

GNEITING, T. (2011) Making and evaluating point forecasts. *J. Amer. Stat. Assoc.*, **106**, 746–762. (Cited on pp. 26, 153, and 156.)

GNEITING, T., BALABDAOUI, F., AND RAFTERY, A. (2007) Probabilistic forecasts, calibration and sharpness. *J. Roy. Stat. Soc.* Ser. B, **69**, 243–268. (Cited on pp. 150 and 153.)

GOFFE, W., FERRIER, G., AND ROGERS, J. (1994) Global optimization of statistical functions with simulated annealing. *J. Econ.*, **60**, 65–99. (Cited on pp. 342 and 392.)

GOLDENBERG, A., ZHENG, A., FEINBERG, S., AND AIROLDI, E. (2009) A survey of statistical networks. See: https://arxiv.org/abs/0912.5410v1. Accessed May 15, 2017. (Cited on pp. 570, 573, 574, 576, 581, 582, and 583.)

GONG, P. (1996) Integrated analysis of spatial data from multiple sources: using evidential reasoning and artificial neural network techniques for geological mapping. *Photogrammetric Eng. Rem. Sensing*, **62**, 513–523. (Cited on p. 585.)

GONZÁLEZ, I. (2014) Statistical analysis of RNA-seq data. Plateforme Bioinformatique – INRA Toulouse, and Plateforme Biostatistique – IMT Université Toulouse III. (Cited on p. 589.)

GONZÁLEZ, J., RODRÍGUEZ-CORTÉS, F., CRONIE, O., AND MATEU, J. (2016) Spatio-temporal point process statistics: a review. *Spatial Stat.*, **18**, 505–544. (Cited on pp. 565, 567, and 568.)

GORDON, K. (1924) Group judgments in the field of lifted weights. *J. Exp. Psych.*, **7**, 398–400. (Cited on p. 451.)

GOWEN, C. AND FONG, S. (2013) Linking RNA measurements and proteomics with genome-scale models. In: Alper, H. (ed.), *Systems Metabolic Engineering*, pp. 429–446. Humana Press, Springer, New York. (Cited on p. 596.)

GREENWOOD, M. (1926) The errors of sampling of the survivorship tables. *Rep. Health and Stat. Subjects* **33**, 1–26. (Cited on p. 214.)

GUHA, S. AND RYAN, L. (2006) Spatio-temporal analysis of areal data and discovery of neighborhood relationships in conditionally autoregressive models. Harvard University Biostatistics Working Paper Series, No. 61. (Cited on pp. 568 and 569.)

GUHA, S., MEYERSON, A., MISHRA, N., MOTWANI, R., AND O'CALLAGHAN, L. (2003). Clustering data streams: theory and practice. *IEEE Trans. Knowl. Data Eng.*, **15**, 515528. (Cited on p. 547.)

GUPTA, S., AND SAXENA, S. (2016) *Real-Time Big Data Analytics.* Packt Publishing. (Cited on p. 537.)

GUTIERREZ-PENA, E., AND WALKER, S. (2001) A Bayesian predictive approach to model selection. *J. Stat. Planning Inf.*, **93**, 259–276. (Cited on p. 72.)

GWALANI, H., MIKLER, R., RAMISETTY-MIKLER, S., AND ONEILL, M. (2017) Collection and integration of multi-spatial and multi-type data for vulnerability analysis in emergency response plans. In: Wohlgemuth, V., Fuchs-Kittowski, F., and Wittmann, J. (eds.), *Advances and New Trends in Environmental Informatics.*, pp. 89–101, Springer International, Switzerland. (Cited on p. 599.)

GYORFI, L. (1984) Adaptive linear procedures under general conditions, *IEEE Trans. Inform. Theory*, **30**, 262–267. (Cited on p. 286.)

GYORFI, L., AND OTTUCSAK, G. (2011) Nonparametric sequential prediction of stationary time series. In: *Machine Learning for Financial Engineering*, pp. 179–226, World Scientific (Imperial College Press). (Cited on p. 285.)

GYORFI, L., AND OTTUCSAK, G. (2012) Nonparametric sequential prediction of stationary time series. In: *Machine Learning for Financial Engineering*, pp. 186–230, World Scientific (Imperial College Press). (Cited on p. 259.)

GYORFI, L., KOHLER, M., KRZYZAK, A., AND WALK, A. (2002) *A Distribution-Free Theory of Nonparametric Regression.* Springer, New York. (Cited on pp. 257, 258, 259, 260, and 282.)

HA, W., GOWDA, P., AND HOWELL, T. (2013) A review of downscaling methods for remote sensing-based irrigation management: part I. *Irrig. Sci.*, **31**, 831–850. (Cited on p. 563.)

HAHN, P., AND CARVALHO, C. (2015) Decoupling shrinkage and selection in Bayesian linear models: a posterior summary perspective. *J. Amer. Stat. Assoc.*, **110**, 435–448. (Cited on p. 603.)

HAHN, P., CARVALHO, C., PUELZ, D., AND HE, J., (2018) Regularization and confounding in linear regression for treatment effect estimation. *Bayesian Anal.*, **13**, pp. 163–182. (Cited on p. 603.)

HAJEK, B. (1988) Cooling schedules for optimal annealing. *Math. Oper. Res.*, **13**, 311–329. (Cited on p. 341.)

HALL, P., MARRON, J. S., NEEMAN, A. (2005) Geometric representation of high dimension, low sample size data. *J. Roy. Stat. Soc.* Ser. B, **67**, 427–444. (Cited on p. 525.)

HALL, P., PARK, B., AND SAMWORTH, R. (2008) Choice of neighbor order in nearest neighbors classification. *Ann. Stat.*, **36**, 2135–2152. (Cited on p. 304.)

HAND, D. AND YU, K. (2001) Idiot's Bayes – not so stupid after all? *Int. Stat. Rev.*, **69**, 385–398. (Cited on p. 454.)

HANNAN, E. (1980) Estimation of the order of an ARMA process. *Ann. Stat.* **8**, 1071–1081. (Cited on p. 134.)

HANNAN, J., AND B. QUINN (1979) The determination of the order of an autoregression. *J. Roy. Stat. Soc.* Ser B, **41**, 190–195. (Cited on p. 321.)

HANNIG, J. (2009) On generalized fiducial inference. *Stat. Sinica*, **19**, 491–544. (Cited on p. 41.)

HANS, C., DOBRA, A., AND WEST, M. (2007) Shotgun stochastic search for large-p regression. *J. Amer. Stat. Assoc.*, **102**, 507–516. (Cited on pp. 345 and 346.)

HARMS, S., TADESSE, T., AND WARDLOW, B. (2009) Algorithm and feature selection for Vegout: a vegetation prediction tool. In: Gama, J. *et al.* (eds.), *Discovery Science, Proc. 12th International Conf., DS 2009*, Lecture Notes in Computer Science, vol. 5808, pp. 107–120, Springer, New York. (Cited on pp. 434 and 435.)

HASSAN, S., KHOSRAVI, A., AND JAAFAR, J. (2013) Bayes model averaging of load demand forecasts from neural networks. In: *Proc. 2013 IEEE International Conf. on Systems, Man, and Cybernetics (SMC)*, pp. 3192–3197. (Cited on p. 460.)

HASTAD, J. (1986) Almost optimal lower bounds for small depth circuits. In *Proc. 18th Annual ACM Symp. on Theory of Computing*, pp. 6–20. (Cited on p. 399.)

HASTAD, J. (1991) On the power of small-depth threshold circuits. *Comp. Complex.*, **1**, 113–129. (Cited on p. 399.)

HASTINGS, W. (1970) Monte Carlo sampling methods using Markov chains and their applications. *Biometrika*, **57**, 97–109. (Cited on p. 344.)

HAUSSLER, D. (1990) Probably approximately correct learning. In: *Proc. 8th National Conf. on Artificial Intelligence*, pp. 1101–1108. Morgan Kaufman. (Cited on p. 600.)

HAY, G., NIEMAN, K., AND GOODENOUGH, D. (1997) Spatial thresholds, image-objects and upscaling: a multi-scale evaluation. *Rem. Sensing Environ.*, **62**, 1–19. (Cited on p. 564.)

HENDERSON, R. AND KEITING, N. (2005) Individual survival time prediction using statistical models. *J. Med. Ethics.*, **31**, 703–706. (Cited on p. 233.)

HENDERSON, R., JONES, M. AND STARE, J. (2001) Accuracy of point predictions in survival analysis. *Stat. Med.*, **20**, 3083–3096. (Cited on p. 233.)

HIGGINS, P. AND THOMPSON, S. (2002) Quantifying heterogeneity in a meta-analysis *Stat. Med.*, **21**, 1539–1558. (Cited on p. 586.)

HINTON, G. AND SALAKHUTDINOV, S. (2006) Reducing the dimensionality of data with neural networks. *Science*, **313**, 504–507. (Cited on p. 404.)

HINTON, G., OSINERO, A., AND TEH, Y.W. (2006) A fast learning algorithm for deep belief nets. *Neural Comp.*, **18**, 1527–1554. (Cited on pp. 400 and 404.)

HJORT, N. (1990) Nonparametric Bayes estimators based on beta processes in models for life history data. *Ann. Stat.*, **18**, 1259–1294. (Cited on pp. 207 and 221.)

HJORT, N. AND CLAESKENS, G. (2003) The focused information criterion. *J. Amer. Stat. Soc.*, **98**, 900–945. (Cited on p. 321.)

HOBBS, J., BRAVERMAN, A., BRYNJARSDÓTTIR, J., CRESSIE, N., FU, D., GRANAT, R. *et al.* (2016) Remote sensing retrievals for atmospheric carbon dioxide: quantifying uncertainty in the presence of nonlinearity and nuisance parameters. Talk given at Joint Statistical Meeting 2016; pdf on request from the authors. (Cited on p. 564.)

HOCHREITER, J. (1991) Untersuchungen zu dynamischen neuronalen Netzen. PhD thesis, Institut für Informatik, Technische Universität München. (Cited on p. 400.)

HOETING, J., MADIGAN, D., RAFTERY, A., AND VOLINSKY, C. (1999) Bayes model averaging: a tutorial. *Stat. Sci.*, **14**, 382–417. (Cited on pp. 56, 454, and 457.)

HOFF, P. (2008) Modeling homophily and stochastic equivalence in symmetric relational data. In: *Advances in Neural Information Processing Systems*, pp. 657–664. (Cited on p. 579.)

HÖHLE, M., PAUL, M., AND HELD, L. (2007) Statistical approaches to the surveillance of infectious diseases for veterinary public health. Technical Report No. 14, Deptartment of Statistics, University of Munich. (Cited on p. 568.)

HOLLAND, P. AND LEINHARDT, S. (1976) Local structure in social networks. *Social Meth.*, **7**, 145. (Cited on pp. 576 and 577.)

HOSMER, D. AND LEMESHOW, S. (1999) *Applied Survival Analysis*. Wiley, New York. (Cited on p. 245.)

HOSMER, D., AND LEMESHOW, S. (2008) *Applied Survival Analysis: Regression Modeling of Time to Event Data*, 2nd edn. John Wiley and Sons, New York. (Cited on p. 222.)

HOSMER, D., AND LEMESHOW S. (2000) *Applied Logistic Regression*, 2nd edn. John Wiley and Sons, New York. (Cited on p. 109.)

HOTHORN, T., HORNIK, K., AND ZEILEIS, A. (2004) Unbiased recursive partitioning: a conditional inference framework. Report No. 8, Department of Statistics and Mathematics, Wirtschaftsuniversität Vienna. (Cited on p. 382.)

HOTHORN, T., BÜHLMANN, P., KNEIB, T., SCHMID, M., AND HOFNER, B. (2010) Model-based boosting 2.0. *J. Mach. Learning Res.*, **11**, 2109–2113. (Cited on p. 481.)

HOY, P. AND QUAIFE, T. (2015) Probabilistic downscaling of remote sensing data with applications for multiscale biogeochemical flux modeling. *PLoS One*, **10**. (Cited on p. 563.)

HUBER, P. (1964). Robust estimation of a location parameter. *Ann. Math. Stat.*, **35**, 73–101. (Cited on p. 338.)

HUBER, P. (1973). Robust regression. *Ann Stat.*, **1**, 799–821. (Cited on p. 338.)

HUBER, P. (1985) Projection pursuit. *Ann. Stat.*, **13**, 435–475. (Cited on p. 444.)

HUFKENS, K., BOGAERT, J., DONG, Q., LU, L., HUANG, C., MA, M. *et al.* (2008) Impacts and uncertainties of upscaling of remote-sensing data validation for a semi-arid woodland. *J. Arid Environ.*, **72**, 1490–1505. (Cited on p. 564.)

HWANG, W., ZHANG, H., AND GHOSAL, S. (2009) FIRST: combining forward iterative selection and shrinkage in high dimensional sparse linear regression. *Stat. Interface*, **2**, 341–348. (Cited on pp. 595 and 596.)

IBRAHIM, J., AND LAUD, P. (1991) On Bayes analysis of generalized linear models using Jeffreys' prior. *J. Amer. Stat. Assoc.*, **86**, 981–986. (Cited on p. 459.)

IBRAHIM, J., CHEN, M., AND SINHA, D. (2001) *Bayes Survival Analysis*. Springer, New York. (Cited on pp. 234, 235, 241, 245, and 354.)

INDYK, P. (2006) Stable distributions, pseudorandom generators, embeddings, and data stream computation. *J. ACM*, **53**, 307–323. (Cited on p. 541.)

INTRATOR, O., AND INTRATOR, N. (2001) Interpreting neural networks results: a simulation study. *Comp. Stat. Data Anal.*, **37**, 373–393. (Cited on p. 399.)

ISHWARAN, H., AND RAO, S. (2005) Spike and slab variable selection: frequentist and Bayesian strategies. *Ann. Stat.*, **33**, 730–773. (Cited on p. 101.)

ISHWARAN, H., KOGAHUR, U., AND RAO, S. (2010) **spikeslab**: prediction and variable selection using spike and slab regression. *The R J.*, **2**, 68–73. (Cited on p. 101.)

ISLAM, M., YAO, X, NIRJON, M., ISLAM, M., AND MURASE, K. (2008) Bagging and boosting negatively correlated networks. *IEEE Syst. Man., Cyber.*, **38**, 771–784. (Cited on p. 469.)

JAMES, G., HASTIE, T. AND SUGAR, C. (2000) Principal component models for sparse functional data. *Biometrika*, **87**, 587–602. (Cited on p. 191.)

JANNACH, D., ZANKER, M., FELFERNIG, A., AND FRIEDRICH, G. (2011) *Recommender Systems: An Introduction*. Cambridge University Press, New York. (Cited on pp. 526, 532, and 535.)

JARA, A., HANSON, T., QUINTANA, F., MUELLER, P., AND ROSNER, G. (2011). DPpackage: Bayesian semi- and nonparametric modeling in R. *J. Stat. Software*, **40**, 1–30. See: http://www.jstat soft.org/v40/i05/. (Cited on p. 298.)

JAYNES, E. (2003) *Probability Theory: The Logic of Science.* Cambridge University Press, Cambridge, UK. (Cited on pp. 34 and 51.)

JIANG, J. (2007) *Linear and Generalized Linear Mixed Models and their Applications.* Springer, New York. (Cited on pp. 162 and 187.)

JOHANNESSON, G., DEVECIGIL, D., AND LAW, D. (2016) Applications of remote sensing in precision agriculture at the climate corporation. Talk given at Joint Statistical Meeting 2016; pdf on request from the authors. (Cited on p. 565.)

JOHNDROW, J., DUNSON, D., AND LUM, K. (2013) Diagonal orthant multinomial probit models. In: *Proc. JMLR Workshop and Conf.*, vol. 31, pp. 29–38. (Cited on p. 24.)

JONES, L. (1987) On a conjecture of Huber concerning the convergence of projection pursuit regression. *Ann. Stat.*, **15**, 880–882. (Cited on p. 444.)

JONES, M., MARRON, J., SHEATHER, S. (1996) A brief survey of bandwidth selection for density estimation. *J. Amer. Stat. Assoc.*, **91**, 401–407. (Cited on p. 263.)

JONES, Z., AND LINDER F. (2015) Exploratory data analysis using random forests. See: http://zmjones.com/static/papers/rfss_manuscript.pdf. Accessed May 15, 2017. (Cited on p. 469.)

JORISSEN, R., LIPTON, L., GIBBS, P., CHAPMAN, M., DESAI, J., JONES, I. *et al.* (2008) DNA copy number alterations underlie gene expression differences between microsatellite stable and unstable colorectal cancers. *Clin. Cancer Res.*, **14**, 8061–8069. (Cited on p. 119.)

KALAI, A. AND SERVEDIO R. (2005) Boosting in the presence of noise. *J. Comp. Syst. Sci.*, **71**, 266–290. (Cited on p. 481.)

KALBFLEISCH, J. (1978) Non-parametric Bayes analysis of survival time data. *J. Roy. Stat. Soc.* Ser. B, **40**, 214–221. (Cited on p. 234.)

KALBFLEISCH, J. AND PRENTICE, R. (1973) Marginal likelihoods based on Cox's regression and life model. *Biometrika*, **60**, 267–278. (Cited on p. 229.)

KAMIMURA, R. (2009) Selective enhancement learning in competitive learning. In: *Proc. IEEE International Joint Conf. on Neural Networks*, pp. 1497–1502. (Cited on p. 399.)

KANG, E., MA, P., BRAVERMAN, A., NGUYEN, H., AND CRESSIE, N. (2016) Statistical downscaling for large spatial data and its applications. Talk given at Joint Statistical Meeting 2016; pdf on request from the authors. (Cited on p. 565.)

KAPLAN, E. AND MEIER, P. (1958). Nonparametric estimation from incomplete observations. *J. Amer. Stat. Assoc.*, **53** 457–481. (Cited on p. 208.)

KASS, R. AND RAFTERY, A. (1995) Bayes factors. *J. Amer. Stat. Assoc.*, **90**, 773–795. (Cited on p. 22.)

KATZ, R. (2002) Stochastic modeling of hurricane damage. *J. Amer. Met. Soc.*, **41**, 754–762. (Cited on p. 152.)

KIM, H.-C., AND GHAHRAMANI, Z. (2012) Bayesian classifier combination. *J. Mach. Learning Res., Proceedings Papers* **22**. See: http://jmlr.csail.mit.edu/proceedings/papers/v22/kim12/kim12.pdf. (Cited on p. 454.)

KIM, H.-C., PANG, S., JE, H.-M., KIM, D., AND BANG, S.-Y. (2003) Constructing support vector machine ensemble[s]. *Pattern Recog.*, **36**, 2757–2767. (Cited on p. 470.)

KIM, J. AND POLLARD, D. (1990) Cube root asymptotics. *Ann. Stat.*, **18**, 191–219. (Cited on p. 495.)

KIM, S., HERAZO-MAYA, J., KANG, D., JUAN-GUARDELA, B., TEDROW, J., MARTINEZ, F. *et al.* (2015) Integrative phenotyping framework (iPF): integrative clustering of multiple omics data identifies novel lung disease subphenotypes. *BMC Genomics*, **16**, 924. (Cited on p. 599.)

KIM, Y. AND LEE, J. (2004) On posterior consistency of survival models. *Ann. Stat.*, **29**, 666–686. (Cited on p. 220.)

KIM, Y. AND RILOFF, E. (2015) Stacked generalization for medical concept extraction from clinical notes. In: *Proc. Biomedical Natural Language Processing Workshop*, pp. 61–70. (Cited on p. 477.)

KIMMELDORF, G. AND WAHBA, G. (1971) Some results on Tchebycheffian spline functions. *J. Math. Anal. Applic.*, **33**, 82–95. (Cited on p. 408.)

KIRK, W., FEINSOD, M., FAVIS, R., KLIMAN, R., AND BARANY, R. (2002) Single nucleotide polymorphism seeking long term association with complex disease. *Nucl. Acids Res.*, **30**, 3295–3311. (Cited on p. 587.)

KITSANTAS, P., HOLLANDER, M., AND LEI, M. (2007) Assessing the stability of classification trees using Florida birth data. *J. Stat. Planning Inf.*, **137**, 3917–3929. (Cited on p. 382.)

KLEIN, J. AND MOESCHBERGER, M. (2003) *Survival Analysis*, 2nd edn. Springer, New York. (Cited on pp. 216, 220, 229, 231, and 241.)

KOENKER, R. (2005) *Quantile Regression*. Cambridge University Press, Cambridge. (Cited on p. 102.)

KOENKER, R. AND BASSET, G. (1978) Regression quantiles. *Econometrika*, **46**, 33–50. (Cited on p. 101.)

KOENKER, R. AND HALLOCK, K. (2001) Quantile regression. *J. Econ. Perspectives*, **15**, 143–156. (Cited on p. 101.)

KOEPKE, H. AND CLARKE, B. (2011) On the limits of clustering in high dimensions via cost functions. *Stat. Anal. Data Mining*, **4**, 30–53. (Cited on pp. 519, 525, and 597.)

KOLACYZK, E. (2009) *Statistical Analysis of Network Data*. Springer, New York. (Cited on p. 572.)

KOLACZYK, E., JU, J., AND GOPAL, S. (2012) Multiscale, multigranular statistical image segmentation. *J. Amer. Stat. Assoc.*, **100**, 1358–1369. (Cited on p. 564.)

KONTKANEN, P. AND MYLLYMAKI, P. (2007) A linear-time algorithm for computing the multinomial stochastic complexity. *Inform. Proc. Lett.*, **103**, 227–233. (Cited on p. 521.)

KOTZ, S. AND NADARAJAH, S. (2004) *Multivariate T-distributions and their Applications*. Cambridge University Press. (Cited on p. 164.)

KRIGE, D. (1951) A statistical approach to some basic mine valuation problems on the Witwatersrand. *J. Chem., Metal. Mining Soc. South Africa*, **52**, 119–139. (Cited on p. 557.)

KRISHNAMOORTHY, K. AND MATHEW, T. (2009) *Statistical Tolerance Regions: Theory, Applications, and Computation*. John Wiley and Sons, Hoboken, NJ. (Cited on p. 250.)

KRUHØFFER, M., JENSEN, J., 2, LAIHO, P., DYRSKJOT, L., SALOVAARA, R., ARANGO, D. *et al.* (2005) Gene expression signatures for colorectal cancer microsatellite status and HNPCC. *Brit. J. Cancer*, **92**, 2240–2248. (Cited on p. 119.)

KUO, L. AND GHOSH, S. (2001) Bayes nonparametric inference for non-homogeneous Poisson processes. North Carolina State University Institute of Statistics Mimeo Series No. 2530. (Cited on p. 151.)

KVAM, P. AND VIDAKOVIC, B. (2007) *Nonparametric Statistics with Applications to Science and Engineering*. John Wiley and Sons, Hoboken, NJ. (Cited on p. 249.)

LACHENBRUCH, P. AND MICKEY, M. (1968) Estimation of error rates in discriminant analysis. *Technometrics*, **10**, 1–11. (Cited on p. 334.)

LAIRD, N. AND WARE, J. (1982) Random-effects models for longitudinal data. *Biometrics*, **38**, 963–974. (Cited on p. 188.)

LAMBERT-LACROIX, S. AND ZWALD, L. (2011). Robust regression through Huber's criterion and the adaptive lasso penalty. *Elec. J. Stat.*, **5**, 1015–1053. (Cited on p. 338.)

LAPLACE, S.-P. (1774) Memoir on the probability of the causes of events. English translation by Stephen Stigler, 1986. *Stat. Sci.*, **1**, 364–378. (Cited on p. 34.)

LAUD, P., AND IBRAHIM, J. (1995) Predictive model selection. *J. Roy. Stat. Soc.* Ser. B, **57**, 247–262. (Cited on p. 355.)

LAVINE, M. (1992) Some aspects of Polya trees for statistical modeling. *Ann. Stat.*, **20**, 1222–1235. (Cited on pp. 290 and 291.)

LAVINE, M (1994) More aspects of Polya tree distributions for statistical modeling. *Ann. Stat.*, **22**, 1161–1176. (Cited on p. 289.)

LE, T. AND CLARKE, B. (2016a) Using the Bayesian Shtarkov solution for prediction. *Comp. Stat. Data Analysis*, **104**, 183–196. (Cited on p. 521.)

LE, T. AND CLARKE, B. (2016b) Model averaging vs. model selection: a Bayes interpretation. In revision for acceptance. (Cited on p. 474.)

LE, T. AND CLARKE, B. (2017) A Bayes interpretation of stacking for $\mathcal{M}$-complete and $\mathcal{M}$-open settings. *Bayes. Anal.*, **12**, 807–829. (Cited on p. 474.)

LECUN, Y., BOTTOU, L., ORR, G., MÜLLER, K.R. (1998) Efficient backprop. In: *Neural Networks: Tricks of the Trade*, pp. 9–50, Lecture Notes in Computer Science, vol. 1524 (Cited on pp. 400 and 403.)

LEE, H. (2001) Model selection for neural network classification. *J. Class.*, **18**, 227–243. (Cited on p. 395.)

LEE, H. (2004) *Bayesian Nonparametrics via Neural Networks*. ASA-SIAM Series on Statistics and Applied Probability. Society for Industrial and Applied Mathematics, Philadelphia. (Cited on p. 395.)

LEE, H. (2005) Default priors for neural network classification. Technical Report No. 2005-16, University of California, Santa Cruz, Department of Applied Mathematics and Statistics. (Cited on p. 395.)

LEE, K., CHAKRABORTY, S., AND SUN, J. (2011) Bayes variable selection in semiparametric proportional hazards models for high dimensional data. *Int. J. Biostat.*, Issue 1, Art. 21. (Cited on pp. 234 and 235.)

LEEB, H., AND PÖTSCHER, B. (2005) Model selection and inference: facts and fiction. *Econ. Theory*, **21**, 21–59. (Cited on p. 428.)

LEEB, H., AND PÖTSCHER, B. (2008) Sparse estimators and the oracle property, or the return of Hodges estimator. *J. Econometrics*, **142**, 201–211. (Cited on p. 428.)

LEY, E. AND STEEL, M. (2009) On the effect of prior assumptions on Bayes model averaging with applications to growth processes. *J. Appl. Econ.*, **24**, 651–674. (Cited on p. 349.)

LI, K. (1987) Asymptotic optimality for C_p, C_l, cross-validation, and generalized cross-validation: Discrete index set. *Ann. Stat.*, **15**, 958–975. (Cited on p. 337.)

LIANG, F. (2005) Bayes neural nets for nonlinear time series forecasting. *Stat. and Comp.*, **15**, 13–29. (Cited on p. 460.)

LIANG, F. AND BARRON, A. (2004) Exact minimax strategies for predictive density estimation, data compression, and model selection. *IEEE Trans. Inform. Theory*, **50**, 2708–2726.

LIANG, F., PAULO, R., MOLINA, G., CLYDE, M., AND BERGER, J. (2008) Mixtures of g-priors for Bayesian variable selection. *J. Amer. Stat. Assoc.*, **103**, 410–423. (Cited on pp. 100 and 349.)

LIANG, W., COLVIN, J., SANSÓ, B., AND LEE, H. (2010) Modeling and anomaly detection for event occurrences following an inhomogeneous spatio-temporal Poisson process. U.C. Santa Cruz Technical Report SOE-10-09. (Cited on p. 567.)

LIANG, G., ZHU, X., AND ZHANG, C. (2011) An empirical study of bagging predictors for different learning algorithms. In: *Proc. 25th AAAI Conf. on Artificial Intelligence*. (Cited on p. 471.)

LIANG, M., LI, Z., CHEN, T., AND ZENG, J. (2015) Integrative data analysis of multi-platform cancer data with a multimodal deep learning approach. *IEEE Trans. Comp. Biol. and Informatics*, **12**, 928–937. (Cited on p. 599.)

LIAO, F., PORPORATO, A., RIDOLFI, L., AND RODRIGUEZ-ITURBE, I. (2001) Plants in water-controlled ecosystem: active role in hydrologic processes and response to water stress II. Probabilistic soil moisture dynamics. *Adv. Water Res.*, **24**, 707–723. (Cited on p. 152.)

LICHTENSTERN, A. (2013) Kriging methods in spatial statistics. Bachelor's thesis, Deptartment of Mathematics, Technische Universität München. (Cited on pp. 558 and 560.)

LIN, X., PITTMAN, J., AND CLARKE, B. (2007) Information conversion, effective samples, and parameter size. *IEEE Trans. Inform. Theory*, **53**, 4438–4456. (Cited on p. 602.)

LINDSTROM, M., AND BATES, D. (1990) Nonlinear mixed effects models for repeated measures data. *Biometrics*, **46**, 673–687. (Cited on p. 194.)

LINK, C. (1984). Confidence intervals for the survival function using Cox's proportional-hazard model with covariates. *Biometrics*, **40**, 601–610. (Cited on p. 232.)

LIU, X. (2001) Kernel smoothing for spatially correlated data. PhD dissertation, Department of Statistics, Iowa State University. (Cited on p. 268.)

LOADER, C. (1999) *Local Regression and Likelihood*. Springer, New York. (Cited on p. 255.)

LOCK, E. AND DUNSON, D. (2013) Bayesian consensus clustering. *Bioinformatics*, **29**, 2610–2616. (Cited on pp. 598 and 599.)

LOFTSGAARDEN, D. AND QUESENBERRY, C. (1965) A nonparametric estimate of a multivariate density function. *Ann. Math. Stat.*, **36**, 1049–1051. (Cited on p. 275.)

LOHR, S. (2010) *Sampling Design and Analysis* 2nd edn. Brooks/Cole, Boston MA. (Cited on p. 572.)

LONG, P. AND SERVEDIO, R. (2010) Random classification noise defeats all convex potential boosters. *Mach. Learning*, **78**, 287–304 (Cited on p. 486.)

LOUPPE, G. (2012) Understanding random forests. PhD dissertation, Deptartment of Electrical Engineering and Computer Science, University of Liège (Cited on p. 468.)

LOVÁSZ, L. (2012) *Large Networks and Graph Limits*. American Mathematical Society, Providence, RI. (Cited on p. 579.)

LU, J., WU, D., MAO, M., WANG, W., AND ZHANG, G. (2015) Recommender system application developments: a survey. *Decision Support Syst.*, **74**, 12–32. (Cited on p. 537.)

LUM, K. AND ISAAC, W. (2016) To predict and serve. *Significance*, **13**, 14–19. (Cited on p. 537.)

LUO, X., STEFANSKI, L., AND BOOS, D. (2007) Tuning variable selection procedures by adding noise. *Technometrics*, **48**, 165–175. (Cited on pp. 343 and 382.)

MACEACHERN, S., AND MUELLER, P. (1998) Estimating mixtures of Dirichlet process models. *J. Comp. and Graph. Stat.*, **7**, 223–238. (Cited on pp. 93 and 94.)

MACK, Y. (1981) Local properties of k-NN regression estimates. *SIAM J. Alg. Disc. Meth.*, **2**, 311–323. (Cited on pp. 282 and 284.)

MACK, Y. AND ROSENBLATT, M. (1979) Multivariate k-nearest neighbor density estimates. *J. Mult. Anal.*, **9**, 1–15. (Cited on pp. 276, 278, 279, and 283.)

MACLIN, R., AND OPITZ, D. (1997) An empirical evaluation of bagging and boosting. In: *Proc. AAAI'97/IAAI'97*, pp. 546–551. (Cited on p. 469.)

MADIGAN, D. AND RAFTERY, A. (1994) Model selection and accounting for model uncertainty in graphical models using Occam's window. *J. Amer. Stat. Assoc.*, **89**, 1535–1546. (Cited on p. 461.)

MALLAT, S. (1989) A theory for multiresolution signal decomposition: the wavelet representation. *IEEE Trans. Pattern Anal. Mach. Intell.*, **11**, 674–693. (Cited on p. 443.)

MANNING, C., RAGHAVAN, P., AND SCHÜTZE, H. (2008) *Introduction to Information Retrieval.* Cambridge University Press. (Cited on p. 533.)

MARCHETTI, Y., NGUYEN, H., BRAVERMAN, A., AND CRESSIE, N. (2016) Spatial data compression via adaptive spatial dispersion clustering. Talk given at Joint Statistical Meeting 2016; pdf on request from the authors. (Cited on p. 565.)

MARIELLA, L., AND TARANTINO, M. (2010) Spatial temporal conditional auto-regressive model: a new autoregressive matrix. *Austrian J. Stat.*, **39**, 223–244. (Cited on p. 569.)

MARIN, J.-M. AND ROBERT, C. (2007) *A Practical Guide to Computational Bayes Statistics.* Springer, New York. (Cited on p. 331.)

MARTÍNEZ-MUÑOZ, G., AND SUÁREZ, A. (2006) Pruning in ordered bagging ensembles. In: *Proc. 23rd International Conf. on Machine Learning*, pp. 609–616. (Cited on p. 470.)

MATHERON, G. (1962) *Traité de géostatistique appliquée.* Editions Technip. (Cited on p. 557.)

MAYR, A., BINDER, H., GEFELLER, O., AND SCHMID, M. (2014) Extending statistical boosting. *Methods Inf. Med.*, **53**, 428–435. (Cited on p. 488.)

MCCARTHY, J. AND JACOB, T. (2009) Who are you? A data mining approach to predicting survey nonrespondents. In: *Proc. 64th Annual Conf. of the American Association for Public Opinion Research*. See: http://www.amstat.org/sections/srms/proceedings/y2009/Files/400007.pdf. Accessed May 15, 2017. (Cited on p. 62.)

MCCULLOUGH, C. (2005) Repeated measures ANOVA, R.I.P.? *Chance*, **18**, 29–33. (Cited on p. 172.)

MCGREGOR, A. (2017) See: http://people.cs.umass.edu/ mcgregor/book/book.html. Forthcoming. (Cited on p. 537.)

MCKAY, C. (2004) Automatic genre classification of MIDI recordings. M. A thesis, Faculty of Music, McGill University. (Cited on p. 26.)

MEASE. D. AND WYNER, A. (2008) Evidence contrary to the statistical view of boosting. *J. Mach. Learning Res.*, **9**, 131–156. (Cited on pp. 485 and 486.)

MEEHL, P. (1954) *Clinical Versus Statistical Prediction: A Theoretical Analysis and a Review of the Evidence.* University of Minnesota Press, Minneapolis. (Cited on p. 361.)

MEHLHORN, K. AND SUN, H. (2014) Streaming algorithms. In: *Proc. Conf. on Great Ideas in Theoretical Computer Science*, Saarland University, Summer 2014. See: http://resources.mpi-inf.mpg.de/departments/d1/teaching/ss14/gitcs/notes3.pdf. Accessed May 15, 2017. (Cited on pp. 541 and 542.)

MEIER, P. (1975) Estimation of a distribution function from incomplete observations. In: *Perspectives in Probability and Statistics*, Gani, J. (ed.), pp. 67–87. Academic Press, London. (Cited on p. 215.)

METROPOLIS, N., ROSENBLUTH, A., ROSENBLUTH, M., TELLER, A., AND TELLER, E. (1953) Equations of state calculations by fast computing machines. *J. Chem. Phys.*, **21**, 1087–1092. (Cited on p. 344.)

MEYER, F. (2012) Recommender systems in industrial contexts. PhD thesis, University of Grenoble. (Cited on p. 530.)

MEYER, M. AND LAUD, P. (2002) Predictive variable selection in generalized linear models. *J. Amer. Stat. Assoc.*, **97**, 859–871. (Cited on p. 184.)

MEYER, S., HELD, L., AND HÖLE, M. (2015) Spatio-temporal analysis of epidemic phenomena using the R package surveillance. See: https://arxiv.org/abs/1411.0416v2. Accessed May 15, 2017. (Cited on p. 568.)

MIKA, S., RÄTSCH, WESTON, J., SCHOLKOPF, B., AND MÜLLER, K.-M. (1999) Fisher discriminant analysis with kernels. *Neural Networks for Signal Proc.*, **9**, 41–48. (Cited on p. 409.)

MILLIKEN, G. AND JOHNSON, D. (1992) *Analysis of Messy Data*. Chapman and Hall, New York (Cited on p. 169.)

MINER, G., DELEN, D., ELDER, J., FAST, A., HILL, T., AND NISBET, R. (2012) *Practical Text Mining and Statistical Analysis for Non-Structured Text Data Applications*. Academic Press, Waltham MA. (Cited on p. 449.)

MINKA, T. (2002) Bayes model averaging is not model combination. MIT Media Lab note (7/6/00). See: http://research-srv.microsoft.com/en-us/um/people/minka/papers/min ka-bma-isnt-mc.pdf. Accessed May 15, 2017. (Cited on p. 454.)

MITCHEL, T. AND BEAUCHAMP. J. (1988) Bayesian variable selection in linear regression. *J. Amer. Stat. Assoc.*, **83**, 1023–1036. (Cited on p. 100.)

MOHAMMADAI, M. AND DAS, S. (2015) SNN: Stacked neural networks. See: https://arxiv.org/abs/1605.08512v1. Accessed May 15, 2017. (Cited on p. 476.)

MONAHAN, J. (1980) A structured approach to ARMA time series models. Part I: Distributional results. Inst. Stat. Mimeo Series, No. 1297 North Carolina State University. (Cited on pp. 140 and 141.)

MONAHAN, J. (1983) Fully Bayesian analysis of ARMA time series models. *J. Econometrics*, **21**, 307–331. (Cited on pp. 139, 140, 141, and 149.)

MONTEITH, K., CARROLL, J., SEPPI, K., AND MARTINIZ, T. (2011) Turning Bayes model averaging into Bayes model combination. In: *Proc. 2011 International Joint Conf. on Neural Networks*, pp. 2657–2663. (Cited on p. 454.)

MOORE, D. AND YACKEL, J. (1977) Large sample properties of nearest neighbor density function estimators. In: Gupta, S. and Moore, D. (eds.), *Statistical Decision Theory and Related Topics*, Academic Press, New York. (Cited on p. 275.)

MORRIS, J. BAGGERLY, K. AND COOMBES, K. (2005) Bayesian shrinkage estimation of the relative abundance of mRNA transcripts using SAGE. *Biometrics*, **59**, 476–486. (Cited on p. 24.)

MORVAI, G. YAKOWITZ, S. AND GYORFI, L. (1996) Nonparametric inference for ergodic, stationary time series, *Ann. Stat.*, **24**, 370–379. (Cited on p. 286.)

MUELLER, P. AND MITRA, R. (2013) Bayesian nonparametric inference: why and how. *Bayes Analysis*, **8**, 269–302. (Cited on p. 93.)

MUKHERJEE, I., AND SCHAPIRE, R. (2013) A theory of multiclass boosting. *J. Mach. Learning Res.*, **14**, 437–497. (Cited on p. 484.)

MÜLLER, M. (1992) Aymptotic properties of model selection procedures in regression analysis. PhD thesis, Humbolt University Berlin. (In German.) (Cited on p. 325.)

MÜLLER, M. (1993) Consistency properties of model selection criteria in multiple linear regression. See: *http://citeseerx.ist.psu.edu/viewdoc/summary?doi=10.1.1.45.256*. Accessed May 15, 2017. (Cited on p. 325.)

MUNRO, J. AND PATERSON, M. (1980) Selection and sorting with limited storage. *Theor. Comp. Sci.*, **12**, 315–323. (Cited on p. 545.)

MURPHY, K. (2007) Conjugate Bayesian analysis of the Gaussian distribution. Technical Report, Department of Computer Science. University of British Columbia. (Cited on pp. 163 and 164.)

MURTAGH, F. (2004) On ultrametricity, data coding, and computation. *J. Classification*, **21**, 167–184. (Cited on pp. 519 and 525.)

MURTAGH, F. (2009) The remarkable simplicity of very high dimensional data: application of model-based clustering. *J. Classification*, **26**, 249–277. (Cited on pp. 116 and 525.)

MURTAGH, F. AND CONTRERAS, P. (2011) Hierarchical clustering for finding symmetries and other patterns in massive, high dimensional datasets. In: Holmes, D. (ed.), *Data Mining: Foundations and Intelligent Paradigms*, pp. 95–130 Springer (Cited on p. 525.)

MUTHUKRISHNAN, S. (2009) Data stream algorithms. Notes from a series of lectures, from *Proc. 2009 Barbados Workshop on Computational Complexity*. (Cited on p. 537.)

NEILSON, M (2012) *Neural Networks and Deep Learning*. Online book, see `http://neuralnet worksanddeeplearning.com/index.html`. (Cited on p. 400.)

NETER, J., WASSERMAN, W., KUTNER, M. (1985) *Applied Linear Statistical Models*, 2nd edn. Richard Irwin, Inc., Holmwood, IL (Cited on pp. 169 and 170.)

NGUYEN, J., AND ZHU, M. (2013) Content-boosted matrix factorization techniques for recommender systems. *Stat. Anal. Data Mining*, **6**, 286–301. (Cited on p. 528.)

NGUYEN, A., YOSINSKI, J., AND CLUNE, J. (2015) Deep neural networks are easily fooled: High Confidence Predictions for Unrecognizable Images. In: *Computer Vision and Pattern Recognition*, IEEE, 2015. (Cited on p. 403.)

OKIMOTO, G., ZEINALZADEH, A., WENSKA, T., LOOMIS, M., NATION, J., FABRE, T. *et al.* (2016) Joint analysis of multiple high-dimensional data types using sparse matrix approximations of rank-1 with applications to ovarian and liver cancer. *BioData Mining*, **9**, 24. (Cited on pp. 594 and 595.)

OLIVE, D. (2013) Asymptotically optimal regression prediction intervals and prediction regions for multivariate data. *Int. J. Stat. Prob.*, **2**. (Cited on p. 95.)

ONORANTE, L., AND RAFTERY, A. (2014) Dynamic model averaging in large model spaces using dynamic Occam's window. Technical Report No. 628, Department of Statistics, University of Washington. (Also see `https://arxiv.org/abs/1410.7799v1`.) (Cited on p. 461.)

ORAVA, J. (2011) *k*-nearest neighbor kernel density estimation, the choice of optimal *k*. *Tatra Mt. Math. Publ.*, **50**, 39–50. (Cited on p. 279.)

PABINGER, S., DANDER, A., FISCHER, M., SNAJDER, R., SPERK, M., EFREMOVA, M. *et al.* (2014) A survey of tools for variant analysis of next-generation genome sequencing data. *Brief. Bioinform.*, **15**, 256–278. (Cited on p. 589.)

PAN, J. AND FANG, K. (2002) *Growth Curve Models*. Springer, New York. (Cited on pp. 203 and 204.)

PAN, W. AND LE, C. (2001) Bootstrap model selection in GLM's. *Biometrics*, **6**, 49–61. (Cited on p. 182.)

PANHARD, X. AND SAMSON, A. (2009) Extension of the SAEM algorithm for nonlinear mixed models with 2 levels of random effects. *Biostatistics*, **10**, 121–135. (Cited on p. 194.)

PARRY, M., DAWID, A. P., AND LAURITZEN, S. (2012) Proper local scoring rules. *Ann. Stat.*, **40**, 561–592. (Cited on p. 26.)

PEARL, J. (2000) *Causality: Models, Reasoning, and Inference*. Cambridge University Press. (Cited on p. 479.)

PELLETIER, M. (1998) Weak convergence rates for stochastic approximation with application to multiple targets and simulated annealing. *Ann. Appl. Prob.*, **8**, 10–44. (Cited on p. 393.)

PEPE, M. (2003) *The Statistical Evaluation of Medical Tests for Classification and Prediction*. Oxford University Press. (Cited on pp. 112 and 221.)

PETERSEN, M., MOLINARO, A., SINISI, S., AND VAN DER LAAN, M. (2007) Cross-validated bagged learning. *J. Multivar. Anal.*, **98**, 1693–1704. (Cited on p. 470.)

PHILIPPE, A. (2006) Bayesian analysis of autoregressive moving average processes with unknown orders. *Comp. Stat. Data Anal.* **51**, 1904–1923. (Cited on p. 141.)

PITTMAN, J., HUANG, E., NEVINS, J., WANG, Q., AND WEST, M. (2004a) Bayesian analysis of binary prediction tree models for retrospectively sampled outcomes. *Biostatistics*, **5**, 587-6-1. (Cited on p. 385.)

PITTMAN, J., HUANG, E., DRESSMAN, H., HORNG, C., CHENG, S., TSOU, M. *et al.* (2004b) Integrated modeling of clinical and gene expression information for personalized prediction of disease outcomes. *Proc. Nat. Acad. Sci. USA*, **101**, 8431–8436. (Cited on p. 385.)

PLECHAWSKA-WOJCIK, M. (2012) A comprehensive analysis of MALDI-TOF spectrometry data. See: `http://cdn.intechopen.com/pdfs/31519/InTech-A_comprehensive_analy sis_of_maldi_tof_spectrometry_data.pdf`. Accessed May 1, 2017. (Cited on p. 590.)

POINCARÉ, H. (1913) *The Foundations of Science*, vol. 1 in the series Science and Education. Trans. G. B. Halsted. The Science Press, Garrison, New York. (Cited on p. 125.)

POLE, A., WEST, M. AND HARRISON, J. (1994) *Applied Bayesian Forecasting and Time Series Analysis*. Chapman and Hall, New York. (Cited on p. 159.)

POLITIS, D. AND VASILIEV, V. (2012) Sequential kernel estimation of a multivariate regression function. In: *Proc. 11th International Conf. 'System Identification and Control Problems'*, V. A. Tapeznikov Institute of Control Sciences, pp. 996–1009. (Cited on p. 286.)

POLLEY, E. AND VAN DER LAAN, M. (2010) Super learner in prediction. University of California at Berkeley Division of Biostatistics Working *Paper Series* No. 266. (Cited on p. 478.)

PRIESTLEY, M., AND CHAO, M. (1972) Nonparametric function fitting. *J. Roy. Stat. Soc.*, **34**, 385–392. (Cited on p. 266.)

PROUST-LIMA, C. AND TAYLOR, J. (2009) Development and validation of a dynamic prognostic tool for prostate cancer recurrence using repeated measures of post-treatment PSA: a joint modeling approach. *Biostatistics*, **10**, 535–549. (Cited on p. 191.)

PUELZ, D., HAHN, P., AND CARVALHO, C. (2017) Variable selection in seemingly unrelated regressions with random predictors. *Bayesian Analysis*, **12**, 969–989. (Cited on p. 603.)

R CORE TEAM (2016) *R: A Language and environment for statistical computing.* R Foundation for Statistical Computing, Vienna, Austria, pp. 2–16. See: `https://www.R-project.org`. (Cited on pp. 104, 116, 195, 234, 237, and 418.)

RAFTERY, A., AND ZHENG, Y. (2003) Performance of Bayes model averaging. *J. Amer. Stat. Assoc.*, **98**, 931–938. (Cited on p. 451.)

RAFTERY, A., MADIGAN, D., AND HOETING, J. (1993) Model selection and accounting for model uncertainty in linear regression models. Technical Report No. 262, Department of Statistics, University of Washington. (Cited on p. 461.)

RAKHLIN, A. AND SRIDHARAN, K. (2014) *Statistical Learning and Sequential Prediction.* See: `http://stat.wharton.upenn.edu/~rakhlin/book_draft.pdf`. (Cited on p. 521.)

RAKOTOMARAOLAHY, P. (2012) Statistical properties of nearest neighbors regression functions estimates for strong mixing processes. See: `https://hal.archives-ouvertes.fr/hal-00725841`. (Cited on p. 286.)

RALAIVOLA, L. AND D'ALECHÉ-BUC, F. (2003) Dynamical modeling with kernels for nonlinear time series prediction. In: Thrun, S., Saul, S., and Schölkopf, B. (eds.), *Advances in Neural Information Processing Systems*, vol. 16, pp. 129–136, MIT Press. (Cited on p. 286.)

RANDLES, R. AND WOLFE, D (1979) *Introduction to the Theory of Nonparametric Statistics.* John Wiley and Sons, New York (Cited on p. 251.)

RAO, C. (1987). Prediction of future observations in growth curve models (with discussion). *Stat. Sci.*, **2**, 434–447. (Cited on pp. 162, 189, and 203.)

RAO, P. (1998) *Statistical Research Methods in the Life Sciences.* Duxbury Press, Pacific Grove, CA. (Cited on p. 169.)

RASMUSSEN, C. AND WILLIAMS, C. (2006) *Gaussian Processes for Machine Learning.* MIT Press, Cambridge, MA. (Cited on pp. 293 and 294.)

REID, S. AND GRUDIC, G. (2009) Regularized linear models in stacked generalization. In: Benediktsson, J. *et al.* (eds.), *Multiple Classifier Systems*, pp. 112–121. Springer-Verlag, Berlin. (Cited on p. 475.)

REINSCH, C. (1967) Smoothing by spline functions. *Numerische Math.*, **10**, 177–183. (Cited on p. 303.)

REISS, R. (1989) *Approximate Distributions of Order Statistics.* Spinger-Verlag, New York. (Cited on pp. 88, 216, and 495.)

REN, S., CAO, X., WEI, Y., AND SUN, J. (2015) Global refinement of random forest. In: *Proc. 2015 IEEE Conf. on Computer Vision and Pattern Recognition*, pp. 723–730. (Cited on p. 468.)

RICCI, F., ROKACH, L., SHAPIRA, AND KANTOR, P. (2011) (eds.) *Recommender Systems Handbook.* Springer, New York. (Cited on pp. 526, 532, and 535.)

RICE, J. AND WU, C. (2001) Nonparametric mixed effects models for unequally sampled noisy curves. *Biometrics* **57**, 253–259. (Cited on p. 191.)

RIDGEWAY, G. (1999) The state of boosting. *Comp. Sci. Stat.*, 172–181. (Cited on p. 488.)

RIDGEWAY, G. (2006) Generalized boosted models: a guide to the gbm package. See: `http://ftp.auckland.ac.nz/software/CRAN/doc/vignettes/gbm/gbm.pdf`. (Cited on p. 488.)

RIDGEWAY, G., MADIGAN, D., AND RICHARDSON, T. (1999) Boosting methodology for regression problems. In: *Proc. Conf. on Artificial Intelligence and Statistics*, pp. 152–161. (Cited on pp. 487 and 488.)

RISSANEN, J. (1996) Stochastic complexity and modeling. *Ann. Stat.*, **14**, 1080–1100. (Cited on pp. 361 and 521.)

RITCHIE, M., HOLZINGER, E., LI, R., PENDERGRASS, S., AND KIM, D. (2015) Methods of integrating data to uncover genotype phenotype interactions. *Nat. Rev. Genetics*, **16**, 85–97. (Cited on pp. 592 and 593.)

ROBINSON, G. K. (1991) that BLUP is a good thing: the estimation of random effects. *Stat. Sci.*, **6**, 15–32. (Cited on p. 186.)

ROCCHIO, J. (1971) Relevance feedback in information retrieval. In: Salton, G. (ed.), *The SMART Retrieval System: Experiments in Automatic Document Processing*, pp. 313–323, Prentice Hall, Upper Saddle River, NJ. (Cited on p. 532.)

ROČKOVÁ, V. AND GEORGE, E. (2016) The spike-and-slab LASSO. *To appear in J. Amer. Stat. Assoc.* See: http://dx.doi.org/10.1080/01621459.2016.1260469. (Cited on p. 429.)

RODRIGUEZ, G. (2005) Competing risks. See: http://data.princeton.edu/pop509/Com petingRisks.pdf. (Cited on pp. 247 and 248.)

ROJAS, R. (2009) Adaboost and the superbowl of classifiers: a tutorial introduction to adaptive boosting. Technical Report, Free University, Berlin. See: http://www.inf.fu-berlin.de/inst/ag-ki/adaboost4.pdf. (Cited on p. 481.)

ROKACH, L. (2009) Taxonomy for characterizing ensemble methods in classification tasks: a review and annotated bibliography. *Comp. Stat. Data Anal.*, **53**, 4046–4072. (Cited on p. 489.)

ROKACH, L. (2010) Ensemble based classifiers. *Artific. Intell. Rev.*, **33**, 1–39. (Cited on p. 473.)

RONCHETTI, E., FIELD, C., AND BLANCHARD, W. (1997). Robust linear model selection by cross-validation. *J. Amer. Stat. Assoc.*, **92**, 1017–1023. (Cited on p. 338.)

ROOS, T. (2008) Monte Carlo estimation of minimax regret with an application to MDL model selection. In: *Proc. 2008 Information Theory Workshop*, pp. 284–288, IEEE Press. (Cited on p. 521.)

ROSIPAL, R., AND TREJO, L. (2001) Kernel partial least squares regression in reproducing kernel Hilbert space. *J. Mach. Learning Res.*, **2**, 97–123. (Cited on p. 409.)

ROSTEK, M. (2010) Quantile maximization in decision theory. *Rev. Econ. Stud.*, **77**, 339–371. (Cited on p. 333.)

ROUSSEEUW, P. (1984) Least median squares regression. *J. Amer. Stat. Assoc.*, **79**, 871–880. (Cited on p. 495.)

RUCKSTUHL, A. (2010) Introduction to nonlinear regression. IDP Institut fur Datenanalyse und Prozessdesign, Zurcher Hochschule fur Angewandte Wissenschaften. See: http://www.idp.zhaw.ch. (Cited on p. 365.)

RUCZINSKI, I. (2003) Logic regression. See: http://biostat.jhsph.edu/~iruczins/pr esentations/ruczinski.04.03.census.pdf. (Cited on p. 445.)

RUCZINSKI, I., KOOPERBERG, C., AND LEBLANC, M. (2003) Logic regression. *J. Critical Globalization Stud.*, **12**, 475–511. (Cited on p. 445.)

RUNGE, C. (1901). Über empirishe Funckionen und die Interpolation zwischen äquidistanten Ordinaten. *Z Math. Phys.*, **46**, 224–243. (Cited on p. 260.)

SABARIAN, M., AND VASCONCELOS, N. (2011) Multiclass boosting: theory and algorithms. In: Shawe-Taylor, J., Zemel, R., Bartlett, P., Pereira, F., and Weinberger, K. (eds.), *Advances in Neural Information Processing Systems*, vol. 24, pp. 2124–2132, Curran Associates, Inc. (Cited on p. 484.)

SAKIA, R. (1992) The Box–Cox transformation technique: a review. *The Statistician*, **41**, 169–178. (Cited on p. 204.)

SASU, A. (2012) k-nearest neighbor prediction algorithm for univariate time series prediction. *Bull. Transilvania Univ. Brasov*, **5**, 147–152. (Cited on p. 286.)

SAVAGE, J. (1954) *The Foundations of Statistics*, revised and enlarged in 1972. Dover, New York. (Cited on pp. 36 and 54.)

SAWYER, S. (2003) The Greenwood and exponential Greenwood confidence intervals in survival analysis. See: http://www.math.wustl.edu/~sawyer/handouts/greenwood.pdf. (Cited on p. 213.)

SCHABENBERGER, O. AND GOTWAY, C. (2005) *Statistical Methods for Spatial Data Analysis*. Chapman and Hall/CRC, Boca Raton, FL. (Cited on p. 557.)

SCHAPIRE, R. (1990) The strength of weak learnability. *Mach. Learning*, **5**, 197–227. (Cited on pp. 480 and 481.)

SCHAPIRE, R., AND SINGER, Y. (1999) Improved boosting algorithms using confidence related predictions. *Mach. Learning*, **37**, 297–336. (Cited on p. 481.)

SCHMIDLI, J., FREI, C., AND VIDALE, P. (2006) Downscaling from GCM precipitations: a benchmark for dynamical and statistical downscaling methods. *Int. J. Climatol.*, **26**, 679–689. (Cited on p. 556.)

SCHMIDT, D. (2007) Minimum message length inference and parameter estimation of autoregressive and moving average models. Technical Report No. 206, Clayton School of Information Technology, Monash University. See: http://www.csse.monash.edu.au/ dschmidt/home.html. (Cited on p. 65.)

SCHMIDT, D., AND MAKALIC, E. (2012) The consistency of MDL for linear regression models with increasing signal to noise ratio. *IEEE. Trans. Sig. Proc.*, **60**, 1508–1510. (Cited on p. 65.)

SCHOLKOPF, B., AND SMOLA, A. (2002) *Learning with Kernels*. MIT Press. (Cited on p. 17.)

SCHOLKOPF, B., SMOLA, A., AND MÜLLER, K.-M. (1998) Nonlinear component analysis as a kernel eigenvalue problem. *Neural Comp.*, **10**, 1299–1319. (Cited on p. 409.)

SCHOLKOPF, B., HERBICH, R., AND SMOLA, A. (2001) A generalized representer theorem. In: *Computational Learning Theory*, Lecture Notes in Computer Science, vol. 2111, pp. 416–426. (Cited on p. 408.)

SCHONLAU, M. (2005) Boosted regression (boosting): an introductory tutorial and a Stata plugin. *Stata J.*, **2005**, 330–354. (Cited on p. 488.)

SCORBUREANU, A. (2009) Productivity and growth: least absolute deviation estimator and bootstrap techniques to predict aggregate production elasticities in the Palestinian manufacturing industry. Paper No. 17 966, Munich Personal RePEc Archive. See: http://mpra.ub.uni-muenchen.de/17966/. (Cited on p. 95.)

SEEGER, M. (2000) Bayesian model selection for support vector machines, gaussian processes and other kernel classifiers. In: Solla, S., Leen, T., and Muller, K.-R. (Eds.), *Advances in Neural Information Processing Systems 12 (NIPS 2000)*, pp. 603–609, MIT Press. (Cited on p. 461.)

SESMERO, M., LEDEZMA, A., AND SANCHIS, A. (2015) Generating ensembles of heterogeneous classifiers using stacked generalization. *WIREs Data Mining Knowl. Discov.*, **5**, 2134. (Cited on p. 475.)

SETHURMAN, J. (1994) A constructive definition of Dirichlet priors. *Stat. Sinica*, **4**, 639–650. (Cited on p. 91.)

SEVERINSKI, C., FOKOUE, E., ZHANG, H., AND CLARKE, B. (2010) *Solutions Manual to Accompany 'Principles and Theory for Data Mining and Machine Learning'*, Springer, New York. (Cited on p. 332.)

SEXTON, R., DORESEY, R., AND JOHNSON, J. (1999) Beyond backpropagation: using simulated annealing for training neural networks. *J. End User Comp.*, **11**, 3–11. (Cited on p. 392.)

SHAHBABA, B., AND JOHNSON, W. (2013) Bayesian nonparametric variable selection as an exploratory tool for finding genes that matter. *Stat. Med.*, **32**, 2114–2126. (Cited on p. 603.)

SHALIZI, C. (2009) Classification and regression trees. See: www.stat.cmu.edu/ ~cshalizi/350/lectures/22/lecture-22.pdf, lecture notes 6 November.

SHALIZI, C. (2016) See: http://www.stat.cmu.edu/~cshalizi/networks/16-1/, lecture notes. (Cited on pp. 574, 575, 578, and 584.)

SHAO, J. (1993) Linear model selection by cross-validation. *J. Amer. Stat. Assoc.*, **88**, 486–495. (Cited on pp. 311, 337, and 579.)

SHAO, J. (1997) An asymptotic theory for model selection. *Stat. Sinica*, **7**, 221–264. (Cited on p. 70.)

SHARKEY, A. (ed.) (1999) *Combining Artificial Neural Networks*. Springer-Verlag, London. (Cited on p. 470.)

SHEN, R., OLSHEN, A., AND LADANYI, M. (2009) Integrative clustering of multiple genomic data types using a joint latent variable model with application to breast and lung cancer subtype analysis. *Bioinformatics*, **25**, 2906–2912. (Cited on p. 597.)

SHEN, R., WANG, S., AND MO, Q. (2013) Sparse integrative clustering of multiple omics data sets. *Ann. Appl. Stat.*, **7**, 269–294. (Cited on pp. 597 and 598.)

SHI, M., WEISS, R. AND TAYLOR, J. (1996) An analysis of pediatric CD4 counts for acquired immune deficiency syndrome using flexible random curves. *J. Roy. Stat. Soc. Ser. C*, **45**, 151–163. (Cited on pp. 191, 192, and 204.)

SHIBATA, R. (1976) Selection of the order of an autoregressive model by Akaike's information criterion. *Biometrika*, **63**, 117–126. (Cited on p. 323.)

SHIBATA, R. (1981) An optimal selection of regression variables. *Biometrika*, **68**, 45–54. (Cited on p. 321.)

SHMUELI, G. (2010) To explain or predict? *Stat. Sci.*, **25**, 289–310. (Cited on p. 359.)

SHORACK, G. (1974) Random means. *Ann. Stat.*, **2**, 661–675. (Cited on p. 90.)

SHTARKOV, Y. (1988) Universal sequential coding of single messages. *Trans. Prob. Inf. Transmi.*, **23**, 317. (Cited on pp. 520 and 521.)

SHUMWAY, R., AND STOFFER, D. (2010) *Time Series Analysis and Its Applications with R Examples*, 3rd edn. Springer-Verlag, New York. (Cited on pp. 129 and 159.)

SHUMWAY, R., AND STOFFER, D. (2013) *Time series analysis and its applications with R examples*, EZ 3rd Ed. Download from `http://www.stat.pitt.edu/stoffer/tsa3/tsa3EZ.pdf` (Cited on p. 130.)

SKOURAS, K., AND DAWID, A. P. (1998) On efficient point prediction systems. *J. Roy. Stat. Soc.*, Ser. B, **60**, 765–780. (Cited on p. 361.)

SKOURAS, K., AND DAWID, A. P. (1999) On efficient probability forecasting systems. *Biometrika*, **86**, 765–784. (Cited on p. 231.)

SMOLA, A., AND SCHOLKOPF, B. (2004) A tutorial on support vector regression. *Stat. Comp.*, **14**, 199–222. (Cited on pp. 420 and 421.)

SMYTH, P., AND WOLPERT, D. (1998) Stacked density estimation. In: *Proc. 10th International Conf. on Neural Information Processing Systems*, pp. 668–674, MIT Press. (Cited on p. 473.)

SMYTH, P., AND WOLPERT, D. (1999) Linearly combining density estimators via stacking. *Mach. Learning*, **36**, 59–83. (Cited on p. 473.)

SNIJDERS, T. (2006) Statistical methods for network dynamics. In: Luchini, S. *et al.* (eds.), *Proc 43rd Scientific Meeting, Italian Statistical Society*, pp. 281–296, Padova (Cited on pp. 583 and 584.)

SNIJDERS, T. (2011) Statistical models for social networks. *Ann. Rev. Soc.*, **37**, 131–153. (Cited on pp. 579 and 581.)

SOBER, E. (2002) Instrumentalism, parsimony, and the Akaike framework. *Phil. Sci.*, **69**, S112–S123. (Cited on p. 307.)

SOLLICH, P. (2002) Bayes methods for support vector machines: evidence and predictive class probabilities. *Mach. Learning*, **46**, 21–52. (Cited on p. 461.)

SONG, L., LANGFELDER, P., AND HORVATH, S. (2013) Random generalized linear models: a highly accurate and interpretable ensemble predictor. *BMC Bioinformatics*, **14**, 5. (Cited on p. 471.)

SPIEGELHALTER, D., BEST, N., CARLIN, B., AND VAN DER LINDE, A. (2002). Bayesian measures of model complexity and fit (with discussion). *J. Roy. Stat. Soc.* Ser. B, **64**, 583–639. (Cited on pp. 321 and 354.)

SPIEGELHALTER, D., BEST, N., CARLIN, B., VAN DER LINDE, A. (2014). The deviance information criterion: 12 years on (with discussion). *J. Roy. Stat. Soc.* Ser. B, **76**, 485–493. (Cited on p. 354.)

SRIHARI, S. (2014) Regularization in neural networks. See: `http://www.cedar.buffalo.edu/~srihari/CSE574/Chap5/Chap5.5-Regularization.pdf`. (Cited on p. 391.)

STATHAKIS, D. (2009) How many hidden layers and nodes? *Int. J. Rem. Sensing*, **30**, 2133–2147. (Cited on p. 399.)

STEFANSKI, L., AND COOK, J. (1995) Simulation-extrapolation: The measurement error jackknife. *J. Amer. Stat. Assoc.*, **90**, 1247–1256. (Cited on p. 382.)

STEYERBERG, E., VICKERS, A., COOK, N., GERDS, T., GONEN, M., OBUCHOWSKI, N. *et al.* (2010) Assessing the performance of prediction models: a framework for traditional and novel measures. *Epidemiology*, **21**, 139–141. (Cited on p. 113.)

STOY, P., WILLIAMS, M., DISNEY, M., PRIETO-BLANCO, A., HUNTLEY, B., BAXTER, R. *et al.* (2009) Upscaling as ecological information transfer: a simple framework with application to Arctic ecosystem carbon exchange. *Landscape Ecol.*, **24**, 971–986. (Cited on p. 564.)

STROBL, C., MALLEY, J., AND TUTZ, G. (2009) An introduction to recursive partitioning: rationale, application and characteristics of classification and regression trees, bagging and random forests. *Psychol. Meth.*, **14**, 323–348. (Cited on p. 382.)

SUN, D. (2015) Analysis of spatial and temporal data. Lecture Notes, Stanford University, Summer 2015. See: http://web.stanford.edu/class/stats253/lectures.html. (Cited on p. 558.)

SUSARLA, V., AND VAN RYZIN, J. (1976) Bayes estimation of survival curves from incomplete observations. *J. Amer. Stat. Assoc.*, **71**, 897–902. (Cited on pp. 92, 219, and 220.)

SZEGEDY, C., ZAREMBA, W., SUTSKEVER, H., BRUNA, J., ERHAN, D., GOODFELLOW, I., *et al.* (2014) Intriguing properties of neural networks. See: https://arxiv.org/pdf/1312.6199.pdf. (Cited on p. 403.)

TADDY, M., AND KOTTAS, A. (2001) Mixture modeling for marked Poisson processes. *Bayes Anal.*, **7**, 335–362. (Cited on p. 151.)

TAKEUCHI, K. (1976) Distribution of information statistics and criteria for adequacy of models. *Math. Sci.*, **153**, 12–18. (Cited on p. 321.)

TAN, P.-N., STEINBACH, M., AND KUMAR, V. (2005) *Introduction to Data Mining*, 1st edn. Pearson PLC, London, UK. (Cited on p. 377.)

TEMPFLI, K., KERLE, N., HUURNEMAN, G. AND JANSSEN, L. (eds.) (2009) *Principles of Remote Sensing*. International Institute for Geo-Information Science and Earth Observation, Enschede, Netherlands. (Cited on p. 563.)

TENORIO, M., AND SHI, W.-T. (1990) Self-organizing networks for optimum supervised learning. *IEEE Trans. Neural Net.*, **1**, 100–110. (Cited on pp. 342 and 393.)

THERNEAU, T. AND GRAMBSCH, P. (2000) *Modeling Survival Data: Extending the Cox Model*. Springer, New York. (Cited on p. 217.)

THIEMICHEN, S., FRIEL, N., CAIMO, A., AND KAUERMANN, G. (2016) Bayesian exponential random graph models with nodal random effects. *Social Networks*, **46**, 11–28. (Cited on p. 577.)

THOMAS, W. AND COOK, R. D. (1990) Assessing influence on predictions from generalized linear models. *Technometrics*, **21**, 59–65. (Cited on p. 184.)

TIBSHIRANI, R. (1996) Regression shrinkage and selection via the LASSO. *J. Roy. Stat. Soc.* Ser. B, **58**, 267–288. (Cited on p. 425.)

TING, K. AND WITTEN, I. (1997) Stacked generalization: when does it work? (Working paper 97/03). Hamilton, New Zealand: University of Waikato, Department of Computer Science. (Cited on p. 475.)

TIPPING, M. (2000) The relevance vector machine. In: Solla, S., Leen, T., and Müller, K.-R. (eds.), *Advances in Neural Information Processing Systems*, vol. 12, pp. 652–658. MIT Press. (Cited on pp. 409, 410, and 412.)

TIPPING, M. (2001) Sparse Bayesian learning and the relevance vector machine. *J. Mach. Learning Res.*, **1**, 211–244. (Cited on pp. 17, 410, 411, and 412.)

TITTERINGTON, D. (2004) Bayes methods for neural networks and related models. *Stat. Sci.*, **19**, 128–139. (Cited on p. 395.)

TSIATIS, A. (1981) A large sample study of Cox's regression model. *Ann. Stat.*, **9**. 93–108. (Cited on p. 230.)

TUKEY, J. (1977) *Exploratory Data Analysis*. Addison-Wesley. (Cited on p. 12.)

TURNBULL, B., GHOSAL, S., AND ZHANG, H. (2013) Iterative selection using orthogonal regression techniques. *Stat. Anal. Data Mining*, **6**, 557–564. (Cited on pp. 595 and 596.)

VALDES, C., BRENNAN, M., DOBRA, A., AND CLARKE, B. (2015) Detecting bacterial genomes in metagenomic samples, using NGS reads. *Stat. Interface*, **8**, pp. 173–185. (Cited on p. 24).

VALENTINI, G. AND DIETTERICH, T. (2003) Low bias bagged support vector machines. In: Fawcett, T., and Mishra, N. (eds.), *Proc. 20th International Conf. on Machine Learning*, pp. 752–759, Washington D.C, AAAI Press. (Cited on p. 470.)

VAN DER LAAN, M. AND ROSE, S. (2011) *Targeted Learning: Causal Inference for Observational and Experimental Data*. Springer, New York. (Cited on pp. 477, 478, and 479.)

VAN DER LAAN, M., DUDOIT, S., AND VAN DER VAART, A. (2003) Unified cross-validation methodology for selection among estimators and a general cross-validated adaptive epsilon-net estimator: finite sample oracle inequalities and examples. University of California at Berkeley Division of Biostatistics Working paper Series No. 130. (Cited on p. 478.)

VAN DER LAAN, M., DUDOIT, S., AND VAN DER VAART, A. (2004) The cross-validated adaptive epsilon-net estimator. University of California at Berkeley Division of Biostatistics Working paper Series No. 142. (Cited on p. 478.)

VAN DER LAAN, M., POLLEY, E., AND HUBBARD, A. (2007) Super learner. University of California at Berkeley Division of Biostatistics Working paper Series No. 222. (Cited on pp. 477 and 478.)

VAN DER VAART, A., DUDOIT, S., AND VAN DER LAAN, M. (2006) Oracle inequalities for multi-fold cross validation. *Stat. Decisions*, **24**, 351–371. (Cited on p. 478.)

VAN ERVEN, T., GRUNEWALD, AND P., DE ROOIJ, S. (2012) Catching up faster by switching sooner: a predictive approach to adaptive estimation with an application to the AIC–BIC dilemma. *J. Roy. Stat. Soc.* Ser. B, **74**, 361–417. (Cited on pp. 134, 149, and 333.)

VAN HOUWELINGEN, H. AND PUTTER, H. (2012) *Dynamic Prediction in Clinical Survival Analysis*. CRC Press, Boca Raton, FL. (Cited on pp. 230, 231, and 245.)

VAPNIK, V. (1998) *Statistical Leraning Theory*. John Wiley and Sons, New York (Cited on p. 78.)

VEHTARI, A. AND OJANEN, J. (2012) A survey of Bayesian predictive methods for model assessment, selection and comparison. *Stat. Surveys*, **6**, 142–228. (Cited on p. 25.)

VERBEKE, G. AND MOLENBERGHS, G. (2009) *Linear Mixed Models for Longitudinal Data*. Springer, New York. (Cited on p. 162.)

VIERSTRAETE, A. (2012) Next generation sequencing for dummies. See: `https://use rs.ugent.be/~avierstr/nextgen/Next_generation_sequencing_web.pdf`. Accessed May 16, 2017. (Cited on p. 589.)

VIVAR, J. AND FERREIRA, M. (2009) Spatiotemporal models for Gaussian areal data. *J. Comp. Graph. Stat.*, **18**, 568–674. (Cited on p. 569.)

VLACHOPOULOU, M., GOSINK, L, PULSIPHER, T., HAFEN, R., ROUNDS, J., ZHOU, N. *et al.* (2015) Net interchange schedule forecasting using Bayes model averaging. In: *Proc. Power & Energy Society General Meeting*, IEEE. See: `https://doi.org/10.1109/PESGM.2015.7285722` (Cited on p. 460.)

VOVK, V. (2012) Combining *p*-values via averaging. See: `https://arxiv.org/pdf/1212.4966`, 1–7. (Cited on p. 24.)

WALLER, L. AND GOTWAY, C. (2004) *Applied Spatial Statistics for Public Health Data*. John Wiley and Sons, New York. (Cited on p. 193.)

WANG, F., ZHOU, C., AND NIE, Y. (2013) Event processing in sensor streams. In: Aggarwal, C. (ed.), *Managing and Mining Sensor Data*, pp. 77–102, Springer, New York. (Cited on p. 550.)

WANG, H., SHAN, H., AND BANERJEE, A. (2011) Bayesian cluster ensembles. *Stat. Anal. Data Mining*, **4**, 54–70. (Cited on p. 598.)

WANG, J., ZENG, H., CHEN, Z., LU, H., TAO, L., MA, W. (2003) ReCoM: reinforcement clustering of multi-type interrelated data objects. In: *Proc. 26th Annual International ACM SIGIR Conf. on Research and Development in Information Retrieval*, pp. 274–281. (Cited on pp. 597 and 599.)

WANG, M. (2013) *Spatial and spatio-temporal point process analysis*. PhD thesis, Department of Biostatistics, Emory University. (Cited on p. 567.)

WAND, M. AND JONES, M. (1995) *Kernel Smoothing*. In: *Monographs on Statistics and Applied Probability*, vol. 60, Chapman and Hall, London. (Cited on p. 263.)

WANG, T., LI, Y., AND CUI, H. (2007) On weighted randomly trimmed means. *J. Sys. Sci. Complexity*, **20**, 1–19. (Cited on p. 90.)

WANG, X., AND DEY, D. (2010) Generalized extreme value regression for binary response data. *Ann. Appl. Stat.*, **4**, 2000–2023. (Cited on p. 495.)

WARD, M., GREENHILL, B., AND BAKKE, C. (2010) The perils of policy by *p*-value: predicting civil conflicts. *J. Peace Res.*, **47**, 363–375. (Cited on p. 361.)

WASSERMAN, L. (2006) *All of Nonparametric Statistics*. Springer, New York. (Cited on pp. 255, 256, 577, and 579.)

WASSERMAN, L. (2010) *All of Statistics: A Concise Course in Statistical Inference*. Springer, New York. (Cited on p. 479.)

WASSERMAN, S., ROBBINS, G., AND STEINLY, D. (2006) Statistical models for networks: a brief review of some recent research. Technical Report 06–01, Deptartment of Statistics, Indiana University. (Cited on pp. 573, 574, and 575.)

WEI, Y. (2015) Integrative analyses of cancer data: a review from a statistical perspective. *Cancer Informatics*, **14(S2)**, 173–181. (Cited on p. 594.)

WEISSNER, M. AND PFEIFER, D. (2014) Health recommender systems: concepts, requirements, technical basics and challenges. *Int. J. Environ. Res. Public Health*, **11**, 2580–2607. (Cited on p. 537.)

WEST, M. AND HARRISON, J. (1997) *Bayesian Forecasting and Dynamic Linear Models*. Springer, New York. (Cited on pp. 139, 159, and 160.)

WESTFALL, P. AND YOUNG, S. (1993) *Resampling-Based Multiple Testing*. John Wiley and Sons, New York. (Cited on p. 23.)

WHITE, H. (1989) Some asymptotic results for learning in single hidden-layer feedforward network models. *J. Amer. Stat. Assoc.*, **84**, 1003–1013. (Cited on pp. 390, 392, and 403.)

WIKLE, C. (2015) An introduction to DSTMs. See: `http://www.stat.osu.edu/~pfc/ MBI/files/OSU_MBI_DSTMs_wikle_rev_2pg.pdf`. (Cited on pp. 558, 560, 561, 562, and 569.)

WILKS, S. (1941) Determination of sample sizes for setting tolerance limits. *Ann. Math. Stat.*, **12**, 91–96. (Cited on pp. 250 and 251.)

WILSON, E. O. (1998) *Consilience*. Vintage Books, a division of Random House Inc., New York. (Cited on p. 524.)

WOLPERT, D. (1992a) Stacked generalization. See: `http://www.machine-learning.mart insewell.com/ensembles/stacking/Wolpert1992.pdf`. (Cited on p. 471.)

WOLPERT, D. (1992b) Stacked generalization. *Neural Networks*, **5**, 241–259. (Cited on p. 471.)

WOLPERT, D. (1992c) On the connection between in-sample testing and generalization error. *Complex. Sys.*, **6**, 47–94. (Cited on p. 523.)

WOLPERT, D. (1996) The lack of a priori distinctions between learning algorithms. *Neural Comp.*, **8**, 1341–1390. (Cited on p. 523.)

WOLPERT, D. (2001) The supervised learning no-free-lunch theorems. See: `http://citeseerx .ist.psu.edu/viewdoc/download?doi10.1.1.99.133&reprep1&typepdf`. (Cited on p. 522.)

WOLPERT, D. AND MACREADY, W. (1996) Combining stacking with bagging to improve a learning algorithm. Technical Report SFI-TR-96-03-123, Santa Fe Institute. (Cited on pp. 474 and 523.)

WOLPERT, D. AND MACREADY, W. (1997) No free lunch theorems for optimization. *IEEE Trans. Evol. Comp.*, **1**, 67–82. (Cited on p. 522.)

WOMACK, A. (2011) Predictive alternatives in Bayes model selection. Electronic Theses and Dissertations paper 381. (Cited on pp. 355 and 356.)

WONG, H. AND CLARKE, B. (2004) Improvement in Bayes prediction in small samples in the presence of model uncertainty. *Can. J. Stat.*, **32**, 269–283. (Cited on pp. 8, 100, 226, 452, and 555.)

WU, H., AND LI, Z.-L. (2009) Scale issues in remote sensing: a review on analysis, processing, and modeling. *Sensors*, **9**, 1768–1793. (Cited on p. 564.)

WU, S., HARRIS, T., AND MCAULEY, K. (2007) The use of simplified or mis-specified models: linear case. *Can. J. Chem. Eng.*, **85**, 386–398. (Cited on p. 361.)

WU, X., LIU, M., AND WU, Y. (2012) In-situ soil moisture sensing: optimal sensor placement and field estimation. *ACM Trans. Sensor Networks*, **8**, 33. (Cited on p. 550.)

XIE, Q. AND BARRON, A. (2000) Asymptotic minimax regret for data compression, gambling, and prediction. *IEEE Trans. Inform. Theory*, **46**, 431–445. (Cited on p. 521.)

XIE, Y. AND AHN, C. (2013) Statistical methods for integrating multiple types of high-throughput data. In: Bang, H. *et al.* (eds.), *Statistical Methods in Molecular Biology*, pp. 511–530, Humana Press, Springer, New York. (Cited on p. 596.)

XU, G., LIANG, F., AND GENTON, M. (2015) A Bayesian spatio-temporal geostatistical model with an auxiliary lattice for large datasets. *Stat. Sinica*, **25**, 61–79. (Cited on p. 561.)

YANG, Y. (2005) Can the strengths of AIC and BIC be shared? A conflict between model identification and regression estimation. *Biometrika*, **92**, 937–950. (Cited on p. 134.)

YANG, Y. (2007) Consistency of cross-validation for comparing regression procedures. *Ann. Stat.*, **35**, 2450–2473. (Cited on p. 337.)

YE, J., ZHU, Z., AND CHENG, H. (2013) What's your next move: user activity prediction in location-based social networks. In: Ghosh, J. *et al.* (eds.), *Proc. 2013 SIAM International Conf. on Data Mining*, pp. 171–179, Society for Industrial and Applied Mathematics. (Cited on p. 447.)

YEN, T. (2011) Majorization-minimization approach to variable selection using spike and slab priors. *Ann. Stat.*, **39**, 1748–1775. (Cited on pp. 100 and 101.)

YOU, D., JIANG, X., CHENG, X., AND WANG, X. (2016) Bayesian kriging modeling for spatiotemporal prediction in squeeze casting. *Int. J. Manu. Technol.*, 1–15. (Cited on p. 561.)

YU, C.-W., AND CLARKE, B. (2010) Asymptotics of Bayes median loss estimation. *J. Mult. Anal.*, **101**, 1950–1958. (Cited on p. 91.)

YU, C.-W., AND CLARKE, B. (2014) Regular, median, and Huber cross-validation: a computational comparison. *Stat. Anal. Data Mining*, **8**, 14–33. (Cited on pp. 338 and 579.)

YU, K. AND MOYEED, R. (2001) Bayesian quantile regression. *Stat. Prob. Lett.* **54**, 437–447. (Cited on pp. 102 and 103.)

YU, Q., MACEACHERN, S., PERUGGIA, M. (2013) Clustered Bayes model averaging. *Bayes Anal.*, **8**, 883–908. (Cited on pp. 461 and 462.)

ZAMAN, F. AND HIROSE, H. (2009) Double SVMbagging: a new double bagging with support vector machines. In: Huang, X., Ao, S., and Castillo, D. (eds.), *Intelligent Automation and Computer Engineering*, Lecture Notes in Electrical Engineering, vol. 52, Springer, Dordrecht. (Cited on p. 470.)

ZARASIZ, G., ELMALI, F., AND OZTURK, A. (2012) Bagging support vector machines for leukemia classification. *Int. J. Comp. Sci.*, **9**, 355–358. (Cited on p. 470.)

ZEILER, M. AND FERGUS, R. (2013) Visualizing and understanding convolutional neural nets. In: *Proc. European Conf. on Computer Vision 2014*, pp. 818–833. (Cited on p. 403.)

ZELLNER, A. (1971) *An Introduction to Bayesian Analysis in Econometrics*. Wiley, New York. (Cited on p. 139.)

ZELLNER, A. (1986) On assessing prior distributions and Bayesian regression analysis with g-prior distributions. In: Goel, P. and Zellner, A. (eds.), *Bayesian Inference and Decision Techniques: Essays in Honor of Bruno de Finetti*, pp. 233–243, North Holland/Elsevier, Amsterdam. (Cited on p. 99.)

ZELLNER, A. (1988) Optimal information processing and Bayes' theorem. *Amer. Stat.*, **42**, 278–280. (Cited on p. 51.)

ZELLNER, A. AND GEISEL, M. (1970) Analysis of distributed lag models with applications to consumption function estimation. *Econometrika* **38**, 865–888. (Cited on p. 139.)

ZELLNER, A. AND WILLIAMS, A. (1973) Bayesian analysis of the Federal Reserve–MIT–Penn model's Almon lag consumption function. *J. Econometrics* **1**, 267–299. (Cited on p. 141.)

ZHANG, C., SUN, S., AND YU, G. (2004) A Bayesian network approach to time series forecasting of short term traffic flows. In: *Proc. 7th International IEEE Conf. on Intelligent Transportation Systems*, pp. 216–221. (Cited on p. 141.)

ZHANG, H. (2002) On estimation and prediction for spatial generalized linear mixed models. *Biometrics*, **58**, 129–136. (Cited on p. 193.)

ZHANG, J. (1999) Developing robust non-linear models through bootstrap aggregated neural networks. *Neurocomp.*, **25**, 93–113. (Cited on p. 470.)

ZHANG, X., MOORE, C., AND NEWMAN, M. (2016) Random graph models for dynamic networks. See: arXiv:1607.07570v1. (Cited on pp. 581 and 582.)

ZHOU, Z.-H., WU, J., AND TANG, W. (2002) Ensembling neural networks: many could be better than all. *Art. Intell.*, **137**, 239–263. (Cited on p. 469.)

ZHU, H., IBRAHIM, J. AND TANG, N. (2011) Bayesian influence analysis: a geometric approach. *Biometrika*, **98**, 307–323. (Cited on p. 65.)

ZIMEK, A., SCHUBERT, E., AND KRIEGEL, H.-P. (2012) A survey on unsupervised outlier detection in high dimensional numerical data. *Stat. Anal. Data Mining*, **5**, 363–387. (Cited on pp. 551 and 554.)

ZINDE-WALSH, V. AND GALBRAITH, J. (1991) Estimation of a linear regression model with stationary $ARMA(p,q)$ errors. *J. Econometrics*, **47**, 333–357. (Cited on pp. 157 and 158.)

ZOU, H. (2006) The adaptive LASSO and its oracle properties. *J. Amer. Stat. Assoc.*, **101**, 1418–1429. (Cited on pp. 424, 425, and 426.)

ZOZUS, M. (2017) *The Data Book: Collection and Management of Research Data*. Chapman and Hall/CRC, Boca Raton, FL. (Cited on p. 80.)

ZREDA, M. (2012) Cosmic-ray hydrometrology: measuring soil moisture with cosmic-ray neutrons. In: Njoku, E. (ed.), *Encyclopedia of Remote Sensing*, Springer-Verlag, Berlin. (Cited on p. 142.) (Cited on p. 57.)

Index

635

Printed in the United States
by Baker & Taylor Publisher Services